Food Safety

Emerging Issues, Technologies, and Systems

Food Safety

Emerging Issues, Technologies, and Systems

Edited by

Steven C. Ricke
Janet R. Donaldson
Carol A. Phillips

AMSTERDAM • BOSTON • HEIDELBERG • LONDON
NEW YORK • OXFORD • PARIS • SAN DIEGO
SAN FRANCISCO • SINGAPORE • SYDNEY • TOKYO
Academic Press is an imprint of Elsevier

Academic Press is an imprint of Elsevier
125 London Wall, London, EC2Y 5AS, UK
525 B Street, Suite 1800, San Diego, CA 92101-4495, USA
225 Wyman Street, Waltham, MA 02451, USA
The Boulevard, Langford Lane, Kidlington, Oxford OX5 1GB, UK

Thank you to the Arkansas Association for Food Protection for allowing Elsevier to publish their logo on the front cover of this book.

British Library Cataloguing in Publication Data
A catalogue record for this book is available from the British Library

Library of Congress Cataloging-in-Publication Data
A catalog record for this book is available from the Library of Congress

ISBN: 978-0-12-800245-2

For information on all Academic Press publications visit our website at http://store.elsevier.com/

Printed and bound in the United States
14 15 16 17 10 9 8 7 6 5 4 3 2 1

Dedication

This book is dedicated to the late Dr. H. Scott Hurd, associate professor of epidemiology at the Iowa State University College of Veterinary Medicine. Dr. Hurd passed away on March 27, 2014, after a courageous battle with cancer.

Dr. Hurd was an internationally renowned epidemiologist who was highly regarded for his expertise in food safety and best practices for antimicrobial usage in food animals. He provided a strong, well-informed voice on how science should influence national policies on animal protein production. In addition to conducting research in food safety and risk assessment, Dr. Hurd also taught veterinary professional and graduate courses in epidemiology, risk assessment, and risk communication. Dr. Hurd also served as deputy and acting undersecretary for food safety at the USDA in 2008.

Dr. Hurd wrote a blog called "Hurd Health" that brought clarity and scientific reason to the issues of antibiotic use, animal health, and animal welfare. He also wrote a blog for Meatingplace.com titled "The gentle vet," and contributed regularly to lay and peer-reviewed scientific publications, helping dispel myths and misinformation about animal agriculture and the food supply chain. Dr. Hurd clearly had a unique ability to effectively communicate science-based information to the public and particularly to consumers.

Dr. Hurd's career was a remarkable example of the land grant mission in action. He was passionate about the outstanding job U.S. livestock producers do in producing a nutritious and safe product. His research had a major influence on animal health and food safety. Dr. Hurd's body of work over his career had a major impact on several of the policies now in place related to food production in the United States. For these and many other reasons, Dr. Hurd will be greatly missed by those involved in U.S. animal agriculture.

Patrick G. Halbur, DVM, MS, PhD

Professor and Chair, Department of Veterinary Diagnostic and Production Animal Medicine,
Executive Director ISU Veterinary Diagnostic Laboratory,
Iowa State University College of Veterinary Medicine

Contents

SECTION 1

DEVELOPMENTS IN FOOD SAFETY TRACKING AND TRACEABILITY

CHAPTER

GLOBAL FOOD SAFETY INITIATIVE: IMPLEMENTATION AND PERSPECTIVES

1

Philip G.Crandall, Corliss A. O'Bryan
Department of Food Science, Center for Food Safety, University of Arkansas, Fayetteville, Arkansas, USA

1 INTRODUCTION

In the 1970s and 1980s British private label brands, including Sainsbury's, Asda, Tesco, and Waitrose, saw a doubling of their market share (Surak and Gombas, 2009). By 2005 private labels accounted for 40% of retail food sales in much of Europe (Surak and Gombas, 2009). However, with this rapid growth came a major problem; before this time the retailers' food product suppliers had ensured that their products met safety and quality standards. With the growth in private labels, retailers now had to assume some of the burden of protecting their own private brands, which inevitably led to an unprecedented proliferation of customer corporate audits and third-party inspections. As a result, international food suppliers were spending a disproportionate amount of their staff time and food safety capital complying with varying audit requirements. This led to widespread efforts to harmonize the dissonance among these competing standards.

Australia developed their Safe Quality Food (SQF) 2000 code in 1994. The British Retail Consortium (BRC) was founded in 1998. In 2003, German retailers organized the International Food Standard (IFS), which was the same year the French government organized the Federation of Commercial and Distribution Companies (FDC). The Dutch Foundation of Food Safety Systems (SCV) developed their own standards in 2004. The Danes had requested ISO to develop an international food safety management standard, which was published in 2005 as ISO 22000. While these efforts represented progress, multinational suppliers still had to comply with multiple audits to sell food products in different countries to differing retailers often with differing standards (Surak and Gombas, 2009).

In 1953, the Food Business Forum was organized in Europe and a U.S. office opened in 1956 (GFSI, 2012). Membership continued to expand over the years and in 2000, CEOs of several global companies came together at The Consumer Goods forum to discuss how best to spend precious food safety resources (GFSI, 2012). This was done against a backdrop of high-profile food recalls, quarantines, and negative publicity for the food industry. There was also an element of "audit fatigue" for manufacturers who had to endure multiple and sometimes contradictory retailer's or third-party inspections, some having more than six audits per year. There was also a lack of agreement on food safety certifications and acceptance among manufacturers. This all combined to form a need for harmonization of national

Food Safety. http://dx.doi.org/10.1016/B978-0-12-800245-2.00001-0

and international safety regulations. They initially agreed on "benchmark" standards by developing a model that determined equivalency between existing food safety schemes. Today, international food safety experts across the entire food supply chain meet in technical working groups, stakeholder conferences, and regional events to share knowledge and promote a harmonized approach to managing food safety across the industry in a noncompetitive environment (GFSI, 2012).

To be certified, a food manufacturer must be audited by a third-party auditor against one of the schemes recognized by GFSI. These third-party auditors belong to a group that provide certification services known as certification bodies (CBs), which are also key stakeholders within GFSI and the technical working groups. GFSI's mantra from its inception has been, "once certified, accepted everywhere." There are several steps essential for an organization to perform to prepare for a GFSI audit.

2 HOW DOES A FOOD MANUFACTURER BEGIN THE PROCESS OF BECOMING GFSI CERTIFIED?

According to Petie (2012) if a customer or company requires that products be produced in accordance with GFSI, the first step should be to thoroughly research all of the recognized schemes. The currently recognized standards are available from the GFSI's website (GFSI's recognized schemes, 2014). Individual schemes are broken down by the type of food product that a company produces to help identify which scheme would be most appropriate for the organization. The categories are shown in Table 1. Within each category, there is a list of suggested schemes that are appropriate to that category of food. For instance, in the Farming of Plants category, GFSI has approved production schemes represented by Primus GFS, Global GAP, or Canada GAP.

Table 1 Categories of Food Product Used to Determine Scheme to Benchmark Against

Category	
Farming	Animals
	Fish
	Plants
	Grains and pulses
Animal conversion	
Preprocessing of plant materials	
Processing	Animal perishable products
	Plant perishable products
	Animal and plant perishable products (mixed)
	Ambient stable products
Production	Feed
	(Bio)chemicals
	Food packaging
Provision of storage and distribution services	

The next step is to visit the websites for each suggested scheme to get copies of details and supporting documents to understand the requirements of that particular scheme. Each scheme differs in matters of process, focus, and how often audits are required. For example, BRC audits the whole system every year, while SQF has an initial document review and facility audit in the first year and then only a facility audit in the years that follow. A scheme should be chosen that fits each individual business, a scheme that is widely recognized by retail customers and has been used by suppliers making similar food products. Additional details on these initial stages to become certified can be found in Mensah and Denyse's (2011) article on implementing food safety management systems in the UK.

Once a scheme and CB have been selected, a company must decide whether to hire an experienced consultant or "go it alone." A search of the internet for "GFSI consultants" returned more than 30,000 hits. Because GFSI auditors are not allowed to provide corrective consulting for the company they are auditing, the company must be prepared with documentation to support their food quality and food safety program before commencing the audit. Wellik (2012) called for "Say what you do, do what you say, and prove it." Written procedures must be developed that address compliance with each specific requirement of the chosen scheme. For example, detailed written procedures would be developed for monitoring the performance of suppliers and the traceability documentation for every ingredient, final product, and service from receipt of raw materials to shipment of finished goods. There must also be documentation the company is following that written procedure, such as production forms or log books; in some cases, electronic databases are used to store these records. The auditor will need to see these documents that prove the system is operating according to the written procedures.

Either with an experienced auditor or going-it-alone, a company will need to select and train their own project team to organize, schedule, budget, and finally enact the chosen scheme. This team will ensure the written procedures meant to comply with the scheme to be audited against are in place and working properly. The project team should prepare for the first audit (Bele, 2011) by performing a gap assessment, which means to look at all current systems in comparison to the requirements of the chosen scheme (Petie, 2012). Hazard and risk assessments should be made and all food safety and quality measures written complete with corrective actions. It is also a good idea to find a CB that will do a preassessment and/or GAP analysis against the scheme that the company has chosen. This will actually function as a mock audit and let the project team know if there are any gaps in the documentation as well as prepare the management and employees for what to expect in a real audit, and to ensure that the nonconformances are understood. At the time of the GFSI audit, all members of the team as well as senior management should be available to meet the auditor(s) and make sure any necessary changes are made.

3 GFSI TODAY

In 2008 the international retailer, Walmart, required that their food suppliers go beyond current USDA and FDA requirements to begin the process of becoming certified under one of the accepted GFSI benchmarked schemes by July 2009 (Crandall et al., 2012). Since that requirement was put in place, there has been a steady progression of additional retailers requiring their suppliers to become certified to a GFSI standard. For example, the BRC has grown to the point that it now certifies more than 15,000 sites around the globe. Wellik (2012) quoted a GFSI consultant as saying that although there was initial resistance to making the investments in capital and staffing in order to become GFSI certified, many clients had experienced large increases in revenue as a result of their certification. Much more than a "check and forget it" list, GFSI certification, when fully supported by management, would require a change in the operating culture of the plant. More than simplistic concern for "damage control," processors will move to "a genuine attempt to do the right thing" from a systems approach (Wellik, 2012). The GFSI is now in its 12th year of constant revision and improvement to its Guidance Document which is now in the Sixth Edition, Issue 3, Version 6.3, which was released in October 2013.

4 IS GFSI CREATING A SAFER FOOD SUPPLY?

Respondents to a survey about implementation of GFSI certification were of the opinion that the certification resulted in a better documented food safety management system, and they also believed the safety of their products was higher (Crandall et al., 2012). Moyers (2012) investigated 342 food and/or ingredient recalls in the United States between May 1, 2011 and April 30, 2012. The most common reason for recall was mislabeling or nonlabeling of allergens, followed by recalls for *Listeria monocytogenes*, *Salmonella*, and foreign material contamination. Nearly 75% of these recalls were made by manufacturers or suppliers that were not certified to any GFSI scheme. Of course any true comparison would have to be done on a company basis with data pre- and postcertification to GFSI, but it might be expected that there would be a trend of fewer recalls as GFSI certifications increase (Moyers, 2012).

5 DOES GFSI REDUCE THE NUMBER OF AUDITS A FOOD MANUFACTURER MUST UNDERGO?

Crandall et al. (2012) surveyed Walmart suppliers about their experiences with GFSI certification; this nationwide survey found that the suppliers did not believe that they had reduced third-party audits, nor did they believe that less money was spent on third-party audits. However, Wellik (2012) quotes a management consultant with a client who found reduction in audits to be a major cost savings from becoming GFSI certified. This consultant stated that one client was initially required to have 17 different audits a year, but with GFSI in place, they had only two different types of audits per year. On the other hand, an informal online poll conducted by the International Food Safety and Quality Network (2014) revealed that, of those who responded, 65% said they had not had a decrease in food safety audits since the implementation of GFSI. It should be noted that in some cases addenda have been added to the GFSI audits to satisfy specific elements of proprietary audits, which should help reduce the number of audits.

6 WHAT ARE THE THRESHOLDS THAT GFSI MUST OVERCOME TO ACHIEVE EVEN MORE WIDESPREAD ACCEPTANCE?

In 2008, Walmart became the first major retailer to require GFSI certification from their suppliers (Labs, 2012). In their survey of suppliers, Crandall et al. (2012) found that the most cited reason for becoming GFSI certified was to satisfy a customer requirement. In 2013, GFSI commissioned McCallum

Layton to do a study of companies who were GFSI certified and once again found that the major reason for GFSI certification was to maintain a relationship with an existing customer (Yiannis, 2014). It seems that customer demand will be the driving force for GFSI; when major retailers require GFSI certification, food companies get on board (Brusco, 2014).

7 HOW IS GFSI EVOLVING?

Private standards have evolved as new research and regulations become available and will continue to do so. If retailers continue to write/determine their own standards, it may be possible to harmonize all of these competing standards using a risk management approach, which will allow retailers to retain their individuality and uniqueness, but allow their manufacturers to comply with globally harmonized standards. The costs of compliance, which are ultimately borne by consumers, are recognized by the owners of the standards and the CBs and there are coordinated efforts to harmonize audit protocols among the competing IFSs. There is a move within some international CBs to promote ISO 22000 encompassing farming and the production of food that may reduce the power of global retailers. However, it is difficult to see retailers giving up their own standards and the control they currently exert as chain captains. Amidst the struggle for private standard dominance, alternative approaches to risk management (e.g., self assessment of risk, independent audits, and risk ranking) may be the way forward in a similar way to how insurance risks are calculated for business. As such, certification of standards could disappear in favor of a more general risk and insurance model for food and farming (Baines and Soon, 2013).

However, the most important element of the GFSI program is that it is a *certification* as opposed to the previous third-party audits. Historically these audits would assess a plant against a standard, identify noncompliances, and *suggest* improvements but did not have the teeth to enforce any changes. In contrast, a major bulwark of the GFSI certification program is the absolute *requirement* to address any noncompliance issues and submit the corrective actions, plus proof that the corrective actions have been implemented, and these have to be accepted by the auditor and CB before a certificate will be issued.

REFERENCES

Baines, R.N., Soon, J.M., 2013. Private food safety and quality standards: facilitating or frustrating fresh produce growers? Laws 2, 1–19.

Bele, P., 2011. Lessons learned from GFSI audits—be prepared with documents and answers. Food Quality April/May, 22–29.

Brusco, J. Global Food Safety Initiative: effect on food manufacturers and processors. Available at: http://www.tracegains.com/blog/global-food-safety-initiative-affect-on-food-manufacturers-and-processors (accessed 28.10.14).

Crandall, P.G., Van Loo, E.J., O'Bryan, C.A., Mauromoustakos, A., Yiannis, F., Dyenson, N., Berdnik, I., 2012. Companies' opinions and acceptance of Global Food Safety Inititiative benchmarks after implementation. J. Food Prot. 75, 1660–1672.

GFSI, 2012. Welcome to the global food safety initiative. http://www.mygfsi.com/ (accessed 08.09.14).

GFSI's recognized schemes, 2014. http://www.mygfsi.com/schemes-certification/recognised-schemes.html (accessed 24.10.14).

International Food Safety and Quality Network, 2014. Member poll. Available at: http://www.ifsqn.com/ (accessed 24.10.14).

Labs, W., 2012. GFSI update: creating a safe and effective worldwide food system. http://www.foodengineeringmag.com/articles/89167-gfsi-update-creating-a-safe-and-effective-worldwide-food-system.

Mensah, L.D., Denyse, J., 2011. Implementation of food safety management systems in the UK. Food Control 22, 1216–1225.

Moyers, S., 2012. Food companies should focus on GFSI compliance while waiting on FSMA. Food Safety Magazine October/November. Available at: http://www.foodquality.com/details/article/2785591/Food_Companies_Should_Focus_on_GFSI_Compliance_While_Waiting_on_FSMA.html (accessed 24.10.14).

Petie, J., 2012. How to prepare for a GFSI audit. http://www.foodprocessing.com/articles/2012/how-to-prepare-gfsi-audit/ (accessed 08.09.14).

Surak, J.G., Gombas, K.L., 2009. GFSI's role in harmonizing food safety standards. Food Safety Magazine June/July, 36–44.

Wellik, R., 2012. Global food safety initiative improves organizational culture, efficiency in food industry. Food Quality April/May. Available at: http://www.foodquality.com/details/article/1721905/Global_Food_Safety_Initiative_Improves_Organizational_Culture_Efficiency_in_Food.html (accessed 01.10.14).

Yiannis, F., 2014. Global food safety conference 2014. Available at: http://www.mygfsi.com/files/Executive_Summary/GFSC_Report_2014.pdf (accessed 06.05.15).

CHAPTER

COMPUTER SYSTEMS FOR WHOLE-CHAIN TRACEABILITY IN BEEF PRODUCTION SYSTEMS

2

Brian Adam[*], Michael Buser[†], Blayne Mayfield[‡], Johnson Thomas[‡], Corliss A. O'Bryan[§], Philip Crandall[§]

Agricultural Economics Department, Oklahoma State University, Stillwater, Oklahoma, USA[*], Biosystems and Agricultural Engineering Department, Oklahoma State University, Stillwater, Oklahoma, USA[†], Computer Science Department, Oklahoma State University, Stillwater, Oklahoma, USA[‡], Food Science Department, University of Arkansas, Fayetteville, Arkansas, USA[§]

1 INTRODUCTION

Product safety is one of the fundamental factors that are making traceability relevant in the food industry. Recent studies have determined that each year in the United States alone 48 million people get sick, 128,000 are hospitalized, and 3000 die of foodborne diseases (Scallan et al., 2011). Only with an efficient tracking system is it possible to have a successful product recall for safety and for effective research into what caused the problems. The need for effective traceability was highlighted by a national outbreak of *Salmonella* illnesses in the spring of 2008 that sickened more than 1300 people across the United States. Initially, investigators at the FDA and the Centers for Disease Control and Prevention (CDC) identified tomatoes as the culprit, and warned the public against consuming them. However, more than a month later, FDA investigators correctly identified the source of the outbreak as peppers from Mexico (CDC, 2008). The delay was partly because of the chaotic record keeping of the growers, distributors, wholesalers, and retailers; in the meantime, the cost to tomato growers in Florida alone was estimated at about $100 million (Jargon, 2008).

Some authors suggest that traceability systems can play a great role in improving food safety (e.g., Fritz and Schiefer, 2009). Traceability systems can help to identify sources of contaminated food and reasons for contamination. This information can assist in minimizing the effect of food contamination and improving quality assurance programs. Resende-Filho and Buhr (2010) show that whole-chain traceability can substantially limit the economic loss of food safety events by enabling firms and regulators to identify sources of any problem while ruling out nonsources.

Moreover, IFT (2010) notes that an effective product tracing system can maintain consumer confidence by providing more rapid resolution to food contamination events. In addition, developing and implementing effective whole-chain traceability systems (WCTS) can help optimize resources and reduce waste. Consumers are also interested in traceability from the standpoint of increased food safety.

Food Safety. http://dx.doi.org/10.1016/B978-0-12-800245-2.00002-2

Several studies have shown that consumers are willing to pay a small premium for traceability, but they are often willing to pay more for traceability when it is attached to another attribute such as enhanced food safety (Dickinson and Bailey, 2002; Hobbs et al., 2005).

However, the motivation for individual segments of the food industry to voluntarily adopt traceability to improve food safety seem to be less clear. Traceability does not directly increase food safety in the manner that production systems such as pathogen reduction and hazard analysis and critical control point do. Instead, it collects information about product characteristics and production practices as the product moves through the supply chain, which itself does not reduce the probability of a food safety crisis. However, the information accumulated in traceability systems could facilitate contractual arrangements between firms in the supply chain to promote food safety. While Pouliot and Sumner (2008) concluded that traceability always increases food safety, Resende-Filho and Hurley (2012) demonstrated that neither voluntary traceability nor mandatory traceability is guaranteed to improve food safety. Whether or not traceability increases food safety, the fact remains that traceability systems of the type used for other food commodities are still far from being fully implemented for livestock and meat products.

2 BENEFITS AND COSTS OF TRACEABILITY IN THE BEEF INDUSTRY

The beef industry is an example of an individual industry that appears to have less motivation to adopt traceability. As such, its unique characteristics merit further investigation. The beef industry is characterized by relatively long production cycles and fragmented supply chains with many relatively small, independent producers. Transactions among these producers involve extensive interaction among individual animals over large geographical areas and high risk of epidemic disease outbreak. Traceability systems in the beef industry could potentially aid animal disease control and reduce international trade restrictions related to animal disease, in addition to improving food safety and increasing consumer confidence (Schnepf, 2009).

According to Schnepf (2009), traceability systems minimize animal health surveillance cost. Furthermore, traceability systems reduce the cost of controlling animal disease outbreaks by quickly identifying the affected animals and minimizing the number of animals to be treated or destroyed. Although industry leaders and government officials had been attempting to develop an animal identification system, discovery in late 2003 of a cow infected with a pathogenic prion that caused bovine spongiform encephalopathy (BSE), also known as mad cow disease, increased interest in these efforts, and a voluntary national animal identification system (NAIS) was created in 2004 (Schroeder and Tonsor, 2012; Crandall et al., 2013). Schroeder and Tonsor (2012) noted that the discovery of the cow had shut down nearly every export market for at least part of 2004, costing the U.S. beef industry $3.2–$4.7 billion in that year alone, according to estimates by Coffey et al. (2005).

However, the voluntary beef traceability system faced much resistance from producers due to cost, lack of confidentiality, and lack of accuracy of the system. The resistance led to low producer participation, and abandonment of the system in 2010 (Schroeder and Tonsor, 2012). Low participation in traceability systems in the U.S. beef industry may cause serious adverse effects in controlling a major animal disease outbreak. In addition, the United States may lose substantial export markets in the time it takes to trace the source and control disease outbreak. According to Schroeder and Tonsor (2012), after the export restrictions arising from the 2003 BSE case it took at least 7 years for the export market to reach approximately the same level it was prior to the BSE case. Some countries may also refuse to accept any imports from the United States if they are not traceable. As a result, U.S. producers may lose competitive advantage in the international beef markets.

Several studies have estimated the economic benefits of a traceability system, particularly through enhanced export activity, and consumers' willingness to pay a premium for traceable beef (e.g., Pendell et al., 2010; Resende-Filho and Buhr, 2006; Lee et al., 2011; Loureiro and Umberger, 2007; Dickinson and Bailey, 2005). For example, the study by Pendell et al. (2010) estimated that the United States could lose a total consumer and producer surplus of $6.65 billion if at least 25% of its untraceable beef product is unacceptable in international trade. Furthermore, they showed that a 1% increase in domestic demand or 34.1% increase in export demand would fully cover the cost and surplus loss of adopting a traceability system with a 90% participation rate.

While most of these potential benefits of traceability would be realized by consumers or the beef industry as a whole, unfortunately, individual producers may not fully realize them. Individual producers participating in a traceability system, particularly small ones, are likely to face increased costs and complicated production activities. Although the benefits of a traceability system can be partially transferred to producers, the level of benefit might not be enough to convince individual beef producers to participate in a traceability system. Although some producers could potentially differentiate their product and receive higher prices to cover the additional costs, as more producers adopted traceability technology, their ability to differentiate their product because of traceability would diminish. Thus, although the industry as a whole might benefit from adopting traceability because of increased reputation and consumer safety, individual firms believed that they were not likely to be able to recover their increased costs through higher prices. Thus, many producers believed that the additional cost of a traceability system may not be covered by the additional revenue that might be obtained from increased willingness to pay for traceable beef.

Moreover, the producers believed that additional cost would greatly affect small-scale producers and cow/calf producers (Schnepf, 2009). Confirming this, Seyoum et al. (2013) summarized research by Blasi et al. (2009) and Butler et al. (2008) that had estimated significant economies of scale in implementing traceability, so that small producers would experience much higher costs per animal than larger producers. The traceability cost per animal is greatly affected by the size of the operation due to a high initial investment cost. A large portion of the traceability system cost is the cost of tagging, which would be undertaken primarily in the cow/calf stage, a stage made up of many small producers who would have difficulty absorbing the cost. Thus, while most of the costs of implementing traceability would be borne by cow/calf producers, the first link in the supply chain and the smallest producers, most of the benefits would accrue to the larger producers and processors further down the supply chain.

In addition to perceived cost inequity, producers raised other concerns as well. While Crandall et al. (2013) suggested that the ability to track individual animal health issues may provide a compelling reason for a beef producer to adopt traceability of individual animals, they noted that livestock producers are concerned about their increased liability, increased trespass on their properties by regulators, and the added expense associated with record keeping. One part of the concern about increased liability is that if an animal disease or food safety event is traced back to a particular producer, that producer would face significant or complete loss, whereas if it could only be traced back to a region or set of producers, at least the misery would be shared misery—all producers in the region would bear the costs. (Of course, a corresponding benefit of participating in a WCTS is that a producer who is not part of an animal disease or food safety event can be more quickly ruled out.) Currently, risk of food safety events is largely borne by firms closer in the supply chain to consumers, such as processors; with traceability, that risk could be shifted back up the supply chain to smaller producers, who typically have less financial resources to absorb the risk.

Another concern was the potential lack of confidentiality of collected business data (Schnepf, 2009). Producers feared that the information collected by the government might not be fully secure. The data stored could be accessed by others and the confidentiality of producers' sensitive information could be compromised.

The accuracy, speed, and technical complexity of the traceability system are also concerns of producers (Bolte et al., 2007). Producers want the traceability system to provide accurate information without affecting the speed of cattle trade. In addition, producers are concerned that the traceability equipment such as the RFID reader, the database, and computer hardware could add complexity to producer activity (Schnepf, 2009).

Surveys of beef producers found that resistance was due to the cost, confidentiality, and [lack of] accuracy of the system (Schroeder and Tonsor, 2012). Beef producers were very reluctant to allow anyone, particularly government agencies and competitors, to have access to information about their animals, and saw very little potential benefit to them as individual producers to make it worth the cost.

3 ADVANCEMENT OF A WCTS

A WCTS in the beef industry has the potential to improve food safety, improve animal disease traceability, improve supply chain management, and enhance value-added opportunities and communication between producers and consumers (both ways, which could increase consumer confidence in food as well as provide valuable feedback to producers to improve their product; Resende-Filho and Buhr (2008) show how a traceability system might accomplish this in a specific application.). However, little of this potential benefit can be realized if a significant number of producers do not participate.

In order to reduce obstacles to beef producers' participation in a traceability system due to concerns about cost, confidentiality, and liability, the USDA funded a multi-institution (Oklahoma State University, University of Arkansas, and the Noble Foundation), multidisciplinary research project to develop a pilot WCTS that would address these concerns. As part of this project, a National Whole Chain Traceability Institute (NWCTI) has been formed. The WCTS technology developed as part of this project forms the backbone of the NWCTI, and is able to resolve the confidentiality concern, in addition to providing a framework to increase value-added opportunities to producers and address liability issues (OSU, 2013).

The WCTS is designed with the goal of an Internet-based, stakeholder-driven traceability and marketing system for agricultural and food products that limits and remedies food safety and animal disease outbreaks and biosecurity breaches, while adding value to the process. The system provides for data input by producers, processors, vendors, and consumers. This data not only provides information to facilitate mitigation, but also provides marketing information, value-added details, cultural and sociological features about the production or handling of specialty crops, quality standards criteria, and a feedback opportunity for consumers to rate or improve product quality.

The key advantage of this system, compared to previously attempted systems, is that those putting information into the system maintain granular privacy control over access to data. In other words, those putting data into the system decide both who can see that information and what pieces (granules) of information they can see. This is critical because the ability to trace food through a supply chain depends on private firms sharing product information with competitors as well as collaborators. As noted above, if they are not assured of privacy control over information, they may refuse to participate in the system, which would severely limit the food safety benefits of the system.

This is in contrast to non-WCTS, such as those using the "one-up, one-back" approach mandated in the Bioterrorism Act of 2002 (U.S. FDA, 2002). Those systems fail to protect privacy and fail to make use of the information in the system to improve supply chain efficiency. Moreover, a firm-by-firm trace-back in the event of a food safety or bioterrorism event is inherently slow with such a system, greatly reducing the effectiveness of intervention. This critical lack of timely information can cause significant economic losses to multiple industries resulting from public uncertainty on the potential for human hazard, affecting industries ultimately found to not be related to outbreaks. Conversely, whole-chain traceability can substantially limit the economic loss of food products that are not directly associated with the specific problem. Moreover, data in a whole-chain system can be used to improve supply chain efficiency and enhance value-added activities.

Figure 1 illustrates the principle of the WCTS in the case of a beef supply chain, the first product for which the traceability system is being developed. Darkened text indicates that the data is visible to

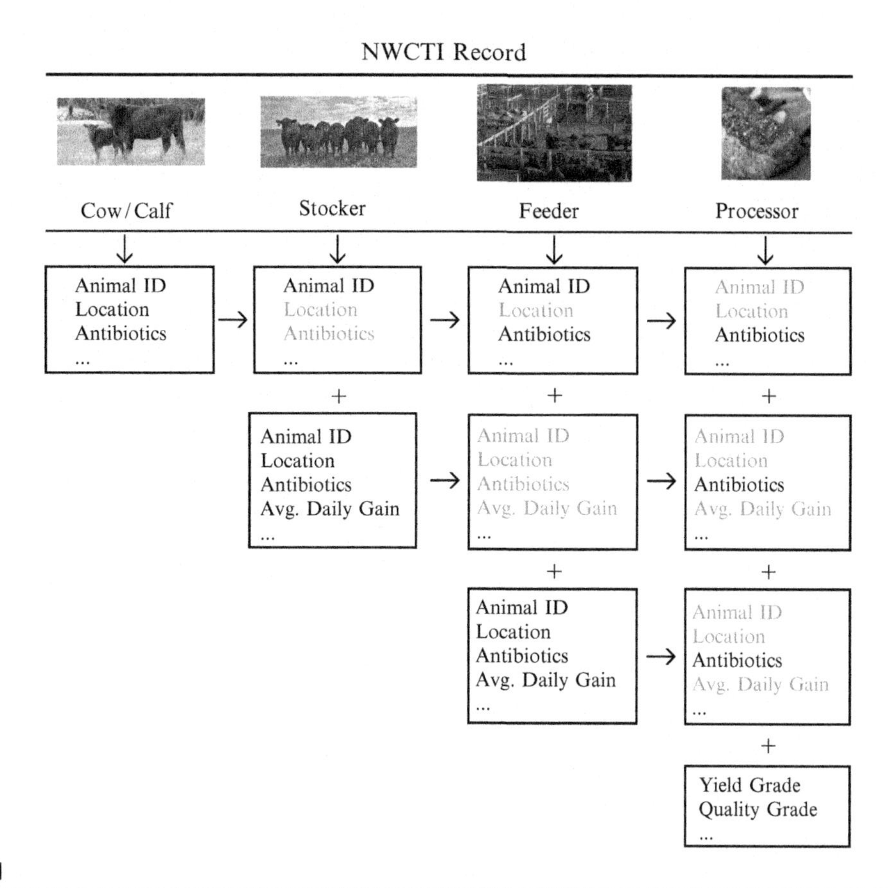

FIGURE 1

Illustration of granular data: information source chooses who sees what information.

Adapted from Buser et al. (2012).

the producer identified at the top of that column in the figure; grayed-out text indicates that the data are in the system but not visible to the producer identified at the top of that column; the boxes on the diagonal indicate information put by the producer identified at the top. In this example, a cow/calf producer might input animal ID, location of the calf, and a record of antibiotics given. The stocker who purchases the calf for further feeding sees only the animal ID that the cow/calf producer has entered, but also adds her own information on location, antibiotics administered, and average daily gain of the calf. The feedlot operator is permitted to see only the animal ID and the record of antibiotics administered by the cow/calf producer, and none of the stocker's information, and puts in his own animal ID, location, antibiotics administered, and average daily gain in the feedlot. The processor who purchases the animal sees only the information on antibiotics, from all three of the previous parties.

Then, after the animal is slaughtered, the processor inputs information on yield grade and quality grade. This information could be shared back up the chain with the cow/calf producer, for example, who might use the information to improve the genetic potential of his herd. This two-way information flow has the potential to greatly improve supply chain efficiency. There may be many kinds of information that would be valuable to someone in the supply chain, and various arrangements could be made to share that information. In the event of a food safety event or disease outbreak, the information needed to track and trace it could be made available to the appropriate agency.

The WCTS is essentially a data management system; as such, it is designed to interface with a range of possible private traceability systems or other data input mechanisms, including independent management software. For example, for the beef supply chain, cow/calf producers could input data from a herd/animal management application software such as Cow Sense™ or CattleMax™.

The information flow is as follows: Initially a producer enters her cow/calf data into CattleMax™ software. Then the producer uploads cattle data and cattle treatment information to the NWCTI server using WCTS iPad™ application. This allows the producer to share her cattle information selectively with other producers. If the producer decides to sell her cattle to a feeder who also participates in WCTS, the producer selects from one of the feeders registered at NWCTI. The cow/calf producer's information about that animal is then sent to the feeder. The feeder would be able to view the animal ID as well as the information the producer expressly chooses to share with the feeder, but none of the other information entered into the NWCTI servers by the cow/calf producer. All information is stored using a secure, open-source database system, further reducing the cost of operation.

The same information transfer principles apply for transactions farther down the supply chain—only information expressly permitted by the firm entering that information is shared with other supply chain participants. Finally, when the processor processes a cattle batch, a unique ID is created for each batch. The processor also enters relevant information, such as yield grade of each batch, into the NWCTI servers. A QR code is created based on the unique ID of each batch, and the QR code is e-mailed to the processor so that it can be printed on beef products.

At the other end of the supply chain, the NWCTI servers can be integrated with traceability systems that interface with consumers. In its current configuration, the NWCTI servers interface with servers operated by *Top 10 Produce, LLC (Top10Produce.org)*, a partner in this project. *Top 10* is a 100% traceable produce brand created to encourage smaller growers to work together to market item-level traceable produce directly to consumers. In *Top10's* relatively short supply chains, consumers can verify the actual location of the grower, as well as other information about the farm, simply by pointing a smartphone at the QR code on a package originating from the farm. Thus, by integrating *Top10*'s system and NWCTI's

WCTS, consumers can instantly get all the information made available to them by all producers in the beef supply chain by pointing their smartphones at the QR codes on the retail beef products.

All of the data in the system will be managed by a trusted third party, further addressing privacy issues, especially those issues concerning state or federal government access to the data. To assure accuracy of the data, it is made immutable after it is approved by the trusted third party. Once the data is made immutable, it cannot be changed. Immutable data can only be appended, much as real estate transactions are appended to the property records rather than changing those records. A change log keeps track of all the changes made to the immutable data and the dates of those changes, and these change logs are archived periodically.

4 SUMMARY

While WCTS have much potential to increase food safety and, in the case of livestock, enhance animal disease control, those benefits are greatly reduced with low producer participation rates. The WCTS technology described here has the potential to resolve some of the major obstacles to participation, especially costs of implementation and operation, loss of confidentiality in business transactions, and increased liability.

REFERENCES

Blasi, D., Brester, G., Crosby, C., Dhuyvetter, K., Freeborn, J., Pendell, D., et al., 2009. Benefit-cost analysis of the national animal identification system. Final report submitted to USDA-APHIS on January 14.

Bolte, K., Dhuyvetter, K., Schroeder, T., Richard, B., 2007. Adopting animal identification systems and services in Kansas auction markets: costs, opportunities, and recommendations. Kansas State University Agricultural Experiment Station and Cooperative Extension Service, MF-2780, April.

Buser, M.D., Adam, B.D., Bowser, T.J., Mayfield, B.E., Thomas, J.P., Crandall, P.G., Ricke, S.C., 2012. Concept of a stakeholder-driven whole-chain traceability system for beef cattle. In: Presented at the Arkansas Association for Food Protection Conference, Fayetteville, AR, September 11–12.

Butler, L.J., Farshid, H., Oltjen, J.W., Caja, G., Evans, J., Velez, V., Bennett, L., Li, C., 2008. Benefits and Costs of Implementing an Animal Identification and Traceability System in California – Beef, Dairy, Sheep and Goats. California Department of Food and Agriculture, Sacramento, CA.

CDC, 2008. Outbreak of Salmonella serotype Saintpaul infections associated with multiple raw produce items – United States. MMWR 57, 929–934.

Coffey, B., Mintert, J., Fox, S., Schroeder, T., Valentin, L., 2005. The economic impact of BSE on the US beef industry: product value losses, regulatory costs, and consumer reactions. Kansas State University Agricultural Experimental Station and Cooperative Extension, Serial No. MF-2678, April.

Crandall, P.G., O'Bryan, C.A., Babu, D., Jarvis, N., Davis, M.L., Buser, M., Adam, B., Marcy, J., Ricke, S.C., 2013. Whole-chain traceability, is it possible to trace your hamburger to a particular steer, a U. S. perspective? Meat Sci. 95 (2), 137–144.

Dickinson, D.L., Bailey, D., 2002. Meat traceability: are US consumers willing to pay for it? J. Agric. Resour. Econ. 27, 348–364.

Dickinson, D.L., Bailey, D., 2005. Experimental evidence on willingness to pay for red meat traceability in the United States, Canada, the United Kingdom, and Japan. J. Agr. Appl. Econ. 37, 537–548.

Fritz, M., Schiefer, G., 2009. Tracking, tracing, and business process interests in food commodities: a multi-level decision complexity. Int. J. Prod. Econ. 117, 317–329.

Hobbs, J.E., Bailey, D., Dickinson, D.L., Haghiri, M., 2005. Traceability in the Canadian red meat market sector: do consumers care? Can. J. Agric. Econ. 53, 47–65.

IFT (Institution of Food Technologists), 2010. Traceability (product tracing) in food systems: an IFT report submitted to the FDA, volume 1: technical aspects and recommendations. Compr. Rev. Food Sci. Food Saf. 1, 92–158.

Jargon, J., 2008. Grocers and restaurants toss out tomatoes – *Salmonella* scare delivers a blow to crop's growers. Wall Street J. (June 10).

Lee, J.Y., Han, D.B., Nayga Jr., R.M., Lim, S.S., 2011. Valuing traceability of imported beef in Korea: an experimental auction approach. Aust. J. Agric. Resour. Econ. 55 (3), 360–373.

Loureiro, M.L., Umberger, W.J., 2007. A choice experiment model for beef: what US consumer responses tell us about relative preferences for food safety, country-of-origin labeling and traceability. Food Policy 32 (4), 496–514.

Oklahoma State University (OSU), 2013. Agricultural Economics Department Fall 2013 Research Update. Accessible at: http://agecon.okstate.edu/files/Ag%20Econ%20Research%202013.pdf, pp. 6–7, 17.

Pendell, D.L., Brester, G.W., Schroeder, T.C., Dhuyvetter, K.C., Tonsor, G.T., 2010. Animal identification and tracing in the United States. Am. J. Agric. Econ. 92, 927–940.

Pouliot, S., Sumner, D.A., 2008. Traceability, liability, and incentives for food safety and quality. Am. J. Agric. Econ. 90, 15–27.

Resende-Filho, M.A., Buhr, B.L., 2006. Economic Evidence of Willingness to Pay for the National Animal Identification System in the US. Presented at International Association of Agricultural Economists Conference, Gold Coast, Australia. pp. 12–18.

Resende-Filho, M.A., Buhr, B.L., 2008. A principal-agent model for evaluating the economic value of a traceability system: a case study with injection-site lesion control in fed cattle. Am. J. Agric. Econ. 90, 1091–1102.

Resende-Filho, M., Buhr, B., 2010. Economics of traceability for mitigation of food recall costs. Munich Personal RePEc Archive (MPRA) Paper No. 27677. Accessible at http://mpra.ub.uni-muenchen.de/27677/, posted December 27.

Resende-Filho, M.A., Hurley, T.M., 2012. Information asymmetry and traceability incentives for food safety. Int. J. Prod. Econ. 139, 596–603.

Scallan, E., Hoekstra, R.M., Angulo, F.J., Tauxe, R.V., Widdowson, M.A., Roy, S.L., 2011. Foodborne illness acquired in the United States—major pathogens. Emerg. Infect. Dis. 17, 7–15.

Schnepf, R., 2009. Animal identification: overview and issues. Congr. Res. Serv. 7-5700.

Schroeder, T., Tonsor, G., 2012. International cattle ID and traceability: competitive implications for the US. Food Policy 37, 31–40.

Seyoum, B., Adam, B.D., Ge, C., Devuyst, E.A., 2013. The value of genetic information in a whole-chain traceability system for beef. In: Selected Paper at the AAEA Annual Meeting in Washington, DC, August 4–6. Available at, http://purl.umn.edu/150458.

U.S. FDA (Food and Drug Administration), 2002. The Bioterrorism Act of 2002.

CHAPTER

TRACKING PATHOGENS IN THE ENVIRONMENT: APPLICATIONS TO FRESH PRODUCE PRODUCTION

3

Kristen E. Gibson

Department of Food Science, Center for Food Safety, University of Arkansas, Fayetteville, Arkansas, USA

1 INTRODUCTION

Exposure to foodborne pathogens results in an estimated 47.8 million illnesses each year in the United States (Scallan et al., 2011a,b). The etiological agents that cause the most foodborne illnesses include human norovirus, nontyphoidal *Salmonella*, *Clostridium perfringens*, and *Camyplobacter* spp. (Scallan et al., 2011b). Additional foodborne pathogens of concern include *Listeria monocytogenes*, Shiga toxin-producing *Escherichia coli*, *Vibrio* spp., Hepatitis A virus, *Toxoplasma gondii*, and *Cyclospora cayetanensis*. The economic cost of foodborne illnesses can be quite significant as well. Scharff (2012) estimated the annual economic cost attributed to foodborne infections to be between US$51 and US$77.7 billion which includes the loss of wages, medical care costs, and loss of quality of life associated with disease.

The primary causative agents of foodborne illnesses may originate from a variety of reservoirs (i.e., environmental, animals, vectors, and humans) and enter the food supply at various nodes along the farm to fork pathway. As indicated by Fu and Li (2014), in order to better understand the human health risk associated with contamination of food as well as to develop strategies to eliminate the contamination, it is essential to determine the source as well as the original reservoir when possible. One approach would be the adoption and application of fecal source tracking (FST) methods that are currently used to determine the origin of nonpoint source fecal pollution in environmental water sources. Briefly, FST is an area of research revolving around the need to investigate the sources of elevated levels of fecal indicator bacteria (FIB) such as generic *E. coli* and enterococci—both used as indicators of the microbial quality of water and in some instances, food (Ailes et al., 2008; Johnston et al., 2006; USEPA, 2002). With respect to water quality, the primary goal of FST has often been to identify the source of contamination and, if possible, to eliminate the source. However, also intrinsic to the goal of FST is to allow for estimation of human health risk due to exposure (ingestion and recreational) to contaminated water sources. In order to do this, a more specific research area known as microbial source tracking (MST) has evolved. Within MST a variety of phenotypic and genotypic methods using fecal indicator organisms and pathogens have been developed to try and determine the dominant sources of fecal contamination (Fu and Li, 2014; Stoeckel and Harwood, 2007). Even though most MST methods have been developed for and applied to contaminated waterways, these methods may also provide the food safety community with new tools that can be broadly applied to tracking foodborne pathogens from farm to fork.

When considering the farm to fork pathway that all food products travel, there are several entry points where fecal microbes and pathogens may be introduced. For raw food products or ingredients,

Food Safety. http://dx.doi.org/10.1016/B978-0-12-800245-2.00003-4

sources of microbes would be soil—including soil amendments and fertilizers—water, and the intestinal tracts of animals (domestic, wild, or food animals) and humans. Fecal organisms may also enter the food supply through the food processing environment including surfaces, food workers/handlers, and air and dust (Fu and Li, 2014). As indicated previously, when MST is included as a part of water resource management, the aim is to determine the contribution of various nonpoint, or diffuse, sources of pollution (e.g., agricultural runoff, food animal manure, and wildlife excreta) and to understand the associated human health risks (Field and Samadpour, 2007). These same issues related to MST in water resource management are also important to investigate within the food supply chain, including the production environment (i.e., farm), the processing environment (i.e., packing plant, cannery, and slaughterhouse), and the food preparation environment. Understanding the sources of microbial contamination at each of these steps and how these microbes gain access will be critical for more precise quantification of food safety hazards and risks (Fu and Li, 2014). A diagram outlining the various sources of fecal contamination in a "generic" food chain was originally depicted by Fu and Li (2014) and has been adapted and modified here as Figure 1. Based on this, the primary purpose of this chapter is to consider how current MST methods used in water quality research may be transitioned for the identification of microbial

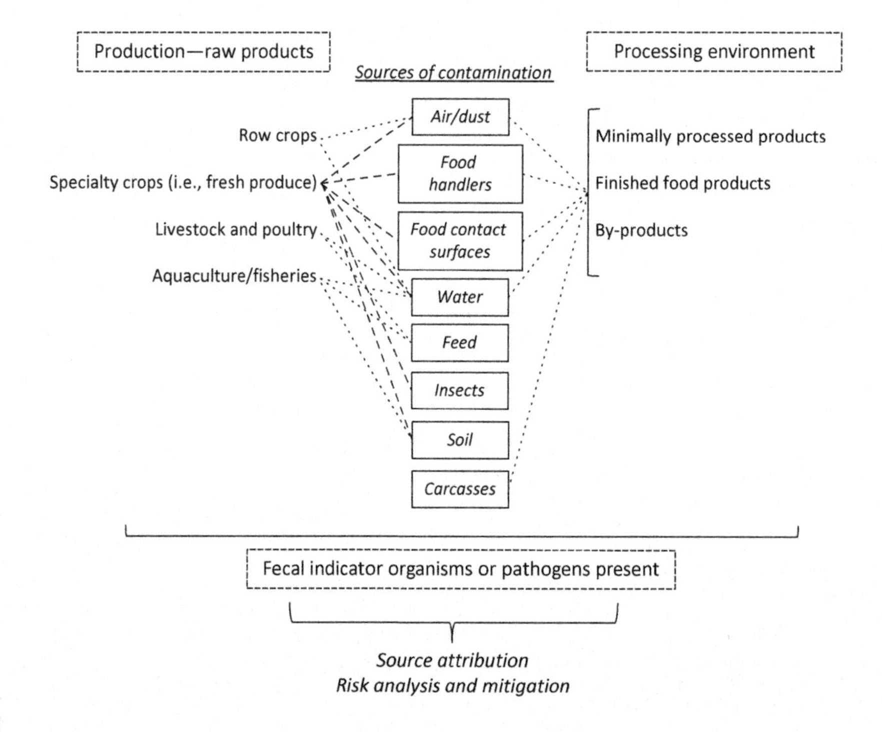

FIGURE 1

Sources of fecal contamination in various production and processing environments that may be addressed by microbial source tracking. Microbial contamination enters the food supply primarily during production and processing: (1) Production environments are the first level of potential microbial contamination and include raw crops, specialty crops, livestock and poultry, or fisheries. Contamination during this preharvest stage is of particular interest especially for foods that are consumed raw or undercooked. (2) Processing environments constitute the second level associated with microbial contamination in the food supply. Contamination during processing can come from a variety of sources such as air/dust, food handlers, food contact surfaces, water, and carcasses contaminated with feces or intestinal contents.

contamination in the food chain and thus increase the safety of the food supply and protect public health. More specifically, the chapter will focus on the fresh produce production environment as a model for adopting and applying MST methods in the food industry as fresh produce is susceptible to a variety of potential contaminants throughout production and postharvest (Berger et al., 2010).

2 MST METHODS OVERVIEW

Traditionally, MST methods are classified into two groups: library dependent (LD) or library independent (LI). LD methods can be either phenotypic or genotypic based. However, over the years, development of MST methods have moved more toward the molecular-based LI methods as LD methods are expensive and time-consuming due to the numerous isolates that are needed to construct robust libraries, and LD methods are often temporally and geographically specific thus limiting their routine use (Stoeckel and Harwood, 2007). Moreover, LD methods are more often specific to bacterial pathogens as opposed to all potential microbial pathogens (i.e., viruses and protozoa). With respect to LI methods, these methods are more efficient and cost-effective when compared to LD methods (Fu and Li, 2014). In this chapter, only molecular-based methods for application to tracing foodborne pathogens and indicators in the fresh produce production environment will be considered.

Techniques developed for and used in MST methods include restriction-based methods, amplification-based methods, and sequencing-based methods. As there are several reviews available that detail the technical nuances of each of the techniques within each of these method categories (Foley et al., 2009; Fu and Li, 2014; Graves, 2011), these methods will not be discussed further except as they apply to specific case studies in the environment or fresh produce industry. Table 1 provides examples of how each of the major foodborne pathogens related to contamination of fresh produce has been studied in various environments—primarily agricultural and urban watersheds—through the application of molecular-based subtyping and MST approaches. Fecal indicators and host-associated fecal markers are also included in Table 1.

Table 1 Examples of MST Application in Agricultural Watersheds and Fresh Produce Production Environments

Target Organism	MST Approach	Location	Application	Primary Findings	References
Pathogens					
Campylobacter spp.	RFLP	Alberta, Canada	To investigate potential sources of contamination in a watershed using bacterial subtyping methods	*Campylobacter* most frequently isolated from fecal, sewage, and water samples with *C. coli* and other species isolated most often from pig and deer	Jokinen et al. (2011)
	PCR	Canada	Occurrence and diversity of thermophilic *Campylobacter* in 23 agricultural watersheds	*Campylobacter* generally more common in watersheds with intensive livestock production	Khan et al. (2014)

Continued

Table 1 Examples of MST Application in Agricultural Watersheds and Fresh Produce Production Environments—cont'd

Target Organism	MST Approach	Location	Application	Primary Findings	References
ST-producing *E. coli*	MLVA, PFGE	Salinas/San Juan Valleys, California	Movement of *E. coli* O157 within the watershed	Recurrence of identical or closely related *E. coli* O157 strains indicate restricted movement in the environment	Cooley et al. (2007)
	PFGE, CGF	Alberta, Canada	To investigate potential sources of contamination in a watershed using bacterial subtyping methods	*E. coli* O157:H7 predominantly isolated from cattle and isolates from cattle, sewage, and water were identical by PFGE patterns	Jokinen et al. (2011)
Listeria spp.	MVLST	Pennsylvania	Determine prevalence of *L. monocytogenes* in mushroom production environment	Analysis of *L. monocytogenes* isolates indicates common origin, possibly horse manure or poultry litter; other virulent strains may be able to colonize production environment	Viswanath et al. (2013)
	sigB PCR, allelic type	New York	Determine diversity of *Listeria* spp. in produce production and natural environments	Temperature and proximity to pastures were strongly associated with isolation of *L. monocytogenes*	Chapin et al. (2014)
Salmonella enterica	PFGE	North Carolina	Determine *Salmonella* source contribution in four different watersheds	Swine production-associated isolates distinctly different	Patchanee et al. (2010)
	Microarray, PFGE	Monterey County, California	Prevalence of *S. enterica* serotypes in the fresh produce production environment	Specific strains persistent in the environment (i.e., water, wildlife, and soil)	Gorski et al. (2011)
	PFGE, serotype PCR	New York, Florida	Incidence of *Salmonella* as well as strain diversity within two regions of the United States	Possibility for PFGE-based source tracking of *Salmonella* in production environments due to region specificity	Strawn et al. (2014)
Host-associated markers					
Bacteriodales	Real-time qPCR	Ontario, Canada	Application of human and bovine *Bacteroidales* markers for MST in water	*Bacteroidales* host markers combined with land use information insists in understanding distribution of source-specific fecal pollution	Lee et al. (2014)

Table 1 Examples of MST Application in Agricultural Watersheds and Fresh Produce Production Environments—cont'd

Target Organism	MST Approach	Location	Application	Primary Findings	References
	Real-time qPCR	Ontario, Canada	Impact of controlled and uncontrolled tile drainage on occurrence of pathogens and host fecal markers in agricultural watersheds	Odds of *Salmonella* spp. presence increased when ruminant marker present and odds of *Arcobacter* spp. presence increased when wildlife marker present	Wilkes et al. (2014)
Host-specific viruses	Real-time RT-qPCR and qPCR	New Zealand	Application of a viral toolbox to distinguish between human and animal fecal pollution	Levels of virus shedding and subsequent detection in environmental water samples were highly variable depending on the virus-host target	Wolf et al. (2010)
	Real-time RT-qPCR and qPCR	Greece, Serbia, and Poland	Presence of host-specific enteric viruses in the leafy green supply chain	Both human and animal enteric viruses were detected in the supply chain with the most positives and diversity in irrigation water	Kokkinos et al. (2012)
Mitochondrial DNA	mtDNA	L'Assomption watershed, Canada	Determine co-linearity of traditional human marker with host-specific mtDNA	Human mtDNA positively correlated with traditional human marker while other sources of mtDNA were not	Villemur et al. (2015)
Fecal indicator bacteria					
Generic *E. coli*	PFGE	Rio Grande River, Texas-Mexico border	Determine genetic relatedness and stability of *E. coli* isolated from irrigation water and sediments	Considerable genetic diversity among *E. coli* isolates may not allow PFGE-based MST to be applied in large watersheds	Lu et al. (2004)
	PFGE, ERIC-PCR	Texas	Evaluation of genotype variation of *E. coli* isolates from surface waters to better understand naturally occurring populations	Discriminatory power of the MST method should be carefully considered for assessing strain diversity and determining the geographic range of identification libraries	Casarez et al. (2007)
Enterococcus spp.	MLST	Spain	Determine how related fresh produce isolates of *E. faecium* are to clinical isolates of *E. faecium*	*E. faecium* isolates from fresh produce are not related to clinically relevant *E. faecium*	Burgos et al. (2013)
	AFLP	Mid-Atlantic region of the United States	Understand diversity of antibiotic resistant *Enterococcus* in tomato production farms	*E. casseliflavus* was most prevalent and is highly associated with plants and aquatic environments; thus, enterococci may not be a suitable fecal indicator	Micallef et al. (2013)

As indicated in Table 1, the primary MST approach that has been applied to studying foodborne pathogen diversity in the environment is pulsed-field gel electrophoresis (PFGE), which is a restriction-based method. Historically, PFGE has been considered the "gold standard" for bacterial subtyping of foodborne pathogens including *Salmonella*, *E. coli*, *Campylobacter*, *Listeria*, *Yersinia*, and *Vibrio* (Foley et al., 2009). The most common application of PFGE is during the identification of multistate foodborne disease outbreaks through the PulseNet program (Swaminathan et al., 2001) followed by utilization in molecular epidemiology studies. However, the discriminatory power (i.e., the average probability that a typing system will assign a different type to two unrelated strains randomly sampled in the microbial population of a given taxon) of PFGE in some instances is not as high for certain foodborne pathogens (e.g., *E. coli* O157:H7 and *Campylobacter*) when compared to other methods such as multilocus sequence typing (MLST) and multilocus variable number tandem repeat analysis (MLVA) (Foley et al., 2009). Even so, application of PFGE for source tracking has enabled a better understanding of the spatial distribution and source contributions of pathogens within agricultural watersheds (Table 1).

For MST using indicators of fecal pollution—as opposed to direct detection of pathogens—a greater variety of methods have been developed. For instance, real-time, quantitative polymerase chain reaction (qPCR) and reverse transcription qPCR (RT-qPCR) have been used for the detection and quantification of host-associated markers including *Bacteroidales* and host-specific viruses while both PFGE and MLST along with other amplification-based methods have been used to characterize generic *E. coli* and *Enterococcus* spp. (Table 1). While the pathogen subtyping methods mainly attempt to elucidate occurrence, prevalence, and diversity with the potential for source attribution, the application of MST to host-associated markers and fecal indicators attempts to correlate these indicators with the presence of frank human pathogens, although often with little success (Harwood et al., 2014).

As part of the overview of MST methods, it is important to identify some of the shortcomings of MST and the key aspects that need to be considered during selection and implementation of one or more of these methods. Currently, the shortcomings of MST—and FST in general—include the increasing variety of methods and target specific assays being utilized of which none have been demonstrated to work quantitatively over time; the continued lack of evidence correlating MST host-associated targets with pathogens; and the lack of any systematic evaluation to assess the value of MST results as they relate to risk characterization. While these shortcomings could be perceived as quite significant, the overall concept of FST and its primary goal are still important to strive for, and thus, there are key aspects to consider during selection and implementation of MST methods in situ. In this regard, the following questions are important to think about:

1. *Under what circumstances have the MST methods correctly identified sources of pollution?*
2. *Are there particular situations when the MST method did not elucidate the source of fecal pollution?*
3. *Have the results of the MST method impacted and/or assisted in determining public health risk?*

First, the decision to apply a specific MST method should be informed primarily by what has been proven to correctly identify sources of pollution, ideally in situ. A study by Boehm et al. (2013) evaluated the sensitivity and specificity of 41 MST methodologies performed by 27 different laboratories. Here, the researchers not only aimed to determine specificity of the assays but also to evaluate the intralaboratory variability. The majority of the assays were PCR- or qPCR-based for detection of molecular markers while there were only three community-based methodologies selected for evaluation including ribosomal RNA-targeted oligonucleotide microarrays (e.g., PhyloChips), universal 16S terminal restriction fragment length polymorphism (TRFLP), and *Bacteroidales* 16S TRFLP. The types of MST methodologies are important to note as most of the MST assays that have been applied in fresh produce production environments are at their core molecular subtyping methods applied to specific foodborne pathogens (Table 1). In addition, the researchers analyzed water samples spiked with either one or two reference fecal samples; thus, not truly applying MST methods to real-world, environmental samples. Regardless, Boehm et al. reported that very few assays met both the predetermined sensitivity and specificity criteria and concluded that field validation studies should be pursued. This study by Boehm et al. (2013) highlights the need for both strict performance criteria and possibly watershed (or foodshed) specific approaches to MST. Harwood and Stoeckel (2011) also indicate the need for robust performance criteria if the results of MST studies are to be applied to risk mitigation.

Along the same lines, the geographic applicability of various MST methods should be known, if available. For instance, to embark on MST efforts within a watershed or fresh produce production region, it is crucial to first identify all relevant sources of human and animal fecal pollution so both the methodologies and effort can be targeted. Along with pollution source profiling, the following information should be known or gathered on specific MST markers (Farnleitner et al., 2011): (1) occurrence in the host population (sensitivity), (2) occurrence in nonhost populations (specificity), (3) expected concentration in pollution sources, (4) persistence and movement in the watershed or foodshed, and (5) limits of detection and performance characteristics of the assay (Wang et al., 2013). Moreover, Farnleitner et al. (2011) urge researchers to design MST studies that integrate the use of multiple tools as opposed to just relying on one method in order to increase the robustness of a given study, as well as the confidence in the results.

Last, a plan for the utilization of results based on the MST study should be in place. As indicated previously, one of the primary goals of MST is to develop intervention strategies to reduce the potential risk to public health. In order to do this, robust science-based tools must be used and the resulting data must be interpreted with care. Teaf et al. (2011) provide an excellent review of the limitations and benefits of MST in the so-called legal arena where MST study results could be used to enforce change or create new regulations.

3 TRACKING FOODBORNE PATHOGENS: WHAT AND WHERE TO TARGET?

As indicated in Figure 1, there are numerous contamination sources that can impact the microbial quality of a variety of raw products as well as processed foods. Of these raw products, specialty crops (i.e., leafy greens, tomatoes, fresh herbs, and in general, fresh produce) can become contaminated through numerous routes of exposure including air and dust, food handlers (i.e., field workers), food contact surfaces (e.g., field harvest containers and harvesting equipment), insects, water (e.g., irrigation, pesticide application, and washing), and soil (e.g., wildlife intrusion and manure). Moreover, the susceptibility of specialty crops to contamination during production is particularly enhanced due to minimal processing during postharvest. These complexities allow specialty crop production to serve as both an ideal, yet complex, example for application of MST in food safety. Based on the number of ways by which hazards can be introduced to fresh produce, it is important to consider which sources to target for tracking important foodborne pathogens using either direct detection of pathogens or fecal indicators.

Of the potential contamination sources in specialty crop production, irrigation water has been repeatedly identified as the major risk factor for pathogen contamination of fresh produce (Gelting et al., 2011; Gorski et al., 2011; Holvoet et al., 2014; Jacobsen and Bech, 2012; Levantesi et al., 2012; Verhoeff-Bakkenes et al., 2011). For instance, Gorski et al. (2011) investigated the prevalence and diversity of *Salmonella* in Monterey County, California—a major produce region—and reported the highest prevalence of *Salmonella* in water samples that were collected and analyzed as compared to wildlife excrement, soil/sediment, cattle feces, and preharvest fresh produce from the same region. Levantesi et al. (2012) also reported that consistent contamination of irrigation waters with *Salmonella* is a common route of crop contamination as illustrated by the increase in *Salmonella* outbreaks related to fresh produce as well as the reportedly high prevalence of *Salmonella* in agricultural waters (Berger et al., 2010; Hanning et al., 2009; Heaton and Jones, 2008). Another study by Holvoet et al. (2014) reported a high prevalence of *Campylobacter* spp. (30.9%) and a much lower prevalence of *Salmonella* (1.4%) in irrigation water samples from lettuce-producing farms in Belgian. While *Campylobacter* is not commonly associated with outbreaks linked to contamination of fresh produce, Verhoeff-Bakkenes et al. (2011) indicate that consumption of raw vegetables and fruits is a risk factor for *Campylobacter* infections. Additional pathogens of concern as well as indicators of fecal pollution have also been reported in agricultural watersheds and irrigation water sources including host-specific *Bacteroidales*, generic *E. coli*, *E. coli* O157, *Enterococcus* spp., human adenovirus and norovirus, bovine polyomavirus, porcine adenovirus, and *L. monocytogenes* (Benjamin et al., 2013; Cooley et al., 2007; Kokkinos et al., 2012; Lee et al., 2014; Lu et al., 2004; Micallef et al., 2013; Wilkes et al., 2011).

A unique aspect of irrigation water is that it can also be impacted by a variety of contamination routes such as soil containing pathogens (Jacobsen and Bech, 2012), agricultural run-off, wastewater discharge, and direct defecation by wildlife and livestock (Steele and Odumeru, 2004). Along the same lines, soil may become contaminated through organic fertilizers; direct deposit of feces by wildlife, domestic, and livestock animals; and transport and deposition of air or dust contaminated with feces (Berger et al., 2010). In addition to soil and irrigation water, fresh produce may become contaminated through contact with field workers (Lynch et al., 2009), dirty equipment (McCollum et al., 2013), and insects (i.e., flies) (Graczyk et al., 2003; Wasala et al., 2013). Therefore, based on the variety of potential contamination sources, the question must arise, "Where should we target MST efforts and what should we be targeting—pathogens or indicators?" Overall, targeting the major risk factor—irrigation water—for fresh produce contamination may be the best approach especially as most MST methods were originally developed for water resource management.

4 MST APPLICATIONS IN FRESH PRODUCE PRODUCTION

This section of the chapter will aim to highlight a few case examples of how various MST approaches can be applied to fresh produce production in order to understand the sources, prevalence, and persistence of pathogens and fecal indicators.

4.1 *E. COLI* O157 IN THE SALINAS AND SAN JUAN VALLEYS IN CALIFORNIA

In the wake of the *E. coli* O157:H7 outbreak related to contaminated spinach in the United States in 2006 (California Food Emergency Response Team, 2007), numerous investigations were spawned that aimed to better understand how a pathogen that was once typically associated with outbreaks related to contaminated ground beef had infiltrated the fresh produce production environment (Rangel et al., 2005; Lynch et al., 2009). More specifically, from 1995 to 2007, 9 of the 22 outbreaks of *E. coli* O157:H7 were traced to the Salinas Valley region on the central coast of California—the major leafy vegetable producer in the United States. In an effort to understand the incidence and factors related to *E. coli* O157:H7 in the major produce production region of California, Cooley et al. (2007) launched an investigation of the Salinas watershed over a 20-month period from January 2005 to August 2006. Additionally, in a companion paper by Jay et al. (2007), samples from feral swine located in the same production region were collected from October to November 2006 to investigate the prevalence of *E. coli* O157:H7 in these animals and to subsequently link related sequences to those found by Cooley et al. (2007) and clinical samples from the spinach outbreak of 2006.

In these studies, the authors used molecular subtyping tools (PFGE and MLVA) to track *E. coli* O157:H7 within the watershed over time and to establish relatedness and potential sources of the environmental isolates. Through application of MLVA, 92 different MLVA types were identified from 1301 *E. coli* O157:H7 isolates analyzed, and 70% of these MLVA types could be grouped into one of eight clusters. The MLVA clusters are of particular interest as four of the clusters were spatially related, meaning the isolates in these clusters were obtained from the same sampling locations or regions on different occasions. For example, strains in one cluster were isolated exclusively from a single creek and its tributaries indicating that if MLVA types from this cluster were detected in fresh produce fields or the produce itself, then the likely source would be this particular creek and its tributaries. However, not all MLVA clusters conformed to these spatial boundaries, which is indicative of the dynamic processes that can occur during the fate and transport of microorganisms in the environment.

Within the same study, the authors compared the discriminatory power of MLVA with PFGE through the selection of strains with identical and closely related MLVA types for further analysis by PFGE. Overall, Cooley et al. (2007) concluded that MLVA is more efficient for the analysis of a large number of isolates and facilitated rapid assessment of relatedness for the purposes of source tracking. In addition, PFGE data often supported the MLVA clusters indicating relatedness with single PFGE profiles associated with multiple MLVA types. However, these results for MLVA as a valuable MST tool may only hold true for *E. coli* O157:H7 relatedness in the Salinas watershed as other studies have suggested that each watershed—or foodshed, in the case of MST in the food production environments—has a unique footprint based on the endemic microbial populations in the environment and in animal populations (Harwood et al., 2014). This study by Cooley et al. (2007) demonstrates the complexity of source tracking as well as the advantage and limitations of applying molecular subtyping tools for this purpose. In the end, the authors posed more questions than answers, calling for an increase in the understanding of factors related to the source and mechanisms of survival and persistence of *E. coli* O157:H7 in fresh produce production environments as well as fate and transport processes that would aid in development of intervention strategies.

4.2 *SALMONELLA* IN THE FRESH PRODUCE PRODUCTION ENVIRONMENT

Along with *E. coli* O157:H7, *Salmonella* is an important foodborne pathogen accounting for an estimated 50% of produce-associated outbreaks in the United States, most notably tomatoes (CDC, 2005, 2007), fresh herbs (Hanning et al., 2009; Pezzoli et al., 2008), and melons (Jackson et al., 2013). One approach to determining reservoirs of *Salmonella* and primary source attribution is through application of bacterial subtyping methods. Studies have shown that some *Salmonella* serovars are routinely associated with certain hosts (e.g., Dublin and Choleraesuis are associated with cattle and swine, respectively) while other serovars (e.g., Typhimurium and Enteriditis) have much broader host ranges (Uzzau et al., 2000). Along with specific host-ranges of *Salmonella* serovars, there may also be a geographical specificity to *Salmonella* serovars. Based on this hypothesis, a study by Strawn et al. (2014) attempted to better understand the potential regional specificity of *Salmonella* strains by analyzing *Salmonella* isolates from two produce-growing areas of the United States.

Through application of PFGE, Strawn et al. (2014) investigated the regional diversity of *Salmonella* associated with fresh produce production environments in New York State and south Florida. The findings of this study support the hypothesis that there are regional differences in *Salmonella* subtypes—information that could actually be advantageous for MST efforts in the food industry. More specifically, the authors reported that two serovars, Cerro and Saphra, were isolated exclusively from New York and south Florida, respectively. In addition, serovar Newport was isolated from both locations though PFGE subtyping indicated that Newport strains may be associated with certain regions. Even though serovars Cerro and Saphra have rarely been associated with human *Salmonella* infections in the United States, the geographical relationship of the serovars could be important if these serovars were identified as the causative agent in an outbreak related to fresh produce as this information could lead to the origin of the contaminated product. In addition, the regional specificity of the *Salmonella* Newport strains could be useful in this regard as well.

While the results of Strawn et al. (2014) seem promising, one major study limitation was the difference in sampling approaches for the two regions with New York samples collected over a 3-year period in diverse production environments while the south Florida samples were collected one day from a single, leafy green farm. Therefore, the geographical specificity of *Salmonella* may be overstated in this study though other studies have also suggested that *Salmonella* subtypes may be more prevalent in certain regions (Gorski et al., 2011, 2013; Haley et al., 2009; Jokinen et al., 2011; Rajabi et al., 2011). These studies highlight the need for future studies that investigate how the type of fresh produce production environment as well as regional factors influences the ecology and serotype diversity of *Salmonellae*. With this information, regionally specific, or even production environment specific, MST approaches can be designed.

4.3 HOST-SPECIFIC ENTERIC VIRUSES IN THE LEAFY GREEN SUPPLY CHAIN

Another approach to MST in the fresh produce supply chain is through detection of host-specific enteric viruses—a technique that has been most often applied to MST for water resource management (Hundesa et al., 2006; Rusinol et al., 2014; Wolf et al., 2010; Wong et al., 2012). The concept of using host-specific enteric viruses follows along with the main assumption of MST, which is that different animal species carry unique, host-specific populations of microorganisms and detection of this unique microorganism indicates that the animal host is impacting environmental sources that contaminate the food supply. Therefore, targeting host-specific enteric viruses can potentially be an effective MST method to distinguish between human, swine, avian, and bovine sources of fecal pollution with a high degree of confidence. The viruses that have been targeted include the human adenoviruses, noroviruses, and polyomaviruses; porcine hepatitis E virus, teschoviruses, adenoviruses, and noroviruses; and bovine noroviruses, polyomaviruses, and enteroviruses (Gibson and Schwab, 2011; Harwood et al., 2013; Jiménez-Clavero et al., 2003, 2005; Kokkinos et al., 2012; Ley et al., 2002; Maluquer de Motes et al., 2004; McQuaig et al., 2006; Hundesa et al., 2006; Wolf et al., 2007).

The selection of specific viruses as markers of fecal contamination must follow certain criteria. These viruses must (1) be specific to the species (i.e., swine, bovine, and human) under investigation, (2) be endemic in the population and excreted in feces at detectable levels, and (3) be stable in the environment (Jiménez-Clavero et al., 2003). If these criteria are not met, then the success of viral source tracking as a mechanism for assessing the burden of animal or human-specific waste streams on the food production environment will be minimal. In general, the porcine and human viruses seem to be extremely host-specific (Harwood et al., 2013) whereas some bovine viruses currently being investigated, such as enteroviruses, are not as species-specific (Jiménez-Clavero et al., 2003, 2005; Maluquer de Motes et al., 2004).

Kokkinos et al. (2012) applied the host-specific virus approach to determine sources of contamination from production, processing, and point-of-sale of fresh leafy greens in three European countries. In this study, the following enteric viruses were targeted: human and porcine adenoviruses, hepatitis A virus, hepatitis E virus, human noroviruses, and bovine polyomaviruses. Predetermined sampling points at production (irrigation water, field worker's hands, toilets, and manure), processing (worker's hands, rinsing water, and knives), and point-of-sale (fresh lettuce, water, and food contact surfaces) were collected in Greece, Serbia, and Poland, and 665 samples were analyzed for enteric viruses by qPCR or RT-qPCR. This leafy green production survey revealed that all target viruses were detected at some point in the supply chain; however, the production samples followed by point-of-sale samples were reported to have the highest percentage of positive samples. Of particular interest is the prevalence of human enteric viruses—human adenoviruses and noroviruses—found in the fresh produce production environment. More specifically, Kokkinos et al. (2012) demonstrated the role field harvesters and seasonal workers could potentially play in dissemination of enteric viruses as the majority of positive samples were human-specific viruses and were detected primarily on hands and toilets as well as at the point-of-sale. Meanwhile, very few samples were positive for porcine or bovine viruses with the exception of irrigation water samples, indicating again that irrigation water is likely a primary source of contamination within fresh produce production, not only for viral pathogens but also for bacterial pathogens of animal origin. A companion study by Maunula et al. (2013) involving the European berry fruit supply chain reported similar findings, highlighting the persistence of human enteric viruses in fresh produce production as well as potential control points, including improved worker hygiene.

While this study by Kokkinos et al. (2012) highlights the usefulness of enteric viruses for MST in the leafy green supply chain, it is important to note some of the limitations of using host-specific viruses. First, efficient recovery of enteric viruses from environmental and food samples has historically been very difficult (Julian and Schwab, 2012; Stals et al., 2012). For instance, the method described by Kokkinos et al. (2012) for recovery of viruses from water involves using 10 L water samples passed through a glass wool filter for adsorption of the viruses followed by elution. Although the glass wool method is an established protocol for recovering viruses from water samples (Cashdollar et al., 2012), the challenge lies with selecting the appropriate volume of water for recovery of viruses. Previous studies on recovery of viruses from water have used variable volumes ranging from 1 to >100 L depending on the quality of the water resource (i.e., ground water, surface water, standing pond, and agricultural runoff) (Gibson et al., 2012; Lambertini et al., 2008); therefore, 10 L water samples may be sufficient in produce production regions with poor irrigation water quality (e.g., surface ponds, shallow wells subject to surface water intrusion, poor water regulations, etc.), but this volume may not be sufficient in regions that use deep ground water wells or protected irrigation water resources (Gibson and Schwab, 2011).

Another limitation of host-specific virus MST is the application of qPCR and RT-qPCR for detection of viral nucleic acids. More specifically, these PCR-based methods can have varying amplification efficiencies depending on the virus target, the addition of a RT step for RNA viruses, and the effect of the sample matrix (i.e., PCR inhibitors) (Gibson et al., 2012; Stals et al., 2012). Therefore, while very robust and sensitive assays have been developed for select enteric viruses, other assays for different virus targets may not be as stringent; thus, potentially biasing any conclusions about dominant sources of fecal pollution based solely on host-specific enteric viruses. Overall, host-specific viruses do seem promising for MST and, although recovery and detection methods are continuously being developed and optimized, there are no standard methods that can be applied.

4.4 ENTEROCOCCI IN TOMATO PRODUCTION

One of the primary ways to determine the microbial quality of either water or food is through the detection of FIB including *E. coli* and enterococci (i.e. *E. faecium* and *E. faecalis*). Various studies have aimed to potentially utilize FIB for MST in water sources with varying success (Byappanahalli et al., 2012; Casarez et al., 2007; Lu et al., 2004; Wang et al., 2013). As indicated in Table 1, there are only a few published studies that investigated enterococci and *E. coli* in either the environment or fresh produce for MST purposes. Overall, the results from these studies indicate the large diversity of FIB population and provide insight into the future application of FIB for MST purposes.

Here, the study by Micallef et al. (2013) will be considered further because the researchers applied amplified fragment length polymorphism (AFLP) for the investigation of enterococci distribution and relatedness in tomato production farms in the Mid-Atlantic. In this study, 295 samples including tomatoes, leaves, irrigation water (pond and ditch), ground water, and soil were collected and analyzed for the presence of *Enterococcus* spp. by standard culture methods. Enterococci were detected in each sample type with the highest prevalence in "bottom" tomato and leaf samples, irrigation pond water and ditches, and irrigation ditch soil. Seven different *Enterococcus* spp. were identified with *E. casseliflavus* as the predominant species followed by *E. faecalis*. The distribution of species is of interest here because *E. casseliflavus* is naturally associated with plants and was found most often in the tomato and leaf samples as well as the irrigation ditch water and soil. Meanwhile, *E. faecalis* was found most often in pond water and is indicative of fecal pollution as well as being associated with human infections.

Determination of genetic relatedness using AFLP was only applied to *E. faecalis* isolates. The authors concluded that *E. faecalis* genetic fingerprints differed over time on a single farm indicating the dynamic process that microbial populations can undergo in the environment. More specifically, the study reveals that *E. faecalis* is able to disperse efficiently throughout the environment with very few habitat-related barriers. Overall, due to the high prevalence of a naturally occurring, plant-associated enterococci species (*E. casseliflavus*), Micallef et al. (2013) suggest that FIB such as enterococci are not ideal choices for MST, which is supported by previous studies related to source apportionment using *E. coli* and *Enterococcus* spp. (Wang et al., 2013).

5 LIMITATIONS AND CHALLENGES OF MST

Although the previous sections have indicated some of the limitations and challenges in MST, a few additional challenges will be discussed here in more detail. One of the first limitations of MST is application of methods that first require isolation of bacterial colonies through traditional culture methods.

This primarily applies to the studies that seek to determine the prevalence and distribution of specific pathogens in the environment (Cooley et al., 2007; Gorski et al., 2011; Strawn et al., 2014). For instance, in Strawn et al. (2014), the authors used a generic enrichment step followed by a variety of selection schemes for the isolation of *Salmonella*, and depending on the isolation media combination, different serovars were detected. This is problematic if one is trying to establish the microbial footprint of a specific environment yet the results vary depending on the selection media of choice. Moreover, issues may also arise if there is a significant portion of the pathogen population of interest that occurs in a viable but nonculturable state in the environment—most notably *Enterococcus* spp. and *Vibrio* spp. (Fakruddin et al., 2013; Goh et al., 2011; Nowakowska and Oliver, 2013). The need for isolation of pure cultures—both time- and labor-intensive as well as costly—is one of the primary reasons why most of the recent watershed-based MST studies have focused on targeting host-specific markers that can be detected by PCR or qPCR (Harwood et al., 2014; Stewart et al., 2013), though these methods also have their own limitations (Gibson et al., 2012; Girones et al., 2010). However, some molecular subtyping tools such as MLVA can be directly applied to environmental and food samples, but this has not been performed routinely (Chen et al., 2011).

Another application of MST is for source attribution of infectious disease outbreaks, primarily foodborne or waterborne outbreaks. As defined by Pires et al. (2009), source attribution is "the partitioning of human disease burden of one or more foodborne infections to specific sources" including animal reservoirs and vehicles of transmission such as food and water. Basically, this means that if certain subtypes of pathogens are almost always found in a specific animal reservoir then one can attribute a foodborne illness or outbreak caused by the same subtype to this specific animal reservoir. Though it may seem easy to establish this linear path for disease transmission, the dynamic processes of pathogens in the environment is much more complex. An example of how MST has been applied to source attribution was discussed by Batz et al. (2005) in which a Danish surveillance study collected and analyzed over 2 million samples for *Salmonella* from food and animal sources and then the *Salmonella* isolates were subtyped using PFGE. The *Salmonella* PFGE patterns were then quantitatively compared to isolates from human infections and a model was developed to attribute animal-food sources to human salmonellosis (Hald et al., 2004). As indicated by Batz et al. (2005), while this modeling approach to source attribution based on the molecular subtyping is novel, the proportional attribution of human cases based on this alone does not discern causation. In addition, while a particular food animal may be the reservoir for a specific *Salmonella* subtype, the model cannot say what the actual food vehicle was because fecal contamination from the food-animal can enter the environment and then become dispersed into different nodes along the farm to fork continuum. It is also important to highlight that while source attribution of *Salmonella* based on host-specific subtypes may have been moderately successful, this approach may not hold true for other foodborne pathogens such as *Campylobacter* or *Listeria*—both pathogens with somewhat homogenous distributions across reservoirs (Batz et al., 2005).

Additional challenges of MST include legal issues that may arise if MST is used to build evidence against or to implicate a farmer, food-animal grower, or company as the cause of fecal pollution. As indicated previously, Teaf et al. (2011) provides an overview of the challenges of using MST with the intent of pursuing legal actions if the results indicate a specific pollution source. Briefly, the main conclusion by Teaf et al. (2011) is that even if performance criteria is met for a specific MST method and accepted by the scientific community, this does not mean that this same method and resulting data are sufficient to satisfy any applicable legal criteria. Therefore, researchers and those who are contracted to perform sampling and analyses to determine the origins of diffuse pollution sources should have a working knowledge and understanding of acceptable requirements for both legal and technical audiences.

6 CONCLUSIONS AND FUTURE DIRECTIONS: WHERE DO WE GO FROM HERE?

Based on the information that has been presented in this chapter, it is clear that MST application for the purposes of enhancing food safety through development of targeted risk mitigation strategies could play a crucial role in future strategies developed to keep our food supply safe from farm to fork. However, there are some obvious limitations and challenges that need to be addressed prior to widespread application of MST in food production environments. Stewart et al. (2013) provides valuable insights into some of these limitations as well recommendations for future method development. In addition, the authors highlight the parameters that are critical for the evaluation and comparison of MST methods and studies, including sample preparation, detection of the MST target (i.e., instrumentation, control for PCR inhibitors or competitive inhibition, and defined limits of detection), and quantification of MST target. However, at the heart of the issue is that, to date, there is no consensus among researchers with respect to the best host-specific markers or indicators for MST. Therefore, researchers working in MST need to move toward the development of a more standardized approach that utilizes not just one method, but a toolbox of MST methods tailored to specific regions, watersheds, or even production systems. Overall, for MST to be truly useful, we need to understand that no single method will be capable of adequately addressing the goal of protecting the food supply and public health; thus, we should step back and approach MST with a toolbox mentality as well as with the mindset that MST may better conform to application at the local, foodshed level as opposed to on a national or international scale.

REFERENCES

Ailes, E.C., Leon, J.S., Jaykus, L.-A., Johnston, L.M., Clayton, H.A., Blanding, S., Kleinbaum, D.G., Backer, L.C., Moe, C.L., 2008. Microbial concentrations on fresh produce are affected by postharvest processing, importation, and season. J. Food Prot. 71, 2389–2397.

Batz, M.B., Doyle, M.P., Morris, G., Painter, J., Singh, R., Tauxe, R.V., Taylor, M.R., Danilo, M.A., Wong, L.L., Food Attribution Working Group, 2005. Attributing illness to food. Emerg. Infect. Dis. 11, 993–999.

Benjamin, L., Atwill, E.R., Jay-Russel, M., Cooley, M., Carychao, D., Gorski, L., Mandrell, R.E., 2013. Occurrence of generic *Escherichia coli*, *E. coli* O157 and *Salmonella* spp. in water and sediment from leafy green produce farms and streams on the Central California coast. Int. J. Food Microbiol. 165, 65–76.

Berger, C.N., Sodha, S.V., Shaw, R.K., Griffin, P.M., Pink, D., Hand, P., Frankel, G., 2010. Fresh fruit and vegetables as vehicles for the transmission of human pathogens. Environ. Microbiol. 12, 2385–2397.

Boehm, A.B., Werfhorst, V.D., Griffith, J.F., Holden, P.A., Jay, J.A., Shanks, O.C., Wang, D., Weisberg, S.B., 2013. Performance of forty-one microbial source tracking methods: a twenty-seven lab evaluation study. Water Res. 47, 6812–6828.

Burgos, M.J., Aguayo, M.C., Pulido, R.P., Gálvez, A., López, R.L., 2013. Multilocus sequence typing and antimicrobial resistance in *Enterococcus faecium* isolates from fresh produce. Antonie Van Leeuwenhoek 105, 413–421.

Byappanahalli, M.N., Nevers, M.B., Korajkic, A., Staley, Z.R., Harwood, V.J., 2012. Enterococci in the environment. Microbiol. Mol. Biol. Rev. 76, 685–706.

California Food Emergency Response Team, 2007. Investigation of an *Escherichia coli* O157:H7 outbreak associated with Dole pre-packaged spinach. http://www.cdph.ca.gov/pubsforms/Documents/fdb%20eru%20Spnch%20EC%20Dole032007wph.PDF.

Casarez, E.A., Pillai, S.D., Di Giovanni, G.D., 2007. Genotype diversity of *Escherichia coli* isolates in natural waters determined by PFGE and ERIC-PCR. Water Res. 41, 3643–3648.

Cashdollar, J.L., Brinkman, N.E., Griffin, S.M., McMinn, B.R., Rhodes, E.R., Varughese, E.A., Grimm, A.C., Parshionikar, S.U., Wymer, L., Fout, G.S., 2012. Development and evaluation of EPA method 1615 for detection of enterovirus and norovirus in water. Appl. Environ. Microbiol. 79, 215–223.

CDC, 2005. Outbreaks of *Salmonella* infections associated with eating Roma tomatoes – United States and Canada, 2004. Morb. Mortal. Wkly. Rep. 54, 325–328.

CDC, 2007. Multistate Outbreaks of *Salmonella* infections associated with raw tomatoes eaten in restaurants – United States, 2005–2006. Morb. Mortal. Wkly. Rep. 56, 909–911.

Chapin, T.K., Nightingale, K.K., Worobo, R.W., Wiedmann, M., Strawn, L.R., 2014. Geographical and meteorological factors associated with isolation of *Listeria* spp. in New York State produce production and natural environments. J. Food Prot. 77, 1919–1928.

Chen, S., Li, J., Saleh-Lakha, S., Allen, V., Odumeru, J., 2011. Multiple-locus variable number of tandem repeat analysis (MLVA) of *Listeria monocytogenes* directly in food samples. Int. J. Food Microbiol. 148, 8–14.

Cooley, M., Carychao, D., Crawford-Miksza, L., Jay, M.T., Myers, C., Rose, C., Keys, C., Farrar, J., Mandrell, R.E., 2007. Incidence and tracking of *Escherichia coli* O157:H7 in a major produce production region in California. PLoS One 2, e1159.

Fakruddin, M., Mannan, K.S.B., Andrews, S., 2013. Viable but nonculturable bacteria: food safety and public health perspective. ISRN Microbiol. http://dx.doi.org/10.1155/2013/703813.

Farnleitner, A.D., Reischer, G.H., Stadler, H., Kollanur, D., Sommer, R., Zerobin, W., Blöschl, G., Barrella, K.M., Truesdale, J.A., Casarez, E.M., Di Giovanni, G.D., 2011. Agricultural and rural watersheds. In: Hagedorn, C., Blanch, A.R., Harwood, V.J. (Eds.), Microbial Source Tracking: Methods, Applications, and Case Studies. Springer, New York, pp. 399–432.

Field, K.G., Samadpour, M., 2007. Fecal source tracking, the indicator paradigm, and managing water quality. Water Res. 41, 3517–3538.

Foley, S.L., Lynne, A.M., Nayak, R., 2009. Molecular typing methodologies for microbial source tracking and epidemiological investigations of Gram-negative bacterial foodborne pathogens. Infect. Genet. Evol. 9, 430–440.

Fu, L., Li, J., 2014. Microbial source tracking: a tool for identifying sources of microbial contamination in the food chain. Crit. Rev. Food Sci. Nutr. 54, 699–707.

Gelting, R.J., Baloch, M.A., Zarate-Bermudez, M., Selman, C., 2011. Irrigation water issues potentially related to the 2006 multistate *E. coli* O157:H7 outbreak associated with spinach. Agric. Water Manag. 98, 1395–1402.

Gibson, K.E., Schwab, K.J., 2011. Detection of bacterial indicators and human and bovine enteric viruses in surface water and groundwater sources potentially impacted by animal and human wastes in Lower Yakima Valley, Washington. Appl. Environ. Microbiol. 77, 355–362.

Gibson, K.E., Schwab, K.J., Spencer, S.K., Borchardt, M.A., 2012. Measuring and mitigating inhibition during quantitative real time PCR analysis of viral nucleic acid extracts from large-volume environmental water samples. Water Res. 46, 4281–4291.

Girones, R., Ferrús, M.A., Alonso, J.L., Rodriguez-Manzano, J., Calgua, B., de Abreu Corrêa, A., Hundesa, A., Carratala, A., Bofill-Mas, S., 2010. Molecular detection of pathogens in water – the pros and cons of molecular techniques. Water Res. 44, 4325–4339.

Goh, S.G., Gin, K.Y.H., Panda, N.R., 2011. Predictive models for the occurrence of viable but non-culturable (VBNC) bacteria under various environmental stresses. Water Pract. Technol. 6, 1–17. http://dx.doi.org/10.2166/wpt.2011.076.

Gorski, L., Parker, C.T., Liang, A., Cooley, M.B., Jay-Russell, M., Gordus, A.G., Atwill, E.R., Mandrell, R.E., 2011. Prevalence, distribution, and diversity of *Salmonella enterica* in a major produce region of California. Appl. Environ. Microbiol. 77, 2734–2748.

Gorski, L., Jay-Russell, M.T., Liang, A.S., Walker, S., Bengson, Y., Govoni, J., Mandrell, R.E., 2013. Diversity of pulsed-field gel electrophoresis pulso-types, serovars, and antibiotic resistance among *Salmonella* isolates from wild amphibians and reptiles in the California central coast. Foodborne Pathog. Dis. 10, 540–548.

Graczyk, T.K., Grimes, B.H., Knight, R., da Silva, A.J., Pieniazek, N.J., Veal, D.A., 2003. Detection of *Cryptosporidium parvum* and *Giardia lamblia* carried by synanthropic flies by combined fluorescent in situ hybridization and a monoclonal antibody. Am. J. Trop. Med. Hyg. 68, 228–232.

Graves, A.K., 2011. Food safety and implications for microbial source tracking. In: Hagedorn, C., Blanch, A.R., Harwood, V.J. (Eds.), Microbial Source Tracking: Methods, Applications, and Case Studies. Springer, New York, pp. 585–608.

Hald, T., Vose, D., Wegener, H.C., Koupeev, T., 2004. A Bayesian approach to quantify the contribution of animal-food sources to human salmonellosis. Risk Anal. 24, 255–269.

Haley, B.J., Cole, D.J., Lipp, E.K., 2009. Distribution, diversity, and seasonality of waterborne *Salmonellae* in a rural watershed. Appl. Environ. Microbiol. 75, 1248–1255.

Hanning, I.B., Nutt, J.D., Ricke, S.C., 2009. Salmonellosis outbreaks in the United States due to fresh produce: sources and potential intervention measures. Foodborne Pathog. Dis. 6, 635–648.

Harwood, V.J., Stoeckel, D.M., 2011. Performance criteria. In: Hagedorn, C., Blanch, A.R., Harwood, V.J. (Eds.), Microbial Source Tracking: Methods, Applications, and Case Studies. Springer, New York, pp. 7–30.

Harwood, V.J., Boehm, A.B., Sassoubre, L.M., Vijayavel, K., Stewart, J.R., Fong, T., Caprais, M.P., Converse, R.R., Diston, D., Ebdon, J., Fuhrman, J.A., Gourmelon, M., Gentry-Shields, J., Griffith, J.F., Kashian, D.R., Noble, R.T., Taylor, H., Wicki, M., 2013. Performance of viruses and bacteriophages for fecal source determination in a multi-laboratory, comparative study. Water Res. 47, 6929–6943.

Harwood, V.J., Staley, C., Badgley, B.D., Borges, K., Korajkic, A., 2014. Microbial source tracking markers for detection of fecal contamination in environmental waters: relationships between pathogens and human health outcomes. FEMS Microbiol. Rev. 38, 1–40.

Heaton, J.C., Jones, K., 2008. Microbial contamination of fruit and vegetables and the behaviour of enteropathogens in the phyllosphere: a review. J. Appl. Microbiol. 104, 613–626.

Holvoet, K., Sampers, I., Seynnaeve, M., Uyttendaele, M., 2014. Relationships among hygiene indicators and enteric pathogens in irrigation water, soil and lettuce and the impact of climatic conditions on contamination in the lettuce primary production. Int. J. Food Microbiol. 171, 21–31.

Hundesa, A., Maluquer de Motes, C., Bofill-Mas, S., Albinana-Gimenez, N., Girones, R., 2006. Identification of human and animal adenoviruses and polyomaviruses for determination of sources of fecal contamination in the environment. Appl. Environ. Microbiol. 72, 7886–7893.

Jackson, B.R., Griffin, P.M., Cole, D., Walsh, K., Chai, S.J., 2013. Outbreak-associated *Salmonella enterica* serotypes and food commodities, United States, 1998–2008. Emerg. Infect. Dis. 19, 1239–1244.

Jacobsen, C.S., Bech, T.B., 2012. Soil survival of *Salmonella* and transfer to freshwater and fresh produce. Food Res. Int. 45, 557–566.

Jay, M.T., Cooley, M., Carychao, D., Wiscomb, G.W., Sweitzer, R.A., Crawford-Miksza, L., Farrar, J.A., Lau, D.K., O'Connell, J., Millington, A., Asmundson, R.V., Atwill, E.R., Mandrell, R.E., 2007. *Escherichia coli* O157:H7 in feral swine near spinach fields and cattle, central California coast. Emerg. Infect. Dis. 13, 1908–1911.

Jiménez-Clavero, M.A., Fernández, C., Ortiz, J.A., Pro, J., Carbonell, G., Tarazona, J.V., Roblas, N., Ley, V., 2003. Teschoviruses as indicators of porcine fecal contamination of surface water. Appl. Environ. Microbiol. 69, 6311–6315.

Jiménez-Clavero, M.A., Escribano-Romero, E., Mansilla, C., Gómez, N., Córdoba, L., Roblas, N., Ponz, F., Ley, V., Súiz, J., 2005. Survey of bovine enterovirus in biological and environmental samples by a highly sensitive real-time reverse transcription-PCR. Appl. Environ. Microbiol. 71, 3536–3543.

Johnston, L.M., Jaykus, L.-A., Moll, D., Anciso, J., Mora, B., Moe, C.L., 2006. A field study of the microbiological quality of fresh produce of domestic and Mexican origin. Int. J. Food Microbiol. 112, 83–95.

Jokinen, C., Edge, T.A., Ho, S., Koning, W., Laing, C., Mauro, W., Medeiros, D., Miller, J., Robertson, W., Taboada, E., Thomas, J.E., Topp, E., Ziebell, K., Gannon, V.P.J., 2011. Molecular subtypes of *Campylobacter* spp., *Salmonella enterica*, and *Escherichia coli* O157:H7 isolated from faecal and surface water samples in the Oldman River watershed, Alberta, Canada. Water Res. 45, 1247–1257.

Julian, T.R., Schwab, K.J., 2012. Challenges in environmental detection of human viral pathogens. Curr. Opin. Virol. 2, 78–83.

Khan, I.U.H., Gannon, V., Jokinen, C.C., Kent, R., Koning, W., Lapen, D.R., Medeiros, D., Miller, J., Neumann, N.F., Phillips, R., Schreier, H., Topp, E., van Bochove, E., Wilkes, G., Edge, T.A., 2014. A national investigation of the prevalence and diversity of thermophilic *Campylobacter* species in agricultural watersheds in Canada. Water Res. 61, 243–252.

Kokkinos, P., Kozyra, I., Lazic, S., Bouwknegt, M., Rutjes, S., Willems, K., Moloney, R., Roda Husman, A.M., Kaupke, A., Legaki, E., D'Agostino, M., Cook, N., Rzeżutka, A., Petrovic, T., Vantarakis, A., 2012. Harmonised investigation of the occurrence of human enteric viruses in the leafy green vegetable supply chain in three European countries. Food Environ. Virol. 4, 179–191.

Lambertini, E., Spencer, S.K., Bertz, P.D., Loge, F.J., Kieke, B.A., Borchardt, M.A., 2008. Concentration of enteroviruses, adenoviruses, and noroviruses from drinking water by use of glass wool filters. Appl. Environ. Microbiol. 74, 2990–2996.

Lee, D., Lee, H., Trevors, J.T., Weir, S.C., Thomas, J.L., Habash, M., 2014. Characterization of sources and loadings of fecal pollutants using microbial source tracking assays in urban and rural areas of the Grand River Watershed, Southwestern Ontario. Water Res. 53, 123–131.

Levantesi, C., Bonadonna, L., Briancesco, R., Grohmann, E., Toze, S., Tandoi, V., 2012. *Salmonella* in surface and drinking water: occurrence and water-mediated transmission. Food Res. Int. 45, 587–602.

Ley, V., Higgins, J., Fayer, R., 2002. Bovine enteroviruses as indicators of fecal contamination. Appl. Environ. Microbiol. 68, 3455–3461.

Lu, L., Hume, M.E., Sternes, K.L., Pillai, S.D., 2004. Genetic diversity of *Escherichia coli* isolates in irrigation water and associated sediments: implications for source tracking. Water Res. 38, 3899–3908.

Lynch, M.F., Tauxe, R.V., Hedberg, C.W., 2009. The growing burden of foodborne outbreaks due to contaminated fresh produce: risks and opportunities. Epidemiol. Infect. 137, 307–315.

Maluquer de Motes, C., Clemente-Casares, P., Hundesa, A., Martín, M., Girones, R., 2004. Detection of bovine and porcine adenoviruses for tracing the source of fecal contamination. Appl. Environ. Microbiol. 70, 1448–1454.

Maunula, L., Kaupke, A., Vasickova, P., Söderberg, K., Kozyra, I., Lazic, S., van der Poel, W.H.M., Bouwknegt, M., Rutjes, S., Willems, K.A., Moloney, R., D'Agostino, M., de Roda Husman, A.M., von Bonsdorff, C., Rzezutka, A., Pavlik, I., Petrovic, T., Cook, N., 2013. Tracing enteric viruses in the European berry fruit supply chain. Int. J. Food Microbiol. 167, 177–185.

McCollum, J.T., Cronquist, A.B., Silk, B.J., Jackson, K.A., O'Connor, K.A., Cosgrove, S., Gossack, J.P., Parachini, S.S., Jain, N.S., Ettestad, P., Ibraheem, M., Cantu, V., Joshi, M., DuVernoy, T., Fogg Jr., N.W., Gorny, J.R., Mogen, K.M., Spires, C., Teitell, P., Joseph, L.A., Tarr, C.L., Imanishi, M., Neil, K.P., Tauxe, R.V., Mahon, B.E., 2013. Multistate outbreak of listeriosis associated with cantaloupe. N. Engl. J. Med. 369, 944–953.

McQuaig, S.M., Scott, T.M., Harwood, V.J., Farrah, S.R., Lukasik, J.O., 2006. Detection of human-derived fecal pollution in environmental waters by use of a PCR-based human polyomavirus assay. Appl. Environ. Microbiol. 72, 7567–7574.

Micallef, S.A., Goldstein, R.E.R., George, A., Ewing, L., Ben, D.T., Boyer, M.S., Joseph, S.W., Sapkota, A.R., 2013. Diversity, distribution and antibiotic resistance of *Enterococcus* spp. recovered from tomatoes, leaves, water and soil on U.S. Mid-Atlantic farms. Food Microbiol. 36, 465–474.

Nowakowska, J., Oliver, J.D., 2013. Resistance to environmental stresses by *Vibrio vulnificus* in the viable but nonculturable state. FEMS Microbiol. Ecol. 84, 213–222.

Patchanee, P., Molla, B., White, N., Line, D.E., Gebreyes, W.A., 2010. Tracking *Salmonella* contamination in various watersheds and phenotypic and genotypic diversity. Foodborne Pathog. Dis. 7, 1113–1120.

Pezzoli, L., Elson, R., Little, C.L., Yip, H., Fisher, I., Yishai, R., Anis, E., Valinsky, L., Biggerstaff, M., Patel, N., Mather, H., Brown, D.J., Coia, J.E., van Pelt, W., Nielsen, E.M., Ethelberg, S., de Pinna, E., Hampton, M.D., Peters, T., Threlfall, J., 2008. Packed with *Salmonella*—investigation of an international outbreak of *Salmonella* Senftenberg infection linked to contamination of prepacked basil in 2007. Foodborne Pathog. Dis. 5, 661–668.

Pires, S.M., Evers, E.G., van Pelt, W., Ayers, T., Scallan, E., Angulo, F.J., Havelaar, A., Hald, T., 2009. Attributing the human disease burden of foodborne infections to specific sources. Foodborne Pathog. Dis. 6, 417–424.

Rajabi, M., Jones, M., Hubbard, M., Rodrick, G., Wright, A.C., 2011. Distribution and genetic diversity of *Salmonella enterica* in the Upper Suwannee River. Int. J. Microbiol. http://dx.doi.org/10.1155/2011/461321 (Article ID 461321).

Rangel, J.M., Sparling, P.H., Crowe, C., Griffin, P.M., Swerdlow, D.L., 2005. Epidemiology of *Escherichia coli* O157:H7 outbreaks, United States, 1982–2002. Emerg. Infect. Dis. 11, 603–609.

Rusinol, M., Fernandez-Cassi, X., Hundesa, A., Vieira, C., Kern, A., Eriksson, I., Ziros, P., Kay, D., Miagostovich, M., Vargha, M., Allard, A., Vantarakis, A., Wyn-Jones, P., Bofill-Mas, S., Girones, R., 2014. Application of human and animal viral microbial source tracking tools in fresh and marine waters from five different geographical areas. Water Res. 59, 119–129.

Scallan, E., Griffin, P., Angulo, F., Tauxe, R., 2011a. Foodborne illness acquired in the United States – unspecified agents. Emerg. Infect. Dis. 17, 16–22.

Scallan, E., Hoekstra, R., Angulo, F., Tauxe, R., 2011b. Foodborne illness acquired in the United States – major pathogens. Emerg. Infect. Dis. 17, 7–15.

Scharff, R.L., 2012. Economic burden from health losses due to foodborne illness in the United States. J. Food Prot. 75, 123–131.

Stals, A., Baert, L., Van Coillie, E., Uyttendaele, M., 2012. Extraction of food-borne viruses from food samples: a review. Int. J. Food Microbiol. 153, 1–9.

Steele, M., Odumeru, J., 2004. Irrigation water as source of foodborne pathogens on fruit and vegetables. J. Food Prot. 67, 2839–2849.

Stewart, J.R., Boehm, A.B., Dubinsky, E.A., Fong, T.T., Goodwin, K.D., Griffith, J.F., Noble, R.T., Shanks, O.C., Vijayavel, K., Weisberg, S.B., 2013. Recommendation following a multi-laboratory comparison of microbial source tracking methods. Water Res. 47, 6829–6838.

Stoeckel, D., Harwood, V., 2007. Performance, design, and analysis in microbial source tracking studies. Appl. Environ. Microbiol. 73, 2405–2415.

Strawn, L.K., Danyluk, M.D., Worobo, R.W., Wiedmann, M., 2014. Distributions of *Salmonella* subtypes differ between two U.S. produce-growing regions. Appl. Environ. Microbiol. 80, 3982–3991.

Swaminathan, B., Barrett, T.J., Hunter, S.B., Tauxe, R.V., PulseNet Task Force, C.D.C., 2001. PulseNet: the molecular typing network for foodborne bacterial disease surveillance, United States. Emerg. Infect. Dis. 7, 382–389.

Teaf, C.M., Garber, M.M., Harwood, V.J., 2011. Use of microbial source tracking in the legal arena: benefits and challenges. In: Hagedorn, C., Blanch, A.R., Harwood, V.J. (Eds.), Microbial Source Tracking: Methods, Applications, and Case Studies. Springer, New York, pp. 301–312.

USEPA, 2002. Method 1600: Enterococci in Water by Membrane Filtration Using Membrane-Enterococcus Indoxyl-B-D-Glucoside Agar (mEI). Office of Water. U.S. Environmental Protection Agency, Washington, DC.

Uzzau, S., Brown, D.J., Wallis, T., Rubino, S., Leori, G., Bernard, S., Casadesus, J., Platt, D.J., Olsen, J.E., 2000. Host adapted serotypes of *Salmonella enterica*. Epidemiol. Infect. 125, 229–255

Verhoeff-Bakkenes, L., Jansen, H.A.P.M., Veld, P.H., Beumer, R.R., Zwietering, M.H., van Leusden, F.M., 2011. Consumption of raw vegetables and fruits: a risk factor for *Campylobacter* infections. Int. J. Food Microbiol. 144, 406–412.

Villemur, R., Imbeau, M., Vuong, M.N., Masson, L., Payment, P., 2015. An environmental survey of surface waters using mitochondrial DNA from human, bovine and porcine origin as fecal source tracking markers. Water Res. 69, 143–153.

Viswanath, P., Murugesan, L., Knabel, S.J., Verghese, B., Chikthimmah, N., LaBorde, L.F., 2013. Incidence of *Listeria monocytogenes* and *Listeria* spp. in a small-scale mushroom production facility. J. Food Prot. 76, 608–615.

Wang, D., Farnleitner, A.H., Field, K.G., Green, H.C., Shanks, O.C., Boehm, A.B., 2013. *Enterococcus* and *Escherichia coli* fecal source apportionment with microbial source tracking genetic markers – is it feasible? Water Res. 47, 6849–6861.

Wasala, L., Talley, J.L., DeSilva, U., Fletcher, J., Wayadande, A., 2013. Transfer of *Escherichia coli* O157:H7 to spinach by house flies, *Musca domestica* (Diptera: Muscidae). Phytopathology 103, 373–380.

Wilkes, G., Edge, T.A., Gannon, V.P.J., Jokinen, C., Lyautey, E., Neumann, N.F., Ruecker, N., Scott, A., Sunohara, M., Topp, E., Lapen, D.R., 2011. Associations among pathogenic bacteria, parasites, and environmental and land use factors in multiple mixed-use watersheds. Water Res. 45, 5807–5825.

Wilkes, G., Brassard, J., Edge, T.A., Gannon, V., Gottschall, N., Jokinen, C.C., Jones, T.H., Khan, I.U.H., Marti, R., Sunohara, M.D., Topp, E., Lapen, D.R., 2014. Long-term monitoring of waterborne pathogens and microbial source tracking markers in paired agricultural watersheds under controlled and conventional tile drainage management. Appl. Environ. Microbiol. 80, 3708–3720.

Wolf, S., Williamson, W.M., Hewitt, J., Rivera-Aban, M., Lin, S., Ball, A., Scholes, P., Greening, G.E., 2007. Sensitive multiplex real-time reverse transcription-PCR assay for the detection of human and animal noroviruses in clinical and environmental samples. Appl. Environ. Microbiol. 73, 5464–5470.

Wolf, S., Hewitt, J., Greening, G.E., 2010. Viral multiplex quantitative PCR assays for tracking sources of fecal contamination. Appl. Environ. Microbiol. 76, 1388–1394.

Wong, K., Fong, T., Bibby, K., Molina, M., 2012. Application of enteric viruses for fecal pollution source tracking in environmental waters. Environ. Int. 45, 151–164.

CHAPTER

APPLICATION OF MOLECULAR METHODS FOR TRACEABILITY OF FOODBORNE PATHOGENS IN FOOD SAFETY SYSTEMS

4

Steven C. Ricke[*,†,‡,§], **Turki M. Dawoud**[*,†], **Young Min Kwon**[*,†,§]

Cell and Molecular Biology Program, University of Arkansas, Fayetteville, Arkansas, USA[*]*, Center for Food Safety, University of Arkansas, Fayetteville, Arkansas, USA*[†]*, Department of Food Science, University of Arkansas, Fayetteville, Arkansas, USA*[‡]*, Department of Poultry Science, University of Arkansas, Fayetteville, Arkansas, USA*[§]

1 INTRODUCTION

Food safety continues to be an important component of food production both economically as well as from a public health standpoint. By the latest estimates, foodborne diseases cost billions of dollars with thousands of hospitalizations occurring each year in the United States alone (Scallan et al., 2011; Scharff, 2012; Batz et al., 2014). While the relative hierarchy of which foodborne pathogen is the annually most prominent varies from year to year, the major foodborne pathogens remain fairly constant. Typically, this list will include *Salmonella* spp., *Staphylococcus aureus*, shiga toxin producing *Escherichia coli*, *Listeria monocytogenes*, *Campylobacter*, and norovirus, among others. While their ability to cause disease ranges from mild to life threatening and the length of incubation for symptoms varies as well, they are all perceived as unacceptable public health hazards. Consequently, there are considerable efforts to not only limit their occurrence in food production and systems but also to assess their epidemiological patterns of dissemination. As public demand for accountability of causes and sources of foodborne disease increases, it becomes ever more critical to accurately trace the pathways that the particular causative organisms have traversed over the course of the outbreak. The ability to track or have traceability capacity for a particular foodborne pathogen is becoming a critical component in overall food safety systems for all types of food commodities whether they are shipped internationally or sold in local farmers' markets.

While in theory tracking foodborne pathogens would seemingly be fairly intuitive, in practice there are several difficulties. Certainly most of the foodborne organisms exhibit biological variations among isolates and strains that can challenge certain approaches for tracking their activities. For example, the genus *Salmonella* is comprised of over 2500 serovars, some of which are more associated with foodborne disease than others, thus making it critical to be able to distinguish particular isolates

Food Safety. http://dx.doi.org/10.1016/B978-0-12-800245-2.00004-6

(Foley et al., 2013; Li et al., 2013; Ricke, 2014). In addition, the physiological status of any given food-borne pathogen after environmental exposure to a variety of stresses either of an intrinsic or extrinsic nature including sanitation disinfectants, antimicrobial agents, fluctuations in temperature, moisture, and oxygen can impede recovery of detectable organisms. Changing regulatory demands can also impact how detection of organisms is conducted. For example, now that several of the pathogenic *E. coli* are considered adulterants, detection limits, and sensitivities are critical for the development of any new identification methods (Wang et al., 2013). Finally, food production and processing systems are considerably complex with a wide range of potentially different sampling matrices to contend with during the farm to fork supply chain. In this review, the complications associated with food production and food safety systems will be discussed along with the needs for further development of traceability and tracking methods. This will be followed by the potential role played by both the historical evolution of technology from basic genetics to the more recent developments in molecular sequencing methods and capabilities.

2 COMPLEXITY OF FOOD PRODUCTION SYSTEMS

Food production continues to become more global with not only increases in economy of scale of production practices occurring but also increases in the sheer volumes of processed food generated and the transportation distances these food products must travel as trade commodities (Ollinger et al., 2005; Windhorst, 2009; Crandall et al., 2012). Obviously, economics is a major driver for the growth in food production global trade along with the advances in food processing and storage technology that have enabled more cost-effective and longer-term retention of food quality attributes. This trend is expected to only continue to increase as the world human population continues growing and the concomitant demand for food that can be sustainably produced attempts to match this demand (Beddington et al. 2012). Given the level of food waste along with finite resources for cultivation of row crops and livestock grazing, sustainability, and achieving environmental balance between greenhouse emissions and other inputs is looming as a prominent global issue (Marsh and Bugusu, 2007; Beddington et al., 2012; Dunkley and Dunkley, 2013; Hoekstra and Wiedmann, 2014). Consequently, life cycle assessment (LCA) approaches have emerged to conduct a more complete quantitative assessment of all inputs and outputs for agricultural systems (Guinée et al., 2011; Halog and Manik, 2011; Castellini et al., 2012; Leinonen et al., 2012; Stoessel et al., 2012; Hellweg and Milà i Canals, 2014; Pelletier et al., 2014).

However, other factors besides LCA have also come into play that are not necessarily economic or sustainable in origin but still impact the means by which food is produced and traded as a commodity. A classic example is the egg industry, where the political emphasis on animal welfare reforms has shifted to cage-free systems that, in turn, has had not only interstate consequences in the United States but international implications (De Reu et al., 2008; Windhorst, 2008, 2009; Mench et al., 2011). Special social and cultural requirements for certain types of meat processing are proving to be economically important as well (Bonne and Verbeke, 2008; Verbeke et al., 2013). Likewise, the popularity of organically produced foods and natural sources of food and food animals along with the growth in local farmers' markets has added to the complicated mosaic of food production with further emphasis on the origins of where food is produced and how it is marketed (Winter and Davis, 2006; O'Bryan et al., 2008, 2014; Ricke et al., 2012; Claeys et al., 2013). At the far end of the food production spectrum is the growing interest in meat substitutes and meat produced from *in vitro* technologies such as tissue culture (Verbeke et al., 2015).

As is the case for the food production sector, the trade commodity sector grows in conjunction with consumer demand for diversified sources and production systems, and accountability becomes ever more critical. However, incidences of food fraud, the potential for food bioterrorism, and ongoing foodborne disease outbreaks have eroded that public trust and placed a premium on accountability (Roberts, 2006; Ricke, 2010; Harrison, 2015). Developing that trust through certification, accurate labeling, and other traceability approaches are critical but still need refinement and remain challenging (Gellynck et al., 2006; Verbeke et al., 2010; Crandall et al., 2012, 2013). Not surprisingly, interest has increased in developing technologies for identifying sources and origins of different commodities of the food production supply chain. This has spawned the concept of traceability technologies that offer up the capability of tracking various components that comprise the final completed food product marketed at retail (Crandall et al., 2013; Angus-Lee, 2015). Such abilities require several sophisticated tools, including not only the data processing and analyses capacity but also the fundamental data that is to be collected for data processing. For some data, such as detailed labeling information, this has progressed substantially over time, but even here issues with the presentation of information remain (Gellynck et al., 2006; Horvath, 2012). Likewise, as discussed in the next section, tracking capabilities for foodborne pathogens in food systems are becoming more reliable but more still remains to be done.

3 FOODBORNE PATHOGENS AND THE POTENTIAL ROLE OF TRACEABILITY

The biological nature of food commodities presents a challenge both in terms of the complex composition and to some degree the sheer volume of material. For some aspects such as foodborne pathogens, tremendous effort and progress have been made using molecular methods to not only detect but also actually quantitate these organisms in foods. This has led to the ability to follow these organisms both as a means to not only track a particular organism to its point of origin but also add much more in-depth epidemiological and risk appraisal on a case-by-case basis. Much less has been done on the actual food material itself and this will be critical in the future as authenticity will be paramount to prevent fraudulent labeling for specific types of foods such as those supposedly generated from certified organic sources. Consequently, it will be important to implement biologically based methods and technologies that can generate data that can easily be processed and incorporated into large electronic data sets for advanced statistical analyses. However, there is already a precedent for this in the food industry with the work undertaken over a number of years to develop better detection approaches for foodborne pathogens in response to public health and regulatory demands.

The historical emphasis on food safety and detection of foodborne pathogens has opened up the opportunity to take advantage of the tremendous advances made in microbial genetics to not only improve detection sensitivity and shorten detection times but also has created the potential for considerable precision. With the explosive advent of molecular-based technologies in the past 30 years, much more is now understood not only about the genetic capabilities of foodborne pathogens but also how they behave in the environment in terms of properties such as the dissemination of antibiotic resistance (Beatson and Walker, 2014). This has led to appreciating not only the characterized general stress responses of some of the foodborne pathogens but also virulence gene expression during foodborne disease pathogenesis as well as factors impacting colonization of food animals (Cotter and Hill, 2003; Ricke, 2003a,b, 2014; Dunkley et al., 2009; Horrocks et al., 2009; Lungu et al., 2009; Soni et al., 2011; Spector and Kenyon, 2012; Callaway et al., 2013; Ferreira et al., 2014). As more has become known, several improvements in food safety have become possible including more optimal intervention strategies in processing as well as in live animal production. In addition, this has led to the development of much more sophisticated risk assessment modeling that not only can better assess where risk priorities should be implemented but also assess the potential impact of proposed interventions. Probably one of the important advances has been made in the improved capabilities of DNA sequencing foodborne pathogen genomes and the subsequent use of this data to not only construct improved detection platforms but also directly use the sequence data to precisely track dissemination of the organism and ultimately achieve complete traceability back to the origin of the outbreak caused by that particular strain (O'Flaherty and Klaenhammer, 2011; Walker and Beatson, 2012; Gerner-Smidt et al., 2013; Park et al., 2014a; Vongkamjn and Weidmann, 2015). In the following sections, the historical progress on understanding of basic genetic principles and the subsequent development of molecular-based applications and sequencing technologies will be discussed. This will be followed by addressing the potential of molecular-based technologies for being a viable component of food traceability systems and the possible future directions that could further merge traceability/tracking data sets with molecular data.

4 DISCOVERY OF DNA AND DEVELOPMENT OF GENETIC FOUNDATIONAL PRINCIPLES

The establishment of **d**eoxyribo**n**ucleic **a**cid (DNA) as the essential program code for organism replication and the emergence of the field of molecular biology has a long history, which has been well covered in a wide range of textbook monographs and reviews (Stent and Calendar, 1978; Judson, 1979; Freifelder, 1983; Sharp, 2014). However, in order to appreciate both the unifying and versatile nature of DNA and its associated entities as a potential common denominator for traceability and tracking in food safety systems requires at least a chronological overview to understand how this came about. This timeline-based understanding is critical because not only does this serve as the basis for the molecular methods currently used for detection and quantitation of foodborne pathogens but it has also fueled evolving molecular technologies for developing functional genomics on both pathogens as well as their respective hosts.

In the middle of the 1860s, Gregor Mendel described his breeding experiments with peas for establishing a simple and essential law of hereditary traits transmission, which later was named "Mendel's laws" (Miko, 2008). Not long after this, Ernst Haeckel proposed that an element inside the nucleus could be accountable for the hereditary traits transmission. This genetic material was first isolated by the Swiss physician biochemist, Johann Friedrich Miescher in 1869. It was an unknown substance, neither protein nor lipid, and it was rich in phosphorous and deficient in sulfur. Because he isolated it from the nuclei of the white blood cells, he named it "nuclein" and by 1871, he published his first paper with others describing this finding (Dahm, 2005, 2008). In 1882, Walther Flemming described chromosomes and investigated their behavior during cell division as "mitosis stages" (Choudhuri, 2003; Dahm, 2005).

In 1889, Richard Altmann performed experiments to successfully separate DNA from proteins. Due to its acidic manner, he named the substance "Nucleïnsäure" or "nucleic acid" (Dahm, 2008; Portin, 2014). Between 1891 and 1893, Albrecht Kossel discovered the primary building blocks of nuclein, the purine (guanine and adenine) and pyrimidine bases (thymine and cytosine), one pentose sugar, along with phosphoric acid, and confirmed that they were limited to the nucleus (Jones, 1953; Dahm, 2008;

Portin, 2014). Theodor Boveri and Walter Sutton suggested in 1902 that the heredity units are located on the chromosomes with "the chromosome theory of inheritance" (Portin, 2014). In 1909, Wilhelm Johannsen utilized "gene" for the units of heredity. In 1928, Frederick Griffith proposed "the transforming principle" that allows transfer of detectable properties from one type of bacteria to another (Downie, 1972). Important contributions were also made by Phoebus Levene in the study of nucleic acid. Levene discovered in 1909 that the carbohydrate "pentose" of the yeast nucleic acid is ribose and thus became known as ribonucleic acid (RNA). Levene and his colleagues required 20 years to identify the D-ribose (deoxyribose) that later came to be known as deoxyribonucleic acid (DNA). Levene proposed a "tetranucleotide hypothesis," postulating an equal quantity of the DNA four bases (A, T, G, and C) from any source (Van Slyke and Jacobs, 1945; Simoni et al., 2002).

The refutability of this hypothesis was tested by Erwin Chargaff and others between 1949 and 1951. They determined that the adenine (A) and thymine (T) were equivalent in their amount and that guanine (G) was equivalent of the cytosine (C) content, which led them to conclude that total purine and total pyrimidine were in fact equal, and that the ratio of purine and pyrimidine was close to 1. Subsequently, this was referred to as "Chargaff's rule." They also discovered that individuals of a particular species possessed the same composition of DNA (the relative nucleic acid base ratios), but distinguishable differences occurred between different species (Kresge et al., 2005b). Doubts were expressed by the research community that DNA was actually the genetic material as they were considered small and not structurally complex, so the bulk of the research attention was directed toward proteins as the better genetic candidates for chromosomal association, due to their complexity of structure as well as the high degree of compositional diversity among the different proteins (Judson, 1979).

In 1941, George Beadle and Edward Tatum proposed "the one-gene-one-enzyme hypothesis," demonstrating that a gene is responsible for enzyme production (Choudhuri, 2003; Horowitz, 1995). Although experiments by Frederick Griffith were the first of their kind to demonstrate that DNA was the genetic material, these findings were not recognized at the time. In 1944, a publication by Oswald Avery, Colin MacLeod, and Maclyn McCarty confirmed that DNA was a transformable substance by injecting a purified DNA from the heat-killed type-S virulent cells to a living type R (rough coat) non-virulent pneumococci through transformation (Avery et al., 1944; Steinman and Moberg, 1994). This finding was later supported by Alfred Hershey and Martha Chase in their classic transduction experiment (Stent and Calendar, 1978; Judson, 1979; Chambers et al., 1994; Choudhuri, 2003).

The eventual discovery of DNA secondary structure and its corresponding function paved the way for answering numerous questions regarding cellular information storage and the replication of DNA. James Watson and Francis Crick were the first to assemble the structure of DNA as an accurate representative model. They accomplished this by collecting and extrapolating a combination of several findings on DNA from Chargaff's rule as well as X-ray crystallography structural analyses by Rosalind Franklin, Raymond Gosling, and Maurice Wilkins. The final constructed structure, the double helix, was possible when Jerry Donohue offered some additional suggestions for proper final configuration of the DNA structure (Watson and Crick, 1953; Judson, 1979; Maddox, 2003).

In 1956, DNA polymerase was discovered by Arthur Kornberg as the enzyme responsible for replicating DNA and in 1958, Kornberg and colleagues published two papers explaining their findings (Kresge et al., 2005a). In the late 1950s, Francis Crick proposed "the central dogma" that information is essentially a transcript from DNA to RNA, subsequently translated to protein and further suggested that three bases in the DNA always coded for one amino acid in a protein (Crick, 1963; Morange, 2005; Thieffry and Sarkar, 1998). The DNA replication mechanism was first described in 1958 by Matthew

Meselson and Franklin Stahl (Meselson and Stahl, 1958; Holmes, 2001). In the early 1960s, Holley, Khorana, and Nirenberg elucidated the genetic code and its function in protein synthesis (Nirenberg, 2004; Söll and RajBhandary, 2006). In the late 1960s, enzymes were utilized to digest the DNA for the first time by Werner Arber, Hamilton Smith, and Daniel Nathans (Nathans, 1979; Roberts, 2005; Smith, 1979). By 1972, Paul Berg employed restriction enzymes for DNA recombination (Mulligan and Berg, 1980; Okayama and Berg, 1982, 1983).

5 GENOMICS AND THE EVOLUTION OF MOLECULAR BIOLOGY

Genome is a term that refers to the genes and chromosomes that comprise the complete set of nuclear material in an organism. Generally, most genomes are of the nucleic acid polymer, DNA, with a few exceptions of some viruses containing RNA as their genome (Gorbalenya et al., 1989). By the mid-1980s, the term "genomics" came into common use as a means to describe the newly emerging scientific field of sequencing, mapping, and analyzing genomes (Maxam and Gilbert, 1977; Hieter and Boguski, 1997; McKusick, 1997). The path to the genomics era was paved by numerous discoveries from the beginning of the second half of the twentieth century (Stent and Calendar, 1978; Judson, 1979). After Watson and Crick elucidated the structure of DNA, it opened the way to tremendous advancements in molecular biology with breakthroughs elucidating the molecular fundamental foundation of the gene functions through gene structure, gene expression, DNA replication, protein synthesis, and other mechanisms (Stent and Calendar, 1978; Judson, 1979; Choudhuri, 2003; Dahm, 2008).

In the middle of the 1960s, Robert Holley with his research group was the first to sequence and determine the structure of alanine tRNA from the yeast, *Saccharomyces cerevisiae*. This finding was the first to generate the sequence of an RNA (Holley, 1965; Holley et al., 1965a,b). In parallel, Frederick Sanger, who pioneered nucleic acid sequencing, along with his group was able to sequence the 5S ribosomal RNA from *E. coli* in the late 1960s (Brownlee and Sanger, 1967, Brownlee et al., 1968). In 1972, Fiers's research group deciphered the nucleotide sequence of a gene coding for a protein from bacteriophage MS2, and subsequently, they reported the first complete genomic sequence of all the genes of the RNA bacteriophage MS2 containing over 3000 nucleotides in its sequence (Min Jou et al., 1972; Fiers et al., 1976).

In 1973, Sanger published a paper describing a procedure for using DNA polymerase I to determine the DNA nucleotide sequence of bacteriophage fl (Sanger et al., 1973). Shortly afterward, Sanger and Coulson (1975) described a simple and rapid method known as "plus and minus" to verify DNA sequences with DNA polymerase I from *E. coli* and T4 DNA polymerase from bacteriophage (Sanger and Coulson, 1975). Following these initial findings, Sanger and colleagues reported the sequence of a bacteriophage genome and described a method for DNA sequencing with chain-terminating inhibitors (Sanger et al., 1977a,b; Brenner, 2014; Roe, 2014). In 1978, Sanger and Coulson developed a method that allowed for visualization of the DNA sequence with better rate and resolution of DNA fragment bands by using thin acrylamide gels (Sanger and Coulson, 1978). Within a couple of years, mitochondrial DNA of human cells was also sequenced by the Sanger research group (Anderson et al., 1981).

In the early 1980s, another innovation was developing that further revolutionized the field of molecular biology, when Kary Mullis developed a simple polymerase-based chain reaction using DNA polymerase from *E. coli* to amplify the desired DNA fragment(s) (Saiki et al., 1985). However, his method still had one limitation involving a heating step to denature the DNA, thus inactivating the

DNA polymerase, and necessitating the addition of new DNA polymerase after each cycle (Bartlett and Stirling, 2003). This was eventually overcome by employing a thermostable polymerase known as Taq DNA polymerase that was first isolated from the extreme thermophile *Thermus aquaticus* (Saiki et al., 1988). Collins and Weissman (1984) achieved a further breakthrough by employing a technique known as "chromosome jumping" for positional cloning of DNA fragments to use for disease gene identification without knowing their functions beforehand.

6 EMERGENCE OF SEQUENCING AS A PRACTICAL TOOL FOR MOLECULAR APPLICATIONS

By the mid 1980s, the interest in sequencing the human genome started to gather momentum mainly led by the U.S. Department of Energy (DOE) and the Office of Health and Environmental Research (OHER). In 1983, GenBank was established as the database of DNA sequences, and by 1990, the basic local alignment search tool (BLAST) was introduced to search the database for query sequence homologous. By 1986, the DOE initiated human genome sequencing (Choudhuri, 2003). Two years later, the National Institute of Health (NIH) joined the DOE collaborating in human genome sequencing research, and by 1989, NIH named their human genome research office as the National Center for Human Genome Research (NCHGR). In October 1990, the Human Genome Project (HGP) led by the United States became a collaborative enterprise performed through the International Human Genome Consortium in several research centers from five countries besides the United States, including the United Kingdom, France, China, Japan, and Germany. In 1997, NCHGR became the National Human Genome Research Institute (NHGRI) and DOE established the Joint Genome Institute (JGI) (Choudhuri, 2006).

In 1998, another human genome sequencing project was initiated in parallel with HGP by Celera Genomics, a private company, to complete sequencing of the human genome ahead of the HGP targets within a 3-year timeline (Venter et al., 2001). In 2003, the HGP announced the completion of the project. Throughout the path to the completion of the HGP, genomes of several species were sequenced in a concerted effort to understand some of the fundamental mechanisms and identify potential conserved genes that would cross align over various species (Collins et al., 2003). In 1995, the genome of *Haemophilus influenza* became the first completely sequenced free-living organism, and sequencing of bacterial genomes has continued at an ever-accelerating pace. Numerous genome sequencing projects have either been completed or are in the process of being completed for many model organisms. With this massive quantity of databases, researchers turned their attention from just studying several genes to more comprehensive systematic studies of the interactions of biological systems, structure, and functions. In late 2003, this transition necessitated the establishment of the **ENC**yclopedia **O**f **DNA** **E**lements (ENCODE) Project, a large-scale international project consortium. The aim of this project was to identify the functional characteristics in genomics analysis (ENCODE Project Consortium, 2004, 2012; Ecker, 2012).

As these sequencing projects progressed, the use of numerous molecular technologies became more routine, to the point that applications for other purposes were possible. Certainly, fundamental understanding of microorganisms was rapidly established for all of the major foodborne pathogens. Evolutionary relationships among isolates within genera, species, and serovars also became a reality and opened the door for understanding how new isolates could emerge and become prominent in a specific outbreak. Once these data were established, molecular tools for detection and other diagnostic purposes could be much more readily developed and applied. Along with the laboratory tools, the corresponding computer program-based search engines were developed that included sequence data from various microorganisms. This allowed not only for comparative analyses among characterized and newly acquired isolates but also for the design of molecular probes and primers useful for distinguishing foodborne pathogens in food matrices based on the PCR and other molecular-based assays. While comprehensive reviews provide much more detailed descriptions and should be referred to for in-depth background (Kwon and Ricke, 2011; O'Flaherty and Klaenhammer, 2011; Douglas et al., 2013; Gerner-Smidt et al., 2013; Park et al., 2014a; Vongkamjn and Weidmann, 2015), the following sections will describe some of the fundamentals of these approaches and how they have been used for traceability/tracking of foodborne pathogens.

7 PROFILING FOODBORNE PATHOGENS USING GEL ELECTROPHORESIS

Once the application of restriction enzymes to cleave DNA at specific nucleotide sites became routine and numerous enzymes were commercially made available, the concept of using gel electrophoresis to differentiate bacterial chromosomal DNA as a function of the restriction enzyme digest-generated banding patterns emerged. One of the initial applications of this principle that became extensively used for a variety of microbial ecosystems to characterize microbial community diversity was denatured gradient gel electrophoresis. It is based on the PCR amplification of 16S rRNA, which generates a range of product sizes from unique sequences representative of each microorganism. These amplified products can be separated on a gradient electrophoresis gel where the gradient is created by the concentration of denaturant starting at the lowest concentration and increasing with the highest concentration at the bottom of the gel (Hanning and Ricke, 2011). The forward primer consists of a 40 bp sequence referred to as a GC clamp, which is designed to keep part of the PCR-generated product double stranded and, therefore, confined within the migration of the gel while the remainder of the 16S rRNA is single stranded (Hanning and Ricke, 2011). The key to separation is the GC content with the higher GC content strands migrating a greater distance than those with higher A-T content (Hanning and Ricke, 2011). The bands separated in the gel can be stained, viewed, and dendrograms of relatedness in a microbial community generated, and depending on the application, individual bands in the gel can be isolated and sequenced to identify individual organisms. This approach has been used to characterize microbial populations in several food-related ecosystems including the gastrointestinal tracts of food animals and the microbiota associated with food processing and retail (Hume et al., 2003; Ricke et al., 2004; Dunkley et al., 2007; Handley et al., 2010; Koo et al., 2012; Mertz et al., 2014; Park et al., 2014b).

In a similar fashion, the ability to discern restriction enzyme generated fragments has been used for developing restriction fragment length polymorphism (RFLP) mapping approaches for typing unique strains of bacteria (Manfreda and De Cesare, 2005; Gerner-Smidt et al., 2013). The key to successful implementation of this approach for molecular subtyping was to reduce the number of restriction fragments generated via application of less frequently digesting restriction enzymes (cleave fewer sites on the target DNA) and use pulsed-field gel electrophoresis (PFGE) to separate the resulting fragments (de Boer and Beumer, 1999; Goering, 2010; Gerner-Smidt et al., 2013). Pulsed gel electrophoresis apparatuses were designed to accommodate large-sized DNA molecules (several Mb) by periodically altering the electrical field as a function of time interval gradients and forcing the DNA fragments to traverse in different orientations in the gel with net mobility being size independent (Herschleb et al., 2007; Goering, 2010; Gerner-Smidt et al., 2013). Thus, choice of molecular size standards, appropriate range of molecular sizes for visualization, and consistent placement in gels are all critical technical considerations along with the appropriate use of computer-assisted data analyses (Goering, 2010). While RFLP-based typing methods have become widely practiced and accepted, Gerner-Smidt et al. (2013) has pointed out there are limitations due to variations among laboratories as well as reproducibility, in part due to the fact that bands of apparently similar size may not only be difficult to separate on the gel but also not come from the same location of the chromosome, thus containing differing genetic content. In addition, Phillips (1999) has noted that the interpretation of the original PFGE band pattern criteria may need modification for some organisms, such as pathogenic *E. coli*, to achieve complete discrimination.

8 PROFILING FOODBORNE PATHOGENS USING DNA SEQUENCE-BASED PROFILING

As genomic sequencing of more foodborne pathogen isolates occurs and the corresponding genome sequences are added to accessible data libraries, the ability to devise the means to do DNA sequence-based comparative typing analyses has become more routine. As has been discussed much more extensively in a number of previous reviews, PCR assays have historically been developed for routine application and marketed as commercial detection systems for foodborne pathogens both as single primer set systems as well as multiplex-based primer sets for use in a wide range of food and animal feed matrices (Hill, 1996; Hanna et al., 2005; Maciorowski et al., 2005; Jarquin et al., 2009; Kuchta et al., 2014; Park et al. 2014a). Further development has led to quantitative PCR- and RNA-based assays to estimate viable pathogen cell numbers and have also been used to determine individual gene expression levels (Wang et al., 2013; Kuchta et al., 2014; Park et al., 2014a). More recent developments in alternative amplification approaches have been described in detail elsewhere (Kuchta et al., 2014) and include Plexor, an alternative real-time PCR chemistry based on quenching during amplification; loop-mediated isothermal amplification; helicase-dependent amplification; isothermal amplification with continuous luminometry; and nucleic acid sequence-based amplification. Kuchta et al. (2014) has also suggested that digital PCR (dPCR) based on endpoint PCR represents a means to achieve better quantification by conducting 15,000-20,000 parallel low-volume PCR reactions in a series of diluted template DNA solutions. Following the PCR assay, they (Kuchta et al., 2014) suggest calculating absolute quantities via the counting of positive versus negative amplification outcomes at the optimal dilution level.

For subtyping, the classic approach has involved using PCR to directly amplify repetitive palindromic elements (rep-PCR) known to occur in strains of a particular organism (Barco et al., 2013; Gerner-Smidt et al., 2013). To achieve more discrimination, combining PCR-based amplification and restriction amplified fragments (AFLP) offers the opportunity to use PCR-specific amplification either of already restriction enzyme digested fragments or for generation of restriction fragments that fit a RFLP profile such as those identified with the O antigen genes in pathogenic *E. coli* (DebRoy et al., 2011; Gerner-Smidt et al., 2013). Multilocus sequence typing (MLST) can be used to distinguish isolates based on subtle nucleotide differences in a select set of housekeeping gene fragments ranging from 450 to 500 bp, applied to much shorter 1-10 bp repeating sequence units called variable number tandem repeats (VNTR) or if multiple locus VNTR (referred to as multiple locus variable number of tandem repeats analysis or MLVA) are needed to achieve discernment (Gerner-Smidt et al., 2013). Foley et al. (2006) concluded from comparing subtyping methods on 128 *Salmonella enterica* serovar Typhimurium isolates from various food animal sources that while PFGE, MLST, and rep-PCR were the best at discernment, no relationship was detected among the typing methods and suggested that combining methods was the optimal approach at least for identification of closely clustering *S.* Typhimurium. Since then some improvements have been made on MLVA methodology to enhance resolution. For example, an MLVA approach combined with flow-through gel electrophoresis has been shown to have sufficient discernment and reproducibility for tracking *S. aureus* spatial and temporal contamination in cheese and meat processing plants (Rešková et al., 2013). Keeratipibul et al. (2015) combined MLV with high-resolution melting analysis (MLV-HRMA) and demonstrated that they could differentiate *S.* Typhimurium isolates from various environments in Thailand based on allelic variants in five tandem repeat loci.

Other sequence sites on foodborne pathogens may offer more unique targets for more precise subtyping among closely related isolates. The clustered regularly interspaced short palindromic repeat (CRISPR) regions in bacteria have been defined by Bhaya et al. (2011) as originating from foreign DNA that is incorporated between host short DNA repeats and thus serve as sort of a "nonself" recognition site for the host cell. These CRISPR regions in turn serve as bacterial adaptive immune monitoring systems that act in concert with CRISPR-associated host proteins (Cas) to recognize and potentially neutralize or "silence" foreign incoming DNA via transcription of small RNAs (Horvath and Barrangou, 2010; Bhaya et al., 2011; Wiedenheft et al., 2012; Hynes and Molneau, 2014). Bhaya et al. (2011) has suggested that both the temporal and spatial hypervariable nature of the CRISPR spacers offer epidemiological opportunities for genotyping of nearly identical bacterial strains both within populations and as a function of time. This has been put into practice with *Salmonella* where CRISPR loci were used along with multivirulence locus sequence typing (CRISPR-MVLST) for the subtyping characterization of *Salmonella* populations from different sources (Liu et al., 2011a,b; DiMarzio et al., 2013). Wang et al. (2013) has suggested that polymorphisms of the CRISPR regions in pathogenic *E. coli* may also prove to be unique targets for PCR. As more becomes known about CRISPR systems and more regions are identified for additional foodborne pathogens, there appear to be opportunities to fully develop subtyping approaches that combine these sequences with other typing methodologies and/or in some cases use them alone (Barco et al., 2013).

9 DNA MICROARRAYS

In the early 2000s, as more genomes were sequenced, efforts were undertaken to use the resulting genomic information to develop methods for simultaneously screening an entire genome of a particular organism. This led to the generation of glass DNA microarray slides, which essentially consist either of cross-linked oligonucleotides or PCR products representing individual genes, the detection of which is essentially based on the level of hybridization between the single-stranded DNA probes attached to the surface versus the target DNA (Lucchini et al., 2001; Sirsat et al., 2010; Douglas et al., 2013; Ricke et al., 2013). As would be expected, the potential for quantitation of gene expression in a presentation of the entire genome for a particular foodborne organism was highly attractive for assessing responses to environmental cues commonly encountered (Kuipers et al., 1999; Ricke et al., 2013). Consequently, a wide range of studies have been conducted examining foodborne pathogen organisms such as *Salmonella* responses to potential stressors including organic acids, mild heat shock, the presence of potentially antimicrobial food matrix substrates, and antibiotics, among others

(Dowd et al., 2007; Milillo et al., 2011; Sirsat et al., 2011a; Chalova et al., 2012; Ricke et al., 2013). The advantage of using microarrays for these types of questions was not just in the ability to assess entire genomes but also in the capacity to determine gene relatedness via quantitation of individual gene level responses and subsequent generation of Venn diagrams to identify shared gene responses to seemingly different environmental conditions (Sirsat et al., 2010; Ricke et al., 2013).

This same property of being able to present large sets of genes in a DNA microarray format also lends itself well for bacterial identification and subsequent subtyping (Goldschmidt, 2006; Rasooly and Herold, 2008; Gerner-Smidt et al., 2013). Consequently, DNA microarrays offered opportunities for more detailed genetic analyses than previously developed detection and typing methods. This may be particularly true when comparing the genomic profiles of closely related organisms such as the multitude of *Salmonella* serovars. For example, Chan et al. (2003) conducted genome hybridization of serovars and strains of both *S. enterica* (subspecies I and IIIa) and *Salmonella bongori* using a *S. enterica* serovar Typhimurium spotted DNA microarray as the comparative template to demonstrate the presence of genomic differences that contribute to the level of host adaptation by a particular serovar. Likewise, Alvarez et al. (2003) used comparative genomic hybridization microarrays to demonstrate that *S. enterica* serovar California isolated from Spanish feed mills was missing 23 gene clusters present in *S.* Typhimurium strain LT2. Huehn et al. (2010) used DNA microarrays consisting of oligonucleotide probes for seven different marker groups (pathogenicity, antimicrobial resistance, serotyping, fimbriae, DNA mobility, metabolism, and prophages) previously identified from the research literature for virulotyping and antimicrobial resistance typing of over 500 *S. enterica* subsp. *enterica* strains making up the five major European serovars (Enteritidis, Typhimurium, Infantis, Virchow, and Hadar). They observed that strains based on virulence genes tended to group together within a serovar except for some within serovar strain variation in mobile genetic elements such as those encoding prophages or plasmids.

Using microarrays directly as a typing tool in a practical setting may be more difficult. When Koyuncu et al. (2011) assessed DNA microarray platforms for the potential to identify and type *Salmonella* in artificially contaminated animal feeds they observed good agreement with DNA aliquots extracted from the selective enrichment (modified semisolid Rappaport Vassiliadus) plates bacterial migration boundaries but not with the samples taken directly after centrifugation of the buffered peptone water. Sirsat et al. (2011b) reported that high inocula levels of *S.* Typhimurium were needed to directly recover sufficient genetic material for consistent microarray analyses of *Salmonella*-spiked chicken breast meat. Specifically, amplifying the pathogen target DNA in the sample using large-scale multiplex PCR has been suggested by Palka-Santini et al. (2009) as a potential means to improve fluorescence-based microarray detection sensitivity. More recently, different approaches to immobilization techniques have been proposed by Nimse et al. (2014) not only as a means to better optimize probe positioning on the surface for improved hybridization efficiency but also to potentially offer up the possibility of regeneration of the surface-immobilized probes for DNA microarray reuse.

10 GENOMICS AND NEXT GENERATION SEQUENCING TECHNOLOGIES

Over the first decade of the twenty-first century, new molecular platforms with sophisticated processing capability for whole genome sequencing technologies and capacities rapidly advanced (Mardis, 2008a,b; Von Bubnoff, 2008; MacLean et al., 2009; Metzker, 2010; Messing, 2014). This is evident in not only the explosive number of bacterial isolates that are now sequenced but the speed and low cost involved in doing the sequencing and laboratory processing (Medini et al., 2005; O'Flaherty and Klaenhammer, 2011; Blow, 2013; Messing, 2014; Park et al., 2014a). Some key technological breakthroughs occurred to make this possible. Historically, sequencing platforms initially involved electrophoretic separation of terminated DNA chains by classic electrophoretic-based separation (Metzker, 2010; O'Flaherty and Klaenhammer, 2011; Grada and Weinbrecht, 2013; Messing, 2014). In contrast, the more recent generation of sequencing platforms are characterized by flow cell sequencing that essentially are conducted as repetitive cycles of nucleotide extension run in parallel on a massive number of clonally amplified template molecules (Holt and Jones, 2008). This results in a stepwise determination of DNA sequences that has the advantage of allowing an extensive parallel sequencing approach for a large number of template fragments; hence the name, massively parallel sequencing or next-generation DNA sequencing (NGS) (Metzker, 2010; Kwon and Ricke, 2011).

As the technology evolved these massively parallel-sequencing platforms became commercialized by different companies (see Ku and Roukos, 2013; Metzker, 2010; Reis-Filho, 2009; Schuster, 2008; Glenn, 2011; O'Flaherty and Klaenhammer, 2011; Quail et al., 2012; Park et al., 2014a for detailed descriptions and comparisons). They differ in their template preparation and sample(s) processing as well as cost per nonmultiplexed sample and instrument run time (Glenn, 2011; Quail et al., 2012; Grada and Weinbrecht, 2013). The amplification step in the 454 FLX Roche system and ABI SOLiD system is an emulsion PCR (emPCR) bead-based enrichment method with one DNA molecule per bead and clonal amplification occurs in microreactors in an emulsion, but are different in sequencing strategy. The 454 FLX Roche System uses pyro-sequencing technique that sequences by synthesis (Margulies et al., 2005; Rothberg and Leamon, 2008) whereas the ABI SOLiD system utilizes ligation sequencing procedures (Park et al. 2014a). The Ilumina genome analyzer involves a bridge amplification step and reverse terminator sequencing strategy with one DNA molecule per cluster (Holt and Jones, 2008; Glenn, 2011; Quail et al., 2012; Grada and Weinbrecht, 2013; Park et al., 2014a). A system known as Helicos Heliscope with no amplification uses a sequencing strategy with a single molecule sequence and is also now being applied to eukaryotic genomes as well (Metzker, 2010; Thompson and Steinmann, 2010; O'Flaherty and Klaenhammer, 2011; Orlando et al., 2015).

These technologies have rapidly progressed from next-generation sequencing or second-generation sequencing platforms to third-generation sequencing and to fourth-generation sequencing (Podolak, 2010; Liang et al., 2014). The main factors for improving and developing new high-throughput sequencing systems include sequencing rate, read lengths, simple sample preparation, and lower analysis cost (Glenn, 2011; Perkel, 2011; Ku and Roukos, 2013; Liang et al., 2014). It is also anticipated that improvements will continue to be made in the fundamental understanding of genomics that could eventually be incorporated into even more efficient and less costly sequencing platforms. For example, the series of studies by Jacqueline Barton's group that have led them to speculate that perhaps DNA repair enzymes may use electron flow to communicate along the chromosome with the genomic DNA template, thus serving as a sort of an electronic "livewire" and concomitantly allowing for much faster repair rates (Service, 2014). This of course sets up the potential premise that if DNA sequences behave in such a fashion this property could somehow be incorporated into further genome analysis.

Other trends are emerging as well. For example, as sequencing technologies continue to develop and become commercialized, Messing (2014) has pointed out that actual sequencing efforts have evolved away from being primarily being focused on large research centers to becoming more decentralized in numerous laboratories. As Blow (2013) has concluded, this is not just because of decreased costs for reagents and equipment but also because the development of innovative methods and techniques help maximize output and applications for the small laboratory user. This in turn has led to more applications related to foodborne pathogens, not only to generate sequence data for assembling fundamental baseline information but also to use in tracking and understanding evolution of strains either as a specific property such as antibiotic resistance, or host preference or origin, and dissemination during an outbreak (Gerner-Smidt et al., 2013; Beatson and Walker, 2014; Park et al., 2014a; Vongkamjn and Weidmann, 2015).

Whole genome sequencing of isolates has already been shown to have had utility in foodborne disease outbreaks for identifying and characterizing the causative isolate of the particular foodborne pathogen involved (den Bakker et al., 2011; Gilmour et al., 2011; Lienau et al., 2011; Hoffmann et al. 2012, 2013; Muniesa et al., 2012; Barco et al., 2013; Gerner-Smidt et al., 2013; Wain et al., 2013; Vongkamjn and Weidmann, 2015). Next-generation sequencing can also be used to improve on existing typing methods. For example, whole genome sequencing offers the opportunity to identify single-nucleotide polymorphisms, which are DNA sequences that differ at a single base-pair between strains and this could be used for comparative purposes (Boxrud, 2010; Gerner-Smidt et al., 2013). In addition, massively parallel sequencing has been used to develop a high-resolution molecular typing method, MLST-seq, that is based on combining a PCR-based target enrichment with next-generation sequencing to simultaneously analyze multiple target gene sequences (Singh et al., 2012, 2013).

As expected, sequencing has started to have an impact over an increasing range of food safety applications and approaches for conducting research on foodborne pathogens. Intuitively, one of the more immediate applications is to use sequence data to assess the effectiveness of previous methodologies for detecting and comparing isolates. Whole genome sequencing has been used to evaluate and compare typing methods for several of the foodborne pathogens. For example, Pendleton et al. (2013) compared whole genome sequencing of *Campylobacter jejuni* isolates that had originated from a wide range of sources including pasture flock-raised poultry with two of the methods more typically used for *Campylobacter*: PFGE typing and *flaA* gene sequencing. Sequencing of the *C. jejuni* isolates was done on a 454 Life Sciences GS FLX Titanium series sequencer (Roche Diagnostics, Branford, CT, USA). They concluded that sequencing was the most discriminatory and revealed unique genotyping characteristics including genome size, percent G-C content, and identification of genomic rearrangements. In addition, they noted that the genome of *Campylobacter* isolates may possess a certain level of plasticity with genomic rearrangements perhaps being fairly common, thus rendering whole genome sequencing more advantageous as a typing and tracking tool than the more conventional typing approaches.

Genome plasticity may be an important factor that impacts typing in other foodborne pathogens as well. Dowd et al. (2010) used a combination of microarray expression analyses, quantitative PCR, and a Roche Applied Science 454 GS-FLX sequencer to assemble draft genomes of two separate genetic lineages of *E. coli* O157:H7 isolated from cattle, one of which was more was more pathogenic to human hosts than the other due to an increased production of shiga toxin 2. Based on these analyses they concluded that transposon rearrangements had occurred in these isolates from the less pathogenic lineage and essentially had disrupted the corresponding genes resulting in repression of expression. Other factors such as cell population density has also been shown to be a factor in *E. coli* causing plasticity in antibiotic resistance mutation rates (Kraŝovec, et al., 2014). It is conceivable that as other foodborne pathogens are delineated via sequencing of more strains and isolates that such changes in the genome may account for other phenotypic differences among both distinct species, serovars, and even subtle differences in strains (Milillo et al., 2012; Foley et al., 2013). Sorting out phenotypic differences as they relate to genetic strain differences is critical not only for subtyping purposes but also because of the practical issues that impact food safety protocols, such as differing growth rates in enrichment and selective culture media and differences in stress responses to the presence of various interventions (González-Gil et al., 2012; Gorski, 2012; Lianou and Koutsoumanis, 2013a,b).

As more sequencing is done, some of the issues regarding the seemingly complex nature and evolutionary relationships of foodborne pathogen diversity may start to be answered and, in turn, can be taken into account for tracking isolates within and across food production environments during the sequential phases of an outbreak. Certainly, comparative genomics has revealed differences in nontyphoidal versus the more systemic infective typhoidal *Salmonella* species with typhoidal isolates having exclusive protein families associated with them (Zou et al., 2014). Differences in genetics and subsequent evolution appear to occur for nontyphoidal *Salmonella* as well and may be attributable to several factors. For example, based on sequencing of *Salmonella* bacteriophages isolated from dairy farms, Moreno Switt et al. (2013) suggested that some of the evolution associated with *Salmonella* host specificity could be linked to *Salmonella* phage tail fiber and tailspike diversity. This is consistent with studies conducted by Fricke et al. (2011) where they used comparative genomics among sequenced *S. enterica* to conclude that sublineages evolved by two mechanisms; namely, loss of metabolic functions resulting in functional reduction, and acquiring horizontally transferred phage and plasmid DNA that added virulence and resistance functions. However, there may be other controlling factors as they suggested that the presence of CRISPR, which serves as an immune-like defense mechanism against these horizontally transferred genetic elements that could, in turn, impact any short-term phenotype changes and long-term sublineage evolution that might occur.

Developing a better understanding of the genetics of foodborne pathogen evolution has practical significance as it is important to establish which factors are the most critical for controlling not just the pathway but the types of changes. Some factors may be less critical, but other factors could have immediate impacts that matter. For example, increases in properties such as antibiotic resistance or development of simultaneous resistance to multiple antibiotics would have critical public health and clinical consequences. As part of this discovery process it will also be important to know what type of impact the environmental pressures associated with food production and processing have in contributing to these changes. Once more of this becomes known it may be possible to design manageable strategies that serve to reduce food safety risk and still be cost-effective.

11 CONCLUSIONS AND FUTURE DIRECTIONS

Molecular methods have undergone one of the most remarkable development periods in the past half-century not only in the history of scientific research, but also in the concomitant implementation of commercial applications. Numerous fields of biology are now being infiltrated to different degrees by molecular biology and this is continuing to increase. Certainly, foodborne pathogen research has been heavily influenced by the application of molecular approaches all the way from achieving a more in-depth fundamental understanding of a particular organism to the development of better detection systems for strains that commonly contaminate foods. However, food safety systems in general are also being impacted by molecular technology. This is manifested in a number of ways, some of which are more obvious than others. For example, the pace of development and commercialization of molecular based detection technologies can be to some extent driven by food safety regulatory policy, depending on the foodborne pathogen and the perceived threat. This pace is set to some extent by the availability of current commercially relevant technology, the level of federal government and other sources of funding such as agricultural commodity groups to support the development of more advanced and/or applicable technology, and the criteria for approval as a standard, universally acceptable protocol (Ricke, 2010; Shaw et al., 2010; Fields, 2014; Kennedy, 2014).

In addition, innovation will probably benefit from the trend toward convergence of multidisciplinary efforts and corresponding infrastructure that evolves both in academic institutions and in entrepreneurial partnerships with industry (Sharp, 2014). Finally, gaining a better understanding of pathogen evolution will also need to be considered (Raskin et al., 2006). Other impacts of molecular biology that are starting to influence food safety include not only understanding the functional genomics of organisms in the gastrointestinal tract as well as in food processing environments but also, in turn, developing more optimal interventions and control measures. This will include identifying novel targets for new antibiotic development and vaccines or designing unique multiple intervention approaches for different types of food production systems such as those that are organic based (Hughes, 2003; Ricke, 2003b, 2014; Ricke et al., 2005, 2012; Sirsat et al., 2009, 2010).

Certainly, current technological capabilities can have an impact as well, particularly when attempting to track an ongoing foodborne pathogen outbreak. This is most clearly evident in the rapid advances made toward routine whole genome sequencing and the efforts to use it as a tool for identifying specific isolates and their corresponding origins during a disease outbreak, including those caused by foodborne pathogens (Walker and Beatson, 2012; Gerner-Smidt et al., 2013; Vongkamjn and Weidmann, 2015). As sequencing becomes more sophisticated, it is anticipated that the applications that have utility for food safety and foodborne pathogens will only expand and potentially include more integration of omics approaches including metagenomics, transcriptomics, epigenomics, and proteomics (De Keersmaecker et al., 2006; O'Flaherty and Klaenhammer, 2011; Winder et al., 2011; Baker and Dick, 2013; Park et al., 2013).

However, there remain challenges with incorporating rapid sequencing into routine applications for food safety. First of all, devising a means of not only interpreting the data but just the sheer logistics of managing libraries of sequences is a daunting challenge. Consequently, given the vast amounts of genomic information that can now be generated, there is new emphasis on developing more efficient methods to not only simultaneously sequence multiple samples but also to recover individual data sets afterward. The concept of DNA sequence labeling has been suggested as a means to identify each sample by ligating a short, known sequence or "barcode" such that samples can be pooled during sequencing yet each individual sequencing sample be easily discerned during data processing (Andersen et al., 2014). Peikon et al. (2014) has taken an intriguing approach on this by developing a novel method for uniquely tagging single cells *in vivo* with a genetic "barcode" that in turn can be recovered during DNA sequencing. They have suggested that the utility of this approach is the potential ability to track cells in fairly complex networks and populations of organisms. It would not be difficult to consider using such an approach in food systems for tracking either foodborne pathogens and/or bacterial cell members in the indigenous microbiomes associated with a food matrix or gastrointestinal ecosystem.

A different concept that may also have an impact of foodborne pathogen tracking involves identifying DNA sequences already present in an organism that could serve as potential standardized identification "barcodes." As described by Lebonah et al. (2014), this essentially involves identifying a short DNA sequence from a standard section of the bacterial genome and using this as a reference "barcode" for comparing unknown isolates against a reference catalogue of known barcode sequences of previously identified organisms (Hebert et al., 2003; Savolainen et al., 2005; Hajibabaei et al., 2007; Lebonah et al., 2014). Given the rapidly expanding discovery of diverse microorganisms, the logistics for a barcoding approach may prove complicated. However, this concept may have more immediate merit for traceability of specific sources of foods and ingredients and, at least with fish, there have been efforts to develop DNA identification systems for seafoods to address issues of fraud and authentication of fish and seafood species (Rasmussen and Morrissey, 2008; Teletchea, 2009). However, as pointed out by Rasmussen and Morrissey (2008), the genetic database for different species of fish is critical for any development and implementation of DNA identification methods. This will be true for other food-animal, vegetable, and crop species as well, but as more large-scale genome sequencing projects are completed this will change. The much more difficult aspect will be to determine whether there are any distinguishable DNA markers uniquely associated with foods from animals and plants grown under specific environmental settings such as organic production.

The other issue with the rapidly increasing sequencing database is twofold: processing and finalizing sequence information itself and attempting to align this data with more traditional subtyping methods currently used by networks such as U.S. PulseNet and PulseNet International (Gerner-Smidt et al. 2013; Vongkamjn and Weidmann, 2015). As Gerner-Smidt et al. (2013) have pointed out, before whole genome sequencing can become a routine typing method its feasibility still needs to be assessed from the standpoint of validation with defined standards that provide a baseline for natural variation in the sequence of a targeted organism. Likewise, Gerner-Smidt et al. (2013) noted that challenges remain with computational processing of these large data sets of short sequence reads as well as storage, quality control, and transfer capabilities. At least for the near future, it appears that subtyping and identification with PFGE will remain the standard and there is some merit in this. For example, a meta-analysis conducted by Zou et al. (2013a) of over 45,000 PFGE profiles of 32 *Salmonella* serotypes from human sources led them to conclude that there may be possible PFGE markers for serotype identification. This data was subsequently used as part of the development of a bioinformatics software program for enhanced PFGE data retrieval, interpretation, and serotype identification (Zou et al., 2013b). The ability to accomplish this type of analysis with large PFGE data sets suggests that as sequencing-based bioinformatics progresses and the corresponding software becomes more available, similar adaptations of whole genome data can occur.

Further developments in molecular technology for tracking in food safety systems will no doubt continue if not accelerate at an even faster pace. One of the keys to this is the further advancement of fundamental understanding of foodborne pathogens based on in-depth analyses of subcellular processes in single cells. In addition, intercellular interactions among cell populations in ecosystems such as biofilms may help to explain some of differences observed in subpopulations. This, in turn, may also (along with other factors) contribute to the strain-to-strain variation that appears to impact phenotype responses. Whole cell genome sequences will continue to be critical to discern the minor genotype variations that may account for some of the differences. In addition, as bioinformatics becomes more advanced, there should be opportunities to develop mathematical models that have the ability to merge divergent modeling systems for risk assessment and perhaps even environmental LCA analysis. Accomplishing this would lead to truly "systems" approaches to the assessment of food production on a scale that would allow decisions to be made based on a wide range of varying impact factors. Therefore, one could be in a position to predict most, if not all, of the potential outcomes from an environmental, risk, economical standpoint, or any other system models, when only a single factor such as a new intervention agent is introduced.

REFERENCES

Alvarez, J., Porwollik, S., Laconcha, I., Gisakis, V., Vivanco, A.B., Gonzalez, I., Echenagusia, S., Zabala, N., Blackmer, F., McClelland, M., Rementeria, A., Garaizar, J., 2003. Detection of a *Salmonella enterica* serovar California strain spreading in Spanish feed mills and genetic characterization with DNA microarrays. Appl. Environ. Microbiol. 69, 7531–7534.

Andersen, J.D., Pereira, V., Pietroni, C., Mikkelsen, M., Johansen, P., Børsting, C., Morling, N., 2014. Next-generation sequencing of multiple individuals per barcoded library by deconvolution of sequenced amplicons using endonuclease fragment analysis. Biotechniques 57, 91–94.

Anderson, S., Bankier, A.T., Barrell, B.G., de Bruijn, M.H.L., Coulson, A.R., Drouin, J., Eperon, I.C., Nierlich, D.P., Roe, B.A., Sanger, F., Schreier, P.H., Smith, A.J.H., Staden, R., Young, I.G., 1981. Sequence and organization of the human mitochondrial genome. Nature 290, 457–465.

Angus-Lee, H., 2015. The evolution of traceability in the meat & poultry industry. Food Saf. Mag. 20 (42), 44–47.

Avery, O.T., MacLeod, C.M., McCarty, M., 1944. Studies on the chemical nature of the substance inducing transformation of pneumococcal types induction of transformation by a desoxyribonucleic acid fraction isolated from pneumococcus type III. J. Exp. Med. 79, 137–158.

Baker, B.J., Dick, G.J., 2013. Omic approaches in microbial ecology: charting the unknown. Microbe 8, 353–360.

Barco, L., Barrucci, F., Olsen, J.E., Ricci, A., 2013. *Salmonella* source attribution based on microbial subtyping. Int. J. Food Microbiol. 163, 193–203.

Bartlett, J.M.S., Stirling, D., (Eds.) 2003. A short history of the polymerase chain reaction. In: PCR Protocols—Methods in Molecular Biology, vol. 226. Humana Press, Totowa, NJ, pp. 3–6.

Batz, M., Hoffmann, S., Morris Jr., J.G., 2014. Disease-outcome trees, EQ-5D scores, and estimated annual losses of quality-adjusted life years (QALYs) for 14 foodborne pathogens in the United States. Foodborne Pathog. Dis. 11, 395–402. http://dx.doi.org/10.1089/fpd.2013.1658.

Beatson, S.A., Walker, M.J., 2014. Tracking antibiotic resistance—sequencing plasmids can reveal the transmission of resistance among bacteria from patients in a clinical setting. Science 345, 1454–1455.

Beddington, J.R., Asaduzzaman, M., Clark, M.E., Fernández Bremauntz, A., Guillou, M.D., Howlett, D.J.B., Jahn, M.M., Lin, E., Mamo, T., Negra, C., Nobre, C.A., Scholes, R.J., Van Bo, N., Wakhungu, J., 2012. What next for agriculture after Durban? Science 335, 289–290.

Bhaya, D., Davison, M., Barrangou, R., 2011. CRISPR-Cas systems in bacteria and archaea: versatile small RNAs for adaptive defense and regulation. Ann. Rev. Genet. 45, 273–297.

Blow, N., 2013. A sequencer in every lab. Biotechniques 55, 284. http://dx.doi.org/10.2144/000114107.

Bonne, K., Verbeke, W., 2008. Muslim consumer trust in halal meat status and control in Belgium. Meat Sci. 79, 113–123.

Boxrud, D., 2010. Advances in subtyping methods of foodborne disease pathogens. Curr. Opin. Biotechnol. 21, 137–141.

Brenner, S., 2014. Frederick Sanger (1918-2013). Science 343, 262.

Brownlee, G.G., Sanger, F., 1967. Nucleotide sequences from the low molecular weight ribosomal RNA of *Escherichia coli*. J. Mol. Biol. 23, 337–353.

Brownlee, G.G., Sanger, F., Barrell, B.G., 1968. The sequence of 5 s ribosomal ribonucleic acid. J. Mol. Biol. 34, 379–412.

Callaway, T.R., Edrington, T.S., Loneragan, G.H., Carr, M.A., Nisbet, D.J., 2013. Shiga toxin-producing *Escherichia coli* (STEC) ecology in cattle and management based options for reducing fecal shedding. Agric. Food Anal. Bacteriol. 3, 39–69.

Castellini, C., Boggia, A., Paolotti, L., Thoma, G.J., Kim, D.S., 2012. Environmental impacts and life cycle analysis of organic meat production and processing. In: Ricke, S.C., Van Loo, E.J., Johnson, M.G., O'Bryan, C.A. (Eds.), Organic Meat Production and Processing. Wiley Scientific/IFT, New York, NY, pp. 113–136 (Chapter 7).

Chalova, V.I., Hernandez-Hernandez, O., Muthaiyan, A., Sirsat, S.A., Natesan, S., Sanz, M.L., Javier Moreno, F., O'Bryan, C.A., Crandall, P.G., Ricke, S.C., 2012. Growth and transcriptional response of *Salmonella* Typhimurium LT2 to glucose-lysine-based Maillard reaction products generated under low water activity conditions. Food Res. Int. 45, 1044–1053.

Chambers, D.A., Reid, K.B., Cohen, R.L., 1994. DNA: the double helix and the biomedical revolution at 40 years. FASEB J. 8, 1219–1226.

Chan, K., Baker, S., Kim, C.C., Detweiler, C.S., Dougan, G., Falkow, S., 2003. Genomic comparison of *Salmonella enterica* serovars and *Salmonella bongori* by use of an *S. enterica* serovar Typhimurium DNA microarray. J. Bacteriol. 185, 553–563.

Choudhuri, S., 2003. The path from nuclein to human genome: a brief history of DNA with a note on human genome sequencing and its impact on future research in biology. Bull. Sci. Technol. Soc. 23, 360–367.

Choudhuri, S., 2006. Some major landmarks in the path from nuclein to human genome 1. Toxicol. Mech. Methods 16 (2–3), 137–159.

Claeys, W.L., Cardoen, S., Daube, G., De Block, J., Dewettinck, K., Dierick, K., De Zutter, L., Huyghebaert, A., Imberechts, H., Thiange, P., Vandenplas, Y., Herman, L., 2013. Raw or heated cow milk consumption: review of risks and benefits. Food Control 31, 251–262.

Collins, F.S., Weissman, S.M., 1984. Directional cloning of DNA fragments at a large distance from an initial probe: a circularization method. Proc. Natl. Acad. Sci. U.S.A. 81, 6812–6816.

Collins, F.S., Morgan, M., Patrinos, A., 2003. The human genome project: lessons from large-scale biology. Science 300, 286–290.

Cotter, P.D., Hill, C., 2003. Surviving the acid test: responses of Gram-positive bacteria to low pH. Microbiol. Mol. Biol. Rev. 67, 429–453.

Crandall, P., Van Loo, E.J., O'Bryan, C.A., Mauromoustakos, A., Yiannas, F., Dyenson, N., Berdnik, I., 2012. Companies' opinions and acceptance of global food safety initiative benchmarks after implementation. J. Food Prot. 75, 1660–1672.

Crandall, P.G., O'Bryan, C.A., Babu, D., Jarvis, N., Davis, M.L., Buser, M., Adam, B., Marcy, J., Ricke, S.C., 2013. Whole-chain traceability, is it possible to trace your hamburger to a particular steer, a U.S. perspective. Meat Sci. 95, 137–144.

Crick, F.H.C., 1963. On the genetic code. Science 139, 461–464.

Dahm, R., 2005. Friedrich Miescher and the discovery of DNA. Dev. Biol. 278, 274–288.

Dahm, R., 2008. Discovering DNA: Friedrich Miescher and the early years of nucleic acid research. Hum. Genet. 122, 565–581.

de Boer, E., Beumer, R.R., 1999. Methodology for detection and typing of foodborne microorganisms. Int. J. Food Microbiol. 50, 119–130.

De Keersmaecker, S.C.J., Thijs, I.M.V., Vanderleyden, J., Marchal, K., 2006. Integration of omics data: how well does it work for bacteria? Mol. Microbiol. 62, 1239–1250.

De Reu, K., Messens, W., Heyndrickx, M., Rodenburg, T.B., Uyttendaele, M., Herman, L., 2008. Bacterial contamination of table eggs and the influence of housing systems. Worlds Poult. Sci. J. 64, 5–19.

DebRoy, C., Roberts, E., Fratamico, P.M., 2011. Detection of O antigens in *Escherichia coli*. Anim. Health Res. Rev. 12, 169–185.

den Bakker, H.C., Switt, A.I.M., Cummings, C.A., Hoelzer, K., Degoricija, L., Rodriguez-Rivera, L.D., Wright, E.M., Fang, R., Davis, M., Root, T., 2011. A whole-genome single nucleotide polymorphism-based approach to trace and identify outbreaks linked to a common *Salmonella enterica* subsp. *enterica* serovar Montevideo pulsed-field gel electrophoresis type. Appl. Environ. Microbiol. 77, 8648–8655.

DiMarzio, M., Shariat, N., Kariyawasam, S., Barrangou, R., Dudley, E.G., 2013. Antibiotic resistance in *Salmonella enterica* serovar Typhimurium associates with CRISPR sequence type. Antimicrob. Agents Chemother. 57, 4282–4289.

Douglas, G.L., Pfeiler, E., Duong, T., Klaenhammer, T.R., 2013. Genomics and proteomics of foodborne microorganisms. In: Doyle, M.P., Buchanan, R.L. (Eds.), Food Microbiology—Fundamentals and Frontiers, fourth ed. American Society for Microbiology Press, Washington, DC, pp. 975–995 (Chapter 39).

Dowd, S.E., Killinger-Mann, K., Blanton, J., San Francisco, M., Brashears, M., 2007. Positive adaptive state: microarray evaluation of gene expression in *Salmonella enterica* Typhimurium exposed to nalidixic acid. Foodborne Pathog. Dis. 4, 187–200.

Dowd, S.E., Crippen, T.L., Sun, Y., Gontcharova, V., Youn, E., Muthaiyan, A., Wolcott, R.D., Callaway, T.R., Ricke, S.C., 2010. Microarray analysis and draft genomes of two *Escherichia coli* O157:H7 lineage II cattle isolates FRIK966 and FRIK2000 investigating lack of Shiga toxin expression. Foodborne Pathog. Dis. 7, 763–773.

Downie, A.W., 1972. Pneumococcal transformation—a backward view. Fourth Griffith Memorial Lecture. J. Gen. Microbiol. 73, 1–11.

Dunkley, C.S., Dunkley, K.D., 2013. Greenhouse gas emissions from livestock and poultry. Agric. Food Anal. Bacteriol. 3, 17–29.

Dunkley, K.D., McReynolds, J.L., Hume, M.E., Dunkley, C.S., Callaway, T.R., Kubena, L.F., Nisbet, D.J., Ricke, S.C., 2007. Molting in *Salmonella* Enteritidis challenged laying hens fed alfalfa crumbles. II. Fermentation and microbial ecology response. Poult. Sci. 86, 2101–2109.

Dunkley, K.D., Callaway, T.R., Chalova, V.I., McReynolds, J.L., Hume, M.E., Dunkley, C.S., Kubena, L.F., Nisbet, D.J., Ricke, S.C., 2009. Foodborne *Salmonella* ecology in the avian gastrointestinal tract. Anaerobe 15, 26–35.

Ecker, J.R., 2012. ENCODE explained—serving up a genome feast. Nature 489, 52–53.

ENCODE (ENCyclopedia Of DNA Elements), 2004. ENCODE Project Consortium. Science 306, 636–640.

ENCODE Project Consortium, 2012. An integrated encyclopedia of DNA elements in the human genome. Nature 489, 57–74.

Ferreira, V., Wiedmann, M., Teixeira, P., Stasiewicz, M.J., 2014. *Listeria monocytogenes* persistence in food-associated environments: epidemiology, strain characteristics, and implications for public health. J. Food Prot. 77, 150–170.

Fields, S., 2014. Would Fred Sanger get funded today? Genetics 197, 435–439.

Fiers, W., Contreras, R., Duerinck, F., Haegeman, G., Iserentant, D., Merregaert, J., Min Jou, W., Molemans, F., Raeymaekers, A., Van Den Berghe, A., Volckaert, G., Ysebaert, M., 1976. Complete nucleotide sequence of bacteriophage MS2 RNA: primary and secondary structure of the replicase gene. Nature 260, 500–507.

Foley, S.L., White, D.G., McDermott, P.F., Walker, R.D., Rhodes, B., Fedorka-Cray, P.J., Simjee, S., Zhao, S., 2006. Comparison of subtyping methods for differentiating *Salmonella enterica* serovar Typhimurium isolates obtained from food animal sources. J. Clin. Microbiol. 44, 3569–3577.

Foley, S.L., Johnson, T.J., Ricke, S.C., Nayak, R., Danzelsen, J., 2013. *Salmonella* pathogenicity and host adaptation in chicken-associated serovars. Microbiol. Mol. Biol. Rev. 77, 582–607.

Freifelder, D., 1983. Molecular Biology—A Comprehensive Introduction to Prokaryotes and Eukaryotes. Jones and Bartlett Publishers, Inc., Boston, MA, 979 pp.

Fricke, W.F., Mammel, M.K., McDermott, P.F., Tartera, C., White, D.G., LeClerc, J.E., Ravel, J., Cebula, T.A., 2011. Comparative genomics of 28 *Salmonella enterica* isolates: evidence for CRISPR-mediated adaptive sublineage evolution. J. Bacteriol. 193, 3556–3568.

Gellynck, X., Verbeke, W., Vermeire, B., 2006. Pathways to increase consumer trust in meat as a safe and wholesome food. Meat Sci. 74, 161–171.

Gerner-Smidt, P., Hyytia-Trees, E., Barrett, T.J., 2013. Molecular source tracking and molecular subtyping. In: Doyle, M.P., Buchanan, R.L. (Eds.), Food Microbiology—Fundamentals and Frontiers, fourth ed. American Society for Microbiology Press, Washington, DC, pp. 1059–1077 (Chapter 43).

Gilmour, M.W., Graham, M., Van Domslaar, G., Tyler, S., Kent, H., Trout-Yakel, K.M., Larios, O., Allen, V., Lee, B., Nadon, C., 2011. High-throughput genome sequencing of two *Listeria monocytogenes* clinical isolates during a large foodborne outbreak. BMC Genomics 11, 120.

Glenn, T.C., 2011. Field guide to next-generation DNA sequencers. Mol. Ecol. Resour. 11, 759–769.

Goering, R.V., 2010. Pulsed field gel electrophoresis: a review of application and interpretation in the molecular epidemiology of infectious disease. Infect. Genet. Evol. 10, 866–875.

Goldschmidt, M.C., 2006. The use of biosensor and microarray techniques in the rapid detection and identification of salmonellae. J. Assoc. Off. Anal. Chem. Int. 89, 530–537.

González-Gil, F., Le Bolloch, A., Pendleton, S., Zhang, N., Wallis, A., Hanning, I., 2012. Expression of *hilA* in response to mild acid stress in *Salmonella enterica* is serovar and strain dependent. J. Food Sci. 77, M292–M297.

Gorbalenya, A.E., Koonin, E.V., Donchenko, A.P., Blinov, V.M., 1989. Two related superfamilies of putative helicases involved in replication, recombination, repair and expression of DNA and RNA genomes. Nucleic Acids Res. 17, 4713–4730.

Gorski, L., 2012. Selective enrichment media bias the types of *Salmonella enterica* strains isolated from mixed strain cultures and complex enrichment broths. PLoS One 7, e34722. http://dx.doi.org/10.1371/journal.pone.0034722.

Grada, A., Weinbrecht, K., 2013. Next-generation sequencing: methodology and application. J. Invest. Dermatol. 133, e11. http://dx.doi.org/10.1038/jid.2013.248.

Guinée, J.B., Heijungs, R., Huppes, G., Zamagni, A., Masoni, P., Buonamici, R., Ekvall, T., Rydberg, T., 2011. Life cycle assessment: past, present, and future. Environ. Sci. Technol. 45, 90–96.

Hajibabaei, M., Singer, G.A.C., Hebert, P.D.N., Hickey, D.A., 2007. DNA barcoding: how it complements taxonomy, molecular phylogenetics and population genetics. Trends Genet. 23, 167–172.

Halog, A., Manik, Y., 2011. Advancing integrated systems modelling framework for life cycle sustainability assessment. Sustainability 3, 469–499. http://dx.doi.org/10.3390/su3020469.

Handley, J.A., Hanning, I., Ricke, S.C., Johnson, M.G., Jones, F.T., Apple, R.O., 2010. Temperature and bacterial profile of post chill poultry carcasses stored in processing combo held at room temperature. J. Food Sci. 75, M515–M520.

Hanna, S.E., Conner, C.J., Wang, H.H., 2005. Real-time polymerase chain reaction for the food microbiologist: technologies, applications, and limitations. J. Food Sci. 70, R49–R53.

Hanning, I., Ricke, S.C., 2011. Prescreening methods of microbial populations for the assessment of sequencing potential. In: Kwon, Y.M., Ricke, S.C. (Eds.), Methods in Molecular Microbiology 733—High-Throughput Next Generation Sequencing: Methods and Applications. Springer Protocols, Humana Press, New York, NY, pp. 159–170.

Harrison, S., 2015. Intercepting food fraud before it hits the shelves. Food Qual. Saf. 21, 27–28.

Hebert, P.D.N., Cywinska, A., Ball, S.L., DeWaard, J.R., 2003. Biological identifications through DNA barcodes. Proc. R. Soc. B Biol. Sci. 270, 313–321.

Hellweg, S., Milà i Canals, L., 2014. Emerging approaches, challenges and opportunities in life cycle assessment. Science 344, 1109–1113.

Herschleb, J., Ananiev, G., Schwartz, D.C., 2007. Pulsed-field gel electrophoresis. Nat. Protoc. 2, 677–684.

Hieter, P., Boguski, M., 1997. Functional genomics: it's all how you read it. Science 278, 601–602.

Hill, W.E., 1996. The polymerase chain reaction: applications for the detection of foodborne pathogens. Crit. Rev. Food Sci. Nutr. 36, 123–173.

Hoekstra, A.Y., Wiedmann, T.O., 2014. Humanity's unsustainable environmental footprint. Science 344, 1114–1117.

Hoffmann, M., Zhao, S., Luo, Y., Li, C., Folster, J.P., Whichard, J., Allard, M.W., Brown, E.W., McDermott, P.F., 2012. Genome sequences of five *Salmonella enterica* serovar Heidelberg isolates associated with a 2011 multistate outbreak in the United States. J. Bacteriol. 194, 3274–3275.

Hoffmann, M., Luo, Y., Lafon, P.C., Timme, R., Allard, M.W., McDermott, P.F., Brown, E.W., Zhao, S., 2013. Genome sequences of *Salmonella enterica* serovar Heidelberg isolates isolated in the United States from a multistate outbreak of human *Salmonella* infections. Genome Announc. 1 (1), e00004–e00012.

Holley, R.W., 1965. Structure of an alanine transfer ribonucleic acid. J. Am. Med. Assoc. 194, 868–871.

Holley, R.W., Everett, G.A., Madison, J.T., Zamir, A., 1965a. Nucleotide sequences in the yeast alanine transfer ribonucleic acid. J. Biol. Chem. 240, 2122–2128.

Holley, R.W., Apgar, J., Everett, G.A., Madison, J.T., Marquisee, M., Merrill, S.H., Zamir, A., 1965b. Structure of a ribonucleic acid. Science 147, 1462–1465.

Holmes, F.L., 2001. Meselson, Stahl, and the Replication of DNA: A History of "the Most Beautiful Experiment in Biology". Yale University Press, New Haven, CT.

Holt, R.A., Jones, S.J., 2008. The new paradigm of flow cell sequencing. Genome Res. 18, 839–846.

Horowitz, N.H., 1995. One-gene-one-enzyme: remembering biochemical genetics. Protein Sci. 4, 1017–1019.

Horrocks, S.M., Anderson, R.C., Nisbet, D.J., Ricke, S.C., 2009. Incidence and ecology of *Campylobacter* in animals. Anaerobe 15, 18–25.

Horvath, J.C., 2012. How can better food labels contribute to true choice? Minn. J. Law Sci. Technol. 13, 359–383.

Horvath, P., Barrangou, R., 2010. CRISPR/Cas, the immune system of bacteria and archaea. Science 327, 167–170. http://dx.doi.org/10.1126/science.1179555.

Huehn, S., La Ragione, R.M., Anjurn, M., Saunders, M., Woodward, M.J., Bunge, C., Helmuth, R., Hauser, E., Guerra, B., Beutlich, J., Brisabois, A., Peters, T., Svensson, L., Madajczak, G., Litrup, E., Imre, A., Herrera-Leon, S., Mevious, D., Newell, D.G., Malorny, B., 2010. Virulotyping and antimicrobial resistance typing of *Salmonella enterica* serovars relevant to human health in Europe. Foodborne Pathog. Dis. 7, 523–535.

Hughes, D., 2003. Exploiting genomics, genetics and chemistry to combat antibiotic resistance. Nat. Rev. Genet. 4, 432–441.

Hume, M.E., Kubena, L.F., Edrington, T.S., Donskey, C.J., Moore, R.W., Ricke, S.C., Nisbet, D.J., 2003. Poultry digestive microflora diversity as indicated by denaturing gradient gel electrophoresis. Poult. Sci. 82, 1100–1107.

Hynes, A.P., Molneau, S., 2014. CRISPR-Cas systems: making the cut. Microbe 9, 204–210.

Jarquin, R., Hanning, I., Ahn, S., Ricke, S.C., 2009. Development of rapid detection and genetic characterization of *Salmonella* in poultry breeder feeds. Sensors 9, 5308–5323.

Jones, M.E., 1953. Albrecht Kossel, a biographical sketch. Yale J. Biol. Med. 26, 80.

Judson, H.F., 1979. The Eighth Day of Creation—The Makers of the Evolution in Biology. Simon and Schuster, New York, NY, 686 pp.

Keeratipibul, S., Silamat, P., Phraephaisarn, C., Srisitthinam, D., Takahashi, H., Chaturongkasumrit, Y., Vesaratchavest, M., 2015. Genotyping of *Salmonella enterica* serovar Typhimurium isolates by multilocus variable number of tandem repeat high-resolution melting analysis (MLV-HRMA). Foodborne Pathog. Dis. 12, 8–20.

Kennedy, D., 2014. Building agricultural research. Science 346, 13.

Koo, O.K., Mertz, A.W., Akins, E.L., Sirsat, S.A., Neal, J.A., Morawicki, R., Crandall, P.G., Ricke, S.C., 2012. Analysis of microbial diversity on deli slicers using polymerase chain reaction and denaturing gradient gel electrophoresis technologies. Lett. Appl. Microbiol. 56, 111–119.

Koyuncu, S., Andersson, G., Vos, P., Häggblom, P., 2011. DNA microarray for tracing *Salmonella* in the feed chain. Int. J. Food Microbiol. 145, 518–522.

Kraŝovec, R., Belavkin, R.V., Aston, J.A.D., Channon, A., Aston, E., Rash, B.M., Kadirvel, M., Forbes, S., Knight, C.G., 2014. Mutation rate plasticity in rifampicin resistance depends on *Escherichia coli* cell–cell interactions. Nat. Commun. 5, 3742. http://dx.doi.org/10.1038/ncomms4742.

Kresge, N., Simoni, R.D., Hill, R.L., 2005a. Arthur Kornberg's discovery of DNA polymerase I. J. Biol. Chem. 280, e46.

Kresge, N., Simoni, R.D., Hill, R.L., 2005b. Chargaff's rules: the work of Erwin Chargaff. J. Biol. Chem. 280, e21.

Ku, C.S., Roukos, D.H., 2013. From next-generation sequencing to nanopore sequencing technology: paving the way to personalized genomic medicine. Expert Rev. Med. Dev. 10, 1–6.

Kuchta, T., Knutsson, R., Fiore, A., Kudirkiene, E., Höhl, A., Tomic, D.H., Gotcheva, V., Pöpping, B., Scaramagli, S., To Kim, A., Wagner, M., De Medici, D., 2014. A decade with nucleic acid-based microbiological methods in safety control of foods. Lett. Microbiol. 59, 263–271.

Kuipers, O.P., de Jong, A., Holsappel, S., Bron, S., Kok, J., Hamoen, L.W., 1999. DNA-microarrays and food technology. Antonie Van Leeuwenhoek 76, 353–355.

Kwon, Y.M., Ricke, S.C. (Eds.), 2011. High-throughput next generation sequencing: methods and applications. Methods in Molecular Microbiology, vol. 733. Springer Protocols, Humana Press, New York, 308 pp.

Lebonah, D.E., Dileep, A., Chandrasekhar, K., Sreevani, S., Sreedevi, B., Kumari, J.P., 2014. DNA barcoding on bacteria: a review. Adv. Biol. 2014, 9. http://dx.doi.org/10.1155/2014/541787 (Article ID 541787).

Leinonen, I., Williams, A.G., Wiseman, J., Guy, J., Kyriazakis, I., 2012. Predicting the environmental impacts of chicken systems in the United Kingdom through a life cycle assessment: egg production systems. Poult. Sci. 91, 26–40.

Li, H., Wang, H., D'Aoust, J.-Y., Maurer, J., 2013. *Salmonella* species. In: Doyle, M.P., Buchanan, R.L. (Eds.), Food Microbiology—Fundamentals and Frontiers, fourth ed. American Society for Microbiology Press, Washington, DC, pp. 225–261 (Chapter 10).

Liang, L., Wang, Q., Ågren, H., Tu, Y., 2014. Computational studies of DNA sequencing with solid-state nanopores: key issues and future prospects. Front. Chem. 2, 5. http://dx.doi.org/10.3389/fchem.2014.00005.

Lianou, A., Koutsoumanis, K.P., 2013a. Strain variability of the behavior of foodborne bacterial pathogens: a review. Int. J. Food Microbiol. 167, 310–321.

Lianou, A., Koutsoumanis, K.P., 2013b. Evaluation of the strain variability of *Salmonella enterica* acid and heat resistance. Food Microbiol. 34, 259–267.

Lienau, E.K., Strain, E., Wang, C., Zheng, J., Ottesen, A.R., Keys, C.E., Hammack, T.S., Musser, S.M., Brown, E.W., Allard, M.W., 2011. Identification of a salmonellosis outbreak by means of molecular sequencing. N. Engl. J. Med. 364, 981–982.

Liu, F., Barrangou, R., Gerner-Smidt, P., Ribot, E.M., Knabel, S.J., Dudley, E.G., 2011a. Novel virulence gene and clustered regularly interspaced short palindromic repeat (CRISPR) multilocus sequence typing scheme for subtyping of the major serovars of *Salmonella enterica* subsp. *enterica*. Appl. Environ. Microbiol. 77, 1946–1956.

Liu, F., Kariyawasam, S., Jayarao, B.M., Barrangou, R., Gerner-Smidt, P., Ribot, E.M., Knabel, S.J., Dudley, E.G., 2011b. Subtyping *Salmonella enterica* serovar Enteritidis isolates from different sources by using sequence typing based on virulence genes and clustered regularly interspaced short palindromic repeats (CRISPRs). Appl. Environ. Microbiol. 77, 4520–4526.

Lucchini, S., Thompson, A., Hinton, J.C.D., 2001. Microarrays for microbiologists. Microbiology 147, 1403–1414.

Lungu, B., Ricke, S.C., Johnson, M.G., 2009. Growth, survival, proliferation and pathogenesis of *L. monocytogenes* under low oxygen or anaerobic conditions: a review. Anaerobe 15, 7–17.

Maciorowski, K.G., Pillai, S.D., Jones, F.T., Ricke, S.C., 2005. Polymerase chain reaction detection of foodborne *Salmonella* spp. in animal feeds. Crit. Rev. Microbiol. 31, 45–53.

MacLean, D., Jones, J.D., Studholme, D.J., 2009. Application of 'next-generation' sequencing technologies to microbial genetics. Nat. Rev. Microbiol. 7, 287–296.

Maddox, B., 2003. The double helix and the 'wronged heroine'. Nature 421, 407–408.

Manfreda, G., De Cesare, A., 2005. Campylobacter and Salmonella in poultry and poultry products: hows and whys of molecular typing. Worlds Poult. Sci. J. 61, 185–197.

Mardis, E.R., 2008a. The impact of next generation sequencing technology on genetics. Trends Genet. 24, 133–141.

Mardis, E.R., 2008b. Next-generation DNA sequencing methods. Annu. Rev. Genomics Hum. Genet. 9, 387–402.

Margulies, M., Egholm, M., Altman, W.E., Attiya, S., Bader, J.S., Bemben, L.A., Berka, J., Braverman, M.S., Chen, Y.J., Chen, Z., Dewell, S.B., Du, L., Fierro, J.M., Gomes, X.V., Godwin, B.C., He, W., Helgesen, S., Ho, C.H., Irzyk, G.P., Jando, S.C., Alenquer, M.L., Jarvie, T.P., Jirage, K.B., Kim, J.B., Knight, J.R.,

Lanza, J.R., Leamon, J.H., Lefkowitz, S.M., Lei, M., Li, J., Lohman, K.L., Lu, H., Makhijani, V.B., McDade, K.E., McKenna, M.P., Myers, E.W., Nickerson, E., Nobile, J.R., Plant, R., Puc, B.P., Ronan, M.T., Roth, G.T., Sarkis, G.J., Simons, J.F., Simpson, J.W., Srinivasan, M., Tartaro, K.R., Tomasz, A., Vogt, K.A., Volkmer, G.A., Wang, S.H., Wang, Y., Weiner, M.P., Yu, P., Begley, R.F., Rothberg, J.M., 2005. Genome sequencing in microfabricated high-density picolitre reactors. Nature 437, 376–380.

Marsh, K., Bugusu, B., 2007. Food packaging—roles, materials, and environmental issues. J. Food Sci. 72, R39–R55.

Maxam, A.M., Gilbert, W., 1977. A new method for sequencing DNA. Proc. Natl. Acad. Sci. U.S.A. 74, 560–564.

McKusick, V.A., 1997. Genomics: structural and functional studies of genomes. Genomics 45, 244.

Medini, D., Donati, C., Tettelin, H., Masignani, V., Rappuoli, R., 2005. The microbial pan-genome. Curr. Opin. Genet. Dev. 15, 589–594.

Mench, J.A., Sumner, D.A., Rosen-Molina, J.T., 2011. Sustainability of egg production in the United States—the policy and market context. Poult. Sci. 90, 229–240.

Mertz, A.W., Koo, O.K., O'Bryan, C.A., Morawicki, R., Sirsat, S.A., Neal, J.A., Crandall, P.G., Ricke, S.C., 2014. Microbial ecology of meat slicers as determined by denaturing gradient gel electrophoresis. Food Control 42, 242–247.

Meselson, M., Stahl, F.W., 1958. The replication of DNA in *Escherichia coli*. Proc. Natl. Acad. Sci. U.S.A. 44, 671–682.

Messing, J., 2014. Microbiology spurred massively parallel genomic sequencing and biotechnology. Microbe 9, 271–277.

Metzker, M.L., 2010. Sequencing technologies—the next generation. Nat. Rev. Genet. 11, 31–46.

Miko, I., 2008. Gregor Mendel and the principles of inheritance. Nat. Educ. 1 (1), 134.

Milillo, S.R., Martin, E., Muthaiyan, A., Ricke, S.C., 2011. Immediate reduction of *Salmonella enterica* serotype Typhimurium following exposure to multiple-hurdle treatments with heated, acidified organic acid salt solutions. Appl. Environ. Microbiol. 77, 3765–3772.

Milillo, S.R., Friedly, E.C., Saldivar, J.C., Muthaiyan, A., O'Bryan, C.A., Crandall, P.G., Johnson, M.G., Ricke, S.C., 2012. A review of the ecology, genomics and stress response of *Listeria innocua* and *Listeria monocytogenes*. Crit. Rev. Food Sci. Nutr. 52, 712–725.

Min Jou, W., Haegeman, G., Ysebaert, M., Fiers, W., 1972. Nucleotide sequence of the gene coding for the bacteriophage MS2 coat protein. Nature 237, 82–88. http://dx.doi.org/10.1038/237082a0.

Morange, M., 2005. What history tells us I. The operon model and its legacy. J. Biosci. 30, 313–316.

Moreno Switt, A.I., Orsi, R.H., den Bakker, H.C., Vongkamjan, K., Altier, C., Wiedmann, M., 2013. Genomic characterization provides new insight into *Salmonella* phage diversity. BMC Genomics 14, 481. http://www.biomedcentral.com/1471-2164/14/481.

Mulligan, R.C., Berg, P., 1980. Expression of a bacterial gene in mammalian cells. Science 209, 1422–1427.

Muniesa, M., Jens, A., Hammerl, J.A., Hertwig, S., Appel, B., Brüssow, H., 2012. Shiga toxin-producing *Escherichia coli* O104:H4: a new challenge for microbiology. Appl. Environ. Microbiol. 78, 4065–4073.

Nathans, D., 1979. Restriction endonucleases, simian virus 40, and the new genetics. Science 206, 903–909.

Nimse, S.B., Song, K., Sonawane, M.D., Sayyed, D.R., Kim, T., 2014. Immobilization techniques for microarray: challenges and applications. Sensors 14, 22208–22229. http://dx.doi.org/10.3390/s141222208.

Nirenberg, M., 2004. Historical review: deciphering the genetic code—a personal account. Trends Biochem. Sci. 29, 46–54.

O'Bryan, C.A., Crandall, P.G., Ricke, S.C., 2008. Organic poultry pathogen control from farm to fork. Foodborne Pathog. Dis. 5, 709–720.

O'Bryan, C.A., Crandall, P.G., Davis, M.L., Kostadini, G., Gibson, K.E., Alali, W.Q., Jaroni, D., Ricke, S.C., Marcy, J.A., 2014. Mobile poultry processing units: a safe and cost-effective poultry processing option for the small-scale farmer in the United States. Worlds Poult. Sci. J. 70, 787–802.

O'Flaherty, S., Klaenhammer, T.R., 2011. The impact of omic technologies on the study of food microbes. Annu. Rev. Food Sci. Technol. 2, 16.1–16.19.

Okayama, H., Berg, P., 1982. High-efficiency cloning of full-length cDNA. Mol. Cell. Biol. 2, 161–170.

Okayama, H., Berg, P., 1983. A cDNA cloning vector that permits expression of cDNA inserts in mammalian cells. Mol. Cell. Biol. 3, 280–289.

Ollinger, M., MacDonald, J.M., Madison, M., 2005. Technological change and economies of scale in U.S. poultry processing. Am. J. Agric. Econ. 87, 116–129.

Orlando, L., Ginolhac, A., Raghavan, M., Vilstrup, J., Rasmussen, M., Magnussen, K., Steinmann, K.E., Kapranov, P., Thompson, J.F., Zazula, G., Froese, D., Moltke, I., Shapiro, B., Hofreiter, M., Al-Rasheid, K.A.S., Gilbert, M.T.P., Willerslev, E., 2015. True single-molecular DNA sequencing of a pleistocene horse bone. Genome Res. 21, 1705–1719.

Palka-Santini, M., Cleven, B.E., Eichinger, L., Martin Krönke, M., Krut, O., 2009. Large scale multiplex PCR improves pathogen detection by DNA microarrays. BMC Microbiol. 9, 1. http://dx.doi.org/10.1186/1471-2180-9-1.

Park, S.H., Hanning, I., Perrota, A., Bench, B.J., Alm, E., Ricke, S.C., 2013. Modifying the gastrointestinal ecology in alternatively raised poultry and the potential for molecular and metabolomic assessment. Poult. Sci. 92, 546–561.

Park, S.H., Aydin, M., Khatiwara, A., Dolan, M.C., Gilmore, D.F., Bouldin, J.L., Ahn, S., Ricke, S.C., 2014a. Current and emerging technologies for rapid detection and characterization of *Salmonella* in poultry and poultry products. Food Microbiol. 38, 250–262.

Park, S.H., Gibson, K., Almeida, G., Ricke, S.C., 2014b. Assessment of gastrointestinal microflora in pasture raised chickens fed two commercial prebiotics. J. Prob. Health 2, 122. http://dx.doi.org/10.4172/2329-8901.1000122.

Peikon, I.D., Gizatullina, D.L., Zador, A.M., 2014. In vivo generation of DNA sequence diversity for cellular barcoding. Nucleic Acids Res. 42, e127. http://dx.doi.org/10.1093/nar/gku604.

Pelletier, N., Ibarburu, M., Xin, H., 2014. Comparison of the environmental footprint of the egg industry in the United States in 1960 and 2010. Poult. Sci. 93, 241–255.

Pendleton, S., Hanning, I., Biswas, D., Ricke, S.C., 2013. Evaluation of whole genome sequencing as a genotyping tool for *Campylobacter jejuni* in comparison with pulsed-field gel electrophoresis and *flaA* typing. Poult. Sci. 92, 573–580.

Perkel, J., 2011. Making contact with sequencing's fourth generation. Biotechniques 50, 93–95.

Phillips, C.A., 1999. The epidemiology, detection and control of *Escherichia coli* O157. J. Sci. Food Agric. 79, 1367–1381.

Podolak, E., 2010. Sequencing's new race. Biotechniques 48, 105–111.

Portin, P., 2014. The birth and development of the DNA theory of inheritance: sixty years since the discovery of the structure of DNA. J. Genet. 93, 293.

Quail, M.A., Smith, M., Coupland, P., Otto, T.D., Harris, S.B., Connor, T.R., Bertoni, A., Swerdlow, H.P., Gu, Y., 2012. A tale of three next generation sequencing platforms: comparison of Ion Torrent, Pacific Biosciences and Illumina MiSeq sequencers. BMC Genomics 13, 341. http://www.biomedcentral.com/1471-2164/13/341.

Raskin, D.M., Seshadri, R., Pukatzki, S.U., Mekalanos, J.J., 2006. Bacterial genomics and pathogen evolution. Cell 124, 703–714.

Rasmussen, R.S., Morrissey, M.T., 2008. DNA-based methods for the identification of commercial fish and seafood species. Compr. Rev. Food Sci. Food Saf. 7, 280–295.

Rasooly, A., Herold, K.E., 2008. Food microbial pathogen detection and analysis using DNA microarray technologies. Foodborne Pathog. Dis. 5, 531–550.

Reis-Filho, J.S., 2009. Next-generation sequencing. Breast Cancer Res. 11 (Suppl. 3), S12.

Rešková, Z., Koreňová, J., Kuchta, T., 2013. Effective application of multiple locus variable number of tandem repeats analysis to tracing *Staphylococcus aureus* in food-processing environment. Lett. Appl. Microbiol. 58, 376–383.

Ricke, S.C., 2003a. The gastrointestinal tract ecology of *Salmonella* Enteritidis colonization in molting hens. Poult. Sci. 82, 1003–1007.

Ricke, S.C., 2003b. Perspectives on the use of organic acids and short chain fatty acids as antimicrobials. Poult. Sci. 82, 632–639.

Ricke, S.C., 2010. Future prospects for advancing food safety research in food animals. In: Ricke, S.C., Jones, F.T. (Eds.), Perspectives on Food Safety Issues of Food Animal Derived Foods. University of Arkansas Press, Fayetteville, AR, pp. 335–350 (Chapter 24).

Ricke, S.C., 2014. Application of molecular approaches for understanding foodborne *Salmonella* establishment in poultry production. Adv. Biol. 2014, 25. http://dx.doi.org/10.1155/2014/813275 (Article ID 813275).

Ricke, S.C., Hume, M.E., Park, S.Y., Moore, R.W., Birkhold, S.G., Kubena, L.F., Nisbet, D.J., 2004. Denaturing gradient gel electrophoresis (DGGE) as a rapid method for assessing gastrointestinal tract microflora responses in laying hens fed similar zinc molt induction diets. J. Rapid Methods Autom. Microbiol. 12, 69–81.

Ricke, S.C., Kundinger, M.M., Miller, D.R., Keeton, J.T., 2005. Alternatives to antibiotics: chemical and physical antimicrobial interventions and foodborne pathogen response. Poult. Sci. 84, 667–675.

Ricke, S.C., Van Loo, E.J., Johnson, M.G., O'Bryan, C.A. (Eds.), 2012. Organic Meat Production and Processing. Wiley Scientific/IFT, New York, NY, 444 pp.

Ricke, S.C., Khatiwara, A., Kwon, Y.M., 2013. Application of microarray analysis of foodborne *Salmonella* in poultry production: a review. Poult. Sci. 92, 2243–2250.

Roberts, R.J., 2005. How restriction enzymes became the workhorses of molecular biology. Proc. Natl. Acad. Sci. U.S.A. 102, 5905–5908.

Roberts, M.T., 2006. Role of regulation in minimizing terrorist threats against the food supply: information, incentives, and penalties. Minn. J. Law Sci. Technol. 8, 199–223.

Roe, B.A., 2014. Frederick Sanger (1918–2013). Genome Res. 24 (4), xi–xii.

Rothberg, J.M., Leamon, J.H., 2008. The development and impact of 454 sequencing. Nat. Biotechnol. 26, 1117–1124.

Saiki, R.K., Scharf, S., Faloona, F., Mullis, K.B., Horn, G.T., Erlich, H.A., Arnheim, N., 1985. Enzymatic amplification of beta-globin genomic sequences and restriction site analysis for diagnosis of sickle cell anemia. Science 230, 1350–1354.

Saiki, R.K., Gelfand, D.H., Stoffel, S., Scharf, S.J., Higuchi, R., Horn, G.T., Mullis, K.B., Erlich, H.A., 1988. Primer-directed enzymatic amplification of DNA with a thermostable DNA polymerase. Science 239, 487–491.

Sanger, F., Coulson, A.R., 1975. A rapid method for determining sequences in DNA by primed synthesis with DNA polymerase. J. Mol. Biol. 94, 441–448.

Sanger, F., Coulson, A.R., 1978. The use of thin acrylamide gels for DNA sequencing. FEBS Lett. 87, 107–110.

Sanger, F., Donelson, J.E., Coulson, A.R., Kössel, H., Fischer, D., 1973. Use of DNA polymerase I primed by a synthetic oligonucleotide to determine a nucleotide sequence in phage f1 DNA. Proc. Natl. Acad. Sci. U.S.A. 70, 1209–1213.

Sanger, F., Air, G.M., Barrell, B.G., Brown, N.L., Coulson, A.R., Fiddes, C.A., Hutchison, C.A., Slocombe, P.M., Smith, M., 1977a. Nucleotide sequence of bacteriophage phi X174 DNA. Nature 265, 687–695.

Sanger, F., Nicklen, S., Coulson, A.R., 1977b. DNA sequencing with chain-terminating inhibitors. Proc. Natl. Acad. Sci. U.S.A. 74, 5463–5467.

Savolainen, V., Cowan, R.S., Vogler, A.P., Roderick, G.P., Lane, R., 2005. Towards writing the encyclopaedia of life: An introduction to DNA barcoding. Philos. Trans. R. Soc. B Biol. Sci. 360, 1805–1811.

Scallan, E., Hoekstra, R.M., Angulo, F.J., Tauxe, R.V., Widdowson, M.-A., Roy, S.L., Jones, J.L., Griffin, P.M., 2011. Foodborne illness acquired in the United States—major pathogens. Emerg. Infect. Dis. 17, 7–15.

Scharff, R.L., 2012. Economic burden from health losses due to foodborne illness in the United States. J. Food Prot. 75, 123–131.

Schuster, S.C., 2008. Next-generation sequencing transforms today's biology. Nat. Methods 5, 16–18.

Service, R.F., 2014. Live wire—do cells use electricity to repair DNA? Jacqueline Barton aims to find out. Science 346, 1284–1287.

Sharp, P.A., 2014. Meeting global challenges: discovery and innovation through convergence: integrate biology, physics, engineering, and social science to innovate. Science 346, 1468–1471.

Shaw, A., Stevenson, L., Macklin, K., 2010. Improving research at universities to benefit the poultry industry. J. Appl. Poult. Res. 19, 307–311.

Simoni, R.D., Hill, R.L., Vaughan, M., 2002. The structure of nucleic acids and many other natural products: Phoebus Aaron Levene. J. Biol. Chem. 277, e11.

Singh, P., Foley, S.L., Nayak, R., Kwon, Y.M., 2012. Multilocus sequence typing of *Salmonella* strains by high-throughput sequencing of selectively amplified target genes. J. Microbiol. Methods 88, 127–133.

Singh, P., Foley, S.L., Nayak, R., Kwon, Y.M., 2013. Massively parallel sequencing of enriched target amplicons for high-resolution genotyping of *Salmonella* serovars. Mol. Cell. Probes 27, 80–85.

Sirsat, S.A., Muthaiyan, A., Ricke, S.C., 2009. Antimicrobials for pathogen reduction in organic and natural poultry production. J. Appl. Poult. Res. 18, 379–388.

Sirsat, S.A., Muthaiyan, A., Dowd, S.E., Kwon, Y.M., Ricke, S.C., 2010. The potential for application of foodborne *Salmonella* gene expression profiling assays in postharvest poultry processing. In: Ricke, S.C., Jones, F.T. (Eds.), Perspectives on Food Safety Issues of Food Animal Derived Foods. University of Arkansas Press, Fayetteville, AR, pp. 195–222.

Sirsat, S.A., Burkholder, K.M., Muthaiyan, A., Dowd, S.E., Bhunia, A.K., Ricke, S.C., 2011a. Effect of sublethal heat stress on *Salmonella* Typhimurium virulence. J. Appl. Microbiol. 110, 813–822.

Sirsat, S.A., Muthaiyan, A., Ricke, S.C., 2011b. Optimization of RNA extraction method for transcriptome studies of *Salmonella* inoculated on commercial raw chicken breast samples. BMC Res. Notes 4 (60), 1–7.

Smith, H.O., 1979. Nucleotide sequence specificity of restriction endonucleases. Science 205, 455–462.

Söll, D., RajBhandary, U.L., 2006. The genetic code—thawing the 'frozen accident'. J. Biosci. 31, 459–463.

Soni, K.A., Nannapeneni, R., Tasari, T., 2011. The contribution of transcriptomics and proteomics to elucidating adaptation stress responses of *Listeria monocytogenes*. Foodborne Pathog. Dis. 8, 843–852.

Spector, M.P., Kenyon, W.J., 2012. Resistance and survival strategies of *Salmonella enterica* to environmental stresses. Food Res. Int. 45, 455–481.

Steinman, R.M., Moberg, C.L., 1994. The experiment that transformed biology: discovering the genetic role of DNA. J. Exp. Med. 179, 381–384.

Stent, G.S., Calendar, R., 1978. Molecular Genetics—An Introductory Narrative. W.H. Freeman and Company, San Francisco, CA, 773 pp.

Stoessel, F., Juraske, R., Pfister, S., Hellweg, S., 2012. Life cycle inventory and carbon and water foodprint of fruits and vegetables: application to a Swiss retailer. Environ. Sci. Technol. 46, 3253–3262.

Teletchea, F., 2009. Molecular identification methods of fish species: reassessment and possible applications. Rev. Fish Biol. Fish. 19, 265–293.

Thieffry, D., Sarkar, S., 1998. Forty years under the central dogma. Trends Biochem. Sci. 23, 312–316.

Thompson, J.F., Steinmann, K.E., 2010. Single molecule sequencing with a HeliScope genetic analysis system. Curr. Protoc. Mol. Biol. (Suppl. 92), 7.10.1–7.10.14. http://dx.doi.org/10.1002/0471142727.

Van Slyke, D.D., Jacobs, W., 1945. In: Biographical Memoir of Phoebus Aaron Theodor Levene, vol. 23. National Academy of Sciences, Washington, DC, pp. 75–86.

Venter, J.C., Adams, M.D., Myers, E.W., Li, P.W., Mural, R.J., Sutton, G.G., et al., 2001. The sequence of the human genome. Science 291, 1304–1351.

Verbeke, W., Pérez-Cueto, F.J.A., de Barcellos, M.D., Krystallis, A., Grunert, K.G., 2010. European citizen and consumer attitudes and preferences regarding beef and pork. Meat Sci. 84, 284–292.

Verbeke, W., Rutsaert, P., Bonne, K., Vermeir, I., 2013. Credence quality coordination and consumers' willingness to pay for certified halal labelled meat. Meat Sci. 95, 790–797.

Verbeke, W., Marcu, A., Rutsaert, P., Gaspar, R., Seibt, B., Fletcher, D., Barnett, J., 2015. 'Would you eat cultured meat?': Consumers' reactions and attitude formation in Belgium, Portugal and the United Kingdom. Meat Sci. 102, 49–58.

Von Bubnoff, A., 2008. Next-generation sequencing: the race is on. Cell 132, 721–723.

Vongkamjn, K., Weidmann, M., 2015. Starting from the bench—prevention and control of foodborne and zoonotic diseases. Prev. Vet. Med. 118 (2–3), 189–195. http://dx.doi.org/10.1016/j.prevetmed,2014.11.004.

Wain, J., Keddy, K.H., Hendriksen, R.S., Rubino, S., 2013. Using next generation sequencing to tackle non-typhoidal *Salmonella* infections. J. Infect. Dev. Ctries. 7, 001–005.

Walker, M.J., Beatson, S.A., 2012. Outsmarting outbreaks. Science 338, 1161–1162.

Wang, F., Yang, Q., Kase, J.A., Meng, J., Clotilde, L.M., Lin, A., Ge, B., 2013. Current trends in detecting non-O157 shiga toxin-producing *Escherichia coli* in food. Foodborne Pathog. Dis. 10, 1–13.

Watson, J.D., Crick, F.H.C., 1953. A structure for deoxyribose nucleic acid. Nature 421, 397–398.

Wiedenheft, B., Sternberg, S.H., Doudna, J.A., 2012. RNA-guided genetic silencing systems in bacteria and archaea. Nature 482, 331–338.

Winder, C.L., Dunn, W.B., Goodacre, R., 2011. TARDIS-based microbial metabolomics: time and relative differences in systems. Trends Microbiol. 19, 315–322.

Windhorst, H.-W., 2008. A projection of the regional development of egg production until 2015. Worlds Poult. Sci. J. 64, 356–376.

Windhorst, H.-W., 2009. Recent patterns of egg production and trade: a status report on a regional basis. Worlds Poult. Sci. J. 65, 685–708.

Winter, C.K., Davis, S.F., 2006. Organic foods. J. Food Sci. 71, R117–R124.

Zou, W., Chen, H.-C., Hise, K.B., Tang, H., Foley, S.L., Meehan, J., Lin, W.-J., Nayak, R., Xu, J., Fang, H., Chen, J.J., 2013a. Meta-analysis of pulsed-field gel electrophoresis fingerprints based on a constructed *Salmonella* database. PLoS One 8, e59224. http://dx.doi.org/10.1371/journal.pone.0059224.

Zou, W., Tang, H., Zhao, W., Meehan, J., Foley, S.L., Lin, W.-J., Chen, H.-J., Fang, H., Nayak, R., Chen, J.J., 2013b. Data mining tools for *Salmonella* characterization: application to gel-based fingerprinting analysis. BMC Bioinf. 14 (Suppl. 14), S15. http://www.biomedcentral.com/1471-2105/14/S14/S15.

Zou, Q.-H., Li, R.-Q., Liu, G.-R., Liu, S.-L., 2014. Comparative genomic analysis between typhoidal and non-typhoidal *Salmonella* serovars reveals typhoid-specific protein families. Infect. Genet. Evol. 26, 295–302.

CHAPTER

5

A DESCRIPTIVE TOOL FOR TRACING MICROBIOLOGICAL CONTAMINATIONS

M. Matt*, M.G. Andersson†, G.C. Barker‡, J.H. Smid§, F. Tenenhaus-Aziza¶, A. Pielaat‖

Austrian Agency for Health and Food Safety (AGES), Innsbruck, Austria, National Veterinary Institute (SVA), Uppsala, Sweden†, Institute of Food Research (IFR), Norwich, UK‡, Institute for Risk Assessment Sciences (IRAS), Utrecht University, Utrecht, The Netherlands§, Centre National Interprofessional de l'Economie Laitière (CNIEL), Paris, France¶, National Institute for Public Health and the Environment (RIVM), Bilthoven, The Netherlands‖*

1 INTRODUCTION

Biotracing is an emerging concept within food quality systems. The closely related notion of traceability has been implemented in feed and food production by EU regulation EC/178/2002 since 2002. This regulation defines traceability as "the ability to trace and follow food, feed, and ingredients through all stages of production, processing and distribution." Various systems have been established to support the tracing of food components throughout the food chain with the main focus on food ingredients. Tracing questions arise frequently and some of them strictly refer to biotracing, considering undesirable agents as the main target of the trace. Following an earlier suggestion (Barker et al., 2009), biotraceability is the ability to use downstream information to point to materials, processes, or actions within a particular food chain that can be identified as the source of undesirable agents. The significance and relevance of biotracing tasks have been demonstrated by several recently reported outbreaks: for example, the EHEC outbreak in Germany, starting in May 2011 (Hyde, 2011); the *Listeria monocytogenes* outbreak in Austria in 2009/2010 (Fretz et al., 2010); and the *Salmonella heidelberg* outbreak in ground turkey in August 2011 and chicken in October 2013 in the United States (CDC, 2013). For the EHEC outbreak, epidemiological investigations ultimately linked the cases to bean sprouts; however, the identification of the source proved to be difficult as both producer and supply chains are very complex. Similar challenges have been reported earlier for the same products in the United Kingdom in 2010 (Cleary et al., 2010), where *Salmonella bareilly* was the undesirable agent.

The fundamental distinctions between tracing and biotracing are manifest in the object that is traced: the product itself, respectively, an undesirable agent in the product together with the connection of chain characteristics (e.g. feed and food) with biological processes and their dynamical structures (Hoorfar et al., 2011). In the context of biotracing, the agents of interest are generally microbial pathogens or their products in different food chains. The basis for biotracing within a food processing chain (including feed) is the combination of chain knowledge (as a conceptual model) and the formulation of

Food Safety. http://dx.doi.org/10.1016/B978-0-12-800245-2.00005-8

microbiological dynamics (as a mathematical model) into a domain model. Such a model represents the relation among chain processes, microbial hazards, and the environment in a qualitative or quantitative way and includes an integral representation of uncertainties and/or variability.

Similarly, microbiological risk assessment (MRA) uses a domain model in estimating public health risks, for evaluating intervention strategies, and for revealing data gaps for future analysis. However, MRA differs from biotracing in three essential ways. First, the end point of concern is different. A traditional MRA aims to assess the consumer risk or to evaluate the costs and benefits of proposed intervention measures, whereas the endpoint for a biotracing analysis is dominated by posterior beliefs concerning the sources of contamination. Secondly, the methodology is different: parameters in a MRA model are usually static in the sense that whenever new information becomes available an updated MRA model will generate new results. In contrast, a biotracing model will give an update of the output based on changing evidence. Finally, as a consequence of the previous point and in contrast to a typical MRA, biotracing systems are able to include longitudinal data and, as such, generate outputs based on historical "experience."

Another scientific tool for the identification of sources in the context of food safety is called source attribution (Pires et al., 2009). Attribution methodologies assign the human cases directly to their probable sources in a broader sense without following the (causal) process in the food chain (e.g., environmental pathways). In this way, attribution assigns sources and cases in average terms rather than as one-to-one. In the case of attribution, the potential pathways of undesirable agents in the chain are not considered and, by definition, multiple potential food chains are considered simultaneously to evaluate the relative source strengths. Another important difference concerns outcomes. Source attribution reveals priority setting for future policy and interventions, whereas biotracing aims to promote direct intervention possibilities. In this context, source attribution is complimentary to biotracing (Barker et al., 2009).

1.1 HISTORY OF BIOTRACING

The idea of using evidence from a scene to identify the source of an observed failure or other adverse event has existed for centuries. This strategy underlies most criminal investigations where evidence concerns traces or marks and the event concerns the guilt of an individual but it is also the principle that underlies modern epidemiology. The ability to gather evidence, in many forms, has increased rapidly and investigations of outbreaks of disease are often able to identify the likely source of the infectious agent from pooled case information. This process was extended during the anthrax letter investigations in the United States and spawned a developing field called microbial forensics (Keim et al., 2008). Inference relating to sources, based on evidence, is central to the microbial source-tracking techniques increasingly used to explore episodes of environmental pollution. With an increasing ability to gather and store detailed information relating to food and food chains a biotracing process, or food chain forensics, is a logical development.

1.2 THE OUTLOOK FOR BIOTRACING

Even though the underlying ideas of the biotracing models are generally similar, the scope of the models and, with that, the levels of detail vary strongly. First, the numerous pathogens and various food products require different modeling approaches because of their respective biological behavior and processing features. Additionally, the onset for biotracing can vary from case to case. As a consequence, the identification of "the one and only biotracing system" is not possible. Nevertheless, a general systematic procedure can help to overcome the obstacles shared between the different applications and shed light on basic contemplations when building future biotracing systems. The aim of this chapter is to reveal the generic parts of a scheme for biotracing in food and feed chains.

2 DESCRIPTIVE BIOTRACING TOOL

The purpose of a biotracing system is to make inferences about the most likely source of contamination based on a combination of endpoint observations and specific knowledge about the food chain and/or the supply chain in the specific situations under investigation. In this chapter, we describe elements of a general biotracing tool through the review of different domain models and propose a systematic approach. This provides a comprehensive framework and, therefore, a common terminology of keywords that are essential. Different levels of abstraction and slightly divergent realizations lead to a diversity in approaches to the topic. Terminology is listed in Table 1 to alleviate a number of semantic difficulties. The specified technical terms are suggested in the context of biotracing and do not contradict existing guidelines of MRA; for example, "Principles and guidelines for the conduct of MRA, CAC/GL30" (FAO/WHO, 2002).

Biotracing, biotraceability, or biotrace are equivalent expressions but the interpretation allows two distinct applications. First, the scientific process of constructing a tracing model for an agent is called biotracing (comparable to some extent with risk assessment). However, the practical process of converting information into decisions and actions based on such a model is called biotracing (comparable to some extent with the risk management process). It is essential to realize the existence of both

Table 1 Common Terminology Used for the Descriptive Tool of Biotracing (Defined Words in Bold)

Term	Definition
Biotraceability = biotracing = biotrace	The ability to use downstream information to point to materials, processes or actions within a particular production chain that can be identified as the **source** of undesirable **agents**
Source	The origin of an undesirable **agent**
Agent	Microorganisms and their products
Trigger	Information that initiates **biotracing** activity
Conceptual model	Collection, identification, and organization of data, definitions, sources, and other information about the agent/matrix combination, including their causal relationships
Domain	A subset of the **conceptual model** relevant to investigate the objectives of biotracing
Domain model	A mathematical representation of the **domain**, which facilitates analysis and support for decision making regarding **biotracing**
Core biotrace	Primary scheme for source-level inference based on a **domain model**
Operational biotracing	Implementation of the **domain model** in order to support interventions and decisions by a manager

meanings for biotracing, in a similar way to the benefits achieved in risk analysis by a strict distinction between risk assessment and risk management.

The descriptive, general concept consists of three different but meshing elements of biotracing, which are explained in more detail in the following sections.

2.1 THE TRIGGER

It is important to distinguish between the stimulus for constructing a biotracing model and the signal for operating an existing biotracing model. In the first case, a ***trigger*** indicates that the identification of different sources is possible based on particular endpoint observations (trigger 1 in Figure 1). This trigger 1 could be the detection of *Salmonella* in livestock on a specific farm, for example, triggering investigations on possible feed sources.

For an existing biotracing model, a trigger (trigger 2 in Figure 1) starts a process of investigation aimed at finding the source of contamination or to support a decision tool for prioritizing source investigations. This trigger might be an observed deviation from the expected behavior (anomaly detection) at any point in the processing chain. Such a deviation might be an increase in the pathogen count, change in temperature, or a change in an agent-related marker; for example, alkaline phosphatase activity as an indicator for the quality of a pasteurization process (Barker and Goméz-Tomé, 2013). Another example

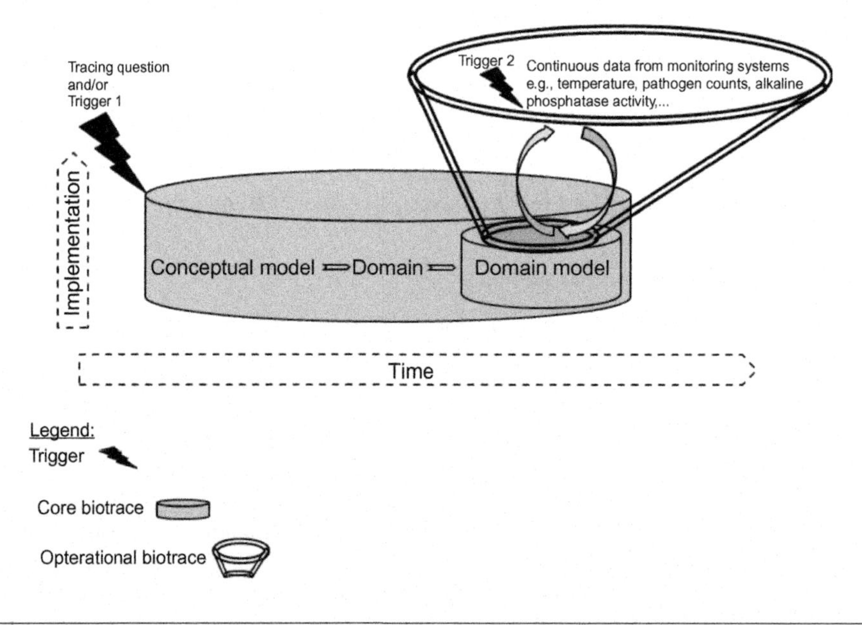

FIGURE 1

Biotracing at a glance: tracing question and/or trigger 1 starts the process of core biotracing. The conceptual model is the basis for the domain and the developed domain model. This double lined flare reflects the possibility for many monitoring systems to be integrated in a domain model. The implementation phase of the resulting operational biotracing system is reflected by the location of the ellipse (double lined). The permanent update of information is indicated by the circulating arrows.

Table 2 Triggers and Tracing Questions Within the Reviewed Core Biotraces

Pathogen/Matrix	Trigger	Tracing Question
Salmonella in feed	*Salmonella* in livestock on the farm *Salmonella* in feed and the feed processing environment	How can we combine information on processes, production flow, and sampling results to improve the identification or exclusion of possible sources for *Salmonella*?
Salmonella in pork	Increased levels of *Salmonella* on the carcass toward the end of slaughter; different distributions of serovars on the equipment compared to the incoming pigs	Can we infer the dominant source (i.e., skin or feces of incoming pigs or house flora) given (concentration) data of *Salmonella* on the exterior and the cut section of a specific carcass after meat inspection?
Staphylococcus aureus enterotoxins in milk	Alkaline phosphatase activity measurements	What is the relative strength for sources of *S. aureus* enterotoxins: toxin formation prior to pasteurization, unmonitored failure of pasteurization, or post process contamination?
L. monocytogenes in cheese	Detection of *L. monocytogenes* at any stage during production	What are the most probable sources of contamination given the sample analysis results of cheese at the end of the process, and/or during manufacturing?

are *Salmonella*-positive samples from the interior and exterior of pig carcasses as they may indicate different sources that trigger the biotracing for *Salmonella* in the pork chain (Smid et al., 2012). Different triggers associated with the examples in this chapter are summarized in Table 2.

2.2 MODELING PRACTICE

The overall aim of biotracing modeling is to drive the implementation of an *operational biotracing* system. Beginning with a *conceptual model*, a vast amount of information is gathered about the pathogen and the food/feed chain of concern. The next step involves refining the essentials, which results in the delineation of a *domain* (a connected set of variables and their dependencies relevant to the specified biotracing question). The implementation of the information about the domain into mathematical expressions provides a *domain model*, which includes the probability pathways leading from sources to observed events (causality) and a scheme for inverting pathways (inference) that can be used to give trace information about undesirable agents in the food chain structure.

Embedded in the operational biotrace is the *core biotrace*, which is the primary, scientific inferential process. Its practical applications involving contributory systems and activities like confirmation and updating, and implementation is the ultimate step to the *operational biotracing* system.

2.2.1 Conceptual model

A conceptual model consist of general information about the agent and matrix of concern and about the process steps without explanation of causal relations or any constraints. It provides an important overall picture and creates broad situation awareness. Gathering data about the agent is the most common start. Prerequisites for bacterial growth, reduction and/or elimination, and possible toxin production are identified. Prevalence and concentration of the pathogen in different stages of production is relevant information. Additionally, specific microbial behavior associated with the matrix or production environment such as biofilm formation or bacterial clustering have to be considered. Potential factors such as nonhomogeneous distribution of bacteria and/or the measures of analytical test performance (like sensitivity and specificity) impact greatly on further analysis. An accurate process description can help to identify possible sources, unintended events, and routes of transmission. Special attention should be paid to the detection of possible unintended events and failures and the assessment of their consequences. Literature research and expert opinions guide the task and, finally, the organization of the collected information and the linkage to causality completes the conceptual model. An elaborated conceptual model of the feed chain gave extraordinary insights regarding biotracing (Binter et al., 2011) (see also Section 3.1). Most conceptual models are not published, although they form the basis of the domain and, consequently, the domain model.

2.2.2 Domain

The domain is characterized by the extraction of relevant information from the *conceptual model* for statistical inference. It is a connected (closed) subset of information and relationships within the boundaries of the specified pathogen/chain combination. Focusing on the trigger and the tracing question(s) (see also Table 2) assists in setting the boundaries for the domain and often results in a reasonable reduction of information. The decisions about which trigger and/or tracing question(s) are considered necessary to induce a biotracing investigation are crucial for the further process. The significance of setting boundaries (scope) is reflected by the consequences: the constructed domain model will be able to answer exclusively questions about information included in the domain.

2.2.3 Domain model

Information about the process links and nodes, including bacterial behavior and specific process knowledge, is used for the mathematical representation of the domain. Different mathematical implementations of a domain can be found in the literature (e.g., Smid et al., 2011). Probabilistic graphical models and particularly Bayesian belief network (BBN) modeling is often applied to convert "forward, that is, tracking" or causal models into "backward, that is, tracing" or inferential domain models (Barker, 2004; Smid et al., 2010).

A *BBN*—also known as a belief network or a causal probabilistic network—is a compact and intuitive graphical representation that represents variables and their conditional dependencies (Cowell, 1999; Jensen, 2007). It provides a consistent approach to probabilistic inference based on combining prior information with observations using Bayes's rule. A BBN is an efficient knowledge representation scheme making it possible to integrate diverse types of information including data and expert opinions into a single domain model.

Biotracing is based on probabilistic modeling of biological systems and in this context two phenomena have to be distinguished: *variability and uncertainty*. Both are represented by probability distributions, but their nature is fundamentally different. Variability is an effect of (biological) variation within a population and is a function of the system. It is not reducible through further measurement. In contrast, uncertainty is a lack of knowledge regarding the true value of a quantity and, therefore, can be reduced by further measurements or additional information.

In the biotracing models reviewed below, the distribution of the processing parameters, general biological parameters, and specific microbiological parameters express variability and uncertainty.

The processing parameters are closely related to the procedures of food production and may differ (be variable) between states, regions, or at company level. General biological data of food chain systems, like a herd size or the weight of a pig, can be gathered and distributions (in the form of probability density functions) describing the variability can be fitted to the available data. Specific microbiological parameters such as growth curves or biofilm forming capacities, and so on, may vary from strain to strain but are assumed to be constant for one strain in particular conditions. Therefore, variability distributions for specific microbiological parameters may be extracted from the scientific literature.

The *importance of uncertainty* is demonstrated in the work of Smid et al. (2010) and may be examined at http://www.ifr.ac.uk/safety/SalmonellaPredictions. In this work, different levels of parameter uncertainty for the growth parameter lead to substantial differences in the discrimination between sources, although the same endpoint observation was entered. This demonstrates the relevance of information quality in biotracing.

In general, *reduction of uncertainty* is part of many investigations, and reducing the uncertainty associated with parameters within a food chain makes an important contribution to improved source identification. Process-related experiments give improved insight in, for example, transfer rates and the amount of transmitted bacteria. Research on the microbiological behavior of pathogens, such as the ability to produce biofilms, is important for future tracing projects. Discovering the main physical location of specific serotypes could change the prior belief about different sources within the chain, and including results of target-oriented sampling reduces uncertainty about loads. This process ensures that research to lower the detection limit or to improve sensitivity associated with refined extraction methods and other molecular tools make major contributions to biotracing in addition to detailed knowledge about the variability within the system.

2.3 IMPLEMENTATION ASPECTS

The *core biotrace* is part of an *operational biotrace*, which leads to the implementation scheme. The construction and evaluation of a domain model reflects the core biotrace and implementation of such a domain model for decision making is called operational biotracing, which is elaborated in the next sections.

2.3.1 Core biotrace

The conceptual model, the domain, and the domain model form the basis of the core biotrace (an inferential process, see Figure 1, gray cylinder). An update of the prior information in a domain model based on new, or additional, data is included in the core biotrace (Pielaat, 2011). When the domain model is complete, the robustness of the model can be evaluated by sensitivity analysis, such as the parameter sensitivity described by Smid et al. (2011). Data quality can be investigated by testing different scenarios and this analysis might, again, and reveal continued data gaps. The domain model can be updated using adaptation (i.e., posterior information developed in response to evidence leading to improved priors); in this way, the core biotrace initiates a progression toward improved biotracing applications.

2.3.2 Operational biotrace

The operational biotrace (Figure 1, ellipse, double lined) is the ultimate target of biotracing and can be seen as a wider "operational" management scheme that is affordable, accepted, progressive, and fits alongside existing practice. The purpose of contributory systems such as surveillance systems, alert systems, and information systems is in their contribution to a user-friendly application and to constant updates of the domain model (Barker et al., 2009).

The distinction between implementation of core and operational biotracing is of a practical nature. The core biotrace is the primary scientific work, whereas the operational biotrace is linked to several monitoring systems and puts the tracing model into practice for future scientific-based source identification.

At least two different users for operational biotracing are conceivable: public health authorities and industry. Most public health surveillance systems generate a vast amount of information. Filtering and implementing most relevant data into a domain model are prerequisites for future applications. An implemented operational biotracing system in governmental public health institutes for specific pathogens can augment epidemiologic outbreak investigations and, therefore, support rapid identification of the source within a food chain. The dimension of an operational biotracing system for public health purposes would be the entire food/feed chain coupled with established tracing procedures. A first attempt could follow the detection of an outbreak, and would involve food/feed chain scenarios for a particular undesirable agent. Outbreak detection systems (ODS) are installed in several European countries like Denmark, the United Kingdom, Germany, the Netherlands, and Sweden (Hulth et al., 2010), as well as Austria (Fuchs et al., 2012). These systems give a signal if the number of reported cases is rising. A further step for an entire biotracing system would assist an epidemiologist in tracing the source; for example, with a model including serotype information of all food samples combined with the outbreak data. Therefore, a complex and entire picture of food/feed production and cooperation of microbiological laboratories would be a necessary. In industry, biotracing is more likely to focus on one commodity and one process chain. The safety of a specific food or feed product is maintained and improved by implementing the information of (in many cases existing) monitoring systems in a domain model on a continuous basis. As most of the (risk) management systems used by the food industry (HACCP, ISO 9001, BRC, IFS) postulate important principles like "measurement, analysis, and continuous improvement" they give a good basis for an operational biotracing system. In an industry context, a biotracing system should provide support for less-intrusive intervention and for preemptive, targeted actions based on in-line monitors, which are additional to established control processes (e.g., Nicolau et al., 2013).

3 EXAMPLES OF BIOTRACING MODELS

The following section gives an overview of biotracing models for several pathogen/matrix combinations. Comparison of different biotracing schemes gives an impression of the diversity of the biotracing concept and how it can be used to trace undesired agents. *Salmonella* in feed serves as an example for a published conceptual model. *Salmonella* in pork, *S. aureus* toxins in milk, and *Listeria* in cheese are further published biotracing examples. The last chapter deals with a complementary, black box concept: source attribution with *Campylobacter* as an example for its distinction from biotracing.

3.1 CONCEPTUAL MODEL: *SALMONELLA* IN FEED

Feed is a potential source for *Salmonella* spp. infection in livestock- and feed-related outbreaks have been reported for pigs, poultry, and cattle (Jones, 2011). Industrially produced feed materials and compound feed are distributed to numerous follow-up operators, which means contaminated material at one place can spread *Salmonella* to several other places in the chain (Binter et al., 2011). Feed materials are bought globally, which means that new serotypes with unpredictable virulence can be introduced with imported feed materials, which may fill ecological niches during mitigating, currently relevant, serotypes (Crump et al., 2002).

The characteristics of processing and distribution pathways for feed were described and visualized in a conceptual model (Binter et al., 2011). The main stages of the feed chain, including transport, were identified and discussed, in relation to hygienic management. To get a complete picture, the information in scientific literature was complemented using interviews with subject experts from reference laboratories, authorities, and industry.

Existing prevalence data on *Salmonella* in feed materials cannot be used directly for the development of a domain model because feed is a highly dispersed matrix. Nonstandardized sampling schemes, origin of the samples, and batch sizes add to the complexity of the system. Sampling for *Salmonella* is not trivial due to the vast amount of feed materials delivered in bulk and the uneven distribution of the pathogen. Additionally, chain dynamics and logistics are subject to short storage capacities, centralized distribution, and international transport routes. Recent detection methods are often not rapid enough to prevent introduction of *Salmonella* to feed processing plants, unless chain logistics, sampling, and detection methods are highly coordinated.

Although thermal treatments are implemented in feed mills, thermal barrier failure and bypass events can subsequently result in growth of *Salmonella* in nonhost environments such as the cooler. Once *Salmonella* is introduced to a farm, sanitation of livestock can be difficult to achieve. Infectious cycles start to build up (Berends et al., 1998; Davies et al., 2004) and cross-contamination scenarios can mask the identification of the ultimate source. Countries with a low herd prevalence report feed to be the most significant source of introducing *Salmonella* to a farm (Wierup and Häggblom, 2010). Difficulties with respect to source tracing at the farm level are mainly related to sampling questions and the comparison of isolates by type matching. The counts of culturable *Salmonella* in feed and feed ingredients are generally low, resulting in a low probability of detection even when the lot in question has introduced *Salmonella* to a farm or feedmill (Binter et al., 2011). This fact may be related to the presence of a large proportion of nonculturable cells (Schelin et al., 2014). The low detection probability in feed materials in combination with the possible presence of multiple *Salmonella* strains in one lot hamper inferences about the source of contamination. To date, the highest probability of isolating strains for type matching is the sampling of dust from the processing line of the feed operators.

Beyond the conceptual model, tracing of *Salmonella* in feed was identified as a process analogous to a forensic investigation, because of a specific tracing investigation per incident or case. Hypotheses about contamination could be identified based on circumstances in the case including origin of feed materials and the history of suppliers. The subsequent domain model can be used to combine expert knowledge with evidence in the form of sampling and typing results to estimate the value of evidence in relation to different hypotheses on contamination. In Andersson and Aspan (2013), we present a proof of concept for how a biotracing model can be used to combine expert knowledge with evidence in the form of sampling and typing results to estimate the value of evidence in relation to different hypotheses on contamination. Whereas the complexity of the feed chain makes it impossible to produce a general domain model for *Salmonella* in feed, the domain for a specific tracing investigation can be represented as a BBN. In this framework, MPRM models (Nauta, 2002) describing specific stages in feed production can be applied in nodes of the BBN in order to estimate the probability of obtaining the observed results given alternative hypotheses of source and, thereby, contribute to produce transparent and well-underpinned statements. In other nodes, the probability of obtaining a particular result would be based on the empirical data, and in this context reference databases for sampling results and genotype distributions would be essential for the interpretation of results during an investigation (Andersson and Aspan, 2013; Sjödin et al., 2013).

3.2 OPERATIONAL BIOTRACING: *SALMONELLA* IN THE PORK SLAUGHTERHOUSE

Observational and epidemiological studies on *Salmonella* in the pork chain may be used to provide insight into potential contamination sites as well as into possible risk factors. Some authors indicate that feed is a source of contamination, which may introduce *Salmonella* into pig farms, where it can multiply (Crump et al., 2002), whereas others emphasize that slaughter hygiene is a determinative factor for carcass contamination (Botteldoorn et al., 2003; Swanenburg et al., 2001). It has been stated that the major sources of *Salmonella* in the slaughterhouse are fecal leakage and subsequent cross-contamination (Borch et al., 1996); for example, through contact surfaces and handling by workers. Persistent slaughterhouse-specific strains, which suggest the existence of house flora, have been found (Swanenburg et al., 2001). The main question for biotracing of *Salmonella* in pork in the slaughterhouse is: "Can we infer the dominant source (i.e., skin or feces of incoming pigs or house flora) given concentration data of *Salmonella* on the exterior and the cut section of a specific carcass after meat inspection?" (Smid et al., 2011). Elevated *Salmonella* concentrations on a carcass may in this case trigger (trigger 1 in Figure 1) the process of biotracing. To answer the biotracing question, an inferential model was developed from a published tracking model through the implementation of a BBN using the HUGIN software platform (Hugin Expert A/S, Aalborg, Denmark).

The EFSA tracking model (EFSA, 2009) describes the microbial behavior of *Salmonella* in the pork slaughter chain in the most extensive way. Qualitative weighting of different models for biotracing purposes resulted in choosing the EFSA model as a basis for the domain. The construction of the core biotracing model is described by Smid et al. (2012). In short, the model was obtained by reducing the amount of variables in the forward MRA model using sensitivity analysis and implementing the mathematical equations into a BBN. Updating the model with sampling data (van Hoek et al., 2012) showed that, for the particular slaughterhouse under investigation, house flora on one of the carcass splitters was an important source of contamination. The constructed core biotracing model is a proof of concept that biotracing in this chain is possible and can form the basis of a future operational biotracing system in a pork slaughterhouse.

One apparent characteristic of this food chain is the discrete, individual objects under investigation; namely, a discrete number of carcasses and the possibility of multiple, simultaneously acting sources. This is an obvious difference to the following milk chain model, where liquid milk is mixed and diluted and where only one source is likely to be responsible for each contamination event.

3.3 *S. AUREUS* ENTEROTOXINS IN PASTEURIZED MILK

Small-scale dairying, where operational margins are very tight and process variability is large, provides an ideal opportunity for implementation of a simple biotracing scheme. In particular, biotracing in relation to hazards that are associated with *S. aureus* in milk provides a good example of domain modeling and of the inference that can be achieved from a systematic combination of evidence. Staphylococcal food poisoning is caused by ingesting toxins produced in food by enterotoxigenic *S. aureus*. Preformed enterotoxins are highly heat resistant and pasteurization has little effect on concentrations (Le Loir et al., 2003). The hazard of milk-borne disease has been decreased over the last decades due to improved hygienic production and heat treatment, particularly during large-scale production, but at the farm scale, often desirable for consumers, controls remain difficult to achieve. In particular, it is apparent that the cooling step after milk collection, the primary heat process (pasteurization), and the hygiene associated with postprocessed milk are all potential sources of problems in relation to *S. aureus*. A biotracing scheme aims to combine information from three observations concerning filler tank milk to provide posterior beliefs about the most likely source of a hazard. In ideal conditions, the biotracing results could preempt disruptive interventions, such as a recall, and stimulate in-line remedial actions. The filler tank measurements include the toxin levels, enzymatic levels (e.g., alkaline phosphatase), and cell populations, and posterior beliefs are expressed as likelihood ratios for the three identified sources. The model presented by Barker and Goméz-Tomé (2013) illustrates that combinations of evidence can be used consistently in the source-tracing process as well as parameter and evidence sensitivities. All the elements of the biotrace exist in current control systems but the biotrace provides added value by combining the information and makes advantageous use of the added sensitivity in modern fluorometric phosphatase tests.

The model is constructed as a modular domain and implemented as a BBN and it supports traditional risk assessment as far as the filler tank. In biotracing mode, it is able to point to regimes in which distinct failures might lead to distinct patterns for associated hazards.

3.4 *L. MONOCYTOGENES* IN CHEESE

L. monocytogenes is ubiquitous and a pathogenic bacterium causing foodborne disease. Human infections with *L. monocytogenes* are rare but important due to the high mortality rate in high-risk groups (Rebagliati et al., 2009); namely, the young, old, pregnant, and immunocompromised. The microorganisms have been detected in different types of food, most commonly in ready-to-eat food, cheese, raw milk, and other dairy products (Rebagliati et al., 2009) but also in cheeses made from pasteurized milk. In that last case, contaminations generally arise from the plant environment and thus one possible tracing question is: "What are the most probable sources of contamination given the sample analysis results of cheese at the end of the process, and/or during the manufacturing?"

Simulation results issued from a simplified version of a (tracking) quantitative microbiological risk assessment (QMRA) model for *L. monocytogenes* in soft cheese made from pasteurized milk (Tenenhaus-Aziza et al., 2014) were used to build a BBN with the HUGIN software package.

The tracking QMRA model simulates the dispersion and the evolution of all the colonies in space and time following an initial contamination at a given process step by *L. monocytogenes*. This primo-contamination occurs after pasteurization, because pathogenic bacteria are not supposed to be present in pasteurized milk. Each process step is modeled as a module and different physiochemical conditions are considered leading to destruction, partial destruction, stress, or growth of the bacteria. The concentration of *L. monocytogenes* in milk was counted as CFU/ml, whereas in the other compartments (cheese, equipment, and environment surfaces) the contamination is given by the total number of bacterial cells. The emerging lag time, resulting from the process (e.g., hygienic operations induce stress on cells, which can prevent growth) is expressed as the quantity of work to characterize the physiological state of the bacteria. Cross-contamination is taken into account as well (Aziza et al., 2006). The inputs of the models are recontamination scenario parameters and process parameters; the outputs of the model are the prevalence and the distribution of the *Listeria* concentrations of a batch at each step of the process. Additionally, the probability of detection of *L. monocytogenes* in the processing plant environment and the probability of detection of a contaminated lot based on specific microbiological sampling plans are estimated. Several recontamination scenarios were simulated with the tracking QMRA model, from which inputs and outputs were extracted to build the BBN. The probability for each recontamination scenario regarding results of sampling plans lead to one application of operational biotracing.

3.5 SOURCE ATTRIBUTION OF *CAMPYLOBACTER*

The relative importance of different reservoirs for *Campylobacter* has been discussed by many authors (e.g., Mullner et al., 2009; Wilson et al., 2008), and poultry has been identified as the main source for human campylobacteriosis in most countries. A top-level tracing question for *Campylobacter* source attribution could be: "What is the contribution of different sources to human Campylobacteriosis?"

A harmonized nomenclature and different methods for attribution of human illness to specific sources has been proposed by Pires et al. (2009). The microbial subtyping approach may be used for answering this particular source-attribution question. The methodology used consists of comparing *Campylobacter* subtypes from suspected sources of human disease to those found in humans. Different source-attribution models using microbial subtyping are described; for example, the Dutch model (van Pelt et al., 1999), the (modified) Hald model (Mullner et al., 2009), the island model (Wilson et al., 2008), combining Bayesian clustering (BAPS software) with phylogenetic analysis (de Haan et al., 2010). The first is a simple frequentist model; the latter are based on Bayesian inference.

These examples show that source attribution partly fits in the concept of biotracing. The main difference is that source attribution only gives insight into the relative contribution of different sources about the food safety problem but cannot provide intervention options within the food chain.

4 CONCLUSIONS AND DISCUSSION

Improving food safety is a major concern of health authorities and industry. Therefore, interdisciplinary science is providing tools for progressive management strategies. Regulations on a national and international level require high standards, and risk analysis is one tool that contributes to maintaining these standards and supporting decision making. Some guidelines, for example, "Principles and guidelines for the conduct of MRA, CAC/GL30," form the basic framework for MRA and, consequently, are a central basis for core biotracing. In addition to the MRA framework, advances for tracing purposes require a complementary and more specific terminology (Table 1) and a compatible inference scheme.

Biotracing is an emerging discipline within microbial risk analysis and approaches to this topic are different. Comparing the genesis of several models serves as a basis for constructing a general, descriptive, and biotracing concept. However, other ways and methods to achieve biotracing might exist as well as different views on applications and operations. Therefore, the proposed scheme makes no claim to be complete. A common terminology, however, is crucial to aid the development of a general descriptive concept (Figure 1). The trigger, modeling practice, and implementation aspects are part of the biotracing tool. Stakeholders have to be involved in the process of developing a conceptual model and a proper, hypothesis-driven, biotracing question. Necessarily modelers and intended users have to share situation awareness about the basic assumptions on the system, which is the result of a multidirectional exchange of knowledge and ideas. Referring to the output of the tracing model, the process of decision making should be considered at an early stage, as decision making is included into the ultimate biotracing tool.

The first attempts of biotracing using mathematical inference proved the applicability of the principle. Still, the models developed so far are primarily academic and refer to core biotracing. Further attempts in reducing uncertainty and the continuous integration of data are necessary to come to operational biotracing. The specific adaptation to an individual processing plant and the continuous use of management and monitoring information (e.g., temperature time-series) will provide a progressive management tool benefiting from its preventive character with respect to food and feed safety. Biotracing should be complementary to existing management tools such as HACCP and should provide additional benefits and advantages such as preemptive interventions. Maybe some of the outbreaks mentioned in the introduction could have benefited from biotracing. For example, a biotracing tool in the responsible cheese factory in Austria might have identified the source of *Listeria* in the specific process earlier and, therefore, the outbreak could have been avoided if corrective actions had been set. Besides this outbreak, the investigation process of, for example, *Salmonella* or EHEC in different production chains as observed in Germany or the United States could have been accelerated.

Joint situation awareness is created when decisions, which should be supported by the system, are defined by stakeholders with the input of modelers providing a scientific basis for decision making. The processes of action have to be linked to decisions and modelers have to recognize how decision making under uncertainty works in the specific chain, which is a link to the field of management. Uncertainty is commonly described with confidence intervals about a point estimate or by using distributions of possible outcomes. A decision maker using a biotracing tool might be confronted with an unfamiliar representation of uncertainty: the result of the tracing model might point out more than one source. A biotracing model will assign a posterior probability to each source. However, when considering multiple possible sources, posterior probabilities reflect both the observed evidence and the structure of the knowledge domain (i.e., the relationship between individual sources). In such cases, an objective decision process becomes crucial. Expressing the outcome as a value of evidence (likelihood ratio) that is not strongly dependent on prior belief is often used for supporting decision making.

It is apparent that uncertainty plays an important role in biotracing systems and this places emphasis on the collection of data and identifying data gaps. If the amount of uncertainty is too high, biotraceability might be unachievable. Typing data might be used as supplemental information for mitigation of uncertainty. Its importance is beyond doubt, although the implementation is not trivial and methodological approaches to this problem are still under development. Additionally, the discriminatory power of typing is characteristic for each pathogen and typing method. The value of this additional information depends on the distribution of different types in distinct sources and on the (sometimes unknown) number of possible sources.

Including microbial variability, on the other hand, is a prerequisite for successful biotracing, as individual cases cannot be traced back without variability in the system. The impact of variability in biotracing is different than MRA, as source-level microbial diversity adds additional insight if it is combined with process-related knowledge. In this sense, a high amount of variability represents a key factor for focusing on, and exclusion of, possible sources. The variability over sources enhances the power of matching that is successful in many forensic activities that concern tracing. However, variability is reduced by several production-related processes like mixing or pooling, often associated with "bow tie-like" production chains, and these create serious limitations to biotracing possibilities.

Future biotracing studies should facilitate ongoing communication between modelers and decision makers. The latter should keep in mind that source identification via biotracing is often uncertain and there is a cost involved in reducing this uncertainty. Mathematical inference can be hampered by uncertainty, lack of variability (no discrimination possible between sources), or by too much variability. The increasing amount of collected data from surveillance, process controls, audits, and so on, combined with increasing sensitivity, throughput of experimental techniques, and "rapid methods" provide strong drivers for the steady progression of biotracing as a component of improved food safety management.

ACKNOWLEDGMENT

The work was supported by the European Union funded Integrated Project BIOTRACER (contract 036272) under the 6th RTD Framework and the Austrian Agency for Health and Food Safety (AGES).

REFERENCES

Andersson, G., Aspan, A., 2013. Application of forensic evaluation of evidence to the tracing of *Salmonella*. In: Presented at the Symposium *Salmonella* and Salmonellosis I3S, Saint-Malo, France.

Aziza, F., Mettler, E., Daudin, J.-J., Sanaa, M., 2006. Stochastic, compartmental, and dynamic modeling of cross-contamination during mechanical smearing of cheeses. Risk Anal. 26, 731–745.

Barker, G.C., 2004. Application of Bayesian Belief Network models to food safety science. In: Bayesian Statistics and Quality Modelling in the Agrifood Production Chain. Kluwer, Dordrecht, pp. 117–130.

Barker, G.C., Goméz-Tomé, N., 2013. A risk assessment model for enterotoxigenic *Staphylococcus aureus* in pasteurized milk: a potential route to source-level inference. Risk Anal. 33, 249–269.

Barker, G.C., Gomez, N., Smid, J., 2009. An introduction to biotracing in food chain systems. Trends Food Sci. Technol. 20, 220–226.

Berends, B.R., Van Knapen, F., Mossel, D.A., Burt, S.A., Snijders, J.M., 1998. Impact on human health of *Salmonella* spp. on pork in The Netherlands and the anticipated effects of some currently proposed control strategies. Int. J. Food Microbiol. 44, 219–229.

Binter, C., Straver, J.M., Häggblom, P., Bruggeman, G., Lindqvist, P.-A., Zentek, J., Andersson, M.G., 2011. Transmission and control of Salmonella in the pig feed chain: a conceptual model. Int. J. Food Microbiol. 145 (Suppl. 1), S7–S17.

Borch, E., Nesbakken, T., Christensen, H., 1996. Hazard identification in swine slaughter with respect to foodborne bacteria. Int. J. Food Microbiol. 30, 9–25.

Botteldoorn, N., Heyndrickx, M., Rijpens, N., Grijspeerdt, K., Herman, L., 2003. Salmonella on pig carcasses: positive pigs and cross contamination in the slaughterhouse. J. Appl. Microbiol. 95, 891–903.

CDC, 2013. *Salmonella* | Multistate Outbreak of Multidrug-Resistant *Salmonella* Heidelberg Infections Linked to Foster Farms Brand Chicken | Oct. 2013 | CDC [WWW Document]. URL http://www.cdc.gov/salmonella/heidelberg-10-13/ (accessed 10.22.13).

Cleary, P., Browning, L., Coia, J., Cowden, J., Fox, A., Kearney, J., Lane, C., Mather, H., Quigley, C., Syed, Q., Tubin-Delic, D., outbreak control team, 2010. A foodborne outbreak of Salmonella Bareilly in the United Kingdom, 2010. Euro Surveill 15 (48), pii: 19732.

Cowell, R.G., 1999. Probabilistic Networks and Expert Systems. Springer, New York.

Crump, J.A., Griffin, P.M., Angulo, F.J., 2002. Bacterial contamination of animal feed and its relationship to human foodborne illness. Clin. Infect. Dis. 35, 859–865.

Davies, P.R., Scott Hurd, H., Funk, J.A., Fedorka-Cray, P.J., Jones, F.T., 2004. The role of contaminated feed in the epidemiology and control of *Salmonella enterica* in pork production. Foodborne Pathog. Dis. 1, 202–215.

de Haan, C.P., Kivistö, R.I., Hakkinen, M., Corander, J., Hänninen, M.-L., 2010. Multilocus sequence types of Finnish bovine *Campylobacter jejuni* isolates and their attribution to human infections. BMC Microbiol. 10, 200.

EFSA, 2009. Quantitative Microbial Risk Assessment on *Salmonella* in Slaughter and Breeder Pigs (No. CFP/EFSA/BIOHAZ/2007/01).

FAO/WHO, 2002. Principles and Guidelines for Incorporating Microbiological Risk Assessment in the Development of Food Safety Standards, Guidelines and Related Texts, a Joint FAO/WHO Consultation, Kiel, Germany, 18–22 March 2002. http://www.fao.org/docrep/006/y4302e/y4302e00.ht.

Fretz, R., Pichler, J., Sagel, U., Much, P., Ruppitsch, W., Pietzka, A.T., Stöger, A., Huhulescu, S., Heuberger, S., Appl, G., Werber, D., Stark, K., Prager, R., Flieger, A., Karpísková, R., Pfaff, G., Allerberger, F., 2010. Update: multinational listeriosis outbreak due to "Quargel", a sour milk curd cheese, caused by two different *L. monocytogenes* serotype 1/2a strains, 2009–2010. Euro Surveill. 15, 19477–19480.

Fuchs, R., Scheriau, S., Kopacka, I., Kornschober, C., Schmid, D., 2012. Outbreak Detection System (ODS): Praktische Umsetzung eines Tools zur Erkennung von Krankheitsausbrüchen. In: F.J. Conraths, (Ed.), Früherkennung und Überwachung neu auftretender Krankheiten. DACH Epidemiologietagung, Gießen, p. 4.

Hoorfar, J., Wagner, M., Jordan, K., Barker, G.C., 2011. Biotracing: a novel concept in food safety integrating microbiology knowledge, complex systems approaches and probabilistic modelling. In: Food Science, Technology and Nutrition. In: Woodhead Publishing Series, vol. 196, pp. 377–390.

Hulth, A., Andrews, N., Ethelberg, S., Dreesman, J., Faensen, D., van Pelt, W., Schnitzler, J., 2010. Practical usage of computer-supported outbreak detection in five European countries. Euro Surveill. 2010;15(36), pii 19658.

Hyde, R., 2011. Germany reels in the wake of *E. coli* outbreak. Lancet 377, 1991.

Jensen, F.V., 2007. Bayesian networks and decision graphs. In: Information Science and Statistics, second ed. Springer, New York.

Jones, F.T., 2011. A review of practical *Salmonella* control measures in animal feed. J. Appl. Poult. Res. 20, 102–113.

Keim, P., Pearson, T., Okinaka, R., 2008. Microbial forensics: DNA Fingerprinting of *Bacillus anthracis* (Anthrax). Anal. Chem. 80, 4791–4800.

Le Loir, Y., Baron, F., Gautier, M., 2003. *Staphylococcus aureus* and food poisoning. Genet. Mol. Res. 2, 63–76.

Mullner, P., Jones, G., Noble, A., Spencer, S.E.F., Hathaway, S., French, N.P., 2009. Source attribution of food-borne zoonoses in New Zealand: a modified Hald model. Risk Anal. 29, 970–984.

Nauta, M.J., 2002. Modelling bacterial growth in quantitative microbiological risk assessment: is it possible? Int. J. Food Microbiol. 73, 297–304.

Nicolau, A.I., Barker, G.C., Aprodu, I., Wagner, M., 2013. Relating the biotracing concept to practices in food safety. Food Control 29, 221–225.

Pielaat, A., 2011. The data supply chain for tracing *Salmonella* in pork production. Int. J. Food Microbiol. 145, S66–S67.

Pires, S.M., Evers, E.G., van Pelt, W., Ayers, T., Scallan, E., Angulo, F.J., Havelaar, A., Hald, T., Med-Vet-Net Workpackage 28 Working Group, 2009. Attributing the human disease burden of foodborne infections to specific sources. Foodborne Pathog. Dis. 6, 417–424.

Rebagliati, V., Philippi, R., Rossi, M., Troncoso, A., 2009. Prevention of foodborne listeriosis. Indian J. Pathol. Microbiol. 52, 145–149.

Schelin, J., Andersson, G., Vigre, H., Norling, B., Häggblom, P., Hoorfar, J., Rådström, P., Löfström, C., 2014. Evaluation of pre-PCR processing approaches for enumeration of *Salmonella enterica* in naturally contaminated animal feed. J. Appl. Microbiol. 116, 167–178.

Sjödin, A., Broman, T., Melefors, Ö., Andersson, G., Rasmusson, B., Knutsson, R., Forsman, M., 2013. The need for high-quality whole-genome sequence databases in microbial forensics. Biosecur. Bioterror. 11 (Suppl. 1), S78–S86.

Smid, J.H., Verloo, D., Barker, G.C., Havelaar, A.H., 2010. Strengths and weaknesses of Monte Carlo simulation models and Bayesian belief networks in microbial risk assessment. Int. J. Food Microbiol. 139 (Suppl. 1), S57–S63.

Smid, J.H., Swart, A.N., Havelaar, A.H., Pielaat, A., 2011. A practical framework for the construction of a biotracing model: application to *Salmonella* in the pork slaughter chain. Risk Anal. 31, 1434–1450.

Smid, J.H., Heres, L., Havelaar, A.H., Pielaat, A., 2012. A biotracing model of *Salmonella* in the pork production chain. J. Food Prot. 75, 270–280.

Swanenburg, M., Urlings, H.A., Snijders, J.M., Keuzenkamp, D., van Knapen, F., 2001. *Salmonella* in slaughter pigs: prevalence, serotypes and critical control points during slaughter in two slaughterhouses. Int. J. Food Microbiol. 70, 243–254.

Tenenhaus-Aziza, F., Daudin, J.-J., Maffre, A., Sanaa, M., 2014. Risk-based approach for microbiological food safety management in the dairy industry: the case of *Listeria monocytogenes* in soft cheese made from pasteurized milk: food safety management in the dairy industry. Risk Anal. 34, 56–74.

Van Hoek, A.H.A.M., de Jonge, R., van Overbeek, W.M., Bouw, E., Pielaat, A., Smid, J.H., Malorny, B., Junker, E., Löfström, C., Pedersen, K., Aarts, H.J.M., Heres, L., 2012. A quantitative approach towards a better understanding of the dynamics of *Salmonella* spp. in a pork slaughter-line. Int. J. Food Microbiol. 153, 45–52.

Van Pelt, W., van de Giessen, A.W., van Leeuwen, W.J., Wannet, W., Henken, A.M., Evers, E.G., 1999. Oorsprong, omvang en kosten van humane salmonellose. Deel 1. Oorsprong van humane salmonellose met betrekking tot varken, rund, kip, ei en overige bronnen. Infectieziekten Bulletin 10, 240–243.

Wierup, M., Häggblom, P., 2010. An assessment of soybeans and other vegetable proteins as source of *Salmonella* contamination in pig production. Acta Vet. Scand. 52, 15.

Wilson, D.J., Gabriel, E., Leatherbarrow, A.J.H., Cheesbrough, J., Gee, S., Bolton, E., Fox, A., Fearnhead, P., Hart, C.A., Diggle, P.J., 2008. Tracing the source of campylobacteriosis. PLoS Genet. 4, e1000203.

CHAPTER

SALMONELLA AND THE POTENTIAL ROLE FOR METHODS TO DEVELOP MICROBIAL PROCESS INDICATORS ON CHICKEN CARCASSES

6

John A. Handley[*], Zhaohao Shi[†,‡], Si Hong Park[†,‡], Turki M. Dawoud[*,†], Young Min Kwon[*,†,§], Steven C. Ricke[*,†,‡,§]

Cell and Molecular Biology Program, University of Arkansas, Fayetteville, Arkansas, USA[*], Center for Food Safety, University of Arkansas, Fayetteville, Arkansas, USA[†], Department of Food Science, University of Arkansas, Fayetteville, Arkansas, USA[‡], Department of Poultry Science, University of Arkansas, Fayetteville, Arkansas, USA[§]

1 INTRODUCTION

Food contamination by foodborne pathogens is considered a major concern because they have been estimated to cause approximately 48 million illnesses in the United States (U.S.) with 128,000 hospitalizations and over 3000 deaths, which means that approximately 15% of the total U.S. population will annually experience a foodborne infection (Scallan et al., 2011). Contaminated food and water with various pathogens have been the main vehicles of infection for diarrheal diseases worldwide (Podewils et al., 2004; O'Ryan et al., 2005; Wilson, 2005; Nath et al., 2006; Sheth and Dwivedi, 2006; Marino, 2007; Zwane and Kremer, 2007; Hanning et al., 2008, 2009b; Horrocks et al., 2009; Schmidt and Cairncross, 2009; Schmidt et al., 2009; Cairncross et al., 2010; Santosham et al., 2010; Milillo et al., 2012). Diarrheal diseases are the second most prominent cause of deaths of children under 5 years old, accounting for 11% of child deaths (Chopra et al., 2013; Farthing et al., 2013; WHO and UNICEF, 2013).

Of the foodborne diseases that have been epidemiologically and economically documented, salmonellosis continues to be one of the more prominent and expensive foodborne diseases (Scharff, 2012). Salmonellosis costs an estimated US$3.3 billion annually for medical care costs and loss of productivity (Hoffmann et al., 2012). Salmonellosis can originate from a wide range of food products, including raw, processed, and even ready-to-eat (RTE) retail food items (Hanning et al., 2009b; Cox et al., 2011; Foley et al., 2011, 2013; Koo et al., 2012). Foodborne salmonellosis originating from poultry products is considered a major problem, and postharvest control is important to prevent systemic *Salmonella* contamination in the processing plant (Crump et al., 2002; Cox et al., 2011; Soria et al., 2011; Finstad

Food Safety. http://dx.doi.org/10.1016/B978-0-12-800245-2.00006-X

et al., 2012). However, assessing when and where *Salmonella* contamination occurs and more importantly which routes of contamination have the greatest impact on introduction and dissemination in the final poultry meat product is an extremely complex question with numerous variables. Consequently, the difficulty lies in the ability to predict which interventions would most likely be effective in preventing postharvest contamination, when they should be applied, and whether certain combinations would be synergistic. This requires appraisal of all aspects of processing and establishing true cause-and-effect relationships of *Salmonella* contamination during these processing steps.

Part of the problem is pulling together the tremendous amounts of empirical data to determine where the data gaps occur and where further research may be needed to provide that data. In this review, methods for identifying the impact of activities, procedures, methodologies, and equipment on *Salmonella* contamination during poultry processing will be discussed along with the potential risk of introduction and dissemination of *Salmonella* beyond processing. Secondly, potential methods for generating bacterial indicator data from the various processing steps that could be used to parallel and therefore potentially predict contamination routes will be examined. Finally, a brief discussion will examine the potential for developing quantitative risk assessment (QRA) models that determine where the highest potential risk(s) for contamination may occur during processing and how this might be used to prioritize optimal intervention strategies that achieve the most effective reductions in *Salmonella* contamination of poultry products.

2 COMMERCIAL POULTRY PROCESSING

Commercial poultry processing is characterized by the sheer large scale and capacity for high-volume generation of meat products. Globally, sources of meat such as broiler chicken and other poultry, as well as turkey and eggs, are among the most highly consumed foods (Kearney, 2010; Rask and Rask, 2011; Harmon, 2013). The average U.S. consumption of chicken per capita is approximately 86 pounds annually (MacDonald, 2008), and by 2010 it surpassed beef (Bentley, 2012). The United States is one of the largest producers of poultry, and its products primarily include chicken meats, turkey meats, and eggs (Harvey, 2012). The U.S. shipments of broiler meat in November 2013 were measured at 639.6 million pounds, with domestic cold storage facilities holding approximately 700 million pounds by the end of 2013, an increase of 7.5% from the same period of 2012 and a major increase (87%) of whole bird stocks of 22 million pounds by the end of November 2013 (Mathews, 2014). In 2014, broiler meat production was estimated at 9.85 billion pounds for the fourth quarter with a yearly total of 38.6 billion pounds, an increase of 2.7% from the previous year (Mathews and Haley, 2015).

Commercial production of chicken meat occurs in three stages: first processing, second processing, and further processing. Figure 1 represents processing of the whole carcass in first processing and Figure 2 represents the portioning of parts in second processing. First processing consists of the steps between receiving the live birds through preparation of the whole bird, which includes the steps of slaughtering, exsanguinating, defeathering, eviscerating, washing, and finally chilling (Owens et al., 2010). In second processing, the whole carcass is portioned into various cuts of meat that include breasts, wings, thighs, legs, and tenders and also can include steps such as deboning, skinning, and seasoning (Owens et al., 2010). The main aspect of second processing is the value additive process by which carcasses are deboned and may be marinade injected. Second processing allows consumers to

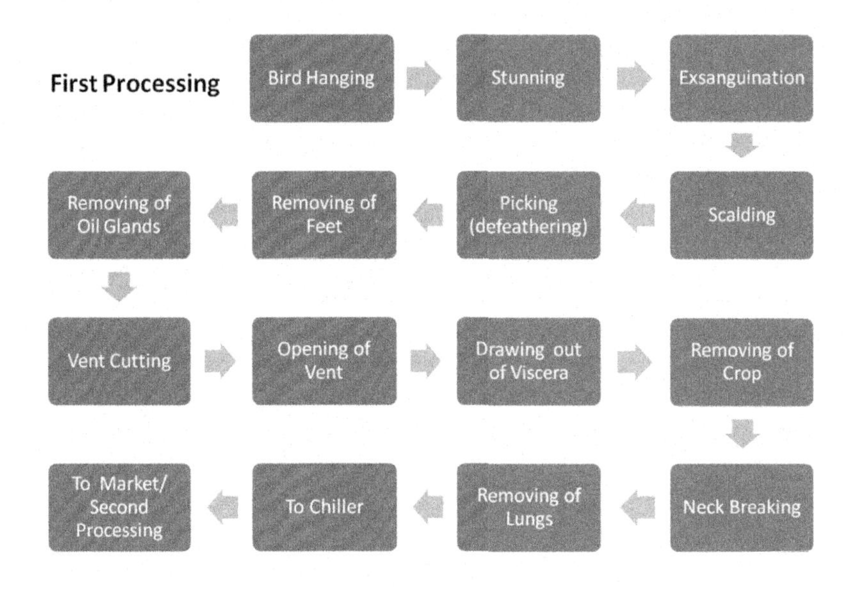

FIGURE 1

First processing of chicken carcass.

Partially adapted from information based on Owens et al. (2010).

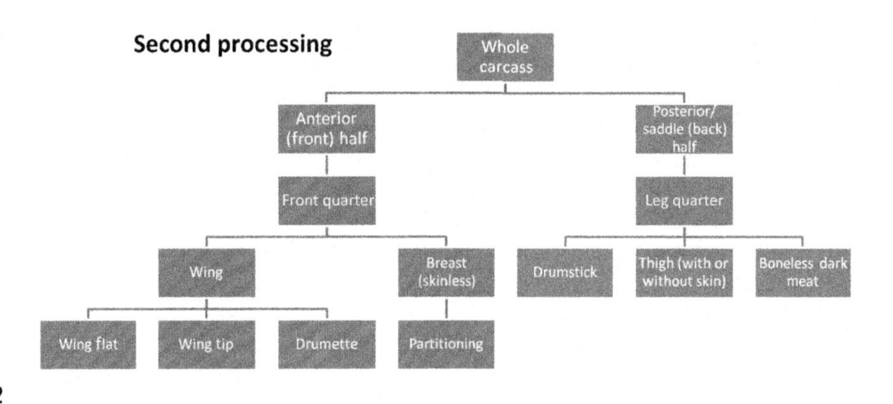

FIGURE 2

Second processing of chicken carcass.

only purchase the parts they desire as well as save time by not having to divide the carcass themselves. The parts may be cut either manually by workers or mechanically by specialized machines depending on the product. In general, second processing begins with the rehanging of the whole carcass out of the chilling stage of first processing. It is then divided in half into anterior (breast, wing, and other parts) and posterior halves (leg, thigh, and other parts). The anterior or front half can subsequently be separated into wings, tenders, and breast meat, which can be further cut down into size according to customer requirements. While this is occurring, the posterior half or saddle is divided into leg quarters, that can be further cut into drumsticks and thighs, which may eventually become deboned dark meat (Owens et al., 2010). All products can either be sent out for further processing or packed and sold as is. Subsequently, the raw meats are either packaged in bulk containers for further processing or retail packaging for consumers, with the meat destined for consumers being crust frozen and transported to distributors prior to selling to customers.

In the third processing stage, cutting, coating, crumbing, and cooking are added to raw chicken meat to produce final RTE foods such as nuggets, chicken patties, breaded tenders or breast strips, and coated chicken meats (Owens et al., 2010). As expected, each of these steps represents particular challenges for prevention of microbial contamination not only due to the complicated nature of the processing steps but also the multitude of environments encountered by the poultry product as it moves through the processing plant and beyond toward its eventual retail or food service destination. The poultry retail aspect is complicated as well because chicken can be marketed not only as whole carcasses but can also be sold as parts of cut up carcasses or as ground meat, each of which can become contaminated and support a microbial contaminated product with a characteristic microbiota that may be somewhat dependent on the type of product.

3 POULTRY PROCESSING AND SOURCES OF MICROBIAL CONTAMINATION

Chickens encounter foodborne pathogens throughout their growing cycle beginning in the hatchery, during stages of grow out, and as live birds being transported to the processing plant. Much of the risk of exposure is subject to a number of factors including contaminated feeds, vectors, and carriers such as insects and aerosols, as well as the physiological status of the bird under certain circumstances like those occurring during times of stress (Pillai and Ricke, 2002; Maciorowski et al., 2004; Park et al., 2008; Dunkley et al., 2009; Finstad et al., 2012; Ricke, 2014). For example, feed removal in laying hens was demonstrated to enhance the virulence properties of *Salmonella* Enteritidis along with concomitant increases in systemic invasion of the internal organs (Durant et al., 1999; Ricke, 2003b; Dunkley et al., 2007; Howard et al., 2012; Ricke et al., 2013a). While different live animal production practices exist for broilers, at least some aspects such as mandatory feed withdrawal prior to transportation to the processing plant do occur and are possible contributors to foodborne pathogen contamination of broilers as well (Ramirez et al., 1997; Byrd et al., 1998a,b). Certainly food safety practices in live birds have become more important as evidenced by the number of intervention measures that have emerged including vaccines, antimicrobials administered in the feed or water, bacteriocins, bacteriophages, probiotics, prebiotics, and synbiotics, to name a few (Joerger, 2003; Patterson and Burkholder, 2003; Ricke, 2003a, 2015; Hume, 2011; Ricke et al., 2012; Siragusa and Ricke, 2012). While these can perform with varying degrees of effectiveness, limits do exist and birds will still become colonized with *Salmonella*. Therefore, interventions at the processing plant remain an important component of an overall *Salmonella* control program for poultry products.

Once live poultry have arrived at the slaughter facility, the slaughter facility offers a new set of potential microbial contaminants that must be overcome. It has been observed that slaughter equipment can be contaminated prior to slaughter activities, but the plucking and scalding areas can have a greater incidence of contamination than the evisceration room (Rasschaert et al., 2007). According to Olsen et al. (2003), some *Salmonella* strains can survive up to 5 days in the commercial slaughter environment despite the daily cleaning and disinfection procedures. The recovery of *Salmonella* strains after sanitation may possibly be due to some acid resistance (Mani-Lopez et al, 2012), or the microbial load may be so high that the sanitation practices are not eliminating 100% of the bacteria. Also, biofilms may be aiding in the survivability of *Salmonella* strains (Van Houdt and Michiels, 2010; Steenackers et al., 2012). Nevertheless, these results indicate that some *Salmonella* strains are surviving in the slaughter environment after sanitation, which reflects how important it is to conduct effective sanitation.

Another mode of contamination comes from bird to bird and/or flock to flock cross-contamination during the kill and evisceration processes. During the transport process, poultry become stressed and excrete more fecal material (Corry et al., 2002). Transportation of chickens to the slaughter facility places the chickens in close proximity to each other, enhancing the potential for bird to bird cross-contamination. Another contributor to contamination is the equipment used to mechanically eviscerate carcasses because every carcass is exposed to the same machinery. However, carcass visual inspection for fecal matter is not necessarily the best indicator of *Salmonella* presence as was demonstrated by Jimenez et al. (2002), who found that 20% of poultry carcasses without visible fecal material were *Salmonella* positive while poultry carcasses with visible fecal material were 20.8% *Salmonella* positive during sampling from evisceration to the postchill process. The results possibly demonstrate how cross-contamination could generally occur by processing equipment during evisceration steps. However, while fecal material on carcasses does not appear to be an appropriate indicator for contamination of *Salmonella*, the fact remains that birds with visible fecal material are required to undergo additional washing steps due to the U.S. Department of Agriculture-Food Safety Inspection Service (USDA-FSIS) zero tolerance regulation for fecal material (Rasekh et al., 2005). It has also been noted that the risk of cross-contamination by pathogens such as *Salmonella* and *Campylobacter* during scalding increases as birds from subsequent flocks are processed during the day (Oosterom et al., 1983; Genigeorgis et al., 1986; Whyte et al., 2004). However, Whyte et al. (2004) observed that levels of contamination in the scalder do not continuously increase during the subsequent passage of birds through the tank. Therefore, incidence of pathogen contamination may rise among all the carcasses slaughtered during the processing day, but the total level of bacteria can be reduced because the microbial load is dispersed among the whole population of carcasses being processed throughout the day. The same is true for the chilling process in that it is a point of cross-contamination in poultry, but if the process is working correctly then the numbers of microorganisms will not increase, but rather decrease. Brewer et al. (1995) demonstrated that the chilling process was unaffected by changes in line speed as long as proportionate changes in chlorinated chiller water were made concurrently.

A possible explanation for ongoing cross-contamination (Lillard, 1989) of *Salmonella* in the processing environment may be that *Salmonella* cells present on the skin are initially entrapped in a water film on the skin and then migrate to the skin itself, where they become entrapped in the ridges and crevices on the skin that become more pronounced after immersion in water. In order to monitor the adequacy of processing, plating by total microbial population enumeration can effectively determine the sanitary condition of the processing abattoir. However, measuring the presence of the *Enterobacteriaceae* family on poultry carcasses may be a better means to indicate inadequate or unhygienic processing, inappropriate handling, or inadequate storage conditions that may result in the presence of pathogens within the family such as *Salmonella* and *Escherichia* (Whyte et al., 2004). Also, USDA-FSIS uses *Escherichia coli* as an indicator organism for fecal contamination because fecal contamination is still considered the primary pathway in which enteric organisms, *Salmonella*, *Campylobacter*, and *E. coli* O157:H7, come in contact with meat and poultry products (Russell, 2000).

4 FOOD SAFETY REGULATIONS FOR POULTRY PROCESSING: PAST AND CURRENT

In order to provide safe products to consumers, meat and poultry firms are required by the USDA-FSIS (USDA-FSIS, 1996, 2010) to follow guidelines for proper slaughter, handling, and storage of raw meat. During raw and fully cooked poultry meat processing, it is imperative that practices such as temperature abuse be prevented to avoid products becoming a food safety hazard. During the late 1990s, the USDA-FSIS adopted what was then considered a new means of self-monitoring for food processors called Hazard Analysis Critical Control Point (HACCP). The Code of Federal Regulation, 9 CFR 417.2 (b)(1), required that every USDA-inspected meat processing establishment develop and implement a written HACCP plan covering each product generated and how they would deal with food safety hazards once an analysis revealed one or more food safety hazards were likely to occur. Therefore, the development of HACCP plans required plants to locate possible points of contamination or locations that could have compromised the safety of a food product.

Several factors such as processing environments, growth conditions, and solidified fecal materials in the broiler house have been demonstrated to influence different bacterial levels and diversity on chickens (Clouser et al., 1995; Russell, 2000; Whyte et al., 2004). Clouser et al. (1995) identified evisceration equipment, chill water, and personnel as potential microbial cross-contamination sources during first processing, and Whyte et al. (2004) also found significant contamination occurring during typical processing procedures. In order to prevent, reduce, or eliminate bacterial levels, plants were allowed to use interventions that would not adulterate the product. Interventions that were available at the time for meat processors to reduce pathogens and microbial load on carcasses included steam pasteurization; steam vacuum; and carcass rinses with chlorine, trisodium phosphate, or organic acids (Hogue et al., 1998; Ricke, 2003a). When choosing such interventions, plant management had to consider the efficiency and effectiveness of the treatment, the cost of the chemical, overall safety to employees and the environment, and finally the best way to apply the treatment in an online processing system (Barbut, 2010). However, the employment of physical and chemical interventions do possibly allow for products to be packaged for consumers that have the potential to be considered safer (Ricke et al., 2005). Since then, more interventions and combinations have been examined that may offer further synergistic potential in chicken meat products (Milillo and Ricke, 2010; Milillo et al., 2011; Ravichandran et al., 2011; Perumalla et al., 2012).

The Food Safety Modernization Act (FSMA) was passed in 2011 to reform aging food safety laws; the USDA-FSIS followed with the release of a "*Salmonella* Action Plan" (USDA-FSIS, 2013) to address the issue of *Salmonella* on meat, which included new testing standards for poultry parts (USDA-FSIS, 2015a). In 2013, a large outbreak of *Salmonella* Heidelberg occurred, causing over 600 cases of illnesses in 29 states across the U.S. Investigations conducted by the U.S. Centers for Disease Control (CDC) along with the FSIS linked the outbreak to Foster Farms brand chicken (CDC, 2014). USDA regulations state that a maximum of 7.5% of chicken carcasses can test positive for *Salmonella* (USDA-FSIS, 2014d), but company representatives reported that they had found 0% prevalence from most of their whole carcass sampling. However, when USDA inspectors tested chicken parts that had been further processed rather than whole carcass testing, they found *Salmonella* on a quarter of the samples analyzed (Charles, 2014). The Foster Farms outbreak and the new regulatory testing standards illustrated a need to look further into what happens between when the whole carcass is prepared during first processing and when the parts are removed and prepared in second processing. This is now being considered a key step for the new testing standards to be applied to and is an area of research focus.

Previous studies looking at *Salmonella* and *Campylobacter* prevalence on retail poultry products indicate that their respective levels can be high. Zhao et al. (2001) found high levels of *Campylobacter* contamination on chicken obtained from local supermarkets in the Washington, DC, area with 70.7% of chicken samples being contaminated, while a study done by Mazengia et al. (2014) investigated retail chicken parts and found *Salmonella* to be present on 10 to 16% of samples. In addition, Cook et al. (2012) identified higher levels of *Campylobacter* on chicken breasts with skin retained compared to skin removed, but observed similar levels of *Salmonella* contamination between the two types of samples. Before the implementation of the new parts-testing requirement by FSIS in January of 2015, previous studies on Quantitative Microbial Risk Assessment (QMRA) focused more on whole carcasses from first processing (Nauta et al., 2005) or on eggs (Latimer et al., 2008) rather than on second processing. However, recently the USDA-FSIS published a risk assessment as well as a cost-benefit analysis based on the implementation of the new poultry parts-testing requirements (USDA-FSIS, 2015a).

5 FOODBORNE PATHOGEN ANALYSES: CULTURAL METHODS

Numerous cultural/selective media-based methodologies have been used over the years for detecting *Salmonella* in poultry products but unfortunately a pre-enrichment step is needed to recover sufficient *Salmonella* from poultry products before PCR, biochemical tests, and selective enrichments can be done, resulting in 5–7 days before *Salmonella* serovars can be confirmed (Sun et al., 2010). For the isolation and identification of *Salmonella*, the USDA-FSIS Microbiology Laboratory Guidebook (MLG) recommends using carcass rinses or swabs. This is followed by the BAX® PCR assay or with selective enrichment with tetrathionate (TT) broth and plating onto selective agar media (USDA-FSIS, 2014b). The population of viable organisms can be estimated probabilistically via the Most Probable Number (MPN) determination technique (USDA-FSIS, 2014a). *Campylobacter* can be quantitatively detected and enumerated based on direct plating or qualitatively detected through enrichment (USDA-FSIS, 2014c). Additional indicator bacteria can be enumerated through aerobic plate counts (USDA-FSIS, 2015b). Once the microorganisms have been detected and isolated, further characterization can be done using Pulsed-Field Gel Electrophoresis (PFGE) for fingerprinting the microorganisms, followed by subtyping and serotyping in specialized laboratories around the U.S. (Nde et al., 2006; Wu et al., 2014).

The traditional MPN method recommended by FSIS protocols has several drawbacks concerning the large amount of time and materials required. Significant numbers of tubes must be individually prepared for even the least time-consuming "three-tube MPN" technique as well as large amounts of media for growth and confirmation plating. Miniaturized most probable number (mMPN) techniques to enumerate *Salmonella* have been examined as a means to streamline the method as well as reduce the volumes required for a comprehensive analyses (Pavic et al., 2010; Berghaus et al., 2013). In their research work, Pavic et al. (2010) determined that the miniaturized MPN method they developed to be statistically similar to traditional MPN methods while saving considerable time and materials primarily with the use of microtiter plates and multichannel pipettes.

6 NUCLEIC ACID-BASED APPROACHES

Nucleic acids based or molecular techniques have become commonly used within the food industry because of the several modifications of the basic polymerase chain reaction (PCR) approach, such as real-time quantitative or multiplex PCR that have made these types of assays more applicable for different food groups or industries. Molecular methodologies enable faster detection and enumeration of specific bacteria compared to culturing methods (Maukonen and Saarela, 2009). The target nucleic acid sequence can be retrieved from DNA, mRNA, or rRNA preparations (Maukonen and Saarela, 2009). The principle theory of PCR is based on the hybridization of DNA or RNA gene fragments by primers, oligonucleotides, or single-stranded DNA fragments, specific to the target gene, and subsequent synthesis of the single-stranded target gene by a thermostable DNA polymerase enzyme (Pillai and Ricke, 1995; Scheu et al., 1998; Maciorowski et al., 2005). PCR has been well established to be able to detect low levels of bacterial cells; Lampel et al. (2000) reported detection of *Shigella flexneri*, *Salmonella* Typhimurium, and *Listeria monocytogenes* as low as 30, 50, and 200 Colony Forming Units (CFU), respectively. Using real-time PCR, Bohaychuk et al. (2007) were able to detect as low as 1 CFU of *Salmonella* in artificially contaminated chicken carcasses. Hanning et al. (2009a) detected *Pseudomonas* species immediately after processing, where traditional culturing methods did not yield results until day 4. This lower detection limit can reduce the amount of preenrichment time, thus becoming closer to a real-time aspect of data collection. However, while PCR targeting DNA can enumerate both viable and nonviable bacterial cellular DNA, the use of mRNA and rRNA as the PCR target will detect only actively growing cells (Hanning et al., 2009a; Maukonen and Saarela, 2009; Park et al., 2014; de Medici et al., 2015). However, rRNA and mRNA do have a shorter half-life compared to DNA and require more labor because they are more difficult to isolate in pure form (Scheu et al., 1998; Ye et al., 2012). Another drawback is the DNA purification process, where losses of targeted DNA can occur. An already low amount of DNA present may be lost with multiple wash steps, preventing identification of a potential pathogen or spoilage organism (Miller et al., 1999; Hanning et al., 2009a). Therefore, DNA purification is relative to the detection limit.

Another nucleic acid-based system, multiplex PCR, enhances PCR technology by offering the ability to detect multiple target genes within a single sample using multiple primers. The ability to detect multiple genes can allow multiple serotypes of a genus to be detected, offering an advantage over cultural detection methods (Levin, 2009; Park et al., 2011, 2014). This technology would allow processors to increase their monitoring of both pathogenic and spoilage organisms in one assay. Hanning et al. (2009a) utilized a multiplex PCR setup to detect five different *Pseudomonas* species to monitor spoilage on poultry carcasses, while Park et al. (2009) were able to identify *Salmonella enterica* serovars Typhimurium, Enteritidis, and Typhi at the same time. Levin (2009) noted that simultaneous detection of *Salmonella* and *L. monocytogenes* posed critical problems due to the fundamental differences between the two genera, primarily the fact that *Salmonella* is Gram-negative and *L. monocytogenes* is Gram-positive. However, Jofré et al. (2005) were able to discriminate multiple *Salmonella* and *L. monocytogenes* isolates by enriching samples in half Buffered Peptone Water (BPW) and half Fraser Broth (HF). Kawasaki et al. (2005) developed a multiplex PCR for the detection of *L. monocytogenes*, *Salmonella*, and *E. coli* O157:H7 in meat samples. Multiplex PCR does have a limit to how many primers can be effective because researchers have noted that more than five or six primers demonstrate cross-reactivity between primers (Jarquin et al., 2009).

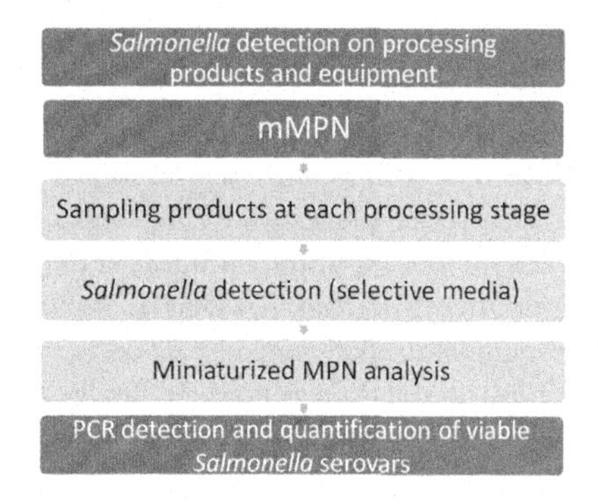

FIGURE 3

Combining PCR with miniaturized Most Probable Number (mMPN) for *Salmonella* detection.

Partially adapted from information based on Pavic et al. (2010) and Berghaus et al. (2013).

In the future, combining mMPN along with confirmation via single PCR using pathogen-specific primers may allow for not only the confirmation of the presence of the respective foodborne pathogen, but perhaps quantitation as well (Figure 3). The advantage here would be that these values could be confirmed via a cultural-based MPN and viable cells could be recovered for further characterization if needed for regulatory or tracking purposes. In addition, more recent developments with species- and strain-specific primers along with the capability of combining these as a multiplex system offers possibilities of not only quantitating total *Salmonella*, but actually differentiating individual isolates (Maciorowski et al., 2005; Kim et al., 2006; Park et al., 2014; Park and Ricke, 2015). As more whole genome sequences are generated for other bacterial foodborne pathogens it is anticipated that similar discernment among species and strains will be possible for these organisms as well.

Assessing individual genetic responses such as stress or virulence genes for a particular foodborne pathogen is now possible as well. Microarray technologies can also be employed within a species (Chan et al., 2003; Ricke et al., 2013b). A microarray approach requires the extraction of mRNA from samples, with the extracted mRNA being used to prepare a radioactively or fluorescently labeled DNA sequence by reverse transcription (Gibson, 2002). Hybridization of DNA sequences to a DNA probe with a known fluorescent label upon excitation from an analyte (Jarquin et al., 2009) yields presence or absence responses by a reader. Microarrays are capable of detecting hundreds to thousands of gene targets on glass slides, polyacrylamide gel pads, microchips, or microsphere beads (Call, 2005). However, there are high costs associated with this procedure for fabrication and equipment, which limits the applicability of the technology, along with difficulty in extracting sufficient quantities of RNA from surfaces such as chicken products (Call, 2005; Jarquin et al., 2009; Sirsat et al., 2011).

7 INDICATOR MICROBIAL ANALYSES

While using PCR-based technology to detect a few select organisms can be very beneficial, it lacks the overall ability to define the community or population diversity within the sample. Therefore, it is also critical to establish nonpathogen bacterial profiles that allow for a more continuous monitoring or oversight of microbial shifts that potentially could occur during processing (Sofos et al., 2013;

Giombelli and Gloria, 2014). There are several approaches for this type of analysis. One of the initial methods involved Denaturing Gradient Gel Electrophoresis (DGGE) that uses a gel with a denaturing agent prepared in a concentration gradient to melt DNA at different steps (Hanning and Ricke, 2011). Differences in DNA sequences cause them to partially melt at different positions along the gel gradient, allowing for the comparison of microbial ecologies by running samples within the same gel. The DGGE methodology is a potential useful tool for initial monitoring of overall microbial profile changes on products and equipment during poultry processing. A DGGE approach has been utilized in the past to document microbiota changes to temperature changes on carcasses after chilling and during storage as well as assessing microbial diversity on retail meat product slicing equipment (Handley et al., 2010; Koo et al., 2013; Mertz et al., 2014).

The DGGE technique is a nucleic acid-based assay that utilizes a PCR product of the same length from either DNA or RNA to compare genera within a given sample or over a period of time (Hanning and Ricke, 2011). The principle strategy for DGGE assessment in microbial diversity is that it uses small nucleic acid sequences, between 200 and 700 base pairs, on an acrylamide gel with a low to high denaturing gradient (Justé et al., 2008), where the denaturing agent is formamide or urea (Muyzer et al., 1993; Muyzer and Smalla, 1998). In order to prevent 100% denaturation, a guanine and cytosine-rich region of 40 bases, termed a GC-clamp, is utilized to prevent the formation of two bands from one organism due to different molecular weights of single-stranded DNA fragments (Justé et al., 2008; Hanning and Ricke, 2011; Bokulich and Mills, 2012).

As the PCR product goes from a lower concentration to a higher concentration of denaturing agent in the acrylamide gel, the hydrogen bonds between the double-stranded DNA begin to break in regions. The areas of DNA that break first are the regions that contain higher adenine (A) and thymine (T) nucleotide content because these two nucleotides are joined by two hydrogen bonds. The regions with higher guanine (G) and cytosine (C) content remain more stable due to the presence of three hydrogen bonds between the two nucleotides. However, with enough denaturing agent the double-stranded DNA will separate into a single strand of DNA, with the exception of the GC-clamp. Once denatured DNA is present, it will come to rest at the location on the gel where the concentration of denaturing agent was great enough to split the double-stranded DNA. Therefore, bands on a polyacrylamide gel indicate PCR product with specific melting points due to the amount of GC present (Muyzer and Smalla, 1998).

The method of DGGE has been used in combination with PCR amplication by many authors to analyze the DNA of microbial populations. Some authors (Randazzo et al., 2002; Ercolini et al., 2003; Sun et al., 2004; Camu et al., 2007) were able to utilize several regions of the 16S rRNA gene, while others (Otawa et al., 2006; Dar et al., 2007) were able to use mRNA. Both mRNA and rRNA PCR products allow for monitoring viable or metabolically active cells during studies over time, whereas DNA can be collected from either viable or nonviable organisms. Koo et al. (2013) and Mertz et al. (2014) determined the genetic ecology within different locations of a deli slicer using DGGE. Handley et al. (2010) utilized the DGGE technique to evaluate the bacterial profile on postchill chicken carcasses at different temperatures and storage periods, and observed a 95% similarity within carcass rinses obtained from the same time period but only an 8% similarity among samples collected over the entire 54 h. They concluded that this increase in diversity over time reflected the considerable changes in microbial populations that occurred as the temperature increased. In comparing DGGE data with the microbial plate count data, it was clear that bacterial population numbers were increasing over time, but proportional shifts in microbial populations could not be detected with petrifilm. Employing a molecular strategy was necessary to detect variations in overall population responses.

While this method has been useful in the detection of discernible shifts in microbial populations, it does have its shortcomings (Hanning and Ricke, 2011). The use of DGGE as an initial monitoring tool in microbial ecology is useful for separating DNA fragments, but it is not ideal for high throughput, as only a small number of samples can be run at one time. Another downfall of this method is that band comigration and multiple copies of the rRNA gene can hinder band purity and increase redundancy of bands for one organism (Bokulich and Mills, 2012). Finally, this is a nonquantitative method, so there is no capability to be able to retrieve information relating to the actual quantitative percentage of an individual organism within the total population (Busté et al., 2008; Bokulich and Mills, 2012).

8 TERMINAL RESTRICTION FRAGMENT LENGTH POLYMORPHISM

Terminal Restriction Fragment Length Polymorphism (TRFLP) is another molecular methodology that differentiates microbial populations based on the terminal restriction fragment size (Justé et al., 2008; Bokulich and Mills, 2012). Fluorescently labeled universal primers are utilized during the amplification of a target gene sequence. Upon completion of PCR, the product is subject to multiple restriction enzymes, as one restriction enzyme would not provide enough resolution in the microbial community (Justé et al., 2008). After restriction enzyme digestion, the sample is subsequently separated using capillary electrophoresis and the 5′ fluorescently labeled terminal fragment is detected (Justé et al., 2008; Bokulich and Mills, 2012). This methodology has high throughput capability and is very sensitive in regards to discriminating microbial communities, but there are limits to the accurate prediction of the microbiota present. A library has to be built, which can be time-consuming, and basic assumptions regarding the microbial community have to be made in order to make accurate predictions of the bacteria present (Bokulich and Mills, 2012). In addition, the method can be pseudoquantitative, as incomplete or nonspecific digestion by the restriction enzyme can lead to an overestimation of the diversity present (Justé et al., 2008). Finally, sequence homology between taxa can lead to an underestimation of the organisms present, but this can be overcome by using a more species-specific gene target (Bokulich and Mills, 2012).

9 DENATURING HIGH PERFORMANCE LIQUID CHROMATOGRAPHY

Initially, Denaturing High Performance Liquid Chromatography (DHPLC) was developed to study single nucleotide polymorphisms and other medical research questions of interest. More recently, this method has begun to be used within the beer and wine industry, cheese processing, and in soil ecology (Ercolini et al., 2008; Wagner et al., 2009; Hutzler et al., 2010). This is a more recently developed approach to microbial community profiling where DNA is separated by utilizing a specific nucleic acid column for DHPLC. The DHPLC approach can use traditional ion-pair reversed phase HPLC systems with minor modifications to separate PCR products of the same length based on guanine (G) and cytosine (C) percent content (Wagner et al., 2009). Like other current microbial profiling methods, the use of a GC-clamp is employed to increase discrimination of the microbial community members (Barlaan et al., 2005). The process works by taking a sample and amplifying a variable region of the 16S rRNA gene for all genomic DNA present via PCR. After amplification, the sample is prepared on an HPLC unit, where it is exposed to partial denaturation conditions between 50 and 70 °C (Premstaller and Oefner, 2003). The sample flows in the liquid phase containing acetonitrile (ACN) and triethylammonium acetate (TEAA) through the stationary phase in an HPLC column containing alkylated nonporous polystyrene/polydivinylbenzene hydrophobic beads. The chemical TEAA has a hydrophilic and hydrophobic region. The positively charged hydrophilic acetate in TEAA will interact with the negatively charged phosphate group of the partially denatured DNA backbone and the hydrophobic portion of TEAA will interact with the hydrophobic beads (Barlaan et al., 2005). As the ACN gradient increases, DNA with lower GC content will be released as double-stranded DNA and a peak will form on the chromatograph with a specific retention time. The retention time can be adjusted to increase peak resolution, and those factors that can be adjusted include the flow rate or gradient of denaturant, the heated column temperature, the PCR primer, and the PCR protocol.

Major advantages of this protocol over the other two methods previously discussed include that this technology automates and yields higher throughput than the process of DGGE with an acrylamide gel (Maukonen and Saarela, 2009) and that one can use fragment lengths of DNA adequate for future DNA sequencing, unlike some of the TRFLP fragments that are generated. By using a fraction collector, individual peaks can be collected for further analysis, such as DNA sequencing. For commercial laboratories, heated column HPLC units with UV detection are readily available. Additionally, it replaces manual handling of gels by laboratory technicians, which decreases variability and increases repeatability. Conversely, this technology is not the future of molecular technology due to the advent of next-generation sequencing (NGS) (Quigley et al., 2011). However, as utilizing sequencing technologies becomes more routine, DHPLC can aid in the differentiation of sequencing until metagenomics can work out current shortcomings and become more cost-effective.

10 NGS FOR METAGENOMICS

This type of organism identification is considered the future of routine molecular analyses. It operates by yielding the genomic DNA sequence of any organism present in the sample. Several sequencing systems have been used, including the Roche 454 pyrosequencing, Applied Biosciences SOLiD technology and ion torrent systems, and Illumina's systems (Justé et al., 2008; Kwon and Ricke, 2011; Bokulich and Mills, 2012; Park et al., 2012; Galimberti et al., 2015). In reassembling a bacterial genome, it is very important to get as much genome sequence coverage as one can because it indicates how many times the genome was completely sequenced to achieve a measure of accuracy and a level of confidence in the genomic sequence. All NGS systems sequence thousands to billions of reads per run, translating numerous observations for robust, saturated diversity and are detected using high-resolution optics or semiconductor chips (Kwon and Ricke, 2011; Bokulich and Mills, 2012). Data obtained through metagenomic analysis are usually reported at the phyla level even though a microbial community may have multiple species present with specialized roles within the community (Foster et al., 2012). As it may be difficult to determine minor genomic variation at the metagenomic level, researchers have to perform targeted speciation techniques to overcome this shortcoming (Foster et al., 2012).

Major issues associated with NGS are computing power, analysis, and data storage because the data generated is thousands to billions of base-pair reads (Gilbert et al., 2011; Bokulich and Mills, 2012). Another issue is the database of genome sequences available. The DNA match is only as good as the database that is inputted (Justé et al., 2008). If the database utilized has been poorly maintained or the initially annotated sequences are incorrect, then the data obtained may not be accurate (Foster et al., 2012; Temperton and Giovannoni, 2012). Therefore, users must be highly selective in the database they use and it is best to use a database that is curated. For prokaryotes, several databases have been identified as being reliable including Greengenes, Silva, and the Ribosomal database project (Bokulich and Mills, 2012). Metagenomics is becoming one of the more commonly used tools to explore gene distribution among populations, but while there is difficulty in processing or contextualizing the data, there are attempts to make progress in this area, including recognizing the need for data standardization for research comparisons (Gilbert et al., 2011; Foster et al., 2012; Knight et al., 2012; Temperton and Giovannoni, 2012).

11 THE POTENTIAL FOR APPLICATION OF INDICATOR ORGANISM PROFILES

Monitoring approaches for assessing overall microbial community populations on carcasses and processing environments and their respective responses to environmental changes in the poultry processing plant has utility for assessing likely pathways for cross-contamination. Environmental impacts such as various intervention steps applied during processing or general shifts in temperature could be assessed in terms of shifts in microbiota on the carcass and their corresponding relatedness to the original microbial population within the respective chicken carcass. In addition, different carcasses within the same batch or across batches could be compared to further isolate consistent patterns in microbiota that are associated with a particular external factor such as a specific temperature change. With this type of information, it may be possible to derive signature or indicator patterns of microbiota that parallel fluctuations in foodborne pathogen levels that correspond to the application of interventions or other mitigating factors. Likewise, the association of a particular microbiota on a chicken carcass with a particular pathogen isolate such as a specific *Salmonella* serovar may be a potential outcome that could be employed for predictions of where problems could occur.

The key point is that the general microbial method for assessing carcass microbial communities could be based on DGGE, some other method along these same lines, or as the cost becomes more reasonable, complete microbiome sequencing (Figure 4). All of these approaches offer a

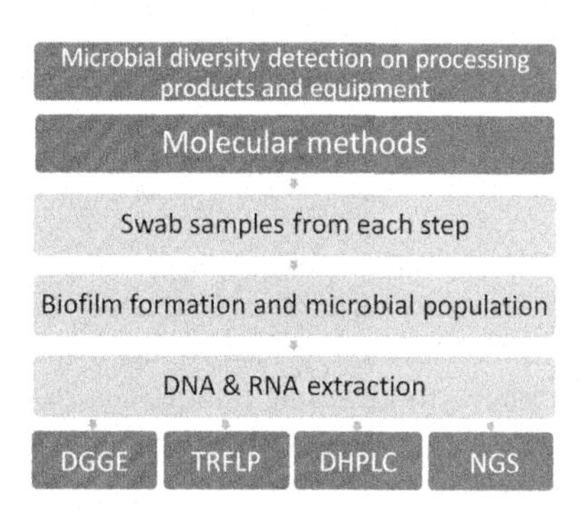

FIGURE 4

Assessment of microbial diversity from carcass sampling.

monitoring method based on the complete microbial profiles and potential indicator microorganisms that occur consistently at a detectable population level. This is critical as relying strictly on detection and quantification of foodborne pathogens during poultry processing may not be sufficiently consistent enough to offer a continuous data baseline for assessing the effectiveness of process control measures and plant cleaning cycles. Consequently, general bacterial loads and the ability to profile and/or quantitate individual bacterial genera, species, and even potential strains would offer a much more complete picture of microbial contamination that occurs over the course of a processing cycle on a regular basis. This ability would allow for a nonpathogen bacterial profile process indicator assessment that would not rely directly on being able to recover a certain level of foodborne pathogen from carcasses as they are being processed all the way to eventual retail or consumer preparation. Depending on how this data correlated with occurrence and frequency of foodborne pathogens such as *Salmonella* during these steps in poultry processing would also allow for development of more detailed assessments that are more quantitative and therefore more predictive of potential problems.

12 THE POTENTIAL FOR APPLICATION OF QRA

The risk associated with the presence of pathogens to the health of consumers is generally determined by conducting Microbial Risk Assessments (MRA) within the process of risk analysis. This has been described in detail elsewhere (Ruzante et al., 2013) and will only be briefly discussed here. According to the guidelines set by the Codex Alimentarius Commission (2012) and the USDA Microbial Risk Assessment Guidelines (USDA-FSIS, 2012), the major steps of a MRA include hazard identification, exposure assessment, hazard characterization, and risk characterization. In hazard identification, the goal is to identify and determine whether the particular microorganism poses a health risk to humans. Exposure assessment estimates the extent of human exposure to the hazard, looking at potential for consumption of the pathogen and the amount consumed. Hazard characterization qualitatively describes the potential harmful effects of an exposure by looking at dose-response relationships between how much pathogen is ingested and the effect on health. Risk characterization describes the previous risk assessment steps in a meaningful manner to aid in making decisions, allowing risk managers to present and discuss possible scenarios. When performing QRAs, statistical models are often implemented to aid in designing scenarios to determine incidence and risk (USDA-FSIS, 2012). Ruzante et al. (2013) offers an overview of types of models and their use. These steps of risk assessment are followed by the implementation of risk management by administrators to select the appropriate response to hazards and risk communication to inform everyone who may become affected (USDA-FSIS, 2012).

QRA offers a commonsense solution by utilizing methodology that involves probability risk assessment, benefit/cost analysis, or systems analysis (Brown, 2002). Practically speaking, this involves incorporating all relevant data available (published in peer reviewed journals and other sources of relevant data) to generate a model that establishes causal linkages between systemic *Salmonella* contamination in poultry carcasses and the frequency and contamination level during processing. The final outcome is a model that predicts which factors determine the introduction and dissemination events that occur during all stages of processing that are most likely to lead to contaminated poultry meat that is of greatest risk to the consumer. Once these events have been identified, then intervention strategies can be designed from a benefit/cost standpoint that will most effectively target these "high risk" events of systemic contamination.

Conducting QRA modeling to assess the impact on *Salmonella* contamination during processing on the production of *Salmonella*-contaminated poultry meat will be an important aspect for meeting future challenges in limiting *Salmonella* contamination. Specific goals for QRA modeling are to statistically estimate the probability of a particular contamination event taking place during processing. The three steps for risk assessment modeling include (1) identifying contamination routes throughout the processing production chain; (2) characterizing the risk probability of final product being contaminated during processing; and (3) recommending the best sampling methods and intervention practices that maximize reduction of *Salmonella* contamination during processing. Once these steps are completed and commercially available, risk analysis software is available to calculate the probabilities of contamination at each step of production under the different conditions and run simulations to characterize the risk throughout the process and to determine the best methods to reduce microbial contamination levels.

13 CONCLUSIONS AND FUTURE DIRECTIONS

Salmonella and other foodborne pathogen contamination before, during, and even after poultry processing continues to be an ongoing problem. Certainly, this is a sustained interest of federal regulatory agencies and, when outbreaks occur, this becomes a heightened concern of the public as well. While a number of interventions and control measures are in place or in the process of being developed, application and overall strategies for implementation remain somewhat unclear. Part of the problem lies in the inability to connect overall microbial contamination with foodborne pathogen frequency and occurrence. This is important because as pathogen control measures improve, it will become much more difficult to consistently generate pathogen data as a means to routinely monitor processing impact on pathogen levels. Certainly, the ability to detect and even in some cases quantitate foodborne pathogens has improved tremendously. However, if pathogen levels are generally lower and more infrequent, bacterial data gaps can still occur during regular cycles of processing. Thus, general bacterial load data consisting of nonpathogen bacterial populations become a much more important source of a continuous microbial data pool to serve as a process indicator for assessing impact of respective control measures.

Historically, petri plate enumeration approaches such as aerobic plate counts served as a constant data source that can be easily collected and provide at least remedial information on bacterial loads. With the recent advent of a multitude of molecular approaches and methodologies, opportunities now exist to generate a much more comprehensive profile of the bacterial populations on chicken carcasses and the surrounding poultry processing plant environment. As the discernment power grows with the development of more cost-effective microbiome sequencing capacity, the potential exists to generate very complete pictures of the makeup of microbial consortia on carcasses during all aspects of processing. This type of data offers the possibility of conducting QRA on poultry processing operations and how they may contribute to foodborne pathogen contamination of poultry meat during processing by identifying signature or indicator organisms that reflect contamination patterns consistent with the corresponding foodborne pathogen of interest. By looking at complete microbial profiles during processing, additional routes of contamination from common procedures may be identified and, subsequently, the best processing practices to reduce their impact may be determined. Examining the processing equipment itself can also be used to link any microbial populations found on equipment with populations occurring on the product to identify where in the processing chain the product was contaminated. Probabilities of microbial contamination from all stages of processing can, in turn, become information to consider when constructing risk models, and simulations may be run using risk analysis software to determine where interventions would be most effective. In future, such combinations of more complete microbial profiles from all aspects of poultry processing, along with models that predict the most likely sources of contamination, will provide a much more precise strategic plan for the most cost-effective applications of intervention steps. Finally, such information could assist in controlling the level of microbial spoilage in the respective poultry products and retain freshness of the end product. Improvements in product quality and extended shelf life would further enhance the cost-benefit for utilizing more sophisticated microbial monitoring approaches.

ACKNOWLEDGMENTS

Some of the research cited in this review was supported by a U.S. Poultry & Egg Association grant to author S.C.R. Author T.M.D. was supported by King Saud University Scholarship, Department of Botany & Microbiology, Science College, Riyadh, Saudi Arabia.

REFERENCES

Barbut, S., 2010. Poultry Products Processing: An Industry Guide. CRC Press, Boca Raton, FL.

Barlaan, E.A., Sugimori, M., Furukawa, S., Takeuchi, K., 2005. Profiling and monitoring of microbial populations by denaturing high-performance liquid chromatography. J. Microbiol. Methods 61, 399–412.

Bentley, J., 2012. U.S. Per Capita Availability of Chicken Surpasses That of Beef. USDA-Economic Research Service (ERS), Amber Waves. Available at: http://www.ers.usda.gov/amber-waves/2012-september/us-consumption-of-chicken.aspx#.VNz79PnF-Dh (accessed 02.12.15).

Berghaus, R.D., Thayer, S.G., Law, B.F., Mild, R.M., Hofacre, C.L., Singer, R.S., 2013. Enumeration of *Salmonella* and *Campylobacter* spp. in environmental farm samples and processing plant carcass rinses from commercial broiler chicken flocks. Appl. Environ. Microbiol. 79, 4106–4114.

Bohaychuk, V., Gensler, G., McFall, M., King, R., Renter, D., 2007. A real-time PCR assay for the detection of *Salmonella* in a wide variety of food and food-animal matrices. J. Food Prot. 70, 1080–1087.

Bokulich, N.A., Mills, D.A., 2012. Next-generation approaches to the microbial ecology of food fermentations. Biochem. Mol. Biol. Rep. 45, 377–389.

Brewer, R.L., James, W.O., Prucha, J.C., Johnston, R.W., Alvarez, C.A., Kelly, W., Bergeron, E.A., 1995. Poultry processing line speeds as related to bacteriologic profile of broiler carcasses. J. Food Sci. 60, 1022–1024.

Brown, M.H., 2002. Quantitative microbiological risk assessment: principles applied to determining the comparative risk of salmonellosis from chicken products. Int. Biodeter. Biodegr. 50, 155–160.

Byrd, J.A., Corrier, D.E., Hume, M.E., Bailey, R.H., Stanker, L.H., Hargis, B.M., 1998a. Incidence of *Campylobacter* in crops of preharvest market-age broiler chickens. Poult. Sci. 77, 1303–1305.

Byrd, J.A., Corrier, D.E., Hume, M.E., Bailey, R.H., Stanker, L.H., Hargis, B.M., 1998b. Effect of feed withdrawal on *Campylobacter* in the crops of market-age broiler chickens. Avian Dis. 42, 802–806.

Cairncross, S., Hunt, C., Boisson, S., Bostoen, K., Curtis, V., Fung, I.C., Schmidt, W.P., 2010. Water, sanitation and hygiene for the prevention of diarrhoea. Int. J. Epidemiol. 39 (Suppl. 1), i193–i205.

Call, D.R., 2005. Challenges and opportunities for pathogen detection using DNA microarrays. Crit. Rev. Microbiol. 31, 91–99.

Camu, N., De Winter, T., Verbrugghe, K., Cleenwerck, I., Vandamme, P., Takrama, J.S., Vancanneyt, M., De Vuyst, L., 2007. Dynamics and biodiversity of populations of lactic acid bacteria and acetic acid bacteria involved in spontaneous heap fermentation of cocoa beans in Ghana. Appl. Environ. Microbiol. 73, 1809–1824.

Centers for Disease Control and Prevention (CDC), 2014. Multistate Outbreak of Multidrug-Resistant *Salmonella* Heidelberg Infections Linked to Foster Farms Brand Chicken. CDC. Available at: http://www.cdc.gov/salmonella/heidelberg-10-13 (accessed 02.12.15).

Chan, K., Baker, S., Kim, C.C., Detweiler, C.S., Dougan, G., Falkow, S., 2003. Genomic comparison of *Salmonella enterica* serovars and *Salmonella bongori* by use of an *S. enterica* serovar Typhimurium DNA microarray. J. Bacteriol. 185, 553–563.

Charles, D., 2014. How Foster Farms is Solving the Case of the Mystery *Salmonella*. The Salt, National Public Radio. Available at: http://www.npr.org/blogs/thesalt/2014/08/28/342166299/how-foster-farms-is-solving-the-case-of-the-mystery-salmonella (accessed 02.12.15).

Chopra, M., Mason, E., Borrazzo, J., Campbell, H., Rudan, I., Liu, L., Black, R.E., Bhutta, Z.A., 2013. Ending of preventable deaths from pneumonia and diarrhoea: an achievable goal. Lancet 381, 1499–1506.

Clouser, C.S., Doores, S., Mast, M.G., Knabel, S.J., 1995. The role of defeathering in the contamination of turkey skin by *Salmonella* species and *Listeria monocytogenes*. Poult. Sci. 74, 723–731.

Codex Alimentarius Commission, 2012. Codex principles and guidelines for the conduct of microbiological risk assessment. Document CAC/GL 30-1999. Available at: http://www.codexalimentarius.org/download/standards/357/CXG_030e_2012.pdf (accessed 02.12.15).

Cook, A., Odumeru, J., Lee, S., Pollari, F., 2012. *Campylobacter*, *Salmonella*, *Listeria monocytogenes*, verotoxigenic *Escherichia coli*, and *Escherichia coli* prevalence, enumeration, and subtypes on retail chicken breasts with and without skin. J. Food Prot. 75, 34–40.

Corry, J.E., Allen, V.M., Hudson, W.R., Breslin, M.F., Davies, R.H., 2002. Sources of *Salmonella* on broiler carcasses during transportation and processing: modes of contamination and methods of control. J. Appl. Microbiol. 92, 424–432.

Cox, N.A., Cason, J.A., Richardson, L.J., 2011. Minimization of *Salmonella* contamination on raw poultry. Ann. Rev. Food Sci. Technol. 2, 75–95.

Crump, J.A., Griffin, P.M., Angulo, F.J., 2002. Bacterial contamination of animal feed and its relationship to human foodborne illness. Clin. Infect. Dis. 35, 859–865.

Dar, S.A., Yao, L., van Dongen, U., Kuenen, J.G., Muyzer, G., 2007. Analysis of diversity and activity of sulfate-reducing bacterial communities in sulfidogenic bioreactors using 16S rRNA and *dsrB* genes as molecular markers. Appl. Environ. Microbiol. 73, 594–604.

De Medici, D., Kuchta, T., Knutsson, R., Angelov, A., Auricchio, B., Barbanera, M., Diaz-Amigo, C., Fiore, A., Kudirkiene, E., Hohl, A., 2015. Rapid methods for quality assurance of foods: the next decade with polymerase chain reaction (PCR)-based food monitoring. Food Anal. Methods 8, 255–271.

Dunkley, K.D., McReynolds, J.L., Hume, M.E., Dunkley, C.S., Callaway, T.R., Kubena, L.F., Nisbet, D.J., Ricke, S.C., 2007. Molting in *Salmonella* Enteritidis-challenged laying hens fed alfalfa crumbles. I. *Salmonella* Enteritidis colonization and virulence gene *hilA* response. Poult. Sci. 86, 1633–1639.

Dunkley, K.D., Callaway, T.R., Chalova, V.I., McReynolds, J.L., Hume, M.E., Dunkley, C.S., Kubena, L.F., Nisbet, D.J., Ricke, S.C., 2009. Foodborne *Salmonella* ecology in the avian gastrointestinal tract. Anaerobe 15, 26–35.

Durant, J.A., Corrier, D.E., Byrd, J.A., Stanker, L.H., Ricke, S.C., 1999. Feed deprivation affects crop environment and modulates *Salmonella* Enteritidis colonization and invasion of leghorn hens. Appl. Environ. Microbiol. 65, 1919–1923.

Ercolini, D., Hill, P.J., Dodd, C.E., 2003. Bacterial community structure and location in Stilton cheese. Appl. Environ. Microbiol. 69, 3540–3548.

Ercolini, D., Frisso, G., Mauriello, G., Salvatore, F., Coppola, S., 2008. Microbial diversity in natural whey cultures used for the production of Caciocavallo Silano PDO cheese. Int. J. Food Microbiol. 124, 164–170.

Farthing, M., Salam, M.A., Lindberg, G., Dite, P., Khalif, I., Salazar-Lindo, E., Ramakrishna, B.S., Goh, K.L., Thomson, A., Khan, A.G., Krabshuis, J., LeMair, A., WGO, 2013. Acute diarrhea in adults and children: a global perspective. J. Clin. Gastroenterol. 47, 12–20.

Finstad, S., O'Bryan, C.A., Marcy, J.A., Crandall, P.G., Ricke, S.C., 2012. *Salmonella* and broiler processing in the United States: relationship to foodborne salmonellosis. Food Res. Int. 45, 789–794.

Foley, S.L., Nayak, R., Hanning, I.B., Johnson, T.J., Han, J., Ricke, S.C., 2011. Population dynamics of *Salmonella enterica* serotypes in commercial egg and poultry production. Appl. Environ. Microbiol. 77, 4273–4279.

Foley, S.L., Johnson, T.J., Ricke, S.C., Nayak, R., Danzeisen, J., 2013. *Salmonella* pathogenicity and host adaptation in chicken-associated serovars. Microbiol. Mol. Biol. Rev. 77, 582–607.

Foster, J.A., Bunge, J., Gilbert, J.A., Moore, J.H., 2012. Measuring the microbiome: perspectives on advances in DNA-based techniques for exploring microbial life. Brief. Bioinform. 13, 420–429.

Galimberti, A., Bruno, A., Mezzasalma, V., De Mattia, F., Bruni, I., Labra, M., 2015. Emerging DNA-based technologies to characterize food ecosystems. Food Res. Int. 69, 424–433.

Genigeorgis, C., Hassuneh, M., Collins, P., 1986. *Campylobacter jejuni* infection on poultry farms and its effect on poultry meat contamination during slaughtering. J. Food Prot. 49, 895–903.

Gibson, G., 2002. Microarrays in ecology and evolution: a preview. Mol. Ecol. 11, 17–24.

Gilbert, J.A., Laverock, B., Temperton, B., Thomas, S., Muhling, M., Hughes, M., 2011. Chapter 12. Metagenomics. In: Kwon, Y.M., Ricke, S.C. (Eds.), High-Throughput Next Generation Sequencing: Methods and Applications, vol. 733. Springer Protocols, Humana Press, New York, pp. 173–183.

Giombelli, A., Gloria, M.B.A., 2014. Prevalence of *Salmonella* and *Campylobacter* on broiler chickens from farm to slaughter and efficiency of methods to remove visible fecal contamination. J. Food Prot. 77, 1851–1859.

Handley, J.A., Hanning, I., Ricke, S.C., Johnson, M.G., Jones, F.T., Apple, R.O., 2010. Temperature and bacterial profile of post chill poultry carcasses stored in processing combo held at room temperature. J. Food Sci. 75, M515–M520.

Hanning, I.B., Ricke, S.C., 2011. Chapter 11. Prescreening of microbial populations for the assessment of sequencing potential. In: Kwon, Y.M., Ricke, S.C. (Eds.), High-Throughput Next Generation Sequencing: Methods and Applications, vol. 733. Springer Protocols, Humana Press, New York, pp. 159–170.

Hanning, I.B., Johnson, M.G., Ricke, S.C., 2008. Precut prepackaged lettuce: a risk for listeriosis? Foodborne Pathog. Dis. 5, 731–746.

Hanning, I.B., Jarquin, R., O'Leary, A., Slavik, M., 2009a. Polymerase chain reaction-based assays for the detection and differentiation of poultry significant *Pseudomonads*. J. Rapid Methods Autom. Microbiol. 17, 490–502.

Hanning, I.B., Nutt, J.D., Ricke, S.C., 2009b. Salmonellosis outbreaks in the United States due to fresh produce: sources and potential intervention measures. Foodborne Pathog. Dis. 6, 635–648.

Harmon, D.E., 2013. Poultry: From the Farm to Your Table, first ed. The Rosen Publishing Group, Inc., New York.

Harvey, D., 2012. Poultry & Eggs. USDA-ERS. Available at: http://www.ers.usda.gov/topics/animal-products/poultry-eggs/.aspx#.Uu9K1z2Qy1J (accessed 24.11.14).

Hoffmann, S., Batz, M.B., Morris Jr., J.G., 2012. Annual cost of illness and quality-adjusted life year losses in the United States due to 14 foodborne pathogens. J. Food Prot. 75, 1292–1302.

Hogue, A.T., White, P.L., Heminover, J.A., 1998. Pathogen Reduction and Hazard Analysis and Critical Control Point (HACCP) systems for meat and poultry. USDA. Vet. Clin. North Am. Food Anim. Pract. 14, 151–164.

Horrocks, S.M., Anderson, R.C., Nisbet, D.J., Ricke, S.C., 2009. Incidence and ecology of *Campylobacter jejuni* and *coli* in animals. Anaerobe 15, 18–25.

Howard, Z.R., O'Bryan, C.A., Crandall, P.G., Ricke, S.C., 2012. *Salmonella* Enteritidis in shell eggs: current issues and prospects for control. Food Res. Int. 45, 755–764.

Hume, M.E., 2011. Historic perspective: prebiotics, probiotics, and other alternatives to antibiotics. Poult. Sci. 90, 2663–2669.

Hutzler, M., Geiger, E., Jacob, F., 2010. Use of PCR-DHPLC (Polymerase Chain Reaction—Denaturing High Performance Liquid Chromatography) for the rapid differentiation of industrial *Saccharomyces pastorianus* and *Saccharomyces cerevisiae* strains. J. Inst. Brew. 116, 464–474.

Jarquin, R., Hanning, I., Ahn, S., Ricke, S.C., 2009. Development of rapid detection and genetic characterization of *Salmonella* in poultry breeder feeds. Sensors 9, 5308–5323.

Jimenez, S., Salsi, M., Tiburzi, M., Pirovani, M., 2002. A comparison between broiler chicken carcasses with and without visible faecal contamination during the slaughtering process on hazard identification of *Salmonella* spp. J. Appl. Microbiol. 93, 593–598.

Joerger, R.D., 2003. Alternatives to antibiotics: bacteriocins, antimicrobial peptides and bacteriophages. Poult. Sci. 82, 640–647.

Jofré, A., Martin, B., Garriga, M., Hugas, M., Pla, M., Rodríguez-Lázaro, D., Aymerich, T., 2005. Simultaneous detection of *Listeria monocytogenes* and *Salmonella* by multiplex PCR in cooked ham. Food Microbiol. 22, 109–115.

Justé, A., Thomma, B., Lievens, B., 2008. Recent advances in molecular techniques to study microbial communities in food-associated matrices and processes. Food Microbiol. 25, 745–761.

Kawasaki, S., Horikoshi, N., Okada, Y., Takeshita, K., Sameshima, T., Kawamoto, S., 2005. Multiplex PCR for simultaneous detection of *Salmonella* spp., *Listeria monocytogenes*, and *Escherichia coli* O157: H7 in meat samples. J. Food Prot. 68, 551–556.

Kearney, J., 2010. Food consumption trends and drivers. Philos. Trans. R. Soc. Lond. Ser. B Biol. Sci. 365, 2793–2807.

Kim, H., Park, S., Lee, T., Nahm, B., Chung, Y., Seo, K., Kim, H., 2006. Identification of *Salmonella enterica* serovar Typhimurium using specific PCR primers obtained by comparative genomics in *Salmonella* serovars. J. Food Prot. 69, 1653–1661.

Knight, R., Jansson, J., Field, D., Fierer, N., Desai, N., Fuhrman, J.A., Hugenholtz, P., Van Der Lelie, D., Meyer, F., Stevens, R., 2012. Unlocking the potential of metagenomics through replicated experimental design. Nat. Biotechnol. 30, 513–520.

Koo, O.K., Sirsat, S.A., Crandall, P.G., Ricke, S.C., 2012. Physical and chemical control of *Salmonella* in ready-to-eat products. Agric. Food Anal. Bacteriol. 2, 56–68.

Koo, O.K., Mertz, A.W., Akins, E.L., Sirsat, S.A., Neal, J.A., Morawicki, R., Crandall, P.G., Ricke, S.C., 2013. Analysis of microbial diversity on deli slicers using polymerase chain reaction and denaturing gradient gel electrophoresis technologies. Lett. Appl. Microbiol. 56, 111–119.

Kwon, Y.M., Ricke, S.C. (Eds.), 2011. High-Throughput Next Generation Sequencing: Methods and Applications. Methods in Molecular Microbiology, vol. 733. Springer Protocols, Humana Press, New York, NY, 308 pp.

Lampel, K.A., Orlandi, P.A., Kornegay, L., 2000. Improved template preparation for PCR-based assays for detection of food-borne bacterial pathogens. Appl. Environ. Microbiol. 66, 4539–4542.

Latimer, H.K., Marks, H.M., Coleman, M.E., Schlosser, W.D., Golden, N.J., Ebel, E.D., Kause, J., Schroeder, C.M., 2008. Evaluating the effectiveness of pasteurization for reducing human illnesses from *Salmonella* spp. in egg products: results of a quantitative risk assessment. Foodborne Pathog. Dis. 5, 59–68.

Levin, R.E., 2009. The use of molecular methods for detecting and discriminating *Salmonella* associated with foods – a review. Food Biotechnol. 23, 313–367.

Lillard, H., 1989. Factors affecting the persistence of *Salmonella* during the processing of poultry. J. Food Prot. 52, 829–832.

MacDonald, J.M., 2008. The Economic Organization of U.S. Broiler Production. In: Economic Information Bulletin: The Economic Organization of U.S. Broiler Production. USDA-ERS. Available at: http://www.ers.usda.gov/publications/eib-economic-information-bulletin/eib38.aspx (accessed 15.02.15).

Maciorowski, K.G., Jones, F.T., Pillai, S.D., Ricke, S.C., 2004. Incidence, sources, and control of foodborne *Salmonella* spp. in poultry feeds. Worlds Poult. Sci. J. 60, 446–457.

Maciorowski, K.G., Pillai, S.D., Jones, F.T., Ricke, S.C., 2005. Polymerase chain reaction detection of foodborne *Salmonella* spp. in animal feeds. Crit. Rev. Microbiol. 31, 45–53.

Mani-Lopez, E., García, H.S., López-Malo, A., 2012. Organic acids as antimicrobials to control *Salmonella* in meat and poultry products. Food Res. Int. 45, 713–721.

Marino, D.D., 2007. Water and food safety in the developing world: global implications for health and nutrition of infants and young children. J. Am. Diet. Assoc. 107, 1930–1934.

Mathews, K., 2014. Holidays and Weather Disrupted Commodity Flows and Increased Price Volatility. USDA-ERS. Available at: http://www.ers.usda.gov/media/1260436/ldpm235.pdf (accessed 02.12.15).

Mathews, K., Haley, M., 2015. Quarterly Hogs and Pigs Report Shows Reduced PEDv Effects. Available at: http://www.ers.usda.gov/media/1737701/ldpm-247.pdf (accessed 13.02.15).

Maukonen, J., Saarela, M., 2009. Microbial communities in industrial environment. Curr. Opin. Microbiol. 12, 238–243.

Mazengia, E., Samadpour, M., Hill, H., Greeson, K., Tenney, K., Liao, G., Huang, X., Meschke, J., 2014. Prevalence, concentrations, and antibiotic sensitivities of *Salmonella* serovars in poultry from retail establishments in Seattle, Washington. J. Food Prot. 77, 885–893.

Mertz, A.W., Koo, O.K., O'Bryan, C.A., Morawicki, R., Sirsat, S.A., Neal, J.A., Crandall, P.G., Ricke, S.C., 2014. Microbial ecology of meat slicers as determined by denaturing gradient gel electrophoresis. Food Control 42, 242–247.

Milillo, S.R., Ricke, S.C., 2010. Synergistic reduction of *Salmonella* in a model raw chicken media using a combined thermal and acidified organic acid salt intervention treatment. J. Food Sci. 75, M121–M125.

Milillo, S.R., Martin, E., Muthaiyan, A., Ricke, S.C., 2011. Immediate reduction of *Salmonella enterica* serotype Typhimurium viability via membrane destabilization following exposure to multiple-hurdle treatments with heated, acidified organic acid salt solutions. Appl. Environ. Microbiol. 77, 3765–3772.

Milillo, S.R., Friedly, E.C., Saldivar, J.C., Muthaiyan, A., O'Bryan, C., Crandall, P.G., Johnson, M.G., Ricke, S.C., 2012. A review of the ecology, genomics, and stress response of *Listeria innocua* and *Listeria monocytogenes*. Crit. Rev. Food Sci. Nutr. 52, 712–725.

Miller, D.N., Bryant, J.E., Madsen, E.L., Ghiorse, W.C., 1999. Evaluation and optimization of DNA extraction and purification procedures for soil and sediment samples. Appl. Environ. Microbiol. 65, 4715–4724.

Muyzer, G., Smalla, K., 1998. Application of denaturing gradient gel electrophoresis (DGGE) and temperature gradient gel electrophoresis (TGGE) in microbial ecology. Antonie Van Leeuwenhoek 73, 127–141.

Muyzer, G., De Wall, E., Uitterlinden, A., 1993. Profiling of complex microbial populations by denaturing gradient gel electrophoresis of 16S ribosomal DNA fragments. Appl. Environ. Microbiol. 59, 695–700.

Nath, K., Bloomfield, S., Jones, M., 2006. Household Water Storage, Handling and Point-of-Use Treatment. International Scientific Forum on Home Hygiene. Available at: http://www.ifh-homehygiene.org/system/files_force/publications/low_res_water_paper.pdf (accessed 02.12.15).

Nauta, M., Der Fels-Klerx, V., Havelaar, A., 2005. A poultry-processing model for quantitative microbiological risk assessment. Risk Anal. 25, 85–98.

Nde, C.W., Sherwood, J.S., Doetkott, C., Logue, C.M., 2006. Prevalence and molecular profiles of *Salmonella* collected at a commercial turkey processing plant. J. Food Prot. 69, 1794–1801.

O'Ryan, M., Prado, V., Pickering, L.K., 2005. A millennium update on pediatric diarrheal illness in the developing world. Semin. Pediatr. Infect. Dis. 16, 125–136.

Olsen, J., Brown, D., Madsen, M., Bisgaard, M., 2003. Cross-contamination with *Salmonella* on a broiler slaughterhouse line demonstrated by use of epidemiological markers. J. Appl. Microbiol. 94, 826–835.

Oosterom, J., Notermans, S., Karman, H., Engels, G., 1983. Origin and prevalence of *Campylobacter jejuni* in poultry processing. J. Food Prot. 46, 339–344.

Otawa, K., Asano, R., Ohba, Y., Sasaki, T., Kawamura, E., Koyama, F., Nakamura, S., Nakai, Y., 2006. Molecular analysis of ammonia-oxidizing bacteria community in intermittent aeration sequencing batch reactors used for animal wastewater treatment. Environ. Microbiol. 8, 1985–1996.

Owens, C.M., Alvarado, C., Sams, A.R. (Eds.), 2010. Poultry Meat Processing, second ed. CRC Press, Boca Raton, FL.

Park, E., Chun, J., Cha, C., Park, W., Jeon, C.O., Bae, J., 2012. Bacterial community analysis during fermentation of ten representative kinds of kimchi with barcoded pyrosequencing. Food Microbiol. 30, 197–204.

Park, S.H., Aydin, M., Khatiwara, A., Dolan, M.C., Gilmore, D.F., Bouldin, J.L., Ahn, S., Ricke, S.C., 2014. Current and emerging technologies for rapid detection and characterization of *Salmonella* in poultry and poultry products. Food Microbiol. 38, 250–262.

Park, S.H., Hanning, I., Jarquin, R., Moore, P., Donoghue, D.J., Donoghue, A.M., Ricke, S.C., 2011. Multiplex PCR assay for the detection and quantification of *Campylobacter* spp., *Escherichia coli* O157: H7, and *Salmonella* serotypes in water samples. FEMS Microbiol. Lett. 316, 7–15.

Park, S.H., Kim, H.J., Cho, W.H., Kim, J.H., Oh, M.H., Kim, S.H., Lee, B.K., Ricke, S.C., Kim, H.Y., 2009. Identification of *Salmonella enterica* subspecies I, *Salmonella enterica* serovars Typhimurium, Enteritidis and Typhi using multiplex PCR. FEMS Microbiol. Lett. 301, 137–146.

Park, S.H., Ricke, S.C., 2015. Development of multiplex PCR assay for simultaneous detection of *Salmonella* genus, *Salmonella* subspecies I, *Salm.* Enteritidis, *Salm.* Heidelberg and *Salm.* Typhimurium. J. Appl. Microbiol. 118, 152–160.

Park, S.Y., Woodward, C.L., Kubena, L.F., Nisbet, D.J., Birkhold, S.G., Ricke, S.C., 2008. Environmental dissemination of foodborne *Salmonella* in preharvest poultry production: reservoirs, critical factors, and research strategies. Crit. Rev. Environ. Sci. Technol. 38, 73–111.

Patterson, J.A., Burkholder, K.M., 2003. Application of prebiotics and probiotics in poultry production. Poult. Sci. 82, 627–631.

Pavic, A., Groves, P., Bailey, G., Cox, J., 2010. A validated miniaturized MPN method, based on ISO 6579: 2002, for the enumeration of *Salmonella* from poultry matrices. J. Appl. Microbiol. 109, 25–34.

Perumalla, A., Hettiarachchy, N.S., Over, K.F., Ricke, S.C., Gbur, E.E., Zhang, J., Davis, B., 2012. Effect of potassium lactate and sodium diacetate combination to inhibit *Listeria monocytogenes* in low and high fat chicken and turkey hotdog model systems. Open Food Sci. J. 6, 16–23.

Pillai, S.D., Ricke, S.C., 1995. Strategies to accelerate the applicability of gene amplification protocols for pathogen detection in meat and meat products. Crit. Rev. Microbiol. 21, 239–261.

Pillai, S.D., Ricke, S.C., 2002. Review/Synthèse Bioaerosols from municipal and animal wastes: background and contemporary issues. Can. J. Microbiol. 48, 681–696.

Podewils, L.J., Mintz, E.D., Nataro, J.P., Parashar, U.D., 2004. Acute, infectious diarrhea among children in developing countries. Semin. Pediatr. Infect. Dis. 15, 155–168.

Premstaller, A., Oefner, P., 2003. Denaturing high-performance liquid chromatography. In: Kwok, P. (Ed.), Single Nucleotide Polymorphisms: Methods and Protocols. In: Methods in Molecular Biology, vol. 212. Springer, New York, pp. 15–35.

Quigley, L., O'Sullivan, O., Beresford, T.P., Ross, R.P., Fitzgerald, G.F., Cotter, P.D., 2011. Molecular approaches to analysing the microbial composition of raw milk and raw milk cheese. Int. J. Food Microbiol. 150, 81–94.

Ramirez, G.A., Sarlin, L.L., Caldwell, D.J., Yezak Jr., C.R., Hume, M.E., Corrier, D.E., Deloach, J.R., Hargis, B.M., 1997. Effect of feed withdrawal on the incidence of *Salmonella* in the crops and ceca of market age broiler chickens. Poult. Sci. 76, 654–656.

Randazzo, C.L., Torriani, S., Akkermans, A.D., de Vos, W.M., Vaughan, E.E., 2002. Diversity, dynamics, and activity of bacterial communities during production of an artisanal Sicilian cheese as evaluated by 16S rRNA analysis. Appl. Environ. Microbiol. 68, 1882–1892.

Rasekh, J., Thaler, A.M., Engeljohn, D.J., Pihkala, N.H., 2005. Food safety and inspection service policy for control of poultry contaminated by digestive tract contents: a review. J. Appl. Poultry Res. 14, 603–611.

Rask, K.J., Rask, N., 2011. Economic development and food production–consumption balance: a growing global challenge. Food Policy 36, 186–196.

Rasschaert, G., Houf, K., De Zutter, L., 2007. Impact of the slaughter line contamination on the presence of *Salmonella* on broiler carcasses. J. Appl. Microbiol. 103, 333–341.

Ravichandran, M., Hettiarachchy, N.S., Ganesh, V., Ricke, S.C., Singh, S., 2011. Enhancement of antimicrobial activities of naturally occurring phenolic compounds by nanoscale delivery against *Listeria monocytogenes*, *Escherichia coli* O157: H7 and *Salmonella typhimurium* in broth and chicken meat system. J. Food Saf. 31, 462–471.

Ricke, S.C., 2003a. Perspectives on the use of organic acids and short chain fatty acids as antimicrobials. Poult. Sci. 82, 632–639.

Ricke, S.C., 2003b. The gastrointestinal tract ecology of *Salmonella* enteritidis colonization in molting hens. Poult. Sci. 82, 1003–1007.

Ricke, S.C., 2014. Application of molecular approaches for understanding foodborne *Salmonella* establishment in poultry production. Adv. Biol. 2014. http://dx.doi.org/10.1155/2014/813275, Article ID 813275, 25 pages.

Ricke, S.C., 2015. Potential of fructooligosaccharide prebiotics in alternative and nonconventional poultry production systems. Poult. Sci. http://dx.doi.org/10.3382/ps/pev049, http://www.hindawi.com/journals/ab/2014/813275/cta/.

Ricke, S.C., Kundinger, M.M., Miller, D.R., Keeton, J.T., 2005. Alternatives to antibiotics: chemical and physical antimicrobial interventions and foodborne pathogen response. Poult. Sci. 84, 667–675.

Ricke, S.C., Hererra, P., Biswas, D., 2012. Bacteriophages for potential food safety applications in organic meat production. In: Ricke, S.C., Van Loo, E.J., Johnson, M.G., O'Bryan, C.A. (Eds.), Organic Meat Production and Processing. Wiley Scientific/IFT, New York, pp. 407–424.

Ricke, S.C., Dunkley, C.S., Durant, J.A., 2013a. A review on development of novel strategies for controlling *Salmonella* Enteritidis colonization in laying hens: fiber-based molt diets. Poult. Sci. 92, 502–525.

Ricke, S.C., Khatiwara, A., Kwon, Y.M., 2013b. Application of microarray analysis of foodborne *Salmonella* in poultry production: a review. Poult. Sci. 92, 2243–2250.

Russell, S.M., 2000. Overview of *E. coli* on poultry. Zootecnica Int. 23, 49–50.

Ruzante, J.M., Whiting, R.C., Dennis, S.B., Buchanan, R.L., 2013. Microbial risk assessment. In: Doyle, M.P., Buchanan, R.L. (Eds.), Food Microbiology: Fundamentals and Frontiers, fourth ed. American Society for Microbiology Press, Washington, DC, pp. 1023–1037.

Santosham, M., Chandran, A., Fitzwater, S., Fischer-Walker, C., Baqui, A.H., Black, R., 2010. Progress and barriers for the control of diarrhoeal disease. Lancet 376, 63–67.

Scallan, E., Hoekstra, R.M., Angulo, F.J., Tauxe, R.V., Widdowson, M., Roy, S.L., Jones, J.L., Griffin, P.M., 2011. Foodborne illness acquired in the United States-major pathogens. Emerg. Infect. Dis. 17, 7–15.

Scharff, R.L., 2012. Economic burden from health losses due to foodborne illness in the United States. J. Food Prot. 75, 123–131.

Scheu, P.M., Berghof, K., Stahl, U., 1998. Detection of pathogenic and spoilage micro-organisms in food with the polymerase chain reaction. Food Microbiol. 15, 13–31.

Schmidt, W., Cairncross, S., 2009. Household water treatment in poor populations: is there enough evidence for scaling up now? Environ. Sci. Technol. 43, 986–992.

Schmidt, W.P., Cairncross, S., Barreto, M.L., Clasen, T., Genser, B., 2009. Recent diarrhoeal illness and risk of lower respiratory infections in children under the age of 5 years. Int. J. Epidemiol. 38, 766–772.

Sheth, M., Dwivedi, R., 2006. Complementary foods associated diarrhea. Indian J. Pediatr. 73, 61–64.

Siragusa, G.R., Ricke, S.C., 2012. Probiotics as pathogen control agents for organic meat production. In: Ricke, S.C., Van Loo, E.J., Johnson, M.G., O'Bryan, C.A. (Eds.), Organic Meat Production and Processing. Wiley Scientific/IFT, New York, pp. 331–349.

Sirsat, S.A., Muthaiyan, A., Ricke, S.C., 2011. Optimization of the RNA extraction method for transcriptome studies of *Salmonella* inoculated on commercial raw chicken breast samples. BMC Res. Notes 4, 60.

Sofos, J.N., Flick, G., Nychas, G.J., O'Bryan, C.A., Ricke, S.C., Crandall, P.G., 2013. Meat, poultry, and seafood In: Doyle, M.P., Buchanan, R.L. (Eds.), Food Microbiology-Fundamentals and Frontiers, fourth ed. American Society for Microbiology Press, Washington, DC, pp. 111–167.

Soria, M.C., Soria, M.A., Bueno, D.J., Colazo, J.L., 2011. A comparative study of culture methods and polymerase chain reaction assay for *Salmonella* detection in poultry feed. Poult. Sci. 90, 2606–2618.

Steenackers, H., Hermans, K., Vanderleyden, J., de Keersmaecker, S.C.J., 2012. *Salmonella* biofilms: an overview on occurrence, structure, regulation and eradication. Food Res. Int. 45, 502–531.

Sun, H.Y., Deng, S.P., Raun, W.R., 2004. Bacterial community structure and diversity in a century-old manure-treated agroecosystem. Appl. Environ. Microbiol. 70, 5868–5874.

Sun, F., Wu, D., Qiu, Z., Jin, M., Li, J., 2010. Development of real-time PCR systems based on SYBR Green for the specific detection and quantification of *Klebsiella pneumoniae* in infant formula. Food Control 21, 487–491.

Temperton, B., Giovannoni, S.J., 2012. Metagenomics: microbial diversity through a scratched lens. Curr. Opin. Microbiol. 15, 605–612.

USDA-FSIS, 1996. Pathogen reduction; hazard analysis critical control point (HACCP) systems. USDA-FSIS, Federal Register 61, 38806–38944.

USDA-FSIS, 2010. New performance standards for *Salmonella* and *Campylobacter* in young chickens and turkey slaughter establishments; new compliance guides. USDA-FSIS, Federal Register 75, 27288.

USDA-FSIS, 2012. Microbial Risk Assessment Guideline: Pathogenic Microorganisms with Focus on Food and Water. Available at: http://www.fsis.usda.gov/wps/wcm/connect/d79eaa29-c53a-451e-ba1c-36a76a6c6434/Microbial_Risk_Assessment_Guideline_2012-001.pdf?MOD=AJPERES (accessed 02.12.15).

USDA-FSIS, 2013. Strategic Performance Working Group *Salmonella* Action Plan. Available at: http://www.fsis.usda.gov/wps/wcm/connect/aae911af-f918-4fe1-bc42-7b957b2e942a/SAP-120413.pdf?MOD=AJPERES (accessed 02.12.15).

USDA-FSIS, 2014a. Microbiological Laboratory Guidebook. Appendix 2.05. Available at: http://www.fsis.usda.gov/wps/wcm/connect/8872ec11-d6a3-4fcf-86df-4d87e57780f5/MLG-Appendix-2.pdf?MOD=AJPERES (accessed 02.12.15).

USDA-FSIS, 2014b. Microbiological Laboratory Guidebook. Document 4.08. Available at: http://www.fsis.usda.gov/wps/wcm/connect/700c05fe-06a2-492a-a6e1-3357f7701f52/MLG-4.pdf?MOD=AJPERES (accessed 02.12.15).

USDA-FSIS, 2014c. Pathogen Reduction – *Salmonella* and *Campylobacter* Performance Standards Verification Testing. Available at: http://www.fsis.usda.gov/wps/wcm/connect/b0790997-2e74-48bf-9799-85814bac9ceb/28_IM_PR_Sal_Campy.pdf?MOD=AJPERES (accessed 02.12.15).

USDA-FSIS, 2014d. Microbiological Laboratory Guidebook. Document 41.03. Available at: http://www.fsis.usda.gov/wps/wcm/connect/0273bc3d-2363-45b3-befb-1190c25f3c8b/MLG-41.pdf?MOD=AJPERES (accessed 02.12.15).

USDA-FSIS, 2015a. Public Health Effects of Raw Chicken Parts and Comminuted Chicken and Turkey Performance Standards. Available at: http://www.fsis.usda.gov/wps/wcm/connect/afe9a946-03c6-4f0d-b024-12aba4c01aef/Effects-Performance-Standards-Chicken-Parts-Comminuted.pdf?MOD=AJPERES (accessed 02.12.15).

USDA-FSIS, 2015b. Microbiological Laboratory Guidebook. Document 3.02. Available at: http://www.fsis.usda.gov/wps/wcm/connect/f2162091-af72-4888-997b-718d6592bcc9/MLG-3.pdf?MOD=AJPERES (accessed 02.12.15).

Van Houdt, R., Michiels, C.W., 2010. Biofilm formation and the food industry, a focus on the bacterial outer surface. J. Appl. Microbiol. 109, 1117–1131.

Wagner, A.O., Malin, C., Illmer, P., 2009. Application of denaturing high-performance liquid chromatography in microbial ecology: fermentor sludge, compost, and soil community profiling. Appl. Environ. Microbiol. 75, 956–964.

WHO, UNICEF, 2013. Ending Preventable Child Deaths from Pneumonia and Diarrhoea by 2025: The Integrated Global Action Plan for Pneumonia and Diarrhoea (GAPPD). World Health Organization. Available at: http://apps.who.int/iris/bitstream/10665/79200/1/9789241505239_eng.pdf?ua=1 (accessed 02.12.15).

Whyte, P., McGill, K., Monahan, C., Collins, J., 2004. The effect of sampling time on the levels of micro-organisms recovered from broiler carcasses in a commercial slaughter plant. Food Microbiol. 21, 59–65.

Wilson, M.E., 2005. Diarrhea in nontravelers: risk and etiology. Clin. Infect. Dis. 41 (Suppl. 8), S541–S546.

Wu, D., Alali, W., Harrison, M., Hofacre, C., 2014. Prevalence of *Salmonella* in neck skin and bone of chickens. J. Food Prot. 77, 1193–1197.

Ye, K., Zhang, Q., Jiang, Y., Xu, X., Cao, J., Zhou, G., 2012. Rapid detection of viable *Listeria monocytogenes* in chilled pork by real-time reverse-transcriptase PCR. Food Control 25, 117–124.

Zhao, C., Ge, B., De Villena, J., Sudler, R., Yeh, E., Zhao, S., White, D.G., Wagner, D., Meng, J., 2001. Prevalence of *Campylobacter* spp., *Escherichia coli*, and *Salmonella* serovars in retail chicken, turkey, pork, and beef from the greater Washington, DC, area. Appl. Environ. Microbiol. 67, 5431–5436.

Zwane, A.P., Kremer, M., 2007. What works in fighting diarrheal diseases in developing countries? A critical review. World Bank Res. Obs. 22, 1–24.

SECTION

NEW STRATEGIES FOR STUDYING FOODBORNE PATHOGEN ECOLOGY

2

CHAPTER

SALMONELLA CONTROL IN FOOD PRODUCTION: CURRENT ISSUES AND PERSPECTIVES IN THE UNITED STATES

7

Steven C. Ricke, Juliany Rivera Calo, Pravin Kaldhone
Department of Food Science, Center for Food Safety, University of Arkansas, Fayetteville, Arkansas, USA

1 INTRODUCTION

The *Salmonella* genus is a member of the Enterobactericeae family and is a motile, nonspore forming, facultative Gram-negative anaerobe that is oxidase negative, capable of fermenting glucose, and reducing nitrates to nitrites (Yan et al., 2003; Pui et al., 2011; Li et al., 2013; Ricke et al., 2013c). Based on the Kauffman-White classification system as well as more recently developed subtyping methods, over 2500 different serotypes have been identified as belonging to the *Salmonella* genus (Pui et al., 2011; Foley et al., 2006, 2013; Li et al., 2013). *Salmonella* spp. are facultative intracellular pathogens that cause gastroenteritis and are one of the more prominent foodborne pathogens that represent a major health risk to humans (Li et al., 2011; Ricke et al., 2013c). Salmonellosis continues to be a recurring public health problem over the past decades with several million cases of *Salmonella* infections in the United States per year that result in several thousand hospitalizations and hundreds of deaths (Tauxe, 1991; Mead et al., 1999; Olsen et al., 2001; Vugia et al., 2004; Scallan et al., 2011; Scharff, 2012). Actual patient-related costs and job loss time due to recovery can lead to multiple burdens on the economy (Scharff, 2012; McLinden et al., 2014). Commonly reported serovars include *Salmonella enterica* Typhimurium, Enteritidis, Newport, and Heidelberg (CDC, 2011).

Despite the inroads made into developing a better understanding of *Salmonella* biology and the means to control it under different food production conditions, food-related outbreaks still occur. Along with the outbreaks, the contamination of *Salmonella* in foods also continues to represent public health concerns beyond just foodborne disease. For example, occurrence of antibiotic resistance in *Salmonella* strains represents a major public health problem. This resistance is a threat to the management and treatment of salmonellosis in both veterinary and human clinical practices. Historically, it was always believed that a potential cause of the selection and proliferation of resistant organisms was the use of clinically important antimicrobial agents in the animal production environment (Jones and Ricke, 2003). Although these types of antimicrobials are now being removed from animal production systems and antimicrobial-free animal production systems such as organic-based operations have become more popular, the end results with regards to *Salmonella* and other foodborne pathogens have not necessarily

Food Safety. http://dx.doi.org/10.1016/B978-0-12-800245-2.00007-1

been clear-cut (Mather et al., 2013; Woolhouse and Ward, 2013; Mollenkopf et al., 2014; Ricke and Rivera Calo, 2015). This may be expected given the complexity of most organisms to both acquire and express antibiotic resistance. Therefore, an increased understanding of the distribution and potential mechanisms of antimicrobial resistance transmission is important to facilitate mitigation strategies that limit salmonellosis caused by drug-resistant strains. Likewise, clinical antimicrobial-free production environments will require development and/or identification of new and/or alternative antimicrobials that will meet the requirements and standards that define these types of operations (Van Loo et al., 2012a,b; Pittman et al., 2012). The focus of this chapter is to examine of the recent trends associated with these factors and potential strategies that offer potential mitigation under these circumstances.

2 *SALMONELLA* AND FOODBORNE SALMONELLOSIS

Salmonella spp. are consistently isolated from a wide range of food animals and food products including fresh produce, fruits, meat products, and retail foods (Olsen et al., 2001; Nayak et al., 2003; Ricke, 2003b; Patrick et al., 2004; Braden, 2006; Boyen et al., 2008; Dunkley et al., 2009; Hanning et al., 2009; Foley et al., 2008, 2011, 2013; Finstad et al., 2012; Howard et al., 2012; Koo et al., 2012; Painter et al., 2013; Pires et al., 2014; Walsh et al., 2014). There are several factors that have contributed to the ongoing prevalence of foodborne *Salmonella* in the food supply. Certainly changes in the nation's food supply make it even more vulnerable to widespread and sustained disease outbreaks, mainly due to highly centralized large-scale food processing systems, characterized by extensive transport and distribution operations that have become prominent both nationally as well as internationally (Ricke, 2010). Consequently, this means that if contamination occurs early in the process it will impact large numbers of people over greater geographical regions, often crossing regional, national, and even continental boundaries (Ricke, 2010). In addition, Akil et al. (2014) have suggested that climate change-induced warming weather patterns may be leading to increased incidence of *Salmonella* outbreaks in geographical regions such as the southern United States.

Salmonellosis in humans is typically zoonotic in origin and is often the result of consumption of undercooked or mishandled poultry products such as eggs, chicken, turkey, milk, as well as beef and pork (Olsen et al., 2001; Li et al., 2013; Ricke et al., 2013c). Clinical manifestations of *Salmonella* infection include enterocolitis, enteric fever, bacteremia, and localized infection (Li et al., 2013; Ricke et al., 2013c). Enterocolitis is manifested clinically as abdominal pain, nausea, vomiting, diarrhea, and headache, typically occurring 12-36 h after bacterial ingestion (Li et al., 2013; Ricke et al., 2013c). In healthy adults, salmonellosis typically results in self-limiting diarrhea that resolves itself in a week. Immunocompromised patients, such as those with HIV or sickle cell anemia, and infants commonly experience symptoms of severe extraintestinal salmonellosis, which has the potential to spread causing bacteremia and possibly life-threatening septicemia (Olsen et al., 2001; Li et al., 2013; Ricke et al., 2013c).

Disease symptoms identified with *Salmonella* infection are associated with the invasion of intestinal cells with attachment of bacteria to intestinal and oropharyngeal epithelia being mediated by type 1 fimbriae or pili, which help to augment colonization (Ernst et al., 1990; Van Asten et al., 2000, 2004; Foley et al., 2013; Li et al., 2013; Ricke et al., 2013c). Pathogenesis in *Salmonella* is expressed via a type III secretion system (TTSS-1) encoded by *Salmonella* pathogenicity island 1 (SPI-1) along with additional SPIs that represent gene clusters involved in all systemic infection aspects associated with its virulence phenotype (Cirillo et al., 1998; Hensel, 2000; Winnen et al., 2008; Foley et al., 2013; Li et al.,

2013; Ricke et al., 2013c). These virulence gene clusters are in turn controlled by a complex array of regulatory genes (Lee et al., 1990; Altier, 2005; Coburn et al., 2007; Foley et al., 2013; Li et al., 2013). The genes in SPI-1 are responsible for secretion of proteins that interact with host cells and facilitate invasion, altering host cell physiology (Coburn et al., 2007; Foley et al., 2013). Typically, TTSS-containing bacteria are able to invade nonphagocytic host cells (Finlay and Falkow, 1997; Hueck, 1998; Schmidt and Hensel, 2004). Thus, TTSS-1 acts as a major virulence factor in the intestinal phase of infection for *Salmonella*, (Finlay and Falkow, 1997; Hueck, 1998; Schmidt and Hensel, 2004; Foley et al., 2013; Li et al., 2013; Ricke et al., 2013c). In addition, to the classical virulence properties associated with *Salmonella*, it also possesses several survival mechanisms once it enters the gut, including acid tolerance that allows it to survive the acidic stomach during transit to the intestines as well as fermentation acids in the lower gut, and *Salmonella* spp. can also induce inflammation for a competitive edge against the native fermentative gut microbiota (Foster and Spector, 1995; Ricke, 2003a; Santos et al., 2009; Winter et al., 2010; Ahmer and Gunn, 2011).

3 *SALMONELLA* SEROVARS AND POULTRY

Contaminated poultry, meat, and eggs are important vehicles of *Salmonella* infections and under the right circumstances the organism can occur in the egg contents via transovarian infection (Guard-Petter, 2001; Gantois et al., 2009; Foley et al., 2011; Howard et al., 2012; Galis et al., 2013). Egg-associated outbreaks, in particular, have continued to receive considerable public attention in the United States due to the more recent investigations such as the salmonellosis outbreak caused by *S.* Enteritidis being traced back to contaminated eggs from Iowa (Kuehn, 2010). Several *Salmonella* serovars have been associated with the consumption of poultry, eggs, and their corresponding food products; thus, they are considered primary sources of *Salmonella* (Guard-Petter, 2001; Hennessy et al., 2004; Finstad et al., 2012; Foley et al., 2011, 2013; Howard et al., 2012; Martelli and Davies, 2012). The emergence and prevalence of *Salmonella* serovars in chickens and eggs have been extensively reported and over time the hierarchy of predominant serovars has varied somewhat each year (CDC, 2009, 2011; Finstad et al., 2012; Foley et al., 2011, 2013; Howard et al., 2012; Martelli and Davies, 2012).

However, the variability in serovar frequency does to appear to have a pattern, as the predominant *Salmonella* serovars associated with poultry and human infections in the United States have shifted over the past several decades. Certainly, *S.* Enteritidis and Typhimurium remain dominant, but lately Heidelberg and Kentucky have been more frequently isolated as well (CDC, 2011; Finstad et al., 2012; Foley et al., 2013). *S.* Enteritidis and *S.* Heidelberg have been identified among the top five serovars responsible for human infections as well as the most frequently detected serovars in poulty over the last 25 years. *S.* Enteritidis is still considered a public health concern for commercial poultry, but a detectable decrease has occurred in chickens in the United States since the mid 1990s. From 1997 to 2006, *S.* Heidelberg supplanted *S.* Enteritidis as the predominant serovar, but by 2007 *S.* Kentucky became the most commonly isolated serovar (Foley et al., 2008, 2011, 2013). This serovar has not been identified as a frequent or common source of human salmonellosis and why it consistently colonizes the chicken ceca remains unresolved, but thus far it has not posed a significant threat to humans unlike the serovars *S.* Typhimurium, *S.* Enteritidis, or *S.* Heidelberg (Foley et al., 2011). Although *S.* Kentucky is not among the most common serovars causing human diseases, a fairly high frequency of multidrug-resistant strains have been isolated from poultry and humans in other parts of the world (Foley et al., 2011).

4 FACTORS THAT IMPACT *SALMONELLA* PERSISTENCE IN POULTRY PRODUCTION

The ongoing persistence of *Salmonella* in poultry environments as well as the other food production system environments is due to a number of factors, including a wide range of insect, wild birds and animal carriers, animal feed as a reservoir, and transfer via aerosols, among others (Pillai and Ricke, 2002; Maciorowski et al., 2004; Park et al., 2008; Ricke, 2014). In the past few years, the colonization of *Salmonella* serovars in chickens has been extensively studied from an experimental perspective but the focus on particular serovars has varied somewhat over the years (Foley et al., 2008, 2011, 2013; Ricke, 2014). In general, colonization of *Salmonella* in poultry is impacted by the age and genetic susceptibility of the birds, bird stress due to overcrowding or underlying illness, level of exposure to the pathogen, competition with gut microbiota for the colonization sites, the specific colonizing *Salmonella* serovar, and whether or not the *Salmonella* strains carry genetic factors that may facilitate the attachment to the gastrointestinal tract of the bird or evade host defenses (Stavric, 1987; Bailey, 1988; Park et al., 2008; Dunkley et al., 2009; Calenge and Beaumont, 2012; Foley et al., 2011, 2013; Ricke, 2014). Certainly, age of bird and dietary intake versus feed removal are critical determinants on susceptibility to colonization (Stavric, 1987; Dunkley et al., 2009; Ricke, 2003b, 2014; Ricke et al., 2013a).

In general, environmental stress has been identified as a consistent determining factor not only in *Salmonella* survival but expression of virulence genes (Humphrey, 2004; Spector and Kenyon, 2012). *Salmonella* spp. encounter numerous stresses during their life cycle, including exposure to acidic pH, dessication, fermentation acids, suboptimal temperatures, oxygen availability, suboptimal osmolarity conditions, and nutritional starvation, just to name a few (Francis et al., 1992; Tartera, and Metcalf, 1993; Foster and Spector, 1995; Schiemann, 1995; Hirshfield et al., 2003; Ricke, 2003a, 2014; Spector and Kenyon, 2012). Some of these factors can have diverse impacts on *Salmonella*, such as exposure to short chain fatty acids (SCFA) that not only can induce acid tolerance responses, but modulate virulence gene expression levels and influence the level of *in vitro* attachment and invasion of epithelial cells (Kwon and Ricke, 1998; Durant et al., 1999b, 2000a,c,d; Lawhon et al., 2002; Ricke, 2003a; Van Immerseel et al., 2004, 2006). This is further complicated by the fact that these responses may also depend upon acid type as well as *Salmonella* serovar and strain (Durant et al., 1999b, 2000a,c,d; Lawhon et al., 2002; Van Immerseel et al., 2004, 2006; González-Gil et al., 2012).

As more has become known, some environmental cues such as starvation have been established to have an impact on *Salmonella* virulence expression (Guiney, 1997; Altier, 2005; Spector and Kenyon, 2012). This is consistent with the suggestion by Rohmer et al. (2011) that nutrient limitation was a primary driver for pathogens to elect to become invasive in their respective host. For example, many bacteria have developed a range of iron chelating compounds to extract iron from their surrounding biological environments where iron is relatively insoluble at neutral pH and often bound to specific iron-binding proteins (Neilands, 1981, 1982; Crosa, 1989; Litwin and Calderwood, 1993; Wooldridge and Williams, 1993). Iron, not surprisingly, is considered a key environmental regulator of virulence expression in pathogens (Mekalanos, 1992). Consequently, in some cases, this has resulted in some fairly distinct host responses such as the evolution of iron transport systems in mammals to sequester iron away from pathogen invaders (Armitage and Drakesmith, 2014; Barber and Elde, 2014). In poultry eggs, the iron sequestering protein ovotransferrin has been identified as a key defensive measure elicited to secure iron away from bacteria that can cause egg rots and has been further characterized via an extensive proteomic examination of egg protein profiles (Tranter and Board, 1982; Mayes and

Takeballi, 1983; Burley and Vadehra, 1989; Ibrahim, 2000; Mann, 2007; Mann et al., 2008; Gantois et al., 2009). In response to this, some pathogens such as *S.* Enteritidis, have in turn evolved high-affinity iron uptake systems (Chart and Rowe, 1993). The importance of available iron in eggs is further reflected in the supplementation of iron to improve recovery of *Salmonella* from egg samples and the recommendation to reduce the available iron concentration in egg washes used to process table eggs (Gast, 1993; Messens et al., 2011; Galis et al., 2013).

The impact of a general starvation or nutrient limitation condition has also been observed for *Salmonella* in the gastrointestinal tract of poultry. In a series of studies, it was demonstrated that feed removal as a means to induce molt in laying hens not only increased systemic infection by *S.* Enteritidis but also could be linked to expression of key virulence regulatory genes (Durant et al., 1999a; Ricke, 2003b; Dunkley et al., 2007b). This has been shown to also impact the crop and cecal indigenous microbiota, both in terms of microbial composition as well as fermentation patterns (Durant et al., 1999a; Dunkley et al., 2007a,c; Ricke et al., 2013a). The dietary connection was demonstrated by feeding a range of high-fiber diets that were not necessarily well utilized by the laying hen but could be fermented by the bird's cecal microbiota both *in vitro* as well as *in vivo* (Woodward et al., 2005; Saengkerdsub et al., 2006; Dunkley et al., 2007a,c; Donalson et al., 2008a,b; Ricke et al., 2013a). Although *in vitro* models are limited that can directly simulate nutrient limitations in the gastrointestinal tract, spent medium has been previously applied to simulate chicken gastrointestinal tract conditions after feed withdrawal (Nutt et al., 2002). Spent media is a growth medium supernatant harvested after growth of selected bacterial species; thus, it represents a medium exhausted of nutrients (Nutt et al., 2002). Spent media has been shown to influence the virulence gene expression patterns of *S.* Typhimurium as well as acid tolerance responses in pathogenic *Escherichia coli* (Durant et al., 2000b; Nutt et al., 2002; Chen et al., 2004b; De Keersmaecker et al., 2005). The filtered supernatant of spent media has the potential to serve as a high-throughput rapid *in vitro* molecular and metabolic screen for identification of specific inhibitory compounds after growth of either specific microorganisms or a microbial consortium from a specific *in vivo* site such as the gastrointestinal tract (Nutt et al., 2002; De Keersmaecker et al., 2005, 2006; Kundinger et al., 2007).

Avian host immune responses also play a role in responding to foodborne *Salmonella* colonization, and this has certainly been exploited with the development of a series of vaccines designed to limit *Salmonella* infection in birds (Kaiser, 2010; Revolledo and Ferreira, 2012; Wang et al., 2013; Ricke, 2014). However, successful efforts to limit *Salmonella* colonization in birds through prior vaccination have been variable (Ricke, 2014). Certainly, some strain behavior variation is to be expected for any pathogen (Lianou and Koutsoumanis, 2013). However, there is some indication that *Salmonella* serovars interact differently with avian host tissues. For example, Kaiser et al. (2000) demonstrated that invasion of primary chick kidney cells by either of the two nonhost specific serovars, *S.* Enteritidis and *S.* Typhimurium, caused an 8-10-fold increase in production of the proinflammatory cytokine IL-6 versus minimal stimulation by the avian host specific serovar *S.* Gallinarum. More recent studies have focused on the use of a macrophage-like immortalized cell line (HD11), originally derived from chicken bone marrow and transformed with the avian myelocytomatosis type MC29 virus (Beug et al., 1979). He et al. (2012) compared *Salmonella* cell invasion and intracellular survival of five different poultry-associated serovars (*S.* Heidelberg, *S.* Enteritidis, *S.* Typhimurium, *S.* Senftenberg, and *S.* Kentucky) to HD11 cells. Their results showed that *S.* Enteritidis was more resistant to intracellular killing, leading them to believe that the intracellular survival ability of this serovar may be related to systemic invasion in chickens (He et al., 2012).

There may even be strain differences within a particular *Salmonella* serovar. Shah et al. (2011) suggested that not all isolates of *S.* Enteritidis recovered from poultry might be equally pathogenic or have similar potential to invade cells and that *S.* Enteritidis pathogenicity was related to both the secretion of T3SS effector proteins and motility. Saeed et al. (2006) reported that, compared to isolates of *S.* Enteritidis recovered from chicken ceca, isolates recovered from eggs or from human clinical cases exhibited a greater adherence and invasiveness of chicken ovarian granulosa cells. In short, both pathogen and host factors may be important keys to determine the mechanisms for the differences in *Salmonella* responses within different cell types (Bueno et al., 2012; Foley et al., 2013; Ricke, 2014). Understanding the ability and mechanisms of this pathogen to attach and invade different cell lines could be helpful for the reduction of *Salmonella* infections by designing more precise vaccine constructs (Ricke, 2014).

5 CLINICAL ANTIMICROBIAL AGENTS

Given the public health concerns regarding foodborne salmonellosis and the potential in some cases for severe or life-threatening circumstances, treatment by clinical administration of antimicrobials is a critical component. Antimicrobial agents are defined as substances produced by organisms or synthetically derived that antagonize the growth of or kill other organisms in high dilution (Foley and Lynne, 2008; FDA, 2002, CDC, 2010a,b). Various antimicrobial agents have different mechanisms of action by which they are classified. Not surprisingly, the mechanisms by which organisms develop resistance varies according to the antimicrobial agent but essentially can be categorized as either intrinsic (bacteria simply lacks a target for the respective antibiotic) or the gene encoding resistance is located on either the bacterial chromosome or plasmid. Two issues make foodborne *Salmonella* antibiotic resistance problematic for public health. First of all, antimicrobial resistance genes can be transferred from one bacterium to another, thus making nonpathogenic bacterial populations potential reservoirs for antibiotic resistance genes (Frye and Jackson, 2013). Secondly, *Salmonella* spp. can express resistance to multiple antibiotics (multiple drug resistance or MDR), which makes their clinical treatment more difficult (Alcaine et al., 2007). Different aspects of antimicrobial response in *Salmonella* response to clinical antimicrobials has been extensively covered elsewhere (Angulo et al., 2000; Alcaine et al., 2007; Foley and Lynne, 2008; Hur et al., 2012; Frye and Jackson, 2013; Ricke and Rivera Calo, 2015) and so will be only briefly touched on here.

Cell wall synthesis inhibitors disrupt the bacteria's ability to synthesize an adequate cell wall structure by disrupting the synthesis and cross-linking of peptidoglycan polymer. Antimicrobial agents that either inhibit the bacterial cell's ability to maintain the peptidoglycan structure or altogether prevent peptidoglycan synthesis include several classes of beta lactams such as penicillin, modified versions of penicillin, namely oxacillin and methicillin, along with cephalosporins and carbapenems (Frye and Jackson, 2013). These antimicrobials function by substituting for one of the components for constructing petidoglycan, thus binding irreversibly to the penicillin binding protein (PBP) enzyme that catalyzes peptidoglycan synthesis. *Salmonella* primarily become resistant to beta lactams by excreting plasmid encoded beta lactamases, which cleave the beta lactam ring structure common to these antibiotics, thus rendering them ineffective (Alcaine et al., 2007; Hur et al., 2012; Frye and Jackson, 2013). Resistance can also be manifested in *Salmonella*, albeit much less commonly, by decreasing porin (OmpC and OmpF) concentration to limit uptake of beta lactams (Alcaine et al., 2007; Frye and Jackson, 2013).

Other classes of antimicrobial agents target protein synthesis within the organism. Protein synthesis inhibitors include the aminoglycosides that inhibit formation of ribosomal initiation complex by binding to the 30S ribosomal unit (Alcaine et al., 2007; Frye and Jackson, 2013). Tetracyclines inhibit protein synthesis by binding to the 30S ribosomal unit (Frye and Jackson, 2013). These agents prevent attachment of aminoacyl-tRNA complex to an acceptor site (A site) in the ribosome (Alcaine et al., 2007). *Salmonella* spp. primarily become resistant to tetracyline by synthesizing aminoglycoside-modifying enzymes including *N*-acetyltransferase that modifies the amino group of aminoglycosides and the *O*-phosphotransferase and *O*-nucleotidyltransferase, which alter the hydroxyl group of aminoglycosides rendering them unable to bind to the respective aminoglycoside (Frye and Jackson, 2013). The phenicol-based antimicrobial chloramphenicol inhibits the formation of peptide bonds by binding to the 50S ribosomal subunit, thus halting peptide transfer activity (Alcaine et al., 2007; Frye and Jackson, 2013). Resistance in *Salmonella* is mediated by either plasmid encoded enzymatic inactivation or via efflux pumps encoded by both chromosomal and plasmid genes (Alcaine et al., 2007).

Antimicrobial agents can also directly impact the bacterial cell at the genomic level such as the synthetically constructed quinolones that interfere with replication by targeting DNA topoisomerase IV and DNA gyrase (Alcaine et al., 2007; Frye and Jackson, 2013). The combination of two quinolone resistance mechanisms has been identified in *Salmonella*, one of which involves a series of point mutations leading to amino acid substitutions in specific subunits of DNA gyrase and DNA topoisomerase IV altering their respective structures, while the second mechanism involves mutations of the regulators for the efflux system that lead to increased efflux activity (Alcaine et al., 2007; Hur et al., 2012).

Metabolic pathways required by bacteria are also targets of antimicrobial activity. For example, folic acid biosynthesis antimicrobials include the sulfonamides that competitively inhibit dihydropteroate synthetase while trimethoprim targets dihydrofolate reductase, enzymes that are both required for folic acid synthesis (Alcaine et al., 2007; Frye and Jackson, 2013). Sulfonamide resistance is acquired by altered synthesis of the enzyme dihydropteroate synthetase leading to an insensitivity to sulfonamide, allowing the bacteria to become resistant to sulfonamides (Frye and Jackson, 2013).

Despite general historical success of clinical antimicrobials against most bacterial pathogens, resistance of these organisms to many of these commonly used antimicrobials has emerged and now presents a major obstacle for continued public heath success against these pathogens (Jones and Ricke, 2003; Aminov, 2010). Likewise, Faleiro (2011) has pointed out that although control of food spoilage and pathogenic bacteria has primarily been accomplished via chemically mediated control measures, the use of synthetic chemicals remains somewhat restrictive because of several undesirable potential outcomes including carcinogenicity, acute toxicity, teratogenicity, and slow degradation periods that could result in environmental issues such as pollution (Faleiro, 2011). In addition, a rising public sentiment for more natural and even entirely organic livestock production systems that avoid usage of any of the classically defined antimicrobials used for production purposes has gained regulatory momentum (Ricke, 2010; Ricke et al., 2012). Consequently, the negative public perception of industrially synthesized food antimicrobials has generated interest in the use of more naturally occurring compounds (Sofos et al., 1998; Faleiro, 2011). Clearly, alternative antimicrobial options are needed.

6 ORGANIC FOODS AND THE DEMAND FOR ALTERNATIVE ANTIMICROBIALS

In its simplest terms, organic food is defined as foods generated using methods that do not incorporate use of synthetic pesticides or chemical fertilizers, do not contain genetically modified organisms, and avoid processing via irradiation, industrial solvents, or chemical food additives (DeSoucey, 2007; Van Loo et al, 2012b). As the U.S. organic food industry evolved in the 1980s and 1990s and expanded commercially more precise definitions were required to avoid fraudulent practices (Van Loo et al, 2012b). National U.S. standards for regulating organically produced farm products were launched by the U.S. Congress in 1990 as the Organic Food Production Act to guarantee consumers a consistent regulatory standard that could transcend interstate commerce in organically generated fresh and processed food (Pittman et al., 2012). By the end of the decade, the USDA issued a final rule on national standards leading to the formation of the National Organic Program (NOP), which allows USDA-accredited agents to certify compliance of all handlers and producers of organic foods to the national standards as defined by the NOP (Pittman et al., 2012).

Given the regulatory rigor associated with organic practices, the obvious question to ask is whether there is economic market demand that warrants development of alternative control measures for food animals such as organically produced foods that would meet the certification requirements. Crandall et al. (2010) collected responses from northwest Arkansas farmers' market customers and reported that 80% would buy more organic products if the price is comparable to that of conventional foods. In a follow-up study, Van Loo et al. (2011) assessed consumers' willingness to pay specifically for organic chicken, and their results demonstrated that consumers were willing to pay a 35% premium for a general organic-labeled chicken breast and 104% for a USDA certified organic-labeled organic chicken breast. In their initial survey, Van Loo et al. (2010) reported that consumers leaned toward buying organic chicken due to their perception that organic chicken is safer, healthier, and has fewer pesticides, antibiotics, and hormones. Crandall et al. (2011) surveyed over 300 consumers to determine their concerns and beliefs about the safety of foods obtained at three Arkansas farmers' markets. Results showed that 36% of the consumer preference for organic foods was that they wanted food free of chemicals; 45% of the biggest safety concerns by consumers were pesticides; and only 2-6% were actually concerned about foodborne pathogens. A total of 76% of the surveyed consumers believed that organic foods were safer than conventional foods (Crandall et al., 2011).

Perceptions on the attributes of organically produced foods continue to be promoted, and a wide range of benefits have been touted over the years. Some of the factors that meet consumers' concerns and thus increase the demand for organic foods include the use of pesticides in farming and the use of growth hormones and antibiotics in livestock production (Magkos et al., 2003; Ricke et al., 2012; Van Loo et al., 2010, 2012a); lower environmental impact and improved animal welfare practices (Magnusson et al., 2003; Van Loo et al., 2010; Ricke et al., 2012); and quality and safety of organic foods (Magkos et al., 2006; Van Loo et al., 2010). However, most of the research on organic foods has concluded that there is no evidence that organic food is safer, healthier, or more nutritious (Magkos et al., 2003; Ricke et al., 2012; Van Loo et al., 2012a). In a summary overview of current published research reports, Van Loo et al. (2012a) concluded that consumers should not assume that chickens labeled as organic, all natural, or free range are any less contaminated with *Salmonella* than conventionally raised chickens. In addition, organic foods are required to meet the same food safety standards as foods that are conventional; there are no stricter food safety requirements (Ricke et al., 2012; Van Loo et al., 2012a). Therefore, a need certainly exists for alternative antimicrobials that meet regulatory requirements but are sufficiently effective to retard spoilage as well as limit foodborne pathogen levels.

7 ALTERNATIVE ANTIMICROBIALS: GENERAL ASPECTS

The emergence of bacterial antibiotic resistance and the negative consumer attitudes toward the more traditional food preservatives have led to an increased interest in the use of compounds that contain potential alternative antimicrobial agents deemed as more "natural" for the control of food spoilage and harmful pathogens (Shelef, 1983; Smith-Palmer et al., 1998; Sofos et al., 1998; Burt, 2004; Nostro et al., 2004; Fisher and Phillips, 2008). Consequently, as Fratianni et al. (2010) have pointed out, there has been an extensive search for potential natural food additive candidates that retain a broad spectrum of antioxidant and antimicrobial activities but still possess the ability to improve the quality and shelf life of perishable foods. Substances that are naturally occurring and thus directly derived from biological systems without alteration or modification in a laboratory setting are recognized as natural antimicrobials (López-Malo et al., 2000; Sirsat et al., 2009; Li et al., 2011). A number of naturally occurring antimicrobial agents are present in animal and plant tissues, where they probably evolved as part of their hosts' defense mechanisms against microbiological invasion and exist as natural ingredients in foods (Sofos et al., 1998; Li et al., 2011).

In general, these natural compounds with animal, plant, or microbiological origins have been employed in order to kill or at least prevent the growth of pathogenic microorganisms in a variety of different applications (Joerger, 2003; Ricke, 2003a; Berghman et al., 2005; Sirsat et al., 2009; Hume, 2011; Li et al., 2011; Muthaiyan et al., 2011; Jacob and Pescatore, 2012; Juneja et al., 2012; Herrera et al., 2013). In organic food production, several biological agents such as bacteriophage, bacteriocins, botanicals, prebiotics, and probiotics have been proposed or in some cases actually implemented for limiting foodborne pathogens either in livestock production or meat processing (O'Bryan et al., 2008b; Sirsat et al., 2009; Ricke et al., 2012; Van Loo et al., 2012b). An ideal antimicrobial would be one that is available in large volumes as a coproduct and one that has GRAS status because it has already been part of the typical human diet for years (Friedly et al., 2009; Nannapaneni et al., 2009b; Chalova et al., 2010; Callaway et al., 2011a). In the following section, botanical sources are examined for their potential to meet this criteria as well as their efficacy against foodborne pathogens such as *Salmonella*.

8 NATURAL ANTIMICROBIALS FROM PLANT SOURCES

Chemicals deployed by plants against bacteria to obtain an advantage in their ecosystem include essential oils (EOs), organic acids, and specific toxins (Friedman et al, 2002; Callaway et al., 2011a). Consequently, there has always been great potential for new antimicrobial discoveries based on collecting and characterizing traditional medicinal plants throughout the world (Lewis and Elvin-Lewis, 1995). These botanicals have been examined as feed additives to improve feed characteristics, increase digestion and performance if the animal is being fed these compounds, or to enhance desirable characteristics of the meat produced (Jacob and Pescatore, 2012). Likewise, based on health, economic, and environmental issues, natural plant extracts have been promoted as a safer choice as alternatives for synthetic antimicrobials (Burt, 2004; Jo et al., 2004; Callaway et al., 2011a; Muthaiyan et al., 2011; Bajpai et al., 2012). Numerous products derived from plant sources (e.g., fruit preparations, vegetable preparations, or extracts and spices) have been used throughout history for the preservation and extension of the shelf life of foods (Billing and Sherman, 1998; Cutter, 2000; McCarthy et al., 2001; Islam et al., 2002; Draughon, 2004; Kim et al., 2011). Of course the classic example is the use of the hop plant in beer fermentation that, when added, possesses antibacterial iso α acids that act as ionophores to control lactic acid bacterial spoilage during the yeast-based fermentation to ethanol (Sakamoto and Konings, 2003; Suzuki et al., 2006; Muthaiyan et al., 2011). As more has become understood about the composition of the hop plant, additional health-promoting benefits for some of the bioactive components such as the hop bitter acids have also been promoted (Van Cleemput et al., 2009).

In addition to the hop plant, a wide range of plant extracts have been investigated for their phytochemical properties and antibacterial effects on both Gram-positive and Gram-negative bacteria including, among others, *Campylobacter*, *Escherichia coli*, *Mycobacterium tuberculosis*, *Pseudomonas*, *Salmonella*, and *Staphylococcus* (Deans and Ritchie, 1987; O'Bryan et al., 2008a; Li et al., 2011; Callaway et al., 2011a; Hume, 2011; Crandall et al., 2012; Van Loo et al., 2012a). Of the various plant compounds that have been examined essential oils (EOs) from various plants have been one of the more thoroughly examined family of compounds for their potential as antimicrobials as well as myriad other potentially beneficial applications (Bakkali et al., 2008; Hardin et al., 2010; Callaway et al., 2011a; Rivera Calo et al., 2015). The term "essential" derives from the fact that these oils are responsible for the distinctive taste and aroma or "essence" that distinguishes the particular plant of origin (Hardin et al., 2010). These naturally occurring antimicrobials have extensive histories of their use in foods and can be identified from various components of the plants including leaves, barks, stems, roots, flowers, and fruits (Rahman and Gray, 2002; Erasto et al., 2004; Zhu et al., 2004). They are widely used in part because they meet the criteria of being readily acceptable from a regulatory standpoint, are already fairly common in human foodstuffs, and possess the capacity to be generated from large-scale industrial sources such as the citrus industry (Hardin et al., 2010). Brenes and Roura (2010) suggested that plant EOs have utility in poultry nutrition for enhancing the production of digestive secretions, stimulating blood circulation, exerting antioxidant properties, and perhaps enhancing immune status. In ruminants, EOs have been touted as beneficial rumen fermentation modifiers for lowering methane production and altering volatile fermentation acids (Calsamiglia et al., 2007; Chaves et al., 2008). Plant EOs have also been widely used for a variety of medicinal and cosmetic purposes (Hardin et al., 2010).

Part of the attraction for the multitude of diversified applications is due to the fact that EOs can be administered in fairly crude or impure extract form and still maintain efficacy when fed in bulk. For example, in a series of studies Callaway et al. (2008, 2011b,c) demonstrated that including orange peel and pulp as dietary components reduced populations of *S.* Typhimurium and *Escherichia coli* O157:H7 in the gut of experimentally inoculated sheep as well as *E. coli* O157:H7 and *Salmonella* Typhimurium either in pure culture or in fermentation with mixed ruminal microorganisms *in vitro*. Obviously, particular forms of the various EOs have potential for specific applications such as antimicrobials against particular organisms. It has been recognized that certain plant extracts, including their EOs and essences possess antimicrobial properties against bacteria, yeast, and molds (Smith-Palmer et al., 1998, 2001; Chao and Young, 2000; Dorman and Deans, 2000; Burt, 2004; Fisher and Phillips, 2006, 2008, 2009; Edris, 2007, Nannapaneni et al., 2008, 2009a,b, Brenes and Roura, 2010; Kim et al., 2011; Shannon et al., 2011a,b; Muthaiyan et al., 2012; Solórzano-Santos and Miranda-Novales, 2012; Jayasena and Jo, 2013; Laird and Phillips, 2013).

The potential use of antimicrobials as a preharvest intervention, before crops or livestock products are sold, would be considered an important part of animal production and food safety management programs. This would be true not only in their use as feed supplements but plant oils may also have merit in other phases of livestock production such as the management of accumulated wastes in large confinement animal operations. However, Varel (2002a) has pointed out that such applications would come with certain limitations such as stopping microbial degradation of waste that would be environmentally counterproductive to one of the primary goals of livestock disposal. Likewise, their cost effectiveness for these types of large-scale operations is not clear (Varel, 2002a). Varel (2002a) countered these disadvantages with some potential merits associated with EOs' application in animal waste management as antimicrobial chemicals. For example, because of the inhibition of microbial fermentation of the corresponding waste, odor, and global warming gas emissions would be reduced while nutrients that would have value in other applications contained in waste could be retained (Varel and Miller, 2001; Varel, 2002a,b). Microbial inhibition would also result in the destruction of pathogenic fecal coliforms and the presence of EOs also could also serve as pesticides for controlling pests such as flies (Varel and Miller, 2001; Varel, 2002a). Finally, phenolic plant oils have the advantage of being stable under these anaerobic conditions and being classified as GRAS (Varel, 2002a,b). It would be interesting if EOs entering the waste stream after being consumed by animals would remain intact and exhibit some of the same properties described by Varel (2002a,b) for those directly added into the waste material.

Among the sources of EOs, citrus EOs have received most of the attention because these compounds are already produced in large scale from the commercial citrus industry, with the United States alone producing over 12 million tons of citrus fruit in 2009 (Robinson, 1999; Hardin et al., 2010). Including citrus extracts and EOs in animal feeds and human foods was reported to be a means to improve human health, be environmentally friendly, serve as natural oxidants, and be economically feasible (Jo et al., 2004; Callaway et al., 2011a). Citrus products and other EOs sources have been proposed as green alternatives to reduce clinical antibiotic use (Callaway et al., 2011a). Not surprisingly, citrus EOs have a long history of being assessed for potential antimicrobial interventions against foodborne pathogens (Subba et al., 1967; Dabbah et al., 1970). Most of the more recent focus for citrus EOs has been on postharvest food production applications to limit foodborne pathogens such as *Salmonella*, and these reports have been reviewed extensively elsewhere (Fisher and Phillips, 2008; Bajpai et al., 2012; Jayasena and Jo, 2013; Rivera Calo et al., 2015).

9 MULTIPLE HURDLE APPROACHES FOR ALTERNATIVE ANTIMICROBIALS

A hurdle intervention is any treatment, physical, chemical, or biological, that is used to reduce or eliminate the presence of pathogens on a food surface (Leistner, 1985; Leistner and Gorris, 1995). The multiple hurdle technology developed by Leistner and Gorris (1995) sought to prevent the survival and regrowth of pathogens in food by using a series of combinations of preservation techniques. The concept of some sort of multiple intervention or multiple hurdle approach that employs administration of several antimicrobials in a food production system is not a new concept. There are several potential advantages to combining interventions, particularly if they differ in their antimicrobial mechanisms. This would allow for potentially lower concentrations being used for each individual hurdle, thus offering cost savings as well as diminishing the opportunity for the respective pathogen to overcome specific interventions due to genetically independent mechanisms of antimicrobial activity (Sirsat et al., 2010).

Thus, instead of using high doses of a single antimicrobial compound, combinations of more than one antimicrobial at lower concentrations could effectively inhibit the contaminants to the same degree, and due to synergism, potentially result in minimal development of antimicrobial resistance (Ricke et al., 2005; Sirsat et al., 2009, 2010; Ricke, 2010; Muthaiyan et al., 2011). Combinations of more than one antimicrobial intervention treatment in lower doses have been often found to increase bactericidal activity more than any single treatment by working synergistically (Leistner and Gorris, 1995; Leistner, 2000; Ricke et al., 2005; Tiwari et al., 2009; Li et al., 2011). Certainly, this would appear to be intuitive for combining unrelated compounds or chemical and physical interventions. For example, synergism has also been reported when heat was combined with organic acids against *S.* Typhimurium (Milillo and Ricke, 2010; Milillo et al., 2011). In at least some cases there appears to be an identifiable mechanistic basis for the synergism. When the synergism between heat and organic acids was examined more closely by Milillo et al. (2011) using microarray transcriptome analyses, the combination of organic acids with thermal exposure (55 °C for 1 min) appeared to visually disrupt the bacterial cell membrane as indicated by electron microscopy and repress genes associated with heat shock responses.

This concept would certainly apply to the application of complex plant-based extracts containing compounds such as EOs along with other plant components where antimicrobial mechanism(s) may vary considerably for individual compounds. Consequently, when used in a combination of more than one component, EOs may work differently than when they are used as independent compounds, and this may explain some of the efficacy differences seen between crude plant extracts and purified individual compounds even when applied in combination. This may be particularly important for preharvest applications where the complexity of the gastrointestinal tract represents a challenge for effective dosage of a specific intervention. For example, Callaway et al. (2011b) concluded that including citrus oils and other products in the ruminant diet could be utilized as a part of an integrated system with the objective of reducing the passage of foodborne pathogens from farm to human consumers.

For plant-based antimicrobials such as EOs there is experimental evidence to support improved performance against foodborne pathogens when compounds are combined either *in vitro* or *in vivo* (Zhou et al., 2007b; Friedly et al., 2009). For example, Friedly et al. (2009) reported synergism when acids were combined with EOs administered against *Listeria monocytogenes*, and this was also observed for several bacteria when terpeneless Valencia orange was combined with dispersing agents (Chalova et al., 2010). More recently, Alali et al. (2013) determined the effect of nonpharmaceuticals (a blend of organic acids, a blend of EOs, lactic acid, and a combination of levulinic acid and sodium dodecyl sulfate) on weight gain, feed conversion ratio, and mortality of broilers and their ability to reduce colonization and fecal shedding of *Salmonella* Heidelberg following *S.* Heidelberg challenge and feed withdrawal. Their results showed that the broilers that received the EOs had significantly increased weight gain and mortality was lower compared to the other treatments. *S.* Heidelberg contamination in crops was significantly lower in challenged and unchallenged broilers that received EOs and lactic acid in drinking water than any of the other treatments (Alali et al., 2013).

Opportunities for synergism also may occur with more closely related compounds such as the various EOs and other similar compounds extracted from plants. Barnhart et al. (1999) studied the potential to eliminate *Salmonella* in the presence of organic matter and reported a synergistic combination of *d*-limonene (DL) and citric acid (CA), when used in a combination of 2% CA with 0.5% DL, completely eliminated detectable *Salmonella* in a poultry feed slurry. Their observations suggested that, in nature, the unknown anti-*Salmonella* action of DL/CA in combination might be bactericidal, rather than bacteriostatic. When Zhou et al. (2007a) compared cinnamaldehyde, thymol, and carvacrol antimicrobial properties against *S.* Typhimurium, the lowest concentrations inhibiting the growth significantly were 200, 400, and 400 mg/L, respectively, when used individually, while the combinations of cinnamaldehyde/thymol, cinnamaldehyde/carvacrol, and thymol/carvacrol, decreased the effective concentrations to 100, 100, and 100 mg/L of the individual compounds, respectively.

10 CONCLUSIONS AND FUTURE PROSPECTS

Foodborne *Salmonella* spp. continues to be problematic as a U.S. public health issue. There are several reasons for this. It is partly due to the fact that food is now being processed in much larger volumes and, at least in some cases, shipped over increasingly greater distances (Ricke, 2010). In addition, antibiotic resistance appears to be an important and consistent trait in pathogens and for foodborne pathogens such as *Salmonella*; resistant strains can be isolated at most stages of production from the farm all the way to the retail shelf (Holmberg et al., 1984; White et al., 2001; Helms et al., 2002; Sørum and L'Abée-Lund, 2002; Johnson et al., 2005; McMahon et al., 2007; Zhao et al., 2006, 2008; Schmidt et al., 2012; CDC, 2013; Ricke and Rivera Calo, 2015). This is a critical concern as many of these antibiotics are clinically important and treatment options for severely ill patients are limited as resistance accumulates and becomes more prevalent (Holmberg et al, 1987; Su et al., 2004; Mølbak, 2005; Zhao et al., 2006, 2008; Rivera Calo et al., 2015). This is further confounded as MDR *Salmonella* strains have emerged in many of these environmental settings as well as resistance to biocides (Poppe et al., 1996; Winokur et al., 2000; Chen et al., 2004a,b; Velge et al., 2005; Zhao et al., 2003, 2007; Brichta-Harhay et al., 2011; Condell et al., 2012a,b; Mather et al., 2013). Finally, Martínez et al. (2015) have suggested that antibiotic resistance genes are sufficiently common in the environment that theoretically, risk would be fairly high that human pathogens can acquire antibiotic resistance genes but there are several biological bottlenecks that may limit actual dissemination of antibiotic resistant genes. However, they also suggest that even though generation of more genomic data currently appears to have made the ability to draw clear-cut conclusions difficult, opportunity does exist to prioritize risk with an identifiable environmental resistome approach.

Deriving specific risk assessment for foodborne *Salmonella* may be difficult to accomplish in practice for various food production systems. Part of the problem is not only the overwhelming number of *Salmonella* serovars that can cause disease but also the variability not just in their respective genetic and phenotypic traits but also the changing nature of the epidemiology associated with each outbreak that occurs. While much less obvious but perhaps equally important may be the strain and isolate differences that occur within a particular *Salmonella* serovar for important pathogenic traits such as the level of infectivity (Poppe et al., 1993; Shah et al., 2011; Heithoff et al., 2012; Foley et al., 2013; Lianou and Koutsoumanis, 2013; Ricke, 2014). As more DNA sequence data is being generated it is becoming clear that minor differences in chromosomal sequence do occur among *Salmonella* strains. How well this matches up with differences in *Salmonella* phenotypic characteristics that impact food production and processing remains to be determined. Fortunately, more precise molecular tools are being developed that go beyond DNA sequencing, such as transcriptomic techniques involving microarrays and more recently developed RNA sequencing and mutagenesis approaches (Ricke et al., 2013b; Ricke, 2014). As more data is generated with these approaches, a better understanding of *Salmonella* responses to various environmental stimuli should provide insight to develop optimal control measures for more effective interventions and better risk assessment modeling both in live animal production as well as during food processing.

The other issue that will need to be addressed with *Salmonella* is the need for development of a so-called next generation of antimicrobials. This demand is driven in part because of the regulatory push to limit clinically important antimicrobials for food-animal production and the consumer enthusiasm for food products generated from natural and organic production systems (Overybye and Barrett, 2005; Aminov, 2010; Benowitz et al., 2010; Ricke et al., 2012; Ricke and Rivera Calo, 2015). As the focus has turned to plants as a source of phytochemicals which may possess antimicrobial properties, a corresponding research explosion has occurred in the past few years to identify such compounds and develop food production applications (McChesney et al., 2007; Ricke, 2010; Ricke et al., 2012). A wide range of experimental approaches and suggested applications have emerged, but several issues remain before use of candidate compounds can become routine. First of all, understanding the mechanism(s) of how some of these compounds function as antimicrobials against a respective pathogen is needed to establish optimal parameters for concentration, appropriate delivery vehicles, and cost-effective large-scale production. Optimizing cost-effective delivery of novel antimicrobials will not only require determination of the most effective concentration dosage but designing carriers that ensure stability of the candidate compound during the required time of exposure while avoiding sensory issues that would negatively impact consumer perception of the food product. The best strategy will probably involve combinations of compounds to create antimicrobial synergism as a multiple hurdle approach. This would not only increase the inhibitory nature of the overall intervention step but may represent a means to lower concentrations of individual compounds, thus avoiding sensory issues.

Finally, particularly for preharvest interventions, it will be critical to establish effectiveness of the compound under fairly anaerobic gastrointestinal conditions where oxygen is limited and organic acid concentration varies in conjunction with fermentation activity. Another aspect that has not been considered is assessing the impact of such compounds on the indigenous gastrointestinal microbiome, as well as any probiotic culture that may administered at the same time. This is critical because most clinical antibiotics tend to be broad-spectrum enough to impact the nonpathogen population in the gastrointestinal tract, which has consequences not only in altering the microbial balance but also creating potential microbial reservoirs for transmitting antimicrobial resistance. Recent advances in microbiome sequencing and characterization offers a means to more precisely evaluate potential impact of such compounds once they reach the gastrointestinal tract (Holmes et al., 2011; Kwon and Ricke, 2011; Gordon, 2012; Haiser and Turnbaugh, 2012; Xu et al., 2013). As more becomes known it may be possible to either design compounds or delivery systems that more specifically target *Salmonella* in the gastrointestinal tract without impacting the rest of the microbiome. However, this requires both an understanding of how resistance in a pathogen such as *Salmonella* develops and application of in-depth knowledge of *Salmonella* capabilities to identify optimal and highly specific targets for a new generation of antimicrobials that either circumvent or avoid expression of resistance altogether. As more becomes known about *Salmonella* from both a genomics as well as a metabolomic standpoint, this should become more feasible.

REFERENCES

Ahmer, B.M.M., Gunn, J.S., 2011. Interaction of *Salmonella* spp. with the intestinal microbiota. Front. Microbiol. 2 (Article 101), 1–9.

Akil, L., Ahmad, H.A., Reddy, R.S., 2014. Effects of climate change on *Salmonella* infections. Foodborne Pathog. Dis. 11, 974–980.

Alali, W.Q., Hofacre, C.L., Mathis, G.F., Faltys, G., Ricke, S.C., Doyle, M.P., 2013. Effect of non-pharmaceutical compounds on shedding and colonization of *Salmonella enterica* serovar Heidelberg in broilers. Food Control 31, 125–128.

Alcaine, S.D., Warnick, L.D., Wiedmann, M., 2007. Antimicrobial resistance in nontyphoidal *Salmonella*. J. Food Prot. 70, 780–790.

Altier, C., 2005. Genetic and environmental control of *Salmonella* invasion. J. Microbiol. 43, 85–92.

Aminov, R.I., 2010. A brief history of the antibiotic era: lessons learned and challenges for the future. Front. Microbiol. 1, 134.

Angulo, F.J., Johnson, K.R., Tauxe, R.V., Cohen, M.L., 2000. Origins and consequences of antimicrobial-resistant nontyphoidal *Salmonella*: implications for the use of fluoroquinolones in food animals. Microb. Drug Resist. 6, 77–83.

Armitage, A.E., Drakesmith, H., 2014. The battle for iron – evolutionary analysis shows how a mammalian iron transport evolves to avoid capture by bacteria. Science 346, 1299–1300.

Bailey, J.S., 1988. Integrated colonization control of *Salmonella* in poultry. Poult. Sci. 67, 928–932.

Bajpai, V.K., Baek, K.-H., Kang, S.C., 2012. Control of *Salmonella* in foods by using essential oils: a review. Food Res. Int. 45, 722–734.

Bakkali, F., Averbeck, S., Averbeck, D., Idaomar, M., 2008. Biological effects of essential oils – a review. Food Chem. Toxicol. 46, 446–475.

Barber, M.F., Elde, N.C., 2014. Escape from bacterial iron piracy through rapid evolution of transferrin. Science 346, 1362–1366.

Barnhart, E.T., Sarlin, L.L., Caldwell, D.J., Byrd, J.A., Corrier, D.E., Hargis, B.M., 1999. Evaluation of potential disinfectants for preslaughter broiler crop decontamination. Poult. Sci. 78, 32–37.

Benowitz, A.B., Hoover, J.L., Payne, D.J., 2010. Antibacterial drug discovery in the age of resistance. Microbe 5, 390–396.

Berghman, L.R., Abi-Ghanem, D., Waghela, S.D., Ricke, S.C., 2005. Antibodies: an alternative for antibiotics? Poult. Sci. 84, 660–666.

Beug, H., von Kirchbach, A., Doderlein, G., Conscience, J.F., Graf, T., 1979. Chicken hematopoietic cells transformed by seven strains of defective avian leukemia viruses display three distinct phenotypes of differentiation. Cell 18, 375–390.

Billing, J., Sherman, P.W., 1998. Antimicrobial functions of spices: why some like it hot. Q. Rev. Biol. 73, 3–49.

Boyen, F., Haesebrouck, F., Maes, D., Immerseel, F., Ducatelle, R., Pasmans, F., 2008. Non-typhoidal *Salmonella* infections in pigs: a closer look at epidemiology, pathogenesis and control. Vet. Microbiol. 130, 1–19.

Braden, C.R., 2006. *Salmonella enterica* serotype Enteritidis and eggs: a national epidemic in the United States. Clin. Infect. Dis. 43, 512–517.

Brenes, A., Roura, E., 2010. Essential oils in poultry nutrition: main effects and modes of action. Anim. Feed Sci. Technol. 158, 1–14.

Brichta-Harhay, D.M., Arthur, T.M., Bosilevac, J.M., Kalchayanand, N., Shackelford, S.D., Wheeler, T.L., Koohmaraie, M., 2011. Diversity of multidrug-resistant *Salmonella enterica* strains associated with cattle at harvest in the United States. Appl. Environ. Microbiol. 77, 1783–1796.

Bueno, S.M., Riquelme, S., Riedel, C.A., Kalergis, A.M., 2012. Mechanisms used by virulent *Salmonella* to impair dendritic cell function and evade adaptive immunity. Immunology 137, 28–36.

Burley, R.W., Vadehra, D.V., 1989. The Avian Egg – Chemistry and Biology. John Wiley and Sons, New York, 472 p.

Burt, S., 2004. Essential oils: their antibacterial properties and potential applications in foods – a review. Int. J. Food Microbiol. 94, 223–253.

Calenge, F., Beaumont, C., 2012. Toward integrative genomics study of genetic resistance to *Salmonella* and *Campylobacter* intestinal colonization in fowl. Front. Genet. 3, 261.

Callaway, T.R., Carroll, J.A., Arthington, J.D., Pratt, C., Edrington, T.S., Anderson, R.C., Galyean, M.L., Ricke, S.C., Crandall, P., Nisbet, D.J., 2008. Citrus products decrease growth of *E. coli* O157:H7 and *Salmonella typhimurium* in pure culture and in fermentation with mixed ruminal microorganisms in vitro. Foodborne Pathog. Dis. 5, 621–627.

Callaway, T.R., Carroll, J.A., Arthington, J.D., Edrington, T.S., Anderson, R.C., Ricke, S.C., Crandall, P., Collier, C., Nisbet, D.J., 2011a. Chapter 17. Citrus products and their use against bacteria: potential health and cost benefits. In: Watson, R., Gerald, J.L., Preedy, V.R. (Eds.), Nutrients, Dietary Supplements, and Nutriceuticals: Cost Analysis versus Clinical Benefits. Humana Press, New York, NY, pp. 277–286.

Callaway, T.R., Carroll, J.A., Arthington, J.D., Edrington, T.S., Anderson, R.C., Rossman, M.L., Carr, M.A., Genovese, K.J., Ricke, S.C., Crandall, P., Nisbet, D.J., 2011b. Orange peel products can reduce *Salmonella* populations in ruminants. Foodborne Pathog. Dis. 8, 1071–1075.

Callaway, T.R., Carroll, J.A., Arthington, J.D., Edrington, T.S., Rossman, M.L., Carr, M.A., Krueger, N.A., Ricke, S.C., Crandall, P., Nisbet, D.J., 2011c. *Escherichia coli* O157:H7 populations in ruminants can be reduced by orange peel product feeding. J. Food Prot. 74, 1917–1921.

Calsamiglia, S., Busquet, M., Cardozo, P.W., Castillejos, L., Ferret, A., 2007. Invited Review: essential oils as modifiers of rumen microbial fermentation. J. Dairy Sci. 90, 2580–2595.

Centers for Disease Control and Prevention, 2010a. Glossary: antibiotic/antimicrobial resistance. http://www.cdc.gov/drugresistance/glossary.html.

Centers for Disease Control and Prevention, 2010b. Antibiotic/antimicrobial resistance. http://www.cdc.gov/drugresistance/glossary.html#antibiotic.

Centers for Disease Control and Prevention (CDC), 2009. Multistate outbreaks of *Salmonella* infections associated with live poultry – United States, 2007. Morb. Mortal. Wkly Rep. 58, pp. 25–29.

Centers for Disease Control and Prevention (CDC), 2011. National *Salmonella* surveillance annual data summary, 2009. Atlanta, GA: US Department of Health and Human Services, CDC.

Centers for Disease Control and Prevention (CDC), 2013. National Antimicrobial Resistance Monitoring System for Enteric Bacteria (NARMS): human isolates final report, 2011. Atlanta, GA: U.S. Department of Health and Human Services, CDC.

Chalova, V.I., Crandall, P.G., Ricke, S.C., 2010. Microbial inhibitory and radical scavenging activities of cold-pressed terpeneless Valencia orange (*Citrus sinensis*) oil in different dispersing agents. J. Sci. Food Agric. 90, 870–876.

Chao, S.C., Young, D.G., 2000. Screening for inhibitory activity of essential oils on selected bacteria, fungi and viruses. J. Essent. Oil Res. 12, 639–649.

Chart, H., Rowe, B., 1993. Iron restriction and the growth of *Salmonella enteritidis*. Epidemiol. Infect. 110, 41–47.

Chaves, A.V., He, M.L., Yang, W.Z., Hristov, A.N., McAllister, T.A., Benchaar, C., 2008. Effects of essential oils on proteolytic, deaminative and methanogenic activities of mixed ruminal bacteria. Can. J. Anim. Sci. 88, 117–122.

Chen, S., Zhao, S., White, D.G., Schroeder, C.M., Lu, R., Yang, H., McDermott, P.F., Ayers, S., Meng, J., 2004a. Characterization of multiple-antimicrobial-resistant *Salmonella* serovars isolated from retail meats. Appl. Environ. Microbiol. 70, 1–7.

Chen, Y., Liming, S.H., Bhagwat, A.A., 2004b. Occurrence of inhibitory compounds in spent growth media that interfere with acid-tolerance mechanisms of enteric pathogens. Int. J. Food Microbiol. 91, 175–183.

Cirillo, D.M., Valdivia, R.H., Monack, D.M., Falkow, S., 1998. Macrophage-dependent induction of the *Salmonella* pathogenicity island 2 type III secretions system and its role in intracellular survival. Mol. Microbiol. 30, 175–188.

Coburn, B., Grassl, G.A., Finlay, B.B., 2007. *Salmonella*, the host and disease: a brief review. Immunol. Cell Biol. 85, 112–118.

Condell, O., Sheridan, A., Power, K.A., Bonilla-Santiago, R., Sergeant, K., Renaut, J., Burgess, C., Fanning, S., Nally, J.E., 2012a. Comparative proteomic analysis of *Salmonella* tolerance to the biocide active agent triclosan. J. Proteomics 75, 4505–4519.

Condell, O., Iversen, C., Cooney, S., Power, K.A., Walsh, C., Burgess, C., Fanning, S., 2012b. Efficacy of biocides used in the modern food industry to control *Salmonella enterica*, and links between biocide tolerance and resistance to clinically relevant antimicrobial compounds. Appl. Environ. Microbiol. 78, 3087–3097.

Crandall, P.G., Friedly, E.C., Patton, M., O'Bryan, C.A., Gurubaramurugeshan, A., Seideman, S., Ricke, S.C., Rainey, R., 2010. Estimating the demand for organic foods by consumers at farmers' markets in Northwest Arkansas. J. Agr. Food Inform. 11, 185–208.

Crandall, P.G., Friedly, E.C., Patton, M., O'Bryan, C.A., Gurubaramurugeshan, A., Seideman, S., Ricke, S.C., Rainey, R., 2011. Consumer awareness of and concerns about food safety at three Arkansas farmers' markets. Food Protect. Trends 31, 156–165.

Crandall, P.G., Ricke, S.C., O'Bryan, C.A., Parrish, N.M., 2012. *In vitro* effects of citrus oils against *Mycobacterium tuberculosis* and non-tuberculous *Mycobacteria* of clinical importance. J. Environ. Sci. Health B 47, 736–741.

Crosa, J.H., 1989. Genetics and molecular biology of siderophore-mediated iron transport in bacteria. Microbiol. Rev. 53, 517–530.

Cutter, C.N., 2000. Antimicrobial effect of herb extracts against *Escherichia coli* O157:H7, *Listeria monocytogenes*, and *Salmonella* Typhimurium associated with beef. J. Food Prot. 63, 601–607.

Dabbah, R., Edwards, V.M., Moats, W.A., 1970. Antimicrobial action of some citrus fruit oils on selected food-borne bacteria. Appl. Microbiol. 19, 27–31.

De Keersmaecker, S.C.J., Marchal, K., Verhoeven, T.L.A., Engelen, K., Vanderleyden, J., Detweiler, C.S., 2005. Microarray analysis and motif detection reveal new targets of the *Salmonella enterica* serovar Typhimurium HilA regulatory protein, including *hilA* itself. J. Bacteriol. 187, 4381–4391.

De Keersmaecker, S.C.J., Verhoeven, T.L.A., Desair, J., Marchal, K., Vanderleyden, J., Nagy, I., 2006. Strong antimicrobial activity of *Lactobacillus rhamnosus* GG against *Salmonella typhimurium* is due to accumulation of lactic acid. FEMS Microbiol. Lett. 259, 89–96.

Deans, S.G., Ritchie, G., 1987. Antimicrobial properties of plant essential oils. Int. J. Food Microbiol. 5, 165–180.

DeSoucey, M., 2007. Organic food. In: Allen, G.J., Albala, K. (Eds.), The Business of Food: Encyclopedia of the Food and Drink Industries. ABC-CLIO Greenwood, Santa Barbara, CA.

Donalson, L.M., Kim, W.K., Chalova, V.I., Herrera, P., McReynolds, J.L., Gotcheva, V.G., Vidanović, D., Woodward, C.L., Kubena, L.F., Nisbet, D.J., Ricke, S.C., 2008a. In vitro fermentation response of laying hen cecal bacteria to combinations of fructooligosaccharide (FOS) prebiotic with alfalfa or a layer ration. Poult. Sci. 87, 1263–1275.

Donalson, L.M., McReynolds, J.L., Kim, W.K., Chalova, V.I., Woodward, C.L., Kubena, L.F., Nisbet, D.J., Ricke, S.C., 2008b. The influence of a fructooligosacharide prebiotic combined with alfalfa molt diets on the gastrointestinal tract fermentation, *Salmonella* Enteritidis infection and intestinal shedding in laying hens. Poult. Sci. 87, 1253–1262.

Dorman, H.J.D., Deans, S.G., 2000. Antimicrobial agents from plants, antibacterial activity of plant volatile oils. J. Appl. Microbiol. 88, 308–316.

Draughon, F.A., 2004. Use of botanicals as biopreservatives in foods. Food Technol. 58, 20–28.

Dunkley, K.D., Dunkley, C.S., Njongmeta, N.L., Callaway, T.R., Hume, M.E., Kubena, L.F., Nisbet, D.J., Ricke, S.C., 2007a. Comparison of in vitro fermentation and molecular microbial profiles of high-fiber feed substrates (HFFS) incubated with chicken cecal inocula. Poult. Sci. 86, 801–810.

Dunkley, K.D., McReynolds, J.L., Hume, M.E., Dunkley, C.S., Callaway, T.R., Kubena, L.F., Nisbet, D.J., Ricke, S.C., 2007b. Molting in *Salmonella* Enteritidis challenged laying hens fed alfalfa crumbles I. *Salmonella* Enteritidis colonization and virulence gene *hilA* response. Poult. Sci. 86, 1633–1639.

Dunkley, K.D., McReynolds, J.L., Hume, M.E., Dunkley, C.S., Callaway, T.R., Kubena, L.F., Nisbet, D.J., Ricke, S.C., 2007c. Molting in *Salmonella* Enteritidis challenged laying hens fed alfalfa crumbles II. Fermentation and microbial ecology response. Poult. Sci. 86, 2101–2109.

Dunkley, K.D., Callaway, T.R., Chalova, V.I., McReynolds, J.L., Hume, M.E., Dunkley, C.S., Kubena, L.F., Nisbet, D.J., Ricke, S.C., 2009. Foodborne *Salmonella* ecology in the avian gastrointestinal tract. Anaerobe 15, 26–35.

Durant, J.A., Corrier, D.E., Byrd, J.A., Stanker, L.H., Ricke, S.C., 1999a. Feed deprivation affects crop environment and modulates *Salmonella enteritidis* colonization and invasion of Leghorn hens. Appl. Environ. Microbiol. 65, 1919–1923.

Durant, J.A., Lowry, V.K., Nisbet, D.J., Stanker, L.H., Corrier, D.E., Ricke, S.C., 1999b. Short-chain fatty acids affect cell-association and invasion of HEp-2 cells by *Salmonella typhimurium*. J. Environ. Sci. Health B 34, 1083–1099.

Durant, J.A., Corrier, D.E., Ricke, S.C., 2000a. Short-chain volatile fatty acids modulate the expression of the *hilA* and *invF* genes of *Salmonella typhimurium*. J. Food Prot. 63, 573–578.

Durant, J.A., Corrier, D.E., Stanker, L.H., Ricke, S.C., 2000b. *Salmonella enteritidis hilA* gene fusion response after incubation in spent media from either *S. enteritidis* or a poultry *Lactobacillus* strain. J. Environ. Sci. Health B 35, 599–610.

Durant, J.A., Lowry, V.K., Nisbet, D.J., Stanker, L.H., Corrier, D.E., Ricke, S.C., 2000c. Late logarithmic *Salmonella typhimurium* HEp-2 cell-association and invasion response to short chain volatile fatty acid addition. J. Food Saf. 20, 1–11.

Durant, J.A., Lowry, V.K., Nisbet, D.J., Stanker, L.H., Corrier, D.E., Ricke, S.C., 2000d. Short-chain fatty acids alter HEp-2 cell association and invasion by stationary growth phase *Salmonella typhimurium*. J. Food Sci. 65, 1206–1209.

Edris, A.E., 2007. Pharmaceutical and therapeutic potentials of essential oils and their individual volatile constituents: a review. Phytother. Res. 4, 308–323.

Erasto, P., Bojase-Moleta, G., Majinda, R.R.T., 2004. Antimicrobial and antioxidant flavonoids from the root wood of *Bolusanthus speciosus*. Phytochemistry 65, 875–880.

Ernst, R.K., Dombroski, D.M., Merrick, J.M., 1990. Anaerobiosis, type 1 fimbriae, and growth phase are factors that affect invasion of HEP-2 cells by *Salmonella typhimurium*. Infect. Immun. 58, 2014–2016.

Faleiro, M.L., 2011. The mode of antibacterial action of essential oil. In: Méndez-Vilas, A. (Ed.), Science Against Microbial Pathogens – Communicating Current Research and Technological Advances. Formatex Research Center, Badajoz, Spain, pp. 1143–1156.

FDA, 2002. The judicious use of antimicrobials for beef producers. Retrieved from: http://www.fda.gov/downloads/AnimalVeterinary/SafetyHealth/AntimicrobialResistance/JudiciousUseofAntimicrobials/UCM095583.pdf (accessed 05.12.12).

Finlay, B.B., Falkow, S., 1997. Common themes in microbial pathogenicity revisited. Microbiol. Mol. Biol. Rev. 61, 136–169.

Finstad, S., O'Bryan, C.A., Marcy, J.A., Crandall, P.G., Ricke, S.C., 2012. *Salmonella* and broiler processing in the United States: relationship to foodborne salmonellosis. Food Res. Int. 45, 789–794.

Fisher, K., Phillips, C.A., 2006. The effect of lemon, orange and bergamot essential oils and their components on the survival of *Campylobacter jejuni, Escherichia coli* O157:H7, *Listeria monocytogenes*, *Bacillus cereus* and *Staphylococcus aureus* in vitro and in food systems. J. Appl. Microbiol. 101, 1232–1240.

Fisher, K., Phillips, C.A., 2008. Potential antimicrobial uses of essential oils in food: is citrus the answer? Trends Food Sci. Technol. 19, 156–164.

Fisher, K., Phillips, C., 2009. The mechanism of action of a citrus oil blend against *Enterococcus faecium* and *Enterococcus faecalis*. J. Appl. Microbiol. 106, 1343–1349.

Foley, S.L., Lynne, A.M., 2008. Food animal-associated *Salmonella* challenges: pathogenicity and antimicrobial resistance. J. Anim. Sci. 86, E173–E187.

Foley, S.L., White, D.G., McDermott, P.F., Walker, R.D., Rhodes, B., Fedorka-Cray, P.J., Simjee, S., Zhao, S., 2006. Comparison of subtyping methods for *Salmonella enterica* serovar Typhimurium from food animal sources. J. Clin. Microbiol. 44, 3569–3577.

Foley, S.L., Lynne, A.M., Nayak, R., 2008. *Salmonella* challenges: prevalence in swine and poultry and potential pathogenicity of such isolates. J. Anim. Sci. 86, E149–E162.

Foley, S., Nayak, R., Hanning, I.B., Johnson, T.L., Han, J., Ricke, S.C., 2011. Population dynamics of *Salmonella enterica* serotypes in commercial egg and poultry production. Appl. Environ. Microbiol. 77, 4273–4279.

Foley, S.L., Johnson, T.J., Ricke, S.C., Nayak, R., Danzelsen, J., 2013. *Salmonella* pathogenicity and host adaptation in chicken-associated serovars. Microbiol. Mol. Biol. Rev. 77, 582–607.

Foster, J.W., Spector, M.P., 1995. How *Salmonella* survive against the odds. Annu. Rev. Microbiol. 49, 145–174.

Francis, C.L., Starnbach, M.N., Falkow, S., 1992. Morphological and cytoskeletal changes in epithelial cells occur immediately upon interaction with *Salmonella typhimurium* grown under low-oxygen conditions. Mol. Microbiol. 6, 3077–3087.

Fratianni, F., De Martino, L., Melone, A., De Feo, V., Coppola, R., Nazzaro, F., 2010. Preservation of chicken breast meat treated with thyme and balm essential oils. J. Food Sci. 75, M528–M535.

Friedly, E.C., Crandall, P.G., Ricke, S.C., Roman, M., O'Bryan, C.O., Chalova, V.I., 2009. *In vitro* antilisterial effects of citrus oil fractions in combination with organic acids. J. Food Sci. 74, M67–M72.

Friedman, M., Henika, P.R., Mandrell, R.E., 2002. Bactericidal activities of plant essential oils and some of their isolated constituents against *Campylobacter jejuni, Escherichia coli, Listeria monocytogenes*, and *Salmonella enterica*. J. Food Prot. 65, 1545–1560.

Frye, J.G., Jackson, C.R., 2013. Genetic mechanisms of antimicrobial resistance identified in *Salmonella enterica, Escherichia coli,* and *Enteroccocus* spp. isolated from U.S. food animals. Front. Microbiol. 4, 135. http://dx.doi.org/10.3389/fmicb.2013.00135.

Galis, A.M., Marcq, C., Marlier, D., Portetelle, D., Van, I., Beckers, Y., Théwis, A., 2013. Control of *Salmonella* contamination of shell eggs—preharvest and postharvest methods: a review. Compr. Rev. Food Sci. Food Saf. 12, 155–181.

Gantois, I., Ducatelle, R., Pasmans, F., Haesebrouck, F., Gast, R., Humphrey, T.J., Van Immerseel, F., 2009. Mechanisms of egg contamination by *Salmonella* Enteritidis. FEMS Microbiol. Rev. 33, 718–738.

Garibaldi, J.A., 1970. Role of microbial iron transport compounds in the bacterial spoilage of eggs. Appl. Microbiol. 20, 558–560.

Gast, R.K., 1993. Recovery of *Salmonella enteritidis* from inoculated pools of egg contents. J. Food Prot. 56, 21–24.

González-Gil, F., Le Bolloch, A., Pendleton, S., Zhang, N., Wallis, A., Hanning, I., 2012. Expression of *hilA* in response to mild acid stress in *Salmonella enterica* is serovar and strain dependent. J. Food Sci. 77, M292–M297.

Gordon, J.I., 2012. Honor thy gut symbionts redux. Science 336, 1251–1253.

Guard-Petter, J., 2001. The chicken, the egg and *Salmonella enteritidis*. Environ. Microbiol. 3, 421–430.

Guiney, D.G., 1997. Regulation of bacterial virulence gene expression by the host environment. J. Clin. Investig. 99, 565–569.

Haiser, H.J., Turnbaugh, P.J., 2012. Is it time for a metagenomic basis of therapeutics? Science 336, 1253–1255.

Hanning, I.B., Nutt, J.D., Ricke, S.C., 2009. Salmonellosis: outbreaks in the United States due to fresh produce: sources and potential intervention measures. Foodborne Pathog. Dis. 6, 635–648.

Hardin, A., Crandall, P.G., Stankus, T., 2010. Essential oils and antioxidants derived from citrus by-products in food protection and medicine: an introduction and review of recent literature. J. Agric. Food Inform. 11, 99–122.

He, H., Genovese, K.J., Swaggerty, C.L., Nisbet, D.J., Kogut, M.H., 2012. A comparative study on invasion, survival, modulation of oxidative burst, and nitric oxide responses of macrophages (HD11), and systemic infection in chickens by prevalent poultry *Salmonella* serovars. Foodborne Pathog. Dis. 9, 1104–1110.

Heithoff, D.M., Shimp, W.R., House, J.K., Xie, Y., Weimer, B.C., Sinsheimer, R.L., Mahan, M.J., 2012. Intraspecies variation in the emergence of hyperinfectious bacterial strains in nature. PLoS Pathog. 8, 1–17.

Helms, M., Vastrup, P., Gerner-Smidt, P., Mølbak, K., 2002. Excess mortality associated with antimicrobial drug-resistant *Salmonella typhimurium*. Emerg. Infect. Dis. 8, 490–495.

Hennessy, T.W., Cheng, L.H., Kassenborg, H., Ahuja, S.D., Mohle-Boetani, J., Marcus, R., Shiferaw, B., Angulo, F.J., 2004. Egg consumption is the principal risk factor for sporadic *Salmonella* serotype Heidelberg infections: a case-control study in FoodNet sites. Clin. Infect. Dis. 38 (Suppl. 3), S237–S243.

Hensel, M., 2000. *Salmonella* pathogenicity island 2. Mol. Microbiol. 36, 1015–1023.

Herrera, P., Aydin, M., Park, S.H., Khatiwara, A., Ahn, S., 2013. Utility of egg yolk antibodies for detection and control of foodborne *Salmonella*. Agric. Food Anal. Bacteriol. 3, 195–217.

Hirshfield, I.N., Terzulli, S., O'Byrne, C., 2003. Weak acids: a panoply of effects on bacteria. Sci. Prog. 86, 245–269.

Holmberg, S.D., Wells, J.G., Cohen, M.L., 1984. Animal-to-man transmission of antimicrobial-resistant *Salmonella*: investigations of US outbreaks, 1971–1983. Science 225, 833–834.

Holmberg, S.D., Solomon, S., Blake, P., 1987. Health and economic impacts of antimicrobial resistance. Rev. Infect. Dis. 9, 1065–1078.

Holmes, E., Li, J.V., Athanasiou, T., Hutan Ashrafian, H., Nicholson, J.K., 2011. Understanding the role of gut microbiome–host metabolic signal disruption in health and disease. Trends Microbiol. 19, 349–359.

Howard, Z.R., O'Bryan, C.A., Crandall, P.G., Ricke, S.C., 2012. *Salmonella* Enteritidis in shell eggs: current issues and prospects for control. Food Res. Int. 45, 755–764.

Hueck, C.J., 1998. Type III protein secretion systems in bacterial pathogens of animals and plants. Microbiol. Mol. Biol. Rev. 62, 379–433.

Hume, M.E., 2011. Historic perspective: prebiotics, probiotics, and other alternatives to antibiotics. Poult. Sci. 90, 2663–2669.

Humphrey, T., 2004. *Salmonella*, stress responses and food safety. Nat. Rev. Microbiol. 2, 504–509.

Hur, J., Jawale, C., Lee, J.H., 2012. Antimicrobial resistance of *Salmonella* isolated from food animals: a review. Food Res. Int. 45, 819–830.

Ibrahim, H.R., 2000. Chapter 7. Ovotransferrin. In: Naidu, A.S. (Ed.), Natural Food Antimicrobial Systems. CRC Press, Boca Raton, FL, pp. 211–226.

Islam, M., Chen, J., Doyle, M.P., Chinnan, M., 2002. Preservative sprays on growth of *Listeria monocytogenes* on chicken luncheon meat. J. Food Prot. 65, 794–798.

Jacob, J., Pescatore, A., 2012. Chapter 21. Gut health and organic acids, antimicrobial peptides, and botanicals as natural feed additives. In: Ricke, S.C., Van Loo, E.J., Johnson, M.G., O'Bryan, C.A. (Eds.), Organic Meat Production and Processing. Wiley Scientific/IFT, New York, NY, pp. 351–378.

Jayasena, D.D., Jo, C., 2013. Essential oils as potential antimicrobial agents in meat and meat products: a review. Trends Food Sci. Technol. 34, 96–108.

Jo, C., Kang, H.J., Lee, M., Lee, N.Y., Byun, M.W., 2004. The antioxidative potential of lyophilized citrus peel extract in different meat model systems during storage at 20°C. J. Muscle Foods 15, 95–107.

Joerger, R.D., 2003. Alternatives to antibiotics: bacteriocins, antimicrobial peptides and bacteriophages. Poult. Sci. 82, 640–647.

Johnson, J.M., Rajic, A., McMullen, L.M., 2005. Antimicrobial resistance of selected *Salmonella* isolates from food animals and food in Alberta. Can. Vet. J. 46, 141–146.

Jones, F.T., Ricke, S.C., 2003. Observations on the history of the development of antimicrobials and their use in poultry feeds. Poult. Sci. 82, 613–617.

Juneja, V.K., Dwivedi, H.P., Yan, X., 2012. Novel natural food antimicrobials. J. Food Sci. Technol. 3, 381–403.

Kaiser, P., 2010. Advances in avian immunology – prospects for disease control: a review. Avian Pathol. 39, 309–324.

Kaiser, P., Rothwell, L., Galyov, E.E., Barrow, P.A., Burnside, J., Wigley, P., 2000. Differential cytokine expression in avian cells in response to invasion by *Salmonella typhimurium*, *Salmonella enteritidis* and *Salmonella gallinarum*. Microbiology 146, 3217–3226.

Kim, S.-Y., Kang, D.-H., Kim, J.K., Ha, Y.-G., Hwang, J.Y., Kim, T., Lee, S.-H., 2011. Antimicrobial activity of plant extracts against *Salmonella typhimurium*, *Escherichia coli* O157:H7, and *Listeria monocytogenes* on fresh lettuce. J. Food Sci. 76, M41–M46.

Koo, O.K., Sirsat, S.A., Crandall, P.G., Ricke, S.C., 2012. Physical and chemical control of *Salmonella* in ready-to-eat products. Agric. Food Anal. Bacteriol. 2, 56–68.

Kuehn, B.M., 2010. *Salmonella* cases traced to egg producers – findings trigger recall of more than 500 million eggs. J. Am. Med. Assoc. 304, 1316.

Kundinger, M.M., Zabala-Diaz, I.B., Chalova, V.I., Kim, W.-K., Moore, R.W., Ricke, S.C., 2007. Characterization of *rsmC* as a potential reference gene for *Salmonella typhimurium* gene expression during growth in spent media. Sens. Instrum. Food Qual. Saf. 1, 99–103.

Kwon, Y.M., Ricke, S.C., 1998. Induction of acid resistance of *Salmonella typhimurium* by exposure to short-chain fatty acids. Appl. Environ. Microbiol. 64, 3458–3463.

Kwon, Y.M., Ricke, S.C. (Eds.), 2011. High-Throughput Next Generation Sequencing: Methods and Applications. Methods in Molecular Microbiology, vol. 733. Springer Protocols, Humana Press, New York, NY, 308 pp.

Laird, K., Phillips, C., 2013. Vapour phase: a potential future use for essential oils as antimicrobials? Lett. Appl. Microbiol. 54, 169–174.

Lawhon, S.D., Maurer, R., Suyemoto, M., Altier, C., 2002. Intestinal short-chain fatty acids alter *Salmonella typhimurium* invasion gene expression and virulence through BarA/SirA. Mol. Microbiol. 46, 1451–1464.

Lee, C.A., Jones, B.D., Falkow, S., 1990. Identification of a *Salmonella typhimurium* invasion locus by selection for hyperinvasive mutants. Proc. Natl. Acad. Sci. U. S. A. 89, 1847–1851.

Leistner, L., 1985. Hurdle technology applied to meat products of the shelf stable product and intermediate moisture food types. In: Simatos, D., Multon, J.L. (Eds.), Properties of Water in Foods in Relation to Quality and Stability. Nijhoff Publishers, Dordrecht, The Netherlands, pp. 309–329.

Leistner, L., 2000. Basic aspects of food preservation by hurdle technology. Int. J. Food Microbiol. 55, 181–186.

Leistner, L., Gorris, L.G.M., 1995. Food preservation by hurdle technology. Trends Food Sci. Technol. 6, 41–46.

Lewis, W.H., Elvin-Lewis, M.P., 1995. Medicinal plants as sources of new therapeutics. Ann. Mo. Bot. Gard. 82, 16–24.

Li, M., Muthaiyan, A., O'Bryan, C.A., Gustafson, J.E., Li, Y., Crandall, P.G., Ricke, S.C., 2011. Use of natural antimicrobials from a food safety perspective for control of *Staphylococcus aureus*. Curr. Pharm. Biotechnol. 12, 1240–1254.

Li, H., Wang, H., D'Aoust, J.-Y., Maurer, J., 2013. Chapter 10 – *Salmonella* species. In: MDoyle, M.P., Buchanan, R.L. (Eds.), Food Microbiology – Fundamentals and Frontiers, fourth ed. American Society for Microbiology Press, Washington, DC, pp. 225–261.

Lianou, A., Koutsoumanis, K.P., 2013. Strain variability of the behavior of foodborne bacterial pathogens: a review. Int. J. Food Microbiol. 167, 310–321.

Litwin, C.M., Calderwood, S.B., 1993. Role of iron in regulation of virulence genes. Clin. Microbiol. Rev. 6, 137–149.

López-Malo, A., Alzamora, S.M., Guerrero, S., 2000. Natural antimicrobials from plants. In: Alzamora, S.M., Tapia, M.S., López-Malo, A. (Eds.), Minimally Processed Fruits and Vegetables: Fundamental Aspects and Applications. Aspen Publication, Gaithersburg, MD, pp. 237–258.

Maciorowski, K.G., Jones, F.T., Pillai, S.D., Ricke, S.C., 2004. Incidence and control of food-borne *Salmonella* spp. in poultry feeds – a review. Worlds Poult. Sci. J. 60, 446–457.

Magkos, F., Arvaniti, F., Zampelas, A., 2003. Putting the safety of organic food into perspective. Nutr. Res. Rev. 16, 211–222.

Magkos, F., Arvaniti, F., Zampelas, A., 2006. Organic food: buying more safety or just peace of mind? A critical review of the literature. Crit. Rev. Food Sci. Nutr. 46, 23–56.

Magnusson, M.K., Arvola, A., Hursti, U.K., Äberg, L., Sjödén, P., 2003. Choice of organic foods is related to perceived consequences for human health and to environmentally friendly behaviour. Appetite 40, 109–117.

Mann, K., 2007. The chicken egg white proteome. Proteomics 7, 3558–3568.

Mann, K., Olsen, J.V., Maček, B., Gnad, F., Mann, M., 2008. Identification of new chicken egg proteins by mass spectrometry-based proteomic analysis. Worlds Poult. Sci. J. 64, 209–218.

Martelli, F., Davies, R.H., 2012. *Salmonella* serovars isolated from table eggs: an overview. Food Res. Int. 45, 745–754.

Martínez, J.L., Coque, T.M., Baquero, F., 2015. What is a resistance gene? Ranking risk in resistomes. Nat. Rev. Microbiol. 13, 116–123. http://dx.doi.org/10.1038/nmicro3399.

Mather, A.E., Reid, S.W.J., Maskell, D.J., Parkhill, J., Fookes, M.C., Harris, S.R., Brown, D.J., Coia, J.E., Mulvey, M.R., Gilmour, M.W., Petrovska, L., de Pinna, E., Kuroda, M., Akiba, M., Izumiya, H., Connor, T.R., Suchard, M.A., Lemey, P., Mellor, D.J., Haydon, D.T., Thomson, N.R., 2013. Distinguishable epidemics of multidrug-resistant *Salmonella typhimurium* DT104 in different hosts. Science 341, 1514–1517.

Mayes, F.J., Takeballi, M.A., 1983. Microbial contamination of the hen's egg: a review. J. Food Prot. 46, 1092–1098.

McCarthy, T.L., Kerry, J.P., Kerry, J.F., Lynch, P.B., Buckley, D.J., 2001. Evaluation of the antioxidant potential of natural food/plant extracts as compared with synthetic antioxidants and vitamin E in raw and cooked pork patties. Meat Sci. 58, 45–52.

McChesney, J.D., Venkataraman, S.K., Henri, J.T., 2007. Plant natural products: back to the future or into extinction? Phytochemistry 68, 2015–2022.

McLinden, T., Sargeant, J.M., Thomas, M.K., Papadopolo, A., Fazil, A., 2014. Association between component costs, study methodologies, and foodborne illness – related factors with the cost of nontyphoidal *Salmonella* illness. Foodborne Pathog. Dis. 11, 718–726.

McMahon, M.A.S., Xu, J., Moore, J.E., Blair, I.S., McDowell, D.A., 2007. Environmental stress and antibiotic resistance in food-related pathogens. Appl. Environ. Microbiol. 73, 211–217.

Mead, P.S., Slutsker, L., Dietz, V., McCaig, L.F., Bresee, J.S., Shapiro, C., Griffin, P.M., Tauxe, R.V., 1999. Food-related illness and death in the United States. Emerg. Infect. Dis. 5, 607–625.

Mekalanos, J.J., 1992. Environmental signals controlling expression of virulence determinants in bacteria. J. Bacteriol. 174, 1–7.

Messens, W., Gittins, J., Leleu, S., Sparks, N., 2011. Chapter 9. Egg decontamination by washing. In: Van Immerseel, F., Nys, Y., Bain, M. (Eds.), Improving the Safety and Quality of Eggs and Egg Products. In: Egg Safety and Nutritional Quality, vol. 2. Woodhead Publishing Limited, Oxford, UK, pp. 163–180.

Milillo, S.R., Ricke, S.C., 2010. Synergistic reduction of *Salmonella* in a model raw chicken media using a combined thermal and organic acid salt intervention treatment. J. Food Sci. 75, M121–M125.

Milillo, S.R., Martin, E., Muthaiyan, A., Ricke, S.C., 2011. Immediate reduction of *Salmonella enterica* serotype Typhimurium viability via membrane destabilization following exposure to multiple-hurdle treatments with heated, acidified organic acid salt solutions. Appl. Environ. Microbiol. 77, 3765–3772.

Mølbak, H., 2005. Human health consequences of antimicrobial drug-resistant *Salmonella* and other foodborne pathogens. Clin. Infect. Dis. 41, 1613–1620.

Mollenkopf, D.F., Cenera, J.K., Bryant, E.M., King, C.A., Kashoma, I., Kumar, A., Funk, J.A., Rajashekara, G., Wittum, T.E., 2014. Organic or antibiotic-free labeling does not impact the recovery of enteric pathogens and antimicrobial-resistant *Escherichia coli* from fresh retail chicken. Foodborne Pathog. Dis. 11, 920–929.

Muthaiyan, A., Limayem, A., Ricke, S.C., 2011. Antimicrobial strategies for limiting bacterial contaminants in fuel bioethanol fermentations. Prog. Energ. Combust. 37, 351–370.

Muthaiyan, A., Muthaiyan, E.M., Natesan, S., Crandall, P.G., Wilkinson, B.J., Ricke, S.C., 2012. Antimicrobial effect and mode of action of terpeneless cold pressed Valencia orange essential oil on methicillin resistant *Staphylococcus aureus* cell lysis. J. Appl. Microbiol. 112, 1020–1033.

Nannapaneni, R., Muthaiyan, A., Crandall, P.G., Johnson, M.G., O'Bryan, C.A., Chalova, V.I., Callaway, T.R., Carroll, J.A., Arthington, J.D., Nisbet, D.J., Ricke, S.C., 2008. Antimicrobial activity of commercial citrus-based natural extracts against *Escherichia coli* O157:H7 isolates and mutant strains. Foodborne Pathog. Dis. 5, 695–699.

Nannapaneni, R., Chalova, V.I., Crandall, P.G., Ricke, S.C., Johnson, M.G., O'Bryan, C.A., 2009a. *Campylobacter* and *Arcobacter* species sensitivity to commercial orange oil fractions. Int. J. Food Microbiol. 129, 43–49.

Nannapaneni, R., Chalova, V.I., Story, R., Wiggins, K.C., Crandall, P.G., Ricke, S.C., Johnson, M.G., 2009b. Ciprofloxacin-sensitive and ciprofloxacin-resistant *Campylobacter jejuni* are equally susceptible to natural orange oil-based antimicrobials. J. Environ. Sci. Health B 44, 571–577.

Nayak, R., Kenney, P.B., Keswani, J., Ritz, C., 2003. Isolation and characterisation of *Salmonella* in a turkey production facility. Br. Poultry Sci. 44, 192–202.

Neilands, J.B., 1981. Microbial iron compounds. Annu. Rev. Biochem. 50, 715–731.

Neilands, J.B., 1982. Microbial envelope proteins related to iron. Annu. Rev. Microbiol. 36, 285–309.

Nostro, A., Blanco, A.R., Cannatelli, M.A., Enea, V., Flamini, G., Morelli, I., Sudano Roccaro, A., Alonzo, V., 2004. Susceptibility of methicillin-resistant staphylococci to oregano essential oil, carvacrol and thymol. FEMS Microbiol. Lett. 230, 191–195.

Nutt, J., Kubena, L., Nisbet, D.J., Ricke, S.C., 2002. Virulence response of a *Salmonella typhimurium hilA:lacZY* fusion strain to spent media from pure cultures of selected bacteria and poultry cecal mixed culture. J. Food Saf. 22, 169–181.

O'Bryan, C.A., Crandall, P.G., Chalova, V.I., Ricke, S.C., 2008a. Orange essential oils antimicrobial activities against *Salmonella* spp. J. Food Sci. 73, M264–M267.

O'Bryan, C.A., Crandall, P.G., Ricke, S.C., 2008b. Organic poultry pathogen control from farm to fork. Foodborne Pathog. Dis. 5, 709–720.

Olsen, S.J., Bishop, R., Brenner, F.W., Roels, T.H., Bean, N., Tauxe, R.V., Slutsker, L., 2001. The changing epidemiology of *Salmonella*: trends in serotypes isolated from humans in the United States, 1987-1997. J. Infect. Dis. 183, 753–761.

Overybye, K.M., Barrett, J., 2005. Antibiotics: where did we go wrong? Drug Discov. Today 10, 45–52.

Painter, J.A., Hoekstra, R.M., Ayers, T., Tauxe, R.V., Braden, C.R., Angulo, F.J., Griffin, P.M., 2013. Attribution of foodborne illnesses, hospitalizations, and deaths to food commodities by using outbreak data, United States, 1998–2008. Emerg. Infect. Dis. 19, 407–415.

Park, S.Y., Woodward, C.L., Kubena, L.F., Nisbet, D.J., Birkhold, S.G., Ricke, S.C., 2008. Environmental dissemination of foodborne *Salmonella* in preharvest poultry production: reservoirs, critical factors and research strategies. Crit. Rev. Environ. Sci. Technol. 38, 73–111.

Patrick, M.E., Adcock, P.M., Gomez, T.M., Altekruse, S.F., Holland, B.H., Tauxe, R.V., Swerdlow, D.L., 2004. *Salmonella* Enteritidis infections, United States, 1985–1999. Emerg. Infect. Dis. 10, 1–7.

Pillai, S.D., Ricke, S.C., 2002. Aerosols from municipal and animal wastes: background and contemporary issues. Can. J. Microbiol. 48, 681–696.

Pires, S.M., Vieira, A.R., Hald, T., Cole, D., 2014. Source attribution of human salmonellosis: an overview of methods and estimates. Foodborne Pathog. Dis. 11, 667–676.

Pittman, H.M., Boling, K.C., Mirus, S.J., 2012. Chapter 3. Regulatory issues in domestically raised and imported organic meats in the United States. In: Ricke, S.C., Van Loo, E.J., Johnson, M.G., O'Bryan, C.A. (Eds.), Organic Meat Production and Processing. Wiley Scientific/IFT, New York, NY, pp. 23–51.

Poppe, C., Demczuk, W., McFadden, K., Johnson, R.P., 1993. Virulence of *Salmonella enteritidis* Phagetypes 4, 8 and 13 and other *Salmonella* spp. for day-old chicks, hens and mice. Can. J. Vet. Res. 57, 281–287.

Poppe, C., McFadden, K.A., Demczuk, W.H., 1996. Drug resistance, plasmids, biotypes and susceptibility to bacteriophages of *Salmonella* isolated from poultry in Canada. Int. J. Food Microbiol. 30, 325–344.

Pui, C.F., Wong, W.C., Chai, L.C., Tunung, R., Jeyaletchumi, P., Noor Hidayah, M.S., Ubong, A., Farinazleen, M.G., Cheah, M.G., Cheah, Y.K., Son, R., 2011. Review article. *Salmonella*: a foodborne pathogen. Int. Food Res. J. 18, 465–473.

Rahman, M.M., Gray, A.I., 2002. Antimicrobial constituents from the stem bark of *Feronia limonia*. Phytochemistry 59, 73–77.

Revolledo, L., Ferreira, A.J.P., 2012. Current perspectives in avian salmonellosis: vaccines and immune mechanisms of protection. J. Appl. Poultry Res. 21, 418–431.

Ricke, S.C., 2003a. Perspectives on the use of organic acids and short chain fatty acids as antimicrobials. Poult. Sci. 82, 632–639.

Ricke, S.C., 2003b. The gastrointestinal tract ecology of *Salmonella* Enteritidis colonization in molting hens. Poult. Sci. 82, 1003–1007.

Ricke, S.C., 2010. Chapter 24. Future prospects for advancing food safety research in food animals. In: Ricke, S.C., Jones, F.T. (Eds.), Perspectives on Food Safety Issues of Food Animal Derived Foods. University of Arkansas Press, Fayetteville, AR, pp. 335–350.

Ricke, S.C., 2014. Application of molecular approaches for understanding foodborne *Salmonella* establishment in poultry production. Advances in Biology 2014. http://dx.doi.org/10.1155/2014/813275, Article ID 813275, 25 pp.

Ricke, S.C., Rivera Calo, C., 2015. Chapter 3. Antibiotic resistance in pathogenic *Salmonella*. In: Chen, C.-Y., Yan, X., Jackson, C. (Eds.), Antimicrobial Resistance and Food Safety: Methods and Techniques. Elsevier Science, Amsterdam, The Netherlands, pp. 37–53.

Ricke, S.C., Kundinger, M.M., Miller, D.R., Keeton, J.T., 2005. Alternatives to antibiotics: chemical and physical antimicrobial interventions and foodborne pathogen response. Poult. Sci. 84, 667–675.

Ricke, S.C., Van Loo, E.J., Johnson, M.G., O'Bryan, C.A. (Eds.), 2012. Organic Meat Production and Processing. Wiley Scientific/IFT, New York, NY, 444 pp.

Ricke, S.C., Dunkley, C.S., Durant, J.A., 2013a. A review on development of novel strategies for controlling *Salmonella* Enteritidis colonization in laying hens: fiber-based molt diets. Poult. Sci. 92, 502–525.

Ricke, S.C., Khatiwara, A., Kwon, Y.M., 2013b. Application of microarray analysis of foodborne *Salmonella* in poultry production: a review. Poult. Sci. 92, 2243–2250.

Ricke, S.C., Koo, O.-K., Foley, S., Nayak, R., 2013c. Chapter 7. *Salmonella*. In: Labbé, R., García, S. (Eds.), Guide to Foodborne Pathogens, second ed. Wiley-Blackwell, Oxford, UK, pp. 112–137.

Rivera Calo, J., Crandall, P.G., O'Bryan, C.A., Ricke, S.C., 2015. Essential oils as antimicrobials in food systems – a review. Food Control 54, 111–119. http://dx.doi.org/10.1016/j.foodcont.2014.12.040.

Robinson, R.K., 1999. Handbook of citrus by-products and processing technology. Food Chem. 71, 155–187.

Rohmer, L., Hocquet, D., Miller, S.I., 2011. Are pathogenic bacteria just looking for food? Metabolism and microbial pathogenesis. Trends Microbiol. 19, 341–348.

Saeed, A.M., Walk, S.T., Arshad, M., Whittam, T.S., 2006. Clonal structure and variation in virulence of *Salmonella* Enteritidis isolated from mice, chickens, and humans. J. Assn. Analyt. Chem. Intl. 89, 504–511.

Saengkerdsub, S., Kim, W.-K., Anderson, R.C., Woodward, C.L., Nisbet, D.J., Ricke, S.C., 2006. Effects of nitrocompounds and feedstuffs on *in vitro* methane production in chicken cecal contents and rumen fluid. Anaerobe 12, 85–92.

Sakamoto, K., Konings, W.N., 2003. Beer spoilage and hop resistance. Int. J. Food Microbiol. 89, 105–124.

Santos, R.I., Raffatellu, M., Bevins, C.L., Adams, L.G., Tükel, C., Tsolis, R.M., Bäumler, A.J., 2009. Life in the inflamed intestine, *Salmonella* style. Trends Microbiol. 17, 498–506.

Scallan, E., Hoekstra, R.M., Angulo, F.J., Tauxe, R.V., Widdowson, M.-A., Roy, S.L., Jones, J.L., Griffin, P.M., 2011. Foodborne illness acquired in the United States—major pathogens. Emerg. Infect. Dis. 17, 7–15.

Scharff, R.L., 2012. Economic burden from health losses due to foodborne illness in the United States. J. Food Prot. 75, 123–131.

Schiemann, D.A., 1995. Association with MDCK epithelial cells by *Salmonella typhimurium* is reduced during utilization of carbohydrates. Infect. Immun. 63, 1462–1467.

Schmidt, H., Hensel, M., 2004. Pathogenicity islands in bacterial pathogenesis. Clin. Microbiol. Rev. 17, 14–56.
Schmidt, J.W., Brichta-Harhay, D.M., Kalchayanand, N., Bosilevac, J.M., Shackelford, S.D., Wheeler, T.L., Koohmaraie, M., 2012. Prevalence, enumeration, serotypes, and antimicrobial resistance phenotypes of *Salmonella enterica* on carcasses at two large United States pork processing plants. Appl. Environ. Microbiol. 78, 2716–2726.
Shah, D.H., Zhou, X., Addwebi, T., Davis, M.A., Orfe, L., Call, D.R., Guard, J., Besser, T.E., 2011. Cell invasion of poultry-associated *Salmonella enterica* serovar Enteritidis isolates is associated with pathogenicity, motility and proteins secreted by the type III secretion system. Microbiology 157, 1428–1445.
Shannon, E.M., Milillo, S.R., Johnson, M.G., Ricke, S.C., 2011a. Efficacy of cold-pressed terpeneless valencia oil and its primary components on inhibition of *Listeria* species by direct contact and exposure to vapors. J. Food Sci. 76, M500–M503.
Shannon, E.M., Milillo, S.R., Johnson, M.G., Ricke, S.C., 2011b. Inhibition of *Listeria monocytogenes* by exposure to a combination of nisin and cold-pressed terpeneless Valencia oil. J. Food Sci. 76, M600–M604.
Shelef, L.A., 1983. Antimicrobial effects of spices. J. Food Saf. 6, 24–29.
Sirsat, S.A., Muthaiyan, A., Ricke, S.C., 2009. Antimicrobials for foodborne pathogen reduction in organic and natural poultry production. J. Appl. Poultry Res. 18, 379–388.
Sirsat, S.A., Muthaiyan, A., Dowd, S.E., Kwon, Y.M., Ricke, S.C., 2010. Chapter 15. The potential for application of foodborne *Salmonella* gene expression profiling assays in postharvest poultry processing. In: Ricke, S.C., Jones, F.T. (Eds.), Perspectives on Food Safety Issues of Food Animal Derived Foods. University of Arkansas Press, Fayetteville, AR, pp. 195–222.
Smith-Palmer, A., Stewart, J., Fyfe, L., 1998. Antimicrobial properties of plant essential oils and essences against five important food-borne pathogens. Lett. Appl. Microbiol. 26, 118–122.
Smith-Palmer, A., Stewart, J., Fyfe, L., 2001. The potential application of plant essential oils as natural food preservatives in soft cheese. Food Microbiol. 18, 463–470.
Sofos, J.N., Beuchat, L.R., Davidson, P.M., Johnson, E.A., 1998. Naturally occurring antimicrobials in food. Regul. Toxicol. Pharmacol. 28, 71–72.
Solórzano-Santos, F., Miranda-Novales, M.G., 2012. Essential oils from aromatic herbs as antimicrobial agents. Curr. Opin. Biotechnol. 23, 136–141.
Sørum, H., L'Abée-Lund, T.M., 2002. Antibiotic resistance in food-related bacteria – a result of interfering with the global web of bacterial genetics. Int. J. Food Microbiol. 78, 43–56.
Spector, M.P., Kenyon, W.J., 2012. Resistance and survival strategies of *Salmonella enterica* to environmental stresses. Food Res. Int. 45, 455–481.
Stavric, S., 1987. Microbial colonization control of chicken intestine using defined cultures. Food Technol. 41, 93–98.
Su, L.H., Chiu, C.H., Chu, C., Ou, J.T., 2004. Antimicrobial resistance in nontyphoid *Salmonella* serotypes: a global challenge. Clin. Infect. Dis. 39, 546–551.
Subba, M.S., Soumithri, T.C., Rao, R.S., 1967. Antimicrobial action of citrus oils. J. Food Sci. 32, 225–227.
Suzuki, K., Iijima, K., Sakamoto, K., Sami, M., Yamashita, H., 2006. A review of hop resistance in beer spoilage bacteria. J. Inst. Brew. 112, 173–191.
Tartera, C., Metcalf, E.S., 1993. Osmolarity and growth phase overlap in regulation of *Salmonella typhi* adherence to and invasion of human intestinal cells. Infect. Immun. 61, 3084–3089.
Tauxe, R.V., 1991. *Salmonella*: a postmodern pathogen. J. Food Prot. 54, 563–568.
Tiwari, B.K., Valdramidis, V.P., O'Donnell, C.P., Muthukumarappan, K., Bourke, P., Cullen, P.J., 2009. Application of natural antimicrobials for food preservation. J. Agric. Food Chem. 57, 5987–6000.
Tranter, H.S., Board, R.G., 1982. The antimicrobial defense of eggs: biological perspective and chemical basis. J. Appl. Biochem. 4, 295–338.
Van Asten, F.J., Hendriks, H.G., Koninkx, J.F., Van der Zeijst, B.A., Gaastra, W., 2000. Inactivation of the flagellin gene of *Salmonella enterica* serotype Enteritidis strongly reduces invasion into differentiated Caco-2 cells. FEMS Microbiol. Lett. 185, 175–179.

Van Asten, F.J., Hendriks, H.G., Koninkx, J.F., van Dijk, J.E., 2004. Flagella-mediated bacterial motility accelerates but is not required for *Salmonella* serotype Enteritidis invasion of differentiated Caco-2 cells. Int. J. Med. Microbiol. 294, 395–399.

Van Cleemput, M., Cattoor, K., De Bosscher, K., Haegeman, G., De Keukeleire, D., Heyerick, A., 2009. Hop (*Humulus lupulus*)-derived bitter acids as multipotent bioactive compounds. J. Nat. Prod. 72, 1220–1230.

Van Immerseel, F., De Buck, J., De Smet, I., Pasmans, F., Haesebrouck, F., Ducatelle, R., 2004. Interactions of butyric acid and acetic acid-treated *Salmonella* with chicken primary cecal epithelial cells in vitro. Avian Dis. 48, 384–391.

Van Immerseel, F., Russell, J.B., Flythe, M.D., Gantois, I., Timbermont, L., Pasmans, F., Haesebrouck, F., Ducatelle, R., 2006. The use of organic acids to combat *Salmonella* in poultry: a mechanistic explanation of the efficacy. Avian Pathol. 35, 182–188.

Van Loo, E., Caputo, V., Nayga Jr., R.M., Meullenet, J.-F., Crandall, P.G., Ricke, S.C., 2010. Effect of organic poultry purchase frequency on consumer attitudes toward organic poultry meat. J. Food Sci. 75, S384–S397.

Van Loo, E.J., Caputo, V., Nayga Jr., R.M., Meullenet, J.-F., Ricke, S.C., 2011. Consumers' willingness to pay for organic chicken breast: evidence from choice experiment. Food Qual. Prefer. 22, 603–613.

Van Loo, E.J., Alali, W., Ricke, S.C., 2012a. Food safety and organic meats. Annu. Rev. Food Sci. Technol. 3, 203–225.

Van Loo, E.J., Ricke, S.C., O'Bryan, C.A., Johnson, M.G., 2012b. Chapter 1. Historical and current perspectives on organic meat production. In: Ricke, S.C., Van Loo, E.J., Johnson, M.G., O'Bryan, C.A. (Eds.), Organic Meat Production and Processing. Wiley Scientific/IFT, New York, NY, pp. 1–9.

Varel, V.H., 2002a. Livestock manure odor abatement with plant-derived oils and nitrogen conservation with urease inhibitors: a review. J. Anim. Sci. 80 (E. Suppl.), E1–E7.

Varel, V.H., 2002b. Carvacrol and thymol reduce swine waste odor and pathogens: stability of oils. Curr. Microbiol. 44, 38–43.

Varel, V.H., Miller, D.N., 2001. Plant-derived oils reduce pathogens and gaseous emissions from stored cattle waste. Appl. Environ. Microbiol. 67, 1366–1370.

Velge, P., Cloeckaert, A., Barrow, P., 2005. Emergence of *Salmonella* epidemics: the problems related to *Salmonella enterica* serotype Enteritidis and multiple antibiotic resistance in other major serotypes. Vet. Res. 36, 267–288.

Vugia, D.J., Samuel, M., Farley, M.M., Marcus, R., Shiferaw, B., Shallow, S., Smith, K., Angulo, F.J., 2004. Invasive *Salmonella* infections in the United States, FoodNet, 1996–1999: incidence, serotype distribution, and outcome. Clin. Infect. Dis. 38 (Suppl. 3), S149–S156.

Walsh, K.A., Bennett, S.D., Mahovic, M., Gould, L.H., 2014. Outbreaks associated with cantaloupe, watermelon, and honeydew in the United States, 1973–2011. Foodborne Pathog. Dis. 11, 945–952.

Wang, S., Kong, Q., Curtiss III, R., 2013. New technologies in developing recombinant attenuated *Salmonella* vaccine vectors. Microb. Pathog. 58, 17–28.

White, D.G., Zhao, S., Sudler, R., Ayers, S., Friedman, S., Chen, S., McDermott, S., Wagner, D.D., Meng, J., 2001. The isolation of antibiotic-resistant *Salmonella* from retail ground meats. N. Engl. J. Med. 345, 1147–1154.

Winnen, B., Schlumberger, M.C., Sturm, A., Schupbach, K., Siebenmann, S., Jenny, P., Hardt, W.D., 2008. Hierarchical effector protein transport by the *Salmonella* Typhimurium SPI-1 type III secretion system. PLoS One 3, http://dx.doi.org/10.1371/journal.pone.0002178, e2178.

Winokur, P.L., Brueggemann, A., DeSalvo, D.L., Hoffmann, L., Apley, M.D., Uhlenhopp, E.K., Pfaller, M.A., Doern, G.V., 2000. Animal and human multidrug-resistant, cephalosporin-resistant *Salmonella* isolates expressing a plasmid-mediated CMY-2 AmpC beta-lactamase. Antimicrob. Agents Chemother. 44, 2777–2783.

Winter, S.E., Thiennimitr, P., Winter, M.G., Butler, B.P., Huseby, D.L., Crawford, R.W., Russell, J.M., Bevins, C.L., Adams, L.G., Tsolis, R.M., Roth, J.R., Bäumler, A.J., 2010. Gut inflammation provides a respiratory electron acceptor for *Salmonella*. Nature 467, 426–429.

Woodward, C.L., Kwon, Y.M., Kubena, L.F., Byrd, J.A., Moore, R.W., Nisbet, D.J., Ricke, S.C., 2005. Reduction of *Salmonella enterica* serovar Enteritidis colonization and invasion by an alfalfa diet during molt in Leghorn hens. Poult. Sci. 84, 185–193.

Wooldridge, K.G., Williams, P.H., 1993. Iron uptake mechanisms of pathogenic bacteria. FEMS Microbiol. Rev. 12, 325–348.

Woolhouse, M.E.J., Ward, M.J., 2013. Sources of antimicrobial resistance. Science 341, 1460–1461.

Xu, X., He, S., Zhang, X., 2013. New food safety concerns associated with gut microbiota. Trends Food Sci. Technol. 34, 62–66.

Yan, S.S., Pendrak, M.L., Abela-Ridder, B., Puderson, J.W., Fedorko, D.P., Foley, S.L., 2003. An overview of *Salmonella* typing public health perspectives. Clin. Appl. Immunol. Rev. 4, 189–204.

Zhao, S., Qaiyumi, S., Friedman, S., Singh, R., Foley, S.L., White, D.G., McDermott, P.F., Donkar, T., Bolin, C., Munro, S., Baron, E.J., Walker, R.D., 2003. Characterization of *Salmonella enterica* serotype Newport isolated from humans and food animals. J. Clin. Microbiol. 41, 5366–5371.

Zhao, S., McDermott, P.F., Friedman, S., Abbott, J., Ayers, S., Glenn, A., Hall-Robinson, E., Hubert, S.K., Harbottle, H., Walker, R.D., Chiller, T.M., White, D.G., 2006. Antimicrobial resistance and genetic relatedness among *Salmonella* from retail foods of animal origin: NARMS retail meat surveillance. Foodborne Pathog. Dis. 3, 106–117.

Zhao, S., McDermott, P.F., White, D.G., Qaiyumi, S., Friedman, S.L., Abbott, J.W., Glenn, A., Ayers, S.L., Post, K.W., Fales, W.H., Wilson, R.B., Reggiardo, C., Walker, R.D., 2007. Characterization of multidrug resistant *Salmonella* recovered from diseased animals. Vet. Microbiol. 123, 122–132.

Zhao, S., White, D.G., Friedman, S.L., Glenn, A., Blickenstaff, K., Ayers, S.L., Abbott, J.W., Hall-Robinson, E., McDermott, P.F., 2008. Antimicrobial resistance in *Salmonella* enterica serovar Heidelberg isolates from retail meats, including poultry, from 2002 to 2006. Appl. Environ. Microbiol. 74, 6656–6662.

Zhou, F., Ji, B., Zhang, H., Jiang, H., Yang, Z., Li, J., Li, J., Yan, W., 2007a. The antibacterial effect of cinnamaldehyde, thymol, carvacrol and their combinations against the foodborne pathogen *Salmonella typhimurium*. J. Food Saf. 27, 124–133.

Zhou, F., Ji, B., Zhang, H., Jiang, H., Yang, Z., Li, J., Ren, Y., Yan, W., 2007b. Synergistic effect of thymol and carvacrol combined with chelators and organic acids against *Salmonella typhimurium*. J. Food Prot. 70, 1704–1709.

Zhu, X., Zhang, H., Lo, R., 2004. Phenolic compounds from the leaf extract of artichoke (*Cynara scolymus* L.) and their antimicrobial activities. J. Agric. Food Chem. 52, 7272–7278.

CHAPTER

LISTERIA AND -OMICS APPROACHES FOR UNDERSTANDING ITS BIOLOGY

8

Janet R. Donaldson*, Kamil Hercik†, Aswathy N. Rai†, Sweetha Reddy†, Mark L. Lawrence†, Bindu Nanduri†, Mariola Edelmann†

Department of Biological Sciences, Mississippi State University, Mississippi State, Mississippi, USA, Department of Basic Sciences, College of Veterinary Medicine, Mississippi State University, Mississippi State, Mississippi, USA†*

1 INTRODUCTION

Listeria is a bacillus-shaped, Gram-positive, facultative anaerobe that is widespread in nature. *Listeria* comprises a diverse genus, with 4 clades and 15 different species (Figure 1) (den Bakker et al., 2014). Of these 15 different species, only *Listeria monocytogenes* and *Listeria ivanovii* are known to cause disease in humans and animals. Listeriosis is a very dangerous disease that is typically contracted by eating food contaminated with *L. monocytogenes.* Approximately 1600 illnesses are due to *L. monocytogenes* annually in the United States (Scallen et al., 2011). This microorganism is responsible for nearly 20% of all foodborne deaths in the United States, making it the third-deadliest foodborne pathogen. Additionally, the hospitalization rate for those that acquire *L. monocytogenes* is 94% (Scallen et al., 2011). Those most susceptible to infections are immunocompromised, the elderly, infants, and pregnant women. *L. monocytogenes* has been the culprit of two of the top three deadliest foodborne bacteria outbreaks in the United States. The largest listeriosis outbreak in the history of the United States occurred in 1985 and was associated with the consumption of contaminated cheese. The outbreak resulted in 52 deaths, including 19 stillbirths and 10 infant deaths. In 2011, another major outbreak of listeriosis occurred in the United States that was associated with the consumption of contaminated cantaloupes and resulted in 33 deaths and one miscarriage.

L. monocytogenes contains at least four genetic lineages (Table 1) (Orsi et al., 2011). Strains from lineages I and II are more frequently isolated than strains from lineage III or IV. This is most likely due to the fact that these strains are more frequently associated with human clinical cases; the majority of human cases of listeriosis are caused by serovar 4b. Although serovars in lineage II, such as serovar 1/2a, are also frequently identified in food isolates, they have not been shown to cause epidemics (Wiedmann et al., 1996). Analyses of the sequenced genomes have been utilized to correlate variations observed within lineages (reviewed in Orsi et al., 2011). However, these studies have all led to the conclusion that strain-to-strain variations cannot be accurately attributed to the genetic lineage. Therefore,

Food Safety. http://dx.doi.org/10.1016/B978-0-12-800245-2.00008-3

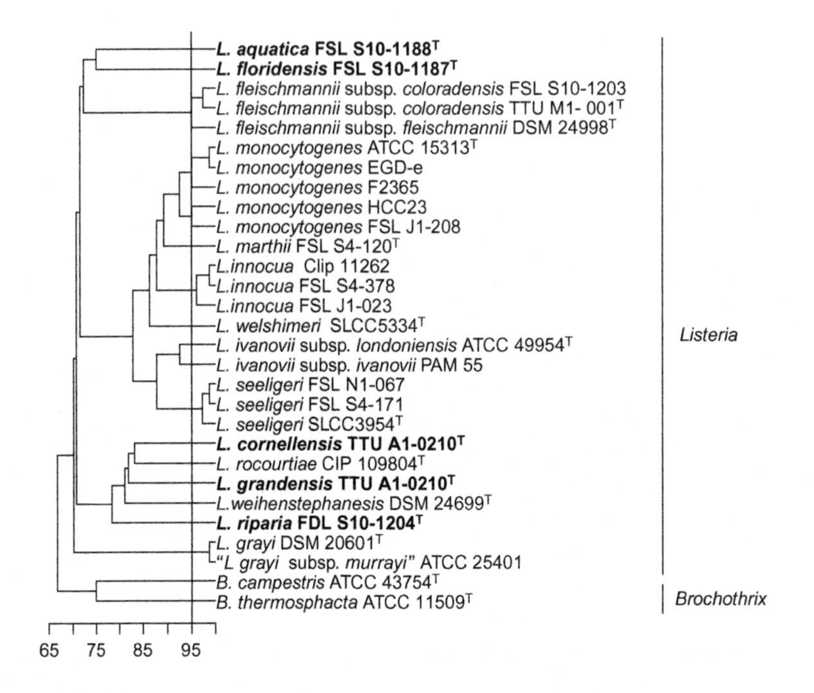

FIGURE 1

UPGMA cluster diagram based on percentage average nucleotide difference. The scale indicates ANIb (%). The gray vertical bar marks 5% average nucleotide difference, or 95% ANIb, a value that has been shown to correlate with the traditional DNA-DNA hybridization value of 70%.

Reproduced from den Bakker et al. (2014). IJSEM, with permission.

Table 1 Genetic Lineages of *Listeria monocytogenes*

Lineage	Serotypes	Genetic Characteristics	Distribution
I	1/2b, 3b, 3c, 4b	Lowest diversity among the lineages	Primarily human isolates
II	1/2a, 1/2c, 3a	Most diverse	Primarily food and food-related isolates, as well as natural environments
III	4a, 4b, 4c	Very diverse	Most isolates from ruminants
IV	4a, 4b, 4c	Few isolates	Most isolates from ruminants

Adapted from Orsi et al. (2011). IJMM with permission.

it is imperative for future studies to include analyses from multiple strains of multiple origins in order to gain a better understanding of the biology of *Listeria.*

In order to expand our knowledge of *Listeria* to be more inclusive for strain variations, "-omics" approaches have been employed. These analyses have sought to identify variations at the transcriptome, proteome, and more recently the metabolome among different serovars of *L. monocytogenes* (Figure 2). This chapter is dedicated to exploring some of the current literature that has utilized -omics technologies to further our understanding of *Listeria.* This has allowed researchers to expand the information obtained from these approaches to applications in food safety and beyond.

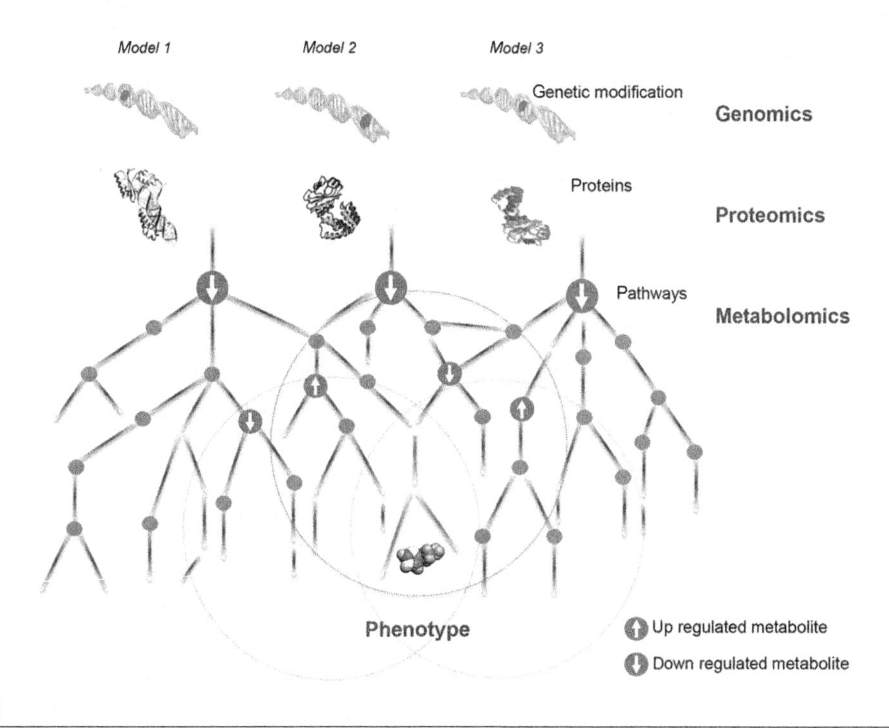

FIGURE 2

The central dogma of biology and the omics cascade. While genes and proteins are subject to regulatory epigenetic processes and posttranslational modifications, respectively, metabolites represent downstream biochemical end products that are closer to the phenotype. Alterations in a single gene (illustrated by blue dots) or a single protein can lead to a cascade of metabolite alterations. In the theoretical schematic shown, up- and down-regulated metabolites are shown in red and unaltered metabolites are shown in grey. Untargeted metabolomics aims at comprehensively profiling metabolites without bias to identify changes that correlate with cellular function or phenotype. By performing meta-analysis, metabolic alterations shared between multiple animal models or multiple genetic modifications may be identified as shown by the superimposed Venn diagram.

Patti et al. (2012), Nature Reviews, Molecular Cell Biology, with permission.

2 TRANSCRIPTOMICS APPROACHES FOR STUDYING *LISTERIA*

Analysis of the transcriptome allows for the identification of responses specific to changing environmental and physiological conditions, as well as different developmental stages in an organism (Fang et al., 2013). The transcriptome also provides information on operon structures and mapping transcriptional start and stop sites (Wang et al., 2009). Transcriptomics has also been used to elucidate the regulatory role of noncoding RNA, which revealed the abundance of antisense RNA in bacteria (Cho et al., 2013; Guell et al., 2011; Passalacqua et al., 2009). Methods that are employed for transcriptome analysis include expressed sequence tags, Northern blots, serial analysis of gene expression (Adams et al., 1991; Velculescu et al., 1995), and whole genome sequencing (Bowman et al., 2008; Chan et al., 2007).

More recently, RNA sequencing (RNASeq) has become more prominent for the analysis of bacterial transcriptomes. This method involves next generation sequencing of cDNA produced from enriched mRNA samples, which generates thousands of gene sequences that are then mapped to the corresponding genes and used for gene expression quantification. To perform operon analysis and gene expression quantifications, computational methods are used, such as TopHaT, Cufflinks, and Scripture (Nagalakshmi et al., 2010; Wang et al., 2009). Differentially expressed genes are typically validated using quantitative PCR.

For food safety pathogens such as *Escherichia coli*, *Salmonella enterica*, *Yersinia enterocolitica*, and *L. monocytogenes*, transcriptomics has been used to elucidate mechanisms of food spoilage, biofilm formation, motility, temperature, and high salt concentration, as well as mechanisms involved in virulence and host-pathogen interactions (Chan et al., 2007; Puttamreddy et al., 2008, 2010; Seo et al., 2007; Thompson et al., 2006). In this section, we summarize current literature that has utilized transcriptomic approaches to advance our understanding of how *L. monocytogenes* responds to various stressors.

2.1 TRANSCRIPTIONAL RESPONSES TO HOST

A large portion of the genome of *L. monocytogenes* is dedicated to the regulation of transcription, with 209 transcriptional regulators having been identified (Glaser et al., 2001). The transcription factor *prfA* regulates many important virulence factors, including the internalins A and B, listeriolysin, phospholipases PlcA and PlcB, Mpl protease, sugar phosphate permease, ActA, and InlC (de las Heras et al., 2011; Scortti et al., 2007). PrfA is regulated by σ^B, which is under the control of other internal and external factors (Kazmierczak et al., 2003; Norton et al., 2001; Rocourt et al., 2003). Temporal transcriptomic analysis was used to identify σ^B-dependent genes at different growth phases in the *L. monocytogenes* strain EGD-e, revealing that 7.6% of the genome was under control of σ^B (105 genes upregulated and 111 genes downregulated) (Hain et al., 2008). The σ^B regulon included genes encoding metabolic functions, transportation, general stress proteins, cell wall proteins, bile resistance proteins, and transcriptional factors (Camejo et al., 2009; Hain et al., 2008).

Survival and replication in the intracellular environment is an essential step in the pathogenesis of listeriosis (Portnoy et al., 2002). In fact, the organism's capacity to survive in the intracellular environment and elicit both innate and adaptive immune responses has made it an important model to elucidate immune mechanisms in mammals (Pamer, 2004; Stavru et al., 2011; Zenewicz and Shen, 2007). Key aspects of intracellular survival are well known, including listeriolysin- and phospholipase-mediated lysis of the vacuole and ActA-mediated mobilization of the actin cytoskeleton. However, many important regulatory mechanisms controlling the expression of certain virulence factors in host cells are still under characterized.

Several studies have been conducted to analyze changes in the gene expression during intracellular growth. In a study using the serotype 1/2a strain EGD-e, it was revealed that genes involved in glycolysis and TCA cycle were downregulated and chaperones were upregulated in response to intracellular growth in macrophages (Chatterjee et al., 2006). EGD-e was also found to have an upregulation of PrfA-regulated genes, genes required for utilizing alternate carbon sources, and chaperones during intracellular replication within the human epithelial cell line Caco-2 (Joseph et al., 2006).

The transcriptome of four *L. monocytogenes* strains from genetic lineages I, II, and III in response to intracellular growth in P388D1 murine macrophages showed strong induction of virulence genes known to be necessary for intracellular survival (*prfA*, *plcA*, *hly*, *mpl*, *actA*, *plcB*, *hpt*, *clpE*, and *bilEA*) (Hain et al., 2012). All strains also had strong induction of genes encoding sugar uptake systems, zinc transporters, spermidine/putrescine ABC transporters, and the nonoxidative branch of the pentose phosphate pathway. The most prominent difference between strains was the induction of flagellar genes during intracellular growth; strains with poorer adaptation to intracellular survival had stronger induction of genes related to motility (Hain et al., 2012). This finding was unexpected, as intracellular growth would not require flagellin and expression could be used to enhance clearance of these strains from the host.

2.2 TRANSCRIPTIONAL RESPONSES TO ENVIRONMENTAL STRESSES

Several studies have characterized the *L. monocytogenes* transcriptome response to stress. During death phase or when exposed to extreme stress, *L. monocytogenes* is able to enter into a long-term survival (LTS) phase (Wen et al., 2009). The transcriptional response of cells entering LTS phase and surviving in this state was analyzed by microarray. During the transition from log phase to LTS phase, 225 genes were identified that were up- or downregulated. These included genes encoding several hypothetical proteins, transport proteins, and ribosomal proteins (Wen et al., 2011).

The SOS response is a conserved mechanism that involves the upregulation of genes involved in DNA repair and replication restart in the presence of damage (reviewed in Cox et al., 2000). The SOS response was confirmed to be active in the *L. monocytogenes* strain EGD-e upon exposure to the DNA damaging agent mitomycin C (van der Veen et al., 2010) and also heat stress (van der Veen et al., 2007). Twenty-nine genes were identified to be part of the SOS regulon in *L. monocytogenes*, including *lexA*, *recA*, *uvrAB*, and *dinB*. Interestingly, this study also identified that the *prfA*-regulated bile exclusion system *bilE* (Sleator et al., 2005) was also regulated by LexA. Other transcriptomic analyses characterizing the listerial stress response indicate the role of cell wall synthesis proteins, transcriptional regulators, and multiple biochemical pathways involved in protecting *L. monocytogenes* during exposure to stressful environments (Durack et al., 2013). Like other bacterial species, *L. monocytogenes* adapts to osmotic and cold stress by changing its cell membrane composition (Sengupta and Chattopadhyay, 2013); thus, several of these studies revealed upregulation of cell wall synthesis genes during stress. When subjected to stress, *L. monocytogenes* also had a decrease in overall metabolic turnover and suppression of genes associated with motility.

L. monocytogenes is capable of growing in temperatures ranging from 1 to 45 °C. Adaptation to different temperatures involves the coordination of complex pathways. Along with the *prfA* and σ^B regulons, an auto-induction system *agrBCDA* plays an important role in gene regulation when *L. monocytogenes* is subjected to temperature stress (Autret et al., 2003). Transcriptome analysis of the EGD-e strain in a temperature shift from 25 to 37 °C revealed that AgrA has synergistic functions with several other regulatory functions, including *prfA*, σ^B, and σ^H (Garmyn et al., 2012).

Certain strains of *L. monocytogenes* are better capable of persisting in food processing environments than others; these strains have been referred to as "persisters." It has been proposed that the use of quaternary ammonium compounds (QACs) in food processing units may contribute to the presence of persisters (Holah et al., 2002). To better understand how these strains are able to tolerate the detergents encountered in the food processing environment, the transcriptome responses of persistent and nonpersistent isolates to a sublethal dose of the QAC benzethonium chloride were analyzed (Casey et al., 2014; Fox et al., 2011). Nearly 600 genes were identified that were upregulated upon exposure to benzethonium chloride, including genes involved in cell wall reinforcement, sugar metabolism, pH regulation, transcription, and biosynthesis of cofactors (Fox et al., 2011). A comparison of the transcriptome of the *L. monocytogenes* persister strain 6179 in the presence and absence of benzethonium chloride showed that genes involved in peptidoglycan biosynthesis, carbohydrate uptake, chemotaxis, and motility were overexpressed (Casey et al., 2014). These studies suggest that the persistence of *L. monocytogenes* cannot be attributed to a single trait, but rather is due to a combination of factors.

Ready-to-eat food is frequently stored at mildly acidic conditions (approximately pH 5) in order to extend the shelf life and control growth of microorganisms. However *L. monocytogenes* is able to survive these conditions (O'Driscoll et al., 1996). To understand mechanisms of survival under acidic pH, the transcriptome of several food isolates of *L. monocytogenes* was compared to EGD at acidic conditions (Bowman et al., 2010; Tessema et al., 2012). It was identified that strain-specific differences were evident based on the isolation source, including differential expression of genes involved in cell membrane maintenance and oxidative stress response (Bowman et al., 2010). Additionally, the transcriptome of the *L. monocytogenes* strain L502 when grown at pH 5 was compared to growth at pH 6.7. In the presence of organic acids, several genes encoding virulence factors, transport proteins, and transcriptional regulators (including σ^L) were highly induced. This upregulation was also observed for HCl, but it was more pronounced in the presence of organic acids, indicating the role of organic acids during stress (Tessema et al., 2012). These studies provided key aspects to common changes to the cell wall that occur between strains during acid stress and that strain-specific differences can be attributed to expression of cell wall maintenance and oxidative stress response.

2.3 NONCODING RNAs

Noncoding RNA molecules (ncRNAs) have been identified as important regulators in metabolism and virulence in *L. monocytogenes* (Guell et al., 2011; Sorek and Cossart, 2010). Initially, 12 ncRNAs were detected in *L. monocytogenes*, 5 of which were not detected in the avirulent strain *L. innocua* and, therefore, were proposed to be involved in regulation of virulence (Mandin et al., 2007). Advances in whole genome tiling arrays and RNAseq have now revealed that *L. monocytogenes* possesses more than 100 different ncRNAs (reviewed in Mellin and Cossart, 2012). Comparing the transcriptome profiles from *L. monocytogenes* and *L. innocua* strains led to the identification of 33 novel small ncRNAs, and 53 novel antisense ncRNAs specific to *L. monocytogenes* (Wurtzel et al., 2012). Another study that utilized ultradeep sequencing of the strain EGD-e identified nine new ncRNAs associated with housekeeping genes, such as *purA*, *fumC*, and *pgi*, indicating the role of ncRNAs in regulating metabolic pathways (Behrens et al., 2014). Though many of these ncRNAs have been experimentally detected, specific roles in regulating transcription/translation have not yet been determined. This is an important aspect that needs to be deciphered for *Listeria*.

3 PROTEOMICS APPROACHES FOR STUDYING *LISTERIA*

Proteomics has become a fundamental part of *modus operandi* in protein characterization and quantification. Proteomics provides information about the identities and quantities of proteins on a large scale, as well as information related to the proteolytic cleavage, posttranslational modification, and even specific mutations in proteins. This information has been used to create proteomic maps specific to different strains of *Listeria*, including maps of the surfaceomes and secretomes. This information can then be applied to mechanistic studies focused on understanding the molecular basis of pathogenesis in *Listeria*, such as mechanisms involved in biofilm formation or bacterial resistance to various environmental stressors. This section focuses on recent literature that has analyzed the proteome in order to further our understanding of the biology of *L. monocytogenes*.

3.1 PROTEOME OF *LISTERIA*

Although the genomes of several strains are available, there is still a need to characterize the proteome of *Listeria*. A partial proteome reference map was created for exponentially grown *L. monocytogenes* strain EGD-e (serotype 1/2a) using two-dimensional polyacrylamide gel electrophoresis (2D-PAGE). This study identified 33 proteins involved in cell envelope and cellular processes, intermediary metabolism, information pathways, and adaptation to atypical conditions (Ramnath et al., 2003). This study also analyzed a partial fractionation of membrane and cytosolic proteins for three serotype 1/2a strains and one serotype 1/2b strain food isolates in comparison to EGD-e. When compared to the profile of EGD-e, 13% of the prominent protein spots identified in the food isolates were not found in EGD-e; even greater variation was observed for minor protein spots identified. Two proteins involved in metabolic pathways in EGD-e were absent in the proteome profile of one of the food isolates, although PCR analysis revealed both genes were present. This suggests that the absence could be due to changes in pI or molecular weight attributed to posttranslational modifications or proteolytic cleavage of these proteins (Ramnath et al., 2003). This reveals the importance of coupling proteomics assays with gene expression analysis, and vice versa.

Another study mapped the proteins of EGD-e grown to either mid-log or stationary phase by 2-DE coupled with matrix-assisted laser desorption/ionization-time of flight-mass spectrometry (MALDI-TOF MS). Thirty-eight proteins were identified as specifically expressed in either the exponential or stationary phase of growth. Among the differentially expressed proteins were ribosomal proteins, proteins involved in cellular metabolism and stress response, and posttranslationally modified listeriolysin (Folio et al., 2004).

In an effort to understand why certain strains lead to epidemics while others cause only isolated cases, the proteome of the *L. monocytogenes* strain EGD-e (serotype 1/2a) was compared to the outbreak strain F2365 (serotype 4b). A total of 1754 proteins were identified from EGD-e and 1427 proteins from F2365; 1077 proteins were common to both, while the remaining were strain specific. Analysis of protein expression profiles indicated that 413 proteins were significantly altered between these two strains, such as proteins involved in cell wall maintenance, flagellar biosynthesis, DNA repair, and stress responses (Donaldson et al., 2009). BLAST searches found that 141 of the expressed EGD-e proteins did not have orthologous protein-encoding genes in F2365 (including bacteriophage proteins and internalin B), and 77% of these were still classified as hypothetical proteins in EGD-e. F2365 had 79 expressed proteins with no ortholog in EGD-e, 62% of which were classified as hypothetical. This suggests that there are a significant number of uncharacterized proteins that might contribute to strain-specific characteristics. The expression of pyruvate kinase and phosphoglyceromutase were elevated in the virulent strains EGD-e and F2365 in comparison to HCC23, while the expression of hydroxymethylglutaryl-CoA synthase and cardiolipin synthetase, which are required for lipid metabolism, were reduced in HCC23 (Donaldson et al., 2011). These studies underpin the fact that changes in metabolism is tightly linked to virulence.

Proteomics approaches have also been used to characterize the transcription regulation of the alternative sigma factors σ^B, σ^C, σ^H, and σ^L (or σ^{54}) (Chaturongakul et al., 2011). To identify σ^B -dependent proteins, a proteomics approach was used that involved 2D-PAGE with isobaric tag (iTRAQ) quantitation and MS identification. In this approach, a total of 38 proteins were identified as σ^B-dependent. Expression of 17 proteins involved in regulation of metabolism and stress-related functions was found to be upregulated by σ^B (Abram et al., 2008). Proteomics has also been used to decipher the regulation of σ^B in the context of protective mechanisms relevant in survival and adaptation to low pH (Wemekamp-Kamphuis et al., 2004). Another proteomic study compared proteomes of the parent strain 10403S and an isogenic Δ*sigB* mutant in combination with meta-analysis of earlier published microarray (Oliver et al., 2010; Ollinger et al., 2009; Raengpradub et al., 2008) and RNASeq (Oliver et al., 2009) data. This comprehensive workflow led to the identification of 134 genes positively regulated by σ^B. This study indicated that σ^B positively regulated pathways pertaining to uptake and biosynthesis of glycine and also energy metabolism (Mujahid et al., 2013).

The role of the alternative σ^{54} factor, which is encoded by *rpoN*, was investigated using comparative proteomics, in which the proteome of EGD-e strain was compared to the isogenic *rpoN* mutant. By using 2D-PAGE, nine proteins with expression modulated by σ^{54} were identified; most of these proteins had functions in carbohydrate, particularly pyruvate, metabolism (Arous et al., 2004). Together, these studies indicate that proteomics can be used to identify novel regulatory mechanisms.

3.2 SECRETOME

Analysis of the secretome provides valuable information on how *Listeria* strains interact with their environment and host. Examples of secretory proteins are internalin A (InlA), InlB, listeriolysin O, phospholipases PlcA and PlcB, and actin assembly inducing protein (ActA). These proteins are first displayed on the bacterial cell surface and then further secreted or directly injected into a host cell (Cabanes et al., 2002). The transport of proteins across the membrane can be performed by seven secretion systems, but is primarily dependent upon the Sec secretion system in *Listeria* (Desvaux and Hebraud, 2006). SecA2 is an ATPase that is a part of Sec translocase in *Listeria* and is responsible for export of a subset of proteins secreted by the Sec pathway (Feltcher and Braunstein, 2012). In order to characterize the Sec-dependent secretome, a *secA2* mutant of the strain EGD-e was constructed and compared to the wild-type strain. Analysis of the secretomes identified 13 novel proteins secreted by the Sec pathway. The functions of these putative secreted proteins were related to cell wall metabolism, adhesion, as well as biofilm formation (Renier et al., 2013).

A study that used signal peptide prediction and in-depth bioinformatics analysis of the secretome in EGD-e in comparison to *L. innocua* CLIP11262 led to annotation of 121 putatively secreted proteins, including 45 novel proteins. This was further supported by an experimental proteomics dataset, where 43 proteins were specifically identified in either EGD-e or *L. innocua.* A larger study aimed at investigating the diversity of *Listeria* compared the cytoplasmic and secreted proteomes of 12 strains belonging to serovars 4b, 1/2a, and 1/2b using 2D-PAGE (Dumas et al., 2008). The hierarchical clustering of these proteomes confirmed the phylogenetic lineage of these strains, which demonstrated that this technique could be used to some extent for phylogenetic studies. The secretome had more pronounced differences between the strains than the cytoplasmic proteome, suggesting that the secretome could be used for serotyping strains (Dumas et al., 2008). Similar results were also found for a study using strains from only serovar 4b (Dumas et al., 2009b). Another comparative secretome study performed on 12 representative strains by 2D-PAGE coupled with MALDI-TOF MS also identified similarities in virulence factors present in the secretome. This study also established the variant secretome of *Listeria*, which could be used in identifying biomarkers related to virulent strains (Dumas et al., 2009a). Together, these proteomic comparisons have suggested that the strongest point of difference between virulent and avirulent strains is connected to the secretome (Trost et al., 2005).

3.3 SURFACEOME

The "surfaceome" includes surface proteins involved in important biological processes, such as bacterial growth, responses to environmental stress, host invasion, and interference with the immune system (reviewed in Bierne and Cossart, 2007). This validates the importance of in-depth characterizations of the surfaceome. To identify surface proteins of the *Listeria* strain LI0521, cells were treated with trypsin, followed by mass spectrometric analysis of the resulting peptides. Nineteen surface proteins were identified. Two of these proteins (lipoprotein and PBPD1) were validated to be surface proteins by immunofluorescence, providing evidence that analysis of the surfaceome can provide valuable information on feasible protein targets for diagnostic assay development (Zhang et al., 2013).

An analysis of cell wall extracts from EGD-e and *L. innocua* CLIP11262 identified 19 and 11 proteins, respectively. Bioinformatic analysis revealed that 20 of these proteins shared an LPXTG motif used for covalent anchoring to the peptidoglycan, while other proteins were enzymes important in peptidoglycan processing or contained an NXZTN motif (Calvo et al., 2005). Another study utilized N-terminal sequencing and mass spectrometry, which lead to the identification of 55 surface proteins of *Listeria*, a majority of which were previously uncharacterized. This study provided novel information on proteins that were previously only thought to be located in the cytoplasm, confirming the importance of analyzing surfaceomes (Schaumburg et al., 2004).

As it was expected that the cell wall proteome would be modified when in contact with host cells, a proteomic study was designed to characterize the cell wall proteome of the *L. monocytogenes* EGD-e strain during intracellular growth within human epithelial cells. The peptidoglycan was isolated and the proteomic analysis identified 53 proteins specifically expressed during intracellular growth, including proteins containing the LXPTG motif, sortase B substrates Lmo2185 and Lmo2186, as well as enzymes related to peptidoglycan metabolism. Interestingly, one of these proteins was ActA, which had not been previously shown to be peptidoglycan-associated. The ActA protein was missing in peptidoglycan of extracellular bacteria grown in nutrient-rich medium, but it was present in bacteria grown in a chemically defined minimal medium (Garcia-del Portillo et al., 2011).

Additional studies have used proteomics as a means to characterize the function of certain cell wall surface proteins. For instance, the immunogenic surface protein IspC is a hydrolase capable of degrading the cell wall peptidoglycan (Wang and Lin, 2007). To study its biological function, the proteome of the isogenic Δ*ispC* mutant was compared to the parental strain LI0521. The deficiency of IspC had a negative effect on the expression of some surface proteins, such as ActA, a putative LPXTG motif-containing internalin, InlC2, and a homologue to the 30kDa flagellin FlaA. This suggests that IspC might have a role in bacterial virulence, especially because the removal of *ispC* was associated with attenuated virulence in mice (Wang and Lin, 2008).

Another study characterizing the function of cell wall surface proteins focused upon the biological function of sortases. Sortases can cleave the C-terminal motif (such as LPXTG) of protein substrates, thereby anchoring them to the cell wall of bacteria. The surfaceome of strains deficient in sortases A and B were analyzed in comparison to the wild-type strain. This study identified 13 unique LPXTG-containing proteins in sortase A-expressing *Listeria*, and two proteins, Lmo2185 and Lmo2186, were unique to sortase B-containing strains. Further analysis suggested that sortase B might recognize NXZTN and NPKXZ sorting motifs (Pucciarelli et al., 2005).

3.4 PROTEOMIC RESPONSE TO HOST

The proteome of two virulent strains (EGD-e and F2365) and one avirulent strain (HCC23) were analyzed during early and late stages of intracellular replication in murine J774.1 macrophages using strong cation exchange (SCX) chromatography followed by reverse phase chromatography coupled with electrospray ionization tandem mass spectrometry. An increase in oxidative stress response proteins and chaperones was only detected in the virulent strains, suggesting that proteins regulating stress response (e.g., superoxide dismutase (SOD) and catalase) are required for long-term intracellular survival (Donaldson et al., 2011).

The proteome of EGD-e was also analyzed during intracellular growth in THP-1 monocytes using two-dimensional difference gel electrophoresis (2D DIGE) followed by MALDI-TOF MS (Van de Velde et al., 2009). This method identified 245 distinct proteins. Sixty-four proteins with reduced expression were involved with stress defense, transport systems, carbon metabolism, pyrimidines synthesis, and D-Ala-D-Ala ligase. Proteins with increased expression included those involved in the synthesis of cell envelope lipids, glyceraldehyde-3-phosphate, pyruvate, and fatty acids. This analysis suggested that the intracellular environment is more favorable for *Listeria* growth than extracellular culture due to differential regulation of metabolic processes (Van de Velde et al., 2009).

Survival of *L. monocytogenes* in bile salt conditions, such as observed within the gallbladder, is considered a crucial virulence factor (Hardy et al., 2004). To determine how *L. monocytogenes* can survive the stressful environment of high concentrations of bile salts, the proteomes of two virulent strains (EGD-e and F2365), and an avirulent strain (HCC23) were analyzed by LTQ mass spectrometry (Payne et al., 2013). In all strains tested, the response to bile salts was associated with differential expression of cell wall-associated proteins (e.g., peptidoglycan-bound internalins), DNA repair proteins, protein-folding chaperones, and oxidative stress-response proteins. However, there were also differences between these strains in the response to bile salts, which included regulation of osmotic stress-response proteins. The glutamate dehydrogenase (GAD) and (p)ppGpp synthetase were both decreased in EGD-e during the bile salt treatment; the protein 1-pyrroline-5-carboxylate reductase (ProC) increased in expression by 5 h postexposure in HCC23. Another interesting group of proteins regulated by the exposure to bile salts of *Listeria* were proteins associated with biofilm formation. The expression of the invasion-associated protein (Iap P60) was decreased in the virulent strain EGD-e following exposure to bile salts (Payne et al., 2013). This protein has been linked to biofilm formation (Monk et al., 2004), therefore suggesting that Iap P60 may promote biofilm formation to enable LTS within the gallbladder (Begley et al., 2009). EGD-e, but not F2365, formed biofilms in the presence of bile salts, suggesting virulent strains may exhibit different mechanisms for survival in the presence of bile salts (Payne et al., 2013).

A study investigating the gastric stress response of three *L. monocytogenes* strains analyzed the proteome of cells following exposure to sublethal conditions of pH and salt in cheese-simulated medium (Melo et al., 2013). This study identified that adapted cells had an increased expression of carbohydrate proteins and different stress proteins, including SOD, cold shock-like protein CspLB, and universal stress proteins, revealing different strategies are used by cells preexposed to salt to survive the stressful environment of the stomach (Melo et al., 2013).

3.5 PROTEOMIC RESPONSES TO ENVIRONMENTAL STRESSES

As described above, *L. monocytogenes* has the ability to survive in hostile environments, such as high salt concentrations, acidic and alkaline conditions, high hydrostatic pressure, and temperatures ranging from 1 to 45 °C. The mechanisms involved in the resistance of *Listeria* to these environments have been extensively studied by proteomic analyses using both gel-based and nongel-based techniques. These studies have expanded our knowledge of transcriptome responses.

Outbreaks of listeriosis are frequently associated with contaminated ready-to-eat products. Many of these products contain high concentrations of salt. High concentrations of salt can be disruptive to the maintenance of osmotic balance in bacteria, but *Listeria* is known to be able to tolerate this stressor (McClure et al., 1989). The adaptation to salt was analyzed in the *L. monocytogenes* strain L028. In this study, bacteria were grown in a rich medium or in a chemically defined medium supplemented with 3.5% NaCl; proteomes were analyzed by 2-DE combined with mass spectrometric identification by MALDI-TOF. This study indicated that osmotic shock was associated with the differential expression of 59 proteins, including upregulation of proteins closely related to Ctc, GbuA, and the 30S ribosomal protein S6, as well as downregulation of metabolic proteins, such as orthologs of AckA and PdhD (Duche et al., 2002).

Analysis of the proteome following exposure to 4 °C identified an increase in the production of proteins related to protein synthesis or folding, chaperones, and regulation of nutrient uptake or oxidative stress (Cacace et al., 2010). The Pta-AckA pathway, which controls the intracellular level of acetyl phosphate (Wolfe, 2005), was also found to be increased following exposure to cold temperatures. Proteins involved in metabolic pathways were also increased in expression, suggesting an increased energy demand at lower temperatures (Cacace et al., 2010).

As *Listeria* is typically exposed to various stressors within the food processing environment, a study were conducted to determine if exposure to low temperatures provides cross protection against salt. Interestingly, bacteria exposed to low temperatures prior to a mild salt stress had increased survival in 3% NaCl (Pittman et al., 2014). An analysis of the differentially expressed proteome demonstrated that bacteria exposed to cold had an increased expression of proteins involved in cell wall maintenance, penicillin binding, and processes related to amino acid metabolism, such as osmolyte, transport, and lipid biosynthesis (Pittman et al., 2014). Additionally, differential expression of proteins involved in the metabolism of alternative carbohydrates were identified, including α-mannosidase, phosphofructokinase, fructose-1-phosphate, sorbitol dehydrogenase, glucose kinase, and 6-phospho-β-glucosidase (Pittman et al., 2014).

The proteome of *Listeria* in response to acidic conditions has also been analyzed using 2D-PAGE and mass spectrometry. In this study, *Listeria* was exposed to a lethal acidic pH and an intermediary nonlethal acidic pH to mimic acid adaptation (Bowman et al., 2012). Both lethal and nonlethal acidic conditions led to regulation of similar proteins, although the lethal acidic pH-induced expression of additional proteins, including ATP synthase, thioredoxin reductase, alcohol dehydrogenase, chaperonin, GroEL, and ferric uptake regulator (Phan-Thanh and Mahouin, 1999). Adaptation to acidic conditions was also analyzed in a phase-dependent manner for the *L. monocytogenes* strain ScottA cultured in near growth-limiting acidic conditions. Many identified proteins showed growth phase-dependent abundance. Acidification caused reductions in growth rate, oxidative stress, activation of pH homeostatic, and protein turnover mechanisms. Proteins involved in sugar phosphotransfer, glycolysis, and cell wall biosynthesis were repressed (Bowman et al., 2012).

The ability of *Listeria* to survive in alkaline conditions has also been analyzed using 2-DE and MALDI-TOF (Giotis et al., 2008; Taormina and Beuchat, 2001). An increase in the expression of the heat shock proteins GroEL and DnaK was observed under alkaline conditions, which the ATP synthase AtpD and D-alanine-D-alanine ligase were decreased. Multidimensional protein identification technology (MudPIT) was used in another study to investigate alkali adaptation of EGD-e (Nilsson et al., 2013). The data suggest that alkaline pH homeostasis was maintained by acidification through the production, retention, and import of polar and charged proteins, peptides, and amino acids. The physiology of adapted cells also corresponded to bacteria growing under anaerobiosis, which might have important implications to food safety in the context of low oxygen tension packaging technology.

Restrictive nutrient conditions can induce bacteria to enter into the stationary phase of growth. Response to this stress has been analyzed by proteomics, which revealed that *Listeria* adjusts to this transition phase by modulating the expression level of over 50% of all proteins (Weeks et al., 2004). Over 88% of these proteins were downregulated at stationary phase in comparison to exponential phase. The chaperone protein GrpE, SOD, and the transcriptional regulators XyIS/AraC and LacI were all increased in expression during this transition (Weeks et al., 2004).

3.6 PROTEOMIC RESPONSE DURING BIOFILM FORMATION

Biofilms are a form of a microbial communities and are proposed to be the preferred growth and survival strategy of many clinically and environmentally important microorganisms (Donlan and Costerton, 2002). Biofilms are embedded and protected by an extracellular polysaccharide matrix, which allow for the efficient colonization of *Listeria* in the food processing environment (Carpentier and Cerf, 1993). The persistence of these biofilms in the environment is a major health concern. In order to better understand how these biofilms are formed, the proteomes of planktonic versus mature biofilm form of a *L. monocytogenes* strain isolated from a food processing plant were investigated by 2-DE and MALDI-TOF MS. The analysis revealed a significant effect on expression of 31 proteins, such as enzymes that are involved in global carbon metabolism, 30S ribosomal proteins, oxidative stress response, DNA repair enzymes, and cellular division factors (Trémoulet et al., 2002).

A similar study comparing proteomes of *Listeria* grown in suspension and cultured as a biofilm in the presence and absence of glucose identified proteins involved in transport, and metabolism of nutrients, amino acids, lipids, and nucleic acids were induced during the carbon starvation of both planktonic and biofilm forms (Helloin et al., 2003). In biofilms, additional proteins were increased in expression to potentially help bacteria with adaptation to a new sessile state (detoxification challenge, protein damage, starvation, etc.). The reduced number of proteins identified in this study was a limiting factor and no major mechanisms of biofilm formation could be identified. However, regulation of several proteins in this study (Helloin et al., 2003) was consistent with the results in the study that was published later (Hefford et al., 2005), which compared proteomes of biofilm and planktonic cells of *L. monocytogenes* strain 568 using 2-DE and MALDI-TOF analysis. Bacteria within the biofilm had an increased expression of 19 proteins involved in stress response, envelope and protein synthesis, biosynthesis, energy generation, and general regulatory functions (Hefford et al., 2005).

3.7 PROTEOMIC-BASED DETECTION OF *L. MONOCYTOGENES*

Mass spectrometry was first used in 1975 as a novel method for the identification of bacteria (Anhalt and Fenselau, 1975). Proteomic profiles for bacteria are based on specific peptide fingerprints obtained by MALDI-TOF and offer an alternative to traditional detection and characterization techniques, such as cultivation, biochemical profile testing, or nucleic acid detection (Giebel et al., 2010; Seng et al., 2009; van Veen et al., 2010). MS analysis was applied on intact bacterial cells to detect several food-borne pathogens of the genera *Escherichia*, *Yersinia*, *Salmonella*, and *Listeria.* Several genus- and species-specific biomarkers were detected. Notably, the method was able to unambiguously distinguish between *L. monocytogenes* and *L. innocua* (Mazzeo et al., 2006). In another study, a rapid and reproducible method for detecting *Listeria* was established using MALDI-TOF to analyze the spectra representative of 146 different strains (Barbuddhe et al., 2008). This method allowed for the identification of specific strains using single colonies. MALDI-TOF was also recently used for the rapid detection of *Listeria* from selective enrichment broths (Jadhav et al., 2014). Additionally, this approach provided reliable detection and identification of *L. monocytogenes* in contaminated food samples after a 24-h incubation in selective enrichment broth, followed by a 6-h secondary enrichment. This method allowed for detection of a very small number of bacteria in medium, which corresponded to 1 colony-forming unit (CFU) per ml. The detection level for *Listeria* from different solid food products (such as cheese or meat) was 10CFU/ml, which validated the potential use of this cost-effective and rapid detection workflow in food samples (Jadhav et al., 2014).

3.8 POSTTRANSLATIONAL MODIFICATIONS OF *LISTERIA* PROTEOMES

Posttranslational modifications of proteins constitute an important regulatory principle. The phosphoproteome of *Listeria* has been studied using phosphopeptide enrichment methods combined with high accuracy mass spectrometry analysis (Misra et al., 2011). In one study, the phosphoenrichment was achieved by a combination of SCX and TiO_2 chromatography, and lactic acid was used to exclude nonphosphopeptide peptides. This method allowed for the detection of 143 different phosphorylation modifications on 112 distinct proteins. Similar to phosphoproteome studies previously conducted in mammalian cells, the majority of phosphorylation modifications were identified on serine residues and the lowest on tyrosine residues. A large proportion of the identified phosphoproteins were associated with virulence, protein translation, stress response, and key metabolic functions, such as glycolysis (Misra et al., 2011).

4 METABOLOMICS APPROACHES FOR STUDYING *LISTERIA*

Analogous to the transcriptome and proteome that catalog expressed genes and proteins, respectively, the metabolome refers to the repertoire of all metabolites that are produced by an organism (Fiehn, 2002). The metabolomics workflow includes metabolic fingerprinting, profiling, data analysis, and integration to model biological systems. Metabolomics has become a critical component of systems biology and has been applied to bacteria, fungi, plants, and animals (Fernie et al., 2004). Metabolomics is a rapidly growing area of research for biomarker discovery and is an attractive avenue for the identification of novel metabolic targets that could be used for diagnostic strategies and potential intervention of infectious diseases. However, much is still not understood in regards to the complexity of the metabolic requirements in various environments and how these vary between strains. Though many transcriptomic and proteomic studies have been conducted on the metabolic profiles of different strains, these approaches can only provide an indirect inference of the metabolic state of the cell. Methodologies and current pipelines used for data analysis and interpretation have been extensively reviewed elsewhere (Fernie et al., 2004; Hollywood et al., 2006; Patti et al., 2012). Therefore, this section is limited to the studies that have been conducted to characterize the metabolome of *Listeria*.

4.1 GENERAL METABOLOME OF *L. MONOCYTOGENES*

The metabolome of *Listeria* has been analyzed using ^{13}C isotoplogue profiling analysis (Fuchs et al., 2012). This technology was key to understand the central carbon metabolism in *Listeria* (Eisenreich et al., 2006, 2010). *Listeria* has an impaired TCA cycle and cannot convert 2-ketoglutarate to oxaloacetate, due to the lack of a 2-oxoglutarate dehydrogenase. Instead, the catabolism of glucose in the *L. monocytogenes* strain EGD-e occurs via the pentose phosphate pathway. Oxaloacetate is a critical precursor for the biosynthesis of threonine and aspartate and in *Listeria* is generated by pyruvate carboxylation (Eisenreich et al., 2010; Fuchs et al., 2012).

4.2 METABOLOMIC RESPONSES TO HOST

To characterize the metabolome of cells in response to *L. monocytogenes* infections, ^{13}C isotopologue profiling of amino acids was used in an infection model using both primary bone marrow macrophages and J774A.1 murine macrophage cells (Gillmaier et al., 2012). The authors concluded that the metabolite profiles of transformed and primary cells varied due to different metabolic requirements of the cell lines used. Unlike primary cells, the J774A.1 cells had an increase of glycolytic activities (as a result of the "Warburg Effect") that may have allowed for more efficient intracellular metabolic activity of *Listeria.* Interestingly, the metabolic profile of EGD-e was similar between both cell lines, indicating that the response was independent of the cell line status.

Changes to the host's metabolome in response to *L. monocytogenes* infections were analyzed using *Drosophila melanogaster* as a model organism (Chambers et al., 2012). When fruit flies were infected with the *L. monocytogenes* strain 10403S, significant shifts in 221 metabolites were observed. Major changes included reduction in beta-oxidation and glycolysis in flies infected with 10403S. Interestingly, the levels of the disaccharide trehalose, the primary sugar in flies, dropped during the initial period of infection, but were restored to normal levels after 48 h. The ability to obtain energy is critical for both host and pathogen and understanding the differences in energy metabolism is critical to design effective intervention strategies.

4.3 METABOLOMIC RESPONSES TO ENVIRONMENTAL STRESSES

As described above, *L. monocytogenes* can tolerate many stressful environments, such as high salt, acidic or alkaline pH, or temperatures ranging from 1 to 45 °C (Cole et al., 1990; Knabel et al., 1990; Liu et al., 2007). Survival in cold temperatures requires continued protein synthesis, maintenance of membrane fluidity, and uptake of intracellular osmolytes. In order to further understanding of the cold response of *Listeria*, metabolomic profiling of the *L. monocytogenes* strain 10403S was performed following growth at 8 °C in comparison to 37 °C (Singh et al., 2011). Among the 103 metabolites identified, the concentrations of 64 were significantly changed upon exposure to cold temperatures. The metabolites identified included amino acids, sugars, polyamines, and organic acids.

An increase in glutamate was also observed following exposure to 8 °C in comparison to growth at 37 °C (Singh et al., 2011). It was suggested that the increase was due to an accumulation of 2-ketoglutarate and that the TCA cycle combined with high levels of glutamate can be considered as a possible metabolic switch for cold adaptation. Glutamate is also the metabolic precursor for the synthesis of glutamine, a critical scavenger for metabolic nitrogenous waste. Catabolism of amino acids also produces ATP, which in turn helps to meet the energy requirements for survival at low temperatures. Among other amino acids with increased concentrations were branched chain amino acids (BCAA), such as isoleucine and leucine. Previous studies in *Bacillus subtilis* showed that BCAAs serve as precursor for the synthesis of anteiso-branched fatty acids, which are critical to maintain membrane fluidity at low temperatures (Klein et al., 1999). Isoleucine and leucine undergo oxidative deamination and decarboxylation reactions that result in the generation of coenzyme-A, an essential precursor for fatty acid synthesis (Klein et al., 1999). In *L. monocytogenes*, synthesis of branched chain fatty acids is essential for replication in the host cytosol (Chatterjee et al., 2006; Joseph et al., 2006; Keeney et al., 2009) and for adaptation to cold temperatures (Zhu et al., 2005).

Compatible solutes, also called osmoprotectants, are low molecular weight compounds that help cells survive in a variety of different stresses (reviewed in Sleator and Hill, 2010). Glycine, betaine, and carnitine are low molecule weight compatible solutes involved in osmotolerance, cryotolerance, and barotolerance in *L. monocytogenes* (Angelidis and Smith, 2003; Dreux et al., 2008; Sleator et al., 2003, 2009; Smiddy et al., 2004). The osmoprotectants betaine and carnitine were detected following exposure to cold temperatures (Singh et al., 2011).

Differences in the concentration of polyamines were also detected in 10403S when grown at 8 °C in comparison to 37 °C. Polyamines serve diverse functions in pathogenic bacteria, including invasion, virulence, modulation of the type III secretome, oxidative stress, heat stress, and biofilm formation (Di Martino et al., 2013). The polyamines putrescine, cadaverine, spermidine, ornithine, and ethanolamine increased following exposure to cold temperatures (Singh et al., 2011).

4.4 METABOLOMICS FOR DETECTION OF *LISTERIA*

Fast and accurate methods for detecting different strains of *Listeria* are needed to ensure food safety. Recently, metabolomics was tested as an option for detecting *Listeria* in a complex food sample of milk (Beale et al., 2014). The study identified specific metabolites were present in samples specifically contaminated with *Listeria*. This proof of concept study provides great promise for the future of metabolomics in rapid detection of pathogens in complex food samples.

5 FUTURE DIRECTIONS FOR *LISTERIA*-OMICS

Delineating mechanisms of virulence and environmental adaptation are critical for the development of new control measures against *Listeria*. In particular, disruption of regulatory mechanisms controlling virulence or survival in the processing plant is an attractive, novel method to control contamination of products. Using various -omics approaches, much has been deciphered in terms of novel aspects of adaptation to stressors encountered within the host and different environmental stressors, particularly those encountered in the food industry. These novel approaches have allowed for the distinction between strains, which will hopefully improve detection strategies in the near future. Another aspect of metabolomics is the discovery of novel small molecules that alter the growth of *L. monocytogenes*; several of these have been identified and screened as potential novel targets to combat this pathogen (Nguyen et al., 2012). The advent of small molecule libraries and repositories will allow for rapid detection strategies as well as allow for the design of novel small molecules that are toxic bactericidal analogs.

Though these -omics approaches have greatly improved our understanding of the biology of *Listeria*, there are still many aspects that need to be explored further. For instance, biofilm formation is very important to the LTS of *Listeria* in food processing facilities. Though the mechanism of biofilm formation has been studied by proteomics and transcriptomics, the secretome has not been analyzed. This could be an important focus of future pursuits based on studies performed on other bacteria (Resch et al., 2006; Toyofuku et al., 2012). Another aspect that has been limited in studies is in analyzing the response of various host cells infected with *Listeria.* Although there have been several studies that analyze host cell response in terms of the changes in protein expression (Lianou and Koutsoumanis, 2013; Reinl et al., 2009; Ribet et al., 2010; Van Troys et al., 2008), as a next step to these explorations we should expect more in-depth proteomic analyses of various host cells infected with different isolates of *Listeria.* Moreover, this analysis would also be useful in the design of new drugs for the treatment of *Listeria*, as well as monitoring the bacterial response and adaptation to currently used antibiotics and pharmacological treatments.

Another aspect that can be greatly improved upon by using -omics approaches includes deciphering strain-to-strain variability of *L. monocytogenes.* Variations are evident in the way that different strains interact with environments and these characteristics have not been able to be attributed to the genetic analysis alone (Lianou and Koutsoumanis, 2013). This suggests that the differences lie in the regulation of gene expression or protein synthesis mechanisms. As "one strain does not fit all," the major disadvantage is the lack of metabolomics data that compares different strains of *Listeria* under different stresses. Future studies should aim at expanding the list of metabolites that are unique to each strain under standardized growth conditions. Additionally, targeted holistic approaches are needed that couple metabolomics with mass spectroscopy or microarray data.

Some of the challenges stated above are not unique to *L. monocytogenes* and are, in fact, a reflection of the current state of the art in -omics technologies. However, the studies reported in the literature clearly attest to the promise of -omics to unearth novel aspects of regulation of this extremely versatile pathogen. The information that can be gained from these studies holds promise for the development of novel strategies to combat infections and prevent contaminated food products.

REFERENCES

Abram, F., Su, W.L., Wiedmann, M., Boor, K.J., Coote, P., Botting, C., Karatzas, K.A., O'Byrne, C.P., 2008. Proteomic analyses of a *Listeria monocytogenes* mutant lacking sigmaB identify new components of the sigmaB regulon and highlight a role for sigmaB in the utilization of glycerol. Appl. Environ. Microbiol. 74, 594–604.

Adams, M.D., Kelley, J.M., Gocayne, J.D., Dubnick, M., Polymeropoulos, M.H., Xiao, H., Merril, C.R., Wu, A., Olde, B., Moreno, R.F., et al., 1991. Complementary DNA sequencing: expressed sequence tags and human genome project. Science 252, 1651–1656.

Angelidis, A.S., Smith, G.M., 2003. Three transporters mediate uptake of glycine betaine and carnitine by *Listeria monocytogenes* in response to hyperosmotic stress. Appl. Environ. Microbiol. 69, 1013–1022.

Anhalt, J.P., Fenselau, C., 1975. Identification of bacteria using mass spectrometry. Anal. Chem. 47, 219–225.

Arous, S., Buchrieser, C., Folio, P., Glaser, P., Namane, A., Hebraud, M., Hechard, Y., 2004. Global analysis of gene expression in an rpoN mutant of *Listeria monocytogenes*. Microbiology 150, 1581–1590.

Autret, N., Raynaud, C., Dubail, I., Berche, P., Charbit, A., 2003. Identification of the agr locus of *Listeria monocytogenes*: role in bacterial virulence. Infect. Immun. 71, 4463–4471.

Barbuddhe, S.B., Maier, T., Schwarz, G., Kostrzewa, M., Hof, H., Domann, E., Chakraborty, T., Hain, T., 2008. Rapid identification and typing of listeria species by matrix-assisted laser desorption ionization-time of flight mass spectrometry. Appl. Environ. Microbiol. 74, 5402–5407.

Beale, D.J., Morrison, P.D., Palombo, E.A., 2014. Detection of *Listeria* in milk using non-targeted metabolic profiling of *Listera monocytogenes*: a proof of concept application. Food Control 42, 343–346.

Begley, M., Kerr, C., Hill, C., 2009. Exposure to bile influences biofilm formation by *Listeria monocytogenes*. Gut Pathog. 1, 11.

Behrens, S., Widder, S., Mannala, G.K., Qing, X., Madhugiri, R., Kefer, N., Abu Mraheil, M., Rattei, T., Hain, T., 2014. Ultra deep sequencing of *Listeria monocytogenes* sRNA transcriptome revealed new antisense RNAs. PLoS One 9, e83979.

Bierne, H., Cossart, P., 2007. *Listeria monocytogenes* surface proteins: from genome predictions to function. Microbiol. Mol. Biol. Rev. 71, 377–397.

Bowman, J.P., Bittencourt, C.R., Ross, T., 2008. Differential gene expression of *Listeria monocytogenes* during high hydrostatic pressure processing. Microbiology 154, 462–475.

Bowman, J.P., Lee Chang, K.J., Pinfold, T., Ross, T., 2010. Transcriptomic and phenotypic responses of *Listeria monocytogenes* strains possessing different growth efficiencies under acidic conditions. Appl. Environ. Microbiol. 76, 4836–4850.

Bowman, J.P., Hages, E., Nilsson, R.E., Kocharunchitt, C., Ross, T., 2012. Investigation of the *Listeria monocytogenes* Scott A acid tolerance response and associated physiological and phenotypic features via whole proteome analysis. J. Proteome Res. 11, 2409–2426.

Cabanes, D., Dehoux, P., Dussurget, O., Frangeul, L., Cossart, P., 2002. Surface proteins and the pathogenic potential of *Listeria monocytogenes*. Trends Microbiol. 10, 238–245.

Cacace, G., Mazzeo, M.F., Sorrentino, A., Spada, V., Malorni, A., Siciliano, R.A., 2010. Proteomics for the elucidation of cold adaptation mechanisms in Listeria monocytogenes. J. Proteomics 73, 2021–2030.

Calvo, E., Pucciarelli, M.G., Bierne, H., Cossart, P., Albar, J.P., Garcia-Del Portillo, F., 2005. Analysis of the Listeria cell wall proteome by two-dimensional nanoliquid chromatography coupled to mass spectrometry. Proteomics 5, 433–443.

Camejo, A., Buchrieser, C., Couve, E., Carvalho, F., Reis, O., Ferreira, P., Sousa, S., Cossart, P., Cabanes, D., 2009. In vivo transcriptional profiling of *Listeria monocytogenes* and mutagenesis identify new virulence factors involved in infection. PLoS Pathog. 5, e1000449.

Carpentier, B., Cerf, O., 1993. Biofilms and their consequences, with particular reference to hygiene in the food industry. J. Appl. Microbiol. 75, 499–511.

Casey, A., Fox, E.M., Schmitz-Esser, S., Coffey, A., McAuliffe, O., Jordan, K., 2014. Transcriptome analysis of *Listeria monocytogenes* exposed to biocide stress reveals a multi-system response involving cell wall synthesis, sugar uptake, and motility. Front. Microbiol. 5, 68.

Chambers, M.C., Song, K.H., Schneider, D.S., 2012. *Listeria monocytogenes* infection causes metabolic shifts in *Drosophila melanogaster*. PLoS One 7, e50679.

Chan, Y.C., Raengpradub, S., Boor, K.J., Wiedmann, M., 2007. Microarray-based characterization of the *Listeria monocytogenes* cold regulon in log- and stationary-phase cells. Appl. Environ. Microbiol. 73, 6484–6498.

Chatterjee, S.S., Hossain, H., Otten, S., Kuenne, C., Kuchmina, K., Machata, S., Domann, E., Chakraborty, T., Hain, T., 2006. Intracellular gene expression profile of *Listeria monocytogenes*. Infect. Immun. 74, 1323–1338.

Chaturongakul, S., Raengpradub, S., Palmer, M.E., Bergholz, T.M., Orsi, R.H., Hu, Y., Ollinger, J., Wiedmann, M., Boor, K.J., 2011. Transcriptomic and phenotypic analyses identify coregulated, overlapping regulons among PrfA, CtsR, HrcA, and the alternative sigma factors sigmaB, sigmaC, sigmaH, and sigmaL in *Listeria monocytogenes*. Appl. Environ. Microbiol. 77, 187–200.

Cho, S., Cho, Y., Lee, S., Kim, J., Yum, H., Kim, S.C., Cho, B.K., 2013. Current challenges in bacterial transcriptomics. Genomics Inform. 11, 76–82.

Cole, M.B., Jones, M.V., Holyoak, C., 1990. The effect of pH, salt concentration and temperature on the survival and growth of *Listeria monocytogenes*. J. Appl. Bacteriol. 69, 63–72.

Cox, M.M., Goodman, M.F., Kreuzer, K.N., Sherratt, D.J., Sandler, S.J., Marians, K.J., 2000. The importance of repairing stalled replication forks. Nature 404, 37–41.

de las Heras, A., Cain, R.J., Bielecka, M.K., Vazquez-Boland, J.A., 2011. Regulation of *Listeria* virulence: PrfA master and commander. Curr. Opin. Microbiol. 14, 118–127.

den Bakker, H.C., Warchocki, S., Wright, E.M., Allred, A.F., Ahlstrom, C., Manuel, C.S., Stasiewicz, M.J., Burrell, A., Roof, S., Strawn, L.K., Fortes, E., Nightingale, K.K., Kephart, D., Wiedmann, M., 2014. *Listeria floridensis* sp. nov., *Listeria aquatica* sp. nov., *Listeria cornellensis* sp. nov., *Listeria riparia* sp. nov. and *Listeria grandensis* sp. nov., from agricultural and natural environments. Int. J. Syst. Evol. Microbiol. 64, 1882–1889.

Desvaux, M., Hebraud, M., 2006. The protein secretion systems in Listeria: inside out bacterial virulence. FEMS Microbiol. Rev. 30, 774–805.

Di Martino, M.L., Campilongo, R., Casalino, M., Micheli, G., Colonna, B., Prosseda, G., 2013. Polyamines: emerging players in bacteria-host interactions. Int. J. Med. Microbiol. 303, 484–491.

Donaldson, J.R., Nanduri, B., Burgess, S.C., Lawrence, M.L., 2009. Comparative proteomic analysis of *Listeria monocytogenes* strains F2365 and EGD. Appl. Environ. Microbiol. 75, 366–373.

Donaldson, J.R., Nanduri, B., Pittman, J.R., Givaruangsawat, S., Burgess, S.C., Lawrence, M.L., 2011. Proteomic expression profiles of virulent and avirulent strains of *Listeria monocytogenes* isolated from macrophages. J. Proteomics 74, 1906–1917.

Donlan, R.M., Costerton, J.W., 2002. Biofilms: survival mechanisms of clinically relevant microorganisms. Clin. Microbiol. Rev. 15, 167–193.

Dreux, N., Albagnac, C., Sleator, R.D., Hill, C., Carlin, F., Morris, C.E., 2008. Glycine betaine improves *Listeria monocytogenes* tolerance to desiccation on parsley leaves independent of the osmolyte transporters BetL, Gbu and OpuC. J. Appl. Microbiol. 104, 1221–1227.

Duche, O., Tremoulet, F., Namane, A., Labadie, J., European Listeria Genome Consortium, 2002. A proteomic analysis of the salt stress response of *Listeria monocytogenes*. FEMS Microbiol. Lett. 215, 183–188.

Dumas, E., Meunier, B., Berdague, J.L., Chambon, C., Desvaux, M., Hebraud, M., 2008. Comparative analysis of extracellular and intracellular proteomes of *Listeria monocytogenes* strains reveals a correlation between protein expression and serovar. Appl. Environ. Microbiol. 74, 7399–7409.

Dumas, E., Desvaux, M., Chambon, C., Hebraud, M., 2009a. Insight into the core and variant exoproteomes of *Listeria monocytogenes* species by comparative subproteomic analysis. Proteomics 9, 3136–3155.

Dumas, E., Meunier, B., Berdague, J.L., Chambon, C., Desvaux, M., Hebraud, M., 2009b. The origin of *Listeria monocytogenes* 4b isolates is signified by subproteomic profiling. Biochim. Biophys. Acta 1794, 1530–1536.

Durack, J., Ross, T., Bowman, J.P., 2013. Characterisation of the transcriptomes of genetically diverse *Listeria monocytogenes* exposed to hyperosmotic and low temperature conditions reveal global stress-adaptation mechanisms. PLoS One 8, e73603.

Eisenreich, W., Slaghuis, J., Laupitz, R., Bussemer, J., Stritzker, J., Schwarz, C., Schwarz, R., Dandekar, T., Goebel, W., Bacher, A., 2006. ^{13}C isotopologue perturbation studies of *Listeria monocytogenes* carbon metabolism and its modulation by the virulence regulator PrfA. Proc. Natl. Acad. Sci. U. S. A. 103, 2040–2045.

Eisenreich, W., Dandekar, T., Heesemann, J., Goebel, W., 2010. Carbon metabolism of intracellular bacterial pathogens and possible links to virulence. Nat. Rev. Microbiol. 8, 401–412.

Fang, G., Passalacqua, K.D., Hocking, J., Llopis, P.M., Gerstein, M., Bergman, N.H., Jacobs-Wagner, C., 2013. Transcriptomic and phylogenetic analysis of a bacterial cell cycle reveals strong associations between gene co-expression and evolution. BMC Genomics 14, 450.

Feltcher, M.E., Braunstein, M., 2012. Emerging themes in SecA2-mediated protein export. Nat. Rev. Microbiol. 10, 779–789.

Fernie, A.R., Trethewey, R.N., Krotzky, A.J., Willmitzer, L., 2004. Metabolite profiling: from diagnostics to systems biology. Nat. Rev. Mol. Cell Biol. 5, 763–769.

Fiehn, O., 2002. Metabolomics – the link between genotypes and phenotypes. Plant Mol. Biol. 48, 155–171.

Folio, P., Chavant, P., Chafsey, I., Belkorchia, A., Chambon, C., Hebraud, M., 2004. Two-dimensional electrophoresis database of *Listeria monocytogenes* EGDe proteome and proteomic analysis of mid-log and stationary growth phase cells. Proteomics 4, 3187–3201.

Fox, E.M., Leonard, N., Jordan, K., 2011. Physiological and transcriptional characterization of persistent and nonpersistent *Listeria monocytogenes* isolates. Appl. Environ. Microbiol. 77, 6559–6569.

Fuchs, T.M., Eisenreich, W., Kern, T., Dandekar, T., 2012. Toward a systemic understanding of *Listeria monocytogenes* metabolism during infection. Front. Microbiol. 3, 23.

Garcia-del Portillo, F., Calvo, E., D'Orazio, V., Pucciarelli, M.G., 2011. Association of ActA to peptidoglycan revealed by cell wall proteomics of intracellular *Listeria monocytogenes*. J. Biol. Chem. 286, 34675–34689.

Garmyn, D., Augagneur, Y., Gal, L., Vivant, A.L., Piveteau, P., 2012. *Listeria monocytogenes* differential transcriptome analysis reveals temperature-dependent Agr regulation and suggests overlaps with other regulons. PLoS One 7, e43154.

Giebel, R., Worden, C., Rust, S.M., Kleinheinz, G.T., Robbins, M., Sandrin, T.R., 2010. Microbial fingerprinting using matrix-assisted laser desorption ionization time-of-flight mass spectrometry (MALDI-TOF MS) applications and challenges. Adv. Appl. Microbiol. 71, 149–184.

Gillmaier, N., Gotz, A., Schulz, A., Eisenreich, W., Goebel, W., 2012. Metabolic responses of primary and transformed cells to intracellular *Listeria monocytogenes*. PLoS One 7, e52378.

Giotis, E.S., Muthaiyan, A., Blair, I.S., Wilkinson, B.J., McDowell, D.A., 2008. Genomic and proteomic analysis of the alkali-tolerance response (AlTR) in *Listeria monocytogenes* 10403S. BMC Microbiol. 8, 102.

Glaser, P., Frangeul, L., Buchrieser, C., Rusniok, C., Amend, A., Baquero, F., Berche, P., Bloecker, H., Brandt, P., Chakraborty, T., Charbit, A., Chetouani, F., Couve, E., de Daruvar, A., Dehoux, P., Domann, E., Dominguez-Bernal, G., Duchaud, E., Durant, L., Dussurget, O., Entian, K.D., Fsihi, H., Garcia-del Portillo, F., Garrido, P., Gautier, L., Goebel, W., Gomez-Lopez, N., Hain, T., Hauf, J., Jackson, D., Jones, L.M., Kaerst, U., Kreft, J., Kuhn, M., Kunst, F., Kurapkat, G., Madueno, E., Maitournam, A., Vicente, J.M., Ng, E., Nedjari, H., Nordsiek, G., Novella, S., de Pablos, B., Perez-Diaz, J.C., Purcell, R., Remmel, B., Rose, M., Schlueter, T., Simoes, N., Tierrez, A., Vazquez-Boland, J.A., Voss, H., Wehland, J., Cossart, P., 2001. Comparative genomics of Listeria species. Science 294, 849–852.

Guell, M., Yus, E., Lluch-Senar, M., Serrano, L., 2011. Bacterial transcriptomics: what is beyond the RNA horizome? Nat. Rev. Microbiol. 9, 658–669.

Hain, T., Hossain, H., Chatterjee, S.S., Machata, S., Volk, U., Wagner, S., Brors, B., Haas, S., Kuenne, C.T., Billion, A., Otten, S., Pane-Farre, J., Engelmann, S., Chakraborty, T., 2008. Temporal transcriptomic analysis of the *Listeria monocytogenes* EGD-e sigmaB regulon. BMC Microbiol. 8, 20.

Hain, T., Ghai, R., Billion, A., Kuenne, C.T., Steinweg, C., Izar, B., Mohamed, W., Mraheil, M.A., Domann, E., Schaffrath, S., Karst, U., Goesmann, A., Oehm, S., Puhler, A., Merkl, R., Vorwerk, S., Glaser, P., Garrido, P., Rusniok, C., Buchrieser, C., Goebel, W., Chakraborty, T., 2012. Comparative genomics and transcriptomics of lineages I, II, and III strains of *Listeria monocytogenes*. BMC Genomics 13, 144.

Hardy, J., Francis, K.P., DeBoer, M., Chu, P., Gibbs, K., Contag, C.H., 2004. Extracellular replication of *Listeria monocytogenes* in the murine gall bladder. Science 303, 851–853.

Hefford, M.A., D'Aoust, S., Cyr, T.D., Austin, J.W., Sanders, G., Kheradpir, E., Kalmokoff, M.L., 2005. Proteomic and microscopic analysis of biofilms formed by *Listeria monocytogenes* 568. Can. J. Microbiol. 51, 197–208.

Helloin, E., Jansch, L., Phan-Thanh, L., 2003. Carbon starvation survival of *Listeria monocytogenes* in planktonic state and in biofilm: a proteomic study. Proteomics 3, 2052–2064.

Holah, J.T., Taylor, J.H., Dawson, D.J., Hall, K.E., 2002. Biocide use in the food industry and the disinfectant resistance of persistent strains of *Listeria monocytogenes* and *Escherichia coli*. J. Appl. Microbiol. 92 (Suppl.), 111S–120S.

Hollywood, K., Brison, D.R., Goodacre, R., 2006. Metabolomics: current technologies and future trends. Proteomics 6, 4716–4723.

Jadhav, S., Sevior, D., Bhave, M., Palombo, E.A., 2014. Detection of *Listeria monocytogenes* from selective enrichment broth using MALDI-TOF mass spectrometry. J. Proteomics 97, 100–106.

Joseph, B., Przybilla, K., Stuhler, C., Schauer, K., Slaghuis, J., Fuchs, T.M., Goebel, W., 2006. Identification of *Listeria monocytogenes* genes contributing to intracellular replication by expression profiling and mutant screening. J. Bacteriol. 188, 556–568.

Kazmierczak, M.J., Mithoe, S.C., Boor, K.J., Wiedmann, M., 2003. *Listeria monocytogenes* sigma B regulates stress response and virulence functions. J. Bacteriol. 185, 5722–5734.

Keeney, K., Colosi, L., Weber, W., O'Riordan, M., 2009. Generation of branched-chain fatty acids through lipoate-dependent metabolism facilitates intracellular growth of *Listeria monocytogenes*. J. Bacteriol. 191, 2187–2196.

Klein, W., Weber, M.H., Marahiel, M.A., 1999. Cold shock response of *Bacillus subtilis*: isoleucine-dependent switch in the fatty acid branching pattern for membrane adaptation to low temperatures. J. Bacteriol. 181, 5341–5349.

Knabel, S.J., Walker, H.W., Hartman, P.A., Mendonca, A.F., 1990. Effects of growth temperature and strictly anaerobic recovery on the survival of *Listeria monocytogenes* during pasteurization. Appl. Environ. Microbiol. 56, 370–376.

Lianou, A., Koutsoumanis, K.P., 2013. Strain variability of the behavior of foodborne bacterial pathogens: a review. Int. J. Food Microbiol. 167, 310–321.

Liu, D., Lawrence, M.L., Ainsworth, A.J., Austin, F.W., 2007. Toward an improved laboratory definition of *Listeria monocytogenes* virulence. Int. J. Food Microbiol. 118, 101–115.

Mandin, P., Repoila, F., Vergassola, M., Geissmann, T., Cossart, P., 2007. Identification of new noncoding RNAs in *Listeria monocytogenes* and prediction of mRNA targets. Nucleic Acids Res. 35, 962–974.

Mazzeo, M.F., Sorrentino, A., Gaita, M., Cacace, G., Di Stasio, M., Facchiano, A., Comi, G., Malorni, A., Siciliano, R.A., 2006. Matrix-assisted laser desorption ionization-time of flight mass spectrometry for the discrimination of food-borne microorganisms. Appl. Environ. Microbiol. 72, 1180–1189.

McClure, P.J., Roberts, T.A., Oguru, P.O., 1989. Comparison of the effects of sodium chloride, pH and temperature on the growth of *Listeria monocytogenes* on gradient plates and in liquid medium. Lett. Appl. Microbiol. 9, 95–99.

Mellin, J.R., Cossart, P., 2012. The non-coding RNA world of the bacterial pathogen *Listeria monocytogenes*. RNA Biol. 9, 372–378.

Melo, J., Schrama, D., Andrew, P.W., Faleiro, M.L., 2013. Proteomic analysis shows that individual *Listeria monocytogenes* strains use different strategies in response to gastric stress. Foodborne Pathog. Dis. 10, 107–119.

Misra, S.K., Milohanic, E., Ake, F., Mijakovic, I., Deutscher, J., Monnet, V., Henry, C., 2011. Analysis of the serine/threonine/tyrosine phosphoproteome of the pathogenic bacterium *Listeria monocytogenes* reveals phosphorylated proteins related to virulence. Proteomics 11, 4155–4165.

Monk, I.R., Cook, G.M., Monk, B.C., Bremer, P.J., 2004. Morphotypic conversion in *Listeria monocytogenes* biofilm formation: biological significance of rough colony isolates. Appl. Environ. Microbiol. 70, 6686–6694.

Mujahid, S., Orsi, R.H., Vangay, P., Boor, K.J., Wiedmann, M., 2013. Refinement of the *Listeria monocytogenes* sigmaB regulon through quantitative proteomic analysis. Microbiology 159, 1109–1119.

Nagalakshmi, U., Waern, K., Snyder, M., 2010. RNA-Seq: a method for comprehensive transcriptome analysis. Curr. Protoc. Mol. Biol, Unit 4.11, 1–13.

Nguyen, U.T., Wenderska, I.B., Chong, M.A., Koteva, K., Wright, G.D., Burrows, L.L., 2012. Small-molecule modulators of *Listeria monocytogenes* biofilm development. Appl. Environ. Microbiol. 78, 1454–1465.

Nilsson, R.E., Ross, T., Bowman, J.P., Britz, M.L., 2013. MudPIT profiling reveals a link between anaerobic metabolism and the alkaline adaptive response of *Listeria monocytogenes* EGD-e. PLoS One 8, e54157.

Norton, D.M., Scarlett, J.M., Horton, K., Sue, D., Thimothe, J., Boor, K.J., Wiedmann, M., 2001. Characterization and pathogenic potential of *Listeria monocytogenes* isolates from the smoked fish industry. Appl. Environ. Microbiol. 67, 646–653.

O'Driscoll, B., Gahan, C.G., Hill, C., 1996. Adaptive acid tolerance response in *Listeria monocytogenes*: isolation of an acid-tolerant mutant which demonstrates increased virulence. Appl. Environ. Microbiol. 62, 1693–1698.

Oliver, H.F., Orsi, R.H., Ponnala, L., Keich, U., Wang, W., Sun, Q., Cartinhour, S.W., Filiatrault, M.J., Wiedmann, M., Boor, K.J., 2009. Deep RNA sequencing of *L. monocytogenes* reveals overlapping and extensive stationary phase and sigma B-dependent transcriptomes, including multiple highly transcribed noncoding RNAs. BMC Genomics 10, 641.

Oliver, H.F., Orsi, R.H., Wiedmann, M., Boor, K.J., 2010. *Listeria monocytogenes* σ^B has a small core regulon and a conserved role in virulence but makes differential contributions to stress tolerance across a diverse collection of strains. Appl. Environ. Microbiol. 76, 4216–4232.

Ollinger, J., Bowen, B., Wiedmann, M., Boor, K.J., Bergholz, T.M., 2009. *Listeria monocytogenes* sigmaB modulates PrfA-mediated virulence factor expression. Infect. Immun. 77, 2113–2124.

Orsi, R.H., den Bakker, H.C., Wiedmann, M., 2011. *Listeria monocytogenes* lineages: genomics, evolution, ecology, and phenotypic characteristics. Int. J. Med. Microbiol. 301, 79–96.

Pamer, E.G., 2004. Immune responses to *Listeria monocytogenes*. Nat. Rev. Immunol. 4, 812–823.

Passalacqua, K.D., Varadarajan, A., Ondov, B.D., Okou, D.T., Zwick, M.E., Bergman, N.H., 2009. Structure and complexity of a bacterial transcriptome. J. Bacteriol. 191, 3203–3211.

Patti, G.J., Yanes, O., Siuzdak, G., 2012. Innovation: metabolomics: the apogee of the omics trilogy. Nat. Rev. Mol. Cell Biol. 13, 263–269.

Payne, A., Schmidt, T.B., Nanduri, B., Pendarvis, K., Pittman, J.R., Thornton, J.A., Grissett, J., Donaldson, J.R., 2013. Proteomic analysis of the response of *Listeria monocytogenes* to bile salts under anaerobic conditions. J. Med. Microbiol. 62, 25–35.

Phan-Thanh, L., Mahouin, F., 1999. A proteomic approach to study the acid response in *Listeria monocytogenes*. Electrophoresis 20, 2214–2224.

Pittman, J.R., Buntyn, J.O., Posadas, G., Nanduri, B., Pendarvis, K., Donaldson, J.R., 2014. Proteomic analysis of cross protection provided between cold and osmotic stress in *Listeria monocytogenes*. J. Proteome Res. 13, 1896–1904.

Portnoy, D.A., Auerbuch, V., Glomski, I.J., 2002. The cell biology of *Listeria monocytogenes* infection: the intersection of bacterial pathogenesis and cell-mediated immunity. J. Cell Biol. 158, 409–414.

Pucciarelli, M.G., Calvo, E., Sabet, C., Bierne, H., Cossart, P., Garcia-del Portillo, F., 2005. Identification of substrates of the *Listeria monocytogenes* sortases A and B by a non-gel proteomic analysis. Proteomics 5, 4808–4817.

Puttamreddy, S., Carruthers, M.D., Madsen, M.L., Minion, F.C., 2008. Transcriptome analysis of organisms with food safety relevance. Foodborne Pathog. Dis. 5, 517–529.

Puttamreddy, S., Cornick, N.A., Minion, F.C., 2010. Genome-wide transposon mutagenesis reveals a role for pO157 genes in biofilm development in *Escherichia coli* O157:H7 EDL933. Infect. Immun. 78, 2377–2384.

Raengpradub, S., Wiedmann, M., Boor, K.J., 2008. Comparative analysis of the sigma B-dependent stress responses in *Listeria monocytogenes* and Listeria innocua strains exposed to selected stress conditions. Appl. Environ. Microbiol. 74, 158–171.

Ramnath, M., Rechinger, K.B., Jansch, L., Hastings, J.W., Knochel, S., Gravesen, A., 2003. Development of a *Listeria monocytogenes* EGDe partial proteome reference map and comparison with the protein profiles of food isolates. Appl. Environ. Microbiol. 69, 3368–3376.

Reinl, T., Nimtz, M., Hundertmark, C., Johl, T., Keri, G., Wehland, J., Daub, H., Jansch, L., 2009. Quantitative phosphokinome analysis of the Met pathway activated by the invasin internalin B from *Listeria monocytogenes*. Mol. Cell. Proteomics 8, 2778–2795.

Renier, S., Chambon, C., Viala, D., Chagnot, C., Hebraud, M., Desvaux, M., 2013. Exoproteomic analysis of the SecA2-dependent secretion in *Listeria monocytogenes* EGD-e. J. Proteomics 80C, 183–195.

Resch, A., Leicht, S., Saric, M., Pasztor, L., Jakob, A., Gotz, F., Nordheim, A., 2006. Comparative proteome analysis of *Staphylococcus aureus* biofilm and planktonic cells and correlation with transcriptome profiling. Proteomics 6, 1867–1877.

Ribet, D., Hamon, M., Gouin, E., Nahori, M.A., Impens, F., Neyret-Kahn, H., Gevaert, K., Vandekerckhove, J., Dejean, A., Cossart, P., 2010. *Listeria monocytogenes* impairs SUMOylation for efficient infection. Nature 464, 1192–1195.

Rocourt, J., BenEmbarek, P., Toyofuku, H., Schlundt, J., 2003. Quantitative risk assessment of *Listeria monocytogenes* in ready-to-eat foods: the FAO/WHO approach. FEMS Immunol. Med. Microbiol. 35, 263–267.

Scallen, E., Hoekstra, R.M., Angulo, F.J., Tauxe, R.V., Widdowson, M., Roy, S.L., Jones, J.L., Griffin, P.M., 2011. Foodborne illness acquired in the United States – major pathogens. Emerg. Infect. Dis. 17, 7–15.

Schaumburg, J., Diekmann, O., Hagendorff, P., Bergmann, S., Rohde, M., Hammerschmidt, S., Jansch, L., Wehland, J., Karst, U., 2004. The cell wall subproteome of *Listeria monocytogenes*. Proteomics 4, 2991–3006.

Scortti, M., Monzo, H.J., Lacharme-Lora, L., Lewis, D.A., Vazquez-Boland, J.A., 2007. The PrfA virulence regulon. Microbes Infect. 9, 1196–1207.

Seng, P., Drancourt, M., Gouriet, F., La Scola, B., Fournier, P.E., Rolain, J.M., Raoult, D., 2009. Ongoing revolution in bacteriology: routine identification of bacteria by matrix-assisted laser desorption ionization time-of-flight mass spectrometry. Clin. Infect. Dis. 49, 543–551.

Sengupta, D., Chattopadhyay, M.K., 2013. Metabolism in bacteria at low temperature: a recent report. J. Biosci. 38, 409–412.

Seo, J., Savitzky, D.C., Ford, E., Darwin, A.J., 2007. Global analysis of tolerance to secretin-induced stress in *Yersinia enterocolitica* suggests that the phage-shock-protein system may be a remarkably self-contained stress response. Mol. Microbiol. 65, 714–727.

Singh, A.K., Ulanov, A.V., Li, Z., Jayaswal, R.K., Wilkinson, B.J., 2011. Metabolomes of the psychrotolerant bacterium *Listeria monocytogenes* 10403S grown at 37 °C and 8 °C. Int. J. Food Microbiol. 148, 107–114.

Sleator, R.D., Hill, C., 2010. Compatible solutes: the key to Listeria's success as a versatile gastrointestinal pathogen? Gut Pathog. 2, 20.

Sleator, R.D., Francis, G.A., O'Beirne, D., Gahan, C.G., Hill, C., 2003. Betaine and carnitine uptake systems in *Listeria monocytogenes* affect growth and survival in foods and during infection. J. Appl. Microbiol. 95, 839–846.

Sleator, R.D., Wemekamp-Kamphuis, H.H., Gahan, C.G., Abee, T., Hill, C., 2005. A PrfA-regulated bile exclusion system (BilE) is a novel virulence factor in *Listeria monocytogenes*. Mol. Microbiol. 55, 1183–1195.

Sleator, R.D., Banville, N., Hill, C., 2009. Carnitine enhances the growth of *Listeria monocytogenes* in infant formula at 7 °C. J. Food Prot. 72, 1293–1295.

Smiddy, M., Sleator, R.D., Patterson, M.F., Hill, C., Kelly, A.L., 2004. Role for compatible solutes glycine betaine and L-carnitine in listerial barotolerance. Appl. Environ. Microbiol. 70, 7555–7557.

Sorek, R., Cossart, P., 2010. Prokaryotic transcriptomics: a new view on regulation, physiology and pathogenicity. Nat. Rev. Genet. 11, 9–16.

Stavru, F., Archambaud, C., Cossart, P., 2011. Cell biology and immunology of *Listeria monocytogenes* infections: novel insights. Immunol. Rev. 240, 160–184.

Taormina, P.J., Beuchat, L.R., 2001. Survival and heat resistance of *Listeria monocytogenes* after exposure to alkali and chlorine. Appl. Environ. Microbiol. 67, 2555–2563.

Tessema, G.T., Moretro, T., Snipen, L., Heir, E., Holck, A., [illegible], K., Axelsson, L., 2012. Microarray-based transcriptome of *Listeria monocytogenes* adapted to sublethal concentrations of acetic acid, lactic acid, and hydrochloric acid. Can. J. Microbiol. 58, 1112–1123.

Thompson, A., Rowley, G., Alston, M., Danino, V., Hinton, J.C., 2006. Salmonella transcriptomics: relating regulons, stimulons and regulatory networks to the process of infection. Curr. Opin. Microbiol. 9, 109–116.

Toyofuku, M., Roschitzki, B., Riedel, K., Eberl, L., 2012. Identification of proteins associated with the *Pseudomonas aeruginosa* biofilm extracellular matrix. J. Proteome Res. 11, 4906–4915.

Trémoulet, F., Duché, O., Namane, A., Martinie, B., 2002. Comparison of protein patterns of Listeria monocytogenes grown in biofilm or in planktonic mode by proteomic analysis. FEMS Microbiol. Lett. 210, 25–31.

Trost, M., Wehmhoner, D., Karst, U., Dieterich, G., Wehland, J., Jansch, L., 2005. Comparative proteome analysis of secretory proteins from pathogenic and nonpathogenic Listeria species. Proteomics 5, 1544–1557.

Van de Velde, S., Delaive, E., Dieu, M., Carryn, S., Van Bambeke, F., Devreese, B., Raes, M., Tulkens, P.M., 2009. Isolation and 2-D-DIGE proteomic analysis of intracellular and extracellular forms of *Listeria monocytogenes*. Proteomics 9, 5484–5496.

van der Veen, S., Hain, T., Wouters, J.A., Hossain, H., de Vos, W.M., Abee, T., Chakraborty, T., Wells-Bennik, M.H., 2007. The heat-shock response of *Listeria monocytogenes* comprises genes involved in heat shock, cell division, cell wall synthesis, and the SOS response. Microbiology 153, 3593–3607.

van der Veen, S., van Schalkwijk, S., Molenaar, D., de Vos, W.M., Abee, T., Wells-Bennik, M.H., 2010. The SOS response of *Listeria monocytogenes* is involved in stress resistance and mutagenesis. Microbiology 156, 374–384.

Van Troys, M., Lambrechts, A., David, V., Demol, H., Puype, M., Pizarro-Cerda, J., Gevaert, K., Cossart, P., Vandekerckhove, J., 2008. The actin propulsive machinery: the proteome of *Listeria monocytogenes* tails. Biochem. Biophys. Res. Commun. 375, 194–199.

van Veen, S.Q., Claas, E.C., Kuijper, E.J., 2010. High-throughput identification of bacteria and yeast by matrix-assisted laser desorption ionization-time of flight mass spectrometry in conventional medical microbiology laboratories. J. Clin. Microbiol. 48, 900–907.

Velculescu, V.E., Zhang, L., Vogelstein, B., Kinzler, K.W., 1995. Serial analysis of gene expression. Science 270, 484–487.

Wang, L., Lin, M., 2007. Identification of IspC, an 86-kilodalton protein target of humoral immune response to infection with *Listeria monocytogenes* serotype 4b, as a novel surface autolysin. J. Bacteriol. 189, 2046–2054.

Wang, L., Lin, M., 2008. A novel cell wall-anchored peptidoglycan hydrolase (autolysin), IspC, essential for *Listeria monocytogenes* virulence: genetic and proteomic analysis. Microbiology 154, 1900–1913.

Wang, Z., Gerstein, M., Snyder, M., 2009. RNA-Seq: a revolutionary tool for transcriptomics. Nat. Rev. Genet. 10, 57–63.

Weeks, M.E., James, D.C., Robinson, G.K., Smales, C.M., 2004. Global changes in gene expression observed at the transition from growth to stationary phase in *Listeria monocytogenes* ScottA batch culture. Proteomics 4, 123–135.

Wemekamp-Kamphuis, H.H., Wouters, J.A., de Leeuw, P.P., Hain, T., Chakraborty, T., Abee, T., 2004. Identification of sigma factor sigma B-controlled genes and their impact on acid stress, high hydrostatic pressure, and freeze survival in *Listeria monocytogenes* EGD-e. Appl. Environ. Microbiol. 70, 3457–3466.

Wen, J., Anantheswaran, R.C., Knabel, S.J., 2009. Changes in barotolerance, thermotolerance, and cellular morphology throughout the life cycle of *Listeria monocytogenes*. Appl. Environ. Microbiol. 75, 1581–1588.

Wen, J., Deng, X., Li, Z., Dudley, E.G., Anantheswaran, R.C., Knabel, S.J., Zhang, W., 2011. Transcriptomic response of *Listeria monocytogenes* during the transition to the long-term-survival phase. Appl. Environ. Microbiol. 77, 5966–5972.

Wiedmann, M., Bruce, J.L., Knorr, R., Bodis, M., Cole, E.M., McDowell, C.I., McDonough, P.L., Batt, C.A., 1996. Ribotype diversity of *Listeria monocytogenes* strains associated with outbreaks of listeriosis in ruminants. J. Clin. Microbiol. 34, 1086–1090.

Wolfe, A.J., 2005. The acetate switch. Microbiol. Mol. Biol. Rev. 69, 12–50.

Wurtzel, O., Sesto, N., Mellin, J.R., Karunker, I., Edelheit, S., Becavin, C., Archambaud, C., Cossart, P., Sorek, R., 2012. Comparative transcriptomics of pathogenic and non-pathogenic Listeria species. Mol. Syst. Biol. 8, 583.

Zenewicz, L.A., Shen, H., 2007. Innate and adaptive immune responses to *Listeria monocytogenes*: a short overview. Microbes Infect. 9, 1208–1215.

Zhang, C.X., Creskey, M.C., Cyr, T.D., Brooks, B., Huang, H., Pagotto, F., Lin, M., 2013. Proteomic identification of *Listeria monocytogenes* surface-associated proteins. Proteomics 13, 3040–3045.

Zhu, K., Ding, X., Julotok, M., Wilkinson, B.J., 2005. Exogenous isoleucine and fatty acid shortening ensure the high content of anteiso-C15:0 fatty acid required for low-temperature growth of *Listeria monocytogenes*. Appl. Environ. Microbiol. 71, 8002–8007.

CHAPTER

CURRENT ISSUES IN FOODBORNE ILLNESS CAUSED BY *STAPHYLOCOCCUS AUREUS*

9

Mark E. Hart

Division of Microbiology (HFT-250), National Center for Toxicological Research, U.S. Food and Drug Administration, Jefferson, Arkansas, USA

1 INTRODUCTION

Scallan et al. (2011a,b), utilizing active, passive, and outbreak surveillance data obtained primarily for 2000-2008, estimated that of the approximately 48 million episodes of domestically acquired, foodborne-related illnesses that occur each year in the United States, 9.4 million are caused by 31 known pathogens. While over half (59%) of these illnesses are caused by viruses (predominantly noroviruses), approximately 13.5% are caused by *Bacillus cereus*, *Clostridium perfringens*, and *Staphylococcus aureus* (Scallan et al., 2011a,b). Foodborne-related illnesses caused by these three bacteria also resulted in an estimated 1522 hospitalizations and 32 deaths per year for the same time period (Scallan et al., 2011a,b). These cases may very well underestimate the true numbers of cases as the etiologies of foodborne illnesses rely primarily on voluntary reporting of cases as the result of outbreaks (Scallan et al., 2011a,b). Nevertheless, based upon the cases reported, it has also been estimated, using a basic cost of illness model, that the total economic cost of these bacteria alone is approximately $523 million, which represents 3% of the estimated $16.8 billion in total health-related costs of domestic foodborne illness with known causes that occur each year in the United States (Scharff, 2012).

Foodborne illness caused by *B. cereus*, *C. perfringens*, and *S. aureus* is primarily due to intoxication with one or more preformed exotoxins produced by these bacteria. In the case of *B. cereus*, a foodborne illness can be one of two varieties: a diarrheal illness caused by one or more toxins, primarily hemolysin BL, nonhemolytic enterotoxin, and cytotoxin K; or an emetic illness caused by the preformed toxin, cereulide (Logan, 2012; Stenfors Arnesen et al., 2008). While *C. perfringens* is capable of producing over 15 different toxins that contribute to a wide variety of diseases in both animals and humans, strains coding for the *C. perfringens* enterotoxin cause almost all foodborne-related illnesses associated with this organism (Lindström et al., 2011). The most formidable of these three is the nonspore-forming *S. aureus*, primarily because of the variety of exotoxins that any

Food Safety. http://dx.doi.org/10.1016/B978-0-12-800245-2.00009-5

one strain can produce, the relative heat and salt tolerance exhibited by most *S. aureus* strains, as well as the heat and protease resistant nature of most, if not all, of the food poisoning enterotoxins (Spaulding et al., 2013).

2 *STAPHYLOCOCCUS AUREUS*

S. aureus is a Gram-positive, facultatively anaerobic bacterium that resides primarily as a commensal colonizer on skin and mucous membranes of both humans and animals. Persistent carriage or colonization rates at any given time can be as high as 30% among healthy human individuals but as high as 90% among health care workers (Waldvogel, 1995). Therefore, it is not surprising that of the estimated two million hospitalizations each year that result in a nosocomial infection, *S. aureus*, along with the coagulase-negative staphylococci (CoNS) are the most common causative agents (Kallen et al., 2010; Klevens et al., 2007a,b; Sievert et al., 2013).

Strains of *S. aureus* are typically differentiated from other species by catalase, coagulase, and DNAse activity and fermentation of mannitol (Kloos and Bannerman, 1995). In the human clinical setting, the production of coagulase is usually used to distinguish between *S. aureus* and all other species, which are typically grouped together as the CoNS (Archer, 1995). However, species other than *S. aureus* subspecies *aureus* and *anaerobius* (De La Fuente et al., 1985; Kloos and Schleifer, 1986) produce coagulase even though their encounter in the human clinical setting is rare (Archer, 1995; Diekema et al., 2001; Huebner and Goldmann, 1999; Kloos and Schleifer, 1986; Kloos and Bannerman, 1994, 1995; Martin de Nicolas et al., 1995). These include *S. intermedius*, found primarily in association with dog, horse, mink, and pigeon (Hájek, 1976); *S. pseudointermedius*, associated primarily with dog and cat (Devriese et al., 2005); and *S. delphini*, a species first characterized in association with dolphin (Varaldo et al., 1988). Other species include *S. schleiferi* subsp. *coagulans*, associated with dog (Igimi et al., 1990); *S. hyicus* subsp. *hyicus*, primarily associated with pigs where it is known to cause infectious exudative epidermitis (Devriese et al., 1978; Kloos and Schleifer, 1986; Kloos and Bannerman, 1995; Saito et al., 1996); and *S. lutrae*, associated with otter (Varaldo et al., 1988).

S. aureus is one of the most resilient of the nonspore-forming bacteria. It is capable of surviving temperatures as high as 60 °C for 30 min and tolerating approximately 10% NaCl (Somerville and Proctor, 2009). This organism has the capacity to produce a wide variety of extracellular proteins, which have been shown to be involved with some aspect of disease and thus, by definition are virulence factors. These proteins are in the form of enzymes (e.g., proteases, nucleases, lipases, and hyaluronidases), immunotoxins (e.g., the superantigenic enterotoxins and toxic shock syndrome toxin-1), and cytotoxins (e.g., hemolysins and leukocidins) (Iandolo, 1990; Smeltzer et al., 2009). In addition, *S. aureus* also produces a large family of cell wall-associated proteins, collectively known as the *mi*crobial *s*urface *c*omponents *r*ecognizing *a*dhesive *m*atrix *m*olecules that promote adherence to various tissues (e.g., fibronectin-binding proteins), acquisition of iron (iron-responsive surface determinants), and evasion of the host's defense systems (e.g., protein A) (Clarke and Foster, 2006; Iandolo, 1990; Smeltzer et al., 2009). Consequently, with such an armament (greater than 30 virulence factors) that any one *S. aureus* strain is capable of producing, the large number of different kinds of diseases this organism is capable of causing is not surprising. These diseases range from skin abscesses to life-threatening endocarditis, meningitis, osteomyelitis, and pneumonia as well as toxemias, such as scalded skin and toxic shock syndrome and foodborne-related illnesses and economically important diseases, such as bovine mastitis and osteomyelitis in commercial broiler chickens (Bradley, 2002; McNamee and Smyth, 2000; Waldvogel, 2000).

Adding to the seriousness of staphylococcal infections is the occurrence of methicillin resistant *S. aureus* (MRSA) among not only hospital-acquired (HA) MRSA isolates but also the relatively recent emergence of community-associated (CA) MRSA in healthy individuals with no apparent classic risk factors (Boyce et al., 1983; Otto, 2012; Stryjewski and Chambers, 2008). Even now these rather simple distinctions are not sufficient as HA-MRSA are found in the community and CA-MRSA are found in hospital settings (David and Daum, 2010). A third significant MRSA type has recently emerged first in Europe among livestock (Armand-Lefevre et al., 2005; Voss et al., 2005) and now in North America (Huijsdens et al., 2006; Khanna et al., 2008; Smith et al., 2009) and Asia (Yu et al., 2008). Isolates of this type are collectively referred to as livestock-associated MRSA (LA-MRSA) (Fluit, 2012; Price et al., 2012).

MRSA strains have intrinsic resistance to all β-lactam antibiotics including flucloxacillin, cephalosporins, and carbepenems (Lindsay, 2013; Ryffel et al., 1994) and tend to acquire multiple other unrelated resistance determinants as well. In the case of most HA-MRSA infections, vancomycin has historically been the drug of choice (Maple et al., 1989). However, resistance to vancomycin has been reported both abroad and in the United States (Hiramatsu et al., 1997; Tenover et al., 1998). HA-MRSA and CA-MRSA accounts for approximately 25% of all bloodstream and lower-respiratory tract infections and almost 40% of all skin and soft tissue infections in the United States (Diekema et al., 2001). While recent reports (2008-2012) by the Active Bacterial Core Surveillance program of the Centers for Disease Control and Prevention (http://www.cdc.gov/abcs/reports-findings/surv-reports.html) indicate that the incidence of *S. aureus* invasive disease has trended down somewhat, the number of deaths in the United States in 2005 was estimated to be 18,650 (Klevens et al., 2007b). By comparison, that same year, approximately 16,000 people in the United States died of AIDS (Klevens et al., 2007b).

Both methicillin-susceptible *S. aureus* (MSSA) and MRSA have been associated with companion and food production animals (Armand-Lefevre et al., 2005; Fitzgerald, 2012; Fluit, 2012; Graveland et al., 2010; Nemati et al., 2008; Wendlandt et al., 2013). Regardless of methicillin resistance, *S. aureus* causes a variety of diseases in economically important livestock, such as cows, pigs, sheep, goats, poultry, and rabbits (Fitzgerald, 2012; Foster, 2012). The most significant of these is intramammary infection of dairy cattle leading to mastitis, which causes a substantial economic loss to the dairy industry worldwide (Bradley, 2002; Foster, 2012; Vanderhaeghen et al., 2010). *S. aureus* is also a major cause of septic arthritis, subdermal abscesses, and bacterial chondronecrosis with osteomyelitis in commercial broiler chickens (McNamee and Smyth, 2000). This organism is also a concern among small ruminants where sheep and goat cheeses are produced (Menzies and Ramanoon, 2001). *S. aureus* also causes mastitis, skin abscesses, and septicemia in rabbits, which is particularly important in countries of the European Union and in China where an extensive rabbit industries exists (Vancraeynest et al., 2006a,b).

Early phenotypic (e.g., biochemical, hemolytic, and coagulase, as well as phage typing) characterization schemes suggested that *S. aureus* strains were host specific (Cuny et al., 2010). With the advent of powerful and sophisticated DNA-based technologies (pulsed-field gel electrophoresis (PFGE), multilocus sequence typing (MLST), *spa*-typing, and others) it appeared that certain sequencing types (ST), as determined by MLST (which focuses on sequence divergence in seven unrelated housekeeping genes within the core genome (Enright et al., 2000)) and *spa* typing (amplification and sequencing of the *spa* gene (Harmsen et al., 2003)), are primarily the isolates found residing on or causing disease within a particular mammalian host (Fitzgerald, 2012; Lindsay, 2010; Lindsay et al., 2006; Musser et al., 1990; Sung et al., 2008). In other words, with exceptions, *S. aureus* strains are host specific. For example, using microarray technology, a large group of *S. aureus* strains from the United Kingdom, comprising strains from cows, horses, sheep, goats, and one camel, along with human hospital and community strains, demonstrated that the majority of lineages found in animals clustered separately from those found in humans, based upon conservation within the core variable regions of each lineage (Lindsay et al., 2006; Sung et al., 2008). Geographical distinctions occur as well. For example, the CA-MRSA strain currently epidemic in the United States is predominantly ST8 (pulse-field type, USA300), whereas in Europe, the predominant ST is ST80 (European) (Kennedy et al., 2008; Tenover et al., 2006; Tristan et al., 2007; Witte et al., 2007a). Notable exceptions, based on the recent data obtained with MLST and whole-genome sequencing, include several STs that appear to have a wider host range. For example, the recently emerged LA-MRSA ST398 strain (Stegger et al., 2014) colonizes pigs, veal calves, dogs, horses, and chickens, but rarely causes disease in these animals (Cuny et al., 2008, 2010; Fitzgerald, 2012; Graveland et al., 2010; Nemati et al., 2008; Witte et al., 2007b). Poultry-associated *S. aureus* ST5 appears to be the result of a human-to-poultry adaptation that occurred approximately 40 years ago (Lowder et al., 2009). ST5 MRSA has also been found as a predominant strain among pig farms, pig farmers of contaminated farms, and veterinary students with limited pig farm exposure on 12 of 40 pig farms examined in Iowa (Frana et al., 2013). This strain, along with ST398, was also the predominant isolate found across a farm-to-retail continuum involving 10 geographically distinct pig farms in Ohio, their associated lairages and abattoirs, and their final retail destinations (Molla et al., 2012).

While MRSA infections in domestic animals were first reported from cases of mastitis in dairy cows in 1972 (Devriese et al., 1972), MRSA infections in domestic animals have been somewhat rare, with only sporadic cases in horses, chickens, dogs, and cats (Gortel et al., 1999; Hartmann et al., 1997; Seguin et al., 1999). MRSA appears to be somewhat rare in staphylococcal food poisoning as well. To date, there has only been one report of a gastrointestinal illness outbreak in humans caused by a MRSA strain (Jones et al., 2002). However, since staphylococcal food poisoning is generally self-limited and usually has a short duration, antibiotic therapy is not typically used. In fact, rarely are outbreak strains even assessed for antibiotic susceptibility so it is quite possible that MRSA strains causing staphylococcal food poisoning are underrepresented.

3 EMERGENCE OF LA-MRSA

The emergence of new MRSA infections among livestock animals and associated farmers (the so-called LA-MRSA) is of great concern as these sources could potentially serve as reservoirs for zoonotic infections in humans (Cuny et al., 2010). An important case in point is the most recent occurrence of human infections with *S. aureus* ST398, a pig-associated strain of MRSA (Rasigade et al., 2010; van der Mee-Marquet et al., 2011). This strain was originally classified as a non-typable (NT)-MRSA swine isolate in the Netherlands and France (Armand-Lefevre et al., 2005; Huijsdens et al., 2006; van Loo et al., 2007a; Voss et al., 2005) by the fact that these strains are resistant to restriction endonuclease DNA digestion with *Sma*I (Bens et al., 2006). This enzyme is typically used in conjunction with PFGE, the "gold standard" for *S. aureus* epidemiological typing (Bannerman et al., 1995; Mulvey et al., 2001; Murchan et al., 2003). Strains of this ST can, however, be typed by PFGE using the *Sma*I isoschizomer, *Cfr*9I (Argudin et al., 2010; Bosch et al., 2010; Molla et al., 2012).

As stated earlier, the ST398 *S. aureus* isolate was not only found among pigs (e.g., in the Netherlands nearly 40% of 509 pigs representing nine different abattoirs were positive for ST398), but was also more prevalent found among pig and cattle farmers than among nonfarming human controls (de Neeling et al., 2007; van Loo et al., 2007a). In addition, examination of NT-MRSA human case isolates in the Netherlands for a 2-year period showed that the frequency increased from 0% in 2002 to greater than 21% in mid-2006 (van Loo et al., 2007a). This NT-MRSA *S. aureus* strain primarily clustered with pig farming and when typed by MLST was found to be ST398 (van Loo et al., 2007a). Carriage rates were found to be high in the Netherlands as well, with as much as 35% of persons associated with pigs and veal calves being carriers of *S. aureus* ST398 (van den Broek et al., 2009).

Transmission of MRSA between animals and humans (e.g., horses and companion animals and individuals associated with their care) is not new, but the MRSA isolates, in most cases, were determined to be of human origin and thus, actually represent an initial human-to-animal transmission (Cefai et al., 1994; Duquette and Nuttall, 2004; Loeffler et al., 2005; Weese et al., 2005). However, the emergence of a MRSA strain (e.g., members of the 398 clonal complex (CC)) that appears to have an animal origin is of concern. Therefore, studies have been conducted to determine the transmissibility of livestock ST398 strains to humans beyond those individuals associated with contaminated farms. By definition a CC is a group of STs "in which every ST shares at least five of seven identical alleles of the core sequence with at least one other ST in the group. The ST that gives rise to each clonal complex (the clonal ancestor) initially will diversify to produce variants that differ at only one of the seven loci" (Day et al., 2001). In this case, ST398 serves as the clonal ancestor for CC398.

While the first report of a mostly livestock-associated ST398 strain occurred among pig farmers in France and was found to be predominantly MSSA (Armand-Lefevre et al., 2005), the vast majority of studies on ST398, have come from the Netherlands (Smith and Pearson, 2011; Wendlandt et al., 2013). These studies in the Netherlands began with the initial report of a MRSA cultured from a preoperative screening of a 6-month-old child admitted to the hospital for thoracic surgery (Voss et al., 2005). It was subsequently shown that the child's parents, who lived on a farm and raised pigs, were colonized with the identical MRSA isolate as determined by *spa* typing and random amplified polymorphic DNA analysis. Most interesting, as it relates to transmissibility, is occurrence of additional MRSA cases in which a veterinarian, his son, and the nurse who treated the son were colonized with the same *spa* type MRSA isolate (Voss et al., 2005). This report, while small in scale, may represent the first report of transmission between pig to veterinarian, veterinarian to family member, and family member to nurse in a hospital setting of a NT-MRSA isolate with *spa* type t108, which is predominantly found in livestock-associated ST398 (Voss et al., 2005).

In a more direct systematic approach to assess ST398 transmissibility, Cuny et al. (2009) sampled the nares of daily pig workers and their nonexposed family members as well as veterinarians who attended to the same pig farms and their nonexposed family members, all from 47 geographically distinct colonized pig farms in Germany. Isolation of MRSA and subsequent typing determined that 86% and 45% of pig farmers and veterinarians, respectively, carried MRSA strains of the CC398 group. In contrast, only 4.3% and 9% of their respective nonexposed family members were found positive for the same CC398 MRSA strain (Cuny et al., 2009). Similar findings were observed in Denmark and Belgium where ST398 MRSA was found among veterinarians in association with livestock but not in small animals (Garcia-Graells et al., 2012). Veterinarians' positive for ST398 MRSA was not restricted to those with recent exposure to pigs, suggesting that veterinarians could potentially become chronic carriers of MRSA ST398 (Garcia-Graells et al., 2012).

A study conducted in the Netherlands where nasal swabs were taken before, directly after, and 24 h after visiting either pig or veal calf farms positive for MRSA by "persons not exposed to livestock on a daily basis," determined that only 17% of the total number of visits resulted in persons positive for MRSA directly after a farm visit (van Cleef et al., 2011). However, 94% of the positive MRSA visitors were MRSA negative after 24 h (van Cleef et al., 2011). In this study, all but one of the MRSA isolates examined were *spa* type t011, t108, and t567, the *spa* types primarily associated with ST398 isolated from pigs (van Cleef et al., 2011). Similar observations conducted in the Netherlands found that persons with transient or short-term exposure to either pigs (van den Broek et al., 2009) or veal calves (Graveland et al., 2010) and who became positive for MRSA, were also negative for the same MRSA strain associated with the farm shortly after their visit or within two days after exposure. In both studies, the predominant MRSA ST found in both animals and humans was ST398.

A more recent study in the United States demonstrated that 22% of 30 veterinary students who were initially MRSA negative became positive after visiting (nasally sampled approximately 3-4 h after exposure) MRSA-positive pig farms in Iowa (Frana et al., 2013). Surprisingly, none of the students were still MRSA positive by 24 h postvisit (Frana et al., 2013). Although *spa* types associated with ST398 were found (15% among pigs), none were detected in students after exposure (Frana et al., 2013). However, the predominant *spa* type most commonly (84%) detected among the students was associated with ST5, suggesting a possible expansion of LA-MRSA to include ST5 (Frana et al., 2013).

Collectively, these studies suggest that the LA-MRSA ST398 strain is mostly confined to livestock, farmers, and the farmers' family members. However, numerous reports indicate that ST398 has caused infections in humans worldwide albeit less frequently with assorted disease syndromes ranging from minor skin infections to more life-threatening bacteremia and pneumonia (Smith and Pearson, 2011). These cases have not only been associated with livestock (Graveland et al., 2010) but have also occurred in cases with no known livestock contact (Bhat et al., 2009; Uhlemann et al., 2012a,b). These observations suggest a broader transmissibility capacity than originally thought. Recent phylogenetic studies have identified a number of genetic differences between livestock- and human-associated ST398 clinical isolates that could potentially account for ST398 human infections in individuals with no known livestock exposure. Current phylogenetic studies have characterized the CC398 group using whole-genome sequencing, single nucleotide polymorphism, and microarray analysis of CC398 MSSA and MRSA strains (McCarthy et al., 2011; Price et al., 2012; Stegger et al., 2013). After whole-genome sequencing of a group of international isolates comprised of 49 MRSA and 40 MSSA, Price et al. (2012) showed that the CC398 group is likely to have originated in humans as a MSSA strain and then later acquired antibiotic resistance from mobile genetic elements (MGEs) coding for methicillin (SCC*mec* IV, V, and VII-like) and tetracycline (*tet*(M)) resistance once it moved into livestock. Only 6 of 89 genomes contained the *lukF-lukS* genes coding for Panton-Valentine leukocidin (PVL) and all these strains were human associated. The PVL genes are carried predominantly by CA-MRSA strains although the exact role that PVL has in human infection remains controversial (David and Daum,

2010; Otto, 2012). In addition, phylogenetically grouped human-associated MSSA strains had also acquired the human immune evasion cluster (IEC) by way of the lysogenic phage, φSa3, while only one livestock-associated isolate was shown to contain the phage (Price et al., 2012). What is significant about the IEC is that it consists of genes that code for the staphylococcal complement inhibitor (*scn*), chemotaxis inhibitory protein (*chp*), staphylokinase (*sak*), and staphylococcal enterotoxin A (*sea*); gene products shown to be important for human adaptation by protecting *S. aureus* from innate immune defense systems (van Wamel et al., 2006). In a second study, 17 *S. aureus* CC398 genomes from Scotland were analyzed alongside 87 previously published human- and livestock-associated CC398 genomes (Ward et al., 2014). In this case, phylogenetic analysis of the core genome indicated that both human- and livestock-associated CC398 emerged at about the same time, with 70% of the livestock-associated CC398 acquiring the *mecA* determinant for methicillin resistance and 99% acquiring the *tet*(M) determinant for tetracycline resistance (Ward et al., 2014). Human-associated CC398 strains were significantly more likely than the livestock-associated CC398 to be lysogenized by φSa3 (coding for human IEC genes), to contain the PVL genes, and be susceptible to methicillin and tetracycline (Ward et al., 2014).

Additional studies (McCarthy et al., 2011, 2012) using microarray analysis also concluded that the core genomes between 32 human- and 44 pig-associated CC398 isolates are quite similar (McCarthy et al., 2011). However, there was considerable variation in the distribution of MGEs among all CC398 strains (McCarthy et al., 2011, 2012). In addition, only human invasive CC398 strains carried φSa3, and while low, PVL carriage was only present in human CC398 strains (McCarthy et al., 2011). This variation may be explained by what has been reported to be a deficiency in the restriction modification systems in these strains, which would allow an increase in horizontal gene transfer (Stegger et al., 2011; Valentin-Domelier et al., 2011; Waldron and Lindsay, 2006). In a second study using a more geographical diverse group of isolates, microarray analysis demonstrated that the φSa3 lysogenic phage was found in most CC398 MSSA strains from humans without pig contact regardless of the year isolated and occurred in strains across all geographical boundaries examined with one exception (McCarthy et al., 2012). Strains from humans, without pig contact and without the φSa3 phage were phylogenetically grouped with pig and human isolates associated with pigs and all were *tet*(M) positive (McCarthy et al., 2012).

Similar findings were observed in a recent bloodstream infection (BSI) survey conducted in France (Valentin-Domelier et al., 2011). In this study, 31 health care institutions were surveyed over a 4-year period. Of the 18 ST398 *S. aureus* strains isolated from BSIs of patients with no livestock exposure, 17 were MSSA. All 18 strains exhibited *spa* types associated with ST398 but none were of the type (t011, t034, and t108) primarily found associated with ST398 strains isolated from pigs. Both PVL genes and the prophage, φSa3, were detected in all 18 strains by PCR (Valentin-Domelier et al., 2011).

The results from these studies clearly represent the diversity that occurs among strains of *S. aureus* and the all-important shift in fitness to accommodate a change from one host to another. Whether *S. aureus* CC398 diverged as two separate clades, one adapted to humans and one adapted to animals (Ward et al., 2014), or whether one (human) led to the second (livestock), with the loss of phage-carried human virulence genes and the acquisition of methicillin and tetracycline resistance determinants (Price et al., 2012), will require additional studies of much larger group of strains from a variety of environments worldwide. As one might expect, the LA-MRSA *S. aureus* CC398 is not alone in the ability to adapt. Recent reports of an increase in human infections by the leading cause of bovine mastitis in Asia, Europe, and North and South America have led to a phylogenetic study indicating that human *S. aureus* CC97 strains originated from livestock and in the process of adapting to the human host acquired either the SCC*mec* IV or V cassette and the temperate phage, φSa3, containing one of two variants of the IEC (Spoor et al., 2013).

It is particularly alarming that the adaptation from one host to another appears to be dependent upon the acquisition or loss of MGEs that code for key elements necessary for survival in new environment (Price et al., 2012; Spoor et al., 2013; Ward et al., 2014). It is quite possible that in the case of the LA-MRSA CC398 strains with resistance to both methicillin and tetracycline found associated with humans in close association with livestock, would eventually acquire, by MGEs and repeated colonization of livestock-associated humans (Wright, 1982), certain genetic traits (additional antibiotic resistance and virulence factors) that would allow *S. aureus* to colonize both hosts and become a more formidable zoonotic agent. As mentioned, this may easily be accomplished because ST398 strains are deficient in one or more restriction modification systems (Stegger et al., 2011; Valentin-Domelier et al., 2011; Waldron and Lindsay, 2006). This adaptation process may have already occurred as PVL-positive ST398 MRSA strains have been isolated from severe cases of community acquired infections (Stegger et al., 2010; Sung et al., 2008). Most ominous is the finding that these PVL-positive ST398 MRSA strains are also lysogenized with φSa3 phage coding for the human-specific virulence factors *sak*, *scn*, and *chp*.

4 *S. AUREUS* AND FOODS

S. aureus is found on the mucous membranes and skin of mammals and birds, in the air, water, raw milk, and other foods, and on numerous environmental surfaces. Therefore, it is not surprising that there is a great deal of concern for the safety of foods, when both humans and animals can serve as a source of the contaminating agent. Recent surveillance data suggest that the concern is warranted; from 22.5% to 64.8% of retail beef, pork, chicken, and turkey meats in five different geographical locations in the United States were contaminated with *S. aureus* (Bhargava et al., 2011; Kelman et al., 2011; O'Brien et al., 2012; Waters et al., 2011). Most alarming is the relatively recent isolation of MRSA from food animals (Bhargava et al., 2011; Buyukcangaz et al., 2013; Jackson et al., 2013; Kelman et al., 2011; Martins et al., 2013; O'Brien et al., 2012; Pu et al., 2009; Waters et al., 2011). The earliest known report, from the Netherlands, found *S. aureus* among nearly 50% of the retail meat samples that included pork and beef; two were MRSA and one of these was ST398 (van Loo et al., 2007b). A second more recent and broader scope study, also in the Netherlands, examined 2217 samples of retail beef, pork, veal, lamb, chicken, turkey, fowl, and game and found that 11.9% of the samples were contaminated with MRSA and the majority (85%) of these strains were *spa* types associated with ST398 (de Boer et al., 2009). This occurred in a country with one of the lowest if not the lowest rate of HA-MRSA in the world, primarily due to an aggressive "search and destroy" policy, which involves prescreening for high-risk MRSA patients and isolating them until they are MRSA free (van Rijen et al., 2008). While a larger number of the reported staphylococcal food poisoning outbreaks can be traced back to a human source harboring *S. aureus* producing certain staphylococcal enterotoxins (SEs), animals have the potential to be a source of primary contamination as well (Wendlandt et al., 2013). For example, *S. aureus* is a significant cause of mastitis in cows, goats, and ewes and where raw milk is used to make cheeses, the potential for contamination exists (Foster, 2012; Jorgensen et al., 2005a,b; Meyrand et al., 1998; Normanno et al., 2007; Vanderhaeghen et al., 2010).

As one might expect, strains from the *S. aureus* CC398 group have been found in retailed meats and milk worldwide (Agerso et al., 2012; de Boer et al., 2009; Feßler et al., 2011, 2012; O'Brien et al., 2012; Wendlandt et al., 2013). Even though strains of the LA-MRSA CC398 group tend to be a homogenous group (Hallin et al., 2011) in that they are comprised of a limited number of *spa* and SCC*mec* types that cannot to be PFGE typed by conventional methods, and though the *tet*(M) gene is present in the vast majority of strains (Vanderhaeghen et al., 2010), very few studies have actually followed the strains from the farm to their final retail destination (Beneke et al., 2011; Broens et al., 2011; Molla et al., 2012). The most recent of these studies determined the prevalence of LA-MRSA in pigs and pork from 10 different farms involving 240 market-aged pigs in Ohio (Molla et al., 2012). Swab samples were collected from these pigs from farm to abattoir to their final retail destination as well. Approximately 3% of the pigs from 50% of the farms were MRSA positive; ST5 and ST398 were the predominant strains detected at all stages along the farm to processing and retail continuum (Molla et al., 2012).

What is most remarkable is that practically all of the LA-MRSA strains characterized to date, particularly the ST398 group, do not appear to code for any of the known SEs (Huber et al., 2011; McCarthy et al., 2011, 2012; Molla et al., 2012; Price et al., 2012; Schijffelen et al., 2010; Ward et al., 2014). However, there is one report where genes for SEs B, K, and Q have been detected by microarray analysis in 4 of 54 MRSA ST398 strains isolated from geographically diverse pig farms in Germany (Kadlec et al., 2009). Like most SEs, these genes (*seb*, *sek*, and *seq*) are coded for by MGEs (see Table 1) and whether or not the acquisition of these virulence determinants by LA-MRSA may eventually pose a threat to humans is not known.

Table 1 Staphylococcal Enterotoxins

Toxin	Molecular Mass (Da)	Genetic Loci	Emetic Activity[a]	References
SEA	27,100	Prophage	+	Betley and Mekalanos (1985) and Borst and Betley (1994)
SEB	28,336	Chromosome, plasmid, pathogenicity island	+	Altboum et al. (1985), Jones and Khan (1986), Shafer and Iandolo (1978), and Shalita et al. (1977)
$SEC_{1\text{-}3}$	~27,500	Plasmid	+	Altboum et al. (1985), Bergdoll et al. (1965), Bohach and Schlievert (1987), Fitzgerald et al. (2001), and Hovde et al. (1990)
SED	26,360	Plasmid	+	Bayles and Iandolo (1989), Casman et al. (1967), and Chang and Bergdoll (1979)
SEE	26,425	Prophage	+	Bergdoll et al. (1971) and Couch et al. (1988)

Continued

Table 1 Staphylococcal Enterotoxins—cont'd

Toxin	Molecular Mass (Da)	Genetic Loci	Emetic Activity[a]	References
SEG	27,043	*enterotoxin gene cluster* (*egc*), chromosome	+	Jarraud et al. (2001) and Munson et al. (1998)
SEH	24,210	Transposon	+	Cusumano et al. (2014), Noto and Archer (2006), Ren et al. (1994), and Su and Wong (1995, 1996)
SEI	24,928	*egc*, chromosome	+	Jarraud et al. (2001) and Munson et al. (1998)
SE*l*J	28,565	Plasmid	NK	Zhang et al. (1998)
SEK	25,539	Pathogenicity island	+	Omoe et al. (2013) and Orwin et al. (2001)
SE*l*L	25,219	Pathogenicity island	+/–	Omoe et al. (2013) and Orwin et al. (2003)
SEM	24,842	*egc*, chromosome	+	Jarraud et al. (2001) and Omoe et al. (2013)
SEN	26,067	*egc*, chromosome	+	Jarraud et al. (2001) and Omoe et al. (2013)
SEO	26,777	*egc*, chromosome	+	Jarraud et al. (2001) and Omoe et al. (2013)
SEP	26,608	Prophage	+	Kuroda et al. (2001), Omoe et al. (2013), and Omoe et al. (2005)
SEQ	25,076	Pathogenicity island	+	Diep et al. (2006), Jarraud et al. (2002), and Omoe et al. (2013)
SER	27,049	Plasmid	+	Omoe et al. (2003)
SES	26,217	Plasmid	+	Ono et al. (2008)
SET	22,614	Plasmid	+	Ono et al. (2008)
SE*l*U	27,192	*egc*, chromosome	NK	Letertre et al. (2003)
SE*l*U_2	26,672	*egc*, chromosome	NK	Thomas et al. (2006)
SE*l*V	24,997	*egc*, chromosome	NK	Thomas et al. (2006)
SE*l*W[b]	17,824	Chromosome	NK	Okumura et al. (2012)
SE*l*X	19,343	Chromosome	NK	Wilson et al. (2011)

[a] *+, positive reaction; +/–, Omoe et al. (2013) demonstrated an emetic response when given to* Macaca fascicularis *while Orwin et al. (2003) did not observe an emetic response when given to* Macaca nemestrina*; NK, not known.*
[b] *Superantigenic properties of the putative SElW have not been determined.*

5 COAGULASE-NEGATIVE STAPHYLOCOCCI

Currently, the enterotoxigenic potential of CoNS species remains inconclusive (Becker et al., 2014; Doyle et al., 2012; Le Loir et al., 2003, Madhusoodanan et al., 2011; Park et al., 2011; Podkowik et al., 2013). A number of studies using PCR and whole genome sequencing have detected enterotoxin and enterotoxin-*like* genes in a number of CoNS species, but very few studies have demonstrated the production of toxins (Doyle et al., 2012; Le Loir et al., 2003; Park et al., 2011; Podkowik et al., 2013). For example, Park et al. (2011) examined the enterotoxin gene profiles of 263 CoNS strains isolated from bovine intramammary infections and determined by multiplex PCR that approximately 31% of the clinical isolates, representing 11 different species, contained one or more genes for enterotoxins (Park et al., 2011). Of the 18 enterotoxin genes examined, one or more genes were amplified from all CoNS species except one, *S. hominis* (Park et al., 2011). Amplified products for genes for staphylococcal enterotoxin B (SEB), SE*l*N, and SE*l*Q were predominantly found in *S. chromogenes*, *S. haemolyticus*, *S. sciuri* subsp. *carnaticus*, *S. simulans*, *S. succinus*, and *S. xylosus* (Park et al., 2011). In contrast, Rosec and Gigaud (2002) examined 332 foodborne staphylococci and none of the CoNS classified species coded for known enterotoxin genes, which included the classical SEA-SEE as well as the relatively novel SEG-SEJ genes. Nevertheless, Bautista et al. (1988) used an enzyme-linked immunosorbent assay (ELISA) to detect the classical enterotoxins SEA, SEC, and SED from CoNS isolated from sheep's milk. These species included *S. cohnii*, *S. epidermidis*, *S. xylosus*, and *S. haemolyticus* (Bautista et al., 1988). However, there has only been one report of a staphylococcal species other than *S. aureus* responsible for a staphylococcal food poisoning outbreak: the coagulase-positive *S. intermedius* (Becker et al., 2001; Khambaty et al., 1994). What may be at the core of the problem, not only in assessing enterotoxin production by CoNS species but also in the detection of SEs outside the classical SEA-SEE group in *S. aureus*, is the lack of available diagnostic reagents for the detection of novel enterotoxin proteins. However, ELISA's have been described for SEH (Su and Wong, 1996) and for SEG and SEI (Omoe et al., 2002). Although a number of the novel toxins SEG-SEI and SER-SET have been shown to be emetic after oral administration to primates (Omoe et al., 2013) among this group only SEH-producing strains of *S. aureus* have been involved with a staphylococcal food poisoning outbreaks (Ikeda et al., 2005; Jorgensen et al., 2005a; Pereira et al., 1991; Su and Wong, 1996).

6 STAPHYLOCOCCAL FOOD POISONING

Staphylococcal food poisoning is an intoxication by one or more enterotoxins produced primarily by *S. aureus* as a preformed toxin ingested from foods contaminated by a sufficient number of *S. aureus* cells as a result of being improperly prepared and/or stored (Le Loir et al., 2003; Schelin et al., 2011). Because *S. aureus* is found primarily on the skin and mucous membranes of warm-blooded animals, it is not surprising that the primary source of food contamination is food handlers (Le Loir et al., 2003; Schelin et al., 2011). However, this organism can also be endemic in processing environments (Borch et al., 1996). Recently, data obtained from the Foodborne Disease Outbreak Surveillance System (http://www.cdc.gov/foodsafety/fdoss/data/annual-summaries/index.html) for the years 1998-2008 were used to determine that, of the 1229 confirmed and suspected foodborne outbreaks caused by *S. aureus*, *C. perfringens*, and *B. cereus* in the United States, *S. aureus* accounted for almost 40% (Bennett et al., 2013). In this surveillance period, cases of vomiting and/or diarrhea with an onset incubation period of 0.5-8 h were confirmed as foodborne *S. aureus* outbreaks when either the organism of the same phage type was isolated from stool or vomitus of two or more ill persons or an enterotoxin was detected in the epidemiologically implicated food or a concentration of 10^5 viable *S. aureus* cells per gram of epidemiologically implicated food, was found (http://www.cdc.gov/outbreaknet/references_resources/guide_confirming_diagnosis.html). Confirmed *S. aureus* foodborne outbreaks exhibited a 4-h median from ingestion to onset, of an illness that typically lasted approximately 15 h. Undoubtedly, this short duration of illness contributes to the belief that staphylococcal food poisoning is significantly underreported and underdiagnosed (Doyle et al., 2012; Scallan et al., 2011b). The most common manifestations exhibited were abdominal cramps (72% of cases), vomiting (87% of cases), and diarrhea (89% of cases). While rare (9% of cases), *S. aureus* was the only etiological agent of the three that was reported to cause fever. Foods commonly implicated (55% of cases) in confirmed *S. aureus* outbreaks included meat and poultry, with pork (67%) and ham (41%) as primary sources (Bennett et al., 2013; Doyle et al., 2012). These foods were most often eaten at restaurants and delicatessens. Errors in food processing and preparation accounted for 93% of the outbreaks. These errors included insufficient time or temperature in the initial cooking of the foods, reheating the prepared foods, or the incubation of prepared foods for several hours at room or outdoor temperatures prior to consumption. Other errors included slow cooling of foods once prepared and either insufficient time or improper temperature during hot and cold holding. In over half of the *S. aureus* food outbreaks, contamination by a food worker was reported. In addition, foodborne outbreaks caused by *S. aureus*, occurred when foods were cross-contaminated by processing equipment or utensils improperly cleaned (67%) or stored in contaminated environments (39%).

Staphylococcal food poisoning is a worldwide concern as most foods can potentially be contaminated with *S. aureus* by human errors in some aspect of food preparation (Doyle et al., 2012; Hennekinne et al., 2012). This important fact, demonstrating the wide variety of contaminated food sources and the worldwide effect this organism has on foodborne illnesses, was presented in a recent review that tabulated a select group of worldwide foodborne outbreaks dating back to the first description of a staphylococcal foodborne illness in 1884 (Hennekinne et al., 2012).

The effective dose of an enterotoxin that causes intoxication in humans is uncertain. However, in a staphylococcal food poisoning outbreak involving more than 800 schoolaged children who consumed half-pint cartons of 2% chocolate milk in a school district in the United States (Evenson et al., 1988), it was estimated that from 94 to 184 ng of staphylococcal enterotoxin A (SEA) was consumed by each student who became ill (Evenson et al., 1988). A more recent study of a staphylococcal food poisoning outbreak involving over 13,000 people in Japan who consumed dairy products manufactured using powdered skim milk obtained from a single location, the outbreak was determined to be caused by intoxication with an estimated dose of SEA between 20 and 100 ng per individual (Asao et al., 2003). One of the earliest studies conducted to determine an effective dose involved feeding partially (approximately 50%) purified SEB to three human volunteers (Raj and Bergdoll, 1969). Onset of symptoms (nausea, abdominal cramping, followed by vomiting and diarrhea) occurred as early as 3 h postingestion of approximately 20-25 μg of enterotoxin (Raj and Bergdoll, 1969). Although only three volunteers were involved, duration, severity, and onset were observed to be dependent upon volunteer weight and amount of toxin ingested (Raj and Bergdoll, 1969). In general, staphylococcal food poisoning outbreaks typically have reported the presence of 1-5 μg of enterotoxin consumed. However, consumption of as little as 0.01 μg or less of enterotoxin has been associated with staphylococcal food poisoning (Gilbert and Wieneke, 1973; Jablonski and Bohach, 1997). The number of actual *S. aureus* cells needed for production of an effective dose of enterotoxin is dependent upon many of the same factors that affect *S. aureus* growth (Schelin et al., 2011). However, it has been reported that an inoculum of as little as 1000 cells/gram of Camembert-type cheeses made from raw goat's milk generated detectable enterotoxin levels by day 6 of the cheesemaking process (Meyrand et al., 1998). SEA levels by day 6 were approximately 0.5 ng per gram of cheese and rose to 3.2 ng per gram of cheese at the end of the cheese ripening process (approximately 41 days) as detected by ELISA (Meyrand et al., 1998).

7 STAPHYLOCOCCAL ENTEROTOXINS

Staphylococcal food poisoning is dependent upon sufficient amounts of SEs and sufficient amounts of SEs, in general, are dependent upon *S. aureus* growth. Obviously, growth is dependent upon a suitable food source. Unfortunately, growth of *S. aureus* may occur in numerous food matrices that encompass a wide array of environmental factors (Schelin et al., 2011). Limitations of *S. aureus* growth are quite broad with respect to some of the more critical factors. For example, *S. aureus* can proliferate within wide limiting ranges of water activity (0.83 ≥ 0.99), NaCl concentration (0-20%), pH (4-10), and temperature (6-48 °C) typical for many foods (Schelin et al., 2011). Staphylococcal enterotoxin production follows a similar pattern, although the ranges tend to be somewhat narrower (Schelin et al., 2011). SE production is not always dependent upon *S. aureus* growth, as SE production by nonreplicating cells has been observed in ham products (Wallin-Carlquist et al., 2010).

The SEs are a part of a family of exotoxins collectively known as the pyrogenic toxin superantigens (PTSAgs) primarily defined by their ability to be pyrogenic, superantigenic, and their ability to substantially enhance endotoxin (Gram-negative lipopolysaccharide) lethality in rabbits by as much as 100,000-fold (Dinges et al., 2000; Marrack and Kappler, 1990; McCormick et al., 2001). To date, this family of PTSAgs includes 25 serologically distinct exoproteins produced by strains of *S. aureus* and 11 exoproteins produced by *Streptococcus pyogenes* (Dinges et al., 2000; Hennekinne et al., 2012). Most of these PTSAgs are remarkably resistant to heat (particularly SEA, which is by far the most commonly associated with foodborne outbreaks), being biologically active after boiling for 1 h and resistant to stomach acid and proteolytic cleavage by gastrointestinal proteases like pepsin (Spaulding et al., 2013).

This entire family of toxins shares similarity both in structure and function and shares a common phylogeny (Balaban and Rasooly, 2000; Dinges et al., 2000; McCormick et al., 2001; Spaulding et al., 2013). These nonglycosylated, low-molecular mass exoproteins are secreted by all human-pathogenic *S. aureus* strains (Spaulding et al., 2013). All, including toxic shock syndrome toxin-1, are potent superantigens as they indiscriminately bind, as intact proteins, to the major histocompatibility II molecules on the surface of antigen-presenting cells and to certain T-cell receptor proteins (Vβ) on the surface of T-cells without processing (Llewelyn and Cohen, 2002). This interaction, which involves as much as 50% of the entire host T-cell population, leads to an uncontrolled expansion of T-cells, which results in a massive production of pro-inflammatory cytokines that is aptly described as a "cytokine storm" (McCormick et al., 2001; Spaulding et al., 2013). This uncontrolled release of cytokines is believed to be the cause of many of the signs and symptoms associated with toxic shock syndrome, namely hypotension, multiorgan failure, and death (McCormick et al., 2001).

In addition to their superantigenic properties, many of the SEs have emetic activity demonstrated by their ability to induce vomiting when ingested by humans or orally administered to primates (Lina et al., 2004). Those SEs without emetic activity in a primate model or whose emetic activity has not been determined are classified as staphylococcal enterotoxin-like (SE*l*) superantigens (Lina et al., 2004). To date, a total of 25 SEs (including subtypes defined by having more than 90% amino acid sequence similarity to one of the SEs or SE*l*s) have been discovered (Table 1). SEs A, B, $C_{1\text{-}3}$, D, and E are referred as the "classical SEs" as they were all identified as a consequence of staphylococcal food poisoning and classified based upon their serological differences (Table 1) (Bergdoll et al., 1959, 1965, 1971; Casman, 1960; Casman et al., 1967). The majority of reported SFP outbreaks are associated with the classical SEs and of these SEA is the most common cause (Cha et al., 2006; Kerouanton et al., 2007; Wieneke et al., 1993). The predominance of SEA in most SFP outbreaks may be due to its extraordinary ability to resist heat and proteolytic digestion. Indeed it has been reported that SEA retained its activity in low-fat milk even after three successive pasteurizations (defined in this case, as 130 °C for 2 or 4 s) (Asao et al., 2003). These classical enterotoxins, along with the staphylococcal toxic shock syndrome toxin-1 (which has no emetic activity), and as many as 11 streptococcal proteins, have been the mainstay for much of what is known about bacterial proteins with superantigenicity (Lina et al., 2004; Spaulding et al., 2013). The remaining SEs and the SE*l*s are termed novel, and apart from SEH; their role in causing staphylococcal food poisoning remains to be determined (Ikeda et al., 2005; Jorgensen et al., 2005a; Su and Wong, 1996).

Most ominous is the fact that all of these SEs and SE*l*s, with perhaps the exception of SE*LX*, are encoded by genes located on MGEs (see Table 1). These elements, which include genomic islands, *S. aureus* pathogenicity islands, staphylococcal cassette chromosome elements, plasmids and prophages, may very well be the contributing elements that promote host adaptation and the overall pathogenic capability of this organism (Argudin et al., 2010).

8 FUTURE POTENTIAL RAMIFICATIONS

The earliest known report demonstrating an association of a food poisoning outbreak with *S. aureus* and a factor secreted by this organism occurred in the 1930s (Hu and Nakane, 2014), yet it is only recently that we have begun to see more clearly the complexity of staphylococcal food poisoning. The need for additional studies into food poisoning has never been more important, given what appears to food industry marketers as a change in consumer food consumption (Zink, 1997). A trend toward more natural foods with less preservatives, salt, sugar, and fat and foods that are less processed continues to grow. In addition, the consumer's need to have more convenient, ready-to-eat foods has caused the food industry to develop new food processing techniques, such as semiprepared, minimally processed, and chilled foods accompanied by less chemical preservatives, while maintaining a minimal production cost. All of this, compounded by the fact that more and more meals by the average consumer are eaten outside the home, may lead to an increase in food poisoning, particularly by *S. aureus*, which is able to grow and express enterotoxins in a wide variety of food sources (Hennekinne et al., 2012). Therefore, it is paramount that a renewed emphasis in foodborne disease research be initiated in light of these changes in consumer consumption to increase our knowledge base of these foodborne pathogens in hopes of maintaining food safety.

ACKNOWLEDGMENTS

The author is indebted to Drs. Carl Cerniglia, Steve Foley, John Sutherland, and William Tolleson for careful evaluation of the manuscript. The views presented in this article do not necessarily reflect those of the U.S. Food and Drug Administration.

REFERENCES

Agerso, Y., Hasman, H., Cavaco, L.M., Pedersen, K., Aarestrup, F.M., 2012. Study of methicillin resistant *Staphylococcus aureus* (MRSA) in Danish pigs at slaughter and in imported retail meat reveals a novel MRSA type in slaughter pigs. Vet. Microbiol. 157, 246–250.

Altboum, Z., Hertman, I., Sarid, S., 1985. Penicillinase plasmid-linked genetic determinants for enterotoxins B and C1 production in *Staphylococcus aureus*. Infect. Immun. 47, 514–521.

Archer, G.L., 1995. *Staphylococcus epidermidis* and other coagulase-negative staphylococci. In: Mandell, G.L., Bennett, J.E., Dolin, R. (Eds.), Principles and Practice of Infectious Diseases. Churchill Livingstone, New York, NY, pp. 1777–1784.

Argudin, M.A., Mendoza, M.C., Rodicio, M.R., 2010. Food poisoning and *Staphylococcus aureus* enterotoxins. Toxins (Basel) 2, 1751–1773.

Armand-Lefevre, L., Ruimy, R., Andremont, A., 2005. Clonal comparison of *Staphylococcus aureus* isolates from healthy pig farmers, human controls, and pigs. Emerg. Infect. Dis. 11, 711–714.

Asao, T., Kumeda, Y., Kawai, T., Shibata, T., Oda, H., Haruki, K., Nakazawa, H., Kozaki, S., 2003. An extensive outbreak of staphylococcal food poisoning due to low-fat milk in Japan: estimation of enterotoxin A in the incriminated milk and powdered skim milk. Epidemiol. Infect. 130, 33–40.

Balaban, N., Rasooly, A., 2000. Staphylococcal enterotoxins. Int. J. Food Microbiol. 61, 1–10.

Bannerman, T.L., Hancock, G.A., Tenover, F.C., Miller, J.M., 1995. Pulsed-field gel electrophoresis as a replacement for bacteriophage typing of *Staphylococcus aureus*. J. Clin. Microbiol. 33, 551–555.

Bautista, L., Gaya, P., Medina, M., Nunez, M., 1988. A quantitative study of enterotoxin production by sheep milk staphylococci. Appl. Environ. Microbiol. 54, 566–569.

Bayles, K.W., Iandolo, J.J., 1989. Genetic and molecular analyses of the gene encoding staphylococcal enterotoxin D. J. Bacteriol. 171, 4799–4806.

Becker, K., Keller, B., von Eiff, C., Bruck, M., Lubritz, G., Etienne, J., Peters, G., 2001. Enterotoxigenic potential of *Staphylococcus intermedius*. Appl. Environ. Microbiol. 67, 5551–5557.

Becker, K., Heilmann, C., Peters, G., 2014. Coagulase-negative staphylococci. Clin. Microbiol. Rev. 27, 870–926.

Beneke, B., Klees, S., Stuhrenberg, B., Fetsch, A., Kraushaar, B., Tenhagen, B.A., 2011. Prevalence of methicillin-resistant *Staphylococcus aureus* in a fresh meat pork production chain. J. Food Prot. 74, 126–129.

Bennett, S.D., Walsh, K.A., Gould, L.H., 2013. Foodborne disease outbreaks caused by *Bacillus cereus*, *Clostridium perfringens*, and *Staphylococcus aureus* in the United States, 1998-2008. Clin. Infect. Dis. 57, 425–433.

Bens, C.C., Voss, A., Klaassen, C.H., 2006. Presence of a novel DNA methylation enzyme in methicillin-resistant Staphylococcus aureus isolates associated with pig farming leads to uninterpretable results in standard pulsed-field gel electrophoresis analysis. J. Clin. Microbiol. 44, 1875–1876.

Bergdoll, M.S., Surgalla, M.J., Dack, G.M., 1959. Staphylococcal enterotoxin. Identification of a specific precipitating antibody with enterotoxin-neutralizing property. J. Immunol. 83, 334–338.

Bergdoll, M.S., Borja, C.R., Avena, R.M., 1965. Identification of a new enterotoxin as enterotoxin C. J. Bacteriol. 90, 1481–1485.

Bergdoll, M.S., Borja, C.R., Robbins, R.N., Weiss, K.F., 1971. Identification of enterotoxin E. Infect. Immun. 4, 593–595.

Betley, M.J., Mekalanos, J.J., 1985. Staphylococcal enterotoxin A is encoded by phage. Science 229, 185–187.

Bhargava, K., Wang, X., Donabedian, S., Zervos, M., de Rocha, L., Zhang, Y., 2011. Methicillin-resistant *Staphylococcus aureus* in retail meat, Detroit, Michigan, USA. Emerg. Infect. Dis. 17, 1135–1137.

Bhat, M., Dumortier, C., Taylor, B.S., Miller, M., Vasquez, G., Yunen, J., Brudney, K., Sanchez, E.J., Rodriguez-Taveras, C., Rojas, R., Leon, P., Lowy, F.D., 2009. *Staphylococcus aureus* ST398, New York City and Dominican Republic. Emerg. Infect. Dis. 15, 285–287.

Bohach, G.A., Schlievert, P.M., 1987. Nucleotide sequence of the staphylococcal enterotoxin C1 gene and relatedness to other pyrogenic toxins. Mol. Gen. Genet. 209, 15–20.

Borch, E., Nesbakken, T., Christensen, H., 1996. Hazard identification in swine slaughter with respect to foodborne bacteria. Int. J. Food Microbiol. 30, 9–25.

Borst, D.W., Betley, M.J., 1994. Phage-associated differences in staphylococcal enterotoxin A gene (*sea*) expression correlate with *sea* allele class. Infect. Immun. 62, 113–118.

Bosch, T., de Neeling, A.J., Schouls, L.M., van der Zwaluw, K.W., Kluytmans, J.A., Grundmann, H., Huijsdens, X.W., 2010. PFGE diversity within the methicillin-resistant *Staphylococcus aureus* clonal lineage ST398. BMC Microbiol. 10, 40.

Boyce, J.M., White, R.L., Spruill, E.Y., 1983. Impact of methicillin-resistant *Staphylococcus aureus* on the incidence of nosocomial staphylococcal infections. J. Infect. Dis. 148, 763.

Bradley, A., 2002. Bovine mastitis: an evolving disease. Vet. J. 164, 116–128.

Broens, E.M., Graat, E.A., van der Wolf, P.J., van de Giessen, A.W., van Duijkeren, E., Wagenaar, J.A., van Nes, A., Mevius, D.J., de Jong, M.C., 2011. MRSA CC398 in the pig production chain. Prev. Vet. Med. 98, 182–189.

Buyukcangaz, E., Velasco, V., Sherwood, J.S., Stepan, R.M., Koslofsky, R.J., Logue, C.M., 2013. Molecular typing of *Staphylococcus aureus* and methicillin-resistant *S. aureus* (MRSA) isolated from animals and retail meat in North Dakota, United States. Foodborne Pathog. Dis. 10, 608–617.

Casman, E.P., 1960. Further serological studies of staphylococcal enterotoxin. J. Bacteriol. 79, 849–856.

Casman, E.P., Bennett, R.W., Dorsey, A.E., Issa, J.A., 1967. Identification of a fourth staphylococcal enterotoxin, enterotoxin D. J. Bacteriol. 94, 1875–1882.

Cefai, C., Ashurst, S., Owens, C., 1994. Human carriage of methicillin-resistant *Staphylococcus aureus* linked with pet dog. Lancet 344, 539–540.

Cha, J.O., Lee, J.K., Jung, Y.H., Yoo, J.I., Park, Y.K., Kim, B.S., Lee, Y.S., 2006. Molecular analysis of *Staphylococcus aureus* isolates associated with staphylococcal food poisoning in South Korea. J. Appl. Microbiol. 101, 864–871.

Chang, H.C., Bergdoll, M.S., 1979. Purification and some physicochemical properties of staphylococcal enterotoxin D. Biochemistry 18, 1937–1942.

Clarke, S.R., Foster, S.J., 2006. Surface adhesins of *Staphylococcus aureus*. Adv. Microb. Physiol. 51, 187–224.

Couch, J.L., Soltis, M.T., Betley, M.J., 1988. Cloning and nucleotide sequence of the type E staphylococcal enterotoxin gene. J. Bacteriol. 170, 2954–2960.

Cuny, C., Strommenger, B., Witte, W., Stanek, C., 2008. Clusters of infections in horses with MRSA ST1, ST254, and ST398 in a veterinary hospital. Microb. Drug Resist. 14, 307–310.

Cuny, C., Nathaus, R., Layer, F., Strommenger, B., Altmann, D., Witte, W., 2009. Nasal colonization of humans with methicillin-resistant *Staphylococcus aureus* (MRSA) CC398 with and without exposure to pigs. PLoS One 4, e6800.

Cuny, C., Friedrich, A., Kozytska, S., Layer, F., Nubel, U., Ohlsen, K., Strommenger, B., Walther, B., Wieler, L., Witte, W., 2010. Emergence of methicillin-resistant *Staphylococcus aureus* (MRSA) in different animal species. Int. J. Med. Microbiol. 300, 109–117.

Cusumano, Z.T., Watson, M.E., Caparon, M.G., 2014. *Streptococcus pyogenes* arginine and citrulline catabolism promotes infection and modulates innate immunity. Infect. Immun. 82, 233–242.

David, M.Z., Daum, R.S., 2010. Community-associated methicillin-resistant *Staphylococcus aureus*: epidemiology and clinical consequences of an emerging epidemic. Clin. Microbiol. Rev. 23, 616–687.

Day, N.P., Moore, C.E., Enright, M.C., Berendt, A.R., Smith, J.M., Murphy, M.F., Peacock, S.J., Spratt, B.G., Feil, E.J., 2001. A link between virulence and ecological abundance in natural populations of *Staphylococcus aureus*. Science 292, 114–116.

de Boer, E., Zwartkruis-Nahuis, J.T., Wit, B., Huijsdens, X.W., de Neeling, A.J., Bosch, T., van Oosterom, R.A., Vila, A., Heuvelink, A.E., 2009. Prevalence of methicillin-resistant *Staphylococcus aureus* in meat. Int. J. Food Microbiol. 134, 52–56.

De La Fuente, R., Suarez, G., Schleifer, K.H., 1985. *Staphylococcus aureus* subsp. anaerobius subsp. nov., the causal agent of abscess disease of sheep. Int. J. Syst. Bacteriol. 35, 99–102.

de Neeling, A.J., van den Broek, M.J.M., Spalburg, E.C., van Santen-Verheuvel, M.G., Dam-Deisz, W.D.C., Boshuizen, H.C., van de Giessen, A.W., van Duijkeren, E., Huijsdens, X.W., 2007. High prevalence of methicillin resistant *Staphylococcus aureus* in pigs. Vet. Microbiol. 122, 366–372.

Devriese, L.A., Van Damme, L.R., Fameree, L., 1972. Methicillin (cloxacillin)-resistant *Staphylococcus aureus* strains isolated from bovine mastitis cases. Zentralbl. Veterinarmed. B 19, 598–605.

Devriese, L.A., Hájek, V., Oeding, P., Meyer, S.A., Schleifer, K.H., 1978. *Staphylococcus hyicus* (Sompolinsky 1953) comb. nov. and *Staphylococcus hyicus* subsp. chromogenes subsp. nov. Int. J. Syst. Bacteriol. 28, 482–490.

Devriese, L.A., Vancanneyt, M., Baele, M., Vaneechoutte, M., De Graef, E., Snauwaert, C., Cleenwerck, I., Dawyndt, P., Swings, J., Decostere, A., Haesebrouck, F., 2005. *Staphylococcus pseudintermedius* sp. nov., a coagulase-positive species from animals. Int. J. Syst. Evol. Microbiol. 55, 1569–1573.

Diekema, D.J., Pfaller, M.A., Schmitz, F.J., Smayevsky, J., Bell, J., Jones, R.N., Beach, M., Group, S.P., 2001. Survey of infections due to *Staphylococcus* species: frequency of occurrence and antimicrobial susceptibility of isolates collected in the United States, Canada, Latin America, Europe, and the Western Pacific region for the SENTRY Antimicrobial Surveillance Program, 1997-1999. Clin. Infect. Dis. 32 (Suppl. 2), S114–S132.

Diep, B.A., Gill, S.R., Chang, R.F., Phan, T.H., Chen, J.H., Davidson, M.G., Lin, F., Lin, J., Carleton, H.A., Mongodin, E.F., Sensabaugh, G.F., Perdreau-Remington, F., 2006. Complete genome sequence of USA300, an epidemic clone of community-acquired methicillin-resistant *Staphylococcus aureus*. Lancet 367, 731–739.

Dinges, M.M., Orwin, P.M., Schlievert, P.M., 2000. Exotoxins of *Staphylococcus aureus*. Clin. Microbiol. Rev. 13, 16–34.

Doyle, M.E., Hartmann, F.A., Lee Wong, A.C., 2012. Methicillin-resistant staphylococci: implications for our food supply? Anim. Health Res. Rev. 13, 157–180.

Duquette, R.A., Nuttall, T.J., 2004. Methicillin-resistant *Staphylococcus aureus* in dogs and cats: an emerging problem? J. Small Anim. Pract. 45, 591–597.

Enright, M.C., Day, N.P., Davies, C.E., Peacock, S.J., Spratt, B.G., 2000. Multilocus sequence typing for characterization of methicillin-resistant and methicillin-susceptible clones of *Staphylococcus aureus*. J. Clin. Microbiol. 38, 1008–1015.

Evenson, M.L., Hinds, M.W., Bernstein, R.S., Bergdoll, M.S., 1988. Estimation of human dose of staphylococcal enterotoxin A from a large outbreak of staphylococcal food poisoning involving chocolate milk. Int. J. Food Microbiol. 7, 311–316.

Feßler, A.T., Kadlec, K., Hassel, M., Hauschild, T., Eidam, C., Ehricht, R., Monecke, S., Schwarz, S., 2011. Characterization of methicillin-resistant *Staphylococcus aureus* isolates from food and food products of poultry origin in Germany. Appl. Environ. Microbiol. 77, 7151–7157.

Feßler, A.T., Olde Riekerink, R.G., Rothkamp, A., Kadlec, K., Sampimon, O.C., Lam, T.J., Schwarz, S., 2012. Characterization of methicillin-resistant *Staphylococcus aureus* CC398 obtained from humans and animals on dairy farms. Vet. Microbiol. 160, 77–84.

Fitzgerald, J.R., 2012. Livestock-associated *Staphylococcus aureus*: origin, evolution and public health threat. Trends Microbiol. 20, 192–198.

Fitzgerald, J.R., Monday, S.R., Foster, T.J., Bohach, G.A., Hartigan, P.J., Meaney, W.J., Smyth, C.J., 2001. Characterization of a putative pathogenicity island from bovine *Staphylococcus aureus* encoding multiple superantigens. J. Bacteriol. 183, 63–70.

Fluit, A.C., 2012. Livestock-associated *Staphylococcus aureus*. Clin. Microbiol. Infect. 18, 735–744.

Foster, A.P., 2012. Staphylococcal skin disease in livestock. Vet. Dermatol. 23, e342–e351.

Frana, T.S., Beahm, A.R., Hanson, B.M., Kinyon, J.M., Layman, L.L., Karriker, L.A., Ramirez, A., Smith, T.C., 2013. Isolation and characterization of methicillin-resistant *Staphylococcus aureus* from pork farms and visiting veterinary students. PLoS One 8, e53738.

Garcia-Graells, C., Antoine, J., Larsen, J., Catry, B., Skov, R., Denis, O., 2012. Livestock veterinarians at high risk of acquiring methicillin-resistant *Staphylococcus aureus* ST398. Epidemiol. Infect. 140, 383–389.

Gilbert, R.J., Wieneke, A.A., 1973. Staphylococcal food poisoning with special reference to the detection of enterotoxin in food. In: Hobbs, B.C., Christian, J.H.B. (Eds.), The Microbiological Safety of Food. Academic Press, New York, pp. 273–285.

Gortel, K., Campbell, K.L., Kakoma, I., Whittem, T., Schaeffer, D.J., Weisiger, R.M., 1999. Methicillin resistance among staphylococci isolated from dogs. Am. J. Vet. Res. 60, 1526–1530.

Graveland, H., Wagenaar, J.A., Heesterbeek, H., Mevius, D., van Duijkeren, E., Heederik, D., 2010. Methicillin resistant *Staphylococcus aureus* ST398 in veal calf farming: human MRSA carriage related with animal antimicrobial usage and farm hygiene. PLoS One 5, e10990.

Hájek, V., 1976. *Staphylococcus intermedius*, a new species isolated from animals. Int. J. Syst. Bacteriol. 26, 401–408.

Hallin, M., De Mendonca, R., Denis, O., Lefort, A., El Garch, F., Butaye, P., Hermans, K., Struelens, M.J., 2011. Diversity of accessory genome of human and livestock-associated ST398 methicillin resistant *Staphylococcus aureus* strains. Infect. Genet. Evol. 11, 290–299.

Harmsen, D., Claus, H., Witte, W., Rothganger, J., Claus, H., Turnwald, D., Vogel, U., 2003. Typing of methicillin-resistant *Staphylococcus aureus* in a university hospital setting by using novel software for *spa* repeat determination and database management. J. Clin. Microbiol. 41, 5442–5448.

Hartmann, F.A., Trostle, S.S., Klohnen, A.A., 1997. Isolation of methicillin-resistant *Staphylococcus aureus* from a postoperative wound infection in a horse. J. Am. Vet. Med. Assoc. 211, 590–592.

Hennekinne, J.-A., De Buyser, M.-L., Dragacci, S., 2012. *Staphylococcus aureus* and its food poisoning toxins: characterization and outbreak investigation. FEMS Microbiol. Rev. 36, 815–836.

Hiramatsu, K., Hanaki, H., Ino, T., Yabuta, K., Oguri, T., Tenover, F.C., 1997. Methicillin-resistant *Staphylococcus aureus* clinical strain with reduced vancomycin susceptibility. J. Antimicrob. Chemother. 40, 135–136.

Hovde, C.J., Hackett, S.P., Bohach, G.A., 1990. Nucleotide sequence of the staphylococcal enterotoxin C3 gene: sequence comparison of all three type C staphylococcal enterotoxins. Mol. Gen. Genet. 220, 329–333.

Hu, D.L., Nakane, A., 2014. Mechanisms of staphylococcal enterotoxin-induced emesis. Eur. J. Pharmacol. 722, 95–107.

Huber, H., Giezendanner, N., Stephan, R., Zweifel, C., 2011. Genotypes, antibiotic resistance profiles and microarray-based characterization of methicillin-resistant *Staphylococcus aureus* strains isolated from livestock and veterinarians in Switzerland. Zoonoses Public Health 58, 343–349.

Huebner, J., Goldmann, D.A., 1999. Coagulase-negative staphylococci: role as pathogens. Annu. Rev. Med. 50, 223–236.

Huijsdens, X.W., van Dijke, B.J., Spalburg, E., van Santen-Verheuvel, M.G., Heck, M.E., Pluister, G.N., Voss, A., Wannet, W.J., de Neeling, A.J., 2006. Community-acquired MRSA and pig-farming. Ann. Clin. Microbiol. Antimicrob. 5, 26.

Iandolo, J.J., 1990. The genetics of staphylococcal toxins and virulence factors. In: Iglewski, B.H., Clark, V.L. (Eds.), Molecular Basis of Bacterial Pathogenesis. Academic Press, Inc., New York, NY, pp. 399–426.

Igimi, S., Takahashi, E., Mitsuoka, T., 1990. *Staphylococcus schleiferi* subsp. coagulans subsp. nov., isolated from the external auditory meatus of dogs with external ear otitis. Int. J. Syst. Bacteriol. 40, 409–411.

Ikeda, T., Tamate, N., Yamaguchi, K., Makino, S., 2005. Mass outbreak of food poisoning disease caused by small amounts of staphylococcal enterotoxins A and H. Appl. Environ. Microbiol. 71, 2793–2795.

Jablonski, L.M., Bohach, G.A., 1997. Staphylococcus aureus. In: Doyle, M.P., Beuchat, L.R., Montville, T.J. (Eds.), Food Microbiology: Fundamentals and Frontiers. American Society for Microbiology, Washington, DC, pp. 353–375.

Jackson, C.R., Davis, J.A., Barrett, J.B., 2013. Prevalence and characterization of methicillin-resistant *Staphylococcus aureus* isolates from retail meat and humans in Georgia. J. Clin. Microbiol. 51, 1199–1207.

Jarraud, S., Peyrat, M.A., Lim, A., Tristan, A., Bes, M., Mougel, C., Etienne, J., Vandenesch, F., Bonneville, M., Lina, G., 2001. *egc*, a highly prevalent operon of enterotoxin gene, forms a putative nursery of superantigens in *Staphylococcus aureus*. J. Immunol. 166, 669–677.

Jarraud, S., Mougel, C., Thioulouse, J., Lina, G., Meugnier, H., Forey, F., Nesme, X., Etienne, J., Vandenesch, F., 2002. Relationships between *Staphylococcus aureus* genetic background, virulence factors, *agr* groups (alleles), and human disease. Infect. Immun. 70, 631–641.

Jones, C.L., Khan, S.A., 1986. Nucleotide sequence of the enterotoxin B gene from *Staphylococcus aureus*. J. Bacteriol. 166, 29–33.

Jones, T.F., Kellum, M.E., Porter, S.S., Bell, M., Schaffner, W., 2002. An outbreak of community-acquired foodborne illness caused by methicillin-resistant *Staphylococcus aureus*. Emerg. Infect. Dis. 8, 82–84.

Jorgensen, H.J., Mathisen, T., Lovseth, A., Omoe, K., Qvale, K.S., Loncarevic, S., 2005a. An outbreak of staphylococcal food poisoning caused by enterotoxin H in mashed potato made with raw milk. FEMS Microbiol. Lett. 252, 267–272.

Jorgensen, H.J., Mork, T., Rorvik, L.M., 2005b. The occurrence of *Staphylococcus aureus* on a farm with small-scale production of raw milk cheese. J. Dairy Sci. 88, 3810–3817.

Kadlec, K., Ehricht, R., Monecke, S., Steinacker, U., Kaspar, H., Mankertz, J., Schwarz, S., 2009. Diversity of antimicrobial resistance pheno- and genotypes of methicillin-resistant *Staphylococcus aureus* ST398 from diseased swine. J. Antimicrob. Chemother. 64, 1156–1164.

Kallen, A.J., Mu, Y., Bulens, S., Reingold, A., Petit, S., Gershman, K., Ray, S.M., Harrison, L.H., Lynfield, R., Dumyati, G., Townes, J.M., Schaffner, W., Patel, P.R., Fridkin, S.K., for the Active Bacterial Core Surveillance (ABCs) MRSA Investigators of the Emerging Infections Program, 2010. Health care-associated invasive MRSA infections, 2005-2008. JAMA 304, 641–648.

Kelman, A., Soong, Y.A., Dupuy, N., Shafer, D., Richbourg, W., Johnson, K., Brown, T., Kestler, E., Li, Y., Zheng, J., McDermott, P., Meng, J., 2011. Antimicrobial susceptibility of *Staphylococcus aureus* from retail ground meats. J. Food Prot. 74, 1625–1629.

Kennedy, A.D., Otto, M., Braughton, K.R., Whitney, A.R., Chen, L., Mathema, B., Mediavilla, J.R., Byrne, K.A., Parkins, L.D., Tenover, F.C., Kreiswirth, B.N., Musser, J.M., DeLeo, F.R., 2008. Epidemic community-associated methicillin-resistant *Staphylococcus aureus*: recent clonal expansion and diversification. Proc. Natl. Acad. Sci. U. S. A. 105, 1327–1332.

Kerouanton, A., Hennekinne, J.A., Letertre, C., Petit, L., Chesneau, O., Brisabois, A., De Buyser, M.L., 2007. Characterization of *Staphylococcus aureus* strains associated with food poisoning outbreaks in France. Int. J. Food Microbiol. 115, 369–375.

Khambaty, F.M., Bennett, R.W., Shah, D.B., 1994. Application of pulsed-field gel electrophoresis to the epidemiological characterization of *Staphylococcus intermedius* implicated in a food-related outbreak. Epidemiol. Infect. 113, 75–81.

Khanna, T., Friendship, R., Dewey, C., Weese, J.S., 2008. Methicillin resistant *Staphylococcus aureus* colonization in pigs and pig farmers. Vet. Microbiol. 128, 298–303.

Klevens, R.M., Edwards, J.R., Richards Jr., C.L., Horan, T.C., Gaynes, R.P., Pollock, D.A., Cardo, D.M., 2007a. Estimating health care-associated infections and deaths in U.S. hospitals, 2002. Public Health Rep. 122, 160–166.

Klevens, R.M., Morrison, M.A., Nadle, J., Petit, S., Gershman, K., Ray, S., Harrison, L.H., Lynfield, R., Dumyati, G., Townes, J.M., Craig, A.S., Zell, E.R., Fosheim, G.E., McDougal, L.K., Carey, R.B., Fridkin, S.K., Active Bacterial Core Surveillance (ABCs) MRSA Investigators, 2007b. Invasive methicillin-resistant *Staphylococcus aureus* infections in the United States. JAMA 298, 1763–1771.

Kloos, W.E., Bannerman, T.L., 1994. Update on clinical significance of coagulase-negative staphylococci. Clin. Microbiol. Rev. 7, 117–140.

Kloos, W.E., Bannerman, T.L., 1995. *Staphylococcus* and *Micrococcus*. In: Murray, P.R., Baron, E.J., Pfaller, M.A., Tenover, F.C., Yolken, R.H. (Eds.), Manual of Clinical Microbiology. ASM Press, Washington, DC, pp. 289–298.

Kloos, W.E., Schleifer, K.H., 1986. Staphylococcus. In: Sneath, P.H.A., Mair, N.S., Sharpe, M.E., Holt, J.G. (Eds.), Bergey's Manual of Systematic Bacteriology. The Williams and Wilkins Co., Baltimore, MD, pp. 1013–1035.

Kuroda, M., Ohta, T., Uchiyama, I., Baba, T., Yuzawa, H., Kobayashi, I., Cui, L., Oguchi, A., Aoki, K., Nagai, Y., Lian, J., Ito, T., Kanamori, M., Matsumaru, H., Maruyama, A., Murakami, H., Hosoyama, A., Mizutani-Ui, Y., Takahashi, N.K., Sawano, T., Inoue, R., Kaito, C., Sekimizu, K., Hirakawa, H., Kuhara, S., Goto, S., Yabuzaki, J., Kanehisa, M., Yamashita, A., Oshima, K., Furuya, K., Yoshino, C., Shiba, T., Hattori, M., Ogasawara, N., Hayashi, H., Hiramatsu, K., 2001. Whole genome sequencing of meticillin-resistant *Staphylococcus aureus*. Lancet 357, 1225–1240.

Le Loir, Y., Baron, F., Gautier, M., 2003. *Staphylococcus aureus* and food poisoning. Genet. Mol. Res. 2, 63–76.

Letertre, C., Perelle, S., Dilasser, F., Fach, P., 2003. Identification of a new putative enterotoxin SEU encoded by the *egc* cluster of *Staphylococcus aureus*. J. Appl. Microbiol. 95, 38–43.

Lina, G., Bohach, G.A., Nair, S.P., Hiramatsu, K., Jouvin-Marche, E., Mariuzza, R., for the International Nomenclature Committee for Staphylococcal Superantigens, 2004. Standard nomenclature for the superantigens expressed by *Staphylococcus*. J. Infect. Dis. 189, 2334–2336.

Lindsay, J.A., 2010. Genomic variation and evolution of *Staphylococcus aureus*. Int. J. Med. Microbiol. 300, 98–103.

Lindsay, J.A., 2013. Hospital-associated MRSA and antibiotic resistance—what have we learned from genomics? Int. J. Med. Microbiol. 303, 318–323.

Lindsay, J.A., Moore, C.E., Day, N.P., Peacock, S.J., Witney, A.A., Stabler, R.A., Husain, S.E., Butcher, P.D., Hinds, J., 2006. Microarrays reveal that each of the ten dominant lineages of *Staphylococcus aureus* has a unique combination of surface-associated and regulatory genes. J. Bacteriol. 188, 669–676.

Lindström, M., Heikinheimo, A., Lahti, P., Korkeala, H., 2011. Novel insights into the epidemiology of *Clostridium perfringens* type A food poisoning. Food Microbiol. 28, 192–198.

Llewelyn, M., Cohen, J., 2002. Superantigens: microbial agents that corrupt immunity. Lancet Infect. Dis. 2, 156–162.

Loeffler, A., Boag, A.K., Sung, J., Lindsay, J.A., Guardabassi, L., Dalsgaard, A., Smith, H., Stevens, K.B., Lloyd, D.H., 2005. Prevalence of methicillin-resistant *Staphylococcus aureus* among staff and pets in a small animal referral hospital in the UK. J. Antimicrob. Chemother. 56, 692–697.

Logan, N.A., 2012. *Bacillus* and relatives in foodborne illness. J. Appl. Microbiol. 112, 417–429.

Lowder, B.V., Guinane, C.M., Ben Zakour, N.L., Weinert, L.A., Conway-Morris, A., Cartwright, R.A., Simpson, A.J., Rambaut, A., Nubel, U., Fitzgerald, J.R., 2009. Recent human-to-poultry host jump, adaptation, and pandemic spread of *Staphylococcus aureus*. Proc. Natl. Acad. Sci. U. S. A. 106, 19545–19550.

Madhusoodanan, J., Seo, K.S., Remortel, B., Park, J.Y., Hwang, S.Y., Fox, L.K., Park, Y.H., Deobald, C.F., Wang, D., Liu, S., Daugherty, S.C., Gill, A.L., Bohach, G.A., Gill, S.R., 2011. An enterotoxin-bearing pathogenicity island in *Staphylococcus epidermidis*. J. Bacteriol. 193, 1854–1862.

Maple, P.A., Hamilton-Miller, J.M., Brumfitt, W., 1989. World-wide antibiotic resistance in methicillin-resistant *Staphylococcus aureus*. Lancet 1, 537–540.

Marrack, P., Kappler, J., 1990. The staphylococcal enterotoxins and their relatives. Science 248, 705–711.

Martin de Nicolas, M.M., Vindel, A., offez-Nieto, J.A., 1995. Epidemiological typing of clinically significant strains of coagulase-negative staphylococci. J. Hosp. Infect. 29, 35–43.

Martins, P.D., de Almeida, T.T., Basso, A.P., de Moura, T.M., Frazzon, J., Tondo, E.C., Frazzon, A.P., 2013. Coagulase-positive staphylococci isolated from chicken meat: pathogenic potential and vancomycin resistance. Foodborne Pathog. Dis. 10, 771–776.

McCarthy, A.J., Witney, A.A., Gould, K.A., Moodley, A., Guardabassi, L., Voss, A., Denis, O., Broens, E.M., Hinds, J., Lindsay, J.A., 2011. The distribution of mobile genetic elements (MGEs) in MRSA CC398 is associated with both host and country. Genome Biol. Evol. 3, 1164–1174.

McCarthy, A.J., van Wamel, W., Vandendriessche, S., Larsen, J., Denis, O., Garcia-Graells, C., Uhlemann, A.C., Lowy, F.D., Skov, R., Lindsay, J.A., 2012. *Staphylococcus aureus* CC398 clade associated with human-to-human transmission. Appl. Environ. Microbiol. 78, 8845–8848.

McCormick, J.K., Yarwood, J.M., Schlievert, P.M., 2001. Toxic shock syndrome and bacterial superantigens: an update. Annu. Rev. Microbiol. 55, 77–104.

McNamee, P.T., Smyth, J.A., 2000. Bacterial chondronecrosis with osteomyelitis ('femoral head necrosis') of broiler chickens: a review. Avian Pathol. 29, 477–495.

Menzies, P.I., Ramanoon, S.Z., 2001. Mastitis of sheep and goats. Vet. Clin. N. Am. Food Anim. Pract. 17, 333–358, vii.

Meyrand, A., Boutrand-Loei, S., Ray-Gueniot, S., Mazuy, C., Gaspard, C.E., Jaubert, G., Perrin, G., Lapeyre, C., Vernozy-Rozand, C., 1998. Growth and enterotoxin production of *Staphylococcus aureus* during the manufacture and ripening of Camembert-type cheeses from raw goats' milk. J. Appl. Microbiol. 85, 537–544.

Molla, B., Byrne, M., Abley, M., Mathews, J., Jackson, C.R., Fedorka-Cray, P., Sreevatsan, S., Wang, P., Gebreyes, W.A., 2012. Epidemiology and genotypic characteristics of methicillin-resistant *Staphylococcus aureus* strains of porcine origin. J. Clin. Microbiol. 50, 3687–3693.

Mulvey, M.R., Chui, L., Ismail, J., Louie, L., Murphy, C., Chang, N., Alfa, M., The Canadian Committee for the Standardization of Molecular Methods, 2001. Development of a Canadian standardized protocol for subtyping methicillin-resistant *Staphylococcus aureus* using pulsed-field gel electrophoresis. J. Clin. Microbiol. 39, 3481–3485.

Munson, S.H., Tremaine, M.T., Betley, M.J., Welch, R.A., 1998. Identification and characterization of staphylococcal enterotoxin types G and I from *Staphylococcus aureus*. Infect. Immun. 66, 3337–3348.

Murchan, S., Kaufmann, M.E., Deplano, A., de Ryck, R., Struelens, M., Zinn, C.E., Fussing, V., Salmenlinna, S., Vuopio-Varkila, J., El Solh, N., Cuny, C., Witte, W., Tassios, P.T., Legakis, N., van Leeuwen, W., van Belkum, A., Vindel, A., Laconcha, I., Garaizar, J., Haeggman, S., Olsson-Liljequist, B., Ransjo, U., Coombes, G., Cookson, B., 2003. Harmonization of pulsed-field gel electrophoresis protocols for epidemiological typing of strains of methicillin-resistant *Staphylococcus aureus*: a single approach developed by consensus in 10 European laboratories and its application for tracing the spread of related strains. J. Clin. Microbiol. 41, 1574–1585.

Musser, J.M., Schlievert, P.M., Chow, A.W., Ewan, P., Kreiswirth, B.N., Rosdahl, V.T., Naidu, A.S., Witte, W., Selander, R.K., 1990. A single clone of *Staphylococcus aureus* causes the majority of cases of toxic shock syndrome. Proc. Natl. Acad. Sci. U. S. A. 87, 225–229.

Nemati, M., Hermans, K., Lipinska, U., Denis, O., Deplano, A., Struelens, M., Devriese, L.A., Pasmans, F., Haesebrouck, F., 2008. Antimicrobial resistance of old and recent *Staphylococcus aureus* isolates from poultry: first detection of livestock-associated methicillin-resistant strain ST398. Antimicrob. Agents Chemother. 52, 3817–3819.

Normanno, G., Corrente, M., La Salandra, G., Dambrosio, A., Quaglia, N.C., Parisi, A., Greco, G., Bellacicco, A.L., Virgilio, S., Celano, G.V., 2007. Methicillin-resistant *Staphylococcus aureus* (MRSA) in foods of animal origin product in Italy. Int. J. Food Microbiol. 117, 219–222.

Noto, M.J., Archer, G.L., 2006. A subset of *Staphylococcus aureus* strains harboring staphylococcal cassette chromosome *mec* (SCC*mec*) type IV is deficient in CcrAB-mediated SCC*mec* excision. Antimicrob. Agents Chemother. 50, 2782–2788.

O'Brien, A.M., Hanson, B.M., Farina, S.A., Wu, J.Y., Simmering, J.E., Wardyn, S.E., Forshey, B.M., Kulick, M.E., Wallinga, D.B., Smith, T.C., 2012. MRSA in conventional and alternative retail pork products. PLoS One 7, e30092.

Okumura, K., Shimomura, Y., Murayama, S.Y., Yagi, J., Ubukata, K., Kirikae, T., Miyoshi-Akiyama, T., 2012. Evolutionary paths of streptococcal and staphylococcal superantigens. BMC Genomics 13, 404.

Omoe, K., Ishikawa, M., Shimoda, Y., Hu, D.L., Ueda, S., Shinagawa, K., 2002. Detection of *seg*, *seh*, and *sei* genes in *Staphylococcus aureus* isolates and determination of the enterotoxin productivities of *S. aureus* isolates Harboring *seg*, *seh*, or *sei* genes. J. Clin. Microbiol. 40, 857–862.

Omoe, K., Hu, D.L., Takahashi-Omoe, H., Nakane, A., Shinagawa, K., 2003. Identification and characterization of a new staphylococcal enterotoxin-related putative toxin encoded by two kinds of plasmids. Infect. Immun. 71, 6088–6094.

Omoe, K., Imanishi, K., Hu, D.L., Kato, H., Fugane, Y., Abe, Y., Hamaoka, S., Watanabe, Y., Nakane, A., Uchiyama, T., Shinagawa, K., 2005. Characterization of novel staphylococcal enterotoxin-like toxin type P. Infect. Immun. 73, 5540–5546.

Omoe, K., Hu, D.L., Ono, H.K., Shimizu, S., Takahashi-Omoe, H., Nakane, A., Uchiyama, T., Shinagawa, K., Imanishi, K., 2013. Emetic potentials of newly identified staphylococcal enterotoxin-like toxins. Infect. Immun. 81, 3627–3631.

Ono, H.K., Omoe, K., Imanishi, K., Iwakabe, Y., Hu, D.L., Kato, H., Saito, N., Nakane, A., Uchiyama, T., Shinagawa, K., 2008. Identification and characterization of two novel staphylococcal enterotoxins, types S and T. Infect. Immun. 76, 4999–5005.

Orwin, P.M., Leung, D.Y., Donahue, H.L., Novick, R.P., Schlievert, P.M., 2001. Biochemical and biological properties of staphylococcal enterotoxin K. Infect. Immun. 69, 360–366.

Orwin, P.M., Fitzgerald, J.R., Leung, D.Y., Gutierrez, J.A., Bohach, G.A., Schlievert, P.M., 2003. Characterization of *Staphylococcus aureus* enterotoxin L. Infect. Immun. 71, 2916–2919.

Otto, M., 2012. MRSA virulence and spread. Cell. Microbiol. 14, 1513–1521.

Park, J.Y., Fox, L.K., Seo, K.S., McGuire, M.A., Park, Y.H., Rurangirwa, F.R., Sischo, W.M., Bohach, G.A., 2011. Detection of classical and newly described staphylococcal superantigen genes in coagulase-negative staphylococci isolated from bovine intramammary infections. Vet. Microbiol. 147, 149–154.

Pereira, J.L., Salzberg, S.P., Bergdoll, M.S., 1991. Production of staphylococcal enterotoxin D in foods by low-enterotoxin-producing staphylococci. Int. J. Food Microbiol. 14, 19–25.

Podkowik, M., Park, J.Y., Seo, K.S., Bystroń, J., Bania, J., 2013. Enterotoxigenic potential of coagulase-negative staphylococci. Int. J. Food Microbiol. 163, 34–40.

Price, L.B., Stegger, M., Hasman, H., Aziz, M., Larsen, J., Andersen, P.S., Pearson, T., Waters, A.E., Foster, J.T., Schupp, J., Gillece, J., Driebe, E., Liu, C.M., Springer, B., Zdovc, I., Battisti, A., Franco, A., Żmudzki, J., Schwarz, S., Butaye, P., Jouy, E., Pomba, C., Porrero, M.C., Ruimy, R., Smith, T.C., Robinson, D.A., Weese, J.S., Arriola, C.S., Yu, F., Laurent, F., Keim, P., Skov, R., Aarestrup, F.M., 2012. *Staphylococcus aureus* CC398: host adaptation and emergence of methicillin resistance in livestock. mBio 3, e00305-11.

Pu, S., Han, F., Ge, B., 2009. Isolation and characterization of methicillin-resistant *Staphylococcus aureus* strains from Louisiana retail meats. Appl. Environ. Microbiol. 75, 265–267.

Raj, H.D., Bergdoll, M.S., 1969. Effect of enterotoxin B on human volunteers. J. Bacteriol. 98, 833–834.

Rasigade, J.P., Laurent, F., Hubert, P., Vandenesch, F., Etienne, J., 2010. Lethal necrotizing pneumonia caused by an ST398 *Staphylococcus aureus* strain. Emerg. Infect. Dis. 16, 1330.

Ren, K., Bannan, J.D., Pancholi, V., Cheung, A.L., Robbins, J.C., Fischetti, V.A., Zabriskie, J.B., 1994. Characterization and biological properties of a new staphylococcal exotoxin. J. Exp. Med. 180, 1675–1683.

Rosec, J.P., Gigaud, O., 2002. Staphylococcal enterotoxin genes of classical and new types detected by PCR in France. Int. J. Food Microbiol. 77, 61–70.

Ryffel, C., Strassle, A., Kayser, F.H., Berger-Bachi, B., 1994. Mechanisms of heteroresistance in methicillin-resistant *Staphylococcus aureus*. Antimicrob. Agents Chemother. 38, 724–728.

Saito, K., Higuchi, T., Kurata, A., Fukuyasu, T., Ashida, K., 1996. Characterization of non-pigmented *Staphylococcus chromogenes*. J. Vet. Med. Sci. 58, 711–713.

Scallan, E., Griffin, P.M., Angulo, F.J., Tauxe, R.V., Hoekstra, R.M., 2011a. Foodborne illness acquired in the United States – unspecified agents. Emerg. Infect. Dis. 17, 16–22.

Scallan, E., Hoekstra, R.M., Angulo, F.J., Tauxe, R.V., Widdowson, M.A., Roy, S.L., Jones, J.L., Griffin, P.M., 2011b. Foodborne illness acquired in the United States—major pathogens. Emerg. Infect. Dis. 17, 7–15.

Scharff, R.L., 2012. Economic burden from health losses due to foodborne illness in the United States. J. Food Prot. 75, 123–131.

Schelin, J., Wallin-Carlquist, N., Cohn, M.T., Lindqvist, R., Barker, G.C., Radstrom, P., 2011. The formation of *Staphylococcus aureus* enterotoxin in food environments and advances in risk assessment. Virulence 2, 580–592.

Schijffelen, M.J., Boel, C.H., van Strijp, J.A., Fluit, A.C., 2010. Whole genome analysis of a livestock-associated methicillin-resistant *Staphylococcus aureus* ST398 isolate from a case of human endocarditis. BMC Genomics 11, 376.

Seguin, J.C., Walker, R.D., Caron, J.P., Kloos, W.E., George, C.G., Hollis, R.J., Jones, R.N., Pfaller, M.A., 1999. Methicillin-resistant *Staphylococcus aureus* outbreak in a veterinary teaching hospital: potential human-to-animal transmission. J. Clin. Microbiol. 37, 1459–1463.

Shafer, W.M., Iandolo, J.J., 1978. Chromosomal locus for staphylococcal enterotoxin B. Infect. Immun. 20, 273–278.

Shalita, Z., Hertman, I., Sarid, S., 1977. Isolation and characterization of a plasmid involved with enterotoxin B production in *Staphylococcus aureus*. J. Bacteriol. 129, 317–325.

Sievert, D.M., Ricks, P., Edwards, J.R., Schneider, A., Patel, J., Srinivasan, A., Kallen, A., Limbago, B., Fridkin, S., The National Healthcare Safety Network (NHSN) Team and Participating NHSN Facilities, 2013. Antimicrobial-resistant pathogens associated with healthcare-associated infections: summary of data reported to the National Healthcare Safety Network at the Centers for Disease Control and Prevention, 2009-2010. Infect. Control Hosp. Epidemiol. 34, 1–14.

Smeltzer, M.S., Lee, C.Y., Harik, N., Hart, M.E., 2009. Molecular basis of pathogenicity. In: Crossley, K.B., Jefferson, K.K., Archer, G.L., Fowler, V.G. (Eds.), Staphylococci in Human Disease, second ed. Wiley-Blackwell, United Kingdom, pp. 65–108.

Smith, T.C., Pearson, N., 2011. The emergence of *Staphylococcus aureus* ST398. Vector Borne Zoonotic Dis. 11, 327–339.

Smith, T.C., Male, M.J., Harper, A.L., Kroeger, J.S., Tinkler, G.P., Moritz, E.D., Capuano, A.W., Herwaldt, L.A., Diekema, D.J., 2009. Methicillin-resistant *Staphylococcus aureus* (MRSA) strain ST398 is present in Midwestern U.S. swine and swine workers. PLoS One 4, e4258.

Somerville, G.A., Proctor, R.A., 2009. The biology of the staphylococci. In: Crossley, K.B., Jefferson, K.K., Archer, G.L., Fowler, V.G. (Eds.), Staphylococci in Human Disease, 2 ed. Wiley-Blackwell, West Sussex, UK, pp. 3–18.

Spaulding, A.R., Salgado-Pabón, W., Kohler, P.L., Horswill, A.R., Leung, D.Y.M., Schlievert, P.M., 2013. Staphylococcal and streptococcal superantigen exotoxins. Clin. Microbiol. Rev. 26, 422–447.

Spoor, L.E., McAdam, P.R., Weinert, L.A., Rambaut, A., Hasman, H., Aarestrup, F.M., Kearns, A.M., Larsen, A.R., Skov, R.L., Fitzgerald, J.R., 2013. Livestock origin for a human pandemic clone of community-associated methicillin-resistant *Staphylococcus aureus*. mBio 4, e00356-13.

Stegger, M., Lindsay, J.A., Sorum, M., Gould, K.A., Skov, R., 2010. Genetic diversity in CC398 methicillin-resistant *Staphylococcus aureus* isolates of different geographical origin. Clin. Microbiol. Infect. 16, 1017–1019.

Stegger, M., Lindsay, J.A., Moodley, A., Skov, R., Broens, E.M., Guardabassi, L., 2011. Rapid PCR detection of *Staphylococcus aureus* clonal complex 398 by targeting the restriction-modification system carrying *sau1-hsdS1*. J. Clin. Microbiol. 49, 732–734.

Stegger, M., Liu, C.M., Larsen, J., Soldanova, K., Aziz, M., Contente-Cuomo, T., Petersen, A., Vandendriessche, S., Jimenez, J.N., Mammina, C., van Belkum, A., Salmenlinna, S., Laurent, F., Skov, R.L., Larsen, A.R., Andersen, P.S., Price, L.B., 2013. Rapid differentiation between livestock-associated and livestock-independent *Staphylococcus aureus* CC398 clades. PLoS One 8, e79645.

Stegger, M., Wirth, T., Andersen, P.S., Skov, R.L., De Grassi, A., Simões, P.M., Tristan, A., Petersen, A., Aziz, M., Kiil, K., Cirković, I., Udo, E.E., del Campo, R., Vuopio-Varkila, J., Ahmad, N., Tokajian, S., Peters, G., Schaumburg, F., Olsson-Liljequist, B., Givskov, M., Driebe, E.E., Vigh, H.E., Shittu, A., Ramdani-Bougessa, N., Rasigade, J.-P., Price, L.B., Vandenesch, F., Larsen, A.R., Laurent, F., 2014. Origin and evolution of European community-acquired methicillin-resistant *Staphylococcus aureus*. mBio 5, e01044-14.

Stenfors Arnesen, L.P., Fagerlund, A., Granum, P.E., 2008. From soil to gut: *Bacillus cereus* and its food poisoning toxins. FEMS Microbiol. Rev. 32, 579–606.

Stryjewski, M.E., Chambers, H.F., 2008. Skin and soft-tissue infections caused by community-acquired methicillin-resistant *Staphylococcus aureus*. Clin. Infect. Dis. 46, S368–S377.

Su, Y.C., Wong, A.C., 1995. Identification and purification of a new staphylococcal enterotoxin, H. Appl. Environ. Microbiol. 61, 1438–1443.

Su, Y.C., Wong, A.C., 1996. Detection of staphylococcal enterotoxin H by an enzyme-linked immunosorbent assay. J. Food Prot. 59, 327–330.

Sung, J.M., Lloyd, D.H., Lindsay, J.A., 2008. *Staphylococcus aureus* host specificity: comparative genomics of human versus animal isolates by multi-strain microarray. Microbiology 154, 1949–1959.

Tenover, F.C., Lancaster, M.V., Hill, B.C., Steward, C.D., Stocker, S.A., Hancock, G.A., O'Hara, C.M., McAllister, S.K., Clark, N.C., Hiramatsu, K., 1998. Characterization of staphylococci with reduced susceptibilities to vancomycin and other glycopeptides. J. Clin. Microbiol. 36, 1020–1027.

Tenover, F.C., McDougal, L.K., Goering, R.V., Killgore, G., Projan, S.J., Patel, J.B., Dunman, P.M., 2006. Characterization of a strain of community-associated methicillin-resistant *Staphylococcus aureus* widely disseminated in the United States. J. Clin. Microbiol. 44, 108–118.

Thomas, D.Y., Jarraud, S., Lemercier, B., Cozon, G., Echasserieau, K., Etienne, J., Gougeon, M.L., Lina, G., Vandenesch, F., 2006. Staphylococcal enterotoxin-like toxins U2 and V, two new staphylococcal superantigens arising from recombination within the enterotoxin gene cluster. Infect. Immun. 74, 4724–4734.

Tristan, A., Bes, M., Meugnier, H., Lina, G., Bozdogan, B., Courvalin, P., Reverdy, M.E., Enright, M.C., Vandenesch, F., Etienne, J., 2007. Global distribution of Panton-Valentine leukocidin-positive methicillin-resistant *Staphylococcus aureus*, 2006. Emerg. Infect. Dis. 13, 594–600.

Uhlemann, A.C., Dumortier, C., Hafer, C., Taylor, B.S., Sanchez, J., Rodriguez-Taveras, C., Leon, P., Rojas, R., Olive, C., Lowy, F.D., 2012a. Molecular characterization of *Staphylococcus aureus* from outpatients in the Caribbean reveals the presence of pandemic clones. Eur. J. Clin. Microbiol. Infect. Dis. 31, 505–511.

Uhlemann, A.C., Porcella, S.F., Trivedi, S., Sullivan, S.B., Hafer, C., Kennedy, A.D., Barbian, K.D., McCarthy, A.J., Street, C., Hirschberg, D.L., Lipkin, W.I., Lindsay, J.A., DeLeo, F.R., Lowy, F.D., 2012b. Identification of a highly transmissible animal-independent *Staphylococcus aureus* ST398 clone with distinct genomic and cell adhesion properties. mBio 3, 1–9.

Valentin-Domelier, A.S., Girard, M., Bertrand, X., Violette, J., François, P., Donnio, P.Y., Talon, D., Quentin, R., Schrenzel, J., van der Mee-Marquet, N., The Bloodstream Infection Study Group of the Réseau des Hygiénistes du Centre (RHC), 2011. Methicillin-susceptible ST398 *Staphylococcus aureus* responsible for bloodstream infections: an emerging human-adapted subclone? PLoS One 6, e28369.

van Cleef, B.A., Graveland, H., Haenen, A.P., van de Giessen, A.W., Heederik, D., Wagenaar, J.A., Kluytmans, J.A., 2011. Persistence of livestock-associated methicillin-resistant *Staphylococcus aureus* in field workers after short-term occupational exposure to pigs and veal calves. J. Clin. Microbiol. 49, 1030–1033.

van den Broek, I.V.F., van Cleef, B.A.G.L., Haenen, A., Broens, E.M., van der Wolf, P.J., van den Broek, M.J.M., Huijsdens, X.W., Kluytmans, J.A., van de Giessen, A.W., Tiemersma, E.W., 2009. Methicillin-resistant *Staphylococcus aureus* in people living and working in pig farms. Epidemiol. Infect. 137, 700–708.

van der Mee-Marquet, N., François, P., Domelier-Valentin, A.S., Coulomb, F., Decreux, C., Hombrock-Allet, C., Lehiani, O., Neveu, C., Ratovohery, D., Schrenzel, J., Quentin, R., The Bloodstream Infection Study Group of the Réseau des Hygiénistes du Centre (RHC), 2011. Emergence of unusual bloodstream infections associated with pig-borne-like *Staphylococcus aureus* ST398 in France. Clin. Infect. Dis. 52, 152–153.

van Loo, I., Huijsdens, X., Tiemersma, E., de Neeling, A., van de Sande-Bruinsma, N., Beaujean, D., Voss, A., Kluytmans, J., 2007a. Emergence of methicillin-resistant *Staphylococcus aureus* of animal origin in humans. Emerg. Infect. Dis. 13, 1834–1839.

van Loo, I.H., Diederen, B.M., Savelkoul, P.H., Woudenberg, J.H., Roosendaal, R., van Belkum, A., Lemmens-den Toom, N., Verhulst, C., van Keulen, P.H., Kluytmans, J.A., 2007b. Methicillin-resistant *Staphylococcus aureus* in meat products, the Netherlands. Emerg. Infect. Dis. 13, 1753–1755.

van Rijen, M.M., Van Keulen, P.H., Kluytmans, J.A., 2008. Increase in a Dutch hospital of methicillin-resistant *Staphylococcus aureus* related to animal farming. Clin. Infect. Dis. 46, 261–263.

van Wamel, W.J., Rooijakkers, S.H., Ruyken, M., van Kessel, K.P., van Strijp, J.A., 2006. The innate immune modulators staphylococcal complement inhibitor and chemotaxis inhibitory protein of *Staphylococcus aureus* are located on beta-hemolysin-converting bacteriophages. J. Bacteriol. 188, 1310–1315.

Vancraeynest, D., Haesebrouck, F., Deplano, A., Denis, O., Godard, C., Wildemauwe, C., Hermans, K., 2006a. International dissemination of a high virulence rabbit *Staphylococcus aureus* clone. J. Vet. Med. B 53, 418–422.

Vancraeynest, D., Hermans, K., Haesebrouck, F., 2006b. Prevalence of genes encoding exfoliative toxins, leucotoxins and superantigens among high and low virulence rabbit *Staphylococcus aureus* strains. Vet. Microbiol. 117, 211–218.

Vanderhaeghen, W., Hermans, K., Haesebrouck, F., Butaye, P., 2010. Methicillin-resistant *Staphylococcus aureus* (MRSA) in food production animals. Epidemiol. Infect. 138, 606–625.

Varaldo, P.E., Kilpper-Bälz, R., Biavasco, F., Satta, G., Schleifer, K.H., 1988. *Staphylococcus delphini* sp. nov., a coagulase-positive species isolated from dolphins. Int. J. Syst. Bacteriol. 38, 436–439.

Voss, A., Loeffen, F., Bakker, J., Klaassen, C., Wulf, M., 2005. Methicillin-resistant *Staphylococcus aureus* in pig farming. Emerg. Infect. Dis. 11, 1965–1966.

Waldron, D.E., Lindsay, J.A., 2006. *Sau1*: a novel lineage-specific type I restriction-modification system that blocks horizontal gene transfer into *Staphylococcus aureus* and between *S. aureus* isolates of different lineages. J. Bacteriol. 188, 5578–5585.

Waldvogel, F.A., 1995. *Staphylococcus aureus* (including toxic shock syndrome). In: Mandell, G.L., Bennett, J.E., Dolin, R. (Eds.), Principles and Practice of Infectious Diseases. Churchill Livingstone, New York, NY, pp. 1754–1777.

Waldvogel, F.A., 2000. *Staphylococcus aureus* (including staphylococcal toxic shock). In: Mandell, G.L., Bennett, J.E., Dolin, R. (Eds.), Principles and Practice of Infectious Diseases, 5 ed. Churchill Livingstone, Philadelphia, pp. 2072–2073.

Wallin-Carlquist, N., Marta, D., Borch, E., Radstrom, P., 2010. Prolonged expression and production of *Staphylococcus aureus* enterotoxin A in processed pork meat. Int. J. Food Microbiol. 141 (Suppl. 1), S69–S74.

Ward, M.J., Gibbons, C.L., McAdam, P.R., van Bunnik, B.A.D., Girvan, E.K., Edwards, G.F., Fitzgerald, J.R., Woolhouse, M.E.J., 2014. Time-scaled evolutionary analysis of the transmission and antibiotic resistance dynamics of *Staphylococcus aureus* clonal complex 398. Appl. Environ. Microbiol. 80, 7275–7282.

Waters, A.E., Contente-Cuomo, T., Buchhagen, J., Liu, C.M., Watson, L., Pearce, K., Foster, J.T., Bowers, J., Driebe, E.M., Engelthaler, D.M., Keim, P.S., Price, L.B., 2011. Multidrug-resistant *Staphylococcus aureus* in US meat and poultry. Clin. Infect. Dis. 52, 1227–1230.

Weese, J.S., Archambault, M., Willey, B.M., Hearn, P., Kreiswirth, B.N., Said-Salim, B., McGeer, A., Likhoshvay, Y., Prescott, J.F., Low, D.E., 2005. Methicillin-resistant *Staphylococcus aureus* in horses and horse personnel, 2000-2002. Emerg. Infect. Dis. 11, 430–435.

Wendlandt, S., Schwarz, S., Silley, P., 2013. Methicillin-resistant *Staphylococcus aureus*: a food-borne pathogen? Annu. Rev. Food Sci. Technol. 4, 117–139.

Wieneke, A.A., Roberts, D., Gilbert, R.J., 1993. Staphylococcal food poisoning in the United Kingdom, 1969-90. Epidemiol. Infect. 110, 519–531.

Wilson, G.J., Seo, K.S., Cartwright, R.A., Connelley, T., Chuang-Smith, O.N., Merriman, J.A., Guinane, C.M., Park, J.Y., Bohach, G.A., Schlievert, P.M., Morrison, W.I., Fitzgerald, J.R., 2011. A novel core genome-encoded superantigen contributes to lethality of community-associated MRSA necrotizing pneumonia. PLoS Pathog. 7, e1002271.

Witte, W., Strommenger, B., Cuny, C., Heuck, D., Nuebel, U., 2007a. Methicillin-resistant *Staphylococcus aureus* containing the Panton-Valentine leucocidin gene in Germany in 2005 and 2006. J. Antimicrob. Chemother. 60, 1258–1263.

Witte, W., Strommenger, B., Stanek, C., Cuny, C., 2007b. Methicillin-resistant *Staphylococcus aureus* ST398 in humans and animals, Central Europe. Emerg. Infect. Dis. 13, 255–258.

Wright, S., 1982. The shifting balance theory and macroevolution. Annu. Rev. Genet. 16, 1–19.

Yu, F., Chen, Z., Liu, C., Zhang, X., Lin, X., Chi, S., Zhou, T., Chen, Z., Chen, X., 2008. Prevalence of *Staphylococcus aureus* carrying Panton-Valentine leukocidin genes among isolates from hospitalised patients in China. Clin. Microbiol. Infect. 14, 381–384.

Zhang, S., Iandolo, J.J., Stewart, G.C., 1998. The enterotoxin D plasmid of *Staphylococcus aureus* encodes a second enterotoxin determinant (*sej*). FEMS Microbiol. Lett. 168, 227–233.

Zink, D.L., 1997. The impact of consumer demands and trends on food processing. Emerg. Infect. Dis. 3, 467–469.

CHAPTER

SHIGA TOXIN-PRODUCING *E. COLI* AND RUMINANT DIETS: A MATCH MADE IN HEAVEN?

10

Whitney L. Crossland*, Todd R. Callaway†, Luis O. Tedeschi*

Department of Animal Science-Ruminant Nutrition, Texas A&M University, Kleberg, College Station, Texas, USA, USDA-ARS Southern Plains Region, College Station, Texas, USA†*

1 INTRODUCTION

The rumen of cattle is colonized by one of the most complex ecosystems of microflora known. The microbial consortium is composed of bacteria, protozoa, and fungi. Bacteria are the most numerous of these microorganisms at approximately 10 billion cells per mL of rumen fluid. This community of microbes shifts and changes as its host delivers the necessary substrates for survival, and in turn utilizes volatile fatty acids for energy; thus, ruminants can harvest energy from feeds containing cellulose that most mammals are unable to utilize. This symbiotic relationship has evolved for approximately 50 million years (Hackmann and Spain, 2010) and despite our best efforts to manipulate the rumen's inhabitants to perform our will, populations continue to reestablish and thrive in the fight for survival.

Although most bacteria play a vital role in ruminant digestion, some are pathogenic to humans. Shiga toxin-producing *Escherichia coli* (STEC) bacteria are part of the natural reservoir in cattle (Karmali et al., 2010) but can have devastating effects causing serious illness (hemorrhagic colitis and hemolytic uraemic syndrome) and death if ingested by humans. STEC causes more than 175,000 human illnesses and costs the American economy more than $1 billion each year (Scallan et al., 2011; Scharff, 2010). Cross-contamination of these bacteria from the ruminant digesta into the food chain is a constant and major area of concern and has significant economic impacts on the cattle industry.

In the United States, the 1993 outbreak of *E. coli* O157:H7 and the loss of human life prompted new government-mandated procedures and an urgent movement toward more stringent postharvest practices with more in-plant control points. The majority of initiatives in food safety research have focused on this phase of the beef production process. While in-plant strategies have been extremely effective at reducing the number of foodborne illnesses, they have not been perfect (Arthur et al., 2007; Barkocy-Gallagher et al., 2003). This is due in large part to indirect routes of human exposure through animal contact, manure amended soil, and manure contaminated runoff (Jay et al., 2007; Keen et al., 2007; Steinmuller, 2006). Therefore, in an effort to reduce these sources of cross-contamination and human illness, researchers have shifted their focus to the potential of preharvest interventions (Callaway et al., 2004a; LeJeune and Wetzel, 2007; Oliver et al., 2008; Sargeant et al., 2007).

Food Safety. http://dx.doi.org/10.1016/B978-0-12-800245-2.00010-1

Pathogen reduction strategies that focus on decreasing live animal shedding could have substantial impact on related human illnesses (Rotariu et al., 2012; Smith et al., 2012). The logic follows that by reducing foodborne pathogenic bacteria in live cattle there would be (1) a reduction in the load of pathogenic bacteria entering the processing plants rendering in-plant strategies more effective; (2) a reduction in the horizontal spreading between infected and noninfected cattle during transport and lairage; (3) a reduction in the amount of pathogenic bacteria in wastewater streams and soil used for food production; and (4) a lower risk for direct exposure of humans via petting zoos, rodeos, and those whose business or employment requires physical handling of cattle (Callaway et al., 2013a). Therefore, recent research has fixated on two main areas in cattle production systems to accomplish this goal: (1) lairage and transport management practices and (2) cattle diet and water management (Callaway et al., 2013a). The proposed initiative is simple and logical; however, finding a clear-cut solution that is applicable to all cattle production systems remains elusive. STEC bacteria are estimated to have emerged some 400-70,000 years ago and have coevolved with their host to be able to endure incredible variances in environment (Law, 2000; Riley et al., 1983; Wick et al., 2005; Zhou et al., 2010). They will not be eliminated easily.

2 STEC, EHEC, VTEC, AND NON-O157:H7 *E. COLI*

Known by the media as the infamous "hamburger bug," *E. coli* O157:H7 has been referred to by a number of different acronyms and researchers have used them interchangeably: Shiga toxin-producing *E. coli*, Enterohemorrhagic *E. coli* (EHEC), and Verotoxin-producing *E. coli* (VTEC). However, *E. coli* bacteria actually belong to a single group that acquired genes from *Shigella* via a gene transfer event (Karmali et al., 2010; Wick et al., 2005) and therefore, for all intents and purposes of this chapter, we will refer to them as STEC.

Given the complex society of bacteria in the gastrointestinal tract of cattle, it would be naïve to think that *E. coli* O157:H7 evolved in isolation and is the only serotype capable of producing toxins. This particular *E. coli* strain was simply the first to be linked with a major foodborne illness outbreak (Bettelheim, 2007; Fremaux et al., 2007). In fact, recent rule making by the USDA/FSIS has included six more "non O157:H7 STEC" serogroups (O26, O45, O103, O111, O121, and O145) that were associated with foodborne illness outbreaks (Bettelheim, 2007; Fremaux et al., 2007). The USDA-FSIS (2012) has recently declared these bacteria, along with O157:H7, as adulterants of beef.

Currently, research operates on the assumption that, for the most part, these non-O157 STEC will react to changes in their environment in a similar fashion as the O157:H7 STEC (Callaway et al., 2013a). However, like the O157:H7, these highly specialized derivatives vary in genotype and phenotype, even within serotypes (Zhang et al., 2007; Bergholz and Whittam, 2007; Free et al., 2012; Fremaux et al., 2007). There is limited research focusing on the non-O157:H7 STECs and their significance to the bionetwork within the gastrointestinal tract of cattle. For that reason, however logical our assumptions may be, strategies that work to deter one STEC may not hold true for another. Yet we know that O157:H7 is highly adaptable to survive in cattle (Garcia-Gonzalez et al., 2010; O'Reilly et al., 2010) and can contaminate ground beef (Bosilevac and Koohmaraie, 2011; Fratamico et al., 2011). The same is true of other STECs (Arthur et al., 2002; Chase-Topping et al., 2012; Joris et al., 2011; Monaghan et al., 2011; Fratamico et al., 2011). The hypothesis that the non-O157:H7 STECs will behave similarly to the O157 is an educated guess and this limitation is acknowledged by researchers (Callaway et al., 2013a). Still, until more time is dedicated to other STECs and research proves otherwise, we must not worry too much about "what might be" and operate on the premise of "what is."

3 ECOLOGY OF THE RUMINANT RESERVOIR AND STEC

The rumen harbors a microbial ecosystem in which a network of bacteria, protozoa, and fungi break down feed ingredients into useable nutrients; fiber and starch are degraded to the ruminant's primary energy source, volatile fatty acids (VFAs); proteins into peptides, amino acids (AA), and nitrogen (N); vitamins and minerals (Owens, 1988; Yokoyama and Johnson, 1988). Feed ingredient type and form entering the rumen determines the microbiological dynamics of digestion and the concentration of end products made available to the animal. Nutrient (dietary) requirements of cattle are determined by their phase of production (growth, gestation, lactation, and maintenance). Ingredient price, availability, practicality, and other factors, determine the contents of the ration. Therefore, the diets of cattle can vary tremendously depending on the aforementioned factors. However, the majority of cattle entering the processing plant have been consuming an energy- and protein-concentrated ration in order to maximize their feed to gain efficiency (Huntington, 1997). After a meal, the fermentation of a grain concentrate (high starch) ration by rumen microbes results in a rapid accumulation of VFA end products causing the typically neutral rumen environment to become more acidic. Most microbes are sensitive to changes in pH, having an optimal growth range, and this decrease alters the microbial profile of the rumen in favor of those who can grow and proliferate the most efficiently in the new environment. Postruminal digestion of nutrients occurs in similar fashion to monogastric processes. Although primary dissimilation of starch occurs in the rumen, portions of undegraded starch can reach the colon where, to a lesser extent, bacterial fermentation can still take place resulting in accumulation of acids in the lower GIT (Huntington, 1997). Research of intestinal environment indicates that low pH and high concentrations of short-chain VFA result in lower STEC populations, due to their toxic effect on the organisms (Hollowell and Wolin, 1965). However, the discovery of acid-resistant *E. coli* strains (Poynter et al., 1986) implies a competitive advantage among different genotypes and that, ultimately, the niche will be filled (Fuller, 1989).

The entirety of the ruminant digestive tract can harbor various genotypes of *E. coli* O157:H7 (Naylor et al., 2003; Smith et al., 2005). This finding emphasizes the fitness of these organisms and their tenacious ability to coevolve with their preferred host. The optimal site of colonization of STEC O157:H7 is proximal to the recto-anal juncture (RAJ) (about 5 cm of total tract length) where it prefers to adhere to mucosal epithelium (Naylor et al., 2005; Davis et al., 2006; Greenquist et al., 2005; Lim et al., 2007). This site is characterized as having a high density of lymphoid follicles (Naylor et al., 2003) and likely contributes to its survival and the fecal-oral lifestyle of *Escherichia coli*. There the organism produces a cytotoxin (Shiga toxin) that, while it does not affect its typical bovine host (who lacks the necessary toxin receptors) (Pruimboom-Brees et al., 2000), can wreak havoc on an unfortunate human host (Karmali et al., 2010; O'Brien et al., 1992). In humans, this cytotoxin is thought to aid the pathogen's ability to adhere to the cellular wall of the intestine, evade host immune response, and cause massive destruction to the brushy border of the microvilli (LeBlanc, 2003; Nataro and Kaper, 1998). In cattle, this site of adherence ensures the organism's survival, by means of fecal shedding, where it can continue to live in manure and soil (Dargatz et al., 1997; Jiang et al., 2002; Maule, 2000; Yang et al., 2010; Bolton, 2011; Semenov et al., 2009; Van Overbeek et al., 2010). In fact, STEC O157:H7 has been reported to survive in manure in typical environmental temperatures for ≥49 days (Wang et al., 1996). The ability to persist in these intermediate environments for long periods of time allows STEC to spread to new hosts in the same pen or herd, via the fecal-oral route (Russell and Jarvis, 2001). It also ensures the pathogen's ability to cross-contaminate beef products, via presence on handling facilities and surfaces, and cause human illness (Ferens and Hovde, 2000). Another major source of concern is the ability of STEC to persist on the hides of cattle. Hides can collect dirt and feces contaminated with STEC and transfer the organism horizontally, especially during transport and lairage (Bach et al., 2004; McGee et al., 2004). Barkocy-Gallagher et al. (2003) found that the number of hides testing positive for *E. coli* O157:H7 is a more accurate predictor of carcass contamination than is fecal prevalence.

Other indirect routes of human illness have been seen to occur from contaminated runoff water from feedlots and dairies (Edwards et al., 2008; Goss and Richards, 2008; Jay et al., 2007) and food crops irrigated with water from cattle farms (Gerba and Smith, 2005; Manshadi et al., 2001; Natvig et al., 2002). More recent research points to pest flies linking contaminated feedlots and human food crops (Berry, 2013). Thus, the persistence of STEC and their unique adaptation abilities further emphasizes the need for management strategies to reduce the pathogen load entering food chain pathways and potentially causing human infection (Hynes and Wachsmuth, 2000; Loneragan and Brashears, 2005; Sargeant et al., 2007).

Therefore, considering the fecal-oral lifestyle of STEC and their ability to adapt to new environments, it may take a combination of practices at the production level to gain relative control and reduce or manage the level of *E. coli* prevalence in a herd. Many different management strategies, feedstuffs, and supplements have been examined for their ability to reduce STEC colonization and shedding. Although the title of this chapter focuses on the diets of ruminants and their effect on STEC, the mechanism is more complicated and, therefore, management strategies will be discussed first.

4 MANAGEMENT STRATEGIES

The efficiency of animal agriculture relies heavily on good animal and health management strategies of the operation. Prevention of animal disease through sanitation and herd health management protocols improve animal performance and thereby increase profitability. While these practices are largely successful at limiting the spread of infectious animal disease they are not necessarily effective at reducing foodborne pathogens. To complicate the matter further, cattle are asymptomatic carriers of STEC pathogens and there is no visual detection method to distinguish infected animals from noninfected animals. Additionally, due to the long time span from birth to slaughter of cattle and all the comingling in between, there is no single strategy proven effective for complete abatement of foodborne pathogen colonization and shedding that can be implemented for all phases of production (Ellis-Iversen et al., 2008; Ellis-Versen and Van Winden, 2008; LeJeune and Wetzel, 2007). Still, there are practices that have been proven effective at reducing horizontal transmission and recirculation of STEC within a herd of cattle (Ellis-Iversen and Watson, 2008).

4.1 IDENTIFYING HIGH-RISK ANIMALS: SUPER SHEDDERS AND SUPER SPREADERS

Studies conclude the prevalence of cattle testing positive for *E. coli* O157:H7 are generally less than 10% of the population but may fluctuate seasonally (Gansheroff and O'Brien, 2000). Although these cattle are clearly the minority, they are suggested to account for the majority (≥96%) of *E. coli* O157:H7 bacteria shed (Matthews et al., 2006b; Omisakin et al., 2003). This finding warrants urgency for a method of identifying these high-risk animals and strategies to isolate them from their herd mates. Current research is focusing on how and why these few animals are the subject of colonization when the larger majority of cattle are not. As previously mentioned, the preferred site of colonization by *E. coli* O157:H7 is the RAJ. Researchers also identified that cattle colonized at this site were implicated with a higher concentration of O157: H7 fecal excretion and did so for a longer period (weeks to months) of time as compared with "noncolonized" cattle (Low et al., 2005; Cobbold et al., 2007; Lim et al., 2007). The noncolonized cattle still shed the bacteria, indicating proliferation elsewhere in the GIT but at a much lower concentration and for a much shorter period of time (a few days) (Davis et al., 2006; Lim et al., 2007). At this point, researchers deduced that RAJ colonization may be intimately involved with the phenomenon of "super shedding" (Callaway et al., 2006; Cobbold et al., 2007; LeJeune and Kauffman, 2006; Lim et al., 2007).

Super-shedding cattle may be defined as individuals who for a period of time yield a much greater concentration of infectious units per gram of feces than the majority of individuals of the same species (Chase-Topping et al., 2008). Researchers differ on the exact threshold, but for the purposes of this chapter, super-shedding cattle may be classified by direct counts of *E. coli* O157 in fecal samples where excretion of *E. coli* O157:H7 is $\geq 10^4$ CFU (colony forming units) per gram of feces (Chase-Topping et al., 2008). Although there are several theories, the exact mechanism of RAJ colonization and super shedding remains elusive, but it has been proposed that the specific bacterial strain plays a role, as well as animal genotype and animal environment (Chase-Topping et al., 2007; Halliday et al., 2006; Dziva et al., 2004; Matthews et al., 2006a; Naylor et al., 2005; Roe et al., 2004).

Because these super-shedding cattle are visibly asymptomatic, it would seem that any management attempts are futile. However, another term being used in current research is "super spreader" (Chase-Topping et al., 2008). It is important to point out here that the terms super shedder and super spreader are sometimes used synonymously in research, although cattle may be distinctly one or the other, or both. Super-spreading cattle are individuals who may have more opportunities to infect other individuals (Chase-Topping et al., 2008). With this train of thought in mind one could make inferences for intervention strategies for the farm/feedlot. A common sense evaluation of risk potential may lead to tactics that decrease in-farm and between-farm transmissions of many bacteria-related diseases. For example, risk potential of an animal may include behavioral factors such as those that increase nose-to-nose contact (i.e., introduction of new animals, accessibility of animals in neighboring pens, establishing dominance within a group). Another important intervention strategy might be identifying factors that increase mounting (females in heat, cattle that have been implanted, and in some countries feedlots with intact bulls). Lastly, and probably the most influential is the source of procurement of animals. Animals purchased from hubs (like sale barns) have many more opportunities to spread or become infected with STEC than those bought from their farm of origin. Feedlots may consider purchasing groups of cattle that have not been mixed with outside animals to decrease their risk potential. Monte Carlo simulations have been used to model animal risk for being positive for STEC O157:H7 upon entrance to the abattoir (Stacey et al., 2007), and there are many parameters to consider.

4.2 HOUSING

As mentioned earlier, STEC O157:H7 can survive for long periods of time in manure, soil, and other organic materials and be transmitted to new hosts or reinfect previous hosts (Jiang et al., 2002; Maule, 2000; Winfield and Groisman, 2003). This can include bedding typically found in dairy facilities, as well as the organic matter in the pen floor of feedlots. Modeling research (and as common sense would have it) has shown that an increase in bedding cleaning frequency would increase the death rate of *E. coli* O157:H7 (Vosough Ahmadi et al., 2006). Additionally, the type of bedding animals are confined to has been shown to have an effect on STEC O157:H7 transmission (Ellis-Iversen et al., 2008; Ellis-Versen and Van Winden, 2008; LeJeune and Kauffman, 2005; Westphal et al., 2011). Dryer bedding, like sand and sawdust, has the propensity to reduce prevalence and transmission from animal to animal possibly due to desiccation and reduced nutrient availability to the bacteria (Callaway et al., 2013a). This supposition is validated by previous research where the presence of urine was shown to increase the growth of *E. coli* O157:H7 on animal bedding as it may potentially provide substrate (urea or D-serine) for growth (Davis et al., 2005).

4.3 FASTING

Cattle are often fasted 12-48 h prior to slaughter to (1) ease the evisceration process, (2) reduce bacterial cross-contamination due to spilled gut fill, and (3) improve dressing percentage of the carcass (Carr et al., 1971). However, this methodology may actually increase the incidence or potential of adulterated beef. Studies have indicated a paradoxical effect of fasting on *E. coli* O157:H7 populations; actually significantly increasing their concentrations throughout the entire GI tract as opposed to those fed a high-concentrate diet or forage diet (Buchko et al., 2000b; Gregory et al., 2000). Moreover, some have found an increased incidence in the colonization of previously noninfected calves when fasted for 48 hours (Cray et al., 1998; Gregory et al., 2000; Kudva et al., 1995). This result may be a confounding effect between diminishing VFA in the GIT and stress-induced immune suppression from shipping.

4.4 VACCINATION

Because STEC O157:H7 does not cause disease in cattle and natural exposure does not provide any immunity, vaccination may not be the first thought in prevention of colonization. However, by targeting the pathogen's physiological characteristics, vaccination series that interrupt the bacterium's metabolic pathways are proving to be successful in reducing the colonization and shedding (Callaway et al., 2014; Walle et al., 2013).

Currently, there are two vaccines with different modes of action against the bacterium: the siderophore vaccine (SRP) disrupts the iron transport receptor of the bacterium and the other, developed from bacterial extract, targets the type III proteins (cytotoxins) secreted by the bacterium (Callaway et al., 2014; Snedeker et al., 2012; Walle et al., 2013; Varela et al., 2013).

A preliminary study of STEC orally challenged calves using a 3-dose SRP vaccination series showed encouraging results and served as the catalyst for finding the optimal vaccination protocol (Thornton et al., 2009). The SRP vaccine has been conditionally approved for cattle use in the United States and has been shown to reduce fecal populations of *E. coli* O157:H7 in cattle by 98% without affecting cattle performance (Thomson et al., 2009). Another study revealed the SRP vaccine reduced in-herd prevalence of *E. coli* O157:H7 by 50% (Fox et al., 2009). In contrast, a study to investigate the passive immunity of vaccinated dams, nonvaccinated dams but vaccinated calves, and a combination of both using a 3-dose series revealed no statistical differences of fecal prevalence at the time of feedlot entry nor the time of slaughter; however, antibody titers were approximately three times greater for calves of vaccinated dams at 90 days of age (Wileman et al., 2011). However, in this study, the researchers repeatedly vaccinated cattle during times of major stress (branding and castration and post-transport to unfamiliar yardage), which is typical protocol for most cattle processing (Wileman et al., 2011). Human studies have revealed that physiological stress level in the 10 days following vaccination shape the long-term antibody response (Miller et al., 2004), which may have profound implications for vaccination timing protocol in ruminants. An even more recent study in a typical feedlot setting utilized a novel 2-dose series SRP vaccination regimen (requiring less animal handling than the 3-dose series), which reduced both prevalence (by 53%) and incidence (by 77%) of super-shedding ($\geq 10^4$ CFU) feedlot cattle (Cull et al., 2012). This finding may further validate the relationship between stress and vaccination efficiency.

Bacterial extract vaccinations have also been utilized for inhibiting type III secretions, which aid the pathogens' ability to colonize the mucosal epithelium. Econiche™ was the first commercially licensed vaccine of its kind to be utilized and preliminary research reported promising results, reducing fecal shedding of STEC O157: H7 of feedlot cattle from 23% to 9% (Moxley et al., 2003; Potter et al., 2004; Van Donkersgoed et al., 2005). Other observations reported reduced fecal shedding from 46% to 14% in feedlot cattle (Ransom et al., 2003). Number of vaccinations in the series has also been reported, with variable results indicating a 3-dose regimen was most efficient (Moxley et al., 2009) but a 2-dose regimen was sufficient and likely more practical for feedlot operators to implement (Smith et al., 2009). Further development of vaccines to include other bacterial extract proteins are being investigated in experimental infection models (McNeilly et al., 2010; Sharma et al., 2010). Antibodies were present in calves whose dams were given vaccines targeting EspA, EspB, Shiga toxin 2, and intimin proteins (Rabinovitz et al., 2012). Multiprotein vaccines have also been shown to reduce fecal shedding in sheep and goat experimentally challenged models (Yekta et al., 2011; Zhang et al., 2012). Probably the most profound study regarding multiprotein vaccines investigated the cross-reactivity of non-O157 STECs, finding, because the STEC share a similar type-III secretion system, a small degree of cross-protection (Asper et al., 2011).

Some researchers have devised chimeric multiprotein (eae, tir, intimin) vaccines that can be expressed in plants that may potentially be utilized as an edible source of immunity (Amani et al., 2010, 2011). Although edible vaccines would be idealis for reducing animal handling, more work would need to be done to ensure passive protection of the proteins so that they aren't degraded during digestion (Callaway et al., 2014).

4.5 WATER TREATMENT

Researchers have demonstrated that cattle water troughs can be reservoirs for the dissemination of *E. coli* O157:H7 and it is suspected that this is a common route of horizontal transmission between animals (LeJeune et al., 2001). Chlorination is often used to treat microbial activity in water sources and it has also been established that the addition of chlorate to *E. coli* cultures kills the bacteria due to nondifferentiation of nitrate and chlorate by the nitrate reductase enzyme, which quickly induces a toxic chlorite effect (Stouthamer, 1969; Stewart, 1988). In vitro populations of *E. coli* O157:H7 and *Salmonella* were quickly reduced when treated with chlorate (Anderson et al., 2000a). Chlorate-treated feed was shown to reduce experimentally inoculated STEC O157:H7 populations in swine and sheep GITs (Anderson et al., 2001; Edrington et al., 2003c). Further studies revealed significant reductions in ruminal, cecal, and fecal *E. coli* O157:H7 populations when soluble chlorate was administered in drinking water of both cattle and sheep (Anderson et al., 2002; Callaway et al., 2002, 2003). These researchers also pointed out that chlorate treatment of drinking water does not adversely affect fermentation (Anderson et al., 2000b) nor antibiotic resistance, nor toxin production by O157 (Callaway et al., 2004b,c). Additionally, lethal-dose levels to the pathogen pose no severe risk for animal toxicity (Callaway et al., 2013a; Fiume, 1995). Pending FDA approval and more practical-setting research, chlorate has been suggested to be supplemented in drinking water or in the last feeding prior to shipment to slaughter (Callaway et al., 2013a).

5 FEEDSTUFFS

Cattle in beef or milk production facilities are typically fed concentrate rations to maximize their production efficiency (Huntington, 1997). The rumen microbes have the ability to utilize dietary starch; however, depending on the grain source and mechanical processing, some starch may reach the colon undegraded where it can undergo microbial fermentation (Huntington, 1997). The grain source and form have been shown to have an effect on fecal *E. coli* populations and shedding (Bach et al., 2005a,b; Berg et al., 2004; Buchko et al., 2000a; Fox et al., 2007; Scott et al., 2000).

It has been demonstrated that steam-flaked grains increased *E. coli* O157:H7 shedding in feces compared to diets composed of dry-rolled grains (Fox et al., 2007). It was theorized that this difference was due to the processing method; dry-rolled grains allowed more undegraded starch to pass to the hind gut for microbial fermentation resulting in greater VFA production in the colon and toxicity to *E. coli* O157:H7 (Fox et al., 2007). Barley is known to ferment more quickly than corn in the rumen. Barley-fed cattle have been shown to have higher fecal pH and lower VFA concentration and have been associated with greater *E. coli* 0157:H7 populations and shedding compared with corn-fed cattle (Berg et al., 2004; Dargatz et al., 1997; Buchko et al., 2000a). Research indicates the rate of fermentation and rumen availability of starch affects postruminal *E. coli* population and shedding. Moreover, it would seem that by feeding ingredients (not necessarily high in starch) that are readily available in the rumen and limiting the amount of by-pass nutrients to the colon, results in an increase in hindgut pH and possibly a more hospitable environment to *E. coli* 0157:H7. An analysis of several dairy feeding and management strategies revealed that heifers fed corn silage significantly increased the risk of *E. coli* shedding as opposed to those consuming nonfermented feeds (Herriott et al., 1998). Ethanol feed byproducts have also been implicated with increased shedding of *E. coli* 0157:H7 (Jacob et al., 2008a,b; Yang et al., 2010). Cattle fed 40% wet corn distiller's grain (WCDG) had a greater fecal *E. coli* population and a much different microbial population in general than did cattle consuming non-WCDG diets (Durso et al., 2012). Another study exposed a trend of increased survival of *E. coli* O157: H7 in feces as levels of distiller's grain (DG) supplementation increased (Varel et al., 2008). A cow-calf operation in Scotland also reported increased shedding when DG was used in the diet (Synge et al., 2003). One study found that feeding a related product (brewer's grain) to cattle increased the odds of shedding *E. coli* O157:H7 by more than sixfold (Dewell et al., 2005). Therefore, it would seem that feed ingredients that undergo processing to maximize rumen availability or exploit specific nutrient content inherently increase the risk of colonization and shedding of *E. coli* O157: H7.

On the opposite end of the spectrum, researchers have found that feeding forage-based diets actually increased the duration of shedding *E. coli* O157:H7 when compared with grain-based diets (60 days vs. 16 days, respectively; Van Baale et al., 2004). A previous study had also reported experimentally challenged cattle fed hay shed STEC O157:H7 significantly longer than grain-fed cattle (42 days vs. 4 days) but that fecal concentrations were similar between diets (Hovde et al., 1999). These results were likely due to differences in VFA production and pH of the GIT. An in vitro study found similar results where rumen fluid from steers consuming a high-forage diet allowed *E. coli* 0157:H7 to proliferate to higher populations than did rumen fluid from grain-fed steers (Tkalcic et al., 2000). In general, however, higher concentrations of *E. coli* (generic and STEC) are found in the feces of grain-fed cattle versus forage-fed cattle (Callaway et al., 2013a). It should be emphasized that STEC are still isolated from cattle solely fed forage and researchers have found no difference in food safety parameters of beef from grass-fed cattle versus grain-fed cattle (Hussein et al., 2003; Thran et al., 2001; Zhang et al., 2010). However, abruptly shifting from grain to hay appears to cause a severe disruption to the rumen microbial ecosystem (Fernando et al., 2010). Generic *E. coli* populations declined 1000-fold within 5 days when cattle fed a 90% concentrate ration were suddenly switched to a 100% good quality hay ration (Diez-Gonzalez et al., 1998). Therefore, researchers proposed that feedlot cattle could be switched from their concentrate rations to hay 5 days prior to slaughter to reduce pathogen load entering the abattoir (Diez-Gonzalez et al., 1998). Researchers following this logic compared naturally infected with O157:H7 grain-fed controls with grain-fed cattle who were later abruptly switched to hay and reported that 52% of controls remained positive for the pathogen while only 18% of cattle in the hay switch group continued to shed O157 (Keen et al., 1999). Conversely, it was found that when cattle were switched from a forage-based diet to a high-grain diet, ruminal and fecal populations of generic *E. coli* concentrations increased (Berry et al., 2006). Other researchers investigated the effects of switching cattle from pasture to hay for 48 h prior to slaughter, which was found to significantly reduce *E. coli* populations throughout the GIT (Gregory et al., 2000). These same researchers also reported that by feeding hay Enterococci populations were increased, which are capable of inhibiting the activity of *E. coli* through competition for nutrients (Gregory et al., 2000).

Mutual results suggest that shifting cattle from high-concentrate rations to a forage ration (specifically hay) could be a powerful method to reduce STEC populations just prior to harvest. However, the mode of action here remains elusive and total reduction results are somewhat inconsistent. Additionally, feasibility of this practice is not advocated by feedlots due to reports of animal weight loss and other logistical issues (Keen et al., 1999; Stanton and Schutz, 2000). Implementation of this concept may require more research using other high-fiber feedstuffs (hulls) (Callaway et al., 2013a).

6 SUPPLEMENTS

There have been several feed supplements proposed to reduce STEC O157:H7 colonization and shedding in cattle. Some treatments are directly antipathogenic while others compete for substrate reducing the severity of the pathogenic population.

6.1 TANNINS, PHENOLICS, AND ESSENTIAL OIL

Plants contain phenolic and polyphenolic compounds (lignin and tannins) that are known to have antimicrobial affects in ruminants (Berard et al., 2009; Cowan, 1999; Hristov et al., 2001; Jacob et al., 2009; Patra and Saxena, 2009). It is theorized that some of these compounds may penetrate biofilms and have an antiquorum-sensing effect (coordinated gene expression), which may play a role in STEC colonization (Edrington et al., 2009; Kociolek, 2009; Sperandio, 2010). Tannins have been demonstrated to significantly inhibit the growth of *E. coli* O157:H7 in vitro and generic *E. coli* populations in cattle (Berard et al., 2009; Cueva et al., 2010; Min et al., 2007; Wang et al., 2009). Other researchers have found that the phenolic acids of lignin have antimicrobial activity against *E. coli* O157:H7 in fecal slurries, and feeding highly lignified forage to cattle showed a reduced period of *E. coli* O157:H7 shedding compared with those fed only corn silage (Wells et al., 2005). Several in vitro studies using various berry extracts have reported inhibition of *E. coli* O157:H7 growth (Caillet et al., 2012; Fullerton et al., 2011; Lacombe et al., 2012).

Brown seaweed (Tasco-14®) is a feed additive that has been included in cattle diets to improve carcass quality characteristics and shelf life, increase antioxidants in vivo, and to improve ruminal fermentation efficiency (Anderson et al., 2006; Braden et al., 2007; Leupp et al., 2005). In vitro studies have indicated that Tasco-14® can reduce populations of *E. coli* and Salmonella (Callaway, unpublished data), and inclusion of Tasco-14® in feedlot cattle rations has been shown to significantly reduce fecal and hide prevalence of *E. coli* O157 (Braden et al., 2004). More recent research has linked this antipathogen activity to the presence of phlorotannins in the brown seaweed (Wang et al., 2009) and also reported the anti-*E. coli* activity of phlorotannins to be significantly greater as compared to terrestrial tannin sources (Min et al., 2007; Wang et al., 2009). Because Tasco-14® is currently available in the marketplace, this is a product that can easily be included in cattle rations; however, the extent of antipathogen activity in vivo is still not clear, so the cost of addition must be weighed carefully by the producer (Callaway et al., 2013a).

Essential oils are most often associated with aromatic compounds in various plants used as spices or extracts (Barbosa et al., 2009). Many of these essential oils exhibit antimicrobial activity (Dusan et al., 2006; Fisher and Phillips, 2006; Kim et al., 1995; Pattnaik et al., 1996; Reichling et al., 2009; Turgis et al., 2009), often by dissolving bacterial membranes (Di Pasqua et al., 2007; Turgis et al., 2009). As a result, many plant products have been used for food preservation and to increase shelf life (Dabbah et al., 1970). It has been proposed that essential oils may be potential modifiers of the ruminal fermentation (Benchaar et al., 2007, 2008; Boadi et al., 2004; Patra and Saxena, 2009) and, after encouraging in vitro studies, reduce *E. coli* O157:H7 in the live animal (Benchaar et al., 2008; Jacob et al., 2009). Some essential oils have been shown to penetrate biofilms and kill *E. coli* O157:H7 (Pérez-Conesa et al., 2011). Citrus fruits contain a variety of compounds, including essential oils and phytophenols that exhibit antimicrobial activity against foodborne pathogens (Friedly et al., 2009; Mkaddem et al., 2009; Nannapaneni et al., 2008; Viuda-Martos et al., 2008). Orange peel and citrus pulp by-products have excellent nutritional characteristics for cattle and have been included as low-cost ration ingredients in dairy and beef cattle rations for many years (Arthington et al., 2002). Research has demonstrated that the addition of >1% orange peel and pulp reduced populations of *E. coli* O157:H7 and *Salmonella* Typhimurium in mixed ruminal fluid fermentations in vitro (Callaway et al., 2008; Nannapaneni et al., 2008). In newly weaned swine feeding orange peel and pulp reduced intestinal populations of diarrheagenic *E. coli* (Collier et al., 2010). One study proposed a mode of action suggesting that limonids from grapefruit may be responsible for inhibiting cytotoxin secretion and intercellular communication by *E. coli* O157:H7 (Vikram et al., 2010). In sheep, researchers demonstrated that feeding of orange peel and citrus pellets (a 50/50 mixture) at levels up to 10% DM reduced artificially inoculated populations of *E. coli* O157:H7 and *Salmonella* Typhimurium (Callaway et al., 2011a,b). However, when follow-up studies were performed using only dried pelleted orange peel, the pathogen-reducing effect disappeared (Farrow et al., 2012), likely due to the inactivation of essential oils (limonene and terpeneless fraction) during the pelleting process. Postmortem applications to beef carcasses has also shown potential for reducing both STEC and *Salmonella* (Pendleton et al., 2012; Pittman et al., 2011). To date, orange peel feeding has not been examined in large-scale feeding studies, but retains promise as a potential on-farm strategy to reduce the burden of pathogens on the farm, reducing environmental contamination and reinfection (Callaway et al., 2013a).

It is evident from the research mentioned here that plant secondary compounds merit further investigation for their practical use and effectiveness at reducing STEC colonization and shedding.

6.2 ANTIBIOTICS

Traditional antibiotics have been used in cattle diets as prophylactics to decrease the risk of subclinical infections as well as to promote growth and feed efficiency (Alexander et al., 2008; Gaskins et al., 2002). Recent research has shown that select antibiotics have a direct effect on intestinal populations of pathogenic bacteria (Elder et al., 2002; Ransom et al., 2003). Neomycin was shown to significantly reduce STEC O157 shedding in feedlot steers (Elder et al., 2002; Ransom et al., 2003). The product used by the researchers was fed for only 3 days and withdrawn 24h prior to slaughter as per label instructions and is intended for the control of colibacillosis (bacterial enteritis); however, after these findings an extra-label use may be approved by the Food and Drug Administration for the control of *E. coli* O157:H7 (Ransom et al., 2003).

However, the use of antibiotics at the subtherapeutic level in animal diets has been scrutinized for the proliferation of antibiotic-resistant bacteria, which may have profound implications on human pathogens (Callaway et al., 2004a; Ghosh and LaPara, 2007; Wichmann et al., 2014). It should be noted that neomycin targets Gram-negative bacteria and is closely related to many antibiotics commonly used to treat several (some life-threatening) human infections (Mingeot-Leclercq et al., 1999). One in vitro study observed increased cytotoxin activity of STEC O157:H7 due to antimicrobial stress (Walterspiel et al., 1992), which was later correlated with developing the complication of HUS in infected children (≤ 13 years) (Slutsker et al., 1998). Hence, although this product would be easily implemented into the finishing phase of cattle, it may not be utilized due to the risk inherent.

6.3 IONOPHORES

Ionophores are antimicrobials already implemented in many phases of beef production to promote growth and feed efficiency (Russell and Strobel, 1989). Although the role of ionophores is the same as antibiotics (to kill or inhibit growth of bacteria), the modes of action are completely separate and unlikely to contribute to antibiotic resistance (Russell and Houlihan, 2003). Ionophores (monensin, lasalocid, laidlomycin propionate) primarily inhibit the metabolism of Gram-positive bacteria and protozoa in the rumen that decrease animal efficiency (methane production) (Guan et al., 2006). Being Gram-negative, STEC O157:H7 would not be inhibited by ionophores but may, in fact, cause concern over allowing the pathogen a competitive advantage. However, both in vitro and in vivo studies have been conducted to find that feeding ionophores have no effect on pathogen populations (Edrington et al., 2003a,b).

6.4 BACTERIOPHAGE

Bacteriophages are naturally occurring viruses that target and kill bacteria and are often found in the GIT of livestock. Phages are specific to individual bacterial strains (Barrow and Soothill, 1997) targeting specific protein receptors on the surface of the bacterium for entry (Lederberg, 1996). Phages commandeer a specific bacterium and utilize intracellular nutrients to replicate until the host bursts, releasing the next generation of phages to repeat the process (Callaway et al., 2013a). Its nature and specificity has led to phage cultivation for use in pathogen elimination (Klieve and Bauchop, 1988; Klieve et al., 1991; Klieve and Swain, 1993). It has been suggested that dosing livestock with bacteriophage may be more effective to eliminate STEC O157:H7 from a mixed microbial population as opposed to less-discriminatory broad-spectrum antibiotics (Johnson et al., 2008; Merril et al., 1996; Summers, 2001). Several in vivo studies have supported this concept when determining phage effects on swine, sheep, and poultry diseases (Smith and Huggins, 1982, 1983; Huff et al., 2002). In cattle, Enteropathogenic *E. coli* (EPEC) causes diarrhea and has similar physiological and ecological characteristics to STEC O157:H7. Studies have reported calves treated orally, prior to pathogen inoculation, with bacteriophage reduced the incidence of developing EPEC diarrhea and splenic colonization (Smith and Huggins, 1983, Smith et al., 1987), indicating that phages could be used as a prophylactic means of reducing the load of foodborne pathogens from entering the food chain (Callaway et al., 2004a). Subsequent studies attempted to isolate and examine bacteriophages to specifically target *E. coli* O157:H7 in cattle and sheep (Kudva et al., 1999; Bach et al., 2003; Callaway et al., 2003). Preliminary in vitro studies were largely successful, supporting the use of bacteriophages to eliminate STEC O157:H7 (Kudva et al., 1999; Bach et al., 2003), but early in vivo studies revealed a less effective response where, although no significant statistical difference was observed, pathogen concentrations tended to decrease (Bach et al., 2003; Callaway et al., 2003). Since then, researchers have isolated several different phages from feedlot cattle (Callaway et al., 2006; Niu et al., 2009, 2012; Oot et al., 2007) and these have been used to reduce populations of STEC O157:H7 strains in experimentally infected animals as "proofs of concept" (Bach et al., 2009; Callaway et al., 2008; Rivas et al., 2010). Research also indicates ruminants with naturally occurring phage are more resistant to *E. coli* O157:H7 colonization (Raya et al., 2006), which has caused confounding results in some intervention studies (Kropinski et al., 2012).

Research studies have clearly earmarked limitations of using bacteriophages. First, the self-limiting effect of phage, in that, once targeted bacteria are eliminated, phage diminish and the animal is not protected from reinfection of the pathogen (Callaway et al., 2013b). Second, due to their specificity, rapid development of bacterial resistance and virulence gene transfer may occur if lytic multiphage cocktails are not implemented (Brabban et al., 2005; Law, 2000; Tanji et al., 2005).

6.5 DIRECT-FED MICROBIALS AND ORGANIC ACIDS

Supplementation with probiotics has received much attention of late to promote gut health of livestock and in an effort to reduce antimicrobial-resistant pathogen strains (Callaway et al., 2004a). The mode of action is by competitive exclusion (CE), where a benign foreign bacterial population is introduced to the intestinal tract in order to reduce incidence of colonization or decrease the existing population of pathogenic bacteria (Fuller, 1989; Nurmi et al., 1992; Steer et al., 2000). Probiotics may be comprised of a single-strain bacterium population, a multistrain bacterium population, or combine several different species of bacteria. The introduction of exogenous bacteria is thought to have an effect on pathogens by three modes of action: (1) competition for binding sites along the GIT epithelium (Collins and Gibson, 1999; Kim et al., 2008); (2) competition for mutual substrate niche; and (3) their production of toxic compounds (VFA, bacteriocins, antibiotics) (Crittenden and Playne, 1999; Nurmi et al., 1992; Steer et al., 2000). Animals with a well-established microbial network in the GIT are more resistant to infections by opportunistic pathogenic bacteria due their ability to occupy and proliferate more efficiently (Fuller, 1989).

Competitive exclusion has been widely used in poultry and swine to prevent intestinal colonization by *Salmonella* (Anderson et al., 1999; Fedorka-Cray et al., 1995; Genovese et al., 2003; Nisbet and Martin, 1993; Nisbet et al., 1996; Nurmi and Rantala, 1973; Nurmi et al., 1992). The ruminant, however, harbors such an immense microbial reservoir and has a longer production cycle compared with other livestock, that CE was not considered to be an effective strategy (Callaway et al., 2004a). Contrary to that notion, researchers demonstrated that CE and other probiotics could in fact reduce pathogen colonization and shedding in cattle (Fox et al., 2009; Reissbrodt et al., 2009; Tkalcic et al., 2003; Zhao et al., 1998, 2003). In vitro studies revealed a generic *E. coli* strain that produced colicins (toxic proteins that reduce competition of other bacterial strains), which, when observed in vivo, actually displaced the established *E. coli* O157:H7 populations in adult cattle (Zhao et al., 1998) and reduced fecal shedding of Shiga toxin-producing *E. coli* O111:NM and O26:H111 in newborn calves (Zhao et al., 2003) and freshly weaned calves (Tkalcic et al., 2003). This finding has profound implications for CE means of counteraction against the recently discovered STEC serogroups.

Direct-fed microbials (DFM) are used in livestock production to maximize feed efficiency and are typically composed of (living or dead) yeast, fungal, or bacterial cultures or end products of fermentation (Callaway et al., 2013a). Many manufacturers of commercially available probiotics and yeasts claim this benefit and suggest their product also reduces *E. coli* O157:H7 colonization, but reports where both objectives were achieved remains to be seen (Keen and Elder, 2000; Swyers et al., 2011). Hence, probiotics to deter STEC colonization specifically without negative impacts on feed efficiency were proposed and several cultures have been shown to have an effect on O157:H7. A mixed culture of *Streptococcus bovis* and *Lactobacillus gallinarum* given to experimentally infected calves was effective at reducing shedding of O157 due to increased VFA concentration in the GIT (Ohya et al., 2000). Another study revealed dosing lambs with a mixed culture of *Streptococcus faecium*, *Lactobacillus acidophilus*, *L. casei*, *L. fermentum*, and *L. plantarum* significantly reduced fecal shedding O157 and increased gain-to-feed ratio, but no significant effects were seen when dosed with *L. acidophilus* monoculture (Lema et al., 2001). In contrast, researchers reported a rumen-derived *L. acidophilus* culture reduced *E. coli* O157:H7 shedding of feedlot cattle by >50% (Brashears and Galyean, 2002; Brashears et al., 2003a,b). Others reported *L. acidophilus* reduced fecal shedding from 45.8% to 13.3% in feedlot cattle (Ransom et al., 2003). Subsequently, *L. acidophilus* combined with Propionibacterium freudenreichii was found to reduce the concentration of *E. coli* O157:H7 in the feces of feedlot cattle from 27% to 16% and the prevalence of hides from 14% to 4% (Elam et al., 2003; Younts-Dahl et al., 2004). Later research using the same DFM combination validated reduced fecal concentrations of O157 (Stephens et al., 2007). Other studies using the monoculture *L. acidophilus* DFM reported no effect on *E. coli* O157 prevalence when using a low-dose formula, indicating that the prevalence and concentration of *E. coli* O157:H7 in the feces is highly dependent on DFM dosage level (Cull et al., 2012). Currently, there are two (different dosage levels) commercially available—Lactobacillus-based DFMs (Bovamine™ and Bovamine Defend™)—that are widely used to increase feed efficiency of feedlot rations.

Organic acids have been used in ruminant nutrition to modify the end products of fermentation by providing a competitive advantage to some members of the microbial ecosystem and/or inhibiting other species (Grilli et al., 2010; Martin and Streeter, 1995; Piva et al., 2007). Some organic acids (such as lactate, acetate, propionate, malate) have been shown to have antimicrobial activity against *E. coli* O157:H7 (Harris et al., 2006; Sagong et al., 2011; Vandeplas et al., 2010; Wolin, 1969). These acids have been used on hide and carcass washes to reduce pathogen populations, but only recently has interest in using organic acids to reduce pathogens in live animals received interest (Callaway et al., 2010; Nisbet et al., 2009).

Preliminary results do show some success in inhibiting pathogens in the lower intestinal tract of animals (unpublished data); however, further research is needed to be able to release the appropriate organic acid and concentration in the appropriate intestinal location to reduce populations of *E. coli* O157:H7 in cattle (Callaway et al., 2013a).

It would seem that using DFM as a means of reducing *E. coli* O157:H7 (and other STEC serotypes) colonization of ruminants is dependent on animal species, phase of production and, obviously, diet. Although the ability to feed organic acids offers a more direct approach at reducing STEC, due to their nature, they are self-limiting. Using DFM may be more sustainable, as populations are established and continue their role in the metabolic pathways, where reduction of pathogens are concerned. Further research should focus on identifying the most beneficial DFM cultures to not only minimize *E. coli* O157:H7 colonization and shedding of ruminants but also to provide the producer with a positive economic incentive for their usage (e.g., increased feed efficiency, value-added program) (Callaway et al., 2013a).

7 CONCLUSIONS

Recent years have seen a dramatic increase in preharvest intervention research with varying results in an effort to understand microbial population and host physiology effects on STEC populations (Callaway et al., 2013a). Undoubtedly, reducing the pathogen load entering the abattoir would render in-house control points more effective. The complex nature of the ruminant reservoir and the length of time required for getting cattle market ready (12-18 months) should emphasize that there is no "magic bullet" that will solve STEC in cattle: thus, a multihurdle approach will need to be implemented (Callaway et al., 2013a). However, for beef suppliers to adopt these solutions into their management practices they must be cost-effective and provide return on investment. Additionally, researchers must be prepared and vigilant for unintended consequences of this change in food safety (Callaway et al., 2013b). When removing a member of the microbial ecosystem, a niche is opened and another opportunistic organism with new virulence factors may fill the vacuum (Callaway et al., 2013b; Kingsley and Bäumler, 2000). After 50 million years of host coevolution and gene transfer, STEC populations will not be easily eliminated. However, the research summarized here provides evidence of many different management and diet manipulation options that can create roadblocks in the biological pathways of the pathogen's life cycle that may inevitably reduce the incidence of foodborne illness.

REFERENCES

Alexander, T.W., Yanke, L.J., Topp, E., Olson, M.E., Read, R.R., Morck, D.W., McAllister, T.A., 2008. Effect of subtherapeutic administration of antibiotics on the prevalence of antibiotic-resistant *Escherichia coli* bacteria in feedlot cattle. Appl. Environ. Microbiol. 74 (14), 4405–4416.

Amani, J., Salmanian, A.H., Rafati, S., Mousavi, S.L., 2010. Immunogenic properties of chimeric protein from espA, eae and tir genes of *Escherichia coli* O157: H7. Vaccine 28 (42), 6923–6929.

Amani, J., Mousavi, S.L., Rafati, S., Salmanian, A.H., 2011. Immunogenicity of a plant-derived edible chimeric EspA, intimin and tir of *Escherichia coli* O157: H7 in mice. Plant Sci. 180 (4), 620–627.

Anderson, R.C., Stanker, L.H., Young, C.R., Buckley, S.A., Genovese, K.J., Harvey, R.B., DeLoach, J.R., Keith, N.K., Nisbet, D.J., 1999. Effect of competitive exclusion treatment on colonization of early-weaned pigs by Salmonella serovar cholerasuis. Swine Health Prod. 12, 155–160.

Anderson, R.C., Buckley, S.A., Kubena, L.F., Stanker, L.H., Harvey, R.B., Nisbet, D.J., 2000a. Bactericidal effect of sodium chlorate on *Escherichia coli* O157:H7 and *Salmonella typhimurium* DT104 in rumen contents in vitro. J. Food Prot. 63, 1038–1042.

Anderson, R.C., Callaway, T.R., Buckley, S.A., Anderson, T.J., Genovese, K.J., Sheffield, C.L., et al., 2000b. Effect of sodium chlorate on porcine gut concentrations of *Escherichia coli* O157:H7 in vivo. In: Proc. Allen D. Leman Swine Conf., Minneapolis, MN. pp. 29.

Anderson, R.C., Callaway, T.R., Buckley, S.A., Anderson, T.J., Genovese, K.J., Sheffield, C.L., Nisbet, D.J., 2001. Effect of oral sodium chlorate administration on *Escherichia coli* O157:H7 in the gut of experimentally infected pigs. Int. J. Food Microbiol. 71, 125–130.

Anderson, R.C., Callaway, T.R., Anderson, T.J., Kubena, L.F., Keith, N.K., Nisbet, D.J., 2002. Bactericidal effect of sodium chlorate on *Escherichia coli* concentrations in bovine ruminal and fecal contents in vivo. Microb. Ecol. Health Dis. 14, 24–29.

Anderson, M.J., Blanton Jr., J.R., Gleghorn, J., Kim, S.W., Johnson, J.W., 2006. Ascophyllum nodosum supplementation strategies that improve overall carcass merit of implanted English crossbred cattle. Asian Australas. J. Anim. Sci. 19, 1514–1518.

Arthington, J.D., Kunkle, W.E., Martin, A.M., 2002. Citrus pulp for cattle. In: Rogers, G., Poore, M. (Eds.), The Veterinary Clinics of North America – Food Animal Practice. W. B. Saunders Company, Philadelphia, PA, pp. 317–328.

Arthur, T.M., Barkocy-Gallagher, G.A., Rivera-Betancourt, M., Koohmaraie, M., 2002. Prevalence and characterization of non-O157 Shiga toxin-producing *Escherichia coli* on carcasses in commercial beef cattle processing plants. Appl. Environ. Microbiol. 68 (10), 4847–4852.

Arthur, T.M., Bosilevac, J.M., Brichta-Harhay, D.M., Guerini, M.N., Kalchayanand, N., Shackelford, S.D., Wheeler, T.L., Koohmaraie, M., 2007. Transportation and lairage environment effects on prevalence, numbers, and diversity of *Escherichia coli* O157:H7 on hides and carcasses of beef cattle at processing. J. Food Prot. 70, 280–286.

Asper, D.J., Karmali, M.A., Townsend, H., Rogan, D., Potter, A.A., 2011. Serological response of Shiga toxin-producing *Escherichia coli* type III secreted proteins in sera from vaccinated rabbits, naturally infected cattle, and humans. Clin. Vaccine Immunol. 18 (7), 1052–1057.

Bach, S.J., McAllister, T.A., Veira, D.M., Gannon, V.J., Holley, R.A., 2003. Effect of bacteriophage DC22 on *Escherichia coli* O157: H7 in an artificial rumen system (Rusitec) and inoculated sheep. Anim. Res. 52 (2), 89–102.

Bach, S.J., McAllister, T.A., Mears, G.J., Schwartzkopf-Genswein, K.S., 2004. Long-haul transport and lack of preconditioning increases fecal shedding of *Escherichia coli* and *Escherichia coli* O157:H7 by calves. J. Food Prot. 67, 672–678.

Bach, S.J., Selinger, L.J., Stanford, K., McAllister, T., 2005a. Effect of supplementing corn- or barley-based feedlot diets with canola oil on faecal shedding of *Escherichia coli* O157:H7 by steers. J. Appl. Microbiol. 98, 464–475.

Bach, S.J., Stanford, K., McAllister, T.A., 2005b. Survival of *Escherichia coli* O157:H7 in feces from corn- and barley-fed steers. FEMS Microbiol. Lett. 252, 25–33.

Bach, S.J., Johnson, R.P., Stanford, K., McAllister, T.A., 2009. Bacteriophages reduce *Escherichia coli* O157: H7 levels in experimentally inoculated sheep. Can. J. Anim. Sci. 89 (2), 285–293.

Barbosa, L.N., Rall, V.L.M., Fernandes, A.A.H., Ushimaru, P.I., da Silva Probst, I., Fernandes, A., 2009. Essential oils against foodborne pathogens and spoilage bacteria in minced meat. Foodborne Pathog. Dis. 6, 725–728.

Barkocy-Gallagher, G.A., Arthur, T.M., Rivera-Betancourt, M., Nou, X., Shackelford, S.D., Wheeler, T.L., Koohmaraie, M., 2003. Seasonal prevalence of shiga toxin-producing *Escherichia coli*, including O157:H7 and non-O157 serotypes, and Salmonella in commercial beef processing plants. J. Food Prot. 66, 1978–1986.

Barrow, P.A., Soothill, J.S., 1997. Bacteriophage therapy and prophylaxis: rediscovery and renewed assessment of potential. Trends Microbiol. 5 (7), 268–271.

Benchaar, C., Chaves, A.V., Fraser, G.R., Wang, Y., Beauchemin, K.A., McAllister, T.A., 2007. Effects of essential oils and their components on in vitro rumen microbial fermentation. Can. J. Anim. Sci. 87, 413–419.

Benchaar, C., Calsamiglia, S., Chaves, A.V., Fraser, G.R., Colombatto, D., McAllister, T.A., Beauchemin, K.A., 2008. A review of plant-derived essential oils in ruminant nutrition and production. Anim. Feed Sci. Technol. 145, 209–228.

Berard, N.C., Holley, R.A., McAllister, T.A., Ominski, K.H., Wittenberg, K.M., Bouchard, K.S., Bouchard, J.J., Krause, D.O., 2009. Potential to reduce *Escherichia coli* shedding in cattle feces by using sainfoin (Onobrychis viciifolia) forage, tested in vitro and in vivo. Appl. Environ. Microbiol. 75, 1074–1079.

Berg, J.L., McAllister, T.A., Bach, S.J., Stillborn, R.P., Hancock, D.D., LeJeune, J.T., 2004. *Escherichia coli* O157:H7 excretion by commercial feedlot cattle fed either barley- or corn-based finishing diets. J. Food Prot. 67, 666–671.

Bergholz, T.M., Whittam, T.S., 2007. Variation in acid resistance among enterohaemorrhagic *Escherichia coli* in a simulated gastric environment. J. Appl. Microbiol. 102 (2), 352–362.

Berry, E., 2013. Effect of proximity to a cattle feedlot on occurrence of *Escherichia coli* O157:H7-positive pest flies in a leafy green crop. In: 2013 Annual Meeting (July 28–31, 2013). International Association for Food Protection.

Berry, E.D., Wells, J.E., Archibeque, S.L., Ferrell, C.L., Freetly, H.C., Miller, D.N., 2006. Influence of genotype and diet on steer performance, manure odor, and carriage of pathogenic and other fecal bacteria. II. Pathogenic and other fecal bacteria. J. Anim. Sci. 84 (9), 2523–2532.

Bettelheim, K.A., 2007. The non-O157 Shiga-toxigenic (verocytotoxigenic) *Escherichia coli*; under-rated pathogens. CRC Cr. Rev. Microbiol. 33 (1), 67–87.

Boadi, D., Benchaar, C., Chiquette, J., Massao, D., 2004. Mitigation strategies to reduce enteric methane emissions from dairy cows: update review. Can. J. Anim. Sci. 84, 319–335.

Bolton, D.J., 2011. Verocytotoxigenic (Shiga toxin-producing) *Escherichia coli*: virulence factors and pathogenicity in the farm to fork paradigm. Foodborne Pathog. Dis. 8 (3), 357–365.

Bosilevac, J.M., Koohmaraie, M., 2011. Prevalence and characterization of non-O157 Shiga toxin-producing *Escherichia coli* isolates from commercial ground beef in the United States. Appl. Environ. Microbiol. 77 (6), 2103–2112.

Brabban, A.D., Hite, E., Callaway, T.R., 2005. Evolution of foodborne pathogens via temperate bacteriophage-mediated gene transfer. Foodborne Pathog. Dis. 2 (4), 287–303.

Braden, K.W., Blanton Jr., J.R., Allen, V.G., Pond, K.R., Miller, M.F., 2004. Ascophyllum nodosum supplementation: a preharvest intervention for reducing *Escherichia coli* O157:H7 and Salmonella spp. in feedlot steers. J. Food Prot. 67, 1824–1828.

Braden, K.W., Blanton Jr., J.R., Montgomery, J.L., Van Santen, E., Allen, V.G., Miller, M.F., 2007. Tasco supplementation: effects on carcass characteristics, sensory attributes, and retail display shelf-life. J. Anim. Sci. 85, 754–768.

Brashears, M.M., Galyean, M.L., 2002. Testing of probiotic bacteria for the elimination of *Escherichia coli* O157: H7 in cattle. Am. Meat Inst. Found. http://www.foodsafetynetwork. ca/articles/263/probiotic_ecoli_cattle. pdf (accessed 05.09.14).

Brashears, M.M., Jaroni, D., Trimble, J., 2003a. Isolation, selection, and characterization of lactic acid bacteria for a competitive exclusion product to reduce shedding of *Escherichia coli* O157: H7 in cattle. J. Food Prot. 66 (3), 355–363.

Brashears, M.M., Galyean, M.L., Loneragan, G.H., Mann, J.E., Killinger-Mann, K., 2003b. Prevalence of *Escherichia coli* O157: H7 and performance by beef feedlot cattle given Lactobacillus direct-fed microbials. J. Food Prot. 66 (5), 748–754.

Buchko, S.J., Holley, R.A., Olson, W.O., Gannon, V.P.J., Veira, D.M., 2000a. The effect of different grain diets on fecal shedding of *Escherichia coli* O157: H7 by steers. J. Food Prot. 63 (11), 1467–1474.

Buchko, S.J., Holley, R.A., Olson, W.O., Gannon, V.P.J., Veira, D.M., 2000b. The effect of fasting and diet on fecal shedding of *Escherichia coli* O157: H7 by cattle. Can. J. Anim. Sci. 80 (4), 741–744.

Caillet, S., Côté, J., Sylvain, J.F., Lacroix, M., 2012. Antimicrobial effects of fractions from cranberry products on the growth of seven pathogenic bacteria. Food Control 23, 419–428.

Callaway, T., 2010. The use of pre-and probiotics to improve food safety in the live animal. In: Meeting Abstract. pp. 83.

Callaway, T.R., Anderson, R.C., Genovese, K.J., Poole, T.L., Anderson, T.J., Byrd, J.A., et al., 2002. Sodium chlorate supplementation reduces *E. coli* O157: H7 populations in cattle. J. Anim. Sci. 80 (6), 1683–1689.

Callaway, T.R., Edrington, T.S., Anderson, R.C., Genovese, K.J., Poole, T.L., Elder, R.O., et al., 2003. *Escherichia coli* O157: H7 populations in sheep can be reduced by chlorate supplementation. J. Food Prot. 66 (2), 194–199.

Callaway, T.R., Anderson, R.C., Edrington, T.S., Genovese, K.J., Bischoff, K.M., Poole, T.L., Jung, Y.S., Harvey, R.B., Nisbet, D.J., 2004a. What are we doing about *Escherichia coli* O157: H7 in cattle? J. Anim. Sci. 82 (13 suppl.), E93–E99.

Callaway, T.R., Anderson, R.C., Edrington, T.S., Jung, Y.S., Bischoff, K.M., Genovese, K.J., Nisbet, D.J., 2004b. Effects of sodium chlorate on toxin production by *Escherichia coli* O157: H7. Curr. Issues Intest. Microbiol. 5 (1), 19–22.

Callaway, T.R., Anderson, R.C., Edrington, T.S., Bischoff, K.M., Genovese, K.J., Poole, T.L., Nisbet, D.J., 2004c. Effects of sodium chlorate on antibiotic resistance in *Escherichia coli* O157: H7. Foodborne Pathog. Dis. 1 (1), 59–63.

Callaway, T.R., Edrington, T.S., Brabban, A.D., Keen, J.E., Anderson, R.C., Rossman, M.L., Nisbet, D.J., 2006. Fecal prevalence of *Escherichia coli* O157, Salmonella, Listeria, and bacteriophage infecting *E. coli* O157: H7 in feedlot cattle in the southern plains region of the United States. Foodborne Pathog. Dis. 3 (3), 234–244.

Callaway, T.R., Carroll, J.A., Arthington, J.D., Pratt, C., Edrington, T.S., Anderson, R.C., Galyean, M.L., Ricke, S.C., Crandall, P., Nisbet, D.J., 2008. Citrus products decrease growth of *E. coli* O157:H7 and *Salmonella typhimurium* in pure culture and in fermentation with mixed ruminal microorganisms in vitro. Foodborne Pathog. Dis. 5, 621–627.

Callaway, T.R., Carroll, J.A., Arthington, J.D., Edrington, T.S., Anderson, R.C., Rossman, M.L., Carr, M.A., Genovese, K.J., Ricke, S.C., Crandall, P., Nisbet, D.J., 2011a. *Escherichia coli* O157:H7 populations in ruminants can be reduced by orange peel product feeding. J. Food Prot. 74, 1917–1921.

Callaway, T.R., Carroll, J.A., Arthington, J.D., Edrington, T.S., Anderson, R.C., Rossman, M.L., Carr, M.A., Genovese, K.J., Ricke, S.C., Crandall, P., Nisbet, D.J., 2011b. Orange peel pellets can reduce *Salmonella* populations in ruminants. Foodborne Pathog. Dis. 8, 1071–1075.

Callaway, T., Edrington, T., Loneragan, G., Carr, M., Nisbet, D., 2013a. Shiga Toxin-producing *Escherichia coli* (STEC) ecology in cattle and management based options for reducing fecal shedding. Agric. Food Anal. Bacteriol. 3, 39–69.

Callaway, T.R., Edrington, T.S., Loneragan, G.H., Carr, M.A., Nisbet, D.J., 2013b. Current and near-market intervention strategies for reducing Shiga toxin-producing *Escherichia coli* (STEC) shedding in cattle. Agric. Food Anal. Bacteriol. 3, 103–120.

Callaway, T.R., Edrington, T.S., Nisbet, D.J., 2014. Meat science and muscle biology symposium: ecological and dietary impactors of foodborne pathogens and methods to reduce fecal shedding in cattle. J. Anim. Sci. 92 (4), 1356–1365.

Carr, T.R., Allen, D.M., Phar, P., 1971. Effect of preslaughter fasting on bovine carcass yield and quality. J. Anim. Sci. 32 (5), 870–873.

Chase-Topping, M.E., McKendrick, I.J., Pearce, M.C., MacDonald, P., Matthews, L., Halliday, J., Woolhouse, M.E., 2007. Risk factors for the presence of high-level shedders of *Escherichia coli* O157 on Scottish farms. J. Clin. Microbiol. 45 (5), 1594–1603.

Chase-Topping, M., Gally, D., Low, C., Matthews, L., Woolhouse, M., 2008. Super-shedding and the link between human infection and livestock carriage of *Escherichia coli* O157. Nat. Rev. Microbiol. 6 (12), 904–912.

Chase-Topping, M.E., Rosser, T., Allison, L.J., Courcier, E., Evans, J., McKendrick, I.J., Pearce, M.C., Handel, I., Caprioli, A., Karch, H., Hanson, M.F., Pollock, K.J., Locking, M.E., Woolhouse, M.E.J., Matthews, L., Low, J.C., Gally, D.L., 2012. Pathogenic potential to humans of bovine *Escherichia coli* O26, Scotland. Emerg. Infect. Dis. 18 (3), 439–448.

Cobbold, R.N., Hancock, D.D., Rice, D.H., Berg, J., Stilborn, R., Hovde, C.J., Besser, T.E., 2007. Rectoanal junction colonization of feedlot cattle by *Escherichia coli* O157: H7 and its association with supershedders and excretion dynamics. Appl. Environ. Microbiol. 73 (5), 1563–1568.

Collier, C.T., Carroll, J.A., Callaway, T.R., Arthington, J.D., 2010. Oral administration of citrus pulp reduces gastrointestinal recovery of orally dosed *Escherichia coli* F18 in weaned pigs. J. Anim. Vet. Adv. 9, 2140–2145.

Collins, M.D., Gibson, G.R., 1999. Probiotics, prebiotics, and synbiotics: approaches for modulating the microbial ecology of the gut. Am. J. Clin. Nutr. 69 (5), 1052s–1057s.

Cowan, M.M., 1999. Plant products as antimicrobial agents. Clin. Microbiol. Rev. 12, 564–582.

Cray, W.C., Casey, T.A., Bosworth, B.T., Rasmussen, M.A., 1998. Effect of dietary stress on fecal shedding of *Escherichia coli* O157: H7 in calves. Appl. Environ. Microbiol. 64 (5), 1975–1979.

Crittenden, R., Playne, M.J., 1999. Prebiotics. In: Handbook of Probiotics and Prebiotics, second ed. pp. 533–581.

Cueva, C., Moreno-Arribas, M.V., Martín-Álvarez, P.J., Bills, G., Vicente, M.F., Basilio, A., Rivas, C.L., Requena, T., Rodríguez, J.M., Bartolomé, B., 2010. Antimicrobial activity of phenolic acids against commensal, probiotic and pathogenic bacteria. Res. Microbiol. 161, 372–382.

Cull, C.A., Paddock, Z.D., Nagaraja, T.G., Bello, N.M., Babcock, A.H., Renter, D.G., 2012. Efficacy of a vaccine and a direct-fed microbial against fecal shedding of *Escherichia coli* O157: H7 in a randomized pen-level field trial of commercial feedlot cattle. Vaccine 30 (43), 6210–6215.

Dabbah, R., Edwards, V.M., Moats, W.A., 1970. Antimicrobial action of some citrus fruit oils on selected food-borne bacteria. Appl. Microbiol. 19, 27–31.

Dargatz, D.A., Wells, S.J., Thomas, L.A., Hancock, D.D., Garber, L.P., 1997. Factors associated with the presence of *Escherichia coli* O157 in feces of feedlot cattle. J. Food Prot. 60 (5), 466–470.

Davis, M.A., Cloud-Hansen, K.A., Carpenter, J., Hovde, C.J., 2005. *Escherichia coli O157: H7* in environments of culture-positive cattle. Appl. Environ. Microbiol. 71, 6816–6822.

Davis, M.A., Rice, D.H., Sheng, H., Hancock, D.D., Besser, T.E., Cobbold, R., Hovde, C.J., 2006. Comparison of cultures from rectoanal-junction mucosal swabs and feces for detection of *Escherichia coli* O157 in dairy heifers. Appl. Environ. Microbiol. 72 (5), 3766–3770.

Dewell, G.A., Ransom, J.R., Dewell, R.D., McCurdy, K., Gardner, I.A., Hill, A.E., Sofos, J.N., Belk, K.E., Smith, G.C., Salman, M.D., 2005. Prevalence of and risk factors for *Escherichia coli* O157 in market-ready beef cattle from 12 U.S. feedlots. Foodborne Pathog. Dis. 2, 70–76.

Di Pasqua, R., Betts, G., Hoskins, N., Edwards, M., Ercolini, D., Mauriello, G., 2007. Membrane toxicity of antimicrobial compounds from essential oils. J. Agric. Food Chem. 55, 4863–4870.

Diez-Gonzalez, F., Callaway, T.R., Kizoulis, M.G., Russell, J.B., 1998. Grain feeding and the dissemination of acid-resistant *Escherichia coli* from cattle. Science 281, 1666–1668.

Durso, L.M., Wells, J.E., Harhay, G.P., Rice, W.C., Kuehn, L., Bono, J.L., Smith, T.P.L., 2012. Comparison of bacterial communities in faeces of beef cattle fed diets containing corn and wet distillers' grain with solubles. Lett. Appl. Microbiol. 55 (2), 109–114.

Dusan, F., Marian, S., Katarina, D., Dobroslava, B., 2006. Essential oils-their antimicrobial activity against *Escherichia coli* and effect on intestinal cell viability. Toxicol. in vitro 20, 1435–1445.

Dziva, F., van Diemen, P.M., Stevens, M.P., Smith, A.J., Wallis, T.S., 2004. Identification of *Escherichia coli* O157:H7 genes influencing colonization of the bovine gastrointestinal tract using signature-tagged mutagenesis. Microbiology 150, 3631–3645.

Edrington, T.S., Callaway, T.R., Varey, P.D., Jung, Y.S., Bischoff, K.M., Elder, R.O., 2003a. Effects of the antibiotic ionophores monensin, lasalocid, laidlomycin propionate and bambermycin on Salmonella and *E. coli* O157: H7 in vitro. Appl. Environ. Microbiol. 94 (2), 207–213.

Edrington, T.S., Callaway, T.R., Bischoff, K.M., Genovese, K.J., Elder, R.O., Anderson, R.C., Nisbet, D.J., 2003b. Effect of feeding the ionophores monensin and laidlomycin propionate and the antimicrobial bambermycin to sheep experimentally infected with *E. coli* O157: H7 and *Salmonella typhimurium*. J. Anim. Sci. 81 (2), 553–560.

Edrington, T.S., Callaway, T.R., Anderson, R.C., Genovese, K.J., Jung, Y.S., McReynolds, J.L., Nisbet, D.J., 2003c. Reduction of *E. coli* O157: H7 populations in sheep by supplementation of an experimental sodium chlorate product. Small Rumin. Res. 49 (2), 173–181.

Edrington, T.S., Farrow, R.L., Sperandio, V., Hughes, D.T., Lawrence, T.E., Callaway, T.R., Anderson, R.C., Nisbet, D.J., 2009. Acyl-homoserine-lactone autoinducer in the gastrointesinal tract of feedlot cattle and correlation to season, *E. coli* O157:H7 prevalence, and diet. Curr. Microbiol. 58, 227–232.

Edwards, A.C., Kay, D., McDonald, A.T., Francis, C., Watkins, J., Wilkinson, J.R., Wyer, M.D., 2008. Farmyards, an overlooked source for highly contaminated runoff. J. Environ. Manage. 87 (4), 551–559.

Elam, N.A., Gleghorn, J.F., Rivera, J.D., Galyean, M.L., Defoor, P.J., Brashears, M.M., Younts-Dahl, S.M., 2003. Effects of live cultures of *Lactobacillus acidophilus* (strains NP45 and NP51) and *Propionibacterium freudenreichii* on performance, carcass, and intestinal characteristics, and *Escherichia coli* strain O157 shedding of finishing beef steers. J. Anim. Sci. 81 (11), 2686–2698.

Elder, R.O., Keen, J.E., Wittum, T.E., Callaway, T.R., Edrington, T.S., Anderson, R.C., Nisbet, D.J., 2002. Intervention to reduce fecal shedding of enterohemorrhagic *Escherichia coli* O157:H7 in naturally infected cattle using neomycin sulfate. J. Anim. Sci. 80 (Suppl. 1), 15 (Abstr.).

Ellis-Iversen, J., Watson, E., 2008. A 7-point plan for control of VTEC O157, Campylobacter jejuni/coli and Salmonella serovars in young cattle. Cattle Pract. 16, 103–106.

Ellis-Iversen, J., Smith, R.P., Van Winden, S., Paiba, G.A., Watson, E., Snow, L.C., Cook, A.J., 2008. Farm practices to control *E. coli* O157 in young cattle – a randomized controlled trial. Vet. Res. 39 (1), 1–12.

Ellis-Versen, J., Van Winden, S., 2008. Control of E-coli O157 (VTEC) by applied management practices. Cattle Pract. 16, 54.

Farrow, R.L., Edrington, T.S., Krueger, N.A., Genovese, K.J., Callaway, T.R., Anderson, R.C., Nisbet, D.J., 2012. Lack of effect of feeding citrus byproducts in reducing Salmonella in experimentally infected weanling pigs. J. Food Prot. 75, 573–575.

Fedorka-Cray, P.J., Kelley, L.C., Stabel, T.J., Gray, J.T., Laufer, J.A., 1995. Alternate routes of invasion may affect pathogenesis of *Salmonella typhimurium* in swine. Infect. Immun. 63 (7), 2658–2664.

Ferens, W.A., Hovde, C.J., 2000. Antiviral activity of shiga toxin 1: suppression of bovine leukemia virus-related spontaneous lymphocyte proliferation. Infect. Immun. 68, 4462–4469.

Fernando, S.C., Purvis, H.T., Najar, F.Z., Sukharnikov, L.O., Krehbiel, C.R., Nagaraja, T.G., Roe, B.A., DeSilva, U., 2010. Rumen microbial population dynamics during adaptation to a high-grain diet. Appl. Environ. Microbiol. 76, 7482–7490.

Fisher, K., Phillips, C.A., 2006. The effect of lemon, orange and bergamot essential oils and their components on the survival of Campylobacter jejuni, *Escherichia coli* O157, Listeria monocyto-genes, Bacillus cereus and Staphylococcus aureus in vitro and in food systems. J. Appl. Microbiol. 101, 1232–1240.

Fiume, M.Z., 1995. Final report on the safety assessment of potassium chlorate. J. Am. Coll. Toxicol. 14, 221–230.

Fox, J.T., Depenbusch, B.E., Drouillard, J.S., Nagaraja, T.G., 2007. Dry-rolled or steam-flaked grain-based diets and fecal shedding of *Escherichia coli* O157 in feedlot cattle. J. Anim. Sci. 85 (5), 1207–1212.

Fox, J.T., Thomson, D.U., Drouillard, J.S., Thornton, A.B., Burkhardt, D.T., Emery, D.A., Nagaraja, T.G., 2009. Efficacy of *Escherichia coli* O157: H7 siderophore receptor/porin proteins-based vaccine in feedlot cattle naturally shedding *E. coli* O157. Foodborne Pathog. Dis. 6 (7), 893–899.

Fratamico, P.M., Bagi, L.K., Cray Jr., W.C., Narang, N., Yan, X., Medina, M., Liu, Y., 2011. Detection by multiplex real-time polymerase chain reaction assays and isolation of Shiga toxin-producing *Escherichia coli* serogroups O26, O45, O103, O111, O121, and O145 in ground beef. Foodborne Pathog. Dis. 8 (5), 601–607.

Free, A.L., Duoss, H.A., Bergeron, L.V., Shields-Menard, S.A., Ward, E., Callaway, T.R., Carroll, J.A., Schmidt, T.B., Donaldson, J.R., 2012. Survival of O157: H7 and non-O157 serogroups of *Escherichia coli* in bovine rumen fluid and bile salts. Foodborne Pathog. Dis. 9 (11), 1010–1014.

Fremaux, B., Delignette-Muller, M.L., Prigent-Combaret, C., Gleizal, A., Vernozy-Rozand, C., 2007. Growth and survival of non-O157: H7 Shiga-toxin-producing *Escherichia coli* in cow manure. J. Appl. Microbiol. 102 (1), 89–99.

Friedly, E.C., Crandall, P.G., Ricke, S.C., Roman, M., O'Bryan, C.A., Chalova, V.I., 2009. In vitro anti- listerial effects of citrus oil fractions in combination with organic acids. J. Food Sci. 74, M67–M72.

Fuller, R., 1989. Probiotics in man and animals. J. Appl. Bacteriol. 66, 365–378.

Fullerton, M., Khatiwada, J., Johnson, J.U., Davis, S., Williams, L.L., 2011. Determination of anti-microbial activity of sorrel (Hibiscus sabdariffa) on *Escherichia coli* O157:H7 isolated from food, veterinary, and clinical samples. J. Med. Food 14, 950–956.

Gansheroff, L.J., O'Brien, A.D., 2000. *Escherichia coli* O157:H7 in beef cattle presented for slaughter in the U. S.: higher prevalence rates than previously estimated. Proc. Natl. Acad. Sci. U. S. A. 97, 2959–2961.

Garcia-Gonzalez, L., Rajkovic, A., Geeraerd, A.H., Elst, K., Van Ginneken, L., Van Impe, J.F., Devlieghere, F., 2010. The development of *Escherichia coli* and *Listeria monocytogenes* variants resistant to high-pressure carbon dioxide inactivation. Lett. Appl. Microbiol. 50 (6), 653–656.

Gaskins, H.R., Collier, C.T., Anderson, D.B., 2002. Antibiotics as growth promotants: mode of action. Anim. Biotechnol. 13 (1), 29–42.

Genovese, K.J., Anderson, R.C., Harvey, R.B., Callaway, T.R., Poole, T.L., Edrington, T.S., Nisbet, D.J., 2003. Competitive exclusion of Salmonella from the gut of neonatal and weaned pigs. J. Food Prot. 66 (8), 1353–1359.

Gerba, C.P., Smith, J.E., 2005. Sources of pathogenic microorganisms and their fate during land application of wastes. J. Environ. Qual. 34 (1), 42–48.

Ghosh, S., LaPara, T.M., 2007. The effects of subtherapeutic antibiotic use in farm animals on the proliferation and persistence of antibiotic resistance among soil bacteria. ISME J. 1 (3), 191–203.

Goss, M., Richards, C., 2008. Development of a risk-based index for source water protection planning, which supports the reduction of pathogens from agricultural activity entering water resources. J. Environ. Manage. 87 (4), 623–632.

Greenquist, M.A., Drouillard, J.S., Sargeant, J.M., Depenbusch, B.E., Shi, X., Lechtenberg, K.F., Nagaraja, T.G., 2005. Comparison of rectoanal mucosal swab cultures and fecal cultures for determining prevalence of *Escherichia coli* O157: H7 in feedlot cattle. Appl. Environ. Microbiol. 71 (10), 6431–6433.

Gregory, N.G., Jacobson, L.H., Nagle, T.A., Muirhead, R.W., Leroux, G.J., 2000. Effect of preslaughter feeding system on weight loss, gut bacteria, and the physico-chemical properties of digesta in cattle. N. Z. J. Agric. Res. 43, 351–361.

Grilli, E., Messina, M.R., Tedeschi, M., Piva, A., 2010. Feeding a microencapsulated blend of organic acids and nature identical compounds to weaning pigs improved growth performance and intestinal metabolism. Livest. Sci. 133, 173–175.

Guan, H., Wittenberg, K.M., Ominski, K.H., Krause, D.O., 2006. Efficacy of ionophores in cattle diets for mitigation of enteric methane. J. Anim. Sci. 84 (7), 1896–1906.

Hackmann, T.J., Spain, J.N., 2010. Invited review: ruminant ecology and evolution: perspectives useful to ruminant livestock research and production. J. Dairy Sci. 93 (4), 1320–1334.

Halliday, J.E.B., Chase-Topping, M.E., Pearce, M.C., Mckendrick, I.J., Allison, L., Fenlon, D., Low, C., Mellor, D.J., Gunn, G.J., Woolhouse, M.E.J., 2006. Herd-level risk factors associated with the presence of phage type 21/28 *E. coli* O157 on Scottish farms. BMC Microbiol. 6, 99.

Harris, K., Miller, M.F., Loneragan, G.H., Brashears, M.M., 2006. Validation of the use of organic acids and acidified sodium chlorite to reduce *Escherichia coli* O157 and *Salmonella Typhimurium* in beef trim and ground beef in a simulated processing environment. J. Food Prot. 69, 1802–1807.

Herriott, D.E., Hancock, D.D., Ebel, E.D., Carpenter, L.V., Rice, D.H., Besser, T.E., 1998. Association of herd management factors with colonization of dairy cattle by Shiga toxin-positive *Escherichia coli* O157. J. Food Prot. 61 (7), 802–807.

Hollowell, C.A., Wolin, M.J., 1965. Basis for the exclusion of *Escherichia coli* from the rumen ecosystem. Appl. Microbiol. 13 (6), 918–924.

Hovde, C.J., Austin, P.R., Cloud, K.A., Williams, C.J., Hunt, C.W., 1999. Effect of cattle diet on *Escherichia coli* O157:H7 acid resistance. Appl. Environ. Microbiol. 65, 3233–3235.

Hristov, A.N., Ivan, M., McAllister, T.A., 2001. In vitro effects of feed oils, ionophores, tannic acid, saponin-containing plant extracts and other bioactive agents on ruminal fermentation and protozoal activity. J. Dairy Sci. 84 (Suppl. 1), 360.

Huff, W., Huff, G., Rath, N., Balog, J., Donoghue, A., 2002. Prevention of Escherichia coli infection in broiler chickens with a bacteriophage aerosol spray. Poult. Sci. 81, 1486–1491.

Huntington, G.B., 1997. Starch utilization by ruminants: from basics to the bunk. J. Anim. Sci. 75 (3), 852–867.

Hussein, H.S., Thran, B.H., Hall, M.R., Kvasnicka, W.G., Torell, R.C., 2003. Verotoxin-producing *Escherichia coli* in culled beef cows grazing rangeland forages. Exp. Biol. Med. 228, 352–357.

Hynes, N.A., Wachsmuth, I.K., 2000. *Escherichia coli* O157: H7 risk assessment in ground beef: a public health tool. In: 4th International Symposium and Workshop on Shiga Toxin (Verocytotoxin)-Producing *Escherichia coli* Infections, Kyoto, Japan. pp. 46.

Jacob, M.E., Fox, J.T., Drouillard, J.S., Renter, D.G., Nagaraja, T.G., 2008a. Effects of dried distillers' grain on fecal prevalence and growth of *Escherichia coli* O157 in batch culture fermentations from cattle. Appl. Environ. Microbiol. 74, 38–43.

Jacob, M.E., Fox, J.T., Narayanan, S.K., Drouillard, J.S., Renter, D.G., Nagaraja, T.G., 2008b. Effects of feeding wet corn distiller's grains with solubles with or without monensin and tylosin on the prevalence and antimicrobial susceptibilities of fecal food-borne pathogenic and commensal bacteria in feedlot cattle. J. Anim. Sci. 86, 1182–1190.

Jacob, M.E., Callaway, T.R., Nagaraja, T.G., 2009. Dietary interactions and interventions affecting *Escherichia coli* O157 colonization and shedding in cattle. Foodborne Pathog. Dis. 6, 785–792.

Jay, M.T., Cooley, M., Carychao, D., Wiscomb, G.W., Sweitzer, R.A., Crawford-Miksza, L., Farrar, J.A., Lau, D.K., O'Connell, J., Millington, A., Asmundson, R.V., Atwill, E.R., Mandrell, R.E., 2007. *Escherichia coli* O157: H7 in feral swine near spinach fields and cattle, central California coast. Emerg. Infect. Dis. 13 (12), 1908.

Jiang, X., Morgan, J., Doyle, M.P., 2002. Fate of *Escherichia coli* O157: H7 in manure-amended soil. Appl. Environ. Microbiol. 68 (5), 2605–2609.

Johnson, R.P., Gyles, C.L., Huff, W.E., Ojha, S., Huff, G.R., Rath, N.C., Donoghue, A.M., 2008. Bacteriophages for prophylaxis and therapy in cattle, poultry and pigs. Anim. Health Res. Rev. 9 (02), 201–215.

Joris, M.A., Pierard, D., De Zutter, L., 2011. Occurrence and virulence patterns of *E. coli* O26, O103, O111 and O145 in slaughter cattle. Vet. Microbiol. 151 (3), 418–421.

Karmali, M.A., Gannon, V., Sargeant, J.M., 2010. Verocytotoxin-producing *Escherichia coli* (VTEC). Vet. Microbiol. 140 (3), 360–370.

Keen, J., Elder, R., 2000. Commercial probiotics are not effective for short-term control of enterohemorrhagic *Escherichia coli* O157 infection in beef cattle. In: 4th International Symposium Works. Shiga Toxin (Verocytotoxin)-producing *Escherichia coli* Infections 92 (Abstr.).

Keen, J.E., Uhlich, G.A., Elder, R.O., 1999. Effects of hay- and grain-based diets on fecal shedding in naturally acquired enterohemorrhagic *E. coli* (EHEC) O157 in beef feedlot cattle. In: Procs 80th Conf. Research Workers in Animal Diseases, Chicago, IL. (Abstr.).

Keen, J.E., Durso, L.M., Meehan, T.P., 2007. Isolation of Salmonella enterica and Shiga-toxigenic *Escherichia coli* O157 from feces of animals in public contact areas of United States zoological parks. Appl. Environ. Microbiol. 73 (1), 362–365.

Kim, J., Marshall, M.R., Wei, C.I., 1995. Antibacterial activity of some essential oil components against five foodborne pathogens. J. Agric. Food Chem. 43, 2839–2845.

Kim, J.G., Taylor, K.W., Hotson, A., Keegan, M., Schmelz, E.A., Mudgett, M.B., 2008. XopD SUMO protease affects host transcription, promotes pathogen growth, and delays symptom development in Xanthomonas-infected tomato leaves. Plant Cell. 20 (7), 1915–1929.

Kingsley, R.A., Bäumler, A.J., 2000. Host adaptation and the emergence of infectious disease: the Salmonella paradigm. Mol. Microbiol. 36 (5), 1006–1014.

Klieve, A.V., Bauchop, T., 1988. Morphological diversity of ruminal bacteriophages from sheep and cattle. Appl. Environ. Microbiol. 54, 1637–1641.

Klieve, A.V., Swain, R.A., 1993. Estimation of ruminal bacteriophage numbers by pulsed-field gel electrophoresis and laser densitometry. Appl. Environ. Microbiol. 59 (7), 2299–2303.

Klieve, A.V., Gregg, K., Bauchop, T., 1991. Isolation and characterization of lytic phages from *Bacterioides ruminicola* ss brevis. Curr. Microbiol. 23 (4), 183–187.

Kociolek, M.G., 2009. Quorum-sensing inhibitors and biofilms. Antiinfect. Agents Med. Chem. 8, 315–326.

Kropinski, A.M., Lingohr, E.J., Moyles, D.M., Ojha, S., Mazzocco, A., She, Y.M., Johnson, R.P., 2012. Endemic bacteriophages: a cautionary tale for evaluation of bacteriophage therapy and other interventions for infection control in animals. J. Virol. 9, 207.

Kudva, I.T., Hatfield, P.G., Hovde, C.J., 1995. Effect of diet on the shedding of *Escherichia coli* O157: H7 in a sheep model. Appl. Environ. Microbiol. 61 (4), 1363–1370.

Kudva, I.T., Jelacic, S., Tarr, P.I., Youderian, P., Hovde, C.J., 1999. Biocontrol of *Escherichia coli* O157 with O157-specific bacteriophages. Appl. Environ. Microbiol. 65 (9), 3767–3773.

Lacombe, A., Wu, V.C.H., White, J., Tadepalli, S., Andre, E.E., 2012. The antimicrobial properties of the lowbush blueberry (*Vaccinium angustifolium*) fractional components against foodborne pathogens and the conservation of probiotic *Lactobacillus rhamnosus*. Food Microbiol. 30, 124–131.

Law, D., 2000. Virulence factors of *Escherichia coli* O157 and other Shiga toxin-producing *E. coli*. J. Appl. Microbiol. 88 (5), 729–745.

LeBlanc, J.J., 2003. Implication of virulence factors in *Escherichia coli* O157:H7 pathogenesis. Clin. Micobiol. Tev. 29, 277–296.

Lederberg, J., 1996. Genetic recombination in *Escherichia coli*: disputation at Cold Spring Harbor, 1946–1996. Genetics 144 (2), 439.

LeJeune, J.T., Kauffman, M.D., 2005. Effect of sand and sawdust bedding materials on the fecal prevalence of *Escherichia coli* O157: H7 in dairy cows. Appl. Environ. Microbiol. 71 (1), 326–330.

LeJeune, J., Kauffman, M., 2006. Bovine *E. coli* O157 supershedders: mathematical myth or meaningful monsters? In: Procs. in Proceedings of the 2006 VTEC Conference. Melbourne, Austalia (Vol. 26).

LeJeune, J.T., Wetzel, A.N., 2007. Preharvest control of *Escherichia coli* O157 in cattle. J. Anim. Sci. 85 (13 Suppl.), E73–E80.

LeJeune, J.T., Besser, T.E., Hancock, D.D., 2001. Cattle water troughs as reservoirs of *Escherichia coli* O157. Appl. Environ. Microbiol. 67 (7), 3053–3057.

Lema, M., Williams, L., Rao, D.R., 2001. Reduction of fecal shedding of enterohemorrhagic *Escherichia coli* O157: H7 in lambs by feeding microbial feed supplement. Small Rumin. Res. 39 (1), 31–39.

Leupp, J.L., Caton, J.S., Soto-Navarro, S.A., Lardy, G.P., 2005. Effects of cooked molasses blocks and fermentation extract or brown seaweed meal inclusion on intake, digestion, and microbial efficiency in steers fed low-quality hay. J. Anim. Sci. 83, 2938–2945.

Lim, J.Y., Li, J., Sheng, H., Besser, T.E., Potter, K., Hovde, C.J., 2007. *Escherichia coli* O157: H7 colonization at the rectoanal junction of long-duration culture-positive cattle. Appl. Environ. Microbiol. 73 (4), 1380–1382.

Loneragan, G.H., Brashears, M.M., 2005. Pre-harvest interventions to reduce carriage of *E. coli* O157 by harvest-ready feedlot cattle. Meat Sci. 71 (1), 72–78.

Low, J.C., McKendrick, I.J., McKechnie, C., Fenlon, D., Naylor, S.W., Currie, C., Gally, D.L., 2005. Rectal carriage of enterohemorrhagic *Escherichia coli* O157 in slaughtered cattle. Appl. Environ. Microbiol. 71 (1), 93–97.

Manshadi, F.D., Gortares, P., Gerba, C.P., Karpiscak, M., Frentas, R.J., 2001. Role of irrigation water in contamination of domestic fresh vegetables. In: Proceedings of the Amer. Soc. Microbiol. Gen. Meeting, Orlando, FL (Vol. 101), p. 561.

Martin, S.A., Streeter, M., 1995. Effect of malate on in vitro mixed ruminal microorganism fermentation. J. Anim. Sci. 73, 2141–2145.

Matthews, L., Low, J.C., Gally, D.L., Pearce, M.C., Mellor, D.J., Heesterbeek, J.A.P., Chase-Topping, M., Naylor, S.W., Shaw, D.J., Reid, S.W.J., Gunn, G.J., Woolhouse, M.E.J., 2006a. Heterogeneous shedding of *Escherichia coli* O157 in cattle and its implications for control. Proc. Natl. Acad. Sci. 103, 547–552.

Matthews, L., McKendrick, I.J., Ternent, H., Gunn, G.J., Synge, B., Woolhouse, M.E.J., 2006b. Super-shedding cattle and the transmission dynamics of *Escherichia coli* O157. Epidemiol. Infect. 134, 131–142.

Maule, A., 2000. Survival of verocytotoxigenic *Escherichia coli* O157 in soil, water and on surfaces. J. Appl. Microbiol. 88 (S1), 71S–78S.

McGee, P., Scott, L., Sheridan, J.J., Earley, B., Leonard, A., 2004. Horizontal transmission of *Escherichia coli* O157: H7 during cattle housing. J. Food Prot. 67 (12), 2651–2656.

McNeilly, T.N., Mitchell, M.C., Rosser, T., McAteer, S., Low, J.C., Smith, D.G., Gally, D.L., 2010. Immunization of cattle with a combination of purified intimin-531, EspA and Tir significantly reduces shedding of *Escherichia coli* O157: H7 following oral challenge. Vaccine 28 (5), 1422–1428.

Merril, C.R., Biswas, B., Carlton, R., Jensen, N.C., Creed, G.J., Zullo, S., Adhya, S., 1996. Long-circulating bacteriophage as antibacterial agents. Proc. Natl. Acad. Sci. U. S. A. 93 (8), 3188–3192.

Miller, G.E., Cohen, S., Pressman, S., Barkin, A., Rabin, B.S., Treanor, J.J., 2004. Psychological stress and antibody response to influenza vaccination: when is the critical period for stress, and how does it get inside the body? Psychosom. Med. 66 (2), 215–223.

Min, B.R., Pinchak, W.E., Anderson, R.C., Callaway, T.R., 2007. Effect of tannins on the in vitro growth of *Escherichia coli* O157:H7 and in vivo growth of generic *Escherichia coli* excreted from steers. J. Food Prot. 70, 543–550.

Mingeot-Leclercq, M.P., Glupczynski, Y., Tulkens, P.M., 1999. Aminoglycosides: activity and resistance, minireview. Antimicrob. Agents Chemother. 43 (4), 727–737.

Mkaddem, M., Bouajila, J., Ennajar, M., Lebrihi, A., Mathieu, F., Romdhane, M., 2009. Chemical composition and antimicrobial and antioxidant activities of mentha (longifolia L. and viridis) essential oils. J. Food Sci. 74, 358–363.

Monaghan, Á., Byrne, B., Fanning, S., Sweeney, T., McDowell, D., Bolton, D.J., 2011. Serotypes and virulence profiles of non-O157 Shiga toxin-producing *Escherichia coli* isolates from bovine farms. Appl. Environ. Microbiol. 77 (24), 8662–8668.

Moxley, R.A., Smith, D., Klopfenstein, T.J., Erickson, G., Folmer, J., Macken, C., Hinkley, S., Potter, A., Finlay, B., 2003. Vaccination and feeding a competitive exclusion product as intervention strategies to reduce the prevalence of *Escherichia coli* O157: H7 in feedlot cattle. In: Proc. 5th Int. Symp. on Shiga Toxin-Producing *Escherichia coli* Infections, Edinburgh, UK, p. 23.

Moxley, R.A., Smith, D.R., Luebbe, M., Erickson, G.E., Klopfenstein, T.J., Rogan, D., 2009. *Escherichia coli* O157: H7 vaccine dose–effect in feedlot cattle. Foodborne Pathog. Dis. 6 (7), 879–884.

Nannapaneni, R., Muthaiyan, A., Crandall, P.G., Johnson, M.G., O'Bryan, C.A., Chalova, V.I., Callaway, T.R., Carroll, J.A., Arthington, J.D., Nisbet, D.J., Ricke, S.C., 2008. Antimicrobial activity of commercial citrus-based natural extracts against *Escherichia coli* O157:H7 isolates and mutant strains. Foodborne Pathog. Dis. 5, 695–699.

Nataro, J.P., Kaper, J.B., 1998. Diarrheagenic *Escherichia coli*. Clin. Microbiol. Rev. 11, 142–201.

Natvig, E.E., Ingham, S.C., Ingham, B.H., Cooperband, L.R., Roper, T.R., 2002. Salmonella enterica serovar Typhimurium and *Escherichia coli* contamination of root and leaf vegetables grown in soils with incorporated bovine manure. Appl. Environ. Microbiol. 68 (6), 2737–2744.

Naylor, S.W., Low, J.C., Besser, T.E., Mahajan, A., Gunn, G.J., Pearce, M.C., McKendrick, I.J., Smith, G.E., Gally, D.L., 2003. Lymphoid follicle-dense mucosa at the terminal rectum is the principal site of colonization of enterohemorrhagic *Escherichia coli* O157: H7 in the bovine host. Infect. Immun. 71 (3), 1505–1512.

Naylor, S.W., Roe, A.J., Nart, P., Spears, K., Smith, D.G., Low, J.C., Gally, D.L., 2005. *Escherichia coli* O157: H7 forms attaching and effacing lesions at the terminal rectum of cattle and colonization requires the LEE4 operon. Microbiology 151 (8), 2773–2781.

Nisbet, D.J., Martin, S.A., 1993. Effects of fumarate, L-malate, and an *Aspergillus oryzae* fermentation extract on D-lactate utilization by the ruminal bacterium Selenomonas ruminantium. Curr. Microbiol. 26, 136.

Nisbet, D.J., Corrier, D.E., Ricke, S.C., Hume, M.E., Byrd, J.A., DeLoach, J.R., 1996. Maintenance of the biological efficacy in chicks of a cecal competitive-exclusion culture against Salmonella by continuous-flow fermentation. J. Food Prot. 59 (12), 1279–1283.

Nisbet, D.J., Callaway, T.R., Edrington, T., Anderson, R.C., Krueger, N., 2009. Effects of the dicarboxylic acids malate and fumarate on E. coli O157: H7 and Salmonella enterica typhimurium populations in pure culture and in mixed ruminal microorganism fermentations. Curr. Microbiol. 58, 488–492.

Niu, Y.D., Johnson, R.P., Xu, Y., McAllister, T.A., Sharma, R., Louie, M., Stanford, K., 2009. Host range and lytic capability of four bacteriophages against bovine and clinical human isolates of Shiga toxin-producing *Escherichia coli* O157: H7. J. Appl. Microbiol. 107 (2), 646–656.

Niu, Y.D., Stanford, K., Kropinski, A.M., Ackermann, H.W., Johnson, R.P., She, Y.M., McAllister, T.A., 2012. Genomic, proteomic and physiological characterization of a T5-like bacteriophage for control of Shiga toxin-producing *Escherichia coli* O157: H7. PLoS One 7 (4), e34585.

Nurmi, E., Rantala, M., 1973. New aspects of Salmonella infection in broiler production. Nature 241, 210–211.

Nurmi, E., Nuotio, L., Schneitz, C., 1992. The competitive exclusion concept: development and future. Int. J. Food Microbiol. 15 (3), 237–240.

O'brien, A.D., Tesh, V.L., Donohue-Rolfe, A., Jackson, M.P., Olsnes, S., Sandvig, K., Lindberg, A.A., Keusch, G.T., 1992. Shiga toxin: biochemistry, genetics, mode of action, and role in pathogenesis. In: Pathogenesis of Shigellosis. Springer, Berlin, Heidelberg, pp. 65–94.

Ohya, T., Marubashi, T., Ito, H., 2000. Significance of fecal volatile fatty acids in shedding of *Escherichia coli* O157 from calves: experimental infection and preliminary use of a probiotic product. J. Vet. Med. Sci. 62 (11), 1151–1155.

Oliver, C.E., Magelky, B.K., Bauer, M.L., Cheng, F.C., Caton, J.S., Hakk, H., et al., 2008. Fate of chlorate present in cattle wastes and its impact on *Salmonella typhimurium* and *Escherichia coli* O157: H7. J. Agric. Food Chem. 56 (15), 6573–6583.

Omisakin, F., MacRae, M., Ogden, I.D., Strachan, N.J.C., 2003. Concentration and prevalence of *Escherichia coli* O157 in cattle feces at slaughter. Appl. Environ. Microbiol. 69 (5), 2444–2447.

Oot, R.A., Raya, R.R., Callaway, T.R., Edrington, T.S., Kutter, E.M., Brabban, A.D., 2007. Prevalence of *Escherichia coli* O157 and O157: H7-infecting bacteriophages in feedlot cattle feces. Lett. Appl. Microbiol. 45 (4), 445–453.

O'Reilly, K.M., Low, J.C., Denwood, M.J., Gally, D.L., Evans, J., Gunn, G.J., Mellor, D.J., Reid, S.W.J., Matthews, L., 2010. Associations between the presence of virulence determinants and the epidemiology and ecology of zoonotic *Escherichia coli*. Appl. Environ. Microbiol. 76 (24), 8110–8116.

Owens, F.N., 1988. Protein metabolism of ruminant animals. In: Church, D.C. (Ed.), The Ruminant Animal. Waveland Press, Prospect Heights, IL, pp. 227–249.

Patra, A.K., Saxena, J., 2009. Dietary phytochemicals as rumen modifiers: a review of the effects on microbial populations. Antonie Van Leeuwenhoek 39, 1–13.

[illegible], [illegible], Subramanyam, V.R., Kole, C., 1996. Antibacterial and antifungal activity of ten essential oils in vitro. Microbios 86, 237–246.

Pendleton, S.J., Crandall, P.G., Ricke, S.C., Goodridge, L., O'Bryan, C.A., 2012. Inhibition of beef isolates of *E. coli* O157: H7 by orange oil at various temperatures. J. Food Sci. 77, M308–M311.

Pérez-Conesa, D., Cao, J., Chen, L., McLandsborough, L., Weiss, J., 2011. Inactivation of *Listeria monocytogenes* and *Escherichia coli* O157:H7 biofilms by micelle-encapsulated eugenol and carvacrol. J. Food Prot. 74, 55–62.

Pittman, C.I., Pendleton, S., Bisha, B., O'Bryan, C.A., Belk, K.E., Goodridge, L., Crandall, P.G., Ricke, S.C., 2011. Activity of citrus essential oils against *Escherichia coli* O157:H7 and Salmonella spp. and effects on beef subprimal cuts under refrigeration. J. Food Sci. 76, M433–M438.

Piva, A., Grilli, E., Fabbri, L., Pizzamiglio, V., Campani, I., 2007. Free versus microencapsulated organic acids in medicated or not medicated diet for piglets. Livest. Sci. 108, 214–217.

Potter, A.A., Klashinsky, S., Li, Y., Frey, E., Townsend, H., Rogan, D., Erickson, G., Finlay, B.B., 2004. Decreased shedding of *Escherichia coli* O157: H7 by cattle following vaccination with type III secreted proteins. Vaccine 22 (3), 362–369.

Poynter, D., Hicks, S.J., Rowbury, R., 1986. Acid resistance of attached organisms and its implications for the pathogenicity of plasmid-bearing *Escherichia coli*. Lett. Appl. Microbiol. 3, 117–121.

Pruimboom-Brees, I.M., Morgan, T.W., Ackermann, M.R., Nystrom, E.D., Samuel, J.E., Cornick, N.A., Moon, H.W., 2000. Cattle lack vascular receptors for *Escherichia coli* O157: H7 Shiga toxins. Proc. Natl. Acad. Sci. 97 (19), 10325–10329.

Rabinovitz, B.C., Gerhardt, E., Tironi Farinati, C., Abdala, A., Galarza, R., Vilte, D.A., Mercado, E.C., 2012. Vaccination of pregnant cows with EspA, EspB, γ-intimin, and Shiga toxin 2 proteins from *Escherichia coli* O157: H7 induces high levels of specific colostral antibodies that are transferred to newborn calves. J. Dairy Sci. 95 (6), 3318–3326.

Ransom, J.R., Belk, K.E., Sofos, J.N., Scanga, J.A., Rossman, M.L., Smith, G.C., Tatum, J.D., 2003. Investigation of on-Farm Management Practices as Pre-Harvest Beef Microbiological Interventions. Natl. Cattlemen's Beef Assoc. Res Fact Sheet, Centennial, CO.

Raya, R.R., Varey, P., Oot, R.A., Dyen, M.R., Callaway, T.R., Edrington, T.S., et al., 2006. Isolation and characterization of a new T-even bacteriophage, CEV1, and determination of its potential to reduce *Escherichia coli* O157: H7 levels in sheep. Appl. Environ. Microbiol. 72 (9), 6405–6410.

Reichling, J., Schnitzler, P., Suschke, U., Saller, R., 2009. Essential oils of aromatic plants with antibacterial, antifungal, antiviral, and cytotoxic properties – an overview. Forsch. Komplementmed. 16, 79–90.

Reissbrodt, R., Hammes, W.P., Dal Bello, F., Prager, R., Fruth, A., Hantke, K., Williams, P.H., 2009. Inhibition of growth of Shiga toxin-producing Escherichia coli by nonpathogenic *Escherichia coli*. FEMS Microbiol. Lett. 290 (1), 62–69.

Riley, L.W., Remis, R.S., Helgerson, S.D., McGee, H.B., Wells, J.G., Davis, B.R., Hebert, R.J., Olcott, E.S., Johnson, L.M., Hargrett, N.T., Blake, P.A., Cohen, M.L., 1983. Hemorrhagic colitis associated with a rare *Escherichia coli* serotype. New Engl. J. Med. 308 (12), 681–685.

Rivas, L., Coffey, B., McAuliffe, O., McDonnell, M.J., Burgess, C.M., Coffey, A., Duffy, G., 2010. In vivo and ex vivo evaluations of bacteriophages e11/2 and e4/1c for use in the control of *Escherichia coli* O157: H7. Appl. Environ. Microbiol. 76 (21), 7210–7216.

Roe, A.J., Naylor, S.W., Spears, K.J., Yull, H.M., Dransfield, T.A., Oxford, M., Mckendrick, I., Porter, M., Woodward, M., Smith, D.G.E., Gally, D.L., 2004. Co-ordinate single cell expression of LEE4- and LEE5-encoded proteins of *Escherichia coli* O157:H7. Mol. Microbiol. 54, 337–352.

Rotariu, O., Ogden, I.D., MacRitchie, L., Forbes, K.J., Williams, A.P., Cross, P., Hunter, C.J., Teunis, P.F.M., Strachan, N.J.C., 2012. Combining risk assessment and epidemiological risk factors to elucidate the sources of human *E. coli* O157 infection. Epidemiol. Infect. 140 (8), 1414–1429.

Russell, J.B., Houlihan, A.J., 2003. Ionophore resistance of ruminal bacteria and its potential impact on human health. FEMS Microbiol. Rev. 27, 65–74.

Russell, J.B., Jarvis, G.N., 2001. Practical mechanisms for interrupting the oral-fecal lifecycle of *Escherichia coli*. J. Mol. Microbiol. Biotechnol. 3 (2), 265–272.

Russell, J.B., Strobel, H.J., 1989. Effect of ionophores on ruminal fermentation. Appl. Environ. Microbiol. 55, 1–6.

Sagong, H.G., Lee, S.Y., Chang, P.S., Heu, S., Ryu, S., Choi, Y.J., Kang, D.H., 2011. Combined effect of ultrasound and organic acids to reduce *Escherichia coli* O157:H7, *Salmonella Typhimurium*, and *Listeria monocytogenes* on organic fresh lettuce. Int. J. Food Microbiol. 145, 287–292.

Sargeant, J.M., Amezcua, M.R., Rajic, A., Waddell, L., 2007. Pre-harvest interventions to reduce the shedding of *E. coli* O157 in the faeces of weaned domestic ruminants: a systematic review. Zoonoses Public Hlth. 54 (6–7), 260–277.

Scallan, E., Hoekstra, R.M., Angulo, F.J., Tauxe, R.V., Widdowson, M.A., Roy, S.L., Jones, J.L., Griffin, P.L., 2011. Foodborne illness acquired in the United States—major pathogens. Emerg. Infect. Dis. 17, 7–15.

Scharff, R.L., 2010. Health-related costs from foodborne illness in the United States. http://www.publichealth.lacounty.gov/eh/docs/ReportPublication/HlthRelatedCostsFromFoodborneIllinessUS.pdf (accessed 2014).

Scott, T., Wilson, C., Bailey, D., Klopfenstein, T., Milton T., Moxley R., Smith, D., Gray, J., Hungerford, L., 2000. Influence of diet on total and acid resistant *E. coli* and colonic pH. Nebraska beef report, pp. 39–41.

Semenov, A.V., van Overbeek, L., van Bruggen, A.H., 2009. Percolation and survival of *Escherichia coli* O157: H7 and Salmonella enterica serovar Typhimurium in soil amended with contaminated dairy manure or slurry. Appl. Environ. Microbiol. 75 (10), 3206–3215.

Sharma, V.K., Bearson, S.M., Bearson, B.L., 2010. Evaluation of the effects of sdiA, a luxR homologue, on adherence and motility of *Escherichia coli O157: H7.* Microbiology 156, 1303–1312.

Slutsker, L., Ries, A.A., Maloney, K., Wells, J.G., Greene, K.D., Griffin, P.M., 1998. A nationwide case-control study of *Escherichia coli* O157: H7 infection in the United States. J. Infect. Dis. 177 (4), 962–966.

Smith, H.W., Huggins, M.B., 1982. Successful treatment of experimental *Escherichia coli* infections in mice using phage: its general superiority over antibiotics. J. Gen. Microbiol. 128 (2), 307–318.

Smith, H.W., Huggins, M.B., 1983. Effectiveness of phages in treating experimental *Escherichia coli* diarrhea in calves, piglets and lambs. J. Gen. Microbiol. 129 (8), 2659–2675.

Smith, H.W., Huggins, M.B., Shaw, K.M., 1987. Factors influencing the survival and multiplication of bacteriophages in calves and in their environment. J. Gen. Microbiol. 133, 1127–1135.

Smith, L., Mann, J.E., Harris, K., Miller, M.F., Brashears, M.M., 2005. Reduction of *Escherichia coli* O157: H7 and *Salmonella* in ground beef using lactic acid bacteria and the impact on sensory properties. J. Food Prot. 68 (8), 1587–1592.

Smith, D.R., Moxley, R.A., Peterson, R.E., Klopfenstein, T.J., Erickson, G.E., Bretschneider, G., Clowser, S., 2009. A two-dose regimen of a vaccine against type III secreted proteins reduced *Escherichia coli* O157: H7 colonization of the terminal rectum in beef cattle in commercial feedlots. Foodborne Pathog. Dis. 6 (2), 155–161.

Smith, K.E., Wilker, P.R., Reiter, P.L., Hedican, E.B., Bender, J.B., Hedberg, C.W., 2012. Antibiotic treatment of *Escherichia coli* O157 infection and the risk of hemolytic uremic syndrome, Minn. Pediatr. Infect. Dis. J. 31 (1), 37–41.

Snedeker, K.G., Campbell, M., Sargeant, J.M., 2012. A systematic review of vaccinations to reduce the shedding of *Escherichia coli* O157 in the feces of domestic ruminants. Zoonoses Public Hlth. 59 (2), 126–138.

Sperandio, V., 2010. SdiA sensing of acyl-homoserine lactones by enterohemorrhagic *E. coli* (EHEC) serotype O157:H7 in the bovine rumen. Gut Microbes 1, 432–435.

Stacey, K.F., Parsons, D.J., Christiansen, K.H., Burton, C.H., 2007. Assessing the effect of interventions on the risk of cattle and sheep carrying *Escherichia coli* O157:H7 to the abattoir using a stochastic model. Prev. Vet. Med. 79 (1), 32–45.

Stanton, T.L., Schutz, D., 2000. Effect of switching from high grain to hay five days prior to slaughter on finishing cattle performance. Colorado State University Research Report. Colorado State University, Fort Collins, CO.

Steer, T., Carpenter, H., Tuohy, K., Gibson, G.R., 2000. Perspectives on the role of the human gut microbiota and its modulation by pro and prebiotics. Nutr. Res. Rev. 13, 229–254.

Steinmuller, N., 2006. Higher incidence of *Escherichia coli* O157: H7 infection in rural counties and possible association with animal contact. In: 44th Annual Meeting. IDSA.

Stephens, T.P., Loneragan, G.H., Chichester, L.M., Brashears, M.M., 2007. Prevalence and enumeration of *Escherichia coli* O157 in steers receiving various strains of Lactobacillus-based direct-fed microbials. J. Food Prot. 70 (5), 1252–1255.

Stewart, V., 1988. Nitrate respiration in relation to facultative metabolism in enterobacteria. Microbiol. Rev. 52 (2), 190.

Stouthamer, A.H., 1969. A genetical and biochemical study of chlorate-resistant mutants of *Salmonella typhimurium*. Antonie Van Leeuwenhoek 35, 505–521.

Summers, W.C., 2001. Bacteriophage therapy. Annu. Rev. Microbiol. 55 (1), 437–451.

Swyers, K.L., Carlson, B.A., Nightingale, K.K., Belk, K.E., Archibeque, S.L., 2011. Naturally colonized beef cattle populations fed combinations of yeast culture and an ionophore in finishing diets containing dried distiller's grains with solubles had similar fecal shedding of *Escherichia coli* O157: H7. J. Food Prot. 74 (6), 912–918.

Synge, B.A., Chase-Topping, M.E., Hopkins, G.F., McKendrick, I.J., Thomson-Carter, F., Gray, D., Rusbridge, S.M., Munro, F.I., Foster, G., Gunn, G.J., 2003. Factors influencing the shedding of verocytotoxin-producing *Escherichia coli* O157 by beef suckler cows. Epidemiol. Infect. 130, 301–312.

Tanji, Y., Shimada, T., Fukudomi, H., Miyanaga, K., Nakai, Y., Unno, H., 2005. Therapeutic use of phage cocktail for controlling *Escherichia coli* O157: H7 in gastrointestinal tract of mice. J. Biosci. Bioeng. 100 (3), 280–287.

Thomson, D.U., Loneragan, G.H., Thornton, A.B., Lechtenberg, K.F., Emery, D.A., Burkhardt, D.T., Nagaraja, T.G., 2009. Use of a siderophore receptor and porin proteins-based vaccine to control the burden of *Escherichia coli* O157: H7 in feedlot cattle. Foodborne Pathog. Dis. 6 (7), 871–877.

Thornton, A., Thomson, D., Loneragan, G., Fox, J., Burkhardt, D., Emery, D., Nagaraja, T., 2009. Effects of a siderophore receptor and porin proteins–based vaccination on fecal shedding of *Escherichia coli O157: H7* in experimentally inoculated cattle. J. Food Protect. 72, 866–869.

Thran, B.H., Hussein, H.S., Hall, M.R., Khaiboullina, S.F., 2001. Shiga toxin-producing *Escherichia coli* in beef heifers grazing an irrigated pasture. J. Food Prot. 64, 1613–1616.

Tkalcic, S., Brown, C.A., Harmon, B.G., Jain, A.V., Mueler, E.P.O., Parks, A., Jacobsen, K.L., Martin, S.A., Zhao, T., Doyle, M.P., 2000. Effects of diet on rumen proliferation and fecal shedding of *Escherichia coli* O157:H7 in calves. J. Food Prot. 63, 1630–1636.

Tkalcic, S., Zhao, T., Harmon, B.G., Doyle, M.P., Brown, C.A., Zhao, P., 2003. Fecal shedding of enterohemorrhagic *Escherichia coli* in weaned calves following treatment with probiotic *Escherichia coli*. J. Food Prot. 66 (7), 1184–1189.

Turgis, M., Han, J., Caillet, S., Lacroix, M., 2009. Antimicrobial activity of mustard essential oil against *Escherichia coli* O157:H7 and *Salmonella typhimurium*. Food Control 20, 1073–1079.

USDA-FSIS, 2012. Shiga toxin-producing *Escherichia coli* in certain raw beef products. Rules and Regulations-Federal Register 77 (105), 31975–31981. Available at: http://www.fsis.usda.gov/OPPDE/rdad/FRPubs/2010-0023FRN.pdf.

Van Baale, M.J., Sargeant, J.M., Gnad, D.P., DeBey, B.M., Lechtenberg, K.F., Nagaraja, T.G., 2004. Effect of forage or grain diets with or without monensin on ruminal persistence and fecal *Escherichia coli* O157:H7 in cattle. Appl. Environ. Microbiol. 70, 5336–5342.

Van Donkersgoed, J., Hancock, D., Rogan, D., Potter, A.A., 2005. *Escherichia coli* O157: H7 vaccine field trial in 9 feedlots in Alberta and Saskatchewan. Can. Vet. J. 46 (8), 724.

Van Overbeek, L.S., Franz, E., Semenov, A.V., De Vos, O.J., Van Bruggen, A.H.C., 2010. The effect of the native bacterial community structure on the predictability of *E. coli* O157: H7 survival in manure-amended soil. Lett. Appl. Microbiol. 50 (4), 425–430.

Vandeplas, S., Dubois Dauphin, R., Beckers, Y., Thonart, P., Thewis, A., 2010. Salmonella in chicken: current and developing strategies to reduce contamination at farm level. J. Food Prot. 73, 774–785.

Varel, V.H., Wells, J.E., Berry, E.D., Spiehs, M.J., Miller, D.N., Ferrell, C.L., Shackelford, S.D., Koohmaraie, M., 2008. Odorant production and persistence of *Escherichia coli* in manure slurries from cattle fed zero, twenty, forty, or sixty percent wet distillers grains with solubles. J. Anim. Sci. 86, 3617–3627.

Varela, N.P., Dick, P., Wilson, J., 2013. Assessing the existing information on the efficacy of bovine vaccination against *Escherichia coli* O157: H7 – a systematic review and meta-analysis. Zoonoses Public Health 60 (4), 253–268.

Vikram, A., Jesudhasan, P.R., Jayaprakasha, G.K., Pillai, B.S., Patil, B.S., 2010. Grapefruit bioactive limonoids modulate *E. coli* O157:H7 TTSS and biofilm. Int. J. Food Microbiol. 140, 109–116.

Viuda-Martos, M., Ruiz-Navajas, Y., Fernndez-Lapez, A.J., Perez-Ãlvarez, J., 2008. Antibacterial activity of lemon (Citrus lemon L.), mandarin (Citrus reticulata L.), grapefruit (Citrus paradisi L.) and orange (Citrus sinensis L.) essential oils. J. Food Saf. 28, 567–576.

Vosough Ahmadi, B., Velthuis, A.G., Hogeveen, H., Huirne, R., 2006. Simulating *Escherichia coli* O157: H7 transmission to assess effectiveness of interventions in Dutch dairy-beef slaughterhouses. Prev. Vet. Med. 77 (1), 15–30.

Walle, K.V., Vanrompay, D., Cox, E., 2013. Bovine innate and adaptive immune responses against *Escherichia coli* O157: H7 and vaccination strategies to reduce fecal shedding in ruminants. Vet. Immunol. Immunopathol. 152 (1), 109–120.

Walterspiel, J.N., Morrow, A.L., Cleary, T.G., Ashkenazi, S., 1992. Effect of subinhibitory concentrations of antibiotics on extracellular Shiga-like toxin. Infection 20 (1), 25–29.

Wang, G., Zhao, T., Doyle, M.P., 1996. Fate of enterohemorrhagic *Escherichia coli* O157: H7 in bovine feces. Appl. Environ. Microbiol. 62 (7), 2567–2570.

Wang, Y., Xu, Z., Bach, S.J., McAllister, T.A., 2009. Sensitivity of *Escherichia coli* to seaweed (Asco- phyllum nodosum) phlorotannins and terrestrial tannins. Asian Australas. J. Anim. Sci. 22, 238–245.

Wells, J.E., Berry, E.D., Varel, V.H., 2005. Effects of common forage phenolic acids on *Escherichia coli* O157:H7 viability in bovine feces. Appl. Environ. Microbiol. 71, 7974–7979.

Westphal, A., Williams, M.L., Baysal-Gurel, F., LeJeune, J.T., Gardener, B.B.M., 2011. General suppression of *Escherichia coli O157: H7* in sand-based dairy livestock bedding. Appl. Environ. Microbiol. 77, 2113–2121.

Wichmann, F., Udikovic-Kolic, N., Andrew, S., Handelsman, J., 2014. Diverse antibiotic resistance genes in dairy cow manure. mBio 5 (2), e01017.

Wick, L.M., Qi, W., Lacher, D.W., Whittam, T.S., 2005. Evolution of genomic content in the stepwise emergence of *Escherichia coli* O157: H7. J. Bacteriol. 187 (5), 1783–1791.

Wileman, B.W., Thomson, D.U., Olson, K.C., Jaeger, J.R., Pacheco, L.A., Bolte, J., et al., 2011. *Escherichia coli* O157: H7 shedding in vaccinated beef calves born to cows vaccinated prepartum with *Escherichia coli* O157: H7 SRP vaccine. J. Food Prot. 74 (10), 1599–1604.

Winfield, M.D., Groisman, E.A., 2003. Role of nonhost environments in the lifestyles of Salmonella and *Escherichia coli*. Appl. Environ. Microbiol. 69 (7), 3687–3694.

Wolin, M.J., 1969. Volatile fatty acids and the inhibition of *Escherichia coli* growth by rumen fluid. Appl. Microbiol. 17, 83–87.

Yang, H.E., Yang, W.Z., McKinnon, J.J., Alexander, T.W., Li, Y.L., McAllister, T.A., 2010. Survival of *Escherichia coli* O157: H7 in ruminal or fecal contents incubated with corn or wheat dried distillers' grains with solubles. Can. J. Microbiol. 56 (11), 890–895.

Yekta, M.A., Goddeeris, B.M., Vanrompay, D., Cox, E., 2011. Immunization of sheep with a combination of intiminγ, EspA and EspB decreases *Escherichia coli* O157: H7 shedding. Vet. Immunol. Immunopathol. 140 (1), 42–46.

Yokoyama, M.G., Johnson, K.A., 1988. Microbiology of the rumen and intestine. In: Church, D.C. (Ed.), The Ruminant Animal: Digestive Physiology and Nutrition. Waveland Press, Englewood Cliffs, NJ, pp. 125–144.

Younts-Dahl, S.M., Galyean, M.L., Loneragan, G.H., Elam, N.A., Brashears, M.M., 2004. Dietary supplementation with Lactobacillus- and Propionibacterium-based direct-fed microbials and prevalence of *Escherichia coli* O157 in beef feedlot cattle and on hides at harvest. J. Food Prot. 67 (5), 889–893.

Zhang, Y., Laing, C., Steele, M., Ziebell, K., Johnson, R., Benson, A.K., Raboada, E., Gannon, V.P., 2007. Genome evolution in major *Escherichia coli* O157: H7 lineages. BMC Genomics 8 (1), 121.

Zhang, Y., Laing, C., Zhang, Z., Hallewell, J., You, C., Ziebell, K., Gannon, V.P., 2010. Lineage and host source are both correlated with levels of Shiga toxin 2 production by *Escherichia coli* O157: H7 strains. Appl. Environ. Microbiol. 76 (2), 474–482.

Zhang, X., Cheng, Y., Xiong, Y., Ye, C., Zheng, H., Sun, H., Xu, J., 2012. Enterohemorrhagic *Escherichia coli* specific enterohemolysin induced IL-1β in human macrophages and EHEC-induced IL-1β required activation of NLRP3 inflammasome. PLoS One 7 (11), e50288.

Zhao, T., Doyle, M.P., Harmon, B.G., Brown, C.A., Mueller, P.E., Parks, A.H., 1998. Reduction of carriage of enterohemorrhagic *Escherichia coli* O157: H7 in cattle by inoculation with probiotic bacteria. J. Clin. Microbiol. 36 (3), 641–647.

Zhao, T., Tkalcic, S., Doyle, M.P., Harmon, B.G., Brown, C.A., Zhao, P., 2003. Pathogenicity of enterohemorrhagic *Escherichia coli* in neonatal calves and evaluation of fecal shedding by treatment with probiotic *Escherichia coli*. J. Food Prot. 66 (6), 924–930.

Zhou, Z., Li, X., Liu, B., Beutin, L., Xu, J., Ren, Y., Feng, L., Lan, R., Reeves, P.R., Wang, L., 2010. Derivation of *Escherichia coli* O157: H7 from its O55: H7 precursor. PLoS One 5 (1), e8700.

CHAPTER

CURRENT PERSPECTIVES ON *CAMPYLOBACTER*

11

S. Pendleton*, D. D'Souza*, S. Joshi*, I. Hanning*,†

Department of Food Science and Technology, University of Tennessee, Knoxville, Tennessee, USA, Department of Genome Sciences and Technology, University of Tennessee, Knoxville, Tennessee, USA†*

1 EMERGENCE OF *CAMPYLOBACTER* AS A FOODBORNE PATHOGEN

Theodor Escherich is credited with the discovery of *Campylobacter* in 1886 (Altekruse et al., 1999). Escherich was examining the cause of diarrhea in newborns and identified *Campylobacter*-like bacteria in their feces. In 1909, McFadyean and Stockman found similar bacteria in the tissues of aborted sheep fetuses. At the time, it was believed to be a related *Vibrio* spp. due to its spiral morphology and referred to as *Vibrio fetus* because of its association with aborted fetuses in sheep and cattle. In 1947, the first case of a human abortion caused by *Vibrio fetus* was reported (Vincent et al., 1947). In 1957, Elisabeth King studied the first human cases of *Vibrio fetus* and "related vibrio" infection (King, 1957). However, it was not until 1972 when a Belgian group added sheep blood, polymixin-B-sulfate, novobiocin, and actidione to fluid-thioglycolate-agar that isolation of "related vibrio" from stool was possible (Dekeyser et al., 1972). Shortly after, in 1973, *Campylobacter jejuni* and *coli* were finally recognized as new species distinct from *Vibrio* (Skirrow, 1977).

Campylobacter has since become one of the leading causes of bacterial gastroenteritis in the United States (Scallan et al., 2011). It causes an estimated 845,024 cases each year and approximately 17% result in hospitalization. Many of the cases go unreported due to the relatively mild symptoms, which include diarrhea, cramping, abdominal pain, and fever. Nausea and vomiting can also occur with this illness. Guillain-Barré, reactive arthritis, and irritable bowl syndrome have been recognized as sequelae of *Campylobacter* infections (WHO, 2012). This means that a relatively simple foodborne illness can ultimately leave an individual with a life-altering condition. These factors make understanding and control of *Campylobacter* a prime focus for research.

2 SPECIES AND STRAIN VARIATION

Understanding *Campylobacter* begins with determining where it can be found in the environment. Its major reservoir is within the gastrointestinal tract of animals, including both domesticated and wild animals. It is most commonly associated with poultry. With chicken consumption in the United States being between 54 and 60 pounds per capita from 2000 to 2012 (USPoultry, 2012), this creates a very large potential for human exposure and infection. *Campylobacter* has also been found in

Food Safety. http://dx.doi.org/10.1016/B978-0-12-800245-2.00011-3

swine and cattle. Even though *E. coli* is generally considered of more concern for the cattle industry than *Campylobacter*, campylobacteriosis has been linked to consumption of raw milk. As raw milk is becoming popular among co-ops and people that believe it offers better nutrition, it is also becoming a more important reservoir for *Campylobacter*. Water is an additional source of infection, though not considered to be as large as that of animals. Within animal reservoirs, *Campylobacter* is not known to cause disease and is considered to be a commensal organism in the chicken gastrointestinal tract.

It has been suggested that the genome of *Campylobacter* could contain host-specific genes or allelic variants, or possibly correlate them to an isolation source (Champion et al., 2005). Analysis of strain and species phylogeny has yielded varying results. Champion and others identified two clades of *Campylobacter*, one associated with livestock and one associated with environmental isolates (nonlivestock). Hepworth and others (2011) evaluated 80 strains of *C. jejuni* from various sources, including water, wild birds, wild animals, chickens, cattle, and humans and found a similar clonal complex distribution (Hepworth et al., 2011). They identified a "water/wildlife" group and a "food-chain-associated" group. A specific clonal group (ST-3704) were also found only within bank voles, indicating a niche specialization and a host-specific sequence type. The bank vole sequence type was unable to colonize a chicken gut, which supports the hypothesis of niche specialization. While these strains could not colonize a chicken gut, it was not determined if they could infect human intestinal cell, and potentially cause human disease. They also found that water and wildlife isolates all clustered together separate from the other clonal complexes, which indicated a divergence of these strains from the more common clonal complexes.

While Hepworth and others found isolates with niche-specific sequence types, a broad study by Gripp and others (2011) investigated the phylogeny of 300 *C. jejuni* and 173 *Campylobacter coli* strains. In the course of their study, they investigated the clonal group ST-21 in depth (Gripp et al., 2011). They had two genetically similar *C. jejuni* strains of ST-21 from two different isolation sources, chicken and human. To determine if there was host specificity for these two isolates, they examined their ability to colonize the chicken gut. Both strains were able to efficiently colonize the chicken gut, though the human isolate to a 1 log lower cell density. Through the course of their study, they were unable to determine host- or isolation-specific genes or phenotypes, which led them to conclude that most dominant sequence types of *Campylobacter* exhibit a "generalist lifestyle." They hypothesized some strains have contingency genes that allow them to adapt to various environments. The ability to adapt to diverse environments would give a distinct advantage over other strains and lends support to the fact that they are the dominant sequence types. Niche specialization, while enhancing the fitness within that environment, limits the organism to adapt to novel environments and conditions.

Campylobacter spp. are naturally competent, allowing acquisition of foreign DNA. It has been demonstrated that simple coculturing of two *C. jejuni* strains is enough to induce transformation (Wassenaar et al., 1995). *Campylobacter* has also been found to exchange genetic information between species (Chan et al., 2008; Meinersmann et al., 2003; Miller et al., 2006; Schouls et al., 2003). Sheppard and colleagues (2008) examined the evolution of *C. coli* and found indications of convergence between a clade of *C. coli* and *C. jejuni* (Sheppard et al., 2008). They hypothesized that if this trend continues, the agriculture-associated *C. coli* clade 1 will eventually be indistinguishable from *C. jejuni* and that there is potential for the other clades to converge as well and result in despeciation. Further research by Sheppard and colleagues (2010, 2013), found that clade 1 was associated with agriculture, whereas clades 2 and 3 were associated with waterfowl and water sources (Sheppard et al., 2010, 2013). Sheppard and colleagues (2010) also found that *C. coli* clade 1 is responsible for all *C. coli* human illness cases (Sheppard et al., 2010).

Their explanation for this is that humans are exposed more to agricultural sources of *C. coli* than water and wildlife sources. Agriculture itself is another explanation for the convergence of *C. coli* clade 1 and *C. jejuni*, as they are exposed to one another more often in the agriculture environment and therefore have more opportunities for horizontal gene transfer.

C. jejuni NCTC 11168 was the first published genome of the *C. jejuni* species (Parkhill et al., 2000). This sequence has since been referred to as NCTC 11168-GS, for genome sequenced, and has become the genomic reference for the species (Thomas et al., 2014). Skirrow originally isolated the strain, in 1977, from a diarrhetic patient (Skirrow, 1977). A colonization study of 11168-GS by Ahmed and others (2002) found that it could not colonize chickens as efficiently as 81116, another strain isolated from a human source (Ahmed et al., 2002). They also determined this inefficiency was inherent to the strain, as it could not be augmented by in vivo passage. Gaynor and others (2004) examined the original Skirrow isolate, NCTC 11168-O (original), in comparison to 11168-GS, and found that 11168-O is a much more efficient colonizer than the sequenced strain (Gaynor et al., 2004). This result and subsequent characterization of the two variant strains indicate that simple storage and passage of *Campylobacter* can cause it to mutate. The same year, Carrillo and others did a similar comparison between the original 11168 strain and the 11168-GS strain (Carrillo et al., 2004). However, the original 11168 strain had been passed regularly as it's original procurement and had since become a variant strain, 11168-V26. The results from their study again indicated that passage of a strain within a laboratory setting not only altered the genetics of the strain enough to make it a variant but also caused it to be less efficient as a colonizer in the chicken gut. Cooper and others (2013) found that motility of 11168-O was reduced by 25% after 10 in vitro passages (Cooper et al., 2013). In a recent study by Thomas and others (2014), an individual working with *C. jejuni* became infected and analysis of the stool sample isolate found that it was indistinguishable from 11168-GS and a variant laboratory passage strain of 11168 (11168-V26) using a 40 gene comparative genomic fingerprinting method (Thomas et al., 2014). Whole genome sequencing found that the individual had been infected with the genome-sequenced variant, 11168-GS. It was also discovered that the strain had undergone 12 indel mutations within contingency genes. Passage of the 11168-GS variant (isolated from the patient) through a mouse caused additional genetic mutations, and two identified SNPs from the human passage were found to have reverted after passage through the mouse. These data indicate that this *C. jejuni* strain has a very flexible genome that allows it to adapt quickly to new environments.

As *Campylobacters* have been shown to adapt so quickly to new environments, it has been suggested that they should be considered a quasi-species, with a quasi-species being defined as one group of related genotypes that contain at least one or more mutations from the parent offspring. In this model, *C. jejuni* has a high mutation rate that allows it to generate multiple variants quickly in an attempt to ensure one subset of the population will succeed in the environment.

3 GENOME

Flexibility in the genome is very important for *Campylobacter*. The genus as a whole has a relatively small genome compared to other pathogenic bacteria. *Salmonella enterica* serovar Typhimurium and *E. coli* O157:H7 have genomes over 4 million base pairs, while *C. jejuni* has a genome between 1.6 and 1.8 million base pairs, depending on the strain (Champion et al., 2008; McClelland et al., 2001; Perna et al., 2001). With less than half the size of the other foodborne pathogens, variability in the genome becomes very important. The first sequencing of *C. jejuni* identified homopolymeric tracts, which are thought to control contingency genes (Wassenaar et al., 2002). These tracts of repeated bases can cause slip-strand mispairing, allowing insertions and deletions (indels) (Levinson and Gutman, 1987). These indels lead to frameshift mutations and can ultimately cause phenotypic changes that can help the bacteria adapt to a new environment. This seems to be the case with *Campylobacter*, as Wassenaar and others (2002) determined that within single colonies polymorphisms were found in these homopolymeric tracts (Wassenaar et al., 2002). A study by Jerome and others (2011) gave further evidence that these homopolymeric tracts play a role in *Campylobacter* adaptation (Jerome et al., 2011). Sequencing their experimental strain of 11168 pre- and postpassage through a murine model illuminated variation within 23 different contingency genes, with the majority focused on homopolymeric tracts.

4 MECHANISMS OF VIRULENCE

While *Campylobacter* is one of the leading causes of bacterial foodborne illness in the United States, campylobacteriosis is generally resolved without the need for antimicrobial treatment (Shandera et al., 1992). Attempts to elucidate the mechanisms of virulence for *Campylobacter* have yet to yield a complete picture. Many bacteria use similar mechanisms of virulence, but *Campylobacter* has thus far proven to use alternate or complex mechanisms. However, the general course of events and a few specifics about each event have been determined.

Campylobacter enters the gut through a contaminated food or water source. It then follows a chemokine gradient, using flagella for motility, to the mucosa, where it adheres to an epithelial cell (Hendrixson and DiRita, 2004; Larson et al., 2008). Once adhered to the epithelial cell surface, the cell's signaling system is used to "hijack" an envelope and enter the cell (Watson and Galán, 2008). Once inside the cell, it can pass through to the lamina propria, where it can interact with the lymphocytes and stimulate a greater inflammatory response (Hu and Kopecko, 2008).

Motility has been found to be a significant factor for *Campylobacter* virulence. In several studies, flagella mutants have reduced colonization potential compared to wild-type strains (Grant et al., 1993; Nachamkin et al., 1993; Wösten et al., 2004). In a study by Yao and colleagues (1994), eight mutants were created with greatly reduced in invasion potential (Yao et al., 1994). Of the eight invasion-deficient mutants, only three had impaired motility. However, those three were the least efficient invaders, being 90-200-fold less efficient than the wild type, compared to 19-35-fold less efficient for the invasion deficient, but motility positive, mutants. Two of the three mutants had fully developed flagella, but were unable to use the flagella for locomotion. These two adhered moderately, but did not invade, indicating the necessity of motility for invasion (Yao et al., 1994). This motility is controlled through a chemotaxis pathway that has yet to be completely defined (Korolik and Ketley, 2008). However, *Campylobacter* has been found to use aspartate, formate, fucose, fumarate, and pyruvate as attractants (Lertsethtakarn et al., 2011).

The flagella has also been shown to function as a type three secretion system (T3SS), which aids in invasion (Konkel et al., 2004). Using the flagella export apparatus, *Campylobacter* secretes CiaB (*Campylobacter* invasion antigen B), FlaA and FlaB (flagellar filament proteins), FlaC (shares homology to FlaA and FlaB), and FspA (Flagellar secreted protein A) (Larson et al., 2008). Most secreted proteins are not well defined, but a variant of FspA, FspA2, induces host cell apoptosis, which could contribute to the severity of the disease (Poly et al., 2007).

Once the bacterium reaches a cell, it must be able to adhere to the surface. Adhesin proteins on the surface of the bacterium facilitate this bonding. There have been a few different potential adhesins identified for *C. jejuni*: CadF (*Campylobacter* adhesion to fibronectin), PEB1, JlpA (*jejuni* lipoprotein A), CapA (*Campylobacter* adhesion protein A), and a MOMP (Major Outer Membrane Protein) PorA (Ashgar et al., 2007; Jin et al., 2001; Konkel et al., 1997; Moser et al., 1997; Pei and Blaser, 1993). Of these adhesins, the most studied is CadF. CadF is a 37 kDa outer membrane protein with a four amino acid binding domain: phenylalanine-arginine-lucine-serine (Konkel et al., 2005). In a study by Monteville, Yoon, and Konkel, CadF was specific for fibronectin and required for adhesion to INT 407 cells (Monteville et al., 2003). It was also determined that *C. jejuni* infection of the intestinal cells resulted in an increased level of tyrosine phosphorylation associated with the focal adhesion signaling molecule paxillin.

Once attached, *Campylobacter* uses host cell signaling, most likely through tyrosine protein kinase-linked pathways, to initiate entry (Biswas et al., 2004; Wooldridge et al., 1996). Rearrangement of the host cytoskeleton is involved in the uptake of a pathogen into the cell (Hu and Kopecko, 2008). Other foodborne pathogens, including *Salmonella* and *Listeria*, use microfilament-dependent pathways (Elsinghorst et al., 1989; Pistor et al., 1994). However, it appears that *Campylobacter* can use either a microfilament-dependent or microtubule-dependent pathway depending on environmental conditions. Hu and Kopecko suggest that under nutrient-restricted conditions (1% fetal bovine serum) *Campylobacter* uses both microfilaments and microtubules, but under nutrient-rich conditions (10-20% fetal bovine serum) *Campylobacter* uses a microtubule specific pathway (Hu and Kopecko, 2008). *Campylobacter* enters the cell in a vacuole, which has markers associated with lipid rafts and caveolae during its early stages, indicating their importance in internalization, but does not follow a caveolae-mediated endocytic pathway (Watson and Galán, 2008). It is thought that *Campylobacter* uses this noncanonical pathway to avoid association with lysosomes, which would ultimately kill the bacterium (Watson and Galán, 2008).

Interaction with host cells can induce proinflammatory cytokines, which are important to the induction of intestinal inflammation (Watson and Galán, 2008). Hickey and others (2000) found that cytolethal distending toxin produced by *C. jejuni* can induce the production of the proinflammatory cytokine interluken-8 by intestinal cells (Hickey et al., 2000). Additionally, once *Campylobacter* traverses the epithelial barrier it can interact with leukocytes, which can induce more proinflammatory cytokines (Hu and Kopecko, 2008). Induction of the inflammatory immune response is most likely what is responsible for the progression of the disease. Intestinal inflammation and apoptosis of the epithelial cells results in diarrhea and an overstimulation of the immune system could lead to the autoimmune diseases known to be sequelae of *Campylobacter* infections.

5 POULTRY PRODUCTION

Typically, birds are reared in large houses containing anywhere from 10,000 to 30,000 chickens. Feed can be delivered via individual hanging feeders or in long trough types of structures. Water is typically delivered by nipple drinking units.

Conventional rearing methods allow the use of antibiotics and coccidiostats. Antibiotics at subtherapeutic levels are delivered in the feed. However, if the birds are ill, antibiotics may be delivered in the water because sick birds will continue to drink but refuse feed. Alternatively, injections can also be given as a treatment for illnesses.

There are many differences between organic and conventional rearing. First, no antibiotics can be utilized in organic productions. If antibiotics are administered, then the products from those birds cannot be sold as organic. Organic rearing also requires that birds have access to outdoors. For this reason, many variations in housing types can be found. Houses may either be static or mobile. Mobile houses offer the advantage of providing fresh grass in a periodic fashion. Because the birds tend to eat and scratch the grass, most of the outdoor area in static housing has no grass. Conventional rearing uses fast-growing birds, but organic birds are usually slow-growing breeds that can take twice as long to grow to market weight.

6 ANTIBIOTIC RESISTANCE

Several studies have been conducted to monitor antibiotic-resistant bacteria in organic or antibiotic-free farms and compare their profiles with the conventional farms. These studies have reported prevalence of antimicrobial resistance in bacteria despite removal of their application. For example, a study reported tetracycline resistance in *C. jejuni* even after 3 years of removal of the antibiotic (Thibodeau et al., 2011). Some other studies reported that similar observations where tetracycline resistance in organic chickens farms with levels ranging from 0.05% to 79% (Cui et al., 2005; Heuer et al., 2001). In a study conducted by Zhou and others, approximately 9-27% of macrolide-resistant *Clostridia* were found on organic farms as compared to 70-80% macrolide-resistant *Clostridia* on conventional farms (Zhou et al., 2009). Observations from these studies indicate that despite the removal of antibiotic pressure from the environment, resistance characteristics persist in the farm microflora.

There are several possible reasons suggested for the prevalence of drug-resistant strains on organic farms. First, organic certification for a chicken starts from the second day and, therefore, there are no restrictions on the breeder facilities that supply eggs to hatcheries, where eggs might have been injected with antibiotics. Additionally, there is a possibility that resistance characteristics could be transferred to other species through horizontal gene transfer mechanisms (Leverstein-van Hall et al., 2002).

The prevalence of antibiotic resistance could also be partially explained by maintenance of a resistance plasmid because it also carries an essential gene (Johnsen et al., 2009). These genes might include those responsible for efficient colonization of the gut, such as resistance to bile or expression of adhesins for attachment (Amábile-Cuevas and Chicurel, 1996; Gunn, 2000; Lin et al., 2003). Some other genes might encode for traits such as resistance to heavy metals found in the farm environment as was the case for *Staphylococcus* from poultry litter that had resistant characteristics for both arsenic and beta-lactamase (Williams et al., 2006). Plasmid stability systems can be linked to the resistance genes and, therefore, resistance might persist despite the antibiotic pressure (Johnsen et al., 2009). This phenomenon occurs due to simultaneously expression of a stable toxin and an unstable antitoxin. If the plasmid is lost, the bacterium can no longer produce the antitoxin and is then killed by the toxin (Gerdes et al., 1986; Schuster and Bertram, 2013).

7 COCCIDOSTAT USAGE AND ARSENIC RESISTANCE

Coccidiosis is a disease found in livestock and other animals caused by coccidian protozoa. Coccidia are a group of spore-forming, single-cellular, and obligate intracellular parasites. In poultry, the two most pathogenic species are *Eimeria necatrix* and *Eimeria tenella* (Stephens, 1965). The sporozoite stage of the parasite, when ingested by the birds, can invade intestinal cells and reproduce within the cells and ultimately cause hemorrhaging. The disease causes malabsorption, diarrhea, and disrupted intestinal function, which can reduce weight gain in the bird (Lillehoj and Lillehoj, 2000). Weight gain is a primary factor in poultry production, which makes coccidiosis a major disease that must be controlled within the poultry industry.

Roxarsone is mixed into the feed and after consumption the chickens excrete roxarsone, which is consequently converted to inorganic arsenic (arsenite and arsenate) by bacteria present in the poultry litter. This is carried out specifically by *Clostridium* spp. under anaerobic conditions (Stolz et al., 2007). Roxarsone in the litter can also be photo degraded, which also converts it to its inorganic forms (Bednar et al., 2003). Because the birds tend to pick at the litter, the inorganic form of arsenic is consumed, which exposes the gut microflora and can accumulate in the body as shown by several studies (Lasky et al., 2004; Nachman et al., 2013; Pizarro et al., 2004). A study conducted by the FDA showed high levels of inorganic arsenic in the liver of birds supplemented with roxarsone (Kawalek et al., 2011). For this reason, the removal of roxarsone from the market was suggested and Pfizer voluntarily withdrew it from the market (FDA, 2011; Nachman et al., 2013).

Campylobacter colonizes the gut of poultry and, thus, they are exposed to the arsenic compounds, which selects for resistance to roxarsone and arsenic (Sapkota et al., 2006). Recently, a four-gene operon and an additional gene in *Campylobacter* were identified, which provide arsenic resistance (Shen et al., 2013; Wang et al., 2009). The operon encodes a putative membrane permease (ArsP), a transcriptional repressor (ArsR), an arsenate reductase (ArsC), and an efflux protein (Acr3), and the separate gene encodes a putative efflux transporter (ArsB) (Shen et al., 2013; Wang et al., 2009; Figure 1.1). It has been determined that *Campylobacter* mediates much of its antibiotic resistance through a multidrug efflux system, CmeABC (Lin et al., 2002). Therefore, the presence of efflux proteins and a membrane permase indicate potential for antibiotic resistance. However, studies have demonstrated that these genes do not confer additional antibiotic resistance (Shen et al., 2013, 2014; Wang et al., 2009). A study by Pendleton et al. (2014) indicated a correlation in increased arsenic resistance in *Campylobacter* isolates from conventional poultry, where roxarsone was predominantly used, when compared to organic poultry (Pendleton, 2014). While these genes are not involved in antibiotic resistance, they do illustrate the effects of a selective pressure and, thus, the need for increased antibiotic stewardship in livestock production.

8 *CAMPYLOBACTER* IN POULTRY

8.1 PREHARVEST

Campylobacter naturally colonizes the chicken gut when birds are 2-3 weeks old. This lag period has been well documented, but an explanation has not yet been proven (Evans and Sayers, 2000; Jacobs-Reitsma et al., 1995; Kalupahana et al., 2013). Currently, there are two widely accepted hypotheses that explain the lag period from a gut developmental or an immunological standpoint. The developmental view argues that *Campylobacter* is a community-dependent organism due to its dependence on secondary metabolites produced by other bacteria (Hanning et al., 2008; Lee and Newell, 2006). Because the gut is sterile at hatch, these metabolites are not available until a microbial community is established, which does not happen until the second week of life (Danzeisen et al., 2011; Lu et al., 2003). Furthermore, gut structures, including crypts, are virtually absent and not well developed until 2 weeks of age (Lumpkins et al., 2010; Sklan, 2001). This is important because the mature gut with well-developed crypts provides a microaerophilic environment that is ideal for oxygen-sensitive *Campylobacter* (Phillips and Lee, 1983). An additional factor that influences the profile of the microbiome and, in turn, histological development is the nutrient content in feed. Chick starter (used for the first 2 weeks) has higher protein content than grower formulations (used for the final 4 weeks) as well as other differences in vitamin and mineral concentrations. Thus, the feed can influence the microbiome profile, which, in turn, can impact *Campylobacter* colonization abilities (Danzeisen et al., 2011).

Alternatively, the immunological view points to the effect of maternal antibodies. Maternal antibodies are taken up by the egg into the yolk during development and can then be passed to the embryo (Sahin et al., 2001). These antibodies are at their peak when the bird is approximately 1-3 days old, and reduce to negligible levels by 10-13 days old (Shawky et al., 1994). *Campylobacter* has been shown to be susceptible to complement-mediated killing that, in the classical pathway, requires antigen-antibody binding (Blaser et al., 1985). Therefore, the presence of *Campylobacter*-specific antibodies passed from the mother to the offspring could prevent the colonization of this bacterium in the early stages of development.

There are a number of strategies currently used to control *Campylobacter*, including biosecurity. This involves controlling the microorganisms with which the chickens come in contact (Newell et al., 2011). If the grower can prevent *Campylobacter* from entering the environment, then the chickens will not have contact with the bacteria and cannot become colonized. While this is straightforward, it is by no means simple. *Campylobacter* is known to persist and have numerous vectors within the environment (Bull et al., 2006; Evans, 1992; Newell and Fearnley, 2003; Pearson et al., 1993). One possible method *Campylobacter* uses to survive in harsh environments is to enter a viable-but-nonculturable (VBNC), dormant state (Snelling et al., 2005). The VBNC phenotype could lead to false-negative environmental tests and could make development of biosecurity strategies difficult (Stern et al., 1994; Tholozan et al., 1999).

Another strategy that is being developed is the vaccination of chickens against *Campylobacter* (Buckley et al., 2010; Hermans et al., 2011; Saxena et al., 2013). The idea is to create an immune response in the bird that will produce antibodies against *Campylobacter* and thus prevent colonization. The 2-3-week lag in colonization of poultry, believed to be due to maternal antibodies, raises the possibility of using the immune system to prevent colonization (Sahin et al., 2001). Some issues with this technique are the time it takes for the bird to mount an immune response and the identification of an appropriate antigen target. Currently, there are few targets that could work for a vaccine. The most common antigen is the flagella, but this is affected by phase variation and post-translational glycocylation, which changes the antigenic sites making a flagellar vaccine ineffective (Scott, 1997).

Finally, bacteriocins are being studied as potential intervention strategies. These are small antimicrobial peptides that bacteria produce to kill other bacteria. They usually have limited specificity, which is an advantage over antibiotics. Because they do not target a specific site, it is much harder for a bacterium to adapt to them and become resistant. A few different bacteriocins active against *Campylobacter* have been identified (Schoeni and Doyle, 1992; Stern et al., 2006; Svetoch et al., 2005). These have been shown to reduce *Campylobacter*, in colonized birds, to levels below the detection limit. Bacteriocins can be incorporated into the feed, making them easy to use by the poultry industry.

8.2 POSTHARVEST

As established previously, *Campylobacter* spp. act as commensal organisms within the chicken gut. Once colonized, *Campylobacter* levels within the bird can reach levels between 10^5 and 10^9 cfu per gram of intestinal contents (Berndtson et al., 1992; Berrang et al., 2000). The bacteria are then shed by the bird and spread rapidly throughout the flock (Lindblom et al., 1986). Given that these bacteria can have a high prevalence within poultry flocks upon time of slaughter, it is important to maintain control measures in order to ensure a safe product for consumers. There are three main control points present in poultry processing: scalding, washing, and chilling.

Scalding takes place immediately following killing and draining of the carcass. The main purpose of this step is to loosen the feathers for removal, but the high temperatures used can also reduce the number of bacteria present on the skin of the carcass (Rosenquist et al., 2006; Yang et al., 2001). While this process does reduce the number of *Campylobacter* on the exterior of the carcass, the majority of the contamination is within the intestinal tract, which is not exposed to this process.

After the carcass is defeathered and the head and feet are removed, the internal organs are removed during evisceration. If the intestines are ruptured during this step, *Campylobacter* and other fecal bacteria can be released to contaminate the rest of the carcass and the equipment being used to preform the evisceration (Bryan and Doyle, 1995). If the equipment becomes contaminated, this can lead to contamination of *Campylobacter*-free birds if they are processed after a *Campylobacter*-positive flock (Genigeorgis et al., 1986). Additionally, studies have illustrated that *Campylobacter* prevalence was the highest after evisceration, which indicates the need for additional control after this step occurs (Figueroa et al., 2009; Guerin et al., 2010; Hue et al., 2010).

Once evisceration is complete, carcasses are washed to eliminate contamination before being chilled. As evisceration can increase the microbial load on the carcass, washing is an important control point. This step can be augmented by the addition of antimicrobial compounds to the wash water to increase effectiveness (Bashor et al., 2004; Hwang and Beuchat, 1995; Li et al., 1997; Park et al., 2002). Some of the chemicals that have been studied for their effectiveness are lactic acid (Li et al., 1997), lactic acid with sodium benzoate (Hwang and Beuchat, 1995), trisodium phosphate (Bashor et al., 2004; Li et al., 1997), acidified sodium chloride (Bashor et al., 2004), sodium bisulfate (Li et al., 1997), cetylpyridinium chloride (Li et al., 1997), sodium chloride (Li et al., 1997), and chlorine with or without electrolyzed water (Park et al., 2002). While these studies demonstrate that methods reduce *Campylobacter*, they do not eliminate it.

After washing, the carcasses are then chilled to reduce the temperature to 40 °F (~4 °C) before packaging to prevent microbial growth. Most processors in the United States use ice-water immersion, but air chilling is also used (Berrang et al., 2008; Sánchez et al., 2002). Analysis of these two methods has yielded varying results. Two studies have indicated that the methods do not differ significantly (Berrang et al., 2008; Rosenquist et al., 2006), while another found that air chilling resulted in significantly less *Campylobacter* (Sánchez et al., 2002). Often, chlorine or chlorine compounds are used in immersion chilling to reduce bacterial contamination (Bashor et al., 2004). However, the presence of organic matter can reduce the effectiveness of chlorine (Wabeck et al., 1968). Therefore, the difference in effectiveness seen by Sánchez and others (2002) could be due to improper maintenance of the available chlorine levels in the chill tank.

If *Campylobacter* is not effectively controlled at the farm or processing plant level, this presents a risk to consumers. It is then possible that consumers who eat improperly prepared poultry products are at risk of *Campylobacter* infection. One recent incident occurred in 2012, where undercooked chicken livers were responsible for a multistate outbreak of campylobacteriosis (CDC, 2013). When incidents like this occur, epidemiological tracking is needed to determine the source and prevent the spread of further infections.

9 EPIDEMIOLOGY

Tracking of disease outbreaks and the foundation of modern epidemiology began with John Snow and the Broad Street cholera outbreak of 1854. Since John Snow's breakthrough investigation, epidemiologists and medical professionals have sought to apply his methods to outbreaks in order to identify the sources and prevent further spread of disease. This has become increasingly difficult as the world population grows. An outbreak of foodborne illness can now be spread across an entire country or multiple countries. This is the best illustrated with the *E. coli* O104:H4 outbreak that was centered in Germany, but affected a number of people in other countries, including five travelers from the United States (CDC, 2011). Determining the source of the outbreak is not only important to stop an outbreak but also determining blame. As legal suits against offending companies have become increasingly common, accurate methods of typing pathogens are very important so that the parties responsible can be correctly identified.

10 TYPING METHODS

10.1 PULSED-FIELD GEL ELECTROPHORESIS

In 1984, Schwartz and Cantor introduced a new type of electrophoresis called pulsed-field gradient gel electrophoresis (Schwartz and Cantor, 1984). The new technique added electrodes perpendicular to the usual electrophoresis layout. They also used a nonuniform charge in one direction to increase resolution. This new method allowed them to achieve what was once impossible, the separation of DNA larger than 750 kb. This finally allowed analysis of whole genomes; and Schwartz and Cantor separated yeast chromosomal DNA to illustrate this ability.

While this new method did allow for the separation of chromosomal DNA, the DNA ran diagonally across the gel and made it difficult to compare samples on the same gel. Chu, Vollrath, and Davis improved this method by adding an additional set of electrodes to the perpendicular plane and putting them at an angle (Chu et al., 1986). They also did away with the gradient aspect of the electrodes. The new "contour-clamped homogeneous electric field (CHEF)" method worked by alternating the electric field pulses between the two orientations, which were 120° apart. This new configuration and homogeneous field allowed the DNA to travel down the gel in a straight line, which allowed for much easier comparison of samples within the same gel.

The ability to separate whole genomes and compare them on a single gel revolutionized epidemiological tracking of pathogens. In the wake of the Jack-in-the-Box outbreak in the western United States, the CDC collaborated with the Association of Public Health Laboratories to find a new method that would allow for more rapid detection of foodborne illness outbreaks across the country. Pulsed-field gel electrophoresis (PFGE) was found to be ideal for this purpose. They developed the PulseNet database to enable laboratories from across the country to upload PFGE patterns and compare them using the Bionumerics software package. Standard protocols have been developed for foodborne pathogens to ensure the uniformity of patterns uploaded for analysis. Due to the ease of comparison and uniformity of the data, PFGE has become the gold standard for epidemiologic typing of bacterial foodborne pathogens. However, PulseNet released an update indicating a move away from PFGE and toward newer sequenced-based approaches to outbreak tracking (Gerner-Smidt et al., 2006).

10.2 MULITLOCUS SEQUENCE TYPING

Mulitlocus sequence typing (MLST), first introduced by Maiden and others (1998), is designed to analyze housekeeping genes that are less susceptible to mutations due to selective pressures (Maiden et al., 1998). Each nucleotide substitution in the gene is treated as a new allele and by combining the allele types of the target housekeeping genes, a sequence type is given for the bacteria. Sequences give a nonsubjective analysis of the bacteria, which allows for simple and direct comparison between strains and laboratories. Housekeeping genes have low mutation rates, which make them better for analysis of strain evolution, but not ideal for epidemiological studies where more recent mutations can give a better indication of relatedness (Foley et al., 2006; Noller et al., 2003). The addition of virulence genes, as in multi-virulence-locus sequence typing (MVLST), or other highly variable genes can give better discriminatory power for typing isolates (Foley et al., 2006; Tankouo-Sandjong et al., 2007; Zhang et al., 2004).

Even though PFGE is the gold standard for epidemiologic studies, the genome of *Campylobacter* has proven to be more complex and unique than can be accurately described by a PFGE pattern. The amount of variation within the genome enhances the discriminatory power of traditional MLST, which focuses solely on the housekeeping genes *aspA*, *glnA*, *gltA*, *glyA*, *pgm*, *tkt*, and *uncA* (Dingle et al., 2001). While the interlaboratory reproducibility is very high for MLST, it requires a sequencer. These machines have been out of the price range for most laboratories until fairly recently, making this typing method valuable but not appropriate for ubiquitous use in epidemiologic studies. As the cost of sequencers, and sequencing in general, has begun to rapidly decline (Wetterstrand, 2014), the use of MLST as an epidemiological tracking tool for *Campylobacter* has become more feasible.

10.3 *flaA* TYPING

While MLST possesses high discriminatory power, it does not always have enough information needed for outbreak tracking. The sequencing of the *flaA* gene, which is a short variable region (SVR) within the genome controlling the main flagellin protein (Meinersmann et al., 1997), can provide additional discriminatory power (Clark et al., 2012; Dingle et al., 2008; Sails et al., 2003). *flaA* typing not only gives approximately the same discriminatory power as serotyping but also allows for the determination of genetic relatedness of isolates at the locus. Single gene typing, while simpler, will inherently be less descriptive than multiple gene typing.

10.4 WHOLE GENOME SEQUENCING

As mentioned previously, the cost of sequencing has dropped drastically and the time needed to sequence a whole genome has also been reduced to a point where whole genome sequencing is becoming a viable option for outbreak tracking. Whole genome sequencing was applied to the *E. coli* O104:H4 strain that caused a foodborne illness outbreak that affected multiple countries (CDC, 2011). With this approach, data were available within a week and expedited tracking and detection efforts (Mellmann et al., 2011).

Whole genome sequencing was utilized to identify and design a strain-specific assay. Furthermore, the data obtained from sequencing allowed identification of specific virulence genes. While this is a breakthrough for epidemiological tracking, it is still in its infancy. Sequencing the whole genome with sufficient coverage and depth allows for the discovery of single nucleotide polymorphisms (SNPs), which can happen during each replication. This could lead to a situation where two isolates are found to be different, but part of the same outbreak. Therefore, determining a limit for relatedness is required. This is something that needs to be established before whole genome sequencing can replace the current typing methods.

Unlike other typing methods, the use of whole genome sequencing allows for additional data mining applications. Having the whole genome sequences gives access to virulence mechanisms, structural variation, resistance determinants, potential vaccine targets, and general biologic functions (Baba et al., 2002; Cole et al., 1998; Fouts et al., 2005; Parkhill et al., 2001). These features can lead to a better understanding of the pathogens and aid in development of new methods and technologies to prevent further outbreaks.

11 CONCLUSIONS

Campylobacter is a genetically flexible organism, which allows it to adapt a variety of hosts and environments. Poultry is a primary source for *Campylobacter* and causes sporadic infections. However, most outbreaks of *Campylobacter* are typically sourced to water, raw milk, or other foods. Epidemiological monitoring and tracking of sporadic infections can be challenging due to a lack of data and specific infection source. Comprehensive typing tools, including whole genome sequencing, can improve epidemiological efforts because they can provide typing information as well as the presence of virulence and resistance factors, vaccine targets, and host-specific determinants. The wealth of data provided by whole genome sequencing can reveal differences between populations and provide information regarding the impact poultry production practices have on genotypic and phenotypic characteristics of *Campylobacter*. This information can then be used to enhance targeted control efforts and reduce the burden of human campylobacteriosis.

REFERENCES

Ahmed, I.H., Manning, G., Wassenaar, T.M., Cawthraw, S., Newell, D.G., 2002. Identification of genetic differences between two *Campylobacter jejuni* strains with different colonization potentials. Microbiology 148, 1203–1212.

Altekruse, S.F., Stern, N.J., Fields, P.I., Swerdlow, D.L., 1999. *Campylobacter jejuni* – an emerging foodborne pathogen. Emerg. Infect. Dis. 5, 28–35.

Amábile-Cuevas, C.F., Chicurel, M.E., 1996. A possible role for plasmids in mediating the cell–cell proximity required for gene flux. J. Theor. Biol. 181, 237–243.

Ashgar, S.S., Oldfield, N.J., Wooldridge, K.G., Jones, M.A., Irving, G.J., Turner, D.P., Ala'Aldeen, D.A., 2007. CapA, an autotransporter protein of *Campylobacter jejuni*, mediates association with human epithelial cells and colonization of the chicken gut. J. Bacteriol. 189, 1856–1865.

Baba, T., Takeuchi, F., Kuroda, M., Yuzawa, H., Aoki, K.-i., Oguchi, A., Nagai, Y., Iwama, N., Asano, K., Naimi, T., Kuroda, H., Cui, L., Yamamoto, K., Hiramatsu, K., 2002. Genome and virulence determinants of high virulence community-acquired MRSA. Lancet 359, 1819–1827.

Bashor, M.P., Curtis, P.A., Keener, K.M., Sheldon, B.W., Kathariou, S., Osborne, J.A., 2004. Effects of carcass washers on *Campylobacter* contamination in large broiler processing plants. Poultry Sci. 83, 1232–1239.

Bednar, A.J., Garbarino, J.R., Ferrer, I., Rutherford, D.W., Wershaw, R.L., Ranville, J.F., Wildeman, T.R., 2003. Photodegradation of roxarsone in poultry litter leachates. Sci. Total Environ. 302, 237–245.

Berndtson, E., Tivemo, M., Engvall, A., 1992. Distribution and numbers of *Campylobacter* in newly slaughtered broiler chickens and hens. Int. J. Food Microbiol. 15, 45–50.

Berrang, M.E., Buhr, R.J., Cason, J.A., 2000. *Campylobacter* recovery from external and internal organs of commercial broiler carcass prior to scalding. Poult. Sci. 79, 286–290.

Berrang, M.E., Meinersmann, R.J., Smith, D.P., Zhuang, H., 2008. The effect of chilling in cold air or ice water on the microbiological quality of broiler carcasses and the population of *Campylobacter*. Poult. Sci. 87, 992–998.

Biswas, D., Niwa, H., Itoh, K., 2004. Infection with *Campylobacter jejuni* induces tyrosine-phosphorylated proteins into INT-407 cells. Microbiol. Immunol. 48, 221–228.

Blaser, M.J., Smith, P.F., Kohler, P.F., 1985. Susceptibility of *Campylobacter* isolates to the bactericidal activity of human serum. J. Infect. Dis. 151, 227–235.

Bryan, F.L., Doyle, M.P., 1995. Health risks and consequences of *Salmonella* and *Campylobacter jejuni* in raw poultry. J. Food Prot. 58, 326–344.

Buckley, A.M., Wang, J., Hudson, D.L., Grant, A.J., Jones, M.A., Maskell, D.J., Stevens, M.P., 2010. Evaluation of live-attenuated *Salmonella* vaccines expressing *Campylobacter* antigens for control of *C. jejuni* in poultry. Vaccine 28, 1094–1105.

Bull, S.A., Allen, V.M., Domingue, G., Jørgensen, F., Frost, J.A., Ure, R., Whyte, R., Tinker, D., Corry, J.E.L., Gillard-King, J., Humphrey, T.J., 2006. Sources of *Campylobacter* spp. colonizing housed broiler flocks during rearing. Appl. Environ. Microbiol. 72, 645–652.

Carrillo, C.D., Taboada, E., Nash, J.H., Lanthier, P., Kelly, J., Lau, P.C., Verhulp, R., Mykytczuk, O., Sy, J., Findlay, W.A., Amoako, K., Gomis, S., Willson, P., Austin, J.W., Potter, A., Babiuk, L., Allan, B., Szymanski, C.M., 2004. Genome-wide expression analyses of *Campylobacter jejuni* NCTC11168 reveals coordinate regulation of motility and virulence by *flhA*. J. Biol. Chem. 279, 20327–20338.

CDC, 2011. Investigation Update: Outbreak of Shiga toxin-producing *E. coli* O104 (STEC O104:H4) Infections Associated with Travel to Germany. Available at: http://www.cdc.gov/ecoli/2011/ecolio104/.

CDC, 2013. Multistate outbreak of *Campylobacter jejuni* infections associated with undercooked chicken livers — Northeastern United States, 2012. MMWR 62, 874–876.

Champion, O.L., Gaunt, M.W., Gundogdu, O., Elmi, A., Witney, A.A., Hinds, J., Dorrell, N., Wren, B.W., 2005. Comparative phylogenomics of the food-borne pathogen *Campylobacter jejuni* reveals genetic markers predictive of infection source. Proc. Natl. Acad. Sci. U. S. A. 102, 16043–16048.

Champion, O.L., Al-Jaberi, S., Stabler, R.A., Wren, B.W., 2008. Comparative genomics of *Campylobacter jejuni*. In: Nachamkin, I., Szymanski, C.M., Blaser, M.J. (Eds.), Campylobacter, third ed. ASM Press, Washington, DC, pp. 63–95.

Chan, K., Elhanafi, D., Kathariou, S., 2008. Genomic evidence for interspecies acquisition of chromosomal DNA from *Campylobacter jejuni* by *Campylobacter coli* strains of a Turkey-associated clonal group (cluster II). Foodborne Pathog. Dis. 5, 387–398.

Chapman, H.D., Johnson, Z.B., 2002. Use of antibiotics and roxarsone in broiler chickens in the USA: analysis for the years 1995 to 2000. Poult. Sci. 81, 356–364.

Chu, G., Vollrath, D., Davis, R.W., 1986. Separation of large DNA molecules by contour-clamped homogeneous electric fields. Science 234, 1582–1585.

Clark, C.G., Taboada, E., Grant, C.C., Blakeston, C., Pollari, F., Marshall, B., Rahn, K., Mackinnon, J., Daignault, D., Pillai, D., Ng, L.K., 2012. Comparison of molecular typing methods useful for detecting clusters of *Campylobacter jejuni* and *C. coli* isolates through routine surveillance. J. Clin. Microbiol. 50, 798–809.

Cole, S.T., Brosch, R., Parkhill, J., Garnier, T., Churcher, C., Harris, D., Gordon, S.V., Eiglmeier, K., Gas, S., Barry, C.E., Tekaia, F., Badcock, K., Basham, D., Brown, D., Chillingworth, T., Connor, R., Davies, R., Devlin, K., Feltwell, T., Gentles, S., Hamlin, N., Holroyd, S., Hornsby, T., Jagels, K., Krogh, A., McLean, J., Moule, S., Murphy, L., Oliver, K., Osborne, J., Quail, M.A., Rajandream, M.A., Rogers, J., Rutter, S., Seeger, K., Skelton, J., Squares, R., Squares, S., Sulston, J.E., Taylor, K., Whitehead, S., Barrell, B.G., 1998. Deciphering the biology of *Mycobacterium tuberculosis* from the complete genome sequence. Nature 393, 537–544.

Cooper, K.K., Cooper, M.A., Zuccolo, A., Joens, L.A., 2013. Re-sequencing of a virulent strain of *Campylobacter jejuni* NCTC11168 reveals potential virulence factors. Res. Microbiol. 164, 6–11.

Cui, S., Ge, B., Zheng, J., Meng, J., 2005. Prevalence and antimicrobial resistance of *Campylobacter* spp. and *Salmonella* serovars in organic chickens from Maryland retail stores. Appl. Environ. Microbiol. 71, 4108–4111.

Danzeisen, J.L., Kim, H.B., Isaacson, R.E., Tu, Z.J., Johnson, T.J., 2011. Modulations of the chicken cecal microbiome and metagenome in response to anticoccidial and growth promoter treatment. PLoS One 6, e27949.

Dekeyser, P., Gossuin-Detrain, M., Butzler, J.P., Sternon, J., 1972. Acute enteritis due to related *Vibrio*: first positive stool cultures. J. Infect. Dis. 125, 390–392.

Dingle, K.E., Colles, F.M., Wareing, D.R., Ure, R., Fox, A.J., Bolton, F.E., Bootsma, H.J., Willems, R.J., Urwin, R., Maiden, M.C., 2001. Multilocus sequence typing system for *Campylobacter jejuni*. J. Clin. Microbiol. 39, 14–23.

Dingle, K.E., McCarthy, N.D., Cody, A.J., Peto, T.E., Maiden, M.C., 2008. Extended sequence typing of *Campylobacter* spp., United Kingdom. Emerg. Infect. Dis. 14, 1620–1622.

Elsinghorst, E.A., Baron, L.S., Kopecko, D.J., 1989. Penetration of human intestinal epithelial cells by *Salmonella*: molecular cloning and expression of *Salmonella* Typhi invasion determinants in *Escherichia coli*. Proc. Natl. Acad. Sci. U. S. A. 86, 5173–5177.

Evans, S.J., 1992. Introduction and spread of thermophilic *Campylobacters* in broiler flocks. Vet. Rec. 131, 574–576.

Evans, S.J., Sayers, A.R., 2000. A longitudinal study of *Campylobacter* infection of broiler flocks in Great Britain. Prev. Vet. Med. 46, 209–223.

FDA, 2011. FDA: Pfizer will voluntarily suspend sale of animal drug 3-nitro. Available at: http://www.fda.gov/NewsEvents/Newsroom/PressAnnouncements/ucm258342.htm.

Figueroa, G., Troncoso, M., Lopez, C., Rivas, P., Toro, M., 2009. Occurrence and enumeration of *Campylobacter* spp. during the processing of Chilean broilers. BMC Microbiol. 9, 94.

Foley, S.L., White, D.G., McDermott, P.F., Walker, R.D., Rhodes, B., Fedorka-Cray, P.J., Simjee, S., Zhao, S., 2006. Comparison of subtyping methods for differentiating *Salmonella enterica* serovar Typhimurium isolates obtained from food animal sources. J. Clin. Microbiol. 44, 3569–3577.

Fouts, D.E., Mongodin, E.F., Mandrell, R.E., Miller, W.G., Rasko, D.A., Ravel, J., Brinkac, L.M., DeBoy, R.T., Parker, C.T., Daugherty, S.C., Dodson, R.J., Durkin, A.S., Madupu, R., Sullivan, S.A., Shetty, J.U., Ayodeji, M.A., Shvartsbeyn, A., Schatz, M.C., Badger, J.H., Fraser, C.M., Nelson, K.E., 2005. Major structural differences and novel potential virulence mechanisms from the genomes of multiple *Campylobacter* species. PLoS Biol. 3, e15.

Gaynor, E.C., Cawthraw, S., Manning, G., MacKichan, J.K., Falkow, S., Newell, D.G., 2004. The genome-sequenced variant of *Campylobacter jejuni* NCTC 11168 and the original clonal clinical isolate differ markedly in colonization, gene expression, and virulence-associated phenotypes. J. Bacteriol. 186, 503–517.

Genigeorgis, C., Hassuneh, M., Collins, P., 1986. *Campylobacter jejuni* infection on poultry farms and its effect on poultry meat contamination during slaughtering. J. Food Prot. 49, 895–903.

Gerdes, K., Rasmussen, P.B., Molin, S., 1986. Unique type of plasmid maintenance function: postsegregational killing of plasmid-free cells. Proc. Natl. Acad. Sci. U. S. A. 83, 3116–3120.

Gerner-Smidt, P., Hise, K., Kincaid, J., Hunter, S., Rolando, S., Hyytia-Trees, E., Ribot, E.M., Swaminathan, B., 2006. PulseNet USA: a five-year update. Foodborne Pathog. Dis. 3, 9–19.

Grant, C.C., Konkel, M.E., Cieplak Jr., W., Tompkins, L.S., 1993. Role of flagella in adherence, internalization, and translocation of *Campylobacter jejuni* in nonpolarized and polarized epithelial cell cultures. Infect. Immun. 61, 1764–1771.

Gripp, E., Hlahla, D., Didelot, X., Kops, F., Maurischat, S., Tedin, K., Alter, T., Ellerbroek, L., Schreiber, K., Schomburg, D., Janssen, T., Bartholomaus, P., Hofreuter, D., Woltemate, S., Uhr, M., Brenneke, B., Gruning, P., Gerlach, G., Wieler, L., Suerbaum, S., Josenhans, C., 2011. Closely related *Campylobacter jejuni* strains from different sources reveal a generalist rather than a specialist lifestyle. BMC Genomics 12, 584.

Guerin, M.T., Sir, C., Sargeant, J.M., Waddell, L., O'Connor, A.M., Wills, R.W., Bailey, R.H., Byrd, J.A., 2010. The change in prevalence of *Campylobacter* on chicken carcasses during processing: a systematic review. Poult. Sci. 89, 1070–1084.

Gunn, J.S., 2000. Mechanisms of bacterial resistance and response to bile. Microbes Infect. 2, 907–913.

Hanning, I., Jarquin, R., Slavik, M., 2008. *Campylobacter jejuni* as a secondary colonizer of poultry biofilms. J. Appl. Microbiol. 105, 1199–1208.

Hendrixson, D.R., DiRita, V.J., 2004. Identification of *Campylobacter jejuni* genes involved in commensal colonization of the chick gastrointestinal tract. Mol. Microbiol. 52, 471–484.

Hepworth, P.J., Ashelford, K.E., Hinds, J., Gould, K.A., Witney, A.A., Williams, N.J., Leatherbarrow, H., French, N.P., Birtles, R.J., Mendonca, C., Dorrell, N., Wren, B.W., Wigley, P., Hall, N., Winstanley, C., 2011. Genomic variations define divergence of water/wildlife-associated *Campylobacter jejuni* niche specialists from common clonal complexes. Environ. Microbiol. 13, 1549–1560.

Hermans, D., Van Deun, K., Messens, W., Martel, A., Van Immerseel, F., Haesebrouck, F., Rasschaert, G., Heyndrickx, M., Pasmans, F., 2011. *Campylobacter* control in poultry by current intervention measures ineffective: urgent need for intensified fundamental research. Vet. Microbiol. 152, 219–228.

Heuer, O.E., Pedersen, K., Andersen, J.S., Madsen, M., 2001. Prevalence and antimicrobial susceptibility of thermophilic *Campylobacter* in organic and conventional broiler flocks. Lett. Appl. Microbiol. 33, 269–274.

Hickey, T.E., McVeigh, A.L., Scott, D.A., Michielutti, R.E., Bixby, A., Carroll, S.A., Bourgeois, A.L., Guerry, P., 2000. *Campylobacter jejuni* cytolethal distending toxin mediates release of interleukin-8 from intestinal epithelial cells. Infect. Immun. 68, 6535–6541.

Hu, L., Kopecko, D., 2008. Cell biology of human host cell entry by *Campylobacter jejuni*. In: Nachamkin, I., Szymanski, C.M., Blaser, M.J. (Eds.), Campylobacter, third ed. ASM Press, Washington, DC, pp. 297–313.

Hue, O., Le Bouquin, S., Laisney, M.-J., Allain, V., Lalande, F., Petetin, I., Rouxel, S., Quesne, S., Gloaguen, P.-Y., Picherot, M., Santolini, J., Salvat, G., Bougeard, S., Chemaly, M., 2010. Prevalence of and risk factors for *Campylobacter* spp. contamination of broiler chicken carcasses at the slaughterhouse. Food Microbiol. 27, 992–999.

Hwang, C.-a., Beuchat, L.R., 1995. Efficacy of a lactic acid/sodium benzoate wash solution in reducing bacterial contamination of raw chicken. Int. J. Food Microbiol. 27, 91–98.

Jacobs-Reitsma, W.F., Van de Giessen, A.W., Bolder, N.M., Mulder, R.W.A.W., 1995. Epidemiology of *Campylobacter* spp. at two Dutch broiler farms. Epidemiol. Infect. 114, 413–421.

Jerome, J.P., Bell, J.A., Plovanich-Jones, A.E., Barrick, J.E., Brown, C.T., Mansfield, L.S., 2011. Standing genetic variation in contingency loci drives the rapid adaptation of *Campylobacter jejuni* to a novel host. PLoS One 6, e16399.

Jin, S., Joe, A., Lynett, J., Hani, E.K., Sherman, P., Chan, V.L., 2001. JlpA, a novel surface-exposed lipoprotein specific to *Campylobacter jejuni*, mediates adherence to host epithelial cells. Mol. Microbiol. 39, 1225–1236.

Johnsen, P.J., Townsend, J.P., Bohn, T., Simonsen, G.S., Sundsfjord, A., Nielsen, K.M., 2009. Factors affecting the reversal of antimicrobial-drug resistance. Lancet Infect. Dis. 9, 357–364.

Kalupahana, R.S., Kottawatta, K.S.A., Kanankege, K.S.T., van Bergen, M.A.P., Abeynayake, P., Wagenaar, J.A., 2013. Colonization of *Campylobacter* spp. in broiler chickens and laying hens reared in tropical climates with low-biosecurity housing. Appl. Environ. Microbiol. 79, 393–395.

Kawalek, J., Carson, M., Conklin, S., Lancaster, V., Howard, K., Ward, J., Farrell, D., Myers, M., Swain, H., Jeanettes, P., Frobish, S., Matthews, S., McDonald, M., 2011. Final report on study 275.30: provide data on various arsenic species present in broilers treated with roxarsone: comparison with untreated birds. Center for Veterinary Medicine/Office of Regulatory Science, U.S. Food and Drug Administration.

King, E.O., 1957. Human infections with *Vibrio fetus* and a closely related *Vibrio*. J. Infect. Dis. 101, 119–128.

Konkel, M.E., Garvis, S.G., Tipton, S.L., Anderson Jr., D.E., Cieplak Jr., W., 1997. Identification and molecular cloning of a gene encoding a fibronectin-binding protein (CadF) from *Campylobacter jejuni*. Mol. Microbiol. 24, 953–963.

Konkel, M.E., Klena, J.D., Rivera-Amill, V., Monteville, M.R., Biswas, D., Raphael, B., Mickelson, J., 2004. Secretion of virulence proteins from *Campylobacter jejuni* is dependent on a functional flagellar export apparatus. J. Bacteriol. 186, 3296–3303.

Konkel, M.E., Christensen, J.E., Keech, A.M., Monteville, M.R., Klena, J.D., Garvis, S.G., 2005. Identification of a fibronectin-binding domain within the *Campylobacter jejuni* CadF protein. Mol. Microbiol. 57, 1022–1035.

Korolik, V., Ketley, J., 2008. Chemosensory signal transduction pathway of *Campylobacter jejuni*. In: Nachamkin, I., Szymanski, C.M., Blaser, M.J. (Eds.), Campylobacter, third ed. ASM Press, Washington, DC, pp. 351–366.

Larson, C., Christensen, J., Pacheco, S., Minnich, S., Konkel, M., 2008. *Campylobacter jejuni* secretes proteins via the flagellar type III secretion system that contribute to host cell invasion and gastroenteritis. In: Nachamkin, I., Szymanski, C.M., Blaser, M.J. (Eds.), Campylobacter, third ed. ASM Press, Washington, DC, pp. 315–332.

Lasky, T., Sun, W., Kadry, A., Hoffman, M.K., 2004. Mean total arsenic concentrations in chicken 1989-2000 and estimated exposures for consumers of chicken. Environ. Health Perspect. 112, 18–21.

Lee, M.D., Newell, D.G., 2006. *Campylobacter* in poultry: filling an ecological niche. Avian Dis. 50, 1–9.

Lertsethtakarn, P., Ottemann, K.M., Hendrixson, D.R., 2011. Motility and chemotaxis in *Campylobacter* and *Helicobacter*. Annu. Rev. Microbiol. 65, 389–410.

Leverstein-van Hall, M.A., Box, A.T.A., Blok, H.E.M., Paauw, A., Fluit, A.C., Verhoef, J., 2002. Evidence of extensive interspecies transfer of integron-mediated antimicrobial resistance genes among multidrug-resistant *Enterobacteriaceae* in a clinical setting. J. Infect. Dis. 186, 49–56.

Levinson, G., Gutman, G.A., 1987. Slipped-strand mispairing: a major mechanism for DNA sequence evolution. Mol. Biol. Evol. 4, 203–221.

Li, Y., Slavik, M.F., Walker, J.T., Xiong, H.U.A., 1997. Pre-chill spray of chicken carcasses to reduce *Salmonella* Typhimurium. J. Food Sci. 62, 605–607.

Lillehoj, H.S., Lillehoj, E.P., 2000. Avian coccidiosis. A review of acquired intestinal immunity and vaccination strategies. Avian Dis. 44, 408–425.

Lin, J., Michel, L.O., Zhang, Q., 2002. CmeABC functions as a multidrug efflux system in *Campylobacter jejuni*. Antimicrob. Agents Chemother. 46, 2124–2131.

Lin, J., Sahin, O., Michel, L.O., Zhang, Q., 2003. Critical role of multidrug efflux pump CmeABC in bile resistance and in vivo colonization of *Campylobacter jejuni*. Infect. Immun. 71, 4250–4259.

Lindblom, G.B., Sjörgren, E., Kaijser, B., 1986. Natural *Campylobacter* colonization in chickens raised under different environmental conditions. J. Hyg. 96, 385–391.

Lu, J., Idris, U., Harmon, B., Hofacre, C., Maurer, J.J., Lee, M.D., 2003. Diversity and succession of the intestinal bacterial community of the maturing broiler chicken. Appl. Environ. Microbiol. 69, 6816–6824.

Lumpkins, B.S., Batal, A.B., Lee, M.D., 2010. Evaluation of the bacterial community and intestinal development of different genetic lines of chickens. Poult. Sci. 89, 1614–1621.

Maiden, M.C., Bygraves, J.A., Feil, E., Morelli, G., Russell, J.E., Urwin, R., Zhang, Q., Zhou, J., Zurth, K., Caugant, D.A., Feavers, I.M., Achtman, M., Spratt, B.G., 1998. Multilocus sequence typing: a portable approach to the identification of clones within populations of pathogenic microorganisms. Proc. Natl. Acad. Sci. U. S. A. 95, 3140–3145.

McClelland, M., Sanderson, K.E., Spieth, J., Clifton, S.W., Latreille, P., Courtney, L., Porwollik, S., Ali, J., Dante, M., Du, F., Hou, S., Layman, D., Leonard, S., Nguyen, C., Scott, K., Holmes, A., Grewal, N., Mulvaney, E., Ryan, E., Sun, H., Florea, L., Miller, W., Stoneking, T., Nhan, M., Waterston, R., Wilson, R.K., 2001. Complete genome sequence of *Salmonella enterica* serovar Typhimurium LT2. Nature 413, 852–856.

Meinersmann, R.J., Helsel, L.O., Fields, P.I., Hiett, K.L., 1997. Discrimination of *Campylobacter jejuni* isolates by *fla* gene sequencing. J. Clin. Microbiol. 35, 2810–2814.

Meinersmann, R.J., E.Dingle, K., Maiden, M.C.J., 2003. Genetic exchange among *Campylobacter* species. Genome Lett. 2, 48–52.

Mellmann, A., Harmsen, D., Cummings, C.A., Zentz, E.B., Leopold, S.R., Rico, A., Prior, K., Szczepanowski, R., Ji, Y., Zhang, W., McLaughlin, S.F., Henkhaus, J.K., Leopold, B., Bielaszewska, M., Prager, R., Brzoska, P.M., Moore, R.L., Guenther, S., Rothberg, J.M., Karch, H., 2011. Prospective genomic characterization of the German enterohemorrhagic *Escherichia coli* O104:H4 outbreak by rapid next generation sequencing technology. PLoS One 6, e22751.

Miller, W.G., Englen, M.D., Kathariou, S., Wesley, I.V., Wang, G., Pittenger-Alley, L., Siletz, R.M., Muraoka, W., Fedorka-Cray, P.J., Mandrell, R.E., 2006. Identification of host-associated alleles by multilocus sequence typing of *Campylobacter coli* strains from food animals. Microbiology 152, 245–255.

Monteville, M.R., Yoon, J.E., Konkel, M.E., 2003. Maximal adherence and invasion of INT 407 cells by *Campylobacter jejuni* requires the CadF outer-membrane protein and microfilament reorganization. Microbiology 149, 153–165.

Moser, I., Schroeder, W., Salnikow, J., 1997. *Campylobacter jejuni* major outer membrane protein and a 59-kDa protein are involved in binding to fibronectin and INT 407 cell membranes. FEMS Microbiol. Lett. 157, 233–238.

Nachamkin, I., Yang, X.H., Stern, N.J., 1993. Role of *Campylobacter jejuni* flagella as colonization factors for three-day-old chicks: analysis with flagellar mutants. Appl. Environ. Microbiol. 59, 1269–1273.

Nachman, K.E., Baron, P.A., Raber, G., Francesconi, K.A., Navas-Acien, A., Love, D.C., 2013. Roxarsone, inorganic arsenic, and other arsenic species in chicken: a U.S.-based market basket sample. Environ. Health Perspect. 121, 818–824.

Newell, D.G., Fearnley, C., 2003. Sources of *Campylobacter* colonization in broiler chickens. Appl. Environ. Microbiol. 69, 4343–4351.

Newell, D.G., Elvers, K.T., Dopfer, D., Hansson, I., Jones, P., James, S., Gittins, J., Stern, N.J., Davies, R., Connerton, I., Pearson, D., Salvat, G., Allen, V.M., 2011. Biosecurity-based interventions and strategies to reduce *Campylobacter* spp. on poultry farms. Appl. Environ. Microbiol. 77, 8605–8614.

Noller, A.C., McEllistrem, M.C., Stine, O.C., Morris Jr., J.G., Boxrud, D.J., Dixon, B., Harrison, L.H., 2003. Multilocus sequence typing reveals a lack of diversity among *Escherichia coli* O157:H7 isolates that are distinct by pulsed-field gel electrophoresis. J. Clin. Microbiol. 41, 675–679.

Park, H., Hung, Y.-C., Brackett, R.E., 2002. Antimicrobial effect of electrolyzed water for inactivating *Campylobacter jejuni* during poultry washing. Int. J. Food Microbiol. 72, 77–83.

Parkhill, J., Achtman, M., James, K.D., Bentley, S.D., Churcher, C., Klee, S.R., Morelli, G., Basham, D., Brown, D., Chillingworth, T., Davies, R.M., Davis, P., Devlin, K., Feltwell, T., Hamlin, N., Holroyd, S., Jagels, K., Leather, S., Moule, S., Mungall, K., Quail, M.A., Rajandream, M.A., Rutherford, K.M., Simmonds, M.,

Skelton, J., Whitehead, S., Spratt, B.G., Barrell, B.G., 2000. Complete DNA sequence of a serogroup A strain of *Neisseria meningitidis* Z2491. Nature 404, 502–506.

Parkhill, J., Dougan, G., James, K.D., Thomson, N.R., Pickard, D., Wain, J., Churcher, C., Mungall, K.L., Bentley, S.D., Holden, M.T.G., Sebaihia, M., Baker, S., Basham, D., Brooks, K., Chillingworth, T., Connerton, P., Cronin, A., Davis, P., Davies, R.M., Dowd, L., White, N., Farrar, J., Feltwell, T., Hamlin, N., Haque, A., Hien, T.T., Holroyd, S., Jagels, K., Krogh, A., Larsen, T.S., Leather, S., Moule, S., O'Gaora, P., Parry, C., Quail, M., Rutherford, K., Simmonds, M., Skelton, J., Stevens, K., Whitehead, S., Barrell, B.G., 2001. Complete genome sequence of a multiple drug resistant *Salmonella enterica* serovar Typhi CT18. Nature 413, 848–852.

Pearson, A.D., Greenwood, M., Healing, T.D., Rollins, D., Shahamat, M., Donaldson, J., Colwell, R.R., 1993. Colonization of broiler chickens by waterborne *Campylobacter jejuni*. Appl. Environ. Microbiol. 59, 987–996.

Pei, Z., Blaser, M.J., 1993. PEB1, the major cell-binding factor of *Campylobacter jejuni*, is a homolog of the binding component in gram-negative nutrient transport systems. J. Biol. Chem. 268, 18717–18725.

Pendleton, S., Yard, C., Heinz, K., Diaz Sanchez, S., Hanning, I., 2014. Impact of Rearing Conditions on Arsenic Resistance in *Campylobacter* spp. International Association of Food Protection, Indianapolis, IN, August 3–6.

Pendleton, S.J., 2014. An Epidemiological Study of *Campylobacter* Populations Reveals a Selective Pressure by Roxarsone. PhD dissertation, University of Tennessee.

Perna, N.T., Plunkett 3rd., G., Burland, V., Mau, B., Glasner, J.D., Rose, D.J., Mayhew, G.F., Evans, P.S., Gregor, J., Kirkpatrick, H.A., Posfai, G., Hackett, J., Klink, S., Boutin, A., Shao, Y., Miller, L., Grotbeck, E.J., Davis, N.W., Lim, A., Dimalanta, E.T., Potamousis, K.D., Apodaca, J., Anantharaman, T.S., Lin, J., Yen, G., Schwartz, D.C., Welch, R.A., Blattner, F.R., 2001. Genome sequence of enterohaemorrhagic *Escherichia coli* O157:H7. Nature 409, 529–533.

Phillips, M.W., Lee, A., 1983. Isolation and characterization of a spiral bacterium from the crypts of rodent gastrointestinal tracts. Appl. Environ. Microbiol. 45, 675–683.

Pistor, S., Chakraborty, T., Niebuhr, K., Domann, E., Wehland, J., 1994. The ActA protein of *Listeria monocytogenes* acts as a nucleator inducing reorganization of the actin cytoskeleton. EMBO J. 13, 758–763.

Pizarro, I., Gomez, M.M., Fodor, P., Palacios, M.A., Camara, C., 2004. Distribution and biotransformation of arsenic species in chicken cardiac and muscle tissues. Biol. Trace Elem. Res. 99, 129–143.

Poly, F., Ewing, C., Goon, S., Hickey, T.E., Rockabrand, D., Majam, G., Lee, L., Phan, J., Savarino, N.J., Guerry, P., 2007. Heterogeneity of a *Campylobacter jejuni* protein that is secreted through the flagellar filament. Infect. Immun. 75, 3859–3867.

Rosenquist, H., Sommer, H.M., Nielsen, N.L., Christensen, B.B., 2006. The effect of slaughter operations on the contamination of chicken carcasses with thermotolerant *Campylobacter*. Int. J. Food Microbiol. 108, 226–232.

Sahin, O., Zhang, Q., Meitzler, J.C., Harr, B.S., Morishita, T.Y., Mohan, R., 2001. Prevalence, antigenic specificity, and bactericidal activity of poultry anti-*Campylobacter* maternal antibodies. Appl. Environ. Microbiol. 67, 3951–3957.

Sails, A.D., Swaminathan, B., Fields, P.I., 2003. Utility of multilocus sequence typing as an epidemiological tool for investigation of outbreaks of gastroenteritis caused by *Campylobacter jejuni*. J. Clin. Microbiol. 41, 4733–4739.

Sánchez, M.X., Fluckey, W.M., Brashears, M.M., McKee, S.R., 2002. Microbial profile and antibiotic susceptibility of *Campylobacter* spp. and *Salmonella* spp. in broilers processed in air-chilled and immersion-chilled environments. J. Food Prot. 65, 948–956.

Sapkota, A.R., Price, L.B., Silbergeld, E.K., Schwab, K.J., 2006. Arsenic resistance in Campylobacter spp. isolated from retail poultry products. Appl. Environ. Microbiol. 72(4), 3069–3071.

Saxena, M., John, B., Mu, M., Van, T.T.H., Taki, A., Coloe, P.J., Smooker, P.M., 2013. Strategies to reduce *Campylobacter* colonisation in chickens. Procedia Vaccinol. 7, 40–43.

Scallan, E., Hoekstra, R.M., Angulo, F.J., Tauxe, R.V., Widdowson, M.A., Roy, S.L., Jones, J.L., Griffin, P.M., 2011. Foodborne illness acquired in the United States – major pathogens. Emerg. Infect. Dis. 17, 7–15.

Schoeni, J.L., Doyle, M.P., 1992. Reduction of *Campylobacter jejuni* colonization of chicks by cecum-colonizing bacteria producing anti-*C. jejuni* metabolites. Appl. Environ. Microbiol. 58, 664–670.

Schouls, L.M., Reulen, S., Duim, B., Wagenaar, J.A., Willems, R.J., Dingle, K.E., Colles, F.M., Van Embden, J.D., 2003. Comparative genotyping of *Campylobacter jejuni* by amplified fragment length polymorphism, multilocus sequence typing, and short repeat sequencing: strain diversity, host range, and recombination. J. Clin. Microbiol. 41, 15–26.

Schuster, C.F., Bertram, R., 2013. Toxin-antitoxin systems are ubiquitous and versatile modulators of prokaryotic cell fate. FEMS Microbiol. Lett. 340, 73–85.

Schwartz, D.C., Cantor, C.R., 1984. Separation of yeast chromosome-sized DNAs by pulsed field gradient gel electrophoresis. Cell 37, 67–75.

Scott, D.A., 1997. Vaccines against *Campylobacter jejuni*. J. Infect. Dis. 176, S183–S188.

Shandera, W.X., Tormey, M.P., Blaser, M.J., 1992. An outbreak of bacteremic *Campylobacter jejuni* infection. Mt. Sinai J. Med. 59, 53–56.

Shawky, S.A., Saif, Y.M., McCormick, J., 1994. Transfer of maternal anti-rotavirus IgG to the mucosal surfaces and bile of turkey poults. Avian Dis. 38, 409–417.

Shen, Z., Han, J., Wang, Y., Sahin, O., Zhang, Q., 2013. The contribution of ArsB to arsenic resistance in *Campylobacter jejuni*. PLoS One 8, e58894.

Shen, Z., Luangtongkum, T., Qiang, Z., Jeon, B., Wang, L., Zhang, Q., 2014. Identification of a novel membrane transporter mediating resistance to organic arsenic in *Campylobacter jejuni*. Antimicrob. Agents Chemother. 58, 2021–2029.

Sheppard, S.K., McCarthy, N.D., Falush, D., Maiden, M.C., 2008. Convergence of *Campylobacter* species: implications for bacterial evolution. Science 320, 237–239.

Sheppard, S.K., Dallas, J.F., Wilson, D.J., Strachan, N.J.C., McCarthy, N.D., Jolley, K.A., Colles, F.M., Rotariu, O., Ogden, I.D., Forbes, K.J., Maiden, M.C.J., 2010. Evolution of an agriculture-associated disease causing *Campylobacter coli* clade: evidence from national surveillance data in Scotland. PLoS One 5, e15708.

Sheppard, S.K., Didelot, X., Jolley, K.A., Darling, A.E., Pascoe, B., Meric, G., Kelly, D.J., Cody, A., Colles, F.M., Strachan, N.J., Ogden, I.D., Forbes, K., French, N.P., Carter, P., Miller, W.G., McCarthy, N.D., Owen, R., Litrup, E., Egholm, M., Affourtit, J.P., Bentley, S.D., Parkhill, J., Maiden, M.C., Falush, D., 2013. Progressive genome-wide introgression in agricultural *Campylobacter coli*. Mol. Ecol. 22, 1051–1064.

Skirrow, M.B., 1977. *Campylobacter* enteritis: a "new" disease. Br. Med. J. 2, 9–11.

Sklan, D., 2001. Development of the digestive tract of poultry. World Poult. Sci. J 57, 415–428.

Snelling, W.J., Moore, J.E., Dooley, J.S.G., 2005. The colonization of broilers with *Campylobacter*. World Poult. Sci. J 61, 655–662.

Stephens, J.F., 1965. Some physiological effects of coccidiosis caused by *Eimeria necatrix* in the chicken. J. Parasitol. 51, 331–335.

Stern, N.J., Jones, D.M., Wesley, I.V., Rollins, D.M., 1994. Colonization of chicks by non-culturable *Campylobacter* spp. Lett. Appl. Microbiol. 18, 333–336.

Stern, N.J., Svetoch, E.A., Eruslanov, B.V., Perelygin, V.V., Mitsevich, E.V., Mitsevich, I.P., Pokhilenko, V.D., Levchuk, V.P., Svetoch, O.E., Seal, B.S., 2006. Isolation of a *Lactobacillus salivarius* strain and purification of its bacteriocin, which is inhibitory to *Campylobacter jejuni* in the chicken gastrointestinal system. Antimicrob. Agents Chemother. 50, 3111–3116.

Stolz, J.F., Perera, E., Kilonzo, B., Kail, B., Crable, B., Fisher, E., Ranganathan, M., Wormer, L., Basu, P., 2007. Biotransformation of 3-nitro-4-hydroxybenzene arsonic acid (roxarsone) and release of inorganic arsenic by *Clostridium* species. Environ. Sci. Technol. 41, 818–823.

Svetoch, E.A., Stern, N.J., Eruslanov, B.V., Kovalev, Y.N., Volodina, L.I., Perelygin, V.V., Mitsevich, E.N., Mitsevich, I.P., Pokhilenko, V.D., Borzenkov, V.N., Levchuk, V.P., Svetoch, O.E., Kudriavtseva, T.Y., 2005. Isolation of *Bacillus circulans* and *Paenibacillus polymyxa* strains inhibitory to *Campylobacter jejuni* and characterization of associated bacteriocins. J. Food Prot. 68, 11–17.

Tankouo-Sandjong, B., Sessitsch, A., Liebana, E., Kornschober, C., Allerberger, F., Hachler, H., Bodrossy, L., 2007. MLST-v, multilocus sequence typing based on virulence genes, for molecular typing of *Salmonella enterica* subsp. enterica serovars. J. Microbiol. Methods 69, 23–36.

Thibodeau, A., Fravalo, P., Laurent-Lewandowski, S., Guevremont, E., Quessy, S., Letellier, A., 2011. Presence and characterization of *Campylobacter jejuni* in organically raised chickens in Quebec. Can. J. Vet. Res. 75, 298–307.

Tholozan, J.L., Cappelier, J.M., Tissier, J.P., Delattre, G., Federighi, M., 1999. Physiological characterization of viable-but-nonculturable *Campylobacter jejuni* cells. Appl. Environ. Microbiol. 65, 1110–1116.

Thomas, D.K., Lone, A.G., Selinger, L.B., Taboada, E.N., Uwiera, R.R., Abbott, D.W., Inglis, G.D., 2014. Comparative variation within the genome of *Campylobacter jejuni* NCTC 11168 in human and murine hosts. PLoS One 9, e88229.

USPoultry, 2012. Economic Data. Available at: https://www.uspoultry.org/economic_data/.

Vincent, R., Dumas, J., Picard, N., 1947. Septicèmie grave au cours de la grossesse, due à un vibrion. Avortement consecutif. Bull. Acad. Natl Med. 131, 90–92.

Wabeck, C.J., Schwall, D.V., Evancho, G.M., Heck, J.G., Rogers, A.B., 1968. *Salmonella* and total count reduction in poultry treated with sodium hypochlorite solutions. Poult. Sci. 47, 1090–1094.

Wang, L., Jeon, B., Sahin, O., Zhang, Q., 2009. Identification of an arsenic resistance and arsenic-sensing system in *Campylobacter jejuni*. Appl. Environ. Microbiol. 75, 5064–5073.

Wassenaar, T.M., Fry, B.N., van der Zeijst, B.A.M., 1995. Variation of the flagellin gene locus of *Campylobacter jejuni* by recombination and horizontal gene transfer. Microbiology 141, 95–101.

Wassenaar, T.M., Wagenaar, J.A., Rigter, A., Fearnley, C., Newell, D.G., Duim, B., 2002. Homonucleotide stretches in chromosomal DNA of *Campylobacter jejuni* display high frequency polymorphism as detected by direct PCR analysis. FEMS Microbiol. Lett. 212, 77–85.

Watson, R., Galán, J., 2008. Interaction of *Campylobacter jejuni* with host cells. In: Nachamkin, I., Szymanski, C.M., Blaser, M.J. (Eds.), Campylobacter, third ed. ASM Press, Washington, DC, pp. 289–296.

Wetterstrand, K., 2014. DNA Sequencing Costs: Data from the NHGRI Genome Sequencing Program (GSP). Available at: https://www.genome.gov/sequencingcosts/.

WHO, 2012. The global view of campylobacteriosis: report of an expert consultation.

Williams, L.E., Detter, C., Barry, K., Lapidus, A., Summers, A.O., 2006. Facile recovery of individual high-molecular-weight, low-copy-number natural plasmids for genomic sequencing. Appl. Environ. Microbiol. 72, 4899–4906.

Wooldridge, K.G., Williams, P.H., Ketley, J.M., 1996. Host signal transduction and endocytosis of *Campylobacter jejuni*. Microb. Pathog. 21, 299–305.

Wösten, M.M., Wagenaar, J.A., van Putten, J.P., 2004. The FlgS/FlgR two-component signal transduction system regulates the *fla* regulon in *Campylobacter jejuni*. J. Biol. Chem. 279, 16214–16222.

Yang, H., Li, Y., Johnson, M.G., 2001. Survival and death of *Salmonella* Typhimurium and *Campylobacter jejuni* in processing water and on chicken skin during poultry scalding and chilling. J. Food Prot. 64, 770–776.

Yao, R., Burr, D.H., Doig, P., Trust, T.J., Niu, H., Guerry, P., 1994. Isolation of motile and non-motile insertional mutants of *Campylobacter jejuni*: the role of motility in adherence and invasion of eukaryotic cells. Mol. Microbiol. 14, 883–893.

Zhang, W., Jayarao, B.M., Knabel, S.J., 2004. Multi-virulence-locus sequence typing of *Listeria monocytogenes*. Appl. Environ. Microbiol. 70, 913–920.

Zhou, Z., Raskin, L., Zilles, J.L., 2009. Macrolide resistance in microorganisms at antimicrobial-free Swine farms. Appl. Environ. Microbiol. 75, 5814–5820.

CHAPTER

ARCOBACTER SPECIES: AN EMERGED OR EMERGING PATHOGEN?

12

Jodie Score, Carol A. Phillips

University of Northampton, Northampton, UK

1 INTRODUCTION

Members of the family *Campylobacteraceae* are motile, Gram-negative, nonsporing, curved, occasionally straight, rods, which may appear as spirals. They include the genera *Campylobacter*, *Sulfurospirillum*, and *Arcobacter*, and are included in the RNA superfamily VI of the Proteobacteria. The original *Arcobacter* classification was proposed by Vandamme et al. (1991a) to describe organisms previously classified as "aerotolerant campylobacters." In 2001, there were four species identified: *Arcobacter butzleri*, *Arcobacter cryaerophilus* (Groups 1A and 1B), *Arcobacter skirrowii*, and *Arcobacter nitrofigilis*. In 2014, 18 species were described, isolated from a variety of sources (Table 1). *A. butzleri*, *A. skirrowii*, and *A. cryaerophilus* have been associated with human illness, of which the most commonly reported species associated with gastroenteritis and bacteremia is *A. butzleri* (Atabay et al., 2014). The pathogenicity and taxonomy of the genus have been reviewed by authors including Mansfield and Forsythe (2000), Phillips (2001), Snelling et al. (2006), Ho et al. (2006), Collado and Figueras (2011), and Shah et al. (2011).

Originally, the "aerotolerant campylobacters" were differentiated from the "true" campylobacters by their ability to grow at 30 °C after primary isolation in an microaerobic environment (Neill et al., 1978). The key distinguishing features of the genus *Arcobacter* differentiating them from *Campylobacter* are the ability to grow at 15 °C but not at 42 °C; the ability to optimally grow aerobically at 30 °C; G + C content of 27-30 mol% and methyl-substituted menaquinone-6 not present as a major isoprenoid quinone (Ursing et al., 1994; Vandamme et al., 1992b).

In 2007, Miller et al. published the complete genome sequence of the human clinical isolate *A. butzleri* RM4018 containing 2,341,251 bp, and in 2010 Pati et al. published the first complete genome sequence of an *Arcobacter* type strain *A. nitrofigilus* (CI^T) containing 3,192,235 bp. These provide the basis for developing an understanding of the genetics and physiology of the arcobacters and elucidating the pathogenic mechanisms, if any, of the species.

2 ISOLATION

The original methods used for *Arcobacter* spp. isolation were mainly based on those developed for *Campylobacter*. Early studies used CIN (cefsulodin-irgasan-novobiocin) agar, selective for *Yersinia* spp., to recover *Arcobacter* from pork meat samples (Collins et al., 1996a) and from a case of human

Food Safety. http://dx.doi.org/10.1016/B978-0-12-800245-2.00012-5

Table 1 *Arcobacter* Species and Their First Described Source

Species	First Isolation Source	References
A. butzleri	Human feces	Kiehlbauch et al. (1991b)
A. cryaerophilus	Bovine abortion fetuses	Neill et al. (1985)
A. skirrowii	Sheep feces	Vandamme et al. (1992b)
A. nitrophilis	Roots of *Spartina alterniflora* Loisel	McClung and Patriquin (1980)
A. cibarius	Chicken meat	Houf et al. (2005)
A. halophilus	Hypersaline lagoon	Donachie et al. (2005)
A. mytili	Mussels	Collado et al. (2009)
A. molluscorum	Mussels	Figueras et al. (2011a)
A. ellisti	Mussels	Figueras et al. (2011b)
A. thereius	Pigs/ducks	Houf et al. (2009)
A. marinus	Seaweed/seawater/starfish	Kim et al. (2010)
A. trophiarium	Porcine feces	De Smet et al. (2011a,b)
A. defluvii	Sewage	Collado and Figueras (2011)
A. bivalviorum	Shellfish	Levican et al. (2012)
A. venerupis	Shellfish	Levican et al. (2012)
A. cloacae	Shellfish/sewage	Levican et al. (2013)
A. suis	Pork meat	Levican et al. (2013)
A. anaerophilus	Estuarine sediment	Sai Jyothsna et al. (2013)

enteritis (Burnens et al., 1992). Many of the original protocols were time-consuming and lacked specificity. For example, using aero-tolerance as a differentiating characteristic between *Arcobacter* spp. and *Campylobacter* spp. was not always successful, as this was not consistently observed on initial isolation (de Boer et al., 1996). Traditionally, identification required for the isolation of a pure culture, which means that the main features of an isolation method should be that it is specific and not too resource- and time-intensive for routine use. Species and strain identification may not always be required in clinical or food monitoring situations, but in epidemiological studies it is essential to further identify strains and clones in order to determine the source(s) of infection. One of the important features of an isolation medium is that it inhibits the growth of competing microflora in situations where a diverse population of both pathogenic and spoilage organisms may occur; for example, in a chilled storage under MAP or vacuum packaging.

Protocols have been developed over the years for isolating *Arcobacter* per se from meats particularly (Table 2). An early procedure was similar to one suggested as a sensitive and rapid method for isolation of *Campylobacters* that included enrichment in an *Arcobacter*-selective broth (ASB), followed by plating onto semisolid *Arcobacter*-selective medium with cefoperazone, trimethoprim, piperacillin, and cycloheximide and incubating at 24 °C. When assessing this protocol for its efficiency in isolating arcobacters from poultry and other meats, recovery rates varied from 0.5% for pork to 24.1% for poultry meat samples (de Boer et al., 1996).

Preenrichment microaerobically at 30 °C in an *Arcobacter* enrichment broth (AEB) containing cefoperazone as the selective agent, combined with filtering onto mCCDA (modified cefoperazone, charcoal, and deoxycholate) agar and incubation aerobically at 30 °C, proved to be effective for recovery

Table 2 Efficiency of Isolation Methods Used to Recover *Arcobacter* from Meat

Meat	Isolation Method	Efficiency (%) (*n*)	References
Pork	Modified CIN/CAT agar	0-89.9 (299)	Collins et al. (1996b)
	Pre-enrichment in ASB, plate onto semi-solid medium, incubate at 24 °C	0.5 (194)	de Boer et al. (1996)
	Enrichment in Preston medium, culture on mCCDA, both at 37 °C microaerobically	3.7 (27)	Zanetti et al. (1996b)
	Enrichment in *Arcobacter* broth (+CAT), incubate at 24 °C microaerobically	76 (54) 29 (21)	Houf et al. (2000) Rivas et al. (2004)
	Enrichment in CAT broth, incubate 37 °C microaerobically, plate onto blood agar with/without filter, incubate at 25 °C or 37 °C	42 (55)	On et al. (2002)
Chicken	Pre-enrichment in ASB, plate onto semi-solid medium, incubate at 24 °C	22.1 (220)	de Boer et al. (1996)
	Enrichment microaerobically at 30 °C in AEB containing cefoperazone, filter onto mCCDA	97 (125)	Lammerding et al. (1996)
	Enrichment in CAT broth, plate onto CAT/blood agar with/without filter, incubate at 25 °C	100 (25)	Atabay and Corry (1997)
	Enrichment in CAT broth, plate onto CAT/blood agar with/without filter, incubate at 30 °C	28(50)	Scullion et al. (2004)
	Enrichment in JM broth, plate onto JM solid medium 30 °C, examine at 48 h	84 (50)	Johnson and Murano (1999b)
	Enrichment in JM broth, plate onto JM solid medium 30 °C, examine at 24 and 48 h	68 (50)	Scullion et al. (2004)
	Bacto Leptaspira medium/enrichment EMJH, subculture onto Karmali agar supplemented with cefoperazone, colistin, amphotericin and 5-fluorouracil at 25 °C aerobically	80.5 (201)	Marinescu et al. (1996b)
	Enrichment in AB supplemented with CAT microaerobically, subculture onto mCIN/CAT agar, incubate at 30 °C aerobically	53 (96)	Gonzalez et al. (2000)
	Enrichment in Preston medium, culture on mCCDA, both at 37 °C microaerobically	0 (32) 52.3 (170)	Zanetti et al. (1996b) Harrab et al. (1998)
	Enrichment in *Arcobacter* broth (+CAT), incubate at 24 °C microaerobically	73 (22)	Rivas et al. (2004)
	Enrichment in AB with five antibiotics, microaerobic incubation at 28 °C, plated on Arcobacter solid medium with five antibiotics at 28 °C	71 (45)	Houf et al. (2004)
	Enrichment in AB with five antibiotics, microaerobic incubation at 28 °C, plated on Arcobacter solid medium with five antibiotics at 30 °C	68 (50)	Scullion et al. (2004)
Turkey	EMJH enrichment at 30 °C, filter onto BHI agar incubate at 30 °C	77 (395)	Manke et al. (1998)
	Enrichment in Preston medium, culture on mCCDA, both at 37 °C microaerobically	0 (30)	Zanetti et al. (1996b)
	Campylobacter enrichment broth (+CAT) microaerobic, 30 °C for 2 days, filter onto blood agar, incubate aerobically for up to 7 days	24 (17)	Atabay and Aydin (2001)

Continued

Table 2 Efficiency of Isolation Methods Used to Recover *Arcobacter* from Meat—cont'd

Meat	Isolation Method	Efficiency (%) (*n*)	References
Duck	Enrichment microaerobically at 30 °C in AEB containing cefoperazone, filter onto mCCDA	80 (10)	Ridsdale et al. (1998)
	Campylobacter enrichment broth (+CAT), microaerobic, 30 °C for 2 days, filter onto blood agar, incubate aerobically for up to 7 days	70 (10)	Atabay and Aydin (2001)
Beef	Pre-enrichment in ASB, plate onto semi-solid medium, incubate at 24 °C	1.5 (68)	De Boer et al. (1996)
	Enrichment in *Arcobacter* broth (+CAT), incubate at 24 °C microaerobically	22 (7)	Rivas et al. (2004)

from raw poultry, with no growth of competing microflora (Lammerding et al., 1996). However, although both mCCDA and CAT (cefoperazone, amphotericin, and teicoplanin) agars supported growth, the latter tended to support a wider range of arcobacters than the former.

A modification of the Lammerding method using enrichment in CAT broth microaerobically, followed by plating on CAT agar and aerobic incubation at 30 °C, does isolate a wider range of arcobacters than the original method, although the results of a small-scale study suggested a lower isolation rate. In that study, 10 out of 10 supermarket chicken carcasses were positive by the original method compared with 8 out of 10 using the modified method. After enrichment in CAT broth, similar recovery rates were obtained substituting blood agar for CAT agar and also using a 0.65 mm rather than a 45 mm pore size filter. An enrichment step is important for maximal isolation because conventional direct plating onto CAT agar may not be successful in recovering *Arcobacter* from poultry samples (Atabay and Corry, 1997).

A commercial enrichment broth, *Arcobacter* broth (AB), was developed that, when supplemented with either CAT or mCCD, was suggested for the isolation of *Arcobacter* spp. or *A. butzleri*, respectively. Importantly, when eight strains of *Campylobacter* spp. were tested, none of these grew in AB, probably because it contains no oxygen-quenching system, such as blood, which neutralizes the effect of atmospheric oxygen (Atabay and Corry, 1998). Another protocol was suggested by Johnson and Murano in 1999 (JM formulation), which included a combination of enrichment and solid media that allowed optimal aerobic growth at 30 °C. On this agar, a deep red color develops around the *Arcobacter* colonies making presumptive positives easier to recognize (Johnson and Murano, 1999a,b). In a comparative study of the de Boer and JM methods for isolating *Arcobacter* spp. from chicken meat, 14 out of 50 broiler chicken samples were positive with the former while 42 were positive with the latter. In the JM method, both the total method time (including PCR confirmation) reduced significantly from 6-9 days to 4 days and specificity improved. Fourteen percent of samples produced plates containing other microflora using the JM formulation compared with 53% in the de Boer method, and 2% were negative for *Arcobacter* spp. The JM method, therefore, showed promise as a routine recovery method as it appeared both rapid and specific, fulfilling the most important criteria for an isolation method. Eifert (2003) found that the JM method was effective at isolating *Arcobacter* spp. from environmental, cloacal, and fecal samples obtained from a poultry house.

Houf et al. (2001) described a revised protocol for the recovery of *Arcobacter* from poultry products using a medium containing five antibiotics (5-fluorouracil, novobiocin, trimethoprim, and teicoplanin or vancomycin) at concentrations that did not affect *Arcobacter* growth. The Houf medium has been used to isolate and directly enumerate the species from retail pork and minced beef (De Smet et al., 2010; Van Driessche and Houf, 2007b).

Scullion et al. (2004), compared three isolation methods including the Houf method, a method described by On et al. (2002) (both modified by using 30 °C and not 25 °C for plate incubation) and the JM method (modified by plating at 24 and 48 h) for the isolation of *Arcobacter* spp. from retail raw poultry, and suggested that the latter protocol was the overall method of choice. Its isolation rate was the same as the Houf method, but it detected both *A. cryaerophilus* and *A. skirrowii*. However, in a similar study comparing five isolation methods, the one that included preincubation in the broth used in the Houf protocol followed by plating onto mCCDA (both incubated at 30 °C aerobically) proved to be the most sensitive and specific, although the JM method was not tested (Merga et al., 2011).

In another comparative study, this time of the Atabay and Corry media, the JM and Houf agars to recover inoculated *A. butzleri* from vacuum-packed chilled storage beef, Ahmed and Balamurugan (2013) reported that the Houf method was the most suitable as a direct plating agar for enumeration of *A. butzleri* for use in meat challenge studies because it effectively suppresses indigenous beef microflora. Fallas-Padilla et al. (2014) compared six different isolation methods for *Arcobacter* from retail chicken breast meat and concluded that a modification of the standard Houf method using Houf selective broth followed by filtration onto blood agar was the most effective, with a sensitivity of 89% and a specificity of 84%.

Therefore, it seems that the Houf and JM methods both have advantages and disadvantages, and which is the most effective depends on various factors such as presence of competing microflora and origin of the samples tested.

3 IDENTIFICATION

Arcobacter species are defined as Gram-negative, oxidase-positive, curved rods that grow aerobically at 25 °C. The main phenotypic tests used for species identification are catalase activity; nitrate reduction; cadmium chloride susceptibility; microaerobic growth at 20 °C; and growth on McConkey agar and in the presence of 3.5% NaCl and 1% glycine (Kiehlbauch et al., 1991a; Vandamme et al., 1992b).

3.1 IDENTIFICATION BY BIOCHEMICAL METHODS

The original "biotyping" identification scheme for *Campylobacter* spp., based on reaction to a small number of biochemical tests was developed by Lior (1984) and later extended by Bolton et al. (1992). In the early 1990s, biochemical tests, protein profiles, and fatty acid profiles were used to differentiate arcobacters (Vandamme et al., 1992b), and serotyping based on the heat-labile antigens was developed (Lior and Woodward, 1991). The combination of serotyping and biotyping was successful in identifying biotypes and serotypes of *A. butzleri* isolates from poultry in France (Marinescu et al., 1996b).

Species identification within *Campylobacteraceae* using standard biochemical tests is problematical because of the low metabolic activity generally demonstrated by Proteobacteria. Therefore, combining a systematic range of biochemical tests with a probabilistic matrix comparing the characteristics of the unknown isolate with those of defined taxa provides a method of species differentiation. However, this procedure is beneficial only if there is a match between the taxa in the scheme and the unknown isolate. Numerical analysis of phenotypic characters together with the use of an extensive probabilistic identification matrix provides a systematic approach to identification and confirmed pathogenic arcobacters as a distinct group from campylobacters (On and Holmes, 1995).

Whole cell protein profiling distinguishes between *Campylobacter*, *Helicobacter*, and *Arcobacter* (Vandamme et al., 1991b). After differentiating *Campylobacteraceae* from other genera, the use of SDS-PAGE of whole cell proteins was effective in determining *Arcobacter* species isolated from poultry at an abattoir, compared with API Campy strips or biotyping by the Bolton method. In the former, 100% of *Arcobacter* isolates were misidentified while the latter does not speciate *A. butzleri* and *A. skirrowii* (Ridsdale et al., 1998).

3.2 IDENTIFICATION BY MOLECULAR METHODS

Nowadays, molecular or genomic techniques are the method of choice for definitive identification of a species or strain. In the classical biotyping and serotyping methods, different strains frequently belong to the same type, making differentiation difficult. All molecular protocols have advantages and disadvantages and some of these are shown in Table 3. Ideally if a molecular typing technique is to be used in a routine laboratory it must be robust, reliable, relatively inexpensive, discriminatory, and, preferably, rapid.

In the 1990s, ribotyping was proved to be reliable for the identification of *Arcobacter* isolates (Kiehlbauch et al., 1991a, 1994). Ribotyping involving the hybridization of *Pvu*II-digested chromosomal DNA or *Cla*I-digested DNA with probes for the highly conserved 16S rRNA gene discriminates between strains of *Campylobacter*, *Helicobacter*, and *Arcobacter* or maximally between *Arcobacter* strains, respectively (Table 3).

Suarez and coworkers (1997) designed several oligonucleotide primers for the rRNA superfamily (R01-R05), one of which was specific for *A. butzleri*. De Oliveria et al. (1999) and Wesley et al. (1995) used two genus-specific 16S rRNA-based oligonucleotide DNA probes and an *A. butzleri* species-specific DNA probe to identify field isolates from animal and human sources. Although these differentiated *A. cryaerophilus* Group 1A from *A. cryaerophilus* Group B, they did not allow the identification of *A. skirrowii*. However, they did differentiate between the two species associated with human illness.

In 1999, a two-step typing scheme based on the PCR-RFLP analysis of the 16S rRNA gene involving digestion with *Dde*I and *Taq*I was developed that produced unique fingerprints for *A. butzleri*, *A. cryaerophilus*, and *A. skirrowii*. However, because it did not differentiate between *C. jejuni* and *C. coli*, it did not gain widespread use in routine laboratory settings (Marshall et al., 1999).

During the 1990s, there were many research groups developing identification methods based on the 23S rRNA gene. One protocol used three endonucleases, *Hpa*II, *Cfo*I, and *Hin*fI, to digest 23S rRNA gene PCR amplicons, resulting in the generation of conserved restriction profiles, allowing differentiation of *A. butzleri* and *A. nitrofigilis*, although it did not distinguish between *A. cryaerophilus* and *A. skirrowii*. When evaluated as a means of identifying 21 field strains that had proved difficult to characterize by phenotypic tests, 17 (86%) were easily assigned to species level, indicating the reliability of the method (Hurtado and Owen, 1997).

Bastyns et al. (1995) developed a genus-specific PCR and three species-specific PCR assays, based on a target sequence comprising the most variable region of 23S rDNA gene that were reliable in identifying

Table 3 Molecular Typing Methods Used to Identify *Arcobacter*

Method	Samples	Advantages	Disadvantages	References
Ribotyping	Type and reference strains	Discriminatory between *Arcobacter* strains	Minimum number of cells required	Kiehlbauch et al. (1994)
Hybridization with 16S rRNA probes for *Arcobacter* spp. and *A. butzleri*	Field isolates from aborted livestock fetuses; reference strains	No cross reaction of *Arcobacter* spp. probe with Helicobacter/*Campylobacter* spp. or of *A. butzleri* probe with *A. skirrowii* or *A. cryaerophilus*; differentiates between *A. cryaerophilus* 1A and 1B	Does not distinguish *A. skirrowii*	Wesley et al. (1995) and de Oliveria et al. (1999)
PCR using 16S rRNA DNA probes specific for *Arcobacter* (as above)	Porcine feces; water	*A. butzleri*, *A. cryaerophilus* and *A. skirrowii* differentiated; no cross reaction with *Helicobacter* or *Campylobacter*; results within 8 h	None described	Harmon and Wesley (1996) and Rice et al. (1999)
PCR-RFLP analysis 16S rRNA gene	Clinical isolates; reference strains	Differentiates all *Arcobacter* spp.	Does not differentiate between *C. jejuni* and *C. coli*	Marshall et al. (1999) and Figueras et al. (2008)
		Updated for 17 *Arcobacter* spp.	More than one digestion required to identify certain species	
PCR-RFLP analysis 23S rRNA gene	Field isolates; reference strains	Relatively inexpensive; differentiates most *Campylobacter* spp.	Does not distinguish between *A. cryaerophilus* and *A. skirrowii*	Hurtado and Owen (1997)
m-PCR using species-specific primers for *A. butzleri* and *C. jejuni*	Artificially contaminated meat, fruit and dairy products	Identifies *C. jejuni* and *A. butzleri* in one assay within 8 h	Not tested for *A. cryaerophilus*	Winters and Slavik (2000)
m-PCR using primers for 16S rRNA genes of *Arcobacter* spp. and *A. butzleri* specific 23S rRNA gene portion	Reference strains, poultry isolates, porcine isolates	Rapid; specific for *A. butzleri*	Does not differentiate non-butzleri species	Harmon and Wesley (1997)
m-PCR-culture using species specific primers	Reference strains; poultry isolates	Sensitive; rapid; distinguishes *A. butzleri*, *A. cryaerophilus* and *A. skirrowii*	None described	Houf et al. (2000)
PCR-culture using *Arcobacter*-specific primers	Reference strains and chicken isolates	Sensitive, rapid	Only identifies *Arcobacter* spp. with no differentiation	Gonzalez et al. (2000)
PCR-hybridization based on glyA gene	Type and reference strains	Low levels of detection of *A. butzleri* (50 pg DNA or 23,000 copies)	Only *A. butzleri* or *A. butzleri*-like strains identified	Al Rashid et al. (2000)
RT PCR	Chicken and waste water isolates	No enrichment required	None described	Gonzalez et al. (2010) and Hausdorf et al. (2013)

reference strains and field isolates of *Arcobacter* spp. The primer combination ARCO1-ARCO2 defined the genus or, together with the species-specific primers ARCO1-BUTZ, ARCO1-CRYAE, and ARCO1-SKIR, allowed identification to species level (Bastyns et al., 1995).

Multiplex-PCR (m-PCR) techniques based on the multiple primer sets for the detection of *Arcobacter* spp. were also developed (Table 3). One protocol used two primers sets: one targeting a section of the 16S rRNA genes (ARCO1 and ARCO2) and the other amplifying a portion of the 23S rRNA genes unique to *A. butzleri* (ARCO-BUTZ). The reliability of this method was demonstrated by analysis of 108 field strains of *Arcobacter* spp., previously characterized to species level by DNA-DNA hybridization, dot-blot hybridization, ribotyping, or by serology. The 1223 bp multiplex PCR product identified all of the isolates as *Arcobacter* spp., while the presence of both the 1223 and 686 bp amplicons identified 66 strains as *A. butzleri* agreeing with results obtained by other methods (Harmon and Wesley, 1997).

Another m-PCR assay also targeting the 16S and 23S rRNA genes, but using five primers, was developed for the simultaneous detection and identification of *A. butzleri*, *A. cryaerophilus*, and *A. skirrowii*. The selected primers amplified a 257-bp fragment from *A. cryaerophilus*, a 401-bp fragment from *A. butzleri* and a 641-bp fragment from *A. skirrowii*. Using reference strains, this m-PCR proved specific for these three *Arcobacter* species. An evaluation of this m-PCR to identify *Arcobacter* spp. from poultry samples, combined with a preenrichment stage in AB with CAT for 24 h, demonstrated that it was a species-specific and rapid technique for detecting and identifying *A. butzleri*, *A. cryaerophilus*, and *A. skirrowii* (Houf et al., 2000). A species-specific PCR assay for *A. butzleri*, *A. cryaerophilus* 1A and 1B, and *A. skirrowii*, amplifying the most variable areas of the 23S rRNA gene, was developed by Kabeya et al. (2003) and was suggested as a simple one-step PCR to identify these three species, although the original study only included 3 reference strains and 10 *Arcobacter* isolates.

Real-time PCR (RT-PCR) has been shown to have better detection rates than conventional PCR without the need for enrichment. This has been shown in various different sample types such as chicken products (Gonzalez et al., 2010), water and salads (Hausdorf et al., 2013), environmental samples from a poultry processing plant (Brightwell et al., 2007), and human stool specimens (de Boer et al., 2013).

To differentiate between the increasing numbers of newly described *Arcobacter* species, revised protocols have been developed. In 2008, there were six species (*A. butzleri*, *A. cryaerophilus*, *A. skirrowii*, *A. cibararius*, *A. nigtrofigilis*, and *A. halophilus*) and these were discriminated by a method using a single digestion with the *Mse*I endonuclease (Figueras et al., 2008). Douidah et al. (2010) developed a m-PCR assay based on the 23S RNA gene that was able to differentiate between *A. butzleri*, *A. cryaerophilus*, *A. skirrowii*, *A. cibarius*, and *A. thereius* in crude lysates as well as DNA extracts, which reduced time to identification and made it more likely to be used in a routine setting.

Recently, the protocol by Figueras et al. (2008) was extended to include the 17 species identified by 2012. The initial digestion by *Mse*I endonuclease identified 10 out of the 17 species, so that further digestion with *Mnl*I was necessary to differentiate other species, with a further digestion with *Bfa*I required to differentiate between *A. defluvii* and *A. suis* and between *A. trophiarum* and atypical *A. cryaerophilus* strains following *Mnl*I digestion (Figueras et al., 2012).

A comparative study of five molecular methods, including the Figueras et al. (2008), Houf et al. (2000), and Kabeya et al. (2003) methods, for identification of *Arcobacter* spp. against the Figueras et al. (2012) method as the standard, concluded that these methods were unreliable and their use has led to misidentification. Therefore, it was suggested that using a more reliable procedure (i.e., the Figueras et al. (2012) protocol) would result in the diversity of the species being redefined (Levican and Figueras, 2013).

3.3 GENOTYPING

Various methods have been used for differentiating between *Arcobacter* strains including amplified fragment length polymorphism (AFLP) analysis, pulsed-field gel electrophoresis (PFGE), enterobacterial repetitive intergenic consensus-PCR (ERIC-PCR), randomly amplified polymorphic-DNA-PCR (RAPD-PCR), and whole genomic sequencing (Table 4). These were reviewed by Gonzalez et al. (2012).

PFGE is considered to be the gold standard of typing methods available, as it is highly discriminatory and reproducible and, although complex, it is the method of choice for epidemiological studies and investigating sources of contamination (Rivas et al., 2004). An example of the information obtained from a PFGE-type identification method is illustrated by a study of isolates from nursing sows and developing pigs on three farms of a farrow-to-finish swine operation. Using PCR and PFGE, the level of genotypic variation revealed suggested that pigs in this operation were colonized by multiple *Arcobacter* parent genotypes that may have undergone genomic rearrangement during successive passages through the animals. Additionally, the level of genotypic diversity seen among *Arcobacter* isolates from individual farms suggested an important role for genotypic phenotyping as a source identification and monitoring tool during outbreaks (Hume et al., 2001). A similar diversity among isolates from a slaughterhouse was detected using PFGE by Ferreira et al. (2012) with 26 different PFGE fingerprints in 23 *A. butzleri* isolates.

In 2009, Miller et al. reported the first multilocus sequencing typing (MLST) scheme for *Arcobacter* spp. MLST is a useful method for determining diversity of isolates, which helps to inform epidemiological studies. For example, in one study in the United Kingdom of 39 isolates, 11 sequence

Table 4 Genotyping Methods for Identification of *Arcobacter* spp. Isolates

Method	Used In	Species	References
AFLP profiling	Epidemiological studies; identification of new species	Highly discriminatory for *A. butzleri* and *A. cryaerophilus*	Scullion et al. (2001)
		Genetically distinct subgroups of *A. cryaerophilus* differentiated	On et al. (2003, 2004)
PFGE	Source identification and monitoring outbreaks	*A. butzleri* and *A. cryaerophilus*	Hume et al. (2001)
			Rivas et al. (2004)
			Shah et al. (2012)
MLST	Strain discrimination; epidemiological studies	*A. butzleri*, *A. cryaerophilus*, *A. skirrowii*, *A. thereius*, *A. cibarius*	Miller et al. (2009)
	Diversity of isolates	*A. butzleri*	Merga et al. (2011, 2013)
RAPD-PCR	Identification of more than one species genotype from same sample	*A. butzleri*, *A. cryaerophilus*, *A. skirrowii*	Houf et al. (2002)
ERIC-PCR	Strain discrimination; epidemiological studies	*A. butzleri*, *A. cryaerophilus*, *A. skirrowii*	Houf et al. (2002)
			Ho et al. (2008)
Whole genomic sequencing	Phenotypic differences	*A. butzleri*	Merga et al. (2013)

types (STs) were present (Merga et al., 2011); while in a later U.K. study in 2013, 104 *A. butzleri* isolates with 43 different STs were identified (Merga et al., 2013). In both cases, the STs were useful in determining a possible common origin for the contamination by *A. butzleri* and suggesting a means of spread. The fact that MLST is valuable in determining a common origin was emphasized in a study in a Danish slaughterhouse by Rasmussen et al. (2013). In this case, high strain variability was reported and cross-contamination was suggested as the reason for the repeated detection of two STs on consecutive production days (with sanitizing in between). MLST analysis of *Arcobacter* spp. isolates from food sources in Spain also showed a wide diversity of genotypes with no correlation between STs and food source, although lateral gene transfer from *A. skirrowii* to *A. butzleri* was reported (Alonso et al., 2014).

Whole-genomic fingerprinting by AFLP is a high-resolution genotyping method. On and Harrington (2001a) showed that numerical analysis of AFLP profiles based on *Bgl*11-*Csp*61 polymorphisms is useful in identifying taxonomic and epidemiological relationships in a range of *Campylobacter* species. Furthermore, they established AFLP as a method for concurrent species identification of the family *Campylobacteraceae*, including four species of *Arcobacter* (On and Harrington, 2001b) and for identification of genetic diversity among different clonal types from distinct geographical areas (On et al., 2004).

A comparison of m-PCR and AFLP-profiling for speciation of *Arcobacter* spp. by Scullion et al. (2001) resulted in a good correlation between the two methods, although three *A. skirrowii* isolates tested gave "atypical" AFLP profiles (i.e., not similar to the type strain) whereas a PCR method confirmed them as *Arcobacter* spp.

AFLP requires a large reference database although it is less expensive, more rapid, and better suited to typing a large number of isolates and easier to perform than other whole-genome techniques such as MLST and PFGE. Once adopted in a routine laboratory, it has the benefits of being robust and highly discriminatory (On et al., 2004).

ERIC-PCR has been used to genotype isolates in many studies and has proved to be a useful technique in a wide range of circumstances. Examples include to determine strain variation in feces from healthy cattle in Belgium (Van Driessche et al., 2005); to pinpoint sources of contamination in a poultry slaughterhouse (Ho et al., 2008); to identify the molecular epidemiology in human acute gastroenteritis (Houf and Stephan, 2007; Kayman et al., 2012); and to investigate heterogeneity in isolates from different foods (Aydin et al., 2007).

Next-generation whole genomic sequencing, especially with the development of high-throughput sequencing methods, allows both genotypic and phenotypic differences between strains to be assessed (Merga et al., 2013). Differentiation between pathogenic and nonpathogenic strains or isolates from the same niche but from different sources will be possible.

4 OCCURRENCE IN ANIMALS AND FOOD

Arcobacters have been associated with reproductive problems in pigs since their first isolation from aborted fetuses and sows with reproductive problems (Neill et al., 1979). Both *A. cryaerophilus* and *A. butzleri* were isolated from sows with reproductive problems, from aborted porcine fetuses in Brazil (de Oliveira et al., 1997), and from aborted porcine fetal tissues in the United States (SchroederTucker et al., 1996). Strains from reproductively impaired sows are often antigenically similar to those also isolated from normal pigs, suggesting that those associated with infertility are opportunistic pathogens colonizing the fetus following placental damage (Wesley, 1994).

As *Helicobacter* and *Arcobacter* are related within the *Campylobacteriace* and *H. pylori* has a role in the etiology of human gastric ulceration, there was the possibility that *Arcobacter* may have a role in the economically important disease in livestock husbandry (i.e., in gastric ulceration in pigs). However, in a study using PCR combined with hybridization with a species-specific DNA probe, there was no statistical difference when isolation rates from specimens with gross gastric pathology were compared with those without, although there was more likelihood of isolating *Arcobacter* from the nonglandular than the glandular region, which made a definitive association difficult to assess (Suarez et al., 1997). Similarly, *Arcobacter* could not be isolated from artificially infected piglets that develop lesions in their gastric mucosa (Wesley et al., 1996). *Arcobacter* spp. are able to colonize neonatal piglets as demonstrated by their recovery from feces and from tissues such as liver, kidney, ileum and brain, although no gross pathology is observed.

Although pork is contaminated by *Arcobacter* spp., the recovery rate varies, which is probably a reflection both of the actual incidence and the isolation methods used (Table 2). In the 1990s, several studies reported a variety of recovery rates. In Italy, 1 out of 27 (3.7%) pork samples was positive for *A. butzleri* (Zanetti et al., 1996b), while in the Netherlands 1 out of 194 (0.5%) was positive (de Boer et al., 1996). However, in a study of ground pork from five processing plants in Iowa, the isolation rates varied from 90% to 0%. The original primary source of the animals, the hygiene practices at the abattoir itself, or both, probably contributed to the variation (Collins et al., 1996b). A similar disparity was reported from a 9-month study of pigs at slaughter from two farms in the United States where the mean prevalence rate was 5% with a range of between 0% and 20% (Harvey et al., 1999).

Similarly, *Arcobacter* spp. has been isolated from poultry, particularly chicken, with varying recovery rates, which again may be an indication of the actual incidence in the animal population or a reflection of the isolation techniques (Table 2). In the only two studies reporting on recovery from eggs no *Arcobacter* were isolated (Lipman et al., 2008; Zanetti et al., 1996b), although with more up-to-date methods and protocols for isolation this might not prove to be the case.

The source of *Arcobacter* contamination in poultry (and other animals) remains unclear. In several studies, arcobacters have been isolated in poultry processing plants from the carcasses and slaughter equipment, but not from the intestinal content (Gilgen et al., 1998). For example, using the Houf method, *Arcobacter* spp. was not isolated from live birds but was from effluent sludge and stagnant water outside rearing sheds (Gude et al., 2005). Similarly, a later study using the same isolation method to examine the presence of arcobacters in and on living chickens of four flocks at slaughter age, no arcobacters were isolated from the intestinal tract or from the skin or feathers of birds. Therefore, it was concluded that arcobacters were not part of the intestinal or skin flora (Van Driessche and Houf, 2007b).

In a study in Iowa, *Arcobacter* spp. was isolated from 14.3% of 405 cloacal swabs, of which 1% were *A. butzleri*, indicating natural infections do occur. As with pigs, variations in isolation rates from poultry occurred dependent on environmental conditions of the originating flock. However, unlike piglets, experimental infections with *A. butzleri* were not established in orally challenged poultry (Eifert et al., 2003), only in inoculated birds and only in the highly inbred Beltsville White strain of turkeys (Wesley et al., 1996; Wesley and Baetz, 1999).

The results of one survey in 1998 of mechanically separated turkey samples from different processing plants suggested that this meat is highly contaminated by *Arcobacter* spp. or *A. butzleri* with rates of 77% and 56%, respectively, being reported. However, similar to the studies on pig processing plants, there was variation in prevalence. One plant yielded 96% contaminated samples (80% were *A. butzleri*) whereas another yielded 44% positive samples of which 59% were *A. butzleri*. Using a PCR-based identification method, diversity in DNA patterns occurred in all plants sampled, suggesting a multiplicity of contaminating sources rather than one major source (Manke et al., 1998). A high prevalence was also reported in a study in 2008 with numbers of *Arcobacter* on turkey skin varying between 1.7 and 2.4 CFU/cm^2; however, the authors did demonstrate that there was a decrease during processing (Atanassova et al., 2008).

Using selective enrichment plus SDS-PAGE confirmation on a small sample (10) of duck carcasses at the abattoir, *A. butzleri* was isolated from two, *A. skirrowii* from one, and *A. cryaerophilus* from five. One out of four flocks investigated was positive for *A. butzleri* and one for *A. cryaerophilus* (Ridsdale et al., 1998).

Infections in cattle have been reported less often than those in either pigs or poultry and, similarly, beef is not contaminated to such an extent as pork or chicken. In one study, only 1.5% (1/68) of minced beef samples were positive for *Arcobacter* compared with 24.1% (53/220) samples of poultry (de Boer et al., 1996). *Arcobacter* spp. have been isolated from bovine fetuses (Neill et al., 1978, 1979) and cattle with mastitis (Wesley, 1994). In 1995, the first isolation of *A. cryaerophilus* was reported in Chile from a bovine abortion using phenotypic characteristics to speciate; which, as discussed previously, is notoriously difficult (Fernandez et al., 1995b). The organism has also been isolated from aborted bovine fetuses from two cattle herds in Germany and an etiological association between *A. cryaerophilus* and serial abortion established in one of the herds (Parvanta, 1999). *A. cryaerophilus* was also the dominant species detected in a study that examined the feces of 276 healthy cattle from three farms in Belgium, with the prevalence ranging from 7.5% to 15%. The isolates showed a large heterogenicity and animals were colonized with more than one genotype, suggesting multiple contamination sources (Van Driessche et al., 2005).

In a small study in Turkey of 50 samples each of cow's milk, water buffalo milk, and village cheese, *A. buzleri* was isolated from 38.8%, 12.5%, and 21.4% of the samples, respectively, with other species including *A. cryaerophilus* and *A. skirrowii* as well as *A. halophilus* and *A. cibarius*, indicating that nonmeat sources could also be a risk to public health. However, another study on goat and sheep farms in Belgium did not isolate any arcobacters from milk samples, but did use a slightly different isolation method (De Smet et al., 2011a). In the same study, arcobacters were isolated from the feces of both male and female goats and sheep. The most concerning factor about the results of the former study was that all of the *A. skirrowii* ($n=12$) and *A. butzleri* ($n=10$) isolates were resistant to the common antibiotics such as cefoperazone, ampicillicin, erythromycin, cloxacillin, tetracycline, and penicillin G (Wesley et al., 2000; Yesilmen et al., 2014).

As well as being recovered from food animals, *Arcobacter* spp. have been isolated from nonfood animals such as pet cats and dogs (Fera et al., 2009; Peterson et al., 2006) but this is not a consistent finding (De Smet et al., 2011a).

5 OCCURRENCE IN WATER

Using a highly specific m-PCR assay, healthy cows have been shown to shed *Arcobacter* spp. in their feces with feeding of alfalfa and the use of individual waterers being protective (Wesley et al., 2000). This latter finding suggests that water probably has a role in the transmission of *Arcobacter* spp. into the human food chain. Many studies have isolated the organism from a range of water sources from different countries. *A. butzleri* has been isolated from canal water in Thailand (Dhamabutra et al., 1992), a drinking water reservoir in Germany (Jacob et al., 1993), brackish water (Maugeri et al., 2000), well water and beach water in the United States (Agidi et al., 2012; Rice et al., 1999), estuarine water in Italy (Fera et al., 2010), and river water and wastewater in Spain (Collado et al., 2010; Gonzalez et al., 2012).

In a 2-year study of water treatment plants in Germany, *A. butzleri* was isolated at all stages of processing with the predominant serotype being serotype 1, although serotypes 17, 19, and 2 were also isolated (Jacob et al., 1998). This pattern corresponds to results of studies also carried out in the 1990s in poultry (Lammerding et al., 1996; Marinescu et al., 1996b) and clinical samples (Vandamme et al., 1992a). Although significant numbers of activated sludge samples (4%) contain *Arcobacter* (Snaidr et al., 1997; Stampi et al., 1998) an early Italian study found that oxygen-activated sludge treatment followed by tertiary treatment with 2 ppm chlorine dioxide reduced the loading of *Arcobacter* spp. (specifically *A. cryaerophilus*) by 99.9% (Stampi et al., 1993).

Land application of sludge, therefore, may provide a means of animal and human *Arcobacter* spp. infection via runoff into the water course as has been suggested for other pathogens (Keene et al., 1994), although drinking water is not a risk factor because of the effectiveness of the treatment process (Collado et al., 2010).

In 2014 there was the first report of the presence of *A. butzleri* (9/9 locations tested) and, to a much lesser extent, *A. cryaerophilus* (1/9 locations tested) in samples of raw, untreated domestic sewage influent in the United Kingdom. Because *A. butzleri* is sensitive to levels of chlorine used in U.K. drinking water (Moreno et al., 2004; Rice et al., 1999) and to BrCl (Zanetti et al., 1996a) it is highly likely that their presence is from the sewage per se and that these organisms are present in the human population (Merga et al., 2014).

Posttreatment contamination of drinking water may pose a risk for further animal or human infection. In a study by Van Driessche and Houf (2008) comparing the survival in drinking water it was demonstrated that, using a direct plating technique, *A. butzleri* survived longer than the other two species at 4, 7, and 20 °C (>100 days compared with 50 days at 4 and 7 °C and 21 days at 20 °C for *A. cryaerophilus* and *A. skirrowii*) and that all strains survived longer in the presence of organic material with *A. butzleri* surviving over 190 days at 4 and 7 °C and 147 days at 20 °C. However, at higher temperatures (52, 56, or 60 °C) survival was a matter of minutes. The longest survival time (30 min) was for *A. butzleri* both in the absence and presence of organic material. Using an enrichment isolation method, culturable cells were isolated for much longer, with *A. butzleri* being detected up to 250 days at 4 °C and at 7 °C in the presence of organic material. Although this was a laboratory-based study and, therefore, may not reflect a real-life situation, it does indicate that *A. butzleri*, particularly, can survive for relatively long periods at low temperatures, representing a possible reservoir of contamination in slaughterhouses and farms.

Fera and coworkers (2004, 2008, 2010) proposed that arcobacters are able to survive in a viable nonculturable form, as has been reported for *C. jejuni* (Cappelier et al., 1999; Rollins and Colwell, 1986; Talibart et al., 2000; Tholozan et al., 1999), and nonculturable forms have been reported as a survival mechanism in nonchlorinated water (Moreno et al., 2004) and in isolates from a spinach processing plant (Hausdorf et al., 2013). Therefore, it remains a possibility as a potential source of contamination into the human food chain. This emphasizes the point that wherever possible, culture techniques should be confirmed by other means such as PCR and FISH.

With the possibility that water is contaminated by arcobacters, food products such as fresh salads and vegetables either irrigated with water or washed with water during processing could also provide a means of entry into the human food chain. The risk in this case is greater than in meat or meat products as these are generally cooked before consumption. In one study of lettuce on retail sale in Spain (Gonzalez and Ferrus, 2011), *A. butzleri* was detected in 20% (10/50) of samples and Hausdorf et al. (2013) demonstrated that a number of *Arcobacter* spp. were able to survive in a spinach processing plant, thus presenting a continuous contamination route.

6 CONTROL

C. jejuni is a well-described human pathogen present in a range of food products. Many interventions have been evaluated for their effectiveness in eliminating this organism. However, these may not be as effective at inactivating or eliminating *Arcobacter* spp., even though the two are closely related. For example, *A. butzleri* is more resistant than *C. jejuni* to irradiation treatment. The average D_{10} value is 0.27 kGy compared with 0.19 kGy for *C. jejuni*. However, in this case, radiation doses currently allowed for foodstuffs in the United States (0.3-7.0 kGy) and the maximum allowed in the United Kingdom for poultry (7 kGy) do provide an effective method of eliminating *A. butzleri* as well as *C. jejuni.*

In foods liable to be contaminated by *Arcobacter* spp. such as pork and poultry, a number of treatments such as organic acid spray or dips and the incorporation of compounds such as nisin or sodium lactate into the product have been demonstrated to be effective in reducing contamination by pathogens. Dipping chicken carcasses inoculated with *A. butzleri* into various organic acids reduces bacterial counts. In one study comparing eight organic acids, benzoic acid proved by a sensory panel to be the most effective in that it reduced counts without affecting the overall acceptability of the meat and skin (Skrivanova et al., 2011). However the reduction was less than 2 log, which is not sufficient in an industrial situation to be economically effective used on its own and would, therefore, have to be combined with other treatments. Such combination treatments can prove not to be altogether the solution they first appear, as *A. butzleri* is able to withstand lethal acid stress following exposure to sublethal stress-adaptation temperatures (Isohanni et al., 2013).

Although continuous exposure to sodium lactate in pure culture is not effective against *A. butzleri,* both citric acid and lactic acid eliminate growth, while nisin (500 IU/ml) reduces survival by approximately 50% (Phillips, 1999). Short-term treatment with both trisodium phosphate and EDTA (alone and in combination with nisin) are effective in reducing survival of *A. butzleri* in pure culture (Phillips and Duggan, 2001). Studies with other Gram-negative bacteria suggest that nisin is less effective in food systems than in culture and this is also the case for *A. butzleri* (Phillips and Long, 2001). Decreasing temperature appears to slow growth but *A. butzleri* still survives both in culture and in chicken at low temperatures (Long and Phillips, 2003; Phillips and Duggan, 2002).

The use of natural antimicrobials for the control of pathogens and spoilage organisms has been suggested by many researchers over recent years. Little work has been reported investigating the use of natural antimicrobials on *Arcobacter* spp., although as the interest in both grows, this will presumably increase. Plant extracts such as essential oils or EOs (Burt, 2004) and plant-produced small molecules such as those produced to fight off infections have been investigated on a range of bacteria and fungi. An example of the latter is resveratrol (3,4′,5 trihydroxystilbene), which is bactericidal to both *A. butzleri* and *A. cryaerophilus* cells, dependent on time of exposure and growth phase, with stationary phase cells being the most resistant. The mode of action is via cell membrane disintegration and efflux pump inhibition combined with a decrease in DNA content and metabolic activity (Ferreira et al., 2014c).

In terms of EOs, bergamot EO in both liquid and vapor form was found to be the most effective against three *Arcobacter* isolates when compared with lemon and orange EOs. As with nisin, the EO was less effective in vitro than on food (Fisher et al., 2007). Although *C. jejuni* and *C. coli* are susceptible to orange oil fractions, *A. butzleri* is not as susceptible as reported by Nannapaneni et al. (2009). When they tested orange oil fractions, *C. jejuni* and *C. coli* were inhibited by all of the tested commercial orange oil fractions, but *A. butzleri* was inhibited by only one.

Biofilms are a common mechanism of persistence for a number of bacterial species and several studies have demonstrated that *A. butzleri* is able to survive on various surfaces including stainless steel, glass, and polypropylene. Extended survival times have been reported for *A. butzleri* cells on polypropylene at high humidity, particularly in soiled conditions (Cervenka et al., 2008). Under chiller conditions, Kjeldgaard et al. (2009) demonstrated that *A. butzleri* can form biofilms. However, in a study in a Portuguese slaughterhouse where 36 isolates were tested for biofilm formation ability in microaerobic conditions at 37 °C, 26 were defined as "weakly adherent" and 5 "moderately adherent" while under aerobic conditions, none showed biofilm-forming ability (Ferreira et al., 2013). A further study by the same research group of six *A. butzleri* isolates (three human and three poultry) demonstrated that biofilm-forming ability varied from weak in two isolates, moderate in three, and strong in one. The latter was a poultry isolate that demonstrated strong adhesion to polystyrene surfaces (Ferreira et al., 2014b). Taken together, the ability of at least some strains under certain conditions to form biofilms may allow persistence within a food processing plant (Hausdorf et al., 2013) and explain why *A. butzleri* has a low incidence in chicken gut but a high prevalence in chicken carcasses (Assanta et al., 2002).

In terms of controlling human and animal infections, antibiotics are the treatment of choice. In the studies that have been reported there is a distinct variation in antibiotic susceptibilities reported with different species and even different isolates of the same species having diverse antibiotic susceptibilities, emphasizing the fact that, before any antibiotic treatment is given, a full antibiotic susceptibility profile should be undertaken.

These differences are illustrated by the following studies that have been reported in the last few years. In a study testing 13 antibiotics against 70 *Arcobacter* spp. isolates, the *A. butzleri* isolates were found to be resistant to amoxycillin + clavulonic acid, nalidixic acid, and ampicillin, at the rate of 20%, 44.28%, and 78.57%, respectively, although all *Arcobacter* isolates were susceptible to gentamycin (Abay et al., 2012).

In a slightly smaller study testing 43 isolates from poultry and environmental sources at a slaughterhouse, again all the isolates were susceptible to gentamycin. One strain was resistant to chloramphenicol and 24 isolates (55.8%) were resistant to cipro-oxacin, while all the isolates presented resistance to multiple antibiotics simultaneously, especially to ampicillin, vancomycin, trimethoprim, piperacillin, cefoperazone, and amoxicillin. It was suggested by the authors of this study that this antibiotic resistance pattern could represent a potential health hazard for humans through food chain contamination (Ferreira et al., 2012).

In another relatively small study of 16 Arcobacter isolates (Unver et al., 2013), all proved to be resistant to cloxacillin, cefazolin, optochin, vancomycin, and fusidic acid, and most were susceptible to oxytetracycline, chloramphenicol, nitrofurantoin, amikacin, enrofloxacin, ofloxacin, erythromycin, ampicillin, sulbactam, and amoxicillin. However, the *A. butzleri* isolates were resistant to chloroamphenical and amoxicillin. All *A. skirrowii* and the majority of *A. cryaerophilus* isolates were susceptible to amoxicillin/clavulanic acid.

A high resistance to ampicillin was also found in isolates from edible bivalve mollusks (Collado et al., 2014), in isolates from milk and cheese (Yesilmen et al., 2014) and in isolates from human gastroenteriritis stool specimens (Kayman et al., 2012). In the first case, there was a high resistance to nalidixic acid as well, and in the second a high resistance to cefoperazone, tetracycline, erythromycin, cloxacillin, and penicillin G was reported. However, the human isolates were susceptible to gentamycin, tetracycline, erythromycin, and ciprofloxacin. Overall the results of these studies suggest gentomycin or ampicillin may be a sensible option as a first-line treatment.

7 *ARCOBACTER* AND HUMAN INFECTIONS

7.1 SYMPTOMS

The common symptoms of *Arcobacter* infection are persistent diarrhea accompanied by stomach cramps and abdominal pain (Lerner et al., 1994; Vandamme et al., 1992a), similar to those of campylobacteriosis, and may be transient in nature making infection rates difficult to assess.

7.2 ASSOCIATION WITH HUMAN DISEASE

The first description of a clinical case of *A. cryaerophilus* 1B human infection was in 1988 (Tee et al., 1988). *Arcobacter* spp. have been isolated from both children and adults with diarrhea (Marinescu et al., 1996a; Taylor and Kiehlbauch, 1991) while *A. butzleri* has been recovered from cases of bacteremia. The bacteremia cases include a neonatal patient (On et al., 1995) and one with liver cirrhosis (Yan et al., 2000), as well as from two with severe diarrhea, both of whom were suffering underlying chronic disease (Lerner et al., 1994). A case of a uremic patient with bacteremia caused by *A. cryaerophilus* 1B suggests that the organism is able to produce invasive infections although, once again, the patient concerned had an underlying illness that may have contributed to the progress of *A. cryaerophilus* infection (Hsueh et al., 1997). This was also the situation in the first reported occasion of *A. skirrowii* being isolated from a case of human diarrhea (Wybo et al., 2004).

In 1996, Marinescu et al. (1996b) reported that biotypes and serotypes of *A. butzleri* isolates from poultry in France were similar to those isolated from human diarrheal cases, indicating that infection may have arisen from a food reservoir. However, using PCR-mediated DNA fingerprinting, a single clone was found to be the cause of an outbreak in an Italian school, indicating person-to-person spread may be a feature of outbreaks (Vandamme et al., 1992a, 1993), while ribotyping identified several different strains in a nonhuman primate colony, suggesting multiple sources of infection rather than a common one (Anderson et al., 1993).

The significance of *Arcobacter* spp. as a cause of human illness is, as yet, still uncertain but evidence increased during the 1990s culminating with, in 2002, the International Commission on Microbiological Specifications for Foods classifying *A. butzleri* a serious hazard to human health. Originally, this uncertainty may have been because standard primary screening procedures used for *Campylobacter* spp. do not allow recovery of *Arcobacter* spp. (Marinescu et al., 1996b) and clinical specimens were not, and are not even now, routinely tested for *Arcobacter* spp. A Belgian study that screened all routine stool specimens submitted for detection of enteropathogens between 1991 and 1994 found less than 0.1% positive for *Arcobacter* spp. (8 of 21,527 samples). However, the isolation technique involved a nonselective medium together with microaerobic conditions, which are not the optimum conditions for recovery. Using the same protocol, no *Arcobacter* isolates were obtained from 879 samples from 468 elderly patients during a 1-year study (Lauwers et al., 1996).

A 5-year study in Belgium between 1995 and 2000 isolated 336 non-*jejuni/coli Campylobacter* species from 179 patients of which 24.6% were *A. butzleri*. The isolation method included enrichment in Brucella broth (containing antibiotic supplement and 5% horse blood) for 24 h at 24 °C followed by incubation on *Arcobacter*-selective medium for 3 days at 25 °C, with both steps under microaerobic conditions (Vandenberg et al., 2001), although the conditions do not necessarily give optimal isolation rates. The best isolation rates result from using the filter method onto mCCDA or CAT agar, which was shown by a reevaluation of isolation methods for *Campylobacter* and related organisms, including *Arcobacter*, carried out in Denmark in 2000. Using this protocol, it was suggested that the prevalence in human fecal samples was underestimated (Engberg et al., 2000).

A further study carried out between July 2002 and December 2003 in France, applied phenotypic and molecular methods to identify *A. butzleri* in the 2855 *Campylobacter*-like strains isolated using standard protocols for *Campylobacter* isolation (but not for *Arcobacter*) in the French surveillance network (Prouzet-Mauleon et al., 2006). Of the 29 confirmed *A. butzleri* isolates (isolation frequency 0.4%, fourth most common *Campylobacter*-like organism isolated), 27 were from fecal samples of patients with gastroenteritis, one from peritonitis pus, and one from a blood culture. The patients' clinical symptoms were typical of *Arcobacter* infection as described by Vandamme et al. (1992a). In the majority of cases for which there was additional information (14 of 15) no other enteropathogen was isolated, emphasizing the role that *A. butzleri* in particular has as a cause of human infection. Similar percentage recoveries (0.4%) of *A. butzleri* were reported from 487 diarrheal stool samples in the Netherlands (de Boer et al., 2013) and from 3287 diarrheal stool samples (0.27%) in Turkey (Kayman et al., 2012).

A more recent study in Portugal, which examined 298 human diarrheal samples collected between September and November 2012, by directly extracting the DNA from the fecal samples, followed by fluorescence resonance energy transfer-PCR (FRET-PCR), detected four (2.2%) positive pediatric samples compared with one (0.9%) of the adult samples. Four out of the total five positive samples were *A. butzleri* and one *A. cryaerophilus* (Ferreira et al., 2014a). It is important to note the difference in recovery rates from studies is probably due to a number of factors, including different isolation methods and source(s) of the samples.

7.3 PATHOGENESIS MECHANISMS

Despite many studies investigating the pathogenicity and virulence of *Arcobacter*, the mechanisms whereby the organism causes human or animal disease and the typical symptoms of watery diarrhea and abdominal pain reported in humans are poorly understood. Toxins, adherence to and/or invasion of epithelial cells, hemoglutinins, and interleukins have all been suggested as possible disease triggers. It is probable that one or more of the so far described mechanisms is responsible, and that these mechanisms are dependent on species or even strain.

An *Arcobacter* hemagglutinin was described by Tsang et al. (1996) as an immunogenic protein of about 20 kDa. It is a lectin-like molecule, binding to erythrocytes via a glycan receptor. In 1997, an early study by Musmanno demonstrated that some *A. butzleri* environmental isolates produced a cytotoxic effect in CHO and Vero cells in vitro, but no cytolethal distending activity was detected nor any invasiveness to Hela and Intestine 407 cells. The former result was confirmed later when Johnson and Murano (2002) screened a range of environmental, animal, and human isolates for the presence of CDT genes using PCR and reported that, although no CDT amplimers were observed, toxicity with Int407 cells was seen, suggesting that *Arcobacter* can produce an entity that is cytotoxic but that acts via a different mechanism from the *Campylobacter* CDT. Ninety-five (95%) of 101 *Arcobacter* isolates in a large study in Mexico produced a virulence mechanism against Vero cells. Thirty-eight induced cell elongation, indicating enterotoxin production, 18 induced formation of vacuoles, and 39 produced both vacuolization and elongation. It was suggested that the vacuolization effect may be related to a vacuolizing toxin and that *Arcobacter* spp. may show cytotoxic effects other than the recognized enterotoxin production (Villarruel-Lopez et al., 2003).

Fernandez et al. (1995a) proposed host cell invasion as a means of pathogenicity; other studies suggest that the mechanism is via adherence, especially to epithelial cells (Carbone et al., 2003; Fernandez et al., 2010). Table 5 summarizes the studies on invasive versus adherence mechanisms of pathogenesis.

Table 5 Studies Reporting Invasion and/or Adhesion Mechanisms of Action for Virulence in *A. butzleri*, *A. cryaerophilus*, and *A. skirrowii*

Species	Source (*n*)	Cell Line	Adhesion	Invasion	References
A. butzleri	Chicken meat (3)	HT-29	2/3	1/3	Karadas et al. (2013)
	Human (3)		2/3	1/3	
	Chicken meat (3)	Caco-2	3/3	3/3	
	Human (3)		3/3	3/3	
A. butzleri	Various environmental (73)	HEp-2	73/73	NT	Fernandez et al. (2010)
A. butzleri[a]	Various environmental sources (12)	Caco-2	3/12	3/12	Levican et al. (2013)
A. cryaerophilus	Various environmental sources (5)	Caco-2	2/5	1/5	
A. butzleri	Sea water (17)	HEp-2	6/17	NT	
		HeLa	6/17	NT	Carbone et al. (2003)
A. butzleri	River water (18)	Int407	1/18	0/18	Musmanno et al. (1997)
		HeLa	1/18	0/18	
A. butzleri[b]	Human (1)	Caco-2	1/1	0/1	Ho et al. (2007)
		IPI-2I	1/1	0/1	
A. cryaerophilus	Field isolates from pig farm (4)	Caco-2	4/4	2/4	
A. skirrowii	Lamb feces (1)	IPI-2I	4/4	1/4	
		Caco-2	2/2	0/2	
	Sow amniotic fluid (1)	IPI-2I	2/2	0/2	
A. butzleri	Human (3)	Caco-2	3/3	3/3	Ferreira et al. (2014b)
	Poultry (3)	Caco-2	3/3	3/3	
A. butzleri	Chicken meat	HEp-2	10/27	4/27	Fallas-Padilla et al. (2014)
A. cryaerophilus			5/9	0/9	

[a] *Also tested 13 other species.*
[b] *Also tested* A. cibarius *strain which also showed adherence to Caco-2/IPI-2I.*

Table 6 Percentage of *Arcobacter* Isolates with Putative Virulence Genes

Species	Biological Source (No. Tested)	*cad*F	*cia*B	*Cj*l349	*irg*A	*hec*A	*hec*B	*mni*N	*pld*A	*tly*A	References
A. butzleri	Cattle (14)	100	100	100	NT	NT	NT	100	100	100	Tabatabaei et al. (2014)
	Sheep (11)	100	100	100	NT	NT	NT	100	100	100	
	Chicken (52)	100	100	100	NT	NT	NT	100	100	100	
	Other[a] (36)	100	100	100	NT	NT	NT	100	100	100	
A. cryaerophilus	Cattle (8)	62.5	100	75	NT	NT	NT	100	62.5	62.5	
	Sheep (7)	42.8	100	42.8	NT	NT	NT	100	14.2	28.5	
	Chicken (12)	41.6	91.6	33.3	NT	NT	NT	83.3	25	25	
	Other[a] (13)	69.2	100	38.5	NT	NT	NT	84.6	30.7	38.4	
A. skirrowii	Cattle (2)	50	100	100	NT	NT	NT	100	0	0	
	Sheep (3)	33.3	66.6	66.6	NT	NT	NT	100	33.3	66.6	
	Chicken (5)	60	80	40	NT	NT	NT	60	0	20	
	Other[a] (5)	60	100	60	NT	NT	NT	80	20	75	
A. butzleri	Human (3)	100	100	100	33.3	100	1	100	100	100	Ferreira et al. (2014a)
	Poultry (3)	100	100	100	66.6	100	100	100	100	100	
A. butzleri	Human (6)	100	100	100	17	7	33	100	100	100	Karadas et al. (2013)
	Chicken (23)	100	100	100	9	9	43	100	100	100	
	Other[b] (13)	100	100	100	30.7	30.4	53.8	100	100	100	
A. butzleri	Human (78)	100	100	100	34.6	20.5	67.9	100	100	100	Douidah et al. (2012)
	Chicken (36)	100	100	100	25	22.2	44.4	100	100	100	
	Pig (33)	100	100	100	92.1	18.2	42.4	100	100	100	
	Cattle (29)	100	100	100	50	58.6	86.2	100	100	100	
	Other[c] (6)	100	100	100	0	0	50	100	100	100	
A. cryaerophilus	Human (22)	63.6	90.9	72.7	4.5	9	36.3	100	59.1	54.2	
	Chicken (34)	14.7	97	32.3	2.9	2.9	0	94.1	2.9	14.7	
	Pig (23)	26	91.3	65.2	4.3	0	4.3	91.3	26.1	60.8	
	Cattle (9)	44.4	88.8	33.3	0	11.1	0	77.7	55.59	44.4	
	Other[c] (11)	45.5	90.9	54.5	0	0	9.1	81.8	0.1	0	
A. skirrowii	Human (1)	100	100	100	0	0	0	100	100	100	
	Chicken (0)	NT	NT	NT	NT	NT	NT	NT	NT	NT	
	Pig (13)	15.3	91	15.3	0	0	23	7.6	5.3	38.5	
	Cattle (23)	17.3	100	8.6	0	0	21.7	34.7	13	4.3	
	Other[c] (1)	100	100	100	0	0	0	0	100	100	

[a] *Processing water/equipment of processing line.*
[b] *Pork/minced meat/water.*
[c] *Sheep/horse/dog.*

Other mechanisms that have been suggested are the production of IL-8 (Ferreira et al., 2014b; Ho et al., 2007), which would explain the local inflammation seen in some human cases and via a epithelial barrier dysfunction by changes in tight junction proteins and induction of epithelial apoptosis (Bucker et al., 2009), which would account for the leak flux type of diarrhea seen in *A. butzleri* infection.

Following the publication of the genomic sequence of *A. butzleri* in 2007 (Miller et al., 2007), various authors have described the presence of virulence-associated genes in *Arcobacter* spp. Developing and then using a PCR assay for nine putative *Campylobacter* virulence genes, Douidah et al. (2012) screened a large number of *Arcobacter* isolates from the three species related to human disease (192 *A. butzleri*, 99 *A. cryaerophilus*, 38 *A. skirrowii*). They found that 14.8% of the *A. butzleri* isolates but none of the *A. cryaerophilus* or *A. skirrowii* isolates carried all nine genes. Distribution patterns of the genes in terms of relationship with their source were different for each gene. For example, for *A butzleri*, six showed no relationship while *hec*A (member of the filamentous hemagglutinin gene family) was more common in cattle than in humans, pigs, and chickens, and *irg*A (iron regulated outer membrane protein) less represented in pig strains than in human and cattle strains. Other studies have confirmed these findings (Table 6). Levican et al. (2013) tested the 17 species described by 2013 for invasion or adherence properties to Caco-2 cells. *A. butzleri*, *A. cryaerophilus*, *A. skirrowii*, and *A. defluvii* were the most invasive species and, using the Douidah et al. (2010) PCR method, all the invasive strains were positive for *cia*B (encoding the putative invasion protein) and, conversely, no virulence genes were detected in strains that showed little or no invasion of Caco-2 cells. Although other putative genes were present, this suggests that the *cia*B gene is probably essential for the determination of virulence.

8 CONCLUSION

The significance of *Arcobacter* spp. as a human pathogen is not fully evaluated at present although, considering its isolation in cases of human and animal illness and from foods of animal origin, it certainly may be added to the ranks of emerging foodborne pathogens. With the development and use of specific and robust isolation, detection, and identification techniques, accurate information will be accumulated on its epidemiology and occurrence, whether in the environment, in food, or in humans.

REFERENCES

Abay, S., Kayman, T., Hizlisoy, H., Aydin, F., 2012. In vitro antibacterial susceptibility of *Arcobacter butzleri* isolated from different sources. J. Vet. Med. Sci. 74, 613–616.

Agidi, S., Lee, C., Lee, J., Marion, J.W., 2012. *Arcobacter* in Lake Erie beach waters: an emerging gastrointestinal pathogen linked with human-associated fecal contamination. Appl. Environ. Microbiol. 78, 5511–5520.

Ahmed, R., Balamurugan, S., 2013. Evaluation of three *Arcobacter* selective agars for selective enumeration of *Arcobacter butzleri* in beef. Food Res. Int. 52, 522–525.

Al Rashid, S.T., Dakuna, I., Louie, H., Ng, D., Vandamme, P., Johnson, W., Chen, V.L., 2000. Identification of *Campylobacter jejuni*, *C. coli*, *C. lari*, *C. upsaliensis*, *Arcobacter butzleri*, and *A. butzleri*-like species based on the *glyA* gene. J. Clin. Microbiol. 38, 1488–1494.

Alonso, R., Girbau, C., Martinez-Malaxetxebarria, I., Fernández-Astorga, A., 2014. Multilocus sequence typing reveals genetic diversity of foodborne *Arcobacter butzleri* isolates in the North of Spain. Int. J. Food Microbiol. 191, 125–128.

Anderson, K.F., Kiehlbauch, J.A., Anderson, D.C., McClure, H.M., Wachsmuth, I.K., 1993. *Arcobacter-(Campylobacter)-butzleri*-associated diarrheal illness in a non human primate population. Infect. Immun. 61, 2220–2223.

Assanta, M.A., Roy, D., Lemay, M., Montpetit, D., 2002. Attachment of *Arcobacter butzleri*, a new waterborne pathogen, to water distribution pipe surfaces. J. Food Prot. 65, 1240–1247.

Atabay, H.I., Aydin, F., 2001. Susceptibility of *Arcobacter butzleri* isolates to 23 antimicrobial agents. Lett. Appl. Microbiol. 33, 430–433.

Atabay, H., Corry, J., 1997. The prevalence of campylobacters and arcobacters in broiler chickens. J. Appl. Microbiol. 83, 619–626.

Atabay, H.I., Corry, J.E.L., 1998. The isolation and prevalence of campylobacters from dairy cattle using a variety of methods. J. Appl. Microbiol. 84, 733–740.

Atabay, H., Corry, J., Ceylan, C., 2014. Bacteria: *Arcobacter*. In: Motarjemi, Y. (Ed.), Encyclopedia of Food Safety. Academic Press, Waltham, pp. 344–347.

Atanassova, V., Kessen, V., Reich, F., Klien, G., 2008. Incidence of *Arcobacter* spp. in poultry: quantitative and qualitative analysis and PCR differentiation. J. Food Prot. 71, 2384–2595.

Aydin, F., Gumussoy, K.S., Atabay, H.I., Ica, T., Abay, S., 2007. Prevalence and distribution of *Arcobacter* species in various sources in Turkey and molecular analysis of isolated strains by ERIC-PCR. J. Appl. Microbiol. 103, 27–35.

Bastyns, K., Cartuyvels, D., Chapelle, S., Vandamme, P., Goossens, H., DeWachter, R., 1995. A variable 23S rDNA region is a useful discriminating target for genus-specific and species-specific PCR amplification in *Arcobacter* species. Syst. Appl. Microbiol. 18, 353–356.

Bolton, F.J., Wareing, D.R.A., Skirrow, M.B., Hutchinson, D.N., 1992. Identification and biotyping of campylobacters. In: Board, R.G., Jones, D., Skinner, F.A. (Eds.), Identification Methods in Applied and Environmental Microbiology. Blackwell Sciences, United Kingdom, pp. 151–161.

Brightwell, G., Mowat, E., Clemens, R., Boerema, J., Pulford, D.J., On, S.L., 2007. Development of a multiplex and real time PCR assay for the specific detection of *Arcobacter butzleri* and *Arcobacter cryaerophilus*. J. Microbiol. Methods 68, 318–325.

Bucker, R., Troeger, H., Kleer, J., Fromm, M., Schulzke, J.D., 2009. *Arcobacter butzleri* induces barrier dysfunction in intestinal HT-29/B6 cells. J. Infect. Dis. 200, 756–764.

Burnens, A.P., Schaad, U.B., Nicolet, J., 1992. Isolation of *Arcobacter butzleri* from a girl with gastroenteritis on Yersinia CIN agar. Med. Microbiol. Lett. 1, 251–256.

Burt, S., 2004. Essential oils: their antibacterial properties and potential applications in food – a review. Int. J. Food Microbiol. 94, 223–253.

Cappelier, J., Magras, C., Jouve, J., Federighi, M., 1999. Recovery of viable but non-culturable *Campylobacter jejuni* cells in two animal models. Food Microbiol. 16, 373–383.

Carbone, M., Maugeri, T.L., Giannone, M., Gugliandolo, C., Midiri, A., Fera, M.T., 2003. Adherence of environmental *Arcobacter butzleri* and *Vibrio* spp. isolates to epithelial cells in vitro. Food Microbiol. 20, 611–616.

Cervenka, L., Kristlova, J., Peskova, I., Vytrasova, J., Pejchalova, M., Brozkova, I., 2008. Persistence of *Arcobacter butzleri* CCUG 30484 on plastic, stainless steel and glass surfaces. Braz. J. Microbiol. 39, 517.

Collado, L., Figueras, M.J., 2011. Taxonomy, epidemiology and clinical relevance of the genus *Arcobacter*. Clin. Microbiol. Rev. 24, 174–192.

Collado, L., Cleenwick, I., Van Trappen, S., De Vos, P., Figueras, M.J., 2009. *Arcobacter mytili* sp. nov. an indoxyl acetate-hydrolysis-negative bacterium isolated from mussels. Int. J. Syst. Evol. Microbiol. 59, 1391–1396.

Collado, L., Kasimir, G., Perez, U., Bosch, A., Pinto, R., Saucedo, G., Huguet, J.M., Figueras, M.J., 2010. Occurrence and diversity of *Arcobacter* spp. along the Llobregat River catchment, at sewage effluents and in a drinking water treatment plant. Water Res. 44, 3696–3702.

Collado, L., Jara, R., Vásquez, N., Telsaint, C., 2014. Antimicrobial resistance and virulence genes of *Arcobacter* isolates recovered from edible bivalve molluscs. Food Control 46, 508–512.

Collins, C.I., Murano, E.A., Wesley, I.V., 1996a. Survival of *Arcobacter butzleri* and *Campylobacter jejuni* after irradiation treatment in vacuum-packaged ground pork. J. Food Prot. 59, 1164–1166.

Collins, C.I., Wesley, I.V., Murano, E.A., 1996b. Detection of *Arcobacter* spp. in ground pork by modified plating methods. J. Food Prot. 59, 448–452.

de Boer, E., Tilberg, J., Woodward, D., Lior, H., Johnson, W., 1996. A selective medium for the isolation of *Arcobacter* from meats. Lett. Appl. Microbiol. 23, 64–66.

de Boer, R., Ott, A., Guren, P., van Zanten, E., van Belkum, A., Kooistra-Smid, A., 2013. Detection of *Campylobacter* species and *Arcobacter butzleri* in stool samples by use of real-time multiplex PCR. J. Clin. Microbiol. 51, 253–259.

de Oliveira, S.J., Baetz, A.L., Wesley, I.V., Harmon, K.M., 1997. Classification of *Arcobacter* species isolated from aborted pig fetuses and sows with reproductive problems in Brazil. Vet. Microbiol. 57, 347–354.

de Oliveria, S.J., Wesley, I.V., Baetz, A.L., Harmon, K.M., Kader, I., de Uzeda, M., 1999. *Arcobacter cryaerophilus* and *Arcobacter butzleri* isolated from preputial fluid of boars and fattening pigs in Brazil. J. Vet. Diagn. Invest. 11, 462–464.

De Smet, S., De Zutter, L., Van Hende, J., Houf, K., 2010. *Arcobacter* contamination on pre- and post-chilled bovine carcasses and in minced beef at retail. J. Appl. Microbiol. 108, 299–305.

De Smet, S., De Zutter, L., Houf, K., 2011a. Small ruminants as carriers of the emerging foodborne pathogen *Arcobacter* on small and medium farms. Small Rumin. Res. 97, 124–129.

De Smet, S., Vandamme, P., De Zutter, L., On, S., Douidah, L., Houf, K., 2011b. *Arcobacter trophiarum* sp. nov. isolated from fattening pigs. Int. J. Syst. Evol. Microbiol. 63, 356–361.

Dhamabutra, N., Kamol-Rathanakul, P., Pienthaweechai, K., 1992. Isolation of campylobacters form the canals in Bangkok. J. Med. Assoc. Thail. 75, 350–363.

Donachie, S.P., Bowman, J.P., On, S.L.W., Alam, M., 2005. *Arcobacter halophilus* sp nov., the first obligate halophile in the genus Arcobacter. Int. J. Syst. Evol. Microbiol. 55, 1271–1277.

Douidah, L., De Zutter, L., Vandamme, P., Houf, K., 2010. Identification of five human and mammal associated *Arcobacter* species by a novel multiplex-PCR assay. J. Microbiol. Methods 80, 281–286.

Douidah, L., De Zutter, L., Baré, J., De Vos, P., Vandamme, P., Vandenburg, O., Vanden Abeele, A., Houf, K., 2012. Occurrence of putative virulence genes in *Arcobacter* species isolated from humans and animals. J. Clin. Microbiol. 50, 735–742.

Eifert, J.D., Castle, R.M., Pierson, F.W., Larsen, C.T., Hackney, C.R., 2003. Comparison of sampling techniques for detection of *Arcobacter butzleri* from chickens. Poult. Sci. 82, 1898–1902.

Engberg, J., On, S.L.W., Harrington, C.S., Gerner-Smith, P., 2000. Prevalence of *Campylobacter*, *Arcobacter*, *Helicobacter* and Sutterella spp. in human fecal samples as estimated by a reevaluation of isolation methods for Campylobacters. J. Clin. Microbiol. 38, 286–291.

Fallas-Padilla, K.L., Rodriguez-Rodriguez, C.E., Jaramillo, H.E., Echandi, M.L.A., 2014. Arcobacter: comparison of isolation methods, diversity and potential pathogenic factors in commercially retailed chicken breast meat from Costa Rica. J. Food Prot. 77, 880–884.

Fera, M.T., Maugeri, T.L., Gugliandolo, C., Beninati, C., Giannone, M., La Camera, E., Carbone, M., 2004. Detection of *Arcobacter* spp. in the coastal environment of the Mediterranean sea. Appl. Environ. Microbiol. 70, 1271–1276.

Fera, M.T., Maugeri, T.L., Gugliandolo, C., La Camera, E., Lentini, V., Favaloro, A., Bonanno, D., Carbone, M., 2008. Induction and resuscitation of viable nonculturable *Arcobacter butzleri* cells. Appl. Environ. Microbiol. 74, 3266–3268.

Fera, M.T., La Camera, E., Carbone, M., Malara, D., Pennisi, M.G., 2009. Pet cats as carriers of *Arcobacter* spp. in Southern Italy. J. Appl. Microbiol. 106, 1661–1665.

Fera, M.T., Gugliandolo, C., Lentini, V., Favaloro, A., Bonanno, D., La Camera, E., Maugeri, T.L., 2010. Specific detection of *Arcobacter* spp. in estuarine waters of Southern Italy by PCR and fluorescent in situ hybridization. Lett. Appl. Microbiol. 50, 65–70.

Fernandez, H., Eller, G., Paillacar, J., Gajardo, T., Riquelme, A., 1995a. Toxigenic and invasive capacities – possible pathogenic mechanisms in *Arcobacter cryaerophilus*. Mem. Inst. Oswaldo Cruz 90, 633–634.

Fernandez, H., Rojas, X., Gajardo, T., 1995b. First isolation in Chile of *Arcobacter cryaerophilus* from a bovine abortion. Arch. Med. Vet. 27, 111–114.

Fernandez, H., Flores, S., Inzunza, F., 2010. *Arcobacter butzleri* strains isolated from different sources display adhesive capacity to epithelial cells in vitro. Acta Sci. Vet. 38, 287–291.

Ferreira, S., Fraqueza, M.J., Queiroz, J., Domingues, F., Oleastro, M., 2012. Genetic diversity and antibiotic resistance of *Arcobacter butzleri* isolated from poultry and slaughterhouse environment in Portugal. In: Central European Symposium on Antimicrobials and Antimicrobial Resistance, 23–26 September 2012 Book of Abstracts 2010: 97. Croatian Microbiological Society, Primošten, Croatia, p. 40.

Ferreira, S., Fraqueza, M.J., Queiroz, J.A., Domingues, F.C., Oleastro, M., 2013. Genetic diversity, antibiotic resistance and biofilm-forming ability of *Arcobacter butzleri* isolated from poultry and environment from a Portuguese slaughterhouse. Int. J. Food Microbiol. 162, 82–88.

Ferreira, S., Júlio, C., Queiroz, J.A., Domingues, F.C., Oleastro, M., 2014a. Molecular diagnosis of *Arcobacter* and *Campylobacter* in diarrhoeal samples among Portuguese patients. Diagn. Microbiol. Infect. Dis. 78, 220–225.

Ferreira, S., Queiroz, J.A., Oleastro, M., Domingues, F.C., 2014b. Genotypic and phenotypic features of *Arcobacter butzleri* pathogenicity. Microb. Pathog. 76, 19–25.

Ferreira, S., Silva, F., Queiroz, J.A., Oleastro, M., Domingues, F.C., 2014c. Resveratrol against *Arcobacter butzleri* and *Arcobacter cryaerophilus*: activity and effect on cellular functions. Int. J. Food Microbiol. 180, 62–68.

Figueras, M.J., Collado, L., Guarro, J., 2008. A new 16S rDNA-RFLP method for the discrimination of the accepted species of *Arcobacter*. Diagn. Microbiol. Infect. Dis. 62, 11–15.

Figueras, M.J., Collado, L., Levican, A., Perez, J., Solsona, M.J., Yustes, C., 2011a. *Arcobacter molluscorum* sp nov., a new species isolated from shellfish. Syst. Appl. Microbiol. 34, 105–109.

Figueras, M.J., Levican, A., Collado, L., Inza, M.I., Yustes, C., 2011b. *Arcobacter ellisii* sp. nov., isolated from mussels. Syst. Appl. Microbiol. 34, 414–418.

Figueras, M.J., Arturo, L., Luis, C., 2012. Updated 16S rRNA-RFLP method for the identification of all currently characterised *Arcobacter* spp. BMC Microbiol. 12, 292.

Fisher, K., Rowe, C., Phillips, C.A., 2007. The survival of three strains of *Arcobacter butzleri* in the presence of lemon, orange and bergamot essential oils and their components in vitro and on food. Lett. Appl. Microbiol. 44, 495–499.

Gilgen, M., Hubner, P., Hofelein, C., Luthy, J., Candrian, U., 1998. PCR-based detection of verotoxin-producing *Escherichia coli* (VTEC) in ground beef. Res. Microbiol. 149, 145–154.

Gonzalez, A., Ferrus, M.A., 2011. Study of *Arcobacter* spp. contamination in fresh lettuces detected by different cultural and molecular methods. Int. J. Food Microbiol. 145, 311–314.

Gonzalez, I., Garcia, T., Antolin, A., Hernandez, P.E., Martin, R., 2000. Development of a combined PCR-culture technique for the rapid detection of *Arcobacter* spp. in chicken meat. Lett. Appl. Microbiol. 30, 207–212.

Gonzalez, A., Suski, J., Ferrus, M.A., 2010. Rapid and accurate detection of *Arcobacter* contamination in commercial chicken products and waste water samples by real-time polymerase chain reaction. Foodborne Pathog. Dis. 7, 327–338.

Gonzalez, I., Garcia, T., Fernandez, S., Martin, R., 2012. Current status on Arcobacter research: an update on DNA-based identification and typing methodologies. Food Anal. Methods 5, 956–968.

Gude, A., Hillman, T.J., Helps, C.R., Allen, V.M., Corry, J.E.L., 2005. Ecology *of Arcobacter* species in chicken rearing and processing. Lett. Appl. Microbiol. 41, 82–87.

Harmon, K.M., Wesley, I.V., 1996. Identification of *Arcobacter* isolates by PCR. Lett. Appl. Microbiol. 23, 241–244.

Harmon, K.M., Wesley, I.V., 1997. Multiplex PCR for the identification of *Arcobacter* and differentiation of *Arcobacter butzleri* from other arcobacters. Vet. Microbiol. 58, 215–227.

Harrab, B., Schwarz, S., Wenzel, S., 1998. Identification and characterisation of *Arcobacter* isolates from broilers by biochemical tests, antimicrobial resistance patterns and plasmid analysis. J. Vet. Med. Ser. B Infect. Dis. Vet. Public Health 45, 87–94.

Harvey, R.B., Anderson, R.C., Young, C.R., Hume, M.E., Genovese, K.J., Ziprin, R.L., Farrington, L.A., Stanker, L.H., Nisbet, D.J., 1999. Prevalence of *Campylobacter*, *Salmonella*, *Arcobacter* species at slaughter in market age pigs. Adv. Exp. Med. Biol. 473, 237–239. In: Paul and Francis (Eds.) Mechanisms in the Pathogenesis of Enteric Diseases 2, Springer.

Hausdorf, L., Neumann, M., Bergmann, I., Sobiella, K., Mundt, K., Frohling, A., Schluter, O., Klocke, M., 2013. Occurrence and genetic diversity of *Arcobacter* spp. in a spinach-processing plant and evaluation of two *Arcobacter*-specific quantitative PCR assays. Syst. Appl. Microbiol. 36, 235–243.

Ho, H.T., Lipman, L.J., Gaastra, W., 2006. *Arcobacter:* what is known and unknown about a potential foodborne zoonotic agent! Vet. Microbiol. 115, 1–13.

Ho, H.T.K., Lipman, L.J.A., Hendricks, H.G.C., Tooten, P.C.J., Ultee, T., Gaastra, W., 2007. Interaction of *Arcobacter* spp. with human and porcine intestinal epithelial cells. FEMS Immunol. Med. Microbiol. 50, 51–58.

Ho, H.T.K., Lipman, L.J.A., Gaastra, W., 2008. The introduction of *Arcobacter* spp. in poultry slaughterhouses. Int. J. Food Microbiol. 125, 223–229.

Houf, K., Stephan, R., 2007. Isolation and characterization of the emerging foodborne pathogen *Arcobacter* from human stool. J. Microbiol. Methods 68, 408–413.

Houf, K., Tutenel, A., De Zutter, L., Van Hoof, J., Vandamme, P., 2000. Development of a multiplex PCR assay for the simultaneous detection and identification of *Arcobacter butzleri*, *Arcobacter cryaerophilus* and *Arcobacter skirrowii*. FEMS Microbiol. Lett. 193, 89–94.

Houf, K., Devriese, L.A., Zutter, L.D., Hoof, J.V., Vandamme, P., 2001. Development of a new protocol for the isolation and quantification of *Arcobacter* species from poultry products. Int. J. Food Microbiol. 71, 189–196.

Houf, K., De Zutter, L., Van Hoof, J., Vandamme, P., 2002. Assessment of the genetic diversity among Arcobacters isolated from poultry products by using two PCR-based typing methods. Appl. Environ. Microbiol. 68, 2172–2178.

Houf, K., Devriese, L.A., Haesebrouck, F., Vandenberg, O., Butzler, J.P., Van Hoof, J., Vandamme, P., 2004. Antimicrobial susceptibility patterns of *Arcobacter butzleri* and *Arcobacter cryaerophilus* strains isolated from humans and broilers. Microb. Drug Resist. Dis. 10, 243–247.

Houf, K., On, S.L.W., Coenye, T., Mast, J., Van Hoof, J., Vandamme, P., 2005. *Arcobacter cibarius* sp nov., isolated from broiler carcasses. Int. J. Syst. Evol. Microbiol. 55, 713–717.

Houf, K., On, S.L.W., Coenye, T., Debruyne, L., Smet, F.D., Vandamme, P., 2009. *Arcobacter thereius* sp. nov. isolated from pigs and ducks. Int. J. Syst. Evol. Microbiol. 59, 2599–2604.

Hsueh, P., Teng, L., Yang, P., Wang, S., Chang, S., Ho, S., Hsieh, W., Luh, K., 1997. Bacteremia caused by *Arcobacter cryaerophilus 1B*. J. Clin. Microbiol. 35, 489–491.

Hume, M.E., Harvey, R.B., Stanker, L.H., Droleskey, R.E., Poole, T.L., Zhang, H.B., 2001. Genotypic variation among *Arcobacter* isolates from a farrow-to-finish swine facility. J. Food Prot. 64, 645–651.

Hurtado, A., Owen, R.J., 1997. A molecular scheme based on 23S rRNA gene polymorphisms for rapid identification of *Campylobacter* and *Arcobacter* species. J. Clin. Microbiol. 35, 2401–2404.

Isohanni, P., Huehn, S., Aho, T., Alter, T., Lyhs, U., 2013. Heat stress adaptation induces cross-protection against lethal acid stress conditions in *Arcobacter butzleri* but not in *Campylobacter jejuni*. Food Microbiol. 34, 431–435.

Jacob, J., Lior, H., Feuerpfeil, I., 1993. Isolation of *Arcobacter-Butzleri* from a drinking-water reservoir in Eastern Germany. Zentralbl. Hyg. Umweltmed. 193, 557–562.

Jacob, J., Woodward, D., Feuerpfeil, I., Johnson, W.M., 1998. Isolation of *Arcobacter butzleri* in raw water and drinking water treatment plants in Germany. Zentralbl. Hyg. Umweltmed. 201, 189–198.

Johnson, L.G., Murano, E.A., 1999a. Development of a new medium for the isolation of *Arcobacter* spp. J. Food Prot. 62, 456–462.

Johnson, L.G., Murano, E.A., 1999b. Comparison of three protocols for the isolation of *Arcobacter* from poultry. J. Food Prot. 62, 610–614.

Johnson, L.G., Murano, E.A., 2002. Lack of a cytolethal distending toxin among *Arcobacter* isolates from various sources. J. Food Prot. 65, 1789–1795.

Kabeya, H., Kobayashi, Y., Maruyama, S., Mikami, T., 2003. One-step polymerase chain reaction-based typing of *Arcobacter* species. Int. J. Food Microbiol. 81, 163–168.

Karadas, G., Sharbati, S., Hanel, I., Messelhausser, U., Glocker, E., Alter, T., Golz, G., 2013. Presence of virulence genes, adhesion and invasion *of Arcobacter butzleri*. J. Appl. Microbiol. 115, 583–590.

Kayman, T., Abay, S., Hizlisoy, H., Atabay, H.I., Diker, K.S., Aydin, F., 2012. Emerging pathogen *Arcobacter* spp. in acute gastroenteritis: molecular identification, antibiotic susceptibilities and genotyping of the isolated arcobacters. J. Med. Microbiol. 61, 1439–1444.

Keene, W.E., McAnulty, J.M., Hoesly, F.C., Williams, L.P., Hedberg, K., Oxman, G.L., Barrett, T.J., Pfaller, M.A., Fleming, D.W., 1994. A swimming-associated outbreak of hemorrhagic colitis caused by *Escherichia coli O157:H7* and *Shigella sonnei*. N. Engl. J. Med. 331, 579–584.

Kiehlbauch, J.A., Plikaytis, B.D., Swaminathan, B., Cameron, D.N., Wachsmuth, I.K., 1991a. Restriction fragment length polymorphisms in the ribosomal genes for species identification and subtyping of aerotolerant *Campylobacter* species. J. Clin. Microbiol. 29, 1670–1676.

Kiehlbauch, J.A., Brenner, D.J., Nicholson, M.A., Baker, C.N., Patton, C.M., Steigerwait, A.G., Wachsmuth, I.K., 1991b. *Campylobacter butzleri* sp. nov. isolated from humans and animals with diarrheal illness. J. Clin. Microbiol. 29, 376–385.

Kiehlbauch, J.A., Cameron, D.N., Wachsmuth, I.K., 1994. Evaluation of ribotyping techniques as applied to *Arcobacter*, *Campylobacter* and *Helicobacter*. Mol. Cell. Probes 8, 109–116.

Kim, H.M., Hwang, C.Y., Cho, B.C., 2010. *Arcobacter marinus* sp. nov. Int. J. Syst. Evol. Microbiol. 60, 531–536.

Kjeldgaard, J., Jorgensen, K., Ingmer, H., 2009. Growth and survival at chiller temperatures of *Arcobacter butzleri*. Int. J. Food Microbiol. 131, 256–259.

Lammerding, A.M., Harris, J.E., Lior, H., Woodward, D.E., Cole, L., Muckle, C.A., 1996. Isolation method for recovery of *arcobacter butzleri* from fresh poultry and poultry products. In: Newell et al. (Eds)., Campylobacters, Helicobacters and Related Organisms. Springer, pp. 329–333.

Lauwers, S., Breynaert, J., van Etterijck, R., Revets, H., Mets, T., 1996. *Arcobacter butzleri* in the elderly in Belgium. In: Newell et al. (Eds) Campylobacters, Helicobacters and Related Organisms. Springer, pp. 515–518.

Lerner, J., Brumberger, V., Preac-Mursic, V., 1994. Severe diarrhoea associated with *A. butzleri*. Eur. J. Clin. Microbiol. Infect. Dis. 13, 660–662.

Levican, A., Figueras, M.J., 2013. Performance of five molecular methods for monitoring *Arcobacter* spp. BMC Microbiol. 13, 220.

Levican, A., Collado, L., Aguilar, C., Yustes, C., Diéguez, A.L., Romalde, J.L., Figueras, M.J., 2012. *Arcobacter bivalviorum* sp. nov. *and Arcobacter venerupis* sp. nov., new species isolated from shellfish. Syst. Appl. Microbiol. 35, 133–138.

Levican, A., Collado, L., Figueras, M.J., 2013. *Arcobacter cloacae* sp. nov. and *Arcobacter suis* sp. nov., two new species isolated from food and sewage. Syst. Appl. Microbiol. 36, 22–27.

Lior, H., 1984. New, extended biotyping scheme for *Campylobacter jejuni*, *Campyobacter coli* and *Campylobacter laridis*. J. Clin. Microbiol. 20, 636–640.

Lior, H., Woodward, D., 1991. A serotyping scheme for *Campylobacter butzleri*. In: Borriello, S.P., Rolfe, R.D., Hardie, J.M. (Eds.), Microbial Ecology in Health and Disease. Proceedings of the IVth Conference on Campylobacters, Helicobacters and Related Organisms. Wiley, p. S93.

Lipman, L., Ho, H., Gaastra, W., 2008. The presence of *Arcobacter* species in breeding hens and eggs from these hens. Poult. Sci. 87, 2404–2407.

Long, C., Phillips, C.A., 2003. The effect of sodium citrate, sodium lactate and nisin on the survival of *Arcobacter butzleri* NCTC 12481 on chicken. Food Microbiol. 20, 495–502.

Manke, T.R., Wesley, I.V., Dickson, J.S., Harmon, K.M., 1998. Prevalence and genetic variability of *Arcobacter* species in mechanically separated turkey. J. Food Prot. 61, 1623–1628.

Mansfield, L.P., Forsythe, S.J., 2000. *Arcobacter butzleri, A-skirrowii and A-cryaerophilus* – potential emerging human pathogens. Rev. Med. Microbiol. 11, 161–170.

Marinescu, M., Collignon, A., Squinazi, F., Derimay, R., Woodward, D., Lior, H., 1996a. Two cases of persistent diarrhoea associated with *Arcobacter* spp. In: Newell, D.G., Ketley, J.M., Feldman, R.A. (Eds.), Campylobacters, Helicobacters and Related Organisms. Plenum Press, New York, pp. 521–523.

Marinescu, M., Collignon, A., Squinazi, F., Woodward, D., Lior, H., 1996b. Biotypes and serogroups of poultry strains of *Arcobacter* spp isolated in France. In: Newell, D.G., Ketley, J.M., Feldman, R.A. (Eds.), Campylobacters, Helicobacters and Related Organisms. Plenum Press, New York, pp. 519–520.

Marshall, S.M., Melito, P.L., Woodward, D.L., Johnson, W.M., Rodgers, F.G., Mulvey, M.R., 1999. Rapid identification of *Campylobacter*, *Arcobacter* and *Helicobacter* isolates by PCR-restriction fragment length polymorphism analysis of the 16S rRNA gene. J. Clin. Microbiol. 37, 4158–4160.

Maugeri, T.L., Gugliandolo, C., Carbone, M., Fera, M.T., 2000. Isolation of *Arcobacter* spp. from a brackish environment. Microbiologica 23, 143–149.

McClung, C.R., Patriquin, D.G., 1980. Isolation of a nitrogen-fixing *Campylobacte*r species from the roots of *Spartina alterniflora Loisel.* Can. J. Microbiol. 26, 881–886.

Merga, J.Y., Leatherbarrow, A.J., Winstanley, C., Bennett, M., Hart, C.A., Miller, W.G., Williams, N.J., 2011. Comparison of *Arcobacter* isolation methods, and diversity of *Arcobacter* spp. in Cheshire, United Kingdom. Appl. Environ. Microbiol. 77, 1646–1650.

Merga, J.Y., Williams, N.J., Miller, W.G., Leatherbarrow, A., Bennett, M., Hall, N., Ashelford, K.E., Winstanley, C., 2013. Exploring the diversity of *Arcobacter butzleri* from cattle in the UK using MLST and whole genome sequencing. PLoS One 8, Article Number: e55240.

Merga, J.Y., Royden, A., Pandey, A.K., Williams, N.J., 2014. *Arcobacter* spp. isolated from untreated domestic effluent. Lett. Appl. Microbiol. 59, 122–126.

Miller, W.G., Parker, C.T., Rubenfield, M., Mendz, G.L., Wösten, M.M.S.M., Ussery, D.W., Stolz, J.F., Binnewies, T.T., Hallin, P.F., Wang, G., Malek, J.A., Rogosin, A., Stanker, L.H., Mandrell, R.E., 2007. The complete genomic sequence and analysis of the epsilonproteobacterium *Arcobacter butzleri*. PLoS One 12, 358.

Miller, W.G., Wesley, I.V., On, S.L.W., Houf, K., Megraud, F., Wang, G., Yee, E., Srijan, A., Mason, C.J., 2009. First multi-locus sequence typing scheme for *Arcobacter* spp. BMC Microbiol. 9, 196.

Moreno, Y., Alonso, J.L., Botella, S., Ferrus, M.A., Hernandez, J., 2004. Survival and injury of *Arcobacter* after artificial inoculation into drinking water. Res. Microbiol. 155, 726–730.

Musmanno, R., Russi, M., Lior, H., Figura, N., 1997. In vitro virulence factors of *Arcobacter butzleri* strains isolated from superficial water samples. Microbiologica 20, 63–68.

Nannapaneni, R., Chalova, V.I., Crandall, P.G., Ricke, S.C., Johnson, M.G., O'Bryan, C.A., 2009. *Campylobacter* and *Arcobacter* species sensitivity to commercial orange oil fractions. Int. J. Food Microbiol. 129, 43–49.

Neill, S.D., Ellis, W.A., O'Brien, J.J., 1978. The biochemical characteristics of *Campylobacter*-like organisms from cattle and pigs. Res. Vet. Sci. 25, 368–372.

Neill, S.D., Ellis, W.A., O'Brien, J.J., 1979. Designation of aerotolerant *Campylobacter*-like organisms from porcine and bovine abortions to the genus *Campylobacter*. Res. Vet. Sci. 29, 180–186.

Neill, S.D., Campbell, J.N., O'Brien, J.J., Weatherup, S.T., Ellis, W.A., 1985. Taxonomic position of *Campylobacter cryaerophila* sp. nov. Int. J. Syst. Bacteriol. 35, 342–356.

On, S.L.W., Harrington, C.S., 2001a. Identification of taxonomic and epidemiological relationships among *Campylobacter* species by numerical analysis of AFLP profiles. FEMS Microbiol. Lett. 193, 161–169.

On, S.L.W., Harrington, C.S., 2001b. Concurrent species identification and high resolution genotyping of members of the family *Campylobacteraceae* by AFLP profiling. Int. J. Med. Microbiol. 291, 87.

On, S.L.W., Holmes, B., 1995. Classification and identification of Campylobacters, Helicobacters and allied taxa by numerical analysis of phenotypic characters. Syst. Appl. Microbiol. 18, 374–390.

On, S.L.W., Stacey, A., Smyth, J., 1995. Isolation of *Arcobacter butzleri* from a neonate with bacteraemia. J. Infect. 31, 133–138.

On, S.L.W., Jensen, T.K., Bille-Hansen, V., Jorsal, S.E., Vandamme, P., 2002. Prevalence and diversity of *Arcobacter* spp. isolated from the internal organs of spontaneous porcine abortions in Denmark. Vet. Microbiol. 85, 159–167.

On, S.L.W., Harrington, C.S., Atabay, H.I., 2003. Differentiation *of Arcobacter species* by numerical analysis of AFLP profiles and description of a novel *Arcobacter* from pig abortions and turkey faeces. J. Appl. Microbiol. 95, 1096–1105.

On, S.L.W., Atabay, H.I., Amisu, K.O., Coker, A.O., Harrington, C.S., 2004. Genotyping and genetic diversity of *Arcobacter butzleri* by amplified fragment length polymorphism (AFLP) analysis. Lett. Appl. Microbiol. 39, 347–352.

Parvanta, M.F., 1999. *Campylobacter cryaerophila* and *Campylobacter fetus subspecies venerealis* as a cause of serial abortions in two cattle herds in North-Rhine-Westfalia. Tierarztl. Umsch. 54, 364–371.

Peterson, J.J., Beecher, G.R., Bhagwat, A.A., Dwyer, J.T., Gebbhardt, S.E., Haytowitz, D.B., Holden, J.M., 2006. Flavanones in grapefruit, lemons and limes: a compilation and review of the data from the analytical literature. J. Food Compos. Anal. 19, S74–S80.

Phillips, C.A., 1999. The effect of citric acid, lactic acid, sodium citrate and sodium lactate, alone and in combination with nisin, on the growth of *Arcobacter butzleri*. Lett. Appl. Microbiol. 29, 424–428.

Phillips, C.A., 2001. Arcobacters as emerging foodborne pathogens. Food Control 12, 1–6.

Phillips, C.A., Duggan, J., 2001. The effect of EDTA and trisodium phosphate, alone and in combination with nisin, on the growth of *Arcobacter butzleri* in culture. Food Microbiol. 18, 547–554.

Phillips, C.A., Duggan, J., 2002. The effect of temperature and citric acid, alone and in combination with nisin, on the growth of *Arcobacter butzleri* in culture. Food Control 13, 463–468.

Phillips, C.A., Long, C., 2001. The survival of *Arcobacter butzleri* on chicken. Int. J. Med. Microbiol. 291, 93.

Prouzet-Mauleon, V., Labadi, L., Bouges, N., Menard, A., Megraud, F., 2006. *Arcobacter butzleri*: underestimated enteropathogen. Emerg. Infect. Dis. 12, 307–309.

Rasmussen, L.H., Kjeldgaard, J., Christensen, J.P., Ingmer, H., 2013. Multilocus sequence typing and biocide tolerance of *Arcobacter butzleri* from Danish broiler carcasses. BMC Res. Notes 13, 322.

Rice, E.W., Rodgers, M.R., Wesley, I.V., Johnson, C.H., Tanner, S.A., 1999. Isolation of *Arcobacter butzleri* from ground water. Lett. Appl. Microbiol. 28, 31–35.

Ridsdale, J.A., Atabay, H.I., Corry, J.E.L., 1998. Prevalence of campylobacters and arcobacters in ducks at the abattoir. J. Appl. Microbiol. 85, 567–573.

Rivas, L., Fegan, N., Vanderlinde, P., 2004. Isolation and characterisation of *Arcobacter butzleri* from meat. Int. J. Food Microbiol. 91, 31–41.

Rollins, D.M., Colwell, R.R., 1986. Viable but non-culturable stage of *Campylobacter jejuni* and its role in survival in the natural aquatic environment. Appl. Environ. Microbiol. 52, 531–538.

Sai Jyothsna, T.A., Rahul, K., Ramaprasad, E.V., Sasikala, C., Ramana, C., 2013. *Arcobacter anaerophilus* sp. nov., isolated from an estuarine sediment and emended description of the genus *Arcobacter*. Int. J. Syst. Evol. Microbiol. 63, 4619–4625.

SchroederTucker, L., Wesley, I.V., Kiehlbauch, J.A., Laron, D.J., Thomas, L.A., Erickson, G.A., 1996. Phenotypic and ribosomal RNA characterization of *Arcobacter* species isolated from porcine aborted fetuses. J. Vet. Diagn. Invest. 8, 186–195.

Scullion, R., On, S.L.W., Madden, R.H., Harrington, C.S., 2001. Comparison of a multiplex PCR assay and AFLP profiling for speciation of *Arcobacter* spp. Int. J. Med. Microbiol. 291, 140.

Scullion, R., Harrington, C.S., Madden, R.H., 2004. A comparison of three methods for the isolation of *Arcobacter* spp. from retail raw poultry in Northern Ireland. J. Food Prot. 67, 799–804.

Shah, A.H., Saleha, A.A., Zunita, Z., Murugaiyah, M., 2011. Arcobacter – an emerging threat to animals and animal origin food products? Trends Food Sci. Technol. 22, 225–236.

Shah, A.H., Saleha, A.A., Zunita, Z., Cheah, Y.K., Murugaiyah, M., Korejo, N.A., 2012. Genetic characterization of *Arcobacter* isolates from various sources. Vet. Microbiol. 160, 355–361.

Skrivanova, E., Molatova, Z., Matenova, M., Houf, K., Marounek, M., 2011. Inhibitory effect of organic acids on arcobacters in culture and their use for control of *Arcobacter butzleri* on chicken skin. Int. J. Food Microbiol. 144, 367–371.

Snaidr, J., Amann, R., Huber, I., Ludwig, W., Schleifer, K.H., 1997. Phylogenic analysis and in situ identification of bacteria in activated sludge. Appl. Environ. Microbiol. 63, 2884–2896.

Snelling, W.J., Matsuda, M., Moore, J.E., Dooley, J.S., 2006. Under the microscope: *Arcobacter*. Lett. Appl. Microbiol. 42, 7–14.

Stampi, S., Varoli, O., Zanetti, F., Deluca, G., 1993. *Arcobacter cryaerophilus* and thermophilic campylobacters in a sewage-treatment plant in Italy – 2 secondary treatments compared. Epidemiol. Infect. 110, 633–639.

Stampi, S., Luca, G.D., Varoli, O., Zanetti, F., 1998. Occurrence, removal and seasonal variation of thermophilic campylobacters and *Arcobacter* in sewage sludge. Zentralbl. Hyg. Umweltmed. 202, 19–27.

Suarez, D.L., Wesley, I.V., Larson, D.J., 1997. Detection of *Arcobacter* species in gastric samples from swine. Vet. Microbiol. 57, 325–336.

Tabatabaei, M., Shirzad Aski, H., Shayegh, H., Khoshbakht, R., 2014. Occurrence of six virulence-associated genes in *Arcobacter* species isolated from various sources in Shiraz, Southern Iran. Microb. Pathog. 66, 1–4.

Talibart, R., Denis, M., Castillo, A., Cappelier, J.M., Ermel, G., 2000. Survival and recovery of viable but noncultivable forms of *Campylobacter* in aqueous microcosm. Int. J. Food Microbiol. 55, 263–267.

Taylor, D.N., Kiehlbauch, J.A., 1991. Isolation of group 2 aerotolerant *Campylobacter* spp. from Thai children with diarrhoea. J. Infect. Dis. 163, 1062–1067.

Tee, W., Baird, M., Dyall-Smith, M., Dwyer, B., 1988. *Campylobacter cryaerophilus* isolated from a human. J. Clin. Microbiol. 26, 2469–2473.

Tholozan, J.L., Cappelier, J.M., Tissier, J.P., Delattre, G., Federighi, M., 1999. Physiological characterization of viable-but-nonculturable *Campylobacter jejuni* cells. Appl. Environ. Microbiol. 65, 1110–1116.

Tsang, R.S.W., Luk, J.M.C., Woodward, D.L., Johnson, W.M., 1996. Immunochemical characterization of a haemagglutinating antigen of *Arcobacter* spp. FEMS Microbiol. Lett. 136, 209–213.

Unver, A., Atabay, H.I., Sahin, M., Celebi, O., 2013. Antimicrobial susceptibilities of various *Arcobacter* species. Turk. J. Med. Sci. 43, 548–552.

Ursing, J.B., Lior, H., Owen, R.J., 1994. Proposal of minimal standards for describing new species of the family *Campylobacteraceae*. Int. J. Syst. Bacteriol. 44, 842–845.

Van Driessche, E., Houf, K., 2007a. Discrepancy between the occurrence of *Arcobacter* in chickens and broiler carcass contamination. Poult. Sci. 86, 744.

Van Driessche, E., Houf, K., 2007b. Characterization of the *Arcobacter* contamination on Belgian pork carcasses and raw retail pork. Int. J. Food Microbiol. 118, 20–26.

Van Driessche, E., Houf, K., 2008. Survival capacity in water of *Arcobacter* species under different temperature conditions. J. Appl. Microbiol. 105, 443–451.

Van Driessche, E., Houf, K., Vangroenweghe, F., De Zutter, L., Van Hoof, J., 2005. Prevalence, enumeration and strain variation of *Arcobacter* species in the faeces of healthy cattle in Belgium. Vet. Microbiol. 105, 149–154.

Vandamme, P., Falsen, E., Rossau, R., Hoste, B., Segers, P., Tygat, R., Delay, J., 1991a. Revision of *Campylobacter*, *Helicobacter* and *Woinella* taxonomy: emendation of generic descriptions and proposal for *Arcobacter gen. nov*. Int. J. Syst. Bacteriol. 41, 88–103.

Vandamme, P., Pot, B., Kersters, K., 1991b. Identification of campylobacters and allied bacteria by means of numerical analysis of cellular protein electrophorograms. In: Borriello, S.P., Rolfe, R.D., Hardie, J.M. (Eds.), Microbial Ecology in Health and Disease. Proceedings of the IVth International Workshop on Campylobacter, Helicobacter and Related Organisms. Wiley, Chichester, UK, p. S101.

Vandamme, P., Pugina, P., Benzi, G., Etterijk, R.V., Vlaes, L., Kersters, K., Butzler, J.P., Lior, H., Lauwers, S., 1992a. Outbreak of recurrent abdominal cramps associated with *Arcobacter butzleri* in an Italian school. J. Clin. Microbiol. 30, 2335–2337.

Vandamme, P., Vancanneyt, M., Pot, B., Mels, L., Hoste, B., Dewettinck, D., Vlaes, L., Borre, C.V.D., Higgins, R., Hommez, J., Kersters, K., Butzler, J.P., Goosens, H., 1992b. Polyphasic taxonomic study of the emended genus *Arcobacter* with *Arcobacter com.nov.* and *Arcobacter skirrowii* sp. nov., an aerotolerant bacterium isolated from veterinary specimens. Int. J. Syst. Bacteriol. 42, 344–356.

Vandamme, P., Giesendorf, B.A.J., Vanbelkum, A., Pierard, D., Lauwers, S., Kersters, K., Butzler, J.P., Goossens, H., Quint, W.G.V., 1993. Discrimination of epidemic and sporadic isolates of *Arcobacter butzleri* by polymerase chain reaction-mediated DNA-fingerprinting. J. Clin. Microbiol. 31, 3317–3319.

Vandenberg, O., Dediste, A., Vlaes, L., Ebraert, A., Retore, P., Douat, N., Vandamme, P., Butzler, J., 2001. Prevalence and clinical features of non-*jejuni/coli Campylobacter* species and related organisms in stool specimens. Int. J. Med. Microbiol. 291, 144.

Villarruel-Lopez, A., Marquez-Gonzalez, M., Garay-Martinez, L.E., Zepeda, H., Castillo, A., De La Garza, L.M., Murano, E.A., Torres-Vitela, R., 2003. Isolation *of Arcobacter* spp. from retail meats and cytotoxic effects of isolates against Vero cells. J. Food Prot. 66, 1374–1378.

Wesley, I.V., 1994. *Arcobacter* infections. In: Geran, G.W. (Ed.), Handbook of Zoonoses. CRC Press, Boca Raton, FL, pp. 180–190.

Wesley, I.V., Baetz, A.L., 1999. Natural and experimental infections of *Arcobacter* in poultry. Poult. Sci. 78, 536–545.

Wesley, I.V., Schroedertucker, L., Baetz, A.L., Dewhirst, F.E., Paster, B.J., 1995. *Arcobacter*-specific and *Arcobacter butzleri*-specific 16s ribosomal RNA based DNA probes. J. Clin. Microbiol. 33, 1691–1698.

Wesley, I.V., Baetz, A.L., Larson, D.J., 1996. Infection of cesarean-derived colostrum-deprived 1-day-old piglets with *Arcobacter butzleri*, *Arcobacter cryaerophilus*, and *Arcobacter skirrowii*. Infect. Immun. 64, 2295–2299.

Wesley, I.V., Wells, S.J., Harmon, K.M., Green, A., Schroeder-Tucker, L., Glover, M., Siddique, I., 2000. Fecal shedding of *Campylobac*ter and *Arcobacter* spp. in dairy cattle. Appl. Environ. Microbiol. 66, 1994–2000.

Winters, D.K., Slavik, M.F., 2000. Multiplex PCR detection of *Campylobacter jejuni* and *Arcobacter butzleri* in food products. Mol. Cell. Probes 14, 95–99.

Wybo, I., Breynaert, J., Lauwers, S., Lindenburg, F., Houf, K., 2004. Isolation of *Arcobac*ter *skirrowii* from a patient with chronic diarrhea. J. Clin. Microbiol. 42, 1851–1852.

Yan, J.J., Wang, W.C., Huang, A.H., Chen, H.M., Jin, Y.T., Wu, J.J., 2000. *Arcobacter butzleri* bacteremia in a patient with liver cirrhosis. J. Formos. Med. Assoc. 99, 166–169.

Yesilmen, S., Vural, A., Erkan, M.E., Yildirim, I.H., 2014. Prevalence and antimicrobial susceptibility of *Arcobacter* species in cow milk, water buffalo milk and fresh village cheese. Int. J. Food Microbiol. 188, 11–14.

Zanetti, F., Stampi, S., Luca, G.D., Varoli, O., Tonelli, E., 1996a. Comparative disinfection of secondary-treated sewage with chlorine dioxide and bromine chloride. Zentralbl. Hyg. Umweltmed. 198, 567–579.

Zanetti, F., Varoli, O., Stampi, S., Luca, G.D., 1996b. Prevalence of thermophilic *Campylobacter* and *Arcobacter butzleri* in food of animal origin. Int. J. Food Microbiol. 33, 315–321.

CHAPTER

NEW INSIGHTS INTO THE EMERGENT BACTERIAL PATHOGEN *CRONOBACTER*

13

Stephen J. Forsythe
School of Science and Technology, Nottingham Trent University, Nottingham, UK

ABBREVIATIONS

CC	clonal complex
COG	clusters of orthologous groups
FAO-WHO	Food and Agricultural Organization-World Health Organization
MLSA	multilocus sequence analysis
MLST	multilocus sequence typing
MLVA	multiple-locus variable-number tandem repeat analysis
NEC	necrotizing enterocolitis
PIF	powdered infant formula
PFGE	pulsed-field gel electrophoresis
rMLST	ribosomal multilocus sequence typing
STs	sequence types
wgMLST	whole genome multilocus sequence typing
cgMLST	core genome multilocus sequence typing

1 BACKGROUND

It almost goes without saying that food safety is important to everyone, but it is especially true for the highly vulnerable members of our society. This chapter concerns the bacterial genus *Cronobacter*, which though initially publicized due to neonatal infection is now recognized as predominantly causing infections in adults.

The *Cronobacter* genus has come to prominence due to its association with severe neonatal infections (necrotizing enterocolitis (NEC), septicaemia, and meningitis) that can be fatal. As neonates are frequently fed reconstituted powdered infant formula (PIF), which is not a sterile product, this potential vector became the focus of attention for reducing infection risk to neonates. It was considered at the time that the number of other exposure routes was limited.

Food Safety. http://dx.doi.org/10.1016/B978-0-12-800245-2.00013-7

Table 1 Summary Timeline of *Cronobacter* Recognition, Control, and Molecular Profiling

Pre-2002 various outbreaks in neonatal intensive care units and sporadic cases reported	
2002	Outbreak at University of Tennessee investigated by CDC and FDA. Fatal case linked to use of powdered formula
	FDA *Cronobacter* (then *E. sakazakii*) detection method; Bacteriological Analytical Manual announced
2004	First FAO-WHO risk assessment meeting on microbiological safety of powdered infant formula
	Development of first chromogenic differential agar; DFI
2006	Second FAO-WHO risk assessment meeting
	ISO detection method announced; ISO/TS 22964
2008	Third FAO-WHO risk assessment meeting
	Revised Codex Alimentarius Commission guidelines for microbiological specifications for powdered infant formula released, to include *Cronobacter* as a named pathogen
***Cronobacter* genus composed of five species defined**	
2009	Open access 7-loci multilocus sequence typing (MLST; 3036bp) scheme established; pubMLST.org/cronobacter/
2010	First whole sequence for *C. sakazakii* published. Strain BAA-894 had been isolated during the University of Tennessee outbreak of 2002
2011	*C. sakazakii* sequence type 4 (clonal lineage 4) association with neonatal meningitis determined
Two new *Cronobacter* species recognized: *C. condimenti* and *C. universalis*	
2012	First pan-genome analysis of *Cronobacter* genus based on 14 genomes published
Revised FDA method of detection; Bacteriological Analytical Manual	
2013	PubMLST.org/cronobacter/ expanded to include:
	(a) Tax-MLST established, extension of MLST to include genotyping of *ompA* and *rpoB*
	(b) Searchable repository for all published *Cronobacter* spp. genomes established
2014	PubMLST.org/cronobacter/ expanded to include:
	(a) rMLST (51 loci; 20kbp) scheme released
	(b) COG-cgMLST(1865 loci; 1.98Mbp) scheme released

Adapted from Forsythe et al. (2014).

The major advances in our knowledge of the *Cronobacter* genus are summarized in Table 1. This chapter addresses many topics concerning *Cronobacter* with an emphasis on our recent findings on the diversity of the organism and similarities with closely related organisms as well as issues of detection and consumer protection.

2 *CRONOBACTER* TAXONOMY AND IDENTIFICATION

Although taxonomy is not a topic, many food microbiologists are practitioners in, it needs to be appreciated that accurate bacterial taxonomy is essential for regulatory control because the detection methods must be based on a thorough understanding of the diversity of the target organism. As members of the *Cronobacter* genus were formerly known as the single-species *Enterobacter sakazakii*, this name was used in publications before mid-2007. Subsequently, it is uncertain which specific *Cronobacter* species were referred to in many pre-2007 publications. Nevertheless, the majority of isolated strains are usually *Cronobacter sakazakii*, and it is probable that this has been the species of major study to date.

Initially the organism was regarded as a pigmented variety of *Enterobacter cloacae*. In 1974, Brenner showed that the pigmented strains had <50% homology with nonpigmented strains, and it was suggested that they should comprise a new species called *E. sakazakii*. This species was distinguishable from *E. cloacae* based on DNA-DNA hybridization, pigment production, biotype assignment, and antimicrobial resistance (Farmer et al., 1980; Izard et al., 1983). DNA-DNA hybridization values were 41% and 54% for *Citrobacter freundii* and *E. cloacae*, respectively, which were used as representatives of the *Citrobacter* and *Enterobacter* genera (Farmer et al., 1980). The results warranted the recognition of *E. sakazakii* as a separate species as they were phenotypically closer to *E. cloacae*. Additional phenotypic analysis led to the description of 15 *E. sakazakii* biogroups, with biotype 1 being the most common (Farmer et al., 1980).

Since the 1980s, bacterial systematics has increasingly used DNA sequencing for its analysis and for determining relatedness. Analysis of both partial 16S rDNA and hsp60 gene sequencing (Iversen et al., 2004c) showed that isolates, then known as *E. sakazakii*, formed at least four distinct clusters and proposed that these may represent different closely related species (Iversen et al., 2004c). The 15 different biogroups fitted into the four clusters, and a 16th biogroup was added in subsequent work (Iversen et al., 2006). However, full taxonomic revision required considerable further analysis for substantiation. The *Cronobacter* genus was defined first in 2007 and revised in 2008 (Iversen et al., 2007, 2008). Differentiation between the newly defined *Cronobacter* species is primarily based on genotypic (DNA-based) analysis and is largely supported by biochemical traits (Table 2).

Initially, the *Cronobacter* genus was composed of *C. sakazakii*, *Cronobacter turicensis*, *Cronobacter muytjensii*, and *Cronobacter dublinensis*. This was quickly revised (Iversen et al., 2008) with the addition of *Cronobacter malonaticus*. This species had originally been described as a subspecies of *C. sakazakii* by Iversen et al. (2007) who could not distinguish *C. sakazakii* and *C. malonaticus* using 16S rDNA-sequence analysis. The *Cronobacter* species were initially differentiated by Iversen et al. (2008) according to 16 *E. sakazakii* biotypes; *C. sakazakii* (biotypes 1-4, 7, 8, 11, and 13); *C. malonaticus* (biotypes 5, 9, and 14); *C. turicensis* (biotypes 16, 16a, and 16b); *C. muytjensii* (biotype 15); and *C. dublinensis* (biotypes 6, 10, and 12).

Table 2 Description of the *Cronobacter* Genus
Cronobacter spp. formerly known as the single-species *Enterobacter sakazakii*
Description: *Cronobacter* genus. Seven recognized species: *C. condimenti*, *C. dublinensis*, *C. malonaticus*, *C. muytjensii*, *C. sakazakii*, *C. turicensis*, and *C. universalis*. Gram-negative, frequently motile, facultatively aerobic rods, members of the *Enterobacteriaceae*
Clinical presentations: Infections in all age groups, particularly neonates, infants <6 months in age, and the elderly. Life-threatening symptoms in primarily low birth weight neonates and infants <6 months (preweaning) in age include meningitis, necrotising enterocolitis, and respiratory infections. Symptoms in adults include bacteraemia, septicaemia, urosepsis, and wound infections
Source: Infections in neonates associated with contaminated reconstituted powdered infant formula. Ubiquitous in nature, particularly plant material. Human carriage
Detection method: Cultivation on chromogenic agar, followed by sequencing of *fusA*
Adapted from Forsythe et al. (2014).

In contrast, Joseph et al. (2012b, 2013a) used strains selected by multilocus sequence analysis (MLSA; Baldwin et al., 2009) as representatives across the genus and, therefore, overcame the preconceived grouping of strains based on phenotyping. These MLSA studies, which will be described in more detail later in this chapter, led to the naming of two new *Cronobacter* species: *Cronobacter universalis* and *Cronobacter condimenti* (Joseph et al., 2011). Due to numerous limitations such as subjectivity and reproducibility, the earlier phenotyping approach to *Cronobacter*, based on 10 biochemical and physiological tests, has been replaced by various DNA-based techniques. Consequently, biotyping no longer has a role in designating the species of *Cronobacter* isolates.

The genus *Cronobacter* is currently composed of *C. condimenti*, *C. dublinensis*, *C. malonaticus*, *C. muytjensii*, *C. sakazakii*, *C. turicensis*, and *C. universalis*. See Table 2 for a fuller description of the genus. A phylogenetic tree that includes near relatives of the genera *Enterobacter*, *Citrobacter*, *Franconibacter*, and *Siccibacter*, based on 7-loci (<3036 nt concatenated length) multilocus sequence typing (MLST), is shown in Figure 1. A detailed table of phenotyping characters differentiating these genera is given in Table 3.

Further detailed phylogenetic analysis based on whole genome sequencing and 7-loci MLST (1036 nt concatenated length) shows that *C. sakazakii*, *C. malonaticus*, *C. turicensis*, and *C. universalis* appear to be more closely related to each other and separated from *C. muytjensii*, *C. dublinensis*, and *C. condimenti* about 41 million years ago (Figure 2; Joseph et al., 2012b). Accepting that such calculations are based on a number of assumptions, it is notable that this corresponds with the Palaeogene period of the Cenozoic era when early flowering plants evolved. This is an interesting observation as it coincides with the suspected natural plant habitat of the organism and opens speculation for the drivers for the evolution of the genus. The feeding of insect larvae on plants could have led to a host adaptation. It is notable that *Cronobacter* are isolated from flies (Pava-Ripoll et al., 2012). The evolution of other members of the *Enterobacteriaceae* family has been estimated by similar MLST studies; hence, they are comparable. The evolution of the distinguishable *Cronobacter* species appears to have occurred over the same period as the divergence of the *Salmonella* species and subspecies, after its split from *Escherichia coli*.

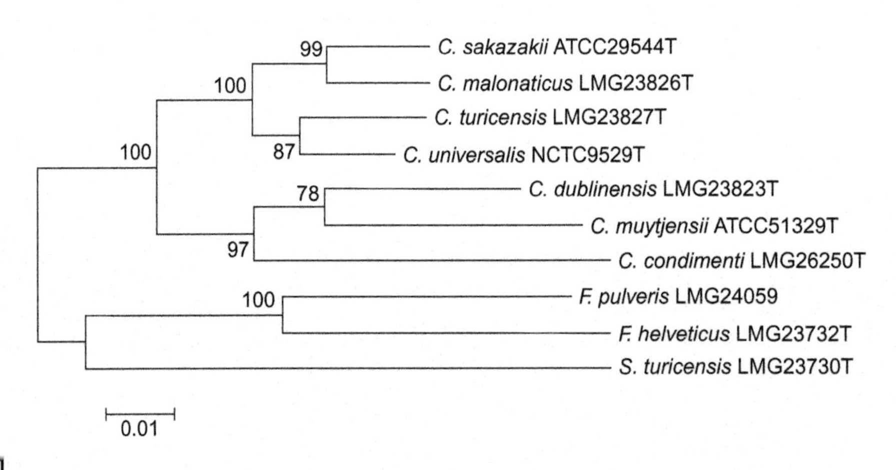

FIGURE 1

Maximum likelihood tree of the seven multilocus sequence typing loci (3036 bp concatenated length) for the members of *Cronobacter* genus and closely related *Franconibacter* and *Siccibacter* genera. The tree was drawn using MEGA5.2 (http://www.megasoftware.net/) with 1000 bootstrap replicates.

Table 3 Phenotypic Characters Differentiating *Siccibacter colletis* sp. nov., and Other Members of the Genera *Siccibacter*, *Franconibacter*, and *Cronobacter*

Characteristic	1	2	3	4	5	6	7	8	9	10	11
Acid phosphatase	−(−)	+(+)	v(−)	+(+)	−	+	+	+	−	−	(−)
***N*-acetyl-ß-glucosaminidase**	+(+)	−(−)	(−)	+(+)	+	+	+	+	+	+	(+)
Motility	+(+)	+(+)	+(+)	+(+)	+(+)	v(+)	+(+)	v(−)	+(+)	+(+)	(−)
Voges-Proskauer	−(−)	−(−)	−(−)	−(−)	+(+)	+(+)	+(+)	+(+)	+(+)	+(+)	(+)
H_2S production	−(−)	−(−)	−(−)	−(−)	−(−)	−(−)	−(−)	−(−)	−(−)	−(−)	(−)
Indole production	−(−)	−(−)	−(−)	−(−)	−(−)	+(+)	−(−)	−(−)	+(+)	+(+)	(+)
Carbon utilization											
Sialic acid	−(−)	+(+)	−(−)	+(+)	+(+)	−(−)	v(−)	−(−)	−(−)	−(−)	(−)
D-Glucose, gas production	−(−)	+(+)	ND	ND	ND	ND	ND	v(−)	ND	ND	(−)
D-Glucose, acid production	+(+)	+(+)	+(+)	+(+)	+(+)	+(+)	+(+)	+(+)	+(+)	+(+)	(+)
Sucrose, acid production	−(−)	−(−)	−(−)	+(+)	+(+)	+(+)	+(+)	+(+)	+(+)	+(+)	(+)
Dulcitol	+(+)	+(+)	+	−	−(−)	−(−)	+(+)	+(+)	+(+)	−(−)	(−)
Malonate	−(−)	−(−)	+(−)	−(−)	−(−)	+(+)	v(+)	+(+)	+(+)	v(+)	(+)
D-melezitose	−(−)	−(−)	−	−	−(−)	+(+)	+(+)	+(+)	−(−)	+(+)	(−)
Inositol	−(−)	−(−)	−(−)	−(−)	v(+)	+(+)	+(+)	+(+)	+(+)	+(+)	(−)
trans-Aconitate	−(−)	(−)	+(+)	+(+)	−(−)	+(+)	−(−)	−(−)	+(+)	+(+)	(−)
Maltitol	−(−)	−(−)	−(−)	v	+(+)	+(+)	+(+)	+(+)	−(−)	+(+)	(−)
D-Arabitol	−(−)	−(−)	−(−)	+(+)	−(−)	−(−)	−(−)	−(−)	−(−)	−(−)	(−)
Lactulose	−(−)	−(−)	−(−)	v	+(+)	v(+)	+(+)	+(+)	+(+)	+(+)	(−)
Putrescine	−(−)	−(−)	+(+)	v	+(+)	+(+)	+(+)	−(−)	+(+)	+(+)	(−)

Taxa: 1, Siccibacter colletis *sp. nov. 1383*^{T} *and 2249; 2,* Siccibacter turicensis *(n = 2); 3,* Franconibacter helveticus *(n = 2); 4,* Franconibacter pulveris *(n = 6); 5, Cronobacter* sakazakii *(n = 163); 6,* Cronobacter malonaticus *(n = 22); 7,* Cronobacter turicensis *(n = 8); 8,* Cronobacter universalis *(n = 4); 9,* Cronobacter muytjensii *(n = 7); 10,* Cronobacter dublinensis *(n = 8); 11,* Cronobacter condimenti *(n = 1). +, Positive; v, 25-75% variable; −, negative; ND, not determined. Key traits for differentiation of* Siccibacter *species are shown in bold. Reactions of the type strains are shown in parentheses.*
Adapted from Jackson et al. (2015).

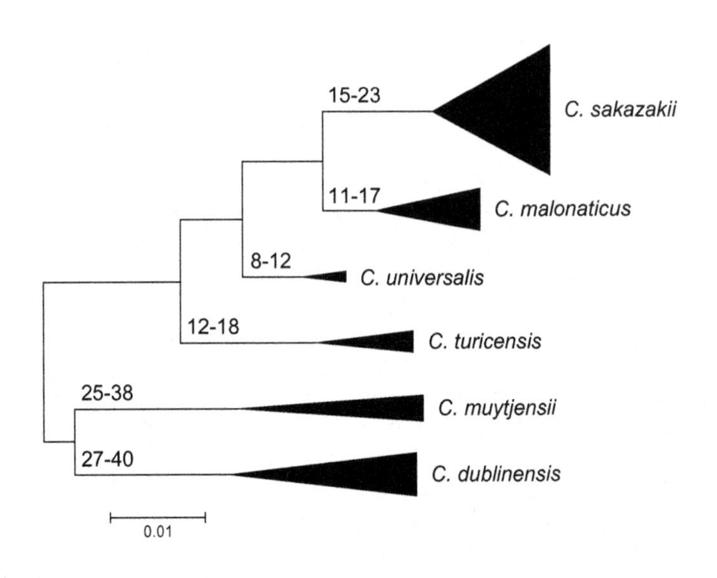

FIGURE 2

Maximum likelihood tree of the *Cronobacter* MLST dataset indicating ranges of the hypothetical divergence dates of each species node measured in millions of years before the present. The tree has been drawn to scale using MEGA5. The bases of the triangles indicate the number of isolates used for the analysis, while the heights indicate the diversity of each branch. *C. condimenti* with the single isolate has been excluded from this analysis.

Joseph and Forsythe (2012).

Distinguishing between *C. sakazakii* and *C. malonaticus* has been problematic, and there are two primary reasons for this. First, the use of biotype profile to designate the species was not totally robust as a few of the biotype index strains were themselves assigned the wrong species (Baldwin et al., 2009). Second, there are seven copies of the rDNA gene in *Cronobacter*, and intrageneric differences can lead to uncertain and inconsistent base calls.

16S rDNA sequences have been traditionally used to determine phylogeny including *Enterobacteriaceae*. However, 16S rDNA sequencing is of limited use for very closely related organisms because of minor differences in the 16S rDNA sequence. The earlier difficulties using 16S rDNA sequencing to distinguish between *C. sakazakii* and *C. malonaticus* were possibly due to polymorphic nucleotide sites and depended on the operator interpretation of the sequencing chromatograms for those loci. Additionally, there is difficulty using biotyping to define the *Cronobacter* species, as some biotype index strains were misassigned their species. Consequently, it is not surprising that numerous confusions from phenotypic and biochemical identification of members of the *Enterobacteriaceae* family as *Cronobacter* spp. have been described, including the following:

1. Fatal case of neonatal sepsis infection in a French neonatal intensive care unit (NICU) by *Cronobacter* spp., reidentified as *E. cloacae* (Caubilla-Barron et al., 2007).
2. NICU outbreak in the United States of *Cronobacter* spp., reidentified as *Enterobacter hormaechei* (Townsend et al., 2008).
3. Mexican infant infections attributed to *C. sakazakii* in infant formula, reidentified as *E. hormaechei* and *Enterobacter* spp. (Flores et al., 2011; Jackson et al., 2015).

4. Quinolone-resistant *Cronobacter* strain, reidentified as *E. hormaechei* (Poirel et al., 2007; Townsend et al., 2008).
5. Oligo-polysaccharide structure for *Cronobacter* strain ZORB 741 reidentified as *Enterobacter ludwiggi* (Szafranek et al., 2005).
6. Zhao et al. (2014) announced the genome of *C. sakazakii* CMCC 45402; however, this strain is in fact *C. malonaticus*.

Contributing to this confusion is that first of all, commercial companies producing phenotyping kits have not been updating their databases inline with the taxonomic revisions; for example, the former *E. sakazakii* Preceptrol™ strain ATCC® 51329 has been reclassified as *C. muytjensii* and should not be confused with *C. sakazakii*. Secondly, the specificity and reliability of some formerly *E. sakazakii* DNA-based PCR probes to the diverse *Cronobacter* genus needs to be reevaluated. Unfortunately, a number of these initial identification and detection methods were based on small numbers of poorly characterized, even misidentified, strains and, therefore, are not necessarily reliable now for their stated purpose.

3 *CRONOBACTER* DETECTION SCHEMES

In 2008, the Codex Alimentarius Commission (CAC) revised their microbiological criteria as applied to PIF with the intended age <6 months by the inclusion of specific testing for *Cronobacter* spp. (CAC, 2008). Hence, a number of methods for the recovery of desiccation-stressed *Cronobacter* cells from this group of products have been developed. A large volume of material needs to be tested as the organism has only been reported at low numbers (<1 cfu/g). The CAC requirement is to test thirty 10 g quantities; therefore, presence/absence testing of PIF is applied rather than direct enumeration. Due to the stressed state of the cells, the initial step involves resuscitation, followed by enrichment, and plating on a differential medium. Early methods for the detection of *Cronobacter* were based on general *Enterobacteriaceae* isolation procedures. However, it is now recognized that these had their limitations due to assumptions regarding upper temperature ranges, pigment production, and identification using phenotyping schemes. Current international detection methods are greatly improved procedures, which are the result of interlaboratory trials, and a greater knowledge of the organism (Chen et al., 2012).

Initial *Cronobacter* detection methods were reminiscent of the stages for *Salmonella* isolation from milk powders. In brief, the steps were (i) preenrichment (225 ml water or BPW + 25 g formula), (ii) enrichment (EE broth), (iii) inoculate VRBG agar, and (iv) pick off five colonies to TSA and identify phenotypically any yellow pigmented colonies. It is now recognized that there are a number of limitations with the above approach. First, there is no initial selection for *Cronobacter*; instead, any *Enterobacteriaceae* could be enriched in EE broth and grow on VRBG agar leading to possible overgrowth of *Cronobacter*. Second, not all *Cronobacter* strains for yellow-pigmented colonies on TSA plates and so could be overlooked. Third, phenotype databases did not adequately cover the genus resulting in conflicting results between commercial kits.

These days improved methods employ chromogenic agars, updated phenotyping databases, along with DNA-based identification and fingerprinting techniques and have considerably greater reliability (Cetinkaya et al., 2013; Jackson et al., 2014). *Cronobacter* has a notable resistance to osmotic stresses, which may be linked to its ecology, and this trait has been used in the design of improved enrichment broths; modified lauryl sulfate broth containing 0.5 M NaCl and *Cronobacter* screening broth with 10% sucrose. The use of chromogenic agar (primarily based on the α-glucosidase reaction) to differentiate

Cronobacter from other *Enterobacteriaceae* present on the plate was a major improvement. The α-glucosidase activity as a test differentiating *Cronobacter* from most other *Enterobacteriaceae* that may occur in the sample (Muytjens et al., 1984) has been incorporated into differential chromogenic agars for *Cronobacter*. An example of such agar is the DFI formulation, named after Patrick *D*ruggan (who designed the agar), *F*orsythe (myself, principal investigator), and Carol *I*versen (my PhD student at the time). The DFI formulated agar will also detect H_2S-producing *Salmonella* as black colonies, whereas *Cronobacter* colonies are blue-green. As well as testing PIF, environmental samples are taken from the production environment as well as from ingredients (especially starches and other plant-derived material). In addition, production facilities and processes are already designed to control enteric pathogens, especially *Salmonella*.

4 *CRONOBACTER* TYPING METHODS

4.1 BIOTYPING

Although it is generally possible to differentiate bacterial species by biochemical profiling, molecular methods are increasingly used as a more rapid and reliable tool to study bacterial genomic diversity and to track sources of infection. Because *Cronobacter* is ubiquitous, typing schemes are required both for epidemiological and environmental investigation. Initially, 15 biogroups of *Cronobacter* were defined with biogroup 1 being the most common (Iversen et al., 2006). However, improved speciation of strains using DNA sequencing-based methods have revealed that the earlier biotyping approach is severely flawed with no more than 50% of strains being correctly assigned to a *Cronobacter* species (Joseph et al., 2013a). This is in part due to the initial use of biotype index strains that were attributed to the wrong *Cronobacter* species (Baldwin et al., 2009). Subsequently, the biotyping approach is now regarded as unreliable for speciation.

4.2 DNA-BASED TYPING METHODS

A number of DNA-based methods for identification, speciation, and profiling have been proposed for *Cronobacter* spp. Initial procedures used plasmid profiling, chromosomal restriction endonuclease analysis, and multilocus enzyme electrophoresis (Clark et al., 1990; Nazarowec-White and Farber, 1999). This was followed by the application of random amplified polymorphic DNA ribotyping, pulsed-field gel electrophoresis (PFGE), and MLVA (multiple-locus variable-number tandem repeat analysis) (Mullane et al., 2008). However, there does not appear to have been a reevaluation of these methods since the taxonomic revisions after 2007, and there is no evidence that they adequately cover the genus. Hence, these methods are not considered in any further details here as they are most likely inappropriate to use.

Proudy et al. (2008) conducted a study using 27 *Cronobacter* spp. strains to evaluate a BOX-PCR technique, a PCR-RFLP sequencing scheme for the flagellin gene *fliC*, and PFGE. Their results showed a comparable discriminatory power of isolates using BOX-PCR and PFGE, though the *fliC* gene did not prove to be a good target to detect strain variability.

PCR probes have been designed for *dnaG*, *rpsU*, and *rpoB* genes Stoop et al., 2009. However, they suffer in that they have not been continued in use or validated against a robust *Cronobacter* strain collection of the seven species. Some require different PCR primer pairs for each species. One exception is *rpoB* allele as opposed to DNA probe sequencing, which has been included in the *Cronobacter* PubMLST database.

4.3 PULSED-FIELD GEL ELECTROPHORESIS

PFGE is used in surveillance and investigations to identify plausible sources of outbreaks. The commonly used epidemiological profiling method of PGFE with two restriction enzymes (*Xba*I and *Spe*I) has been used to investigate neonatal intensive care unit outbreaks (Himelright et al., 2002; Van Acker et al., 2001; Caubilla-Barron et al., 2007). The technique is widely employed and can be used for transnational investigations, as per PulseNet, because the gel results can be electronically analyzed (http://www.cdc.gov/pulsenet/). The method has also been used for source tracing in milk powder (PIF) manufacturing plants (Craven et al., 2010; Jacobs et al., 2011; Reich et al., 2010). However, the method neither speciates isolates nor determine the relatedness of strains (a common misunderstanding). In addition, a proportion of strains do not give profiles and are, therefore, nontypable (Craven et al., 2010).

4.4 PCR O-ANTIGEN SEROGROUPING

The most advanced non-MLST profiling method is serogrouping using PCR (Jarvis et al., 2011, 2013; Sun et al., 2011, 2012). However, the technique suffers from a number of limitations. First, only 17 serogroups have been recognized, which is not very discriminatory given that 7-loci MLST has >300 defined sequence types (STs) across all seven *Cronobacter* species. Second, not all strains give a PCR product, which indicates further unrecognized serogroups exist. Third, there are contradictions in the literature with the same serogroup across more than one species and even genus: *C. sakazakii* O:4=*E. coli* O103, *C. sakazakii* O:3=*C. muytjensii* O:1, and *C. malonaticus* O:1=*C. turicensis* O:1. For comparative purposes, the described serotypes for >300 strains are included in the PubMLST *Cronobacter* database.

4.5 MULTILOCUS SEQUENCE TYPING

Following the association of *Cronobacter* spp. with several publicized fatal outbreaks in NICU of meningitis and NEC, the World Health Organization (WHO) in 2004 requested the establishment of a molecular typing scheme to enable the international control of the organism (Table 4). This section presents the application of next generation sequencing to *Cronobacter*, which has led to the establishment of the *Cronobacter* PubMLST genome and sequence definition database (http://pubmlst.org/cronobacter/) containing over 1000 isolates along with the recognition of specific clonal lineages linked to neonatal meningitis and adult infections. This site also contains the metadata for strains, which have been compiled from various researchers, sources, and countries. The scheme has revealed

Table 4 FAO-WHO (2004) Executive Summary
1. The use of internationally validated detection and molecular typing methods for *Cronobacter* spp. and other relevant microorganisms should be promoted
2. Investigation and reporting of sources and vehicles, including powdered infant formula, of infection by *Cronobacter* spp. and other relevant microorganisms should be promoted
3. Research should be promoted to gain a better understanding of the ecology, taxonomy, virulence and other characteristics of *Cronobacter* spp. and on ways to reduce its levels in reconstituted powdered infant formula

Please note the term "Cronobacter *spp.*" *has been used in place of* "E. sakazakii" *as used in the original text due to changes in taxonomy since the FAO-WHO report.*
Adapted from Forsythe et al. (2014).

stable clones, some of which could be traced over a 50-year period, from a wide range of countries and sources. Typing *Cronobacter* isolates to better understand the diversity of the genus has led to the development of several MLST schemes. The initial scheme used 7-loci (3036 nt), but more recently ribosomal protein-MLST (rMLST) using 53 loci and COG-cgMLST covering ~1/3 genome (1865 loci) have been established. In addition, through the *Cronobacter* PubMLST database, the user can define his or her own scheme. A detailed review of the results obtained through the MLST approach is found in Forsythe et al. (2014). Only the key results are presented here.

The *Cronobacter* 7-loci MLST scheme requires the partial sequence analysis of seven housekeeping genes (Baldwin et al., 2009). These loci were chosen as they are distributed around the *Cronobacter* genome and are not coinherited. Comparing the loci DNA sequences with the *Cronobacter* MLST reference database (http://www.pubMLST.org/cronobacter) generates the 7-digit allele code, and the strain's ST. When concatenated together the seven allele sequences form a 3036 nucleotide length for MLSA and phylogenetic analysis.

The sequencing of protein-coding genes is a useful, more discriminatory alternative to partial 16S rDNA sequencing (ca. 528 nucleotide length), especially as unlinked sequences from multiple protein-coding genes are used. The *Cronobacter* 7-loci MLST analysis is based on seven housekeeping genes; ATP synthase beta chain (*atp*D), elongation factor G (*fus*A), glutaminyl-tRNA synthetase (*gln*S), glutamate synthase large subunit (*glt*B), DNA gyrase subunit B (*gyr*B), translation initiation factor IF-2 (*inf*B), and phosphoenolpyruvate synthase A (*pps*A). The seven sequenced alleles can be concatenated together to give >3000 nucleotide for phylogenetic analysis (Figures 1–3). This is six times the length of the commonly used partial 16S rDNA sequences and has the additional advantage of considerably greater number of variable loci. The initial publication was focused on *C. sakazakii* and *C. malonaticus* due to the reported difficulties in distinguishing between them (Baldwin et al., 2009).

Whole genome sequencing and MLST have supported the formal recognition of the genus *Cronobacter* composed of seven species to replace the former single-species *E. sakazakii*. Hence, MLSA based on housekeeping genes, including *fusA*, has proven to be a useful tool for the taxonomic

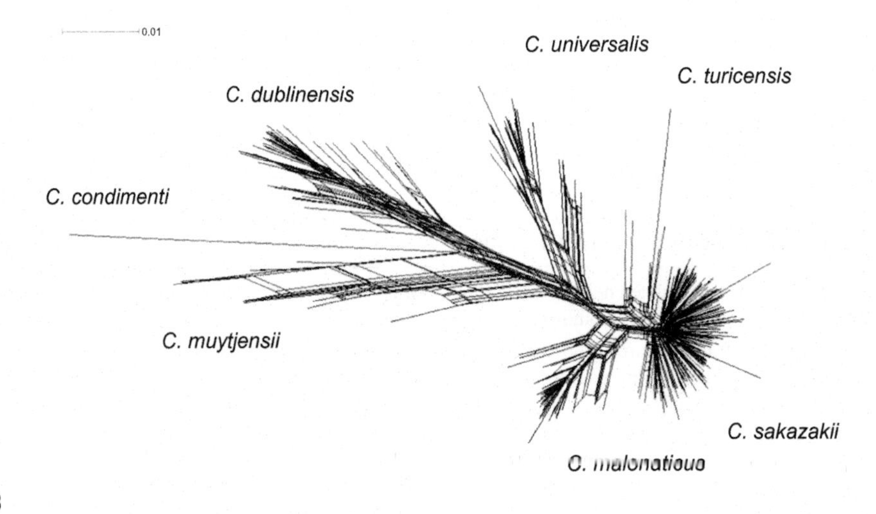

FIGURE 3

7-Loci Splits Network of 297 sequence types from the *Cronobacter* genus.

analysis of *Enterobacteriaceae* and was found to be more effective than phenotyping for *Cronobacter* speciation (Joseph and Forsythe, 2012). Joseph et al. (2011) used 7-loci MLSA for the definitions of two new *Cronobacter* species: *C. universalis* and *C. condimenti.*

The correlation of typing with whole genome phylogeny does not occur with the PCR-serotyping approach. For example, there is a general association between *C. sakazakii* ST4 with *C. sakazakii* serotype O:2. However, not all *C. sakazakii* O:2 are ST4, which is probably due to serotyping not following the phylogeny of the genus; for example, *C. sakazakii* O:4 is the same as *E. coli* O103 (Sun et al., 2012).

There is evidence of homologous recombination (gene conversion) events having occurred in the evolution of the *Cronobacter* genus. This is covered in detail in Joseph et al. (2012b) and Forsythe et al. (2014). Figure 3 shows NeighborNet analysis using Splitstree of the 7-loci alleles to visualize recombination events and evolutionary relationships in the data set. The formation of parallelograms in these figures denotes the possibility of recombination events. The separation of species is not always easy to define, and a large number of recombination events can occur within species, resulting in multiple plausible but only partially consistent taxonomic trees (Figure 3). Whereas the separation of *C. sakazakii* and *C. malonaticus* using 16S rRNA gene sequencing was problematic, the two species are clearly separated using the 7-loci MLST (Figures 1 and 2). The *C. sakazakii* CC4 and *C. malonaticus* CC7 clonal complexes can also be seen as tight clusters. *C. dublinensis* and *C. muytjensii* show wide divergence within their species. Splitstree analysis also reveals a higher diversity in the *C. muytjensii* and *C. dublinensis* species than other *Cronobacter* species, with some individual branches being more genetically distant from the main cluster. This phenomenon has been described as "fuzzy" species in the MLST of *Neisseria* spp. Strain 1330 (ST 98) was one such candidate which can be seen as a lineage branching out from the *C. dublinensis* cluster. Following further phenotypic and DNA-DNA hybridization studies, this strain was confirmed to be a previously unrecognized independent species and was subsequently formally named *C. condimenti* (Joseph et al., 2011). Such recognition of new species in the *Cronobacter* genus is likely as MLST is adopted because the phylogenetic analysis is dynamic and not constrained to an analytical database.

It should be noted that Brady et al. (2013) only used four loci (*atpD*, *gyrB*, *infB*, and *rpoB*) to support their reclassification of *Enterobacter helveticus*, *Enterobacter pulveris*, and *Enterobacter turicensis* to *Cronobacter helveticus*, *Cronobacter pulveris*, and *Cronobacter zurichensis*, respectively. This misclassification was corrected by Stephan et al. (2014) who proposed two new genera containing the species *Franconibacter helveticus*, *Franconibacter pulveris*, and *Siccibacter turicensis* for the same former *Enterobacter* species, based on SNP analysis of whole genomes. Their relatedness to *Cronobacter* is shown in Figure 1.

Investigating 1007 isolate entries in the *Cronobacter* MLST database reveals the temporal, geographic, and source diversity of the organism (Table 5). The earliest isolate (*C. sakazakii* NCIMB 8282) was from dried milk powder in 1950; the genome of which has now been published and is one of 107 genomes that can now be analyzed via the PubMLST database (Masood et al., 2013). *Cronobacter* strains have been isolated from 36 countries and are from clinical (24%), infant formula (15%), food (19%), environmental (15%), and other sources such as water (4%) (Table 5). *Cronobacter* spp. have been frequently isolated from plant material (wheat, rice, herbs, and spices) and various food products (Iversen and Forsythe, 2004). However, due to the history of concern, it is not surprising that a large number of strains are clinical or from PIF in origin. The database includes the MLST profiling of isolates from all reported outbreaks around the world. Because the database now contains over 1000 isolates, an informed understanding of the diversity and sources of the organism can now be obtained. Applying the 7-loci MLST scheme to 1007 strains reveals >300 definable STs (see Tables 5 and 6 which are based on 2014 data).

Table 5 Summary of *Cronobacter* Isolates in the *Cronobacter* PubMLST Database

Species	Number of Strains (%)	Number of STs[a]	Number of Genomes	Earliest Isolate	Countries	Clinical	Infant Formula	Food and Ingredients	Environmental	Other
C. sakazakii	726 (72.1)	155	73	1950	33	19.8[b]	23.8	32.8	16.3	7.3
C. malonaticus	136 (13.5)	53	14	1973	17	36.0	16.2	36.8	7.4	3.7
C. dublinensis	59 (5.9)	44	9	1956	11	5.1	13.6	47.5	6.8	27.1
C. turicensis	41 (4.1)	26	6	1970	12	17.1	12.2	36.6	22.0	12.2
C. muytjensii	35 (3.5)	14	2	1988	10	2.9	28.6	45.7	2.9	20.0
C. universalis	9 (0.9)	5	2	1956	6	11.1	0.0	55.6	11.1	22.2
C. condimenti	1 (0.1)	1	1	2010	1	0	0	100	0	0
Total	1007	298	107		36	20.4[c]	21.6	14.2	35.1	8.7

[a] *Sequence type.*
[b] *Percentage of species total.*
[c] *Percentage of genus total.*
From Forsythe et al. (2014).

Table 6 Profile of the Major Sequence Types in *C. sakazakii* and *C. malonaticus*

Cronobacter Species	Clonal Group	Number of Profiled Isolates[a]	Percentage of Species	Infant (<1 year)				Child (>1 year)	Adult (>18 years)	Unknown	Clinical Total	Infant Formula	Food and Ingredients	Environmental	Others
				Unknown	Meningitis	NEC[b]	Septicaemia								
C. sakazakii (726)a	CC4	195	26.9	4.6[c]	10.0	5.1	4.1	2.1	16.9	2.0	45.1	24.6	8.2	16.9	5.6
	CC1	80	11.0	2.5	3.8	0.0	1.3	1.3	0.0	3.8	12.5	30.0	22.5	31.3	3.8
	CC8	35	4.8	5.7	0.0	0.0	0.0	5.7	0.0	20.0	31.4	17.1	37.1	5.7	8.6
	CC64	28	3.9	0.0	0.0	0.0	0.0	0.0	0.0	3.6	3.6	46.4	35.7	7.1	7.1
	CC45	25	3.4	0.0	0.0	0.0	0.0	0.0	0.0	4.0	4.0	8.0	52.0	32.0	4.0
	CC13	24	3.3	10.5	0.0	0.0	0.0	0.0	0.0	0.0	8.3	29.2	58.3	0.0	4.2
	CC3	19	2.6	13.3	0.0	0.0	0.0	0.0	0.0	0.0	10.5	31.6	36.8	15.8	5.3
	ST12	15	2.1	27.3	0.0	13.3	0.0	0.0	0.0	26.7	60.0	20.0	20.0	0.0	0.0
	CC83	11	1.5	0.0	0.0	0.0	9.1	0.0	0.0	0.0	9.1	18.2	9.1	63.6	0.0
C. malonaticus (136)	CC7	58	42.6	3.4	0.0	0.0	0.0	22.4	15.5	15.5	56.9	12.1	25.9	0.0	5.2

[a] *Number of strains for each species in PubMLST database.*
[b] *Necrotizing enterocolitis.*
[c] *Percentage of strains in each clonal group.*
From Forsythe et al. (2014).

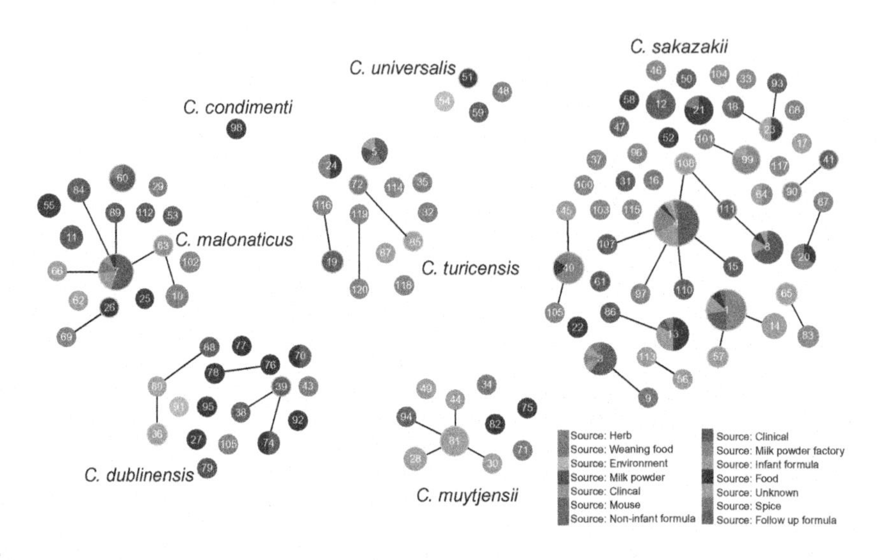

FIGURE 4

GoeBURST analysis of *Cronobacter* STs according to the diversity of their sources of isolation.

GoeBURST analysis shows the formation of clonal complexes among the identified STs for the *Cronobacter* genus (Figure 4), and an uneven distribution of clinical isolates across both the species and STs (Forsythe et al., 2014). Some of these large clonal complexes are especially significant with respect to strain clustering according to their isolation sources. All STs can be analyzed from the open access database, which is updated regularly; however, for convenience only selected ones are considered in detail below. I_A is a measure of the linkage of a population, and when calculated using the MLST sequences, the I_A values for the genus *Cronobacter* were found to be significantly greater than zero (Joseph et al., 2012b). This indicates the presence of linkage disequilibrium or clonality. This analysis has been highly significant for our understanding of both the diversity of the genus and the specificity of clinically relevant strains. Only key clonal lineages are considered in this chapter, and the reader should refer to Joseph et al. (2012b) for a more detailed account.

Clonal complex 1 currently comprises *C. sakazakii* STs 1 and 14. ST1 is a dominant ST consisting of strains isolated from across the world over a period of more than 25 years (Tables 5 and 6). These have been mainly isolated from PIF and clinical cases, and also more recently from milk powder processing factories in Germany and Australia (Craven et al., 2010; Jacobs et al., 2011), apart from a few food isolates. ST14, which is a SLV (*ppsA* allele) of ST1, is isolated from infant formula (Caubilla-Barron et al., 2007). Also linked to this clonal complex is ST57, which is a DLV (alleles *atpD* and *glnS*) to ST1. This is the profile for a PIF isolate from Denmark in 1988, as reported by Muytjens et al. (1988) in their milk powder survey.

Clonal complex 4 comprises *C. sakazakii* STs 4, 15, 97, 107, and 108. This is a key clade with respect to *Cronobacter* spp. epidemiology. ST4 is the most dominant ST in this MLST study with more than 70 isolates, and also the most frequent clinical ST (Table 6; Baldwin et al., 2009). However, the description of "clinical" when referring to the origin of strains is ambiguous. Such strains are not necessarily from the site of infection; that is, conjunctivae swabs from meningitis cases, or even isolates from colonized asymptomatic individuals. Fortunately, sufficient detailed information was available for these clinical *C. sakazakii* strains to reveal what is possibly the most significant finding generated by MLST analysis regarding the epidemiology and trophism of neonatal *Cronobacter* infections. In 2011, Joseph and Forsythe were the first to announce the strong association between neonatal meningitis isolates and the clonal lineage of *C. sakazakii* ST4. This was established using a retrospective study of 41 clinical strains from 1953 to 2008, collected from seven countries. This association was confirmed later in the year by the analysis of a number of highly publicized cases in the United States (CDC 2011; Hariri et al., 2013). Within the 2011 US strains, there were two ST4 SLVs. The cerebrospinal fluid (CSF) strain 1565 (ST107) differed from the ST4 profile in the *fusA* loci by 6/438 nt. Strain 1572 (ST108) isolated from an opened tin of PIF differed in the *fusA* loci by 5/438 nt. These two strains only differ from each other by 1 nucleotide out of 3036 (concatenated length) in the *fusA* loci position 378 (A:T). This level of discrimination is not possible using PFGE.

The reason for the association of one ST out of >300 STs in the genus is unclear, as no particular virulence traits have been determined in strains from the *C. sakazakii*, the clonal lineage of ST4. This clonal lineage is very robust and has been confirmed (Forsythe et al., 2014) using the more discriminatory approaches of ribosomal-MLST (51-loci) and whole genome-MLST (1865 loci). Thus, the clonal complex ST4 represents an evolutionary lineage for the causative agents of neonatal meningitis among the *Cronobacter* isolates (Joseph and Forsythe, 2011; Joseph et al., 2012b; Hariri et al., 2013). Why *C. sakazakii* clonal complex 4 predominates neonatal meningitis cases is presently unclear and could be due to environmental fitness factors as well as virulence traits. It is plausible that adult cases are unreported to date due to the maturity of the blood-brain barrier.

Strains not from the clonal complex 4 have been associated with neonatal meningitis (Table 6). The most well-known being *C. sakazakii* BAA-894 isolated from noninfant formula during the publicized Tennessee outbreak (Himelright et al., 2002). This strain is ST1 and indicates that nonclonal complex 4 strains may on occasion also cause severe brain damage. The additional importance of this strain is that it was the first strain to be fully sequenced (Kucerova et al., 2010, 2011). Also, within the 2011 US isolates that have been profiled is a strain of *C. malonaticus* ST112 (Hariri et al., 2013). The strain had been isolated from the blood of a <1-month-old infant who died from a fatal case of meningitis. This isolate is highly significant, as previously it had been observed that *C. malonaticus* predominates adult infections, and no previous neonatal meningitis cases have been attributed to this species (Joseph and Forsythe, 2011).

C. sakazakii CC4 strains have been isolated from the infant formula (Muytjens et al., 1988) and milk powder-manufacturing plants in Australia (Craven et al., 2010), Germany (Jacobs et al., 2011), Switzerland (Muller et al., 2013), and Ireland (Power et al., 2013); therefore, they may represent a particularly persistent clonal variant resulting in increased neonatal exposure. Sonbol et al. (2013) applied 7-loci MLST to the strains from the studies of Muytjens et al. (1988). Five out of 17 of the *C. sakazakii* isolates were *C. sakazakii* ST4 strains. These had been isolated from PIF samples purchased in Canada, Russia, West Germany, and the Netherlands. Three strains of *C. sakazakii* ST1 were isolated from PIF from the Netherlands and Russia, and two strains of *C. sakazakii* ST3 were from products from Belgium and the Netherlands. Similar analysis of PFGE representatives from the study of five manufacturing plants in Australia (Craven et al., 2010) showed the isolates were primarily *C. sakazakii* (42/52), followed by *C. turicensis* (9/52) and *C. malonaticus* (1/52). The *C. sakazakii* strains were different pulsetypes of 116 isolates from five milk-processing factories. About 12 of these pulse-type representatives were *C. sakazakii* ST4. The *C. sakazakii* ST4 strains had been isolated between 2006 and 2007, from various locations of all five sampled manufacturing plants: tanker bay, factory roofs, milk powder-processing environment, and outside grounds. Two isolates of *C. sakazakii* ST97 were from a tanker bay at one factory. This ST is within clonal complex 4, differing by 1 nucleotide in the *gltB* allele from the ST4 profile. The *C. sakazakii* ST1 strains represented nine pulse types that comprised 33 isolates. These had been isolated from similar milk powder manufacturing areas in three of the five factories sampled. Two strains that could not be profiled using PFGE were *C. sakazakii* STs 3 and 133. A third strain that also could not be profiled using PFGE was *C. turicensis* ST132.

The original study by Jacobs et al. (2011) isolated 81 *E. sakazakii* (they did not refer to *Cronobacter* in their study) strains from one German manufacturing plant, and these were divided into 13 pulse types. In our study, all representative strains of these pulse types were identified as *C. sakazakii*. The strains were primarily in ST1, ST4, and ST99. The *C. sakazakii* ST1 strains were isolated from a roller dryer that had been sampled in 2009. The *C. sakazakii* ST4 strains were isolated from a roller dryer (sampled in 2009) and from a drying tower in 2006. The *C. sakazakii* ST99 strains had been collected from the filter powder and routine testing from two towers in 2006. Additionally, one strain (1530) was ST101. This ST is in clonal complex 10 with ST99. Strain 1530 (ST101) had been isolated from filter powder collected from the same drying tower, as had some of the closely related ST99 strains. Across the three collections, the majority (28 of 39) of STs were identified in *C. sakazakii* compared with only 11 in *C. malonaticus*, *C. turicensis*, and *C. muytjensii*. The main *C. sakazakii* STs were ST4 (24%), ST1 (19%), ST40 (5%), ST99 (5%), and ST3 (5%). *C. sakazakii* ST1 and ST4 were the only STs isolated from all three collections.

A more detailed overall perspective of the STs can be gained from a study of the entries in the PubMLST database (Table 6). Clonal complex 4 (CC4) comprises single and double loci variants of *C. sakazakii* ST4. At the time of writing these strains comprise 19.4% (n=195) of the *Cronobacter* genus database. About 45% (88 of 195) of these are clinical isolates, and 25% (48 of 195) have been isolated from PIF (Table 6). This analysis demonstrates the importance of this clonal lineage with respect to *Cronobacter* spp. epidemiology, as CC4 has been identified as a genetic signature for the *C. sakazakii* meningitic pathovar. The two earlier predictions of the association of *C. sakazakii* CC4 with meningitis (Joseph and Forsythe, 2011; Hariri et al., 2013) are confirmed in Table 6 where, from over a hundred clinical isolates, this clonal lineage is prevalent and is recovered more frequently from infant infections, especially cases of meningitis in 20 of 23 cases. The reason for the association of one clonal complex in the genus with neonatal meningitis is unclear as no particular virulence traits have been determined in *C. sakazakii* CC4 compared to other STs.

It is notable that ca. 25% (n=195) of *C. sakazakii* CC4 isolates were from infant formula and also that ca. 17% of isolates were from environmental sources such as milk powder and infant formula manufacturing plants (Table 6). The table also shows that 40% of *C. malonaticus* strains (n=136) recorded in the database are in clonal complex 7. Strains in this complex have been isolated over the past 30 years. Within this complex, 57% (33 of 58) of strains are clinical in origin and primarily from children and adults (Table 6). Only one reported fatal neonatal meningitis case has been attributed to *C. malonaticus*, though the vehicle of infection is uncertain and not necessarily linked to the consumption of infant formula (Hariri et al., 2013). The database only contains seven *C. malonaticus* CC7 isolates from PIF, indicating a low incidence of this complex complex, and there are no isolates from infant formula or milk powder manufacturing plants.

Table 6 also shows for the first time that *C. sakazakii* ST12 has been associated with cases of NEC (13% of strains) and not neonatal meningitis or septicaemia. Although there are isolates in the database from infant formula, unlike CC4, there have been none from milk powder or infant formula-manufacturing plants. The remaining clonal groups in Table 6 are less associated clinically relevant and are more food and environmental isolates.

Many MLST databases are now implemented using the bacterial isolate genome sequence database (BIGSdb) platform (Jolley and Maiden, 2010; Maiden et al., 2013). This enables a range of MLST-like schemes to be described and applied to whole genome sequenced organisms, in addition to the conventional laboratory-derived seven loci. Hence hierarchical classifications, either predefined in the database or user-defined, can be applied with progressively greater resolution as the number of loci analyzed increases. In addition to the 7-loci MLST scheme, as a part of the *Cronobacter* site, a specific open-access repository for all *Cronobacter* genomes sequenced to date has been established. This enables the scalable analysis of *Cronobacter* genomes, representing all recognized species, for genes of interest and easily accommodates any changes in taxonomy.

Further alternative schemes that researchers can select include genes coding for the ribosomal proteins (rMLST, 53-loci) which has been proposed as a means of integrating microbial genealogy and typing (Jolley et al., 2012). Furthermore, genes tagged as belonging to clusters of orthologous genes can be used to generate COG-MLST with 1865 loci. By producing a phylogenetic network for increasingly large fractions of the genome, rMLST and COG-cgMLST facilitate the identification of STs and species, while illustrating the ambiguities inherent in phylogenetic reconstruction across a genus.

Expanding the multiallelic analysis to the larger rMLST scheme shows the *C. sakazakii* CC4 clonal lineage as more pronounced and separate from the remaining *C. sakazakii* (Figure 5; Forsythe et al., 2014). This is based on 51 of the 53 ribosomal protein sequences due to the absence of two genes (*rpsA* and *rpsQ*) in some strains in the sample group. This reveals the *C. sakazakii* ST4 clonal lineage as defined by just seven loci is robust and is also found using 51 other housekeeping genes.

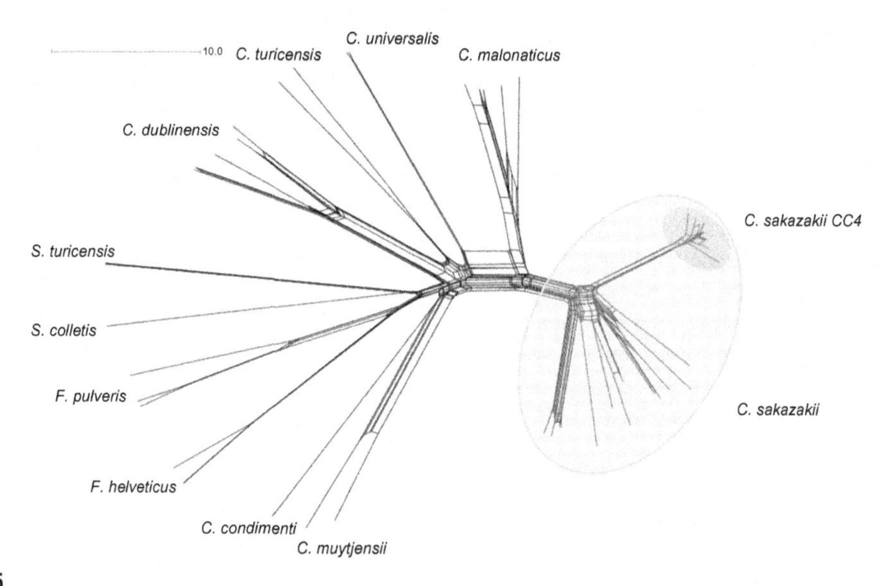

FIGURE 5

rMLST analysis of genomes across the *Cronobacter*, *Franconibacter*, and *Siccibacter* genera.

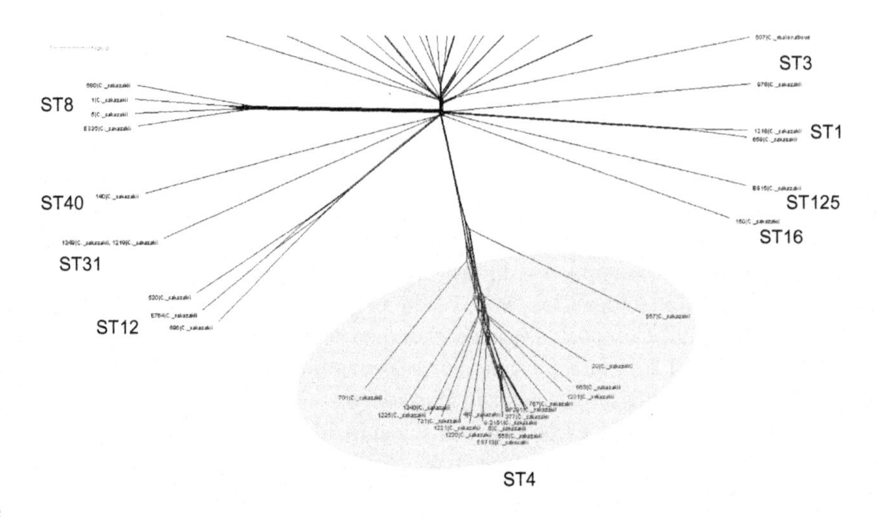

FIGURE 6

COG-cgMLST analysis of *Cronobacter sakazakii* genomes. Seven-loci MLST sequence types highlighted.

Further expansion of the genomic analysis using COG-defined genes of *C. sakazakii* ES15 as the reference genome generated 1865 loci that could be compared across the genus or selected species or strains enables core genome-MLST (cgMLST) analysis (Forsythe et al., 2014). COG loci are identified by their being tagged as belonging to COGs in *C. sakazakii* ES15 (excluding loci named in other analysis schemes). The subsequent figure also highlights the *C. sakazakii* clonal lineage associated with ST 4 (Figure 6). COG-cgMLST analysis can also be used to generate a profile of potential biochemical traits that may be of use for differentiating strains and species using phenotyping. This is particularly useful, as previously the choice of biochemical tests to try has been arbitrary. In addition, the scheme can be used to study a range of physiological traits.

The *Cronobacter* PubMLST database offers a central, open access, reliable sequence-based repository for researchers. It has the capacity to create new analysis schemes "on the fly," and to integrate metadata (source, geographic distribution, clinical presentation). It is also expandable and adaptable to changes in taxonomy, and able to support the development of reliable detection methods of use to industry and regulatory authorities. Using the *Genome Comparator* tool, traits can be identified that distinguish between strains and species. Therefore, it meets the FAO/WHO (2004) request for the establishment of a typing scheme for this emergent bacterial pathogen (Table 4). Whole genome sequencing (considered below) has additionally shown a range of potential virulence and environmental fitness traits that may account for the association of *C. sakazakii* CC4 pathogenicity, and propensity for neonatal CNS.

In summary, the MLST and analysis (MLSA) approaches have

1. Revealed the diversity of the *Cronobacter* genus.
2. Contributed to the recognition of two new species: *C. universalis* and *C. condimenti*.
3. Shown the close relatedness and groupings of the seven species.
4. Shown the evolutionary decent of the genus.
5. Revealed the majority of neonatal meningitis cases are attributable to one clonal lineage: *C. sakazakii* ST4 clonal complex.

6. Formed the basis for future research regarding strain selection for investigating *Cronobacter* virulence and environmental fitness.
7. Established an open access curated database (http://www.pubMLST.org/cronobacter). This is composed of ~1000 7-loci MLST profiled strains that are widely geographically, temporally, and sourced distributed, describing experimental protocols, contains DNA sequences of the 7-loci and 107 genomes for offline analysis, and where users can determine the ST of their strains and undertake further advanced DNA sequence analysis. The database is curated by myself and hosted by the University of Oxford, UK.

4.6 GENOME DESCRIPTION OF *C. SAKAZAKII* AND OTHER *CRONOBACTER* SPP.

A comprehensive description of the 107 *Cronobacter* genomes sequenced to date (December 2014) cannot be covered here. Instead, the reader should consult specific publications. The first *Cronobacter* genome (*C. sakazakii* BAA-894) was published by Kucerova et al. (2010) and included comparative microarray analysis with strains from other *Cronobacter* species recognized at the time. *C. sakazakii* BAA-894 was an isolate from formula in use during the *Cronobacter* outbreak at the University of Tennessee (Himelright et al., 2002). Kucerova et al. (2010) reported that of the 4382 annotated genes, ~55% (2404) were present in all *C. sakazakii* strains, and 43% (1899) were present in the other *Cronobacter* species. An estimated core gene set for *C. sakazakii* species constituted 80.9% (3547) genes and core gene for *Cronobacter* genus includes 75.3% (3301) genes. The vast majority of these shared genes are predicted to encode cellular essential functions such as energy metabolism, biosynthesis, DNA, RNA and protein synthesis, cell division, and membrane transport. Kucerova et al. (2010) also reported 13 regions of genomic variation including three integrated phages, multidrug efflux systems, fimbrial systems, sugar utilization (including sialic acid), and O-antigen. Kucerova et al. (2011) described in more detail several variable regions that are putative virulence factors; that is, fimbriae and multidrug efflux systems, many of which are plasmid borne. Putative virulence traits of particular interest were iron-uptake mechanisms, superoxide dismutase for macrophage survival, haemolysin, flagella, pili, a metalloprotease, an enterotoxin, and plasmid-borne virulence factors such as *Cronobacter* plasminogen activator (cpa) and type six secretion systems (T6SS).

Joseph et al. (2012a) extensively studied the genomic diversity within the *Cronobacter* genus using eleven strains representing the seven species. This covered whole genome phylogeny as well as regions of particular interest: fimbriae, sialic acid utilization, efflux systems, and T6SS. Currently, there are 107 *Cronobacter* genomes in the PubMLST database (http://www.pubmlst.org/cronobacter/), which can be searched using BLAST and used for comparative genomic analysis. The key results are illustrated in detail in Forsythe et al. (2014), and only a few have been selected to present in this chapter (Tables 5 and 6; Figures 3, 5, and 6). The sequenced genomes can be used to help define species boundaries. The previous DNA-DNA hybridization analysis for taxonomic purposes is labor intensive. Alternatively the average nucleotide identity (ANI) can be calculated online. The ANI values for *Cronobacter* and related genera are shown in Table 7 (Jackson et al., 2015). This confirms the separation of *Francoibacter* and *Siccibacter* from *Cronobacter*, in contrast to Brady et al. (2013).

It should be noted that misidentification of strains occurs even with genome sequenced strains. For example, Zhao et al. (2014) announced the genome of *C. sakazakii* CMCC 45402 but, based on the genome sequence, it is definitely *C. malonaticus*. Such misidentification of *C. sakazakii* and *C. malonaticus* may be due to initial identification using phenotyping or 16S rDNA sequence analysis. This has probably also occurred with the mistaken description of *C. sakazakii* serotype O:6, which matches *C. malonaticus* serotype O:2.

Table 7 Average Nucleotide Identity (%) Between *Siccibacter colletis* sp. nov. with Other Members of the Genera *Siccibacter*, *Franconibacter*, and *Cronobacter*

Species	Strain	*S. colletis* 1383^{T}	*S. turicensis* LMG 23730^{T}	*F. helveticus* LMG 23732 ^{T}	*F. pulveris* LMG 24057^{T}	*C. sakazakii* ATCC 29544^{T}	*C. malonaticus* LMG 23826^{T}	*C. universalis* NCTC 9529^{T}	*C. turicensis* LMG 23827^{T}	*C. muytjensii* ATCC 51329^{T}	*C. dublinensis* LMG 23823^{T}
S. colletis	1383^{T}										
S. turicensis	LMG 23730^{T}	87.18									
F. helveticus	LMG 23732 ^{T}	83.4	83.57								
F. pulveris	LMG 24057^{T}	83.32	83.22	88.55							
C. sakazakii	ATCC 29544^{T}	83.58	83.57	84.38	84.49						
C. malonaticus	LMG 23826^{T}	84.19	84.36	84.89	84.96	94.98					
C. universalis	NCTC 9529^{T}	84.42	83.33	85.02	85.11	93.86	94.55				
C. turicensis	LMG 23827^{T}	83.55	83.46	84.08	84.34	92.55	92.62	94.2			
C. muytjensii	ATCC 51329^{T}	83.46	83.31	84.37	84.70	89.6	89.54	90.26	89.87		
C. dublinensis	LMG 23823^{T}	83.92	83.91	84.56	84.77	89.54	89.76	90.44	89.67	82.05	
C. condimenti	NCTC 9529^{T}	83.43	83.49	84.02	84.21	88.13	88.26	88.71	88.18	88.39	88.97

Adapted from Jackson et al. (2015).

5 SOURCES OF *CRONOBACTER* SPP.

5.1 ISOLATION OF *CRONOBACTER* FROM PLANT MATERIALS

Cronobacter spp. has been isolated from a wide range of food and nonfood sources, and there is asymptomatic human carriage (Iversen and Forsythe, 2003; Hochel et al., 2012; Kandhai et al., 2004) (Tables 8 and 9). One probable niche for *Cronobacter* is plant material, as it has been isolated from cereals, wheat, corn, soy, rice, herbs and spices, vegetables, and salads (Friedeman, 2007). Fresh and dried herbs and spices are a productive source of *Cronobacter* strains with a ~30% incidence (Iversen and Forsythe, 2004). In fact, an early patent for a food thickener was material extracted from *Cronobacter* spp. isolated from Chinese tea (Harris and Oriel, 1989; Scheepe-Leberkuhne and Wagner, 1986).

Iversen and Forsythe (2003) hypothesized that a common ecosystem for the *Cronobacter* species might be plants. This was due to physiological features such as the production of a polysaccharide capsule, production of a yellow pigment, and its desiccation resistance (Iversen and Forsythe, 2003). These traits may enable the organism to stick to plant leaves, protect against oxygen radicals from sunlight exposure, and survive dry periods including autumn. They have the ability to survive osmotic stress and desiccation by taking up osmoprotectants including trehalose (via phosphotrasferase system, PTS), glycine, betaine, proline, spermidine, and putrescine, using ABC transporters (Riedel and

Table 8 Survey of Food Products and Food Ingredients from Which *Cronobacter* spp. Have Been Isolated

Food Product or Ingredient	Number of Positive Samples	Total Number of Samples	%
Follow-up formula	10	74	14
Cereal-based follow-up formula	6	100	6
Dry infant food	5	49	10
Dry infant food	22	179	12
Dry infant cereals	2	6	33
Milk powder	3	72	4
Milk powder	2	20	10
Milk powder	3	50	6
Milk-based product	5	20	25
Milk powder and derived products	1	55	2
Starches	40	1389	3
Corn, soy, wheat, and rice	14	78	18
Rice seed			
Rice flour	6	16	38
Fermented bread			
Dry food ingredients	15	66	23
Herbs and spices	40	122	33
Spices	13	21	62
Spices	14	71	20

Continued

Table 8 Survey of Food Products and Food Ingredients from Which *Cronobacter* spp. Have Been Isolated—cont'd

Food Product or Ingredient	Number of Positive Samples	Total Number of Samples	%
Spices and dried herbs	7	26	27
Sprouts and fresh herbs and spices	14	23	61
Dried powdered vegetables	1	50	2
Ready-to-eat salad	19	109	17
Salad	1	15	7
Nuts	2	2	100
Instant soups	2	13	15
Instant soups	6	10	60
Lentils	1	11	9
Vegetables	5	12	42
Vegetables	19	128	15
Courgette			
Cereal products			
Semolina	1	3	33
Sorghum			28
Cereal	8	50	16
Oat flakes	1	10	10
Wheat sprout	1	9	11
Fruit	3	41	7
Tea	3	5	60
Pastries	5	9	56
Lettuce			
Confectionary	3	42	7
Chocolate products	11	37	30
Raw meat	1	64	2
Raw meat	3	15	20
Raw meat (spiced or marinated)	17	48	35
Meat by-products (ready to eat)	9	81	11
Seeds	14	34	41
Desiccated coconut	1	10	10
Coconut biscuits	1	1	100
Dried fish	13	50	26
Sunsik	17	36	47
Tofu	4	11	36
Sour tea			
Cheese	2	62	3
Eggs	1	20	5
Shellfish	3	8	38
Seaweed	3	24	13

Table 9 Survey of Environments from Which *Cronobacter* Has Been Isolated

Source	Details	Positive Samples	Total Sample Size	%
Human body	Mouth			
	Oral rinse	3	27	11
	Pooled plaque	4	27	15
	Oral pharynx	5	1	20
	Breast abscess	*C. malonaticus* type strain		
	Skin	1	116	
	Faeces	1	98	
Environmental	Hospital air			
	Households	5	16	31
	Bathrooms			
	Kitchens	21	78	27
	Chocolate factory	4	9	44
	Cereal factory	4	9	44
	Potato flour factory	4	15	27
	Pasta factory	6	26	23
	Preparation equipment (blender, spoons)			
	Beer mugs			
	Flies			
	Rats			
	Horse nose			
	Soil			
	Rhizosphere			
	Sediment			
	Wetlands			
	Crude oil			
	Cutting fluids			
	Dust			
Liquids	Water			
	Pipelines and biofilm			
	Hydrothermal springs			

Lehner, 2007; Osaili and Forsythe, 2009). About 80% of *Cronobacter* strains produce a nondiffusible, yellow pigment on tryptone soya agar at 25 °C. Pigment production is temperature dependent, and even fewer strains produce it at 37 °C. If the organism colonizes plant material, the yellow carotenoid-based pigmentation may protect it from sunlight-generated oxygen radicals. Schmid et al. (2009) further investigated biochemical traits associated with plant microorganisms with nine strains representing the then recognized five *Cronobacter* species. All strains were able to solubilize mineral phosphate, produce indole acetic acid, and produce siderophores. The strains were also able to endophytically colonize tomato and maize roots. The authors confirmed that plants may be the natural habitat of

Cronobacter spp. and that the rhizosphere might act as a reservoir of the bacterium. The resistance to plant essential oils may be linked to efflux pumps, could contribute to the organism's resistance to osmotic pressure, and can be of use in the design of selective media.

As the organism is probably plant associated, it is not surprising that it can be isolated from a wide range of additional environments including water, soil, as well as a variety of processed foods and fresh produce (Tables 8 and 9). As described later, genomic studies have revealed that an organism may also have a more complex ecology with nonplant ecosystems. The bacterium has been isolated from factories producing milk powder, household vacuum cleaning bags, and also from household utensils used for the reconstitution of PIF (Table 9). Rats and flies may be additional sources of contamination.

The bacterium has been isolated from the hospital environment and clinical samples including hospital air, CSF, blood, bone marrow, sputum, urine, inflamed appendix, neonatal enteral feeding tubes, and conjunctivae. Human carriage has been reported in teeth, saliva, stroke patients, feces, breast milk, and skin (Table 9).

5.2 ISOLATION OF *CRONOBACTER* SPP. FROM PIF, FOLLOW-UP FORMULA, AND WEANING FOODS

Cronobacter was first associated with contaminated PIF by Muytjens et al. (1988) when it was isolated from prepared formula and reconstitution equipment. They reported 52.2% ($n=141$) of PIF samples from 35 countries contained *Enterobacteriaceae*, with 14% containing *Cronobacter* spp. As the organism has undergone various taxonomic revisions since 1988, these strains have been reidentified using DNA-sequence analysis as reported by Sonbol et al. (2013). This much-referenced strain collection was shown to comprise *C. sakazakii* (17/20), *C. malonaticus* (2/20), and *C. muytjensii* (1/20). These strains had been isolated from PIF produced from Australia, Belgium, Canada, Denmark, France, Germany, New Zealand, Russia, the Netherlands, Uruguay, and the United States. One strain identified by Muytjens et al. (1988) as *E. sakazakii* was reidentified as *E. hormaechei* and had been isolated from PIF purchased in India. It should be noted that the much-cited *Cronobacter* outbreak at the University of Tennessee was associated with the use of powdered formula that was not used for its intended target age (noninfants).

The prevalence of *Cronobacter* in PIF has been determined in samples from many countries, and the reported values vary between 2% and 14%. It is notable that there are no published reports of *Cronobacter* spp. in PIF exceeding 1 cell/g. In fact, the likely level is considered to be ~1 bacterial cell/100 g. Hence, the importance of good hygienic practices and temperature control to reduce opportunities for extrinsic bacterial contamination and multiplication. It should be noted that some early surveillance studies used methods that may not have accurately identified *Cronobacter* spp., but the approximate level of <1 cfu/g for the organism in PIF is probably a reasonable estimate.

The plant association of *Cronobacter* may account for physiological traits such as surviving spray drying and prolonged periods in dry materials (i.e., starches), and presence in ingredients that are added to PIF without additional heat treatment (FAO-WHO, 2004, 2006). An international survey for *Cronobacter* and related organisms in PIF, follow-up formula, and infant foods conducted by eight laboratories in seven countries in response to the call for data in preparation for the FAO-WHO (2008) risk assessment analyzed a total of 290 products using a standardized procedure. *Cronobacter* was isolated from 3% ($n=91$) follow-up formulae and 12% ($n=199$) infant foods and drinks (Chap et al., 2009).

The few reported quantitative studies do not show any samples with *Cronobacter* at levels >1 cell/g PIF (Santos et al., 2013). A value of <1 cell in 100g may be more representative, and therefore, large sample volumes (30×10g) are required for testing. There is also the need to consider routes for extrinsic bacterial contamination and multiplication during formula preparation. Caubilla-Barron et al. (2007) reported that *Cronobacter* can persist in PIF for 2 years. This is much longer than most other *Enterobacteriaceae*. It is notable that one of the earliest isolates of *Cronobacter* dates back to dried milk in 1960. Therefore, possibly *Cronobacter* has probably been present in dried milk products for many decades. *Cronobacter* does posses the gene encoding the universal stress protein UspA, which is also found in other closely related *Enterobacteriaceae*. In *E. coli*, the protein is induced following both heat and osmotic shock. Hence, it may be important in the survival of *Cronobacter* during manufacturing processes and the cross-induction of other protection mechanisms. For a fuller review of desiccation survival mechanisms, the reader should refer to Osaili and Forsythe (2009). The isolation of *Cronobacter* from manufacturing plants is covered later with respect to genotyping of isolates using MLST.

6 PHYSIOLOGY AND GROWTH

About 80% of strains produce a nondiffusible, yellow pigment on tryptone soya agar at 25 °C. Pigment production is temperature dependent, and even less strains produce it at 37 °C (Iversen and Forsythe, 2007). The organism probably colonizes plant material and the yellow carotenoid-based pigmentation may protect it from sunlight-generated oxygen radicals (Osaili and Forsythe, 2009). This may explain its frequent isolation from plant ingredients and products. The organism often produces a capsule that has been patented for use as a thickening agent in foods (Harris and Oriel, 1989). This capsular material may facilitate the organism's attachment to plant surfaces. It is known that the organism has been found as a part of the mixed flora biofilm in enteral feeding tubes from NICU of neonates not fed powdered formula (Figure 7; Hurrell et al., 2009b). In laboratory studies, it has been shown that a contaminated feed could cause subsequent feeds to be contaminated due to bacterial attachment and multiplication in the enteral feeding tube (Hurrell et al., 2009a).

Cronobacter can grow over a wide temperature range. The lowest is near refrigeration temperatures (~5 °C), optimal ~37-39 °C, and the maximum growth temperature is 44-47 °C. Growth of *Cronobacter* spp. in fortified breast milk has been compared with breast milk and reconstituted breast milk substitutes by Lenati et al. (2008). At 10 °C, only one in seven of the strains showed growth. Of those where data was obtained, 9 of 23 strains showed a decline or no growth and for the others an average lag time of 5.5 days and generation time of 11.97h. The lag time was significantly longer in breast milk (with or without fortifiers) than in PIF. At 23 °C, the growth rate was slower in fortified breast milk than breast milk or reconstituted PIF; lag time ~4h, generation time ~0.8h. The total growth was lower in fortified breast milk at 10 °C than in other media. The authors hypothesized that this could be *related to the presence of inhibitors in the fortifiers, which may affect E. sakazakii growth.* It should be noted that only a limited number of strains were able to grow in the fortified breast milk at 10 °C and that these were unstressed overnight cultures of *Cronobacter* spp. (akin to recent contamination source), and not stressed (desiccated, stored) cells as would occur with intrinsic contamination. Telang et al. (2005) reported that *C. sakazakii* did not show significant growth at room temperature (22 °C) for over 6h in fortified expressed milk. The inhibition of bacterial and fungal growth in fortified breast milk has been reported by other researchers, though their experimentation was not so detailed as Lenati et al. (2008)

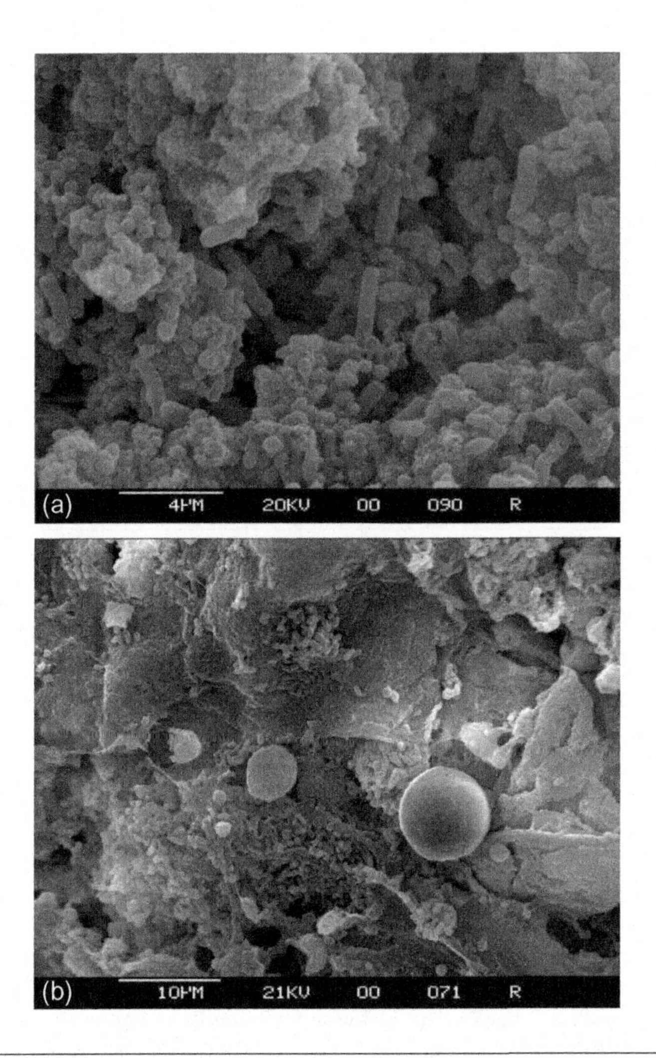

FIGURE 7

(a, b) Electron micrographs of bacteria and fungi attached to enteral feeding tubes from neonates on neonatal intensive care unit.

Hurrell et al. (2007b), copyright retained.

and, therefore, have not been presented here. These researchers also used healthy, unstressed cultures, unlike the condition of microbial cells from intrinsic contamination, which would be injured due to processing, desiccation, and prolonged storage (Caubilla-Barron and Forsythe, 2007).

6.1 CAPSULE AND BIOFILM FORMATION

Cronobacter often produces a heteropolysaccharide capsule composed of glucuronic acid, D-glucose, D-galactose, D-fucose, and D-mannose. Strains from NICU outbreaks produce so much capsular material that on milk agar plates the colonies drip onto the lid of inverted Petri dishes (Caubilla-Barron et al., 2007). This material has been patented for use as a thickening agent in foods (Scheepe-Leberkuhne and Wagner, 1986; Harris and Oriel, 1989). The capsular material, induced under nitrogen-limited conditions, could facilitate the organism's attachment to plant surfaces. Combined with a tolerance to desiccation, this gives the organism an armory to colonize plant material and maybe survive harsh environmental conditions. These traits may also contribute to the organism's presence in starches used in the manufacture of infant formula and persistence during the manufacturing process. The organism attaches to surfaces, forming biofilms that are resistant to cleaning and disinfectant agents, and the organism has been isolated as part of the mixed flora biofilm in enteral feeding tubes of neonates not fed PIF (Hurrell et al., 2009b).

Cronobacter is able to adhere to silicon, latex, and polycarbonate and, to a lesser extent, to stainless steel (Iversen et al., 2004a). Furthermore, *Cronobacter* has been reported to attach and form biofilm on glass and polyvinyl chloride (Lehner et al., 2005). All of these materials are commonly used for infant feeding and food preparation equipment and, if contaminated, may increase the risk of infection. Kim et al. (2007) reported that the ability of *Cronobacter* to form a biofilm is affected by the composition of the media, and that it is enhanced by infant formula components. The infant formula composition can also increase *Cronobacter* resistance to disinfectants, as shown by Kim et al. (2007) who examined the effect of 13 disinfectants commonly used in infant formula preparation areas. Populations of *Cronobacter* cells suspended in water (ca. 7 log CFU/ml) decreased to undetectable levels (<0.3 log CFU/ml) within 1-5 min of treatment with disinfectants, whereas numbers of cells in reconstituted PIF diminished by only 0.02-3.69 log CFU/ml after treatment for 10 min. Furthermore, cells attached to stainless steel were less resistant to disinfectants. It is clear that the ability of *Cronobacter* to attach to surfaces, form biofilms, and resist dry stress conditions contribute to the risk of *Cronobacter* ingestion. Moreover, the composition of PIF has a strong protective effect on the survival of *Cronobacter*.

Cronobacter appears to have the carbon storage regulatory (Csr) system, as evident by the presence of *Csr*A in *Cronobacter* genomes. Although its regulatory role in *Cronobacter* is unknown at present, its role in *E. coli* has been well established. CsrA is an RNA-binding protein that binds to the untranslated leader sequences of target mRNAs and alters their translation and stability. It represses stationary phase processes, including glycogen synthesis and catabolism, gluconeogenesis, and biofilm formation. It also activates glycolysis, motility, and biofilm dispersal. Repression of biofilm formation by *Csr*A involves the synthesis and catabolism of intracellular glycogen. Therefore, biofilm formation in *Cronobacter* is probably linked to central carbon metabolism.

6.2 TEMPERATURE RESPONSE AND THERMAL STRESS

Cronobacter can grow over a wide temperature range of about 5 to 44-47 °C and is strain dependent (Ref?). It should be noted that the *C. sakazakii* type strain (ATCC 29544^T) does not to grow above 42 °C (Nazarowec-White and Farber, 1999), which is the temperature used for the ISO/TS 22964 isolation method. Due to interest in the organism and infant infections, a number of growth and death rates have been determined in reconstituted infant formula. At room temperature (21 °C), *Cronobacter* had a doubling time of 40-94 min. The lowest permissible growth temperature is near refrigeration (~5 °C) and, therefore, the organism may grow following prolonged storage or poor temperature control (Zh Xu et al., 2015).

Decimal reduction times and *z*-values vary considerably between strains; that is, D_{55} 2-49 min, *z*-values 2-14 °C (Caubilla-Barron et al., 2009). Early studies inferred that the organism was very thermotolerant. However, subsequent work clarified that the organism was less thermotolerant than *Listeria monocytogenes* (Nazarowec-White and Farber, 1999). Nevertheless, the organism can survive spray drying, albeit with a considerable reduction in viability, and the surviving cells may be severely damaged (Caubilla-Barron et al., 2009). As stated already, the organism's tolerance to drying has been well noted, and it can survive for 2 years desiccated in infant formula and then rapidly grow on reconstitution (Caubilla-Barron and Forsythe, 2007).

In order to reduce the number of intrinsic bacteria and limit bacterial growth, the FAO-WHO (2004, 2006) expert committees proposed that PIF be reconstituted at temperatures no cooler than 70 °C, and that it be used immediately rather than stored. This is very pertinent because a common feature in a number of outbreaks has been a lack of adequate hygienic preparation and temperature control of the reconstituted infant formula (Himelright et al., 2002; Caubilla-Barron et al., 2007). A second outcome from the FAO-WHO meetings was the production of an online risk model (http://www.mramodels.org/ESAK/default.aspx). The model allows the user to compare the level of risk between different levels of contamination and reconstitution practices. The model was based on growth and death kinetic data for a limited number of *Cronobacter* strains. We extended the risk model to cover all organisms in categories A and B: *Cronobacter* species, *Salmonella*, other named *Enterobacteriaceae*, and *Acinetobacter* spp. It can be accessed from the UK FSA website at http://www.foodbase.org.uk/results.php?f_category_id=&f_report_id=395. The data was generated using casein- and whey-based formulas because the type of formula affects bacterial lag times, growth, and death rates.

As referred to above, the WHO guidelines for hygienic preparation of PIF are aimed at reducing the number of bacteria in the reconstituted product by using hot water and limiting the time available for any survivors to multiply. However, a wider perspective is that neonates are frequently feed via enteral feeding tubes. These tubes are in place for prolonged periods (even several days) to reduce distress to the neonate by the gagging reaction. However, *Cronobacter* and other opportunistic pathogens can attach and colonize these tubes, which are at 37 °C, and at regular intervals receive fresh feed (Hurrell et al., 2009a,b). This scenario is applicable to all neonates with nasogastric tubes, and not only those on reconstituted PIF. In fact, *Cronobacter* and other *Enterobacteriaceae* have been isolated from such tubes in intensive care units from neonates receiving breast milk and various other feeding regimes at levels up to 10^7 cfu per tube (Hurrell et al., 2009a,b). Therefore, hygienic practices and avoidance of temperature abuse are vitally important regardless of the type of feed. Electron micrographs of used enteral feeding tubes from NICU are shown in Figure 7 (Hurrell et al., 2009a). It is evident that there is a considerable biofilm present that is composed of bacteria and fungi. Organisms recovered include *C. sakazakii*, *E. coli*, *E. cloacae*, *E. hormaechei*, *Klebsiella pneumoniae*, *Klebsiella oxytoca*, *Serratia marcensens*, and *Candida* spp. (Hurrell et al., 2009b).

7 *CRONOBACTER* INFECTIONS

Only four *Cronobacter* species are associated with human infections: *C. sakazakii*, *C. malonaticus*, *C. turicensis*, and *C. universalis*. The species can be grouped, with the mostly clinically relevant being Group 1: *C. sakazakii* and *C. malonaticus*, which form the majority of clinical isolates in all age groups, and Group 2: *C. turicensis* and *C. universalis*, which have been less frequently reported. The other three species (*C. dublinensis*, *C. muytjensii*, and *C. condimenti*) are primarily environmental commensals and are probably of little clinical significance. Table 10 shows an age profile of *Cronobacter* isolated

Table 10 Isolation of *Cronobacter* spp. from Throat Swabs of Outpatients According to Age Groups

	Patient Age (Years)								
Year	**<1**	**1-4**	**5-9**	**10-14**	**15-44**	**45-64**	**65-74**	**>75**	**Total**
2005	1	5	1	1(1)	2	0	2	0	12(1)
2006	(1)[a]	1(2)	0	5	2	0	(3)	(2)	8(8)
2007	2	9	0	2	2(1)	(1)	(1)	(1)	15(4)
2008	0	1(1)	2	2(1)	0	0	(2)	(1)	5(5)
2009	(2)	4	1	2	1	(2)	1	(1)	9(5)
2010	0	5	2(1)	0	(2)	1(1)	(2)	2	10(6)
2011	(1)	0	2	0	0	0	0	0	2(1)
Total number of *Cronobacter* isolates	3(4)	25(3)	8(1)	12(2)	7(3)	1(4)	3(8)	2(5)	61(30)
% of isolates/age group	4.9	41	13.1	19.7	11.5	1.6	4.9	3.3	100.0
Number of patients sampled[b]	808	3404	2651	1867	18,223	10,744	3888	3538	45,123
Incidence *Cronobacter* isolates/1000 patients sampled	8.7	8.2	3.4	7.5	0.5	0.5	2.8	2.0	2.0

[a] *Numbers in parenthesis indicate additional isolates from normally sterile sites, sampled according to the clinical presentation of the patient.*
[b] *Total number during period 2005-2011.*
From Holy and Forsythe (2014).

using throat swabs from over 45,000 outpatients during the period 2005-2011. The organism was isolated from every age group, with a higher frequency from children less than 14 years of age.

Members of the *Cronobacter* genus have come to prominence due to its association with severe neonatal infections leading to NEC, septicaemia, and meningitis, which can be fatal. However, *Cronobacter* infections are not unique to neonates. They occur in all age groups, albeit with a greater incidence in the more immunocompromized very young and elderly (FAO-WHO, 2008; Holy and Forsythe, 2014; Patrick et al., 2014). A major difference between the age groups is the severity of infection in neonates. Neonates, particularly those of low birth weight, are the major identified group at risk with a high mortality rate. In the United States, the reported *Cronobacter* infection incidence rate is ~1 per 100,000 infants. This incidence rate increases to 9.4 per 100,000 in infants of very low birth weight (i.e., <1.5 kg, Stoll et al., 2004). The first neonatal deaths attributed to *Cronobacter* were in 1958 (Urmenyi and Franklin, 1961). Symptoms in neonates include NEC, septicaemia, and meningitis. The former is noninvasive (and is multifactorial), whereas in septicaemia and meningitis the organism has attached and invaded presumably through the intestinal epithelial layer. NEC is a common gastrointestinal illness in neonates and can be caused by a variety of bacterial pathogens. It is characterized by ischemia, bacterial colonization of the intestinal tract, and increased levels of proteins in the gastrointestinal lumen. The incidence of NEC is 2-5% of premature infants and 13% in those weighing <1.5 kg at birth. It is 10 times more common in infants fed formula compared with those fed breast milk (Lucas and Cole, 1990). NEC has a high mortality rate: 15-25% of cases (Henry and Moss, 2009). *Cronobacter* has been implicated as a causative agent of NEC, but its role in the pathogenesis of the disease is somewhat unclear. There are reports of *Cronobacter* isolation from babies who developed NEC (van Acker et al., 2001; Caubilla-Barron et al., 2007). This suggests that there is an association between *Cronobacter* occurrence and NEC, although until recently, the organism has not been conclusively proven to cause the disease.

Fatal infant infections have followed cases of NEC, septicemia, and meningitis. Infections in older age groups are principally bacteremia as well as urosepsis and wound infections. NEC is noninvasive (and is multifactorial), whereas in septicemia and meningitis, the organism has attached and invaded presumably through the intestinal epithelial layer. Due to the understandable sensitivity toward neonatal infections, this aspect of the bacterium has been emphasized more than infections in other age groups. NEC is a common gastrointestinal illness in neonates and can be caused by a variety of bacterial pathogens.

In *Cronobacter* meningitis, there is gross destruction of the brain, leading sadly to either death (40% to 80% of cases) or severe neurological retardation (Lai, 2001). The pathogenesis of the meningitis is different from *Neisseria meningitidis* and neonatal meningitic *E. coli* and is similar to that of the closely related bacterium *Citrobacter koseri* Kline et al., 1988. A number of outbreaks of *Cronobacter* spp. have been reported in NICU. *Cronobacter*-related meningitis is characterized by a mortality rate of 40-80% and generally a very poor clinical outcome. The bacterium causes cystic changes, abscesses, fluid collection, brain infarctions, hydrocephalus, necrosis of brain tissue, and liquefaction of white cerebral matter. Patients surviving *Cronobacter*-related meningitis often suffer from severe neurological sequelae, such as hydrocephalus, quadriplegia, and retarded neural development (Lai, 2001). The infection usually arises between the fourth and the fifth days after birth, and it can be fatal within hours to days following the first clinical signs (Muytjens et al., 1983). Compared with patients suffering from *Cronobacter*-induced enterocolitis, infants in whom meningitis developed tend to have normal gestational age and birth weight (Bowen and Braden, 2006).

Epidemiological investigations of outbreaks are generally easier to study than individual cases due to the possibility of identifying a common feature. Hence, there is more documentation of NICU outbreaks associated with *Cronobacter* than cases in the home. Bowen and Braden (2006) have reported that a number of neonatal cases could not be linked to the consumption of reconstituted PIF. Pertinent to this is the recognition of bacterial colonization of the nasogastric enteral feeding tube, and the isolation of *C. sakazakii* from tubes from neonates fed breast milk, and (sterile) ready to feed formula (Hurrell et al., 2009b). Also of possible relevance is the recent MLST analysis that has shown that *Cronobacter* species are clonal; that is, they do not extensively swop DNA within the species and that distinct subgroups exits; some of which are primarily composed of clinical strains (Tables 5 and 6; Baldwin et al., 2009; Joseph et al., 2012b; Forsythe et al., 2014).

Infections in older age groups are principally bacteremias as well as urosepsis and wound infections. *C. malonaticus* ST7 appears to be more associated with adult than neonatal infections (Table 6; Joseph and Forsythe, 2011). Therefore, pathogenicity in humans may be an acquired trait in this genus. Infections caused by *Cronobacter* in adults comprise a wide range of symptoms from conjunctivitis, biliary sepsis, urosepsis, and appendicitis to wound infection and pneumonia. Adult patients at increased risk include those previously treated with antibiotics, immunocompromised and elderly patients, those with medical implants or with acute, chronic, or serious illnesses (Pitout et al., 1997; Patrick et al., 2014).

7.1 SOURCES OF *CRONOBACTER* INFECTION OF NEONATES

While the source of contamination in *Cronobacter*-related outbreaks has not always been confirmed, breast milk substitutes (one group of PIF products) have been epidemiologically or microbiologically established as the source of infection in a number of cases (Muytjens et al., 1983; Clark et al., 1990; van Acker et al., 2001). A strong link between the presence of *Cronobacter* in formula feeding and an outbreak of *Cronobacter* infection was established by the Centers for Disease Control and Prevention in 2002 following the outbreak in a NICU in Tennessee in 2001.

The Tennessee outbreak was investigated by the FDA and the CDC, and their report was published in the CDC's online journal *Emerging Infectious Disease* (Himelright et al., 2002). Using PFGE to genotype isolates, the same strain in the CSF of the neonate who died was also found in a previously unopened tin of powdered formula. According to PFGE profiling, the same strain was recovered from the neonate who died from meningitis, two suspected infections (trachael aspirate positive), and seven infants who were colonized (six fecal positive and one urine positive). The source was attributed to powdered formula. Hence, instead of a generalized nosocomial attributed cause of infection, a source could be named and, therefore, a control measure could be applied. Why one neonate had died from meningitis, while two had respiratory problems and seven were asymptomatic, is unknown. It is often overlooked that the formula in the Tennessee outbreak was a commercial formula product not intended for use by neonates but was used following the instruction of neonatologists. This issue has not been considered further and is generally overlooked by researchers in the topic.

Infections have been directly linked to reconstituted PIF that may have been contaminated intrinsically or during preparation and administration. A common feature in some of these outbreaks is the opportunity for temperature abuse of the prepared feed, which would permit bacterial growth. In reported outbreaks in France and the United States, the neonates were fed using perfusion devices, whereby the reconstituted PIF is slowly pumped at ambient temperature into the neonate through an enteral-feeding tube (Himelright et al., 2002; Caubilla-Barron et al., 2007). Using this procedure there is the possibility of bacterial multiplication in the syringe leading to the ingestion of large numbers of *Cronobacter* by the neonate. The neonate has an immature immune system and a low intestinal microflora density. Consequently, if a large number of *Cronobacter* cells were ingested they would not be outcompeted by the resident intestinal flora. Following invasion of the intestinal cells, the lack of a developed immune system could make the neonate more prone to systemic infection. No infectious dose has been determined for neonates. Animal studies by Pagotto et al. (2003) and Richardson et al. (2009) have used large numbers of *Cronobacter* cells (~10^8) for infection studies. Whether this number is reflective of that necessary for neonates is uncertain, but it does contrast with the number of cells reported in contaminated PIF (<1 cfu/g) and may, therefore, indicate the role of temperature abuse in enabling bacterial multiplication.

Some neonatal infections have been linked to reconstituted breast milk substitutes that may have been contaminated intrinsically or during preparation and administration. It is pertinent to note that the bacterium is isolated from the pharyngeal tract (Table 9; Baltimore et al., 1989; Caubilla-Barron et al., 2007) and has been recovered from the nasogastric feeding tubes of neonates fed breast milk and sterile ready-to-feed formula, not infant formula (Hurrell et al., 2009b). Therefore, wider sources of the organism during an outbreak need to be investigated than just the use of breast milk substitutes. Infants can be colonized by more than one strain of *Cronobacter*, and therefore, multiple isolates need to be characterized in epidemiological investigations (Caubilla-Barron et al., 2007).

Bowen and Bradden have reported that there are a number of neonatal cases that have no links with the ingestion of infant formula (Bowen and Braden, 2006). Therefore, in epidemiological investigations multiple sources should be sampled. Breast milk can contain the bacterium, and the *C. malonaticus* type strain (LMG 23826^T) was isolated from a breast abscess. In some countries, breast milk from mothers with mastitis is still fed to the neonate. The organism has also been isolated from hospital air, human intestines, and throats. So controlling the microbiological content of PIF will not necessarily totally remove the risk of neonate infection by this bacterium.

Factors affecting the probability of infection (which is different from probability of ingestion) relate to the pathogen, the food, and the host. The pathogen may be adapted to survive heat treatment and the acidity of the neonatal stomach. The food may have a high buffering capacity and, hence, protect the bacterium's transit through the stomach, and the host's immune status will affect the likelihood of infection. The infectivity, therefore, can vary according to the organism's history (i.e., prior adaptation to heat and acidity), the host state (healthy or immunocompromized), and the food matrix. In the case of PIF and neonates, the organism will have been stressed during spray drying and storage, the host is immunocompromized and, being a liquid, the initial portion of the feed may pass quickly through the stomach (mild acidic conditions) into the small intestines.

8 VIRULENCE MECHANISMS

Although all *Cronobacter* species are considered potentially pathogenic, to date isolates from infected neonates and adults have been limited to only three species: *C. sakazakii*, *C. malonaticus*, and *C. turicensis* (Kucerova et al., 2010). These species can invade human intestinal cells, replicate in macrophages, and invade the blood-brain barrier. It is known that *Cronobacter* strains and species vary in their virulence (Caubilla-Barron et al., 2007). *In vitro* studies have shown that bacterial attachment and invasion of mammalian intestinal cells, macrophage survival, and serum resistance is comparable with *E. cloacae* and *Cit. freundii*, but less than that for *Salmonella typhimurium* (Townsend et al., 2007). Strains from *C. sakazakii* and *C. malonaticus* showed higher invasion of Caco-2 (human cell line) than other *Cronobacter* species. Similarily, *C. sakazakii* and *C. malonaticus* survive and replicate in macrophages inside phagosomes, whereas *C. muytjensii* dies, and *C. dublinensis* is serum sensitive. Virulence also varies within the *C. sakazakii* species. This was determined from epidemiological studies of an NICU outbreak in France where the clinical outcome of three *C. sakazakii* pulse types varied with only one pulse type causing the three deaths (Caubilla-Barron et al., 2007). The organism can invade human intestinal cells, replicate in macrophages, and invade the blood-brain barrier, and this varies between species (Townsend et al., 2007, 2008). Of particular interest are *C. sakazakii*, *C. malonaticus*, and *C. turicensis*, which are the only *Cronobacter* species that have been isolated from neonatal infections. Based on the clinical outcome of an outbreak in France (Caubilla-Barron et al., 2007) it was proposed that certain variants of *C. sakazakii* were more virulent, and this has been confirmed by MLST (Joseph and Forsythe, 2011; Hariri et al., 2013). However, the basis of the variation in virulence is unknown.

OmpA is produced by *Cronobacter* and has been used as an identification trait (Nair and Venkitanarayanan, 2006). This protein has been extensively studied in *E. coli* K1 as contributes to the organism's serum resistance, adhesion to host cells, and invasion of brain microvascular endothelial cells. It is logical to predict that it also has a role in *Cronobacter* penetrating the blood-brain barrier. However, the mechanism leading to the destruction of the brain cells is unknown and could, in part, be a host response. *Cronobacter* may invade the tissues using pathogenic secretory factors (elastases, glycopeptides, endotoxins, collagenases, and proteases), which increase the permeability of the blood-brain barrier and allow the organism to gain access to the nutrient-rich cerebral matter (Iversen and Forsythe, 2003). Only a limited number of animal studies have been undertaken on *C. sakazakii* (Pagotto et al., 2003; Richardson et al., 2009), but these have confirmed the variation within the species. The majority of animal studies have used *C. muytjensii* ATCC 51329, as this was the Preceptrol™ strain for the former *E. sakazakii* species. However, no clinical cases have been reported for this species and so the relevance of the studies is uncertain (Mittal et al., 2009).

Townsend et al. (2007) showed that *Cronobacter* can attach to intestinal Caco2 cells and survive in macrophages, but the invasion mechanism remains unknown. Kim and Loessner (2008) suggested that the invasion of *Cronobacter* to Caco2 cells may be receptor-mediated, as the bacterial invasion showed characteristics of saturation kinetics. The authors also concluded that bacterial *de novo* protein synthesis was required for invasion. In the same study, pretreatment of Caco2 cells with an actin polymerization inhibitor resulted in decreased invasiveness of *L. monocytogenes* and *S. typhimurium* but enhanced the invasiveness of *Cronobacter*. The authors hypothesized that this enhancement was due to the disruption of tight junction, a membrane-associated structure that acts as a barrier against the molecular exchange between epithelial cells. This was confirmed when the disruption of the tight junction by EGTA significantly increased the invasive properties of *Cronobacter*. They also speculated that frequent lipopolysaccharide (LPS) contamination of PIF that is known to disrupt tight junctions might contribute to the invasiveness of *Cronobacter* (Kim and Loessner, 2008).

Townsend et al. (2008) studied seven *C. sakazakii* strains associated with the largest reported NICU outbreak with the most reported deaths to date. All strains were able to attach and invade intestinal cells Caco2 more than *E. coli* K12 and *Salmonella* Enteritidis, and two *C. sakazakii* strains (767 and 701) associated with fatal cases of meningitis and NEC showed the highest invasion rates. These two strains were also able to replicate within macrophages, while all other strains survived inside macrophages for at least 48 h.

Although Pagotto et al. (2003) described the production of an enterotoxin by some *Cronobacter* strains, the genes encoding the putative toxin have yet to be identified. The *C. sakazakii* type strain ATCC 29544^T showed no enterotoxin production in their study, which confirms that there are significant differences in virulence among *Cronobacter* strains and that some strains may be nonpathogenic. Kothary et al. (2007) characterized a zinc metalloprotease, zpx, which was unique to 135 *Cronobacter* strains tested, which could allow the bacterium to penetrate the blood-brain barrier and cause meningitis. The protein is found in all *Cronobacter* species (Kucerova et al., 2010; Joseph et al., 2012b), although there is some sequence variation (Kothary et al., 2007). Although *C. muytjensii* has not been associated with neonatal infections, one strain (ATCC 51329^T, source unknown) has been used in animal studies to demonstrate its potential to infect neonates (Mittal et al., 2009).

Current knowledge of the virulence and epidemiology of *Cronobacter* spp. is limited. The sequenced genomes of *Cronobacter* species have revealed an array of adhesins, outer-membrane proteins, efflux systems, iron-uptake mechanisms, haemolysins, and type VI secretion systems (Grim et al., 2012, Kucerova et al., 2010, 2011; Joseph et al., 2012a). Other candidate virulence determinants include superoxide dismutase (*sodA*) for macrophage survival (Townsend et al., 2007, 2008), flagella (Cruz et al., 2011), a metalloprotease (Kothary et al., 2007), an enterotoxin (Pagotto et al., 2003), and plasmid-borne virulence factors such as *Cronobacter* plasminogen activator (*cpa*) and type six secretion systems (T6SS) (Franco et al., 2011). The bacteria can attach to intestinal cells and survive in macrophages (Townsend et al., 2007, 2008). OmpA and OmpX possibly have a role in the organism penetrating the blood-brain barrier, though the mechanism leading to the destruction of the brain cells is unknown and could, in part, be a host response (Kim and Loessner, 2008; Kim et al., 2010). Following a multiple-strain *Cronobacter* outbreak at a French NICU, it became apparent that possibly not all *Cronobacter* strains, as profiled using PFGE, were equally virulent (Caubilla-Barron et al., 2007). This observation contributed to the recognition of *C. sakazakii* ST4 as the major clonal lineage associated with neonatal meningitis cases (Joseph and Forsythe, 2011).

Because *Cronobacter* is associated with neonates and infants, the availability of iron in milk or formula could be an important virulence trait. A list of known iron assimilation mechanisms was initially compiled by Kucerova et al. (2010), and their presence in different *Cronobacter* species was evaluated based on the available CGH data. All *Cronobacter* strains examined by CGH possess complete operons for enterobactin synthesis (*entABCDEFS*) and enterobactin receptor and transport (*fepABCDEG*), except *C. dublinensis*, in which *fepE* was absent. All *Cronobacter* species except *C. muytjensii* also possess a complete operon for aerobactin synthesis *iucABCD* and its receptor *iutA*. The operon for salmochelin synthesis is missing in all *Cronobacter* species. The strong genetic similarity between *C. sakazakii* and *Cit. koseri*, as well as urinary pathogenic *E. coli* is evident from the presence of all genes for enterobactin and aerobactin synthesis in these organisms. *C. sakazakii* can cause urinary tract infections, though to date this aspect has not been studied in any detail.

Ten fimbriae clusters were identified in the genomes of *Cronobacter* species (Joseph et al., 2012a). Many fimbriae clusters are common to all species, though there are two interesting exceptions. *C. sakazakii* is the only *Cronobacter* species encoding for β-fimbriae, whereas the genomes of the other species encode for curli fimbriae. This may reflect evolution to the host ecosystem. Also, it has been proposed that the outer-membrane proteins OmpA and OmpX have roles in *Cronobacter* penetrating the blood-brain barrier. The mechanism(s) leading to the destruction of the brain cells in unknown and could in part be a host response.

Type VI secretion system (T6SS) is a newly described system that may be involved in adherence, cytotoxicity, host-cell invasion, growth inside macrophages, and survival within the host. Five putative T6SS clusters have been described for *Cronobacter* (Kucerova et al., 2010; Joseph et al., 2012a). Most T6SS clusters typically encode from 12 to 25 proteins (Filloux et al., 2008) and also encode a ClpV ATPase, which is not found in all *Cronobacter* T6SS regions.

The organism has several plasmid-borne features of interest, including encoding for an outer-membrane protease (*Cronobacter* plasminogen activator, cpa) that has significant identity to proteins belong to the Pla subfamily of omptins. Members of this subfamily of proteins degrade a number of serum proteins, including circulating complement, providing protection from the complement-dependent serum killing (Franco et al., 2011).

C. sakazakii and some *C. turicensis* strains can utilize exogenous sialic acid, and this may have clinical significance (Joseph et al., 2013b). The ability to utilize sialic acid could be a major evolutionary host-adaptation because the compound is found in breast milk, mucin, and gangliosides. Sialic acid is also an ingredient in PIF due to its association with brain development. *C. sakazakii* is also able to grow on the ganglioside GM1 as a sole carbon source.

High levels of heat-stable LPS (endotoxin) in infant formula may enhance the translocation of *Cronobacter* across both the intestines and the blood-brain barrier and, therefore, increase the risk of bacteremia in neonates (Townsend et al., 2007). Levels of endotoxin vary 500-fold in PIF. Levels of endotoxin due to the presence of heat-stable LPS have been measured in infant formula. In rat pups, LPS enhances the translocation of *Cronobacter* across both the intestines and the blood-brain barrier and, therefore, indicates its presence in infant formula could increase the risk of bacteremia in neonates. It has been speculated that frequent LPS contamination of PIF (known to disrupt tight junctions) might contribute to the invasiveness of *Cronobacter* across the blood-brain barrier.

Finally, the organism also has a number of heavy metal resistance factors and biofilm formation, which might enable it to resist disinfectants in food production environments. The organism also encodes for a number of haemolysins.

8.1 ANTIBIOTIC SUSCEPTIBILITY

When an infection by *Cronobacter* occurs, it is essential to provide rapid antibiotic treatment. *Cronobacter*-related infections have been traditionally treated with ampicillin combined with gentamicin or chloramphenicol (Lai, 2001). Initial reports concerning the ability of *Cronobacter* to produce β-lactamases gave conflicting results. The presence of ß-lactamases in *Cronobacter* was reported in a study by Pitout et al. (1997) when all tested strains were positive for Bush group 1 ß-lactamase (cephalosporinase). In 2001, Lai reported increasing β-lactamase production among *Cronobacter* strains. Similarly, Block et al. (2002) reported that all *Cronobacter* isolates tested were β-lactamase positive. However, Stock and Wiedemann (2002) did not find any evidence of β-lactam production in the 35 *Cronobacter* strains tested. The discrepancy in the results might be due to the different selection of strains, the limited number of strains used, as well as differences in the experimental protocol. In 1980, all strains tested by Farmer et al. were susceptible to ampicillin. In 2001, Lai described five cases of *Cronobacter* infection in which one or more of the isolates were resistant to ampicillin and most cephalosporins of the first and the second generations. Kim et al. (2008) reported frequent resistance of *Cronobacter* food isolates to ampicillin and cephalotin. For this reason, the shift to carbapenems or third-generation cephalosporins with an aminoglycoside or trimethoprim with sulfamethoxazole was proposed. This treatment improved the outcome of *Cronobacter* meningitis but may also have caused the increase in resistance to these antimicrobials (Lai, 2001). Unfortunately, if the organism has developed resistance to ampicillin (Muytjens et al., 1983; Lai, 2001), the alternative use of gentamicin is limited as it fails to reach sufficient concentrations in the cerebral spinal fluid (Iversen and Forsythe, 2003).

9 CONTROL MEASURES

9.1 FAO-WHO (2004, 2006, 2008) RISK ASSESSMENTS OF THE MICROBIOLOGICAL SAFETY OF PIF

The International Commission for Microbiological Specifications for Foods (ICMSF) has ranked *Cronobacter* as a "Severe hazard for restricted populations, life threatening or substantial chronic sequelae or long duration" (ICMSF, 2002). To date, the raised awareness of the organism has focused on infant infections and resulted in changes in the microbiological criteria for PIF and reconstitution procedures. In other words, there have been required changes on two sides of the same coin; manufacturing practices and hygienic preparation practices. Such requirements need regulatory enforcement and support, but must be based on robust, reliable information. In order to achieve this, there have been three FAO-WHO risk assessment meetings on the microbiological safety of PIF (FAO-WHO, 2004, 2006, 2008). The key summary is given in Table 1. Those identified as being at high risk of *Cronobacter* infection were neonates (especially low birth weight) for whom their source of nutrition will be limited to breast milk, fortified breast milk, or breast milk replacement. Due to their immature immune system and lack of competing intestinal flora, the hygienic preparation of their feed is essential to reduce the incidence of infections. In order to achieve this, there have been required changes in both manufacturing practices and hygienic preparation practices. Such requirements needed regulatory enforcement and support and must be based on robust, reliable information as reviewed in earlier sections. Prior to the concerns over *Cronobacter*, the CAC microbiolgical criteria for PIF had not been revised since 1979. The CAC guidelines were revised in 2008 for products with an intended use with infants aged 0-6 months.

The risk assessments by FAO-WHO (2004, 2006, 2008) were in part in response to the CAC request for expert advice only for the control of *Cronobacter* spp. in the food chain. The focus on PIF was because it is one of the routes of infection and as such was the one the FAO-WHO and the CAC had scope to improve via guidelines on microbiological criteria and hygienic preparation instructions for the general public to reduce the risk of exposure. As well as testing the final powdered formula, environmental samples are taken in the production environment as well as ingredients. In addition, production facilities and processes are designed to control other enteric pathogens, especially *Salmonella*. PIF are not necessarily sterile but conform to international microbiological specification guidelines.

Although the 1979 CAC criteria did not include *Cronobacter* spp. *per se*, they did specify testing for the general bacterial family *Enterobacteriaceae* or for coliforms. The *Enterobacteriaceae* commonly inhabit the large intestines of animals and humans, as well as soil and are not necessarily harmful. Therefore, their presence can only be used as an indication of the possible presence of serious intestinal pathogens. However, this group of microbes has been hard to specifically define, and the test does not allow for the recovery of bacterial cells that may be injured during processing (i.e., dehydration) and need resuscitation before they can grow on selective agars. Therefore, the current *Enterobacteriaceae* direct count test does not guarantee the absence of enteric pathogens. Hence, there is the need for the presence of *Salmonella* to be tested, which are members of the *Enterobactericaeae* family. Similarly, *Cronobacter* is a genus within the *Enterobacteriaceae* family. However the *Enterobacteriacae* and coliform tests may not always detect *Cronobacter* in PIF samples (Iversen and Forsythe, 2004). This is probably due to the low cell number and injury to the organism during desiccation. Therefore, specific test methods needed to be developed and approved for international usage through the ISO and FDA standards authorities.

The first two FAO-WHO (2004, 2006) meetings to consider *Cronobacter* spp. and other microorganisms in PIF reviewed the published refereed literature concerning bacterial infections attributed to PIF and what was known at that time concerning potential infectious organisms. They categorized bacterial pathogens associated with PIF into three groups.

Category A, Clear evidence of causality;
Salmonella and *Cronobacter* spp.,
Category B, Casuality plausible, but not yet demonstrated;
Enterobacteriaceae: E. coli, Escherichia vulneris, Cit. koseri, E. cloacae, Hafnia alvei, Pantoea agglomerans, K. pneumoniae, and *K. oxytoca*,
Non-*Enterobacteriaceae*: *Acinetobacter* spp.
Category C, Causality less plausible or not yet demonstrated;
Clostridium botulinum, Staphylococcus aureus, L. monocytogenes, and *Bacillus cereus*.

Therefore, only *Salmonella* serovars and *Cronobacter* spp. were attributed to causing neonatal infection through contaminated PIF. Consequently, the focus on the risk assessment of the microbiological safety of PIF was on these two organisms, especially *Cronobacter* spp., as the control of *Salmonella* was already well described; although, the reader should note the FAO-WHO (2004) comments on the potential failure of commonly used detection schemes to recognize the presence of lactose-positive *Salmonella* in PIF.

Because PIF is not produced as a sterile product, the need for good hygienic practice in subsequent preparation before feeding is essential. The ingestion of raised numbers of *Cronobacer* spp. due to temperature abuse after reconstitution is highlighted as well as the uncertain routes of product contamination.

As *Cronobacter* had not been previously recognized as an organism of concern, it was not surprising that initially there were no specific or reliable methods for its recovery or identification. A major improvement was the development of chromogenic agars that enabled the visualization of *Cronobacter* in a mixed culture of other *Enterobacteriaceae*, which was not feasible with MacConkey agar as initially recommended by the FDA (Iversen et al., 2004a). The FAO-WHO (2004) expert committee recommended that research should be promoted to gain a better understanding of the ecology, taxonomy, virulence, and other characteristics of *Cronobacter* (Table 4).

Currently, microbiological criteria for *Cronobacter* spp. are required for infant formulas with an intended target age <6 months. A presence/absence test is applied to large volumes due to the low (<1 cfu/g) incidence of the organism in the product. Although the organism has been recovered from follow-up formulas (infant formulas with intended target age >6 months) and weaning foods, there is currently insufficient epidemiological evidence to support the implementation of criteria for these products. Readers should consult the relevant Codex (2008) documents for details.

Key advice from these FAO-WHO risk assessments was that PIF should be reconstituted with water >70 °C, minimize any storage time by not preparing in advance, and if storage for short periods is necessary then the temperature should be <5 °C. The high water temperature will drastically reduce the number of vegetative bacteria present, and minimizing the storage period will reduce the multiplication of any surviving organisms. These recommendations have been well addressed by the WHO "Guidelines for the safe preparation, storage and handling of powdered infant formula," which are available online and can be downloaded using the URL given in the Reference section. These recommendations have been adopted by some countries, but not all.

Despite the emphasis on *Cronobacter* infections of infants, it should be noted, however, that such neonatal infections are rare, and not all have been associated with reconstituted formula ingestion. Breast milk has been a suspected source in two meningitis cases (Barreira et al., 2003; Stoll et al., 2004). *C. malonaticus*-type strain was isolated from a breast abscess. CDC has reported a number of infections in infants with no known exposure to breast milk substitutes. An infection occurred in the state of Florida in 2011 that was reported be in a breast-fed baby. Similarly, with a case in Israel where there has been a *C. sakazakii* infection without any evident link to PIF exposure. *Cronobacter* species have also been isolated from hospital air, dust, human intestines, and throats. So control of microbiological content of PIF will not necessarily totally remove the risk of neonate infection by this bacterium. Sources of adult infections are unknown at present.

ACKNOWLEDGMENTS

This chapter made use of the *Cronobacter* Multi Locus Sequence Typing website (http://pubmlst.org/cronobacter/) developed by Keith Jolley and sited at the University of Oxford (Jolley and Maiden, 2010). The development of this site has been funded by the Wellcome Trust. The author also thanks the numerous collaborators who have submitted sequences to the *Cronobacter* PubMLST database.

REFERENCES

Baldwin, A., Loughlin, M., Caubilla-Barron, J., Kucerova, E., Manning, G., Dowson, C., Forsythe, S., 2009. Multilocus sequence typing of *Cronobacter sakazakii* and *Cronobacter malonaticus* reveals stable clonal structures with clinical significance which do not correlate with biotypes. BMC Microbiol. 9, 223.

Baltimore, R.S., Duncan, R.L., Shapiro, E.D., Edberg, S.C., 1989. Epidemiology of pharyngeal colonization of infants with aerobic Gram-negative rod bacteria. J. Clin. Microbiol. 27, 91–95.

Barreira, E.R., et al., 2003. Meingite por *Enterobacter sakazakii* em recém-nascido: relato de caso. Pediatria (São Paulo) 25, 65–70.

Block, C., Peleg, O., Minster, N., Bar-Oz, B., Simhon, A., Arad, I., Shapiro, M., 2002. Cluster of neonatal infections in Jerusalem due to unusual biochemical variant of *Enterobacter sakazakii*. Eur. J. Clin. Microbiol. Infect. Dis. 21, 613–616.

Bowen, A.B., Braden, C.R., 2006. Invasive *Enterobacter sakazakii* disease in infants. Emerg. Infect. Dis. 12, 1185–1189.

Brady, C., Cleenwerck, I., Venter, S., Coutinho, T., De Vos, P., 2013. Taxonomic evaluation of the genus *Enterobacter* based on multilocus sequence analysis (MLSA): proposal to reclassify *E. nimipressuralis* and *E. amnigenus* into *Lelliottia* gen. nov. as *Lelliottia nimipressuralis* comb. nov. and *Lelliottia amnigena* comb. nov., respectively, *E. gergoviae* and *E. pyrinus* into *Pluralibacter* gen. nov. as *Pluralibacter gergoviae* comb. nov. and *Pluralibacter pyrinus* comb. nov., respectively, *E. cowanii, E. radicincitans, E. oryzae* and *E. arachidis* into *Kosakonia* gen. nov. as *Kosakonia cowanii* comb. nov., *Kosakonia radicincitans* comb. nov.,

Kosakonia oryzae comb. nov. and *Kosakonia arachidis* comb. nov., respectively, and *E. turicensis*, *E. helveticus* and *E. pulveris* into *Cronobacter* as *Cronobacter zurichensis* nom. nov., *Cronobacter helveticus* comb. nov. and *Cronobacter pulveris* comb. nov., respectively, and emended description of the genera *Enterobacter* and *Cronobacter*. Syst. Appl. Microbiol. 36, 309–319.

Caubilla-Barron, J., Forsythe, S., 2007. Dry stress and survival time of *Enterobacter sakazakii* and other Enterobacteriaceae. J. Food Prot. 70, 2111–2117.

Caubilla-Barron, J., Hurrell, E., Townsend, S., Cheetham, P., Loc-Carrillo, C., Fayet, O., Prere, M.-F., Forsythe, S.J., 2007. Genotypic and phenotypic analysis of *Enterobacter sakazakii* strains from an outbreak resulting in fatalities in a neonatal intensive care unit in France. J. Clin. Microbiol. 45, 3979–3985.

Caubilla-Barron, J., Kucerova, E., Loughlin, M., Forsythe, S.J., 2009. Bacteriocidal Preparation of Powdered Infant Formula. UK Food Standards Agency. Project Code: B13010. http://www.foodbase.org.uk/results.php?f_category_id=&f_report_id=395.

Cetinkaya, E., Joseph, S., Ayhan, K., Forsythe, S.J., 2013. Comparison of methods for the microbiological identification and profiling of *Cronobacter* species from ingredients used in the preparation of infant formula. Mol. Cell. Probes 27, 60–64.

Chap, J., Jackson, P., Siqueira, R., Gaspar, N., Quintas, C., Park, J., Osaili, T., Shaker, R., Jaradat, Z., Hartantyo, S.H.P., Abdullah Sani, N., Estuningsih, S., Forsythe, S.J., 2009. International survey of *Cronobacter sakazakii* and other *Cronobacter* spp. in follow up formulas and infant foods. Int. J. Food Microbiol. 136, 185–188.

Chen, Y., Lampel, K., Hammack, K., 2012. *Cronobacter*. In: Bacteriological Analytical Manual. US Food and Drug Administration, USA (Chapter 29). http://www.fda.gov/Food/FoodScienceResearch/LaboratoryMethods/ucm289378.htm.

Clark, N.C., Hill, B.C., O'Hara, C.M., Steingrimsson, O., Cooksey, R.C., 1990. Epidemiologic typing of *Enterobacter sakazakii* in two neonatal nosocomial outbreaks. Diagn. Microbiol. Infect. Dis. 13, 467–472.

Codex Alimentarius Commission, 2008. Code of Hygienic Practice for Powdered Formulae for Infants and Young Children. CAC/RCP 66-2008. http://www.codexalimentarius.net/download/standards/11026/cxp_066e.pdf.

Craven, H.M., McAuley, C.M., Duffy, L.L., Fegan, N., 2010. Distribution, prevalence and persistence of *Cronobacter* (*Enterobacter sakazakii*) in the nonprocessing and processing environments of five milk powder factories. J. Appl. Microbiol. 109, 1044–1052.

Cruz, A.L., Rocha-Ramirez, L.M., Gonzalez-Pedrajo, B., Ochoa, S.A., Eslava, C., et al., 2011. Flagella from *Cronobacter sakazakii* induced an inflammatory response in human monocytes. Cytokine 56, 95.

Communicable Disease Centre, 2011. CDC Update: Investigation of *Cronobacter* Infections Among Infants in the United States. http://www.cdc.gov/foodsafety/diseases/cronobacter/investigation.html.

Farmer, J.J., Asbury, M.A., Hickman, F.W., Brenner, D.J., 1980. *Enterobacter sakazakii*: a new species of "*Enterobacteriaceae*" isolated from clinical specimens. Int. J. Syst. Bacteriol. 30, 569–584.

Filloux, A., Hachani, A., Bleves, S., 2008. The bacterial type VI secretion machine: yet another player for protein transport across membranes. Microbiology 154, 1570–1583.

Flores, J.P., Medrano, S.A., Sanchez, J.S., Fernandez-Escartin, E., 2011. Two cases of hemorrhagic diarrhea caused by *Cronobacter sakazakii* in hospitalized nursing infants associated with the consumption of powdered infant formula. J. Food Prot. 74, 2177–2181.

Food and Agriculture Organization/World Health Organization (FAO/WHO), 2004. *Enterobacter sakazakii* and other microorganisms in powdered infant formula. In: Meeting Report, MRA Series 6. World Health Organization, Geneva, Switzerland. http://www.who.int/foodsafety/publications/micro/mra6/en/index.html.

Food and Agriculture Organization/World Health Organization (FAO/WHO), 2006. *Enterobacter sakazakii* and *Salmonella* in powdered infant formula. In: Second Risk Assessment Workshop. Meeting Report, MRA Series 10. World Health Organization, Geneva, Switzerland. http://www.who.int/foodsafety/publications/micro/mra10/en/index.html.

Food and Agriculture Organization/World Health Organization (FAO/WHO), 2008. *Enterobacter sakazakii* (*Cronobacter* spp.) in powdered follow-up formula. In: Microbiological Risk Assessment Series 15. Washington. http://www.who.int/foodsafety/publications/micro/mra_followup/en/.

Forsythe, S.J., Dickins, B., Jolley, K.A., 2014. Cronobacter, the emergent bacterial pathogen *Enterobacter sakazakii* comes of age; MLST and whole genome sequence analysis. BMC Genomics 15, 1121.

Franco, A.A., Kothary, M.H., Gopinath, G., Jarvis, K.G., Grim, C.J., Hu, L., Datta, A.R., McCardell, B.A., Tall, B.D., 2011. Cpa, the outer membrane protease of *Cronobacter sakazakii*, activates plasminogen and mediates resistance to serum bactericidal activity. Infect. Immun. 79, 1578–1587.

Friedeman, M., 2007. *Enterobacter sakazakii* in food and beverages (other than infant formula and milk powder). Int. J. Food Microbiol. 116, 1–10.

Grim, C.J., Kothary, M.H., Gopinath, G., Jarvis, K.G., Beaubrun, J.J., McClelland, M., Tall, B.D., Franco, A.A., 2012. Identification and characterization of *Cronobacter* iron acquisition systems. Appl. Environ. Microbiol. 78, 6035–6050.

Hariri, S., Joseph, S., Forsythe, S.J., 2013. *Cronobacter sakazakii* ST4 strains and neonatal meningitis, United States. Emerg. Infect. Dis. 19, 175–177.

Harris, L.S., Oriel, P.J., 1989. Heteropolysaccharide produced by *Enterobacter sakazakii*. United States Patent number 4,806,636.

Henry, M.C., Moss, R.L., 2009. Necrotizing enterocolitis. Annu. Rev. Med. 60, 111–124.

Himelright, I., Harris, E., Lorch, V., Anderson, M., 2002. *Enterobacter sakazakii* infections associated with the use of powdered infant formula—Tennessee, 2001. JAMA 287, 2204–2205.

Hochel, I., Ruzickova, H., Krasny, L., Demnerova, K., 2012. Occurrence of *Cronobacter* spp. in retail foods. J. Appl. Microbiol. 112, 1257–1265.

Holy, O., Forsythe, S.J., 2014. *Cronobacter* species as emerging causes of healthcare-associated infection. J. Hosp. Infect. 86, 169–177.

Holy, O., Petrželová, J., Hanulík, V., Chromá, M., Matoušková, I., Forsythe, S., 2014. Age profiling of *Cronobacter* isolates from hospital patients from the University Hospital Olomouc (Czech Republic). Epidemiol. Mikrobiol. Imunol. 63, 69–72.

Hurrell, E., Kucerova, E., Loughlin, M., Caubilla-Barron, J., Forsythe, S.J., 2009a. Biofilm formation on enteral feeding tubes by *Cronobacter sakazakii*, *Salmonella* serovars and other Enterobacteriaceae. Int. J. Food Microbiol. 136, 227–231.

Hurrell, E., Kucerova, E., Loughlin, M., Caubilla-Barron, J., Hilton, A., Armstrong, R., Smith, C., Grant, J., Shoo, S., Forsythe, S., 2009b. Neonatal enteral feeding tubes as loci for colonisation by members of the *Enterobacteriaceae*. BMC Infect. Dis. 9, 146.

International Commission on Microbiological Specification for Foods (ICMSF), 2002. Microbiological Testing in Food Safety Management, vol. 7. Academic/Plenum Publisher, New York.

Iversen, C., Forsythe, S., 2003. Risk profile of *Enterobacter sakazakii*, an emergent pathogen associated with infant milk formula. Trends Food Sci. Technol. 14, 443–454.

Iversen, C., Forsythe, S.J., 2004. Isolation of *Enterobacter sakazakii* and other Enterobacteriaceae from powdered infant formula milk and related products. Food Microbiol. 21, 771–776.

Iversen, C., Forsythe, S., 2007. Comparison of media for the isolation of *Enterobacter sakazakii*. Appl. Environ. Microbiol. 73, 48–52.

Iversen, C., Lane, M., Forsythe, S.J., 2004a. The growth profile, thermotolerance and biofilm formation of *Enterobacter sakazakii* grown in infant formula milk. Lett. Appl. Microbiol. 38, 378–382.

Iversen, C., Druggan, P., Forsythe, S.J., 2004b. A selective differential medium for *Enterobacter sakazakii*. Int. J. Food Microbiol. 96, 133–139.

Iversen, C., Waddington, M., On, S.L.W., Forsythe, S., 2004c. Identification and phylogeny of *Enterobacter sakazakii* relative to *Enterobacter* and *Citrobacter*. J. Clin. Microbiol. I42, 5368–5370.

Iversen, C., Waddington, M., Farmer, J.J., Forsythe III, S.J., 2006. The biochemical differentiation of *Enterobacter sakazakii* genotypes. BMC Microbiol. 6, 94.

Iversen, C., Lehner, A., Mullane, N., Bidlas, E., Cleenwerck, I., Marugg, J., Joosten, H., 2007. The taxonomy of *Enterobacter sakazakii*: proposal of a new genus *Cronobacter* gen. nov. and descriptions of *Cronobacter sakazakii* comb. nov. *Cronobacter sakazakii* subsp. *sakazakii*, comb. nov., *Cronobacter sakazakii* subsp.

malonaticus subsp. nov., *Cronobacter turicensis* sp. nov., *Cronobacter muytjensii* sp. nov., *Cronobacter dublinensis* sp. nov. and *Cronobacter* genomospecies 1. BMC Evol. Biol. 7, 64.

Iversen, C., Mullane, N., McCardell, B., Tall, B.D., Lehner, A., Fanning, S., Joosten, H., 2008. *Cronobacter* gen. nov., a new genus to accommodate the biogroups of *Enterobacter sakazakii*, and proposal of *Cronobacter sakazakii* gen. nov., comb. nov., *Cronobacter malonaticus* sp. nov., *Cronobacter turicensis* sp. nov., *Cronobacter muytjensii* sp. nov., *Cronobacter dublinensis* sp. nov., *Cronobacter* genomospecies 1, and of three subspecies, *Cronobacter dublinensis* subsp. *dublinensis* subsp. nov., *Cronobacter dublinensis* subsp. *lausannensis* subsp. nov. and *Cronobacter dublinensis* subsp. *lactaridi* subsp. nov. Int. J. Syst. Evol. Microbiol. 58, 1442–1447.

Izard, D., Richar, C., Leclerc, H., 1983. DNA relatedness between *Enterobacter sakazakii* and other members of the genus *Enterobacter*. Ann. Inst. Pasteur Microbiol. 134, 241–245.

Jackson, E.E., Sonbol, H., Masood, N., Forsythe, S.J., 2014. Genotypic and phenotypic characteristics of *Cronobacter* species, with particular attention to the newly reclassified species *C. helveticus*, *C. pulveris*, and *C. zurichensis*. Food Microbiol. 44, 226–235.

Jackson, E.E., Parra Flores, J., Fernandez-Escartin, E., Forsythe, S.J., 2015. Re-evaluation of a suspected *Cronobacter sakazakii* outbreak in Mexico. J. Food Prot. 78, 1191–1196.

Jacobs, C., Braun, P., Hammer, P., 2011. Reservoir and routes of transmission of *Enterobacter sakazakii* (*Cronobacter* spp.) in a milk powder-producing plant. J. Dairy Sci. 94, 3801–3810.

Jarvis, K.G., Grim, C.J., Franco, A.A., Gopinath, G., Sathyamoorthy, V., Hu, L., Lee, C.S., Tall, B.D., 2011. Molecular characterization of *Cronobacter* lipopolysaccharide O-antigen gene clusters and development of serotype-specific PCR assays. Appl. Environ. Microbiol. 77, 4017–4026.

Jarvis, K.G., Yan, Q.Q., Grim, C.J., Power, K.A., Franco, A.A., Hu, L., Gopinath, G., Sathyamoorthy, V., Kotewicz, M.L., Kothary, M.H., Lee, C., Sadowski, J., Fanning, S., Tall, B.D., 2013. Identification and characterization of five new molecular serogroups of *Cronobacter* spp. Foodborne Pathog. Dis. 10, 343–352.

Jolley, K.A., Maiden, M.C., 2010. BIGSdb: scalable analysis of bacterial genome variation at the population level. BMC Bioinform. 11, 595.

Jolley, K.A., Bliss, C.M., Bennett, J.S., Bratcher, H.B., Brehony, C., Colles, F.H., Wimalarathna, H., Harrison, O.B., Sheppard, S.K., Cody, A.J., Maiden, M.C.J., 2012. Ribosomal multilocus sequence typing: universal characterization of bacteria from domain to strain. Microbiology 158, 1005–1015.

Joseph, S., Forsythe, S., 2011. Predominance of *Cronobacter sakazakii* sequence type 4 in neonatal infections. Emerg. Infect. Dis. 17, 1713–1715.

Joseph, S., Forsythe, S.J., 2012. Insights into the emergent bacterial pathogen *Cronobacter* spp., generated by multilocus sequence typing and analysis. Front. Microbiol. 3, 397.

Joseph, S., Cetinkaya, E., Drahovska, H., Levican, A., Figueras, M.J., Forsythe, S.J., 2011. *Cronobacter condimenti* sp. nov., isolated from spiced meat and *Cronobacter universalis* sp. nov., a novel species designation for *Cronobacter* sp. genomospecies 1, recovered from a leg infection, water, and food ingredients. Int. J. Syst. Evol. Microbiol. 62, 1277–1283.

Joseph, S., Desai, P., Ji, Y., Cummings, C.A., Shih, R., Degoricija, L., Forsythe, S.J., 2012a. Comparative analysis of genome sequences covering the seven *Cronobacter* species. PLoS One 7, e49455.

Joseph, S., Sonbol, H., Hariri, S., Desai, P., McClelland, M., Forsythe, S.J., 2012b. Diversity of the *Cronobacter* genus as revealed by multilocus sequence typing. J. Clin. Microbiol. 50, 3031–3039.

Joseph, S., Hariri, S., Forsythe, S.J., 2013a. Lack of continuity between *Cronobacter* biotypes and species as determined using multilocus sequence typing. Mol. Cell. Probes 27, 137–139.

Joseph, S., Hariri, S., Masood, N., Forsythe, S., 2013b. Sialic acid utilization by *Cronobacter sakazakii*. Microb. Inf. Exp. 3, 3.

Kandhai, M.C., Reij, M.W., Gorris, L.G., Guillaume-Gentil, O., van Schothorst, M., 2004. Occurrence of *Enterobacter sakazakii* in food production environments and households. Lancet 363, 39–40.

Kim, K.P., Loessner, M.J., 2008. *Enterobacter sakazakii* invasion in human intestinal Caco-2 cells requires the host cell cytoskeleton and is enhanced by disruption of tight junction. Infect. Immun. 76, 562–570.

Kim, H., Ryu, J.H., Beuchat, L.R., 2007. Effectiveness of disinfectants in killing *Enterobacter sakazakii* in suspension, dried on the surface of stainless steel, and in a biofilm. Appl. Environ. Microbiol. 73, 1256–1265.

Kim, K., Jang, S.S., Kim, S.K., Park, J.H., Heu, S., Ryu, S., 2008. Prevalence and genetic diversity of Enterobacter sakazakii in ingredients of infant foods. Int. J. Food Microbiol. 122, 196–203.

Kim, K., Kim, K.P., Choi, J., Lim, J.A., Lee, J., Hwang, S., Ryu, S., 2010. Outer membrane proteins A (OmpA) and X (OmpX) are essential for basolateral invasion of *Cronobacter sakazakii*. Appl. Environ. Microbiol. 76, 5188–5198.

Kline, M.W., 1988. *Citrobacter* meningitis and brain abscess in infancy: epidemiology, pathogenesis, and treatment. J. Pediatr. 113, 430–434.

Kothary, M.H., McCardell, B.A., Frazar, C.D., Deer, D., Tall, B.D., 2007. Characterization of the zinc-containing metalloprotease encoded by *zpx* and development of a species-specific detection method for *Enterobacter sakazakii*. Appl. Environ. Microbiol. 73, 4142–4151.

Kucerova, E., Clifton, S.W., Xia, X.-Q., Long, F., Porwollik, S., Fulton, L., Fronick, C., Minx, P., Kyung, K., Warren, W., Fulton, R., Feng, D., Wollam, A., Shah, N., Bhonagiri, V., Nash, W.E., Hallsworth-Pepin, K., Wilson, R.K., McClelland, M., Forsythe, S.J., 2010. Genome sequence of *Cronobacter sakazakii* BAA-894 and comparative genomic hybridization analysis with other *Cronobacter* species. PLoS One 5, e9556.

Kucerova, E., Joseph, S., Forsythe, S., 2011. The *Cronobacter* genus: ubiquity and diversity. Qual. Assur. Saf. Crops Food 3, 104–122.

Lai, K.K., 2001. *Enterobacter sakazakii* infections among neonates, infants, children, and adults. Case reports and a review of the literature. Medicine 80, 113–122.

Lehner, A., Riedel, K., Eberl, L., Breeuwer, P., Diep, B., Stephan, R., 2005. Biofilm formation, extracellular polysaccharide production, and cell-to-cell signaling in various *Enterobacter sakazakii* strains: aspects promoting environmental persistence. J. Food Prot. 68, 2287–2294.

Lenati, R.F., O'Connor, D.L., Hébert, K.C., Farber, J.M., Pagotto, F.J., 2008. Growth and survival of *Enterobacter sakazakii* in human breast milk with and without fortifiers as compared to powdered infant formula. Int. J. Food Microbiol. 122, 171–179.

Lucas, A., Cole, T.J., 1990. Breast milk and neonatal necrotising enterocolitis. Lancet 336 (8730), 1519–1523.

Maiden, M.C., van Rensburg, M.J., Bray, J.E., Earle, S.G., Ford, S.A., Jolley, K.A., McCarthy, N.D., 2013. MLST revisited: the gene-by-gene approach to bacterial genomics. Nat. Rev. Microbiol. 11, 728–736.

Masood, N., Moore, K., Farbos, A., Hariri, S., Paszkiewicz, K., Dickins, B., McNally, A., Forsythe, S., 2013. Draft genome sequence of the earliest *Cronobacter sakazakii* sequence type 4 strain NCIMB 8272. Genome Announc. 2, e00585-14.

Mittal, R., Wang, Y., Hunter, C.J., Gonzalez-Gomez, I., Prasadarao, N.V., 2009. Brain damage in newborn rat model of meningitis by *Enterobacter sakazakii*: a role for outer membrane protein A. Lab. Invest. 89, 263–277.

Mullane, N., Healy, B., Meade, J., Whyte, P., Wall, P.G., Fanning, S., 2008. Dissemination of *Cronobacter* spp. (*Enterobacter sakazakii*) in a powdered milk protein manufacturing facility. Appl. Environ. Microbiol. 74, 5913–5917.

Muller, A., Stephan, R., Fricker-Feer, C., Lehner, A., 2013. Genetic diversity of *Cronobacter sakazakii* isolates collected from a Swiss infant formula production facility. J. Food Prot. 76, 883–887.

Muytjens, H.L., Zanen, H.C., Sonderkamp, H.J., Kollee, L.A., Wachsmuth, I.K., Farmer III, J.J., 1983. Analysis of eight cases of neonatal meningitis and sepsis due to *Enterobacter sakazakii*. J. Clin. Microbiol. 18, 115–120.

Muytjens, H.L., van Der Ros-van de Repe, J., van Druten, H.A., 1984. Enzymatic profiles of *Enterobacter sakazakii* and related species with special reference to the alpha-glucosidase reaction and reproducibility of the test system. J. Clin. Microbiol. 20, 684–686.

Muytjens, H.L., Roelofs-Willemse, H., Jaspar, G.H., 1988. Quality of powdered substitutes for breast milk with regard to members of the family *Enterobacteriaceae*. J. Clin. Microbiol. 26, 743–746.

Nair, M.K., Venkitanarayanan, K.S., 2006. Cloning and sequencing of the *ompA* gene of *Enterobacter sakazakii* and development of an *ompA*-targeted PCR for rapid detection of *Enterobacter sakazakii* in infant formula. Appl. Environ. Microbiol. 72, 2539–2546.

Nazarowec-White, M., Farber, J.M., 1999. Phenotypic and genotypic typing of food and clinical isolates of *Enterobacter sakazakii*. J. Med. Microbiol. 48, 559–567.

Osaili, T., Forsythe, S., 2009. Desiccation resistance and persistence of *Cronobacter* species in infant formula. Int. J. Food Microbiol. 136, 214–220.

Pagotto, F.J., Nazarowec-White, M., Bidawid, S., Farber, J.M., 2003. *Enterobacter* sakazakii: infectivity and enterotoxin production *in vitro* and *in vivo*. J. Food Prot. 66, 370–375.

Patrick, M.E., Mahon, B.E., Greene, S.A., Rounds, J., Cronquist, A., Wymore, K., Boothe, E., Lathrop, S., Palmer, A., Bowen, A., 2014. Incidence of *Cronobacter* spp. infections, United States, 2003–2009. Emerg. Infect. Dis. 20, 1520–1523.

Pava-Ripoll, M., Goeriz Pearson, R.E., Miller, A.K., Ziobro, G.C., 2012. Prevalence and relative risk of *Cronobacter* spp., *Salmonella* spp., and *Listeria monocytogenes* associated with the body surfaces and guts of individual filth flies. Appl. Environ. Microbiol. 78, 7891–7902.

Pitout, J.D., Moland, E.S., Sanders, C.C., Thomson, K.S., Fitzsimmons, S.R., 1997. Beta-lactamases and detection of beta-lactam resistance in *Enterobacter* spp. Antimicrob. Agents Chemother. 41, 35–39.

Poirel, L., Nordmann, P., De Champs, C., Eloy, C., 2007. Nosocomial spread of QnrA-mediated quinolone resistance in *Enterobacter sakazakii*. Int. J. Antimicrob. Agents 29, 223–224.

Power, K.A., Yan, Q., Fox, E.M., Cooney, S., Fanning, S., 2013. Genome sequence of *Cronobacter sakazakii* SP291, a persistent thermotolerant isolate derived from a factory producing powdered infant formula. Genome Announc. 1, 13.

Proudy, I., Bougle, D., Coton, E., Coton, M., Leclercq, R., Vergnaud, M., 2008. Genotypic characterization of *Enterobacter sakazakii* isolates by PFGE, BOX-PCR and sequencing of the *fliC* gene. J. Appl. Microbiol. 104, 26–34.

Reich, F., König, R., von Wiese, W., Klein, G., 2010. Prevalence of *Cronobacter* spp. in a powdered infant formula processing environment. Int. J. Food Microbiol. 140, 214–217.

Richardson, A.N., Lambert, S., Smith, M.A., 2009. Neonatal mice as models for *Cronobacter sakazakii* infection in infants. J. Food Prot. 72, 2363–2367.

Riedel, K., Lehner, A., 2007. Identification of proteins involved in osmotic stress response in *Enterobacter sakazakii* by proteomics. Proteomics 7, 1217–1231.

Santos, R.F.S., da Silva, N., Junqueira, V.C.A., Kajsik, M., Forsythe, S., Pereira, J.L., 2013. Screening for *Cronobacter* species in powdered and reconstituted infant formulas and from equipment used in formula preparation in maternity hospitals. Ann. Nutr. Metab. 63, 62–68.

Scheepe-Leberkuhne, M., Wagner, F., 1986. Optimization and preliminary characterization of an exopolysaccharide synthezised by *Enterobacter sakazakii*. Biotechnol. Lett. 8, 695–700.

Schmid, M., Iversen, C., Gontia, I., Stephan, R., Hofmann, A., Hartmann, A., Jha, B., Eberl, L., Riedel, K., Lehner, A., 2009. Evidence for a plant-associated natural habitat for *Cronobacter* spp. Res. Microbiol. 160, 608–614.

Sonbol, H., Joseph, S., McAuley, C., Craven, H., Forsythe, S.J., 2013. Multilocus sequence typing of *Cronobacter* spp. from powdered infant formula and milk powder production factories. Int. Dairy J. 30, 1–7.

Stephan, R., Grim, C.J., Gopinath, G.R., Mammel, M.K., Sathyamoorthy, V., Trach, L.H., Chase, H.R., Fanning, S., Tall, B.D., 2014. Re-examination of the taxonomic status of *Enterobacter helveticus*, *Enterobacter pulveris*, and *Enterobacter turicensis* as members of *Cronobacter* and description of *Siccibacter turicensis* com. nov., *Franconibacter helveticus* comb. nov., and *Franconibacter pulveris* com. nov. Int. J. Syst. Evol. Microbiol 64, 3402–10.

Stock, I., Wiedemann, B., 2002. Natural antibiotic susceptibility of *Enterobacter amnigenus*, *Enterobacter cancerogenus*, *Enterobacter gergoviae* and *Enterobacter sakazakii* strains. Clin. Microbiol. Infect. 8, 564–578.

Stoll, B.J., et al., 2004. *Enterobacter sakazakii* is a rare cause of neonatal septicaemia or meningitis in VLBW infants. J. Pediatr. 144, 821–823.

Stoop, B., Lehner, A., Iversen, C., Fanning, S., Stephan, R., 2009. Development and evaluation of *rpoB* based PCR systems to differentiate the six proposed species within the genus *Cronobacter*. Int. J. Food Microbiol. 136, 165–168.

Sun, Y., Wang, M., Liu, H., Wang, J., He, X., Zeng, J., Guo, X., Li, K., Cao, B., Wang, L., 2011. Development of an O-antigen serotyping scheme for *Cronobacter sakazakii*. Appl. Environ. Microbiol. 77, 2209–2214.

Sun, Y., Wang, M., Wang, Q., Cao, B., He, X., Li, K., Feng, L., Wang, L., 2012. Genetic analysis of the *Cronobacter sakazakii* O4 to O7 O-antigen gene clusters and development of a PCR assay for identification of all *C. sakazakii* O serotypes. Appl. Environ. Microbiol. 78, 3966–3974.

Szafranek, J., Czerwicka, M., Kumirska, J., Paszkiewicz, M., Łojkowska, E., 2005. Repeating unit structure of *Enterobacter sakazakii* ZORB 741 O-polysaccharide. Pol. J. Chem. 79, 287–295.

Telang, S., et al., 2005. Fortifying fresh human milk with commercial powdered human milk fortifiers does not affect bacterial growth during 6 hours at room temperature. J. Am. Diet. Assoc. 105, 1567–1572.

Townsend, S.M., Hurrell, E., Gonzalez-Gomez, I., Lowe, J., Frye, J.G., Forsythe, S., Badger, J.L., 2007. *Enterobacter sakazakii* invades brain capillary endothelial cells, persists in human macrophages influencing cytokine secretion and induces severe brain pathology in the neonatal rat. Microbiology 153, 3538–3547.

Townsend, S., Hurrell, E., Forsythe, S., 2008. Virulence studies of *Enterobacter sakazakii* isolates associated with a neonatal intensive care unit outbreak. BMC Microbiol. 8, 64.

Urmenyi, A.M.C., Franklin, A.W., 1961. Neonatal death from pigmented coliform infection. Lancet 11, 313–315.

van Acker, J., de Smet, F., Muyldermans, G., Bougatef, A., Naessens, A., Lauwers, S., 2001. Outbreak of necrotizing enterocolitis associated with *Enterobacter sakazakii* in powdered milk formula. J. Clin. Microbiol. 39, 293–297.

Zh Xu, Y., Metris, A., Stasinopoulis, D., Forsythe, S.J., Sutherland, J.P., 2015. Effect of heat shock and recovery temperature on variability of single cell lag time of *Cronobacter*. Food Microbiol. 45 (Part B), 195–204, Special Issue on Predictive Modeling in Food.

Zhao, Z., Wang, L., Wang, B., Liang, H., Ye, Q., Zenga, M., 2014. Complete genome sequence of *Cronobacter sakazakii* strain CMCC 45402. Genome Announc. 2, e01139-13.

CHAPTER

NEW AND EMERGING BACTERIAL FOOD PATHOGENS

14

Katie Laird

The School of Pharmacy, De Montfort University, The Gateway, Leicester, UK

1 INTRODUCTION

New and emerging foodborne pathogens present a serious risk to human health, particularly in developing countries. However, increased globalization of the food supply chain means that risks identified in developing countries can spread to more developed countries making identification, prevention, and mitigation of emerging foodborne illness and disease more complex.

An emerging food risk can be defined as an unanticipated risk that may have occurred naturally, accidentally, or deliberately and can have an effect on human and animal health, as well as the environment and the economy (Wentholt et al., 2010).

An example of a new risk that can naturally occur is via genetic transfer between microorganisms, resulting in microbes with different characteristics to those isolated previously. A good example is *Escherichia coli* O157:H7, which originated from its close ancestor O55:H7 through horizontal transfer and recombination. At some point in time, *E. coli* O55:H7 acquired cytotoxic genes from *Shigella dysenteriae* and emerged as *E. coli* O157:H7 (Sheridan and McDowell, 1998).

In recent years, the epidemiology of foodborne pathogens has changed due to the emergence of new microorganisms in food, increases in previously recognized food pathogens, and changes in the foods that they inhabit (Tauxe et al., 2010). One such example is an outbreak of *Yersinia pseudotuberculosis* in Finland that was associated with iceberg lettuce (McCabe-Sellers and Beattie, 2004).

Factors affecting the global emergence of new foodborne pathogens include the introduction of new food types, population growth, culture and religion, infringements and violations of food safety standards, and climate change (Ackerman, 2006). Technological innovation is also a contributory factor; for example, the emergence of *Listeria monocytogenes* with the introduction of refrigeration. Although these factors are well recognized, there remain many obstacles to the efficient management of emerging foodborne pathogens. These include identification, limited access to data, and international networks' lack of data sharing and, importantly, lack of resources to meet substantial costs associated with predictive technologies. The introduction of the Global Outbreak Alert and Response Network (GOARN) by the World Health Organization (WHO) has gone some way toward addressing these problems, with alerts for norovirus in oysters, *E. coli* in spinach, and *Salmonella* sp. in chocolate in recent years (Kleter and Marvin, 2009).

Wentholt et al. (2010) have suggested that six methods could aid in predicting the emergence of new foodborne pathogens. These are early warning systems, risk profiling, risk trending, foresight,

Food Safety. http://dx.doi.org/10.1016/B978-0-12-800245-2.00014-9

vulnerability assessment, and horizon scanning. The importance of evolutionary changes in food-borne pathogens and studies in the changing incidents of these pathogens must also play an important part in the prevention and management of the emergence of new pathogens in food (Kleter and Marvin, 2009).

Situational awareness can be used in the prevention of foodborne pathogens by the use of risk assessment measures such as Hazard Analysis and Critical Control Points (HACCP); information can be collected about the food production situation in real time, and thus, measures can be taken to prevent contaminated foodstuffs being sold commercially (Kleter and Marvin, 2009). HACCP assessments need to be used in combination with detection systems for monitoring hazardous microorganisms in order for the system to be successful. Traditional detection systems have included culture-based methods; however, this process is time and labor intensive. Other detection systems include molecular and immunodetection. However, the need for more rapid testing that still has the reliability of the traditional culturing method has led to the development of biosensors. These are based on molecule trapping to a reactive surface that then uses electrochemical, optical, and acoustic waves to capture the molecule into a measurable signal allowing for real-time analysis (Yeni et al., 2014).

Risk management allows for the data collected from risk assessments to be used in order to devise mitigation strategies. Successful application of this process includes the reduction of *Campylobacter* sp. in foodstuffs by testing flocks 1 week before slaughter and only allowing the distribution of birds that have tested negative (Denmark); the use of water chlorination during poultry slaughtering (United States); and in Sweden lowering the herd prevalence in poultry production (Tauxe et al., 2010).

Hurdle technology is also paramount to the prevention and control of outbreaks of pathogens in the food chain. Elements such as temperature, pH, NaCl, and $NaNO_2$ can be used in combination to prevent further growth of microorganisms (Daskalov, 2006). The microbiological safety of processed food is based on the combination of preservative technologies (hurdles), allowing the nutritional and sensory quality as well as safety of the food to be maintained. Combining hurdles brings about an overall preservation strategy and has shown to be successful on fresh-cut fruit and vegetables (Allende et al., 2006) and meat products (Thomas et al., 2008a,b). By subjecting bacteria to a number of sublethal stresses or hurdles, which act on the same element of the cell, and adding the synergistic effect of combining hurdles, which act simultaneously disturbing several different functions of the cell (cell membrane, DNA, and enzyme systems), an additive inhibitory effect occurs. Whether the effect of the combined technologies results in an additive or synergistic effect on the bacteria, both can lead to cell death (Ross et al., 2003).

A number of pathogens have recently emerged in the food and water chain, including enterohaemorrhagic *E. coli*, *Salmonella typhimurium*, *L. monocytogenes*, and *Campylobacter jejuni*, some of which are multi antibiotic resistant (Duffy et al., 2008). Such microorganisms are well documented. However, it is the more unusual species of foodborne pathogens that may prove to have a greater impact on food safety in the future.

2 *ACINETOBACTOR* SP.

Much of the research conducted on *Acinetobactor* sp. as a foodborne pathogen was carried out over two decades ago, from a range of sources including soil, water, and animals (Gennari et al., 1992; Gennari and Lombardi, 1993). In recent years, *Acinetobacter* sp. has emerged as a hospital-acquired infection (HAI) with significant associated morbidity and mortality rates. This has resulted in much of the published literature focusing on *Acinetobacter* sp. in a health care setting. However, links are starting to be made between food, antibiotic resistance, and human health. The potential association between *Acinetobacter* sp. food contamination and human infection is of particular interest, as *Acinetobacter* sp. is intrinsically resistant to antibiotics.

2.1 PREVALENCE IN FOOD

The published data on *Acintetobacter* sp. as a foodborne pathogen is limited. However, this microorganism has recently been isolated from previously unsuspected food sources, such as fruits and vegetables. The frequency of *Acintetobacter baumannii* was assessed in fresh fruits and vegetables and was found to account for 56% of all strains of *Acintetobacter* sp. isolated from a wide range of fresh fruits, salads, and vegetable products. Many of the sources were ready-to-eat products. Alarmingly, it follows that hospital food could potentially be contaminated and, therefore, responsible for *Acinetobacter* sp. nosocomial infections (Giamarellou et al., 2008).

The widespread use of antibiotics as growth promoters in livestock is associated with the emergence of antibiotic-resistant bacteria in humans, through direct contact with infected animals or consumption of meat products. This relationship is well established; however, information regarding antibiotic-resistant *A. baumannii* in food sources is still limited. About 16 isolates of *A. baumannii* from recently slaughtered pigs and cattle were screened for their susceptibility to 11 antibiotics; all samples (16) were found to be resistant to amoxicillin (MIC >16 mg/L), cefradine (MIC >8 mg/L), and chloramphenicol (MIC >8 mg/L). However, the PFGE profiles of these isolates differed from that of human strains (Hamouda et al., 2011). Metallo-β-lactamase (MBL) genes have also been assessed in Gram-negative bacteria with low susceptibility to imipenem (MIC >8 μg/mL). Isolates from 15 animal farms and one slaughterhouse in China were found to have a MBL phenotype. The $bla_{NDM\text{-}1}$ gene in *E. coli* had been passed on by a plasmid (pAL-01) from *Acinetobacter lwoffii* isolated from chicken. This transfer of antibiotic resistance between bacterial species has implications for food safety due to the potential for contamination by any microorganism carrying the pAL-01 plasmid (Wang et al., 2012).

Raw bulk tank milk (BTM) has also been highlighted as a possible source of *Acinetobacter* sp. In Korea, 2287 dairy farms in six different provinces were screened, and 7.7% BTM milk samples were found to be contaminated with *Acinetobacter* sp., and 32.4% of which were *A. baumannii*. These isolates were found to be resistant to amikacin (2.3%), gentamicin (7.4%), piperacillin (2.3%), and cefotaxime (4%) (Gurung et al., 2013; Tamang et al., 2014).

The transmission routes by which *Acinetobacter* sp. cause HAIs are poorly understood. In order to control this microorganism in the clinical arena, the relationship between food consumption/contamination and infection needs to be fully explored and understood. This is especially important due to the prevalence of multidrug resistance in this genre of bacteria.

3 *AEROMONAS* SP.

Aeromonas sp. are waterborne bacteria isolated from many water sources, including water distribution systems and drinking water. However, the level found in drinking water is low compared to that isolated from food such as meat, fish, vegetables, and processed foods. *Aeromonas* sp. are able to grow at high salt concentrations and at pH 4-10, as well as in low temperatures used in refrigeration when it produces exotoxins. Therefore, it poses a serious food safety issue (Tomas, 2012). This ability of *Aeromonas* sp. to grow and produce toxins at cold temperatures is fundamental to its status as an emerging food pathogen, although associations between *Aeromonas* sp. and foodborne transmission are not well established. Most gastroenteritis cases (>85%) are attributed to *Aeromonas hydrophila* (Daskalov, 2006).

3.1 PREVALENCE IN FOOD

Due to the motility of *Aeromonas* sp., this species is often associated with waterborne transmission, in particular *A. hydrophila*. Many studies in the 1990s established the link between this microorganism and contaminated seafood. In one study, 36.1% of 238 channel catfish fillets were contaminated with *A. hydrophila*, and incidences were higher in the summer months compared to winter (Wang and Silva, 1999). *A. hydrophila* has been isolated from 19%, 28%, 90%, and 22% of fish samples from the United Kingdom, New Zealand, Switzerland, and Taiwan, respectively (Daskalov, 2006). A similar isolation rate of 40% for *A. hydrophila* was found by Davies et al. (2011) who analyzed samples of fresh fish from retail outlets in the United Kingdom, France, Greece, and Portugal. More recently, *A. hydrophila* has been shown to be able to survive on sea bream in modified atmosphere packaging at 4 °C, with growth reaching 1.69-3.34 log cycles at 60/40 atm and 0.12-2.31 log cycles at 0/30 atm, depending on the strain (Provincial et al., 2013). Aeromonads are considered particular foodborne threats when they carry the aerolysin or hemolysin genes; 96% of *Aeromonas veronii* isolated from catfish were found to carry the *aer*A gene and the cytotoxic enterotoxin *act* gene (Nawaz et al., 2010).

The epidemiological relationship between *Aeromonas* sp. found in human diarrheal stools, potable water, rabbit meat, and marine fish has been assessed. The predominant species in clinical samples and water isolates were found to be *Aeromonas caviae* and *Aeromonas media*, whereas the motile strain *Aeromonas salmonicida* was more commonly found in fish and meat. Genetypical relationships were established, and relatedness was observed between the water and clinical isolates, suggesting waterborne transmission of *Aeromonas* sp. (Pablos et al., 2011); Ottaviani et al. (2011) reported very similar results.

Inadequate sanitization of vegetables can lead to contamination of organic vegetables and raw, minimally processed ready-to-eat vegetables. McMahon and Wilson (2001) screened organic vegetables sourced from retail facilities (commercially packed) and organic vegetables for food poisoning microorganisms. *E. coli*, *Salmonella* sp., *Campylobacter* sp., or *Listeria* sp. were not detected, but *Aeromonas* sp. were isolated from 34% of the organic vegetables and 41% of the vegetables that had undergone minimal processing. In a separate study, minimally processed salad stuff (26 types from five producers) were found to also contain *Aeromonas* sp., with 33 isolates of *A. hydrophila* and 12 isolates of *A. caviae* being present (Xanthopoulos et al., 2010). Palú et al. (2006) tested 28 strains of *A. hydrophila* and *A. caviae* isolated from lettuce for their antibiotic susceptibility. Data showed that *A. hydrophila* was resistant to ampicillin/sulbactam and tetracycline in 96.2% and 7.7% of strains, respectively; and *A. caviae* was resistant to ampicillin/sulbactam, tetracycline, and cefoxitin in 55.1%, 10.3%, and 31% of strains, respectively. This increased antibiotic resistance in food isolates of *Aeromonas* sp. in recent years is attributed to plasmid transfer. In addition, the ability of *Aeromonas* sp. to produce toxins is of importance when determining this bacteria's role in food contamination. Of 99 food isolates of *Aeromonas* sp. that were assessed for toxin production, 28% produced enterotoxin, 17.1% showed hemolytic activity, and

72.7% produced cytotoxins. These pathogenic factors are critical for *Aeromonas* sp. to induce food poisoning (Martins et al., 2002). Similar patterns of toxigenicity and multi antibiotic resistance have also been observed in *A. hydrophila* isolates from finfish and prawns from South India; of 225 isolates, 78.4% were hemolysin producers, and all were resistant to bacitracin (Radu et al., 2003).

Aeromonas sp. can be isolated from a diverse range of foods including meat products. However, in the last 20 years, little attention has been paid to meat as a source of *Aeromonas* sp., although a recent study by Fontes et al. (2011) has shown that two-thirds of samples taken from pig carcasses and feces, and dehairing equipment and water in slaughterhouses, contained eight different species of *Aeromonas* sp.: *A. hydrophila, A. salmonicida, Aeromonas bestiarum, A. caviae, A. media, A. veronii, Aeromonas allosaccharophila, Aeromonas simiae*, and *Aeromonas aquariorum. A. simiae*, isolated from healthy monkeys in Mauritius, was first described in 2004. It is believed that the first identification of *A. smiae* was made in 2010 from pig carcasses for human consumption in Portugal (Fontes et al., 2010).

It is the presence of toxin genes that makes *Aeromonas* sp. a food safety threat and endows it with the ability to cause infection in humans. Along with its capacity to inhabit a diverse range of foodstuffs, this makes *Aeromonas* sp. a threat as an emerging foodborne pathogen.

4 *CLOSTRIDIUM DIFFICILE*

Is *Clostridium difficile* a cause of foodborne disease? In recent years the epidemiology of *C. difficile* has changed, and an increased number of cases reported in the community. This has raised concerns about *C. difficile* being a foodborne pathogen, rather than solely a health care associated pathogen. These increased unknown sources of exposure to *C. difficile* have led to greater investigations into food as a potential hub. Multilocus variable number tandem repeat analysis (MLVA) obtained from the Centers for Disease Control and Prevention (CDC) has shown multiple highly related clusters of *C. difficile* isolates from food, animal, and human origins (Marsh, 2013).

In addition to the GI tract of humans, *C. difficile* inhabits a number of environments that could potentially contaminate the food chain, including marine sediment, saltwater, freshwater, soil, root vegetables, and a range of animals from cattle to poultry (Rodriguez-Palacios et al., 2012).

It is the *C. difficile* spores in retail products that are able to survive cooking that makes *C. difficile* a considerable foodborne threat to human health. It has been shown that after heating spores to 71 °C for 30 min, and then reheating to 85 °C for 10 min (the highest validated minimal internal cooking temperature to reduce pathogens from difficult-to-cook foods), that the isolates that survived were of meat origin (Rodriguez-Palacios et al., 2010).

4.1 PREVALENCE IN FOOD

The first food positively identified to contain *C. difficile* was raw ground beef and pork in 1996. Since that time, various retail food products have been found to contain toxigenic strains of *C. difficile*. The ingestion of pork from contaminated pigs can pose a potential food safety threat from *C. difficile*; pigs have been a known source of *C. difficile* for at least two decades. More recently, a 16% mortality rate was observed in 5-day-old piglets due to infection with *C. difficile*. Nondiarrheic piglets producing normal feces are also often found to be TcdA/B positive and, thus, are *C. difficile* carriers; this can be the case for up to 74% of the litter (Songer and Anderson, 2006). In 2011, pigs at a slaughterhouse from

52 farms were sampled for *C. difficile*. Results showed that up to 9% of 677 rectal fecal samples were positive for the microorganism. In total, 61% of herds from each farm were positive for *C. difficile*, and there was no significant difference found between conventional and organic farms. About 16 ribotypes were isolated, with 078 being the most predominant isolate (Keessen et al., 2011). Pork processing environments have also been shown to be positive for *C. difficile*, with 078 again being the most predominant riobtype present. Samples taken from the holding area, manure, and carcasses at poststick, postbleed, and preevisceration had isolation rates for *C. difficile* of 45%, 80%, 10%, 15%, and 15%, respectively (Hawken et al., 2013).

Of 660 raw meat samples obtained from cows, sheep, goats, camels, and buffalos in Iran, 13 samples were found to be positive for *C. difficile*, with the highest prevalence being in buffalo. Overall, 7 of the 13 strains had the toxin genes *fortcdA*, *tcdB*, and *cdtB* (Rahimi et al., 2014). However, published reports vary widely in the types of meat from which *C. difficile* has been isolated. A study in the Netherlands found *C. difficile* in lamb (6.3% of samples) and chicken (2.7% of samples) but not in raw beef (de Boer et al., 2011). In contrast, other studies have shown that retail beef are carriers of *C. difficile* in quantities ranging from 2.4% to 42.4% (Rodriguez-Palacios et al., 2010; Songer and Anderson, 2006). About 3% of Austrian retail ground meat samples were also found to test positive for *C. difficile*, while all samples of raw milk were negative, suggesting that raw milk may not be the source of contamination for *C. difficile* in dairies (Jöbstl et al., 2010).

Seafood may also pose a risk of foodborne *C. difficile*. Of 52 edible mollusks that were sampled, 49% were found to be contaminated with *C. difficile* (Pasquale et al., 2012), and of 119 retail seafood samples in Canada, 4.8% were positive (Metcalf et al., 2011), with the predominant isolate again being the toxigenic 078 in both cases.

A few studies have also been conducted on ready-to-eat vegetables. In France, 2.9% of samples tested were positive for toxigenic *C. difficile* and included lettuce heart, lamb's lettuce salad, and pea sprouts. Spore numbers varied from 3 to 15 cfu per 20 g of salad (Eckert et al., 2013). An isolation rate of 4.5% of *C. difficile* spores from 111 vegetables sampled in Canada has been observed; once again it was the toxigenic strain 078 that was detected (Metcalf et al., 2010).

It is unclear if the source of *C. difficile* in meat products is the gastrointestinal tract of the animal, or whether contamination has occurred from workers, processing equipment, or the slaughterhouse environment. Also, variations in data may result from the different detection methods employed for the isolation of *C. difficile*, making it difficult to compare data across studies or from different countries (Blanco et al., 2013). In addition, as is seen in human cases of *C. difficile* infection, seasonality may play a role in the isolation of bacteria from food (de Boer et al., 2011).

5 CONCLUSION

Understanding the role of new and emerging pathogens in food safety is of paramount importance in preventing and controlling outbreaks of foodborne disease. Predicting new emerging organisms such as *Aeromonas* sp., *Acinetobacter* sp., and *C. difficile* that may pose an increased threat in the food arena in the future, and evaluating the niches they inhabit, how they survive, and their virulence, is the key to ensure the implementation of effective monitoring systems and technologies to prevent growth. This is a particular challenge due to the diversity of the microorganisms emerging in food and their ability to share genetic material, resulting often in antibiotic resistance and the ability to withstand current technologies aimed at their eradication.

REFERENCES

Ackerman, C., 2006. It is hard to predict the future: the evolving nature of threats and vulnerabilities. Rev. Sci. Tech. 25, 353–360.

Allende, A., Tomas-Barberan, F.A., Gil, M.I., 2006. Minimal processing for healthy traditional foods. Trends Food Sci. Technol. 17, 513–519.

Blanco, J.L., Álvarez-Pérez, S., García, M.E., 2013. Is the prevalence of *Clostridium difficile* in animals underestimated? Vet. J. 197, 694–698.

Daskalov, H., 2006. The importance of *Aeromonas hydrophila* in food safety. Food Control 17, 474–483.

Davies, A., Pottage, T., Bennett, A., Walker, J., 2011. Gaseous and air decontamination technologies for *Clostridium difficile* in the healthcare environment. J. Hosp. Infect. 77, 199–203.

de Boer, E., Zwartkruis-Nahuis, A., Heuvelink, A.E., Harmanus, C., Kuijper, E.J., 2011. Prevalence of *Clostridium difficile* in retailed meat in The Netherlands. Int. J. Food Microbiol. 144, 561–564.

Duffy, G., Lynch, O.A., Cagney, C., 2008. Tracking emerging zoonotic pathogens from farm to fork. Meat Sci. 78, 34–42.

Eckert, C., Burghoffer, B., Barbut, F., 2013. Contamination of ready-to-eat raw vegetables with *Clostridium difficile* in France. J. Med. Microbiol. 62, 1435–1438.

Fontes, M.C., Saavedra, M.J., Monera, A., Martins, C., Martínez-Murcia, A., 2010. Phylogenetic identification of *Aeromonas simiae* from a pig, first isolate since species description. Vet. Microbiol. 142, 313–316.

Fontes, M.C., Saavedra, M.J., Martins, C., Martinez-Murcia, A.J., 2011. Phylogenetic identification of *Aeromonas* from pigs slaughtered for consumption in slaughterhouses at north of Portugal. Int. J. Food Microbiol. 146, 118–122.

Gennari, M., Lombardi, P., 1993. Comparative characterization of *Acinetobacter* strains isolated from different foods and clinical sources. Zentralbl. Bakteriol. 279, 553–564.

Gennari, M., Parini, M., Volpon, D., Serio, M., 1992. Isolation and characterization by conventional methods and genetic transformation of *Psychrobacter* and *Acinetobacter* from fresh and spoiled meat, milk and cheese. Int. J. Food Microbiol. 15, 61–75.

Giamarellou, H., Antoniadou, A., Kanellakopoulou, K., 2008. *Acinetobacter baumannii*: a universal threat to public health? Int. J. Antimicrob. Agents 32, 106–119.

Gurung, M., Nam, H.M., Tamang, M.D., Chae, M.H., Jang, G.C., Jung, S.C., Lim, S.K., 2013. Prevalence and antimicrobial susceptibility of *Acinetobacter* from raw bulk tank milk in Korea. J. Dairy Sci. 96, 1997–2002.

Hamouda, A., Findlay, J., Al Hassan, L., Amyes, S.G.B., 2011. Epidemiology of *Acinetobacter baumannii* of animal origin. Int. J. Antimicrob. Agents 38, 314–318.

Hawken, P., Scott Weese, J., Friendship, R., Warriner, K., 2013. Carriage and dissemination of *Clostridium difficile* and methicillin resistant *Staphylococcus aureus* in pork processing. Food Control 31, 433–437.

Jöbstl, M., Heuberger, S., Indra, A., Nepf, R., Köfer, J., Wagner, M., 2010. *Clostridium difficile* in raw products of animal origin. Int. J. Food Microbiol. 138, 172–175.

Keessen, E.C., van den Berkt, A.J., Haasjes, N.H., Hermanus, C., Kuijper, E.J., Lipman, L.J.A., 2011. The relation between farm specific factors and prevalence of *Clostridium difficile* in slaughter pigs. Vet. Microbiol. 154, 130–134.

Kleter, G.A., Marvin, H.J.P., 2009. Indicators of emerging hazards and risks to food safety. Food Chem. Toxicol. 47, 1022–1039.

Marsh, J.W., 2013. Counterpoint: is *Clostridium difficile* a food-borne disease? Anaerobe 21, 62–63.

Martins, L.M., Marquez, R.F., Yano, T., 2002. Incidence of toxic *Aeromonas* isolated from food and human infection. FEMS Immunol. Med. Microbiol. 32, 237–242.

McCabe-Sellers, B.J., Beattie, S.E., 2004. Food safety: emerging trends in foodborne illness surveillance and prevention. J. Am. Diet. Assoc. 104, 1708–1717.

McMahon, M.A.S., Wilson, I.G., 2001. The occurrence of enteric pathogens and *Aeromonas* species in organic vegetables. Int. J. Food Microbiol. 70 (1–2), 155–162.

Metcalf, D.S., Costa, M.C., Dew, W.M.V., Weese, J.S., 2010. *Clostridium difficile* in vegetables, Canada. Lett. Appl. Microbiol. 51, 600–602.

Metcalf, D., Avery, B.P., Janecko, N., Matic, N., Reid-Smith, R., Weese, J.S., 2011. *Clostridium difficile* in seafood and fish. Anaerobe 17, 85–86.

Nawaz, M., Khan, S.A., Khan, A.A., Sung, K., Tran, Q., Kerdahi, K., Steele, R., 2010. Detection and characterization of virulence genes and integrons in *Aeromonas veronii* isolated from catfish. Food Microbiol. 27, 327–331.

Ottaviani, D., Parlani, C., Citterio, B., Masini, L., Leoni, F., Canonico, C., Sabatini, L., Bruscolini, F., Pianetti, A., 2011. Putative virulence properties of *Aeromonas* strains isolated from food, environmental and clinical sources in Italy: a comparative study. Int. J. Food Microbiol. 144, 538–545.

Pablos, M., Huys, G., Cnockaert, M., Rodríguez-Calleja, J.M., Otero, A., Santos, J.A., García-López, M.L., 2011. Identification and epidemiological relationships of *Aeromonas* isolates from patients with diarrhea, drinking water and foods. Int. J. Food Microbiol. 147, 203–210.

Pasquale, V., Romano, V., Rupnik, M., Capuano, F., Bove, D., Aliberti, F., Krovacek, K., Dumontet, S., 2012. Occurrence of toxigenic *Clostridium difficile* in edible bivalve molluscs. Food Microbiol. 31, 309–312.

Palu, A.P., Gomes, L.M., Miguel, M.A.L., Balassiano, I.T., Queiroz, M.L.P., Freitas-Almeida, A.C., de Oliveir, S.S., 2006. Antimicrobial resistance in food and clinical *Aeromonas* isolates. Food Microbiol. 23, 504–509.

Provincial, L., Guillén, E., Alonso, V., Gil, M., Roncalés, P., Beltrán, J.A., 2013. Survival of *Vibrio parahaemolyticus* and *Aeromonas hydrophila* in sea bream (*Sparus aurata*) fillets packaged under enriched CO_2 modified atmospheres. Int. J. Food Microbiol. 166, 141–147.

Radu, S., Ahmad, N., Ling, F.H., Reezal, A., 2003. Prevalence and resistance to antibiotics for *Aeromonas* species from retail fish in Malaysia. Int. J. Food Microbiol. 81, 261–266.

Rahimi, E., Jalali, M., Weese, J.S., 2014. Prevalence of *Clostridium difficile* in raw beef, cow, sheep, goat, camel and buffalo meat in Iran. BMC Public Health 14, 119.

Rodriguez-Palacios, A., Reid-Smith, R.J., Staempfli, H.R., Weese, J.S., 2010. *Clostridium difficile* survives minimal temperature recommended for cooking ground meats. Anaerobe 16, 540–542.

Rodriguez-Palacios, A., LeJeune, J., Hoover, D., 2012. *Clostridium difficle*: an emerging food safety risk. Food Technol. 66, 40–47.

Ross, A.I.V., Griffiths, M.W., Mittal, G.S., Deeth, H.C., 2003. Combining nonthermal technologies to control foodborne microorganisms. Int. J. Food Microbiol. 89, 125–138.

Sheridan, J.J., McDowell, D.A., 1998. Factors affecting the emergence of pathogens on foods. Meat Sci. 49 (Suppl. 1), S151–S167.

Songer, J.G., Anderson, M.A., 2006. *Clostridium difficile*: an important pathogen of food animals. Anaerobe 12, 1–4.

Tamang, M.D., Gurung, M., Nam, H.M., Kim, S.R., Jang, G.C., Jung, S.C., Lim, S.K., 2014. Short communication: genetic characterization of antimicrobial resistance in *Acinetobacter* isolates recovered from bulk tank milk. J. Dairy Sci. 97, 704–709.

Tauxe, R.V., Doyle, M.P., Kuchenmüller, T., Schlundt, J., Stein, C.E., 2010. Evolving public health approaches to the global challenge of foodborne infections. Int. J. Food Microbiol. 139 (Suppl.), S16–S28.

Thomas, R., Anjaneyulu, A.S.R., Kondaiah, N., 2008a. Development of shelf stable pork sausages using hurdle technology and their quality at ambient temperature (37 ± 1 °C) storage. Meat Sci. 79, 1–12.

Thomas, R., Anjaneyulu, A.S.R., Kondaiah, N., 2008b. Effect of hot-boned pork on the quality of hurdle treated pork sausages during ambient temperature (37 ± 1 °C) storage. Food Chem. 107, 804–812.

Tomas, J.M., 2012. The main *Aeromonas* pathogenic factors. ISRN Microbiol. 2012, 256–261.

Wang, C., Silva, J., 1999. Prevalence and characterization of *Aeromonas* species isolated from processed channel catfish. J. Food Prot. 62, 30–34.

Wang, Y., Wu, C., Zhang, Q., Qi, J., Liu, H., Wang, Y., He, T., Ma, L., Lai, J., Shen, Z., Lui, Y., Shen, J., 2012. Identification of New Delhi metallo-β-lactamase 1 in *Acinetobacter lwoffii* of food animal origin. Plos One 8. http://dx.doi.org/10.1371/annotation/e33066ea-f500-4d34-8981-7cfd68497e73.

Wentholt, M.T.A., Fischer, A.R.H., Rowe, G., Marvin, H.J.P., Frewer, L.J., 2010. Effective identification and management of emerging food risks: results of an international Delphi survey. Food Control 21, 1731–1738.

Xanthopoulos, V., Tzanetakis, N., Litopoulou-Tzanetaki, E., 2010. Occurrence and characterization of *Aeromonas hydrophila* and *Yersinia enterocolitica* in minimally processed fresh vegetable salads. Food Control 21, 393–398.

Yeni, F., Acar, S., Polat, Ö.G., Soyer, Y., Alpas, H., 2014. Rapid and standardized methods for detection of foodborne pathogens and mycotoxins on fresh produce. Food Control 40, 359–367.

SECTION

3

NEW DEVELOPMENTS IN FOOD SAFETY EDUCATION—FOOD SYSTEMS AND TRAINING

CHAPTER

FOOD SAFETY AT FARMERS' MARKETS: FACT OR FICTION?

15

Sujata A. Sirsat*, Kristen E. Gibson†, Jack A. Neal*

Conrad N. Hilton College of Hotel and Restaurant Management, University of Houston, Houston, Texas, USA, Center for Food Safety, Department of Food Science, University of Arkansas, Fayetteville, Arkansas, USA†*

1 INTRODUCTION

The number of farmers' markets in the United States has rapidly grown over the last 20 years. The US Department of Agriculture (USDA) listed 1755 markets in 1994 and 8144 in 2013 (USDA AMS, 2013). This statistic demonstrates a 364% increase in farmers' markets in almost two decades. Previous studies have shown that farmers' market consumers are more likely to visit farmers' markets because of the following reasons: to purchase a higher-quality product, economics, to socialize, for entertainment, to buy goods directly from the vendor, and to purchase organic (or naturally grown) fresh fruits and vegetables (Wolf et al., 2005). Similar observations were made by Gao et al. (2012) regarding farmers' market consumers. The results of this study showed that consumers shop at farmers' markets with the belief that local produce is environmentally friendly, to experience the atmosphere, and support local growers. Some markets also feature educational displays, cooking demonstrations, musicians, arts, and seasonal events to involve customers and attract them to the market (Abel et al., 1999).

In 1997, a farmers' market demographic profile analysis showed that shoppers tended to be older, more likely to be married, and more likely to not be employed as compared to nonshoppers (Wolf, 1997). Following this, a 2005 survey reported that that shoppers were more like to be female, married, and to have completed postgraduate education. Additionally, the study showed that the income levels and employment status between shoppers and nonshoppers was similar (Wolf et al., 2005). Schmit (2008) reported that 52% of farmers' market consumers believed it was crucial to buy local produce. The study also showed that many consumers shop for local produce at farmers' markets (61%) and supermarkets (62%). A study by Rainey et al. (2011) obtained similar results. The investigators conducted a comprehensive survey to investigate farmers' market consumers opinions and food safety concerns at three Arkansas farmers' markets. The results showed that the most important reason for consumers to shop at farmers' market is to support local farmers and vendors. Following this, the second reason was the belief that farmers' market produce is fresher. Interestingly, very few shoppers were concerned about harmful bacteria in organic or conventional produce (4% and 11%, respectively).

Food Safety. http://dx.doi.org/10.1016/B978-0-12-800245-2.00015-0

2 FOOD SAFETY AT FARMERS' MARKETS

Farmers' market vendors (i.e., farmers and prepared food workers) and managers need to ensure that appropriate good handling practices (GHPs) and best management practices are followed to ensure that fresh produce commodities sold at the market are safe for human consumption. While few foodborne disease outbreaks have been directly linked to farmers' markets, the majority of foodborne disease outbreaks are never identified or reported. There have been several farmers' market-related foodborne disease outbreaks. For instance, 18 illnesses were reported in 2008 after farmers' market customers consumed contaminated bagged peas sold at an Alaskan farmers' market (Gardner et al., 2011). Following this, in 2010, fresh guacamole and salsa sold at a farmers' market in Iowa was possibly contaminated with *Salmonella* leading to several hospitalizations (Quinlisk, 2010). A produce-related outbreak at an Oregon farmers' market in 2011 was traced to strawberries contaminated with *Escherichia coli* O157:H7 (USDA, 2011). In February 2014, a market vendor in Michigan was convicted under Michigan's Food Law after pleading guilty of willfully misbranding apple cider sold at his booth (Ag and Rural Development, 2014).

The risk for illness related to products sold at farmers' markets may be slightly higher compared to produce at wholesalers as oversight of food safety is largely left up to the market managers who may or may not have the appropriate training or be aware of appropriate resources. Farmers' markets present considerable opportunities for economic development in the local community and in the country as a whole; therefore, it is important that food safety recommendations—and potential regulations—do not stifle the development of local food systems (Rosenberg and Leib, 2012). However, the local economy and vendor sales may be negatively affected if a foodborne disease outbreak is confirmed at a farmers' market. Studies have shown that business planning and training of youth is crucial for the growth of the local food movement (Tropp and Barham, 2008). Interestingly, food safety and handling practices is not considered a high priority (Harrison et al., 2013).

The Produce Safety Project at Georgetown University, an initiative of The Pew Charitable Trusts, released the report *Health-Related Costs from Foodborne Illness in the United States* and estimated the cost of acute foodborne illness in the United States to be $152 billion annually—fresh produce was identified as contributing to 25% of this annual burden (Scharff, 2012). The Centers for Disease Control and Prevention (CDC) estimates that each year roughly 1 out of 6 Americans (or 48 million people) gets sick, 128,000 are hospitalized, and 3000 die from foodborne diseases (Scallan et al., 2011). The Morbidity and Mortality Weekly Report (MMWR) on Foodborne Disease Outbreaks showed that fresh produce was associated with the highest number of illnesses (3377) in 2008 (CDC, 2011). According to The Vegetables and Melons Yearbook (ERS/USDA, 2010), consumption of fresh produce (not including potatoes and melons) in the United States has steadily increased over the past two decades from approximately 120 lbs (farm weight) per capita in 1990 to 140 lbs per capita for 2010. More specifically, leafy vegetable (not including head lettuce) consumption increased from 3.8 to 11.2 lbs per capita in 1990-2010, respectively—nearly a threefold increase (ERS/USDA, 2010). Along with this per capita increase in fresh produce consumption, there has been a strong consumer movement toward supporting local, or regional, food markets and purchasing directly from the producer in venues such as farmers' markets or through memberships to community-supported agriculture programs (Martinez et al., 2010). According to a survey by Mintel in 2009, one in six US consumers makes it a priority to buy local food products, with fruits and vegetables topping the list (Merrett, 2009). In addition to direct purchase of local food products by consumers, the 2012 National Restaurant Survey indicates that sustainable and locally sourced food products (both fresh produce and meat) will occupy a dominant place on restaurant menus (NRA, 2012). The 2007 Census of Agriculture reported that "direct-to-consumer" marketing totaled $1.2 billion in sales compared with $551 million only a decade ago—the majority of these sales going to small farms (Martinez et al., 2010).

As fresh produce continues to be the dominant food commodity implicated in the most foodborne disease outbreaks in the United States, questions concerning agricultural food safety practices related to the production of this commodity are being raised. However, as local food markets increase and smaller farms become the preferable source for many consumers to procure fresh produce, this may make the control of foodborne pathogens associated with fresh produce such as human norovirus, *Salmonella*, *Listeria*, and enterohemorrhagic *E. coli* even more difficult. Many of these small farms are new to agricultural production and have varying field preparation, production, harvest, and postharvest handling practices.

Moreover, on December 21, 2010, the House of Representatives voted 215-144 to pass the Senate version of the Food Safety Modernization Act (FSMA). Included in this bill were several National Sustainable Agricultural Coalition (NSAC) supported amendments that

- Give very small farms and food processing facilities and direct-market farms who sell locally the option of complying with state regulations.
- Provide the FDA with the authority to exempt farms engaged in low or minimal risk processing from new regulatory requirements.
- Reduce unnecessary paperwork and excess regulations required under the preventative control plan.
- Exempt farmers from extensive traceability and record-keeping requirements.

Small farms that market directly to consumers and certain retailers and restaurants with annual sales less than $500,000 are technically exempt from these produce safety standards. However, these small farms are still responsible for either identifying potential hazards associated with the food being produced and implementing and monitoring preventive controls or demonstrating that they will comply with state, county, or other applicable non-Federal food safety laws. This has been a source of great confusion and frustration for many small farmers. In addition, some farmers' markets are starting to require their vendors to meet certain standards for on-farm food safety in order to sell their products (Miller, 2011). It is evident that growers and other food industry processors want to know more about food safety; however, many small farmers are accustomed to complete freedom in how they grow, harvest, and sell their products.

3 COTTAGE FOODS

Cottage foods are home-produced food items that are not potentially hazardous foods and include examples such as baked goods and jams and jellies. In general, these foods do not present a food safety risk if prepared appropriately. As per the FDA Food Code, all food sold to a consumer in any establishment should be prepared in a certified commercial kitchen (Food Code, 2013). However, each state has the primary responsibility regarding their citizens' health and well being. Hence, each state, and not the federal government, has the ability to decide if home-produced food can be sold within the state direct to consumers at venues such as farmers' markets. States that allow cottage food operations are listed in

Table 1. Table 2 lists states that do not allow cottage food operations. Specific cottage food laws may vary among the states that allow the production and sale of these foods directly to consumers. The states that do not allow cottage food production have likely adopted the FDA code that defines and regulates all food establishments (Condra, 2013). The goal of the states that support the cottage food industry is to enhance the local food movement and to improve the economy and general health of the community (Tarr, 2011). However, the perspective of the local and county health department may be different. The sanitarians have the daunting task of inspecting all of the items available at farmers' markets (listed in Table 3). Some examples include but are not limited to produce, meat, eggs, and ready-to-eat (RTE) foods. Hence, adding home-produced cottage foods is an extra responsibility and a potential hazard if

Table 1 States That Allow Cottage Food Operations

Alabama	Maine	Ohio
Alaska	Maryland	Oregon
Arkansas	Michigan	Pennsylvania
California	Minnesota	Rhode Island
Colorado	Mississippi	South Carolina
Delaware	Missouri	South Dakota
Florida	Montana	Tennessee
Georgia	Nebraska	Texas
Illinois	Nevada	Utah
Indiana	New Hampshire	Vermont
Iowa	New Mexico	Virginia
Kentucky	New York	Washington
Louisiana	North Carolina	Wisconsin
Wyoming		

Table 2 States That Do Not Allow Cottage Food Operations

Connecticut	District of Columbia
Hawaii	Idaho
Kansas	New Jersey
North Dakota	Oklahoma
West Virginia	

Table 3 Commodities Sold at Farmers' Markets Across the Unites States

Vegetables	Leafy greens	Meat
Poultry	Candles	Soap
Arts and crafts	Poultry	Ready-to-eat foods
Baked goods	Cottage foods	Fruits

The items may vary depending on the state (USDA, 2014).

the food is not produced properly, and appropriate food safety guidelines are not followed. Hence, it is crucial for states that allow cottage food production to have specific training and education provisions regarding good handling and kitchen practices. Some states such as Texas, Colorado, and California that have passed cottage food laws have online training available for cottage food entrepreneurs who want to sell food items directly to consumers. However, one standardized national resource that can be modified depending on each state's cottage food law would ensure that the educational material is disseminated accurately and effectively.

4 FARMERS' MARKET MICROBIAL RESEARCH

The risk of foodborne illness outbreaks because of fresh produce and prepared foods can be prevented or reduced by following good agriculture practices and GHPs. Water quality, manure, good personal hygiene, equipment sanitation, and tractability have been identified as key areas for decreasing the risk of microbial contamination and, hence, the possibility of a foodborne outbreak (Parker et al., 2012). Along these lines, Zerio-Egli et al. (2014) designed a novel device for small farmers to wash, sanitize, and dry fresh leafy greens on the farm before selling to customers at the farmers' market. To this end, the investigators obtained readily available parts from hardware stores and designed the device using a diluted vinegar solution as the sanitizing wash. Viability studies were performed to test the efficacy of the device in combination with various dilutions of white vinegar (acetic acid) against pathogenic (*Listeria*, *E. coli* O157:H7, and *Salmonella* spp.) and spoilage microorganisms. The data showed that a 1.6% acetic acid solution was effective at reducing coliforms, yeast, mold, and pathogenic bacteria. Following this, detailed information sheets were designed for small farmers with information on how to build the device. The idea was that if the farmers build the device themselves, then they will be more likely to use it (Zerio-Egli et al., 2014).

Harrison et al. (2013) surveyed food safety practices on small- to medium-sized farms and in farmers' markets in Georgia, Virginia, and South Carolina. Their results showed that more than 54% of farmers used manure and of these, 34% use raw manure. Over 27% of farmers do not test water sources for irrigation safety. Over 43% of farmers do not sanitize surfaces that touch produce at the farm and only 33% clean transport containers between uses. The farmers' market survey data revealed that over 42% of managers have no food safety practices in place at the farmers' market and less than 25% of managers sanitize the market surfaces. Overall, these results showed that there is an increasing need to address food safety practices at the farmers' market and upstream on farms to prevent microbial hazards.

To further highlight the need for better farmers' market food safety, a study conducted by Norwood et al. (2014) at four independent farmers' markets in the Houston, Texas, area provided evidence that there is little adherence to basic GHPs at farmers' markets. The investigators observed that RTE foods (e.g., dairy products) were not held at the appropriate temperatures, and some vendors did not use gloves while handling RTE foods. Moreover, many farmers' markets did not have basic facilities including hand-washing stations. However, several of these markets had produce and RTE items available for sampling.

An independent study conducted by Behnke et al. (2012) at Indiana farmers' markets showed that only 9 out of 18 vendors had access to functional hand-washing stations. Even though hand-washing stations and glove use are not required at farmers' markets, implementation would improve food safety behaviors and prevent cross-contamination issues. Teng et al. (2004) reported similar findings while observing cheese vendor practices at farmers' markets. The researchers observed 17 cheese vendors in nine

Ontario farmers' markets. The observations showed that approximately 47% of vendors had issues with refrigeration, 24% did not have hand-washing sinks at their disposal, 41% were not restraining their hair, and in 24% of the cases, cheese was stored adjacent to raw meat products. Research has shown that appropriate hand-washing techniques reduce the risk of foodborne illness significantly (Todd et al., 2010).

Norwood et al. (2014) designed educational videos and information sheets specifically for farmers' market managers and vendors based on their farmers' market observational data. Several of the information sheets were also directed toward farmers' market consumers. For example, the information sheets and videos highlighted the need to (1) wash tote bags to prevent cross-contamination, (2) have more than one vendor present at the booth so that tasks such as produce handling and money handling are separated, and (3) emphasize the need for hand-washing stations at farmers' markets (Norwood et al., 2014).

Multiple studies have performed comparative microbial analyses between items (meat and produce) obtained from farmers' markets and super markets. Overall, the results have shown higher bacterial prevalence on food items purchased from farmers' markets. For instance, Park and Sanders (1992) conducted a study to compare thermotolerant *Campylobacter* in samples obtained from outdoor farmers' markets and supermarkets. The results demonstrated the presence of *Campylobacter jejuni* on six vegetable types obtained from the outdoor farmers' market. All samples obtained from the supermarket were negative for *Campylobacter*. Sirsat and Neal (2013) compared the microbial profiles of farmers' market lettuce and lettuce obtained from grocery stores (bagged and loose leaf). The results showed that lettuce obtained from the farmers' market had significantly higher (10^3 log CFU/g) *E. coli* concentrations compared to the lettuce obtained from the grocery stores (Sirsat and Neal, 2013). Soendjojo (2012) conducted similar analyses to compare microbial counts on leafy greens obtained from the grocery store and farmers' markets. The hypothesis of this study was that farmers' market greens would have fewer bacteria because there is very little time between harvest and selling to the consumers. Moreover, greens sold at the grocery store may have gone through additional handling and processing. However, the results showed that farmers' market Romaine and Bibb lettuce had higher microbial counts (6 log CFU/g) compared to grocery store lettuce (5 log CFU/g). Su (2014) reported similar results while comparing produce items (cilantro, green onions, jalapeños, and serrano peppers) from farmers' markets and supermarkets. The investigators found significantly higher levels of coliform bacteria on produce obtained from the farmers' market.

Bohaychuk et al. (2009) studied the prevalence of *E. coli* on produce (lettuce, spinach, carrots, and green onions) obtained from farmers' markets in Alberta, Canada. The investigators observed approximately 0.48-3 log/g *E. coli* colonies on produce samples. A study performed by Hernández et al. (1997) showed an increased prevalence of rotavirus and hepatitis A virus on lettuce samples obtained from a farmers' market in Costa Rica. Similar conclusions were made by Scheinberg et al. (2013) based on their investigation that demonstrated a higher incidence of *Campylobacter* spp. and *Salmonella* spp. on poultry products obtained from farmers' markets when compared to supermarkets in Pennsylvania. The results of these studies highlight the need for farmers' market vendor food safety training and education.

5 FARMERS' MARKET FOOD SAFETY OUTREACH AND EDUCATION

Food safety recommendations for farmers' markets are available online through extension websites of state land grant universities, state governments, and nonprofit organizations. In addition, a noticeable trend among all of these resources is the format in which they are disseminated—4-40 page brochures/booklets focused on what is allowable or not allowable and only referencing generic safe-handling requirements or website links with no interactive component. As identified by Rosenberg and Leib (2012), there is no national set of model regulations designed specifically for direct marketing venues and such a resource would greatly improve uniformity across farmers' markets with respect to food safety standards. Hence, there is an increasing need for a single, national-based repository for information targeted at market managers and vendors by market size, market type, and location.

A recent study conducted by researchers in British Columbia, Canada, investigated the food safety knowledge of farmers' market vendors, managers, and public health officials (McIntyre et al., 2014). As expected, the investigators found that the health officials had the highest score. In addition, there were no significant differences in food safety knowledge scores between farmers' market managers (79.7%) and vendors (80.7%). However, scores of less than 70% were observed on questions related to recognition and analysis of potentially hazardous foods. Interestingly, McIntyre et al. (2014) and Lee et al. (2010) demonstrated that there was no correlation between food safety training and food safety knowledge.

Similarly, a study by Worsfold et al. (2004) conducted at a UK farmers' market showed that most vendors had had food safety training and rated their hygiene practices very highly. However, less than half of the vendors had a risk management plan and did not consider the possibility that their produce could be contaminated. The consumers who shopped at the farmers' market were also surveyed, and they had no concerns regarding food safety (Worsfold et al., 2004).

Another study carried out by investigators in Pennsylvania used survey results to identify specific training needs that farmers' market managers may have. The results showed that along with help related to enhancing business practices, 64% market managers wanted food safety resources and training materials (Berry et al., 2013). An independent study recommended that training for farmers' market personnel should be different from that offered to restaurant employees (Choi and Almanza, 2012). Training for farmers' market vendors and managers needs to be focused specifically on the food items sold at the farmers' market and good hygiene, handling, and behavior practices.

6 RECOMMENDATIONS

The FSMA, formerly called Senate Bill 510, was signed into law on January 4, 2011. The overarching goal of this act was to respond to safety concerns regarding food produced in the United States and foods imported to the United States from across the world. In addition, the intent of the FSMA was to shift the focus toward preventing rather than managing a crisis or outbreak. While the goal of the FSMA was to help small farmers so that they are exempt from regulation and extensive record keeping, it seems to have created more questions than answers. Several farmers, especially those who wish to sell produce to restaurants and local grocery stores, have very little direction and guidance on best practices. Moreover, one farmer may sell produce at three to four different farmers' markets during a week. Each of these farmers' markets could be in different counties within one state or in different states altogether. Hence, rules and regulations regarding items sold may vary at each market. For instance, some counties may allow the sale of shelled peas while other counties may not. Similarly, some counties may allow market vendors to offer fruit samples to customers while others may not. It has been a challenge for market vendors to follow these dynamic regulations across state

Table 4 Recommendations for Farmers' Market Food Safety (Norwood et al., 2014)

Positive Food Safety Behaviors at Farmers' Markets
Hand-washing stations available for vendors and customers
Vendors display their name and contact information
Designated eating areas are provided and marked clearly
Booths have tent coverings
Vendor uses gloves, tongs, and clean barriers
Food is kept off the ground
If samples are allowed, toothpicks are provided
Antibacterial gels are provided in addition to hand-washing stations
Responsible reuse of tote bags are encouraged (washed and sanitized in a timely manner)
Food is covered and out of direct sunlight

and county lines. Hence, a pilot study by Liou et al. (2014) investigated the guidelines for food safety across 12 different farmers' markets across the Unites States. The results showed an increasing need to standardize food safety instruction, guidelines, and regulations at the farmers' markets. Following this, the investigators designed a unique farmers' market food safety smartphone and tablet application for consumers to be made more aware of safe produce, meat, and RTE food handling at the farmers' market (Liou et al., 2014).

These researchers also performed pilot analyses to identify the best farmers' market vendor booth layout to reduce the amount of microbial cross-contamination at a farmers' market. The investigators designed L-, O-, and U-shaped mock farmers' markets and requested volunteers to "walk through" the market simulating customers' actions. The volunteer's hands were inoculated with a fluorescent compound, and the "contamination" was visually tracked across the markets. The results showed that the U-shaped farmers' market had the least amount of cross-contamination if nonfood vendors (e.g., candles and arts and crafts) were used as "barriers" (Liou et al., 2014).

Some ways to proactively prevent an outbreak are to ensure that farmers' market managers have food safety knowledge. More importantly, the managers should ensure that the market vendors have the required certifications and training. However, the most effective way to ensure good practices is by incorporating food safety practices as a part of regular practices such as harvest, transport, setting up for the market, and storage. Basic recommendations such as separate tasks at vendor booths and hand washing are listed in Table 4.

7 CONCLUSIONS

Locally grown fresh produce and its economic viability will be threatened if a significant foodborne disease outbreak is traced to a farmers' market. In order to prevent this, steps need to be taken to incorporate food safety practices on farms and in farmers' markets across the country. Irrespective of how food is grown (i.e., conventionally, naturally, or organically) employing good food safety practices is the only way to lower the incidence of foodborne disease outbreaks.

REFERENCES

Abel, J., Thomson, J., Maretzki, A., 1999. Extension's role with farmers' markets: working with farmers, consumers, and communities. J. Ext. 37 (5), 150–165.

Ag and Rural Development, 2014. http://www.michigan.gov/mdard/0,4610,7-125-1572_28248-322547-,00.html (accessed 20 September 2014).

Berry, J., Moyer, B., Oberholtzer, L., 2013. Assessing training and information needs for Pennsylvania farmers markets: results from a 2011 survey of market managers. J. NACAA. http://www.nacaa.com/journal/index.php?jid=194 (accessed 25 September 2014).

Behnke, C., Seo, S., Miller, K., 2012. Assessing food safety practices in farmers' markets. Food Protect. Trends 32 (5), 232–239.

Bohaychuk, V.M., Bradbury, R.W., Dimock, R.M., Fehr, G.E., Gensler, R.K., King, R., Romero, B.P., 2009. A microbiological survey of selected Alberta-grown fresh produce from farmers' markets in Alberta, Canada. J. Food Prot. 72 (2), 415–420.

CDC, 2011. Surveillance for foodborne outbreaks – United States, 2008. MMWR 60 (35), 1197–1202.

Choi, J.-K., Almanza, B., 2012. An assessment of food safety risk at fairs and festivals: a comparison of health inspection violations between fairs and festivals and restaurants. Event Manage. 16 (4), 295–303.

Condra, A., 2013. Cottage Food Laws in the United States. http://blogs.law.harvard.edu/foodpolicyinitiative/files/2013/08/FINAL_Cottage-Food-Laws-Report_2013.pdf (accessed 23 September 2014).

ERS/USDA, 2010. The Vegetables and Melon Yearbook. http://www.ers.usda.gov/Publications/VGS/#yearbook (accessed 25 September 2014).

FDA Food Code, 2013. http://www.fda.gov/Food/GuidanceRegulation/RetailFoodProtection/FoodCode/ucm374275.htm (accessed 22 September 2013).

Gao, Z., Swisher, M., Zhao, X., 2012. A new look at farmers' markets: consumer knowledge and loyalty. HortScience 47, 1102–1107.

Gardner, T.J., Fitzgerald, C., Xavier, C., Klein, R., Pruckler, J., Stroika, S., McLaughlin, J.B., 2011. Outbreak of campylobacteriosis associated with consumption of raw peas. Clin. Infect. Dis. 53 (1), 26–32.

Harrison, J.A., Gaskin, J.W., Harrison, M.A., Cannon, J.L., Boyer, R.R., Zehnder, G.W., 2013. Survey of food safety practices on small to medium-sized farms and in farmers markets. J. Food Prot. 76 (11), 1989–1993.

Hernández, F., Monge, R., Jiménez, C., Taylor, L., 1997. Rotavirus and Hepatitis A virus in market lettuce *Latuca sativa* in Costa Rica. Int. J. Food Microbiol. 37 (2), 221–223.

Lee, J.-E., Almanza, B.A., Nelson, D.C., 2010. Food safety at fairs and festivals: vendor knowledge and violations at a regional festival. Event Manage. 14 (3), 215–223.

Liou, L., Sirsat, S.A., Gibson, K.E., Neal, J.A., 2014. Food Safety Guidelines for Farmers' Markets in the United States: A Need for Standardization. International Association of Food Protection, Indianapolis.

Martinez, S., Hand, M., DaPra, M., Pollack, S., Ralston, K., Smith, T., Vogel, S., Clark, S., Lohr, L., Low, S., Newman, C., 2010. Local Food Systems: Concepts, Impacts, and Issues. USDA Economic Research Service, Washington, DC.

McIntyre, L., Herr, I., Kardan, L., Shyng, S., Allen, K., 2014. Competencies of those assessing food safety risks of foods for sale at the farmers' markets in British Columbia, Canada. Food Prot. Trends 34 (5), 331–348.

Merrett, N., 2009. 'Buy local' message requiring promotion push, says Mintel. Br. Food J. 106 (3), 211–227.

Miller, C., 2011. Microcredit and crop agriculture: new approaches, technologies and other innovations to address food insecurity among the poor. http://www.globalmicrocreditsummit2011.org/ (accessed 22 September 2014).

Norwood, H.E., Sirsat, S.A., Neal, J.A., 2014. Farmers' Market Food Safety: Educating While Engaging. International Association of Food Protection, Indianapolis.

NRA, 2012. Chef Survey: What's Hot in 2012. http://www.restaurant.org/News-Research/News/What-s-Hot-in-2012-chef-survey-shows-local-sourcin (accessed 1 September 2014).

Park, C.E., Sanders, G.W., 1992. Occurrence of thermotolerant campylobacters in fresh vegetables sold at farmers' outdoor markets and supermarkets. Can. J. Microbiol. 38 (4), 313–316.

Parker, J.S., Wilson, R.S., LeJeune, T., Doohan, D., 2012. Including growers in the "food safety" conversation: Enhancing the design and implementation of food safety programming based on farm and marketing needs of fresh fruit and vegetable production. Agric. Human Values 29 (3), 303.

Quinlisk, P., 2010. Foodborne Illness and Farmers' Markets. Iowa Department of Public Health—access update. http://publications.iowa.gov/10622/1/hca_newsletter_201008.pdf (accessed 14 September 2014).

Rainey, R., Crandall, P.G., O'Bryan, C.A., Ricke, S.C., Pendleton, S., Seideman, S., 2011. Marketing locally produced organic foods in three metropolitan Arkansas farmers' markets: consumer opinions and food safety concerns. J. Agric. Food Inform. 12 (2), 141–153.

Rosenberg, N., Leib, E.B., 2012. Pennsylvania's Chapter 57 and Its Effects on Farmers Markets. Harvard Food Law and Policy Clinic. http://blogs.law.harvard.edu/foodpolicyinitiative/files/2012/08/PA-FM-FINAL3.pdf (accessed 23 September 2014).

Scallan, E., Hoekstra, R.M., Angulo, F.J., Tauxe, R.V., Widdowson, M.A., Roy, S.L., Griffin, P.M., 2011. Foodborne illness acquired in the United States – major pathogens. Emerg. Infect. Dis. 17 (1), 7–15.

Scharff, R.L., 2012. Economic burden from health losses due to foodborne illness in the United States. J. Food Prot. 75 (1), 123–131.

Scheinberg, J., Doores, S., Cutter, C.N., 2013. A microbiological comparison of poultry products obtained from farmers' markets and supermarkets in Pennsylvania. J. Food Saf. 33 (3), 259–264.

Schmit, J., 2008. Locally Grown Food Sounds Great, but What Does It Mean? USA Today. http://usatoday30.usatoday.com/money/economy/2008-10-27-local-grown-farms-produce_N.htm (accessed 15 September 2014).

Sirsat, S.A., Neal, J.A., 2013. Microbial profile of soil-free *versus* in-soil grown lettuce and intervention methodologies to combat pathogen surrogates and spoilage microorganisms on lettuce. Foods 2 (4), 488–498.

Soendjojo, E., 2012. Is local produce safer? Microbiological quality of fresh lettuce and spinach from grocery stores and farmers' markets. J. Purdue Undergrad. Res. 2 (1), 10.

Su, Y., 2014. Prevalence of *Salmonella, Escherichia coli* O157:H7 and *Shigella* in selected fresh produce from supermarkets, local markets and farmers' markets. In: 2014 Annual Meeting, IAFP.

Tarr, N.W., 2011. Food entrepreneurs and food safety regulation. J. Food Law Policy 7, 35. http://heinonline.org/HOL/Page?handle=hein.journals/jfool7&div=5&g_sent=1&collection=journals#37 (accessed 23 September 2014).

Teng, D., Wilcock, A., Aung, M., 2004. Cheese quality at farmers markets: observation of vendor practices and survey of consumer perceptions. Food Control 15 (7), 579–587.

Tropp, D., Barham, J., 2008. National Farmers Market Summit Proceedings Report. USDA AMS. http://www.ams.usda.gov/AMSv1.0/getfile?dDocName=STELPRDC5066926 (accessed 15 September 2014).

Todd, E.C., Greig, J.D., Michaels, B.S., Bartleson, C.A., Smith, D., Holah, J., 2010. Outbreaks where food workers have been implicated in the spread of foodborne disease. Use of antiseptics and sanitizers in community settings and issues of hand hygiene compliance in health care and food industries. J. Food Prot. 73 (12), 2306–2320.

USDA, 2011. Fresh Strawberries from Washington County Farm Implicated in *E. coli* O157:H7 Outbreak in Oregon. http://www.fda.gov/Safety/Recalls/ucm267667 (accessed 15 September 2014).

USDA, 2014. http://www.ams.usda.gov/AMSv1.0/ams.fetchTemplateData.do?template=TemplateN&leftNav=WholesaleandFarmersMarkets&page=WFMUSDAFarmersMarket&description=USDA%20Farmers%20Market (accessed 22 September 2014).

USDA Agriculture Marketing Service (USDA AMS), 2013. Farmers Markets and Local Food Marketing. http://www.ams.usda.gov/AMSv1.0/ams.fetchTemplateData.do?template=TemplateS&navID=WholesaleandFarmersMarkets&leftNav=WholesaleandFarmersMarkets&page=WFMFarmersMarketGrowth&description=Farmers%20Market%20Growth&acct=frmrdirmkt (accessed 14 September 2014).

Wolf, M.M., 1997. A target consumers profile and positioning for promotion of the direct marketing of fresh produce: a case study. J. Food Distrib. Res. 28 (3), 1–17.

Wolf, M.M., Spittler, A., Ahern, J., 2005. A profile of farmers' market consumers and the perceived advantages of produce sold at farmers' markets. J. Food Distrib. Res. 36 (1), 192–201.
Worsfold, D., Worsfold, P.M., Griffith, C.J., 2004. An assessment of food hygiene and safety at farmers' markets. Int. J. Environ. Health Res. 14 (2), 109–119.
Zerio-Egli, C.N., Sirsat, S.A., Neal, J.A., 2014. Development of a novel economical device to improve post-harvest processing practices on small farms. Food Control 4, 152–158.

CHAPTER

NOVEL APPROACHES FOR RETAIL FOOD SAFETY EDUCATION

16

Jack A. Neal, Sujata A. Sirsat

Conrad N. Hilton College of Hotel and Restaurant Management, Houston, Texas, USA

1 INTRODUCTION

In the United States, over 130 million people eat out every day, totaling over 70 billion meals and snacks each year. Estimates suggest over 76 million people in the United States contract diseases related to foodborne illnesses, and around 75% of foodborne illnesses outbreaks are caused by improper food handling and poor food safety practices (Cho et al., 2013; Almanza and Nesmith, 2004). Despite evidence of a small decline in the number of outbreaks associated with food workers, the total number of outbreaks is still significant, and more efforts must be taken to reduce this. Research addressing risk factors associated with most outbreaks in foodservice establishments shows that many of these occurrences can be avoided (Hertzman and Barrash, 2007).

Even after decades of research, increasing emphasis in the FDA's Model Food Code, and increased spending on food safety, there has not been a significant decrease in foodborne illness outbreaks or food product recalls. According to the US Food and Drug Administration (FDA) website, there have been over 160 food product recalls in 2013 for contamination from both pathogenic microorganisms and physical hazards. These microbial pathogens include *Listeria monocytogenes*, *Escherichia coli* O157:H7, *Salmonella*, and human norovirus (FDA, 2013). Foodborne illness continues to grow as a health concern in the United States and throughout the world. According to the Centers for Disease Control and Prevention (CDC), one in six US citizens get sick each year from foodborne illness (CDC, 2001). In addition, over 128,000 persons are hospitalized, and approximately 3000 die from foodborne illness (Scallan et al., 2011).

Food safety training programs currently available rely heavily on the delivery of information; however, there are suggestions that training programs should also identify the attitudes and behaviors with increased focus on altering the behaviors leading to foodborne illnesses (Egan et al., 2007). The foodservice industry consists of people from a variety of different educational and ethnic backgrounds, and these backgrounds result in diverse beliefs and cultural behaviors about food safety procedures (Neal et al., 2011).

Poor food safety practices and improper food handling cause 97% of all foodborne illnesses in restaurants (Howes et al., 1996; Egan et al., 2007). Some have suggested that there is a correlation between the increase in convenience food consumption in restaurants and grocery stores and the increase of foodborne illnesses in the United States (Cotterchio et al., 1998; Almanza and Nesmith, 2004). With the growth

Food Safety. http://dx.doi.org/10.1016/B978-0-12-800245-2.00016-2

of the foodservice industry, cutbacks on labor, high turnover rates, and language barriers, employees in this industry often receive inadequate training in food safety procedures (Almanza and Nesmith, 2004). However, food safety training and certification is not a new concept.

2 FOOD SAFETY CERTIFICATION

The first mandate requiring a food safety certification was in the state of Washington in the 1950s (Almanza and Nesmith, 2004). Since the introduction of mandatory certification in Washington, over 17 states have passed legislation requiring certifications in the foodservice industry. Some of the states that have passed legislation include California, Connecticut, Florida, Illinois, Louisiana, Mississippi, and Texas (Almanza and Nesmith, 2004). In the late 1980s and early 1990s, there was an increased concern about the rise in foodborne illnesses and the lack of effective food safety training. As a result, the University of Massachusetts Cooperative Extension established a food safety training program titled "Food Handling Is a Risky Business" (Fritiz et al., 1989). The program focused on four objectives: (1) gaining knowledge in proper food handling practices and becoming aware of foods carrying a higher risk for foodborne illnesses; (2) adopting food safety practices for preparation, processing, handling, and storage; (3) increasing knowledge of current food safety regulations; and (4) acquiring knowledge of how food handling affects an organization and population. Prior to this food safety training program, the first two objectives (acquisition of knowledge in proper food handling, and the adoption of food safety practices) were commonly taught; however, this was the first time the additional objectives addressing the risks and regulations were incorporated into a food safety training program (Fritiz et al., 1989). Prior to the more formal food safety workshops created, food safety had been seen as a form of "good manners" and "work rules," but unrelated to regulations and accountability from the foodservice organizations. Employees were trained through nonmandatory seminars hosted by local health departments or by employers (Bryan, 1988). The food safety training programs typically consisted of a 1-h lesson that included one of the three "top learning" methods used at the time: filmstrips (videotape), lectures, and demonstrations (simulation). The program would conclude with a group discussion (interaction) on various food safety scenarios, potentially leading to the risk of foodborne illnesses (Fritiz et al., 1989).

Many state and local health regulators have mandated that managers receive formal food safety training, which is described as the successful completion of a recognized training program, accredited by the American National Standards Institute-Conference for Food Protection (ANSI-CFP, 2012). This certification training may be administered through online or face-to-face instruction, and successful completion of a final exam with a passing grade of 75 or higher. Unfortunately, Olson and Carbone (2011) reported that the existing food safety certification exam does not sufficiently measure food safety procedures, responsibilities, and material obtained by those taking the exam. They suggested that there was a correlation between exam scores and the level of education obtained by the test taker. Because many of the positions within the foodservice industry are entry level, often requiring fewer skills, many employees have a high school education or less. The study concluded the pass rate for a test taker with less than a high school degree was 36.4%, and test takers with a high school degree or equivalent had a pass rate of 78.8%. Taking these poor success rates and educational achievement into consideration, the exam was analyzed, and word usage was tested. The study found that, by slightly modifying the construction of a question and the language used, the pass rate increased by 10% for those with less than a high school degree (Olson and Carbone, 2011).

The content of the exam is intended to cover the top five risk factors associated with foodborne illnesses: food from unsafe sources, inadequate cooking, improper holding temperatures, contaminated equipment, and poor personal hygiene (Hertzman and Barrash, 2007). An example of what the exam covers in personal hygiene includes possessing the knowledge of proper hand-washing methods (including time, water temperature, and the use of appropriate soap). Additionally, personal hygiene also refers to properly keeping one's self groomed (including wearing hairnets, having little to no facial hair, and keeping one's fingernails clean) (Cogan et al., 2002). Another example of material that is covered in the exam is managing controls, which is the use of gloves when necessary, maintaining proper food temperature and heating times, ensuring that expired inventory is disposed of, and ensuring the establishment's employees are properly trained in all areas (Cogan et al., 2002).

3 THE LEADING RISK FACTORS IN FOODBORNE ILLNESS OUTBREAKS

In the mid-1990s, the FDA developed the National Food Retail Steering Committee with representation from the Center for Food Safety and Applied Nutrition (CFSAN), Office of Regulatory Affairs (ORA), Division of Federal/State Relations (DFSR), Division of Human Resource Development (DHRD), and the Interstate Travel Program (ITP) Field Team. This group was tasked with measuring trends in the occurrence of food preparation practices and employee behaviors at the retail and foodservice level that are believed to most commonly contribute to foodborne illness outbreaks. In 2011, the FDA Trend Analysis Report on the occurrence of foodborne illness risk factors in selected institutional foodservice, restaurant, and retail food store facility types (1998-2008) was published. While the level of foodborne illness associated with this segment of the industry would have been an excellent program performance indicator, unfortunately the occurrence of foodborne illness is often unreported making tracking the incidence of foodborne illness unreliable as a program performance measure. Instead, the FDA selected the occurrence of foodborne illness risk factors as the performance indicator.

Findings from the 10-year study were identified from nine facility types from three different segments of the retail and foodservice industry:

Restaurants:

- fast food,
- full service,

Retail food stores:

- deli departments/stores,
- meat and poultry/markets/departments,
- seafood markets/departments,
- produce markets/departments,

Institutional foodservice:

- Hospitals,
- Nursing homes,
- Elementary schools (K-5).

The focus of this discussion will be on the first two segments of the retail and foodservice industry.

The leading risk factors identified were (1) improper holding temperature, (2) contaminated equipment, (3) poor personal hygiene, (4) food from unsafe source, (5) inadequate cooking, and (6) other/chemicals (FDA, 2011). While the data suggested that specific control of certain foodborne illness risk factors improved in most facility types, compliance with important requirements of the FDA Food Code needs further improvement to adequately reduce or prevent foodborne illness outbreaks. For example, there were statistically significant improvements in the improper holding/time and temperature risk factors and poor personal hygiene risk factor for fast food restaurants, full service restaurants, meat and poultry/markets/departments, and produce markets/departments; however, in many facility types the compliance percentages remained low for many of the risk factors, implying that a greater emphasis is needed in training and behaviors to control for these risk factors. In addition, in the proper, adequate handwashing data item, hand-washing practices were observed to be out of compliance approximately once in three out of four full service restaurants and almost half of retail delis (see Table 1).

Even though there were improvements for the risk factors poor personal hygiene and improper holding/time and temperature, these two risk factors had generally lower in compliance percentages compared to the other risk factors. This clearly demonstrates the continued need to ensure that effective protocols and procedures, training, and monitoring are both developed and implemented.

The five foods linked with a majority of foodborne illness outbreaks are produce, seafood, poultry, beef, and eggs (Dewaal et al., 2006). Linking particular foods to specific foodborne illness outbreaks allows for higher inspection of those foods and those areas. For example, 40% of seafood outbreaks are linked to improper holding temperatures, 40% of produce outbreaks are related to poor personal hygiene, and an additional 30% of produce-linked outbreaks are associated with cross-contamination. Inadequate cooking accounts for more than 40% of beef-related outbreaks (Dewaal et al., 2006).

Table 1 Foodborne Illness Risk Factors with Statistically Significant Improvement, 1998–2008.

Industry Segment	Facility Type	Foodborne Illness Risk Factor(s) with Statistically Significant Improvement
Restaurants	Fast food	• Improper holding/time and temperature • Poor personal hygiene
	Full service	• Contaminated equipment/protection from contamination • Improper holding/time and temperature • Poor personal hygiene
Retail food stores	Deli departments/stores	• Poor personal hygiene
	Meat and poultry markets/departments	• Improper holding/time and temperature • Poor personal hygiene
	Seafood markets/departments	• Poor personal hygiene
	Produce markets/departments	• Improper holding/time and temperature • Poor personal hygiene

http://www.fda.gov/downloads/Food/GuidanceRegulation/RetailFoodProtection/FoodborneIllnessRiskFactorReduction/UCM369245.pdf, p. 23.

The most challenging risk factor associated with outbreaks to identify is cross-contamination because it can occur at various stages during food production (Allos et al., 2004). The most common reasons for outbreaks in restaurants are infected foodservice employees and employees handling food with improperly washed hands (Hedberg et al., 2006). By identifying the risk factors and foods associated with foodborne illness outbreaks, researchers and educators can develop task-specific training materials aimed at changing employee behaviors.

4 THEORIES OF ADULT LEARNING

Multiple models, assumptions and principles, and theories and explanations have been created to explain the adult learning knowledge base. The three major theories include androgony, self-directed learning, and transformational learning. Knowles (1980) popularized the concept of androgony, which is described as the art and science of helping adults learn; versus pedagogy, which is the art and science of teaching children to learn. Andragogy poses a set of assumptions about adult learners, namely that the adult learner

- Moves from dependency to increasing self-directedness as he or she matures and directs his or her own learning.
- Relies on previous life experiences to help with learning.
- Becomes ready to learn when he or she experiences a need to learn in order to cope more satisfyingly with real-life tasks or problems.
- Wants to apply new information immediately and wants the information to be problem-focused and applicable.
- Is motivated by internal rather external factors.

It is important for adult learners to understand *why* they are learning something, and effective trainers and educators need to explain clearly their reasons for teaching specific skills. In regards to food safety, trainers must clarify how certain behaviors, such as hand washing, prevent or contribute to the risk of foodborne illness. Adults also learn by doing; therefore, rather than memorizing content, it is better for training and educational materials to focus on task-specific behaviors. As adult learners tend to be problem solvers and learn best when information is of immediate use, effective instruction involves the learner in solving job-related issues and problems.

Self-directed learning is a process in which an individual takes the initiative, without help of others in planning, carrying out, and evaluating their own learning experience (Knowles, 1975). Over 30 years ago, both Tough (1971) and Cross (1981) reported that approximately 70% of all adult learning is self-directed and approximately 90% of all adults conduct at least one self-directed learning project per year. These statements were made prior to the digital age, and with the relatively unlimited access to information from the Internet, these numbers have only increased. What makes learning "self-directed" is determined by who makes decisions (in this case, the learner) concerning content, methods, resources, and evaluation of learning. Specifically, the learner takes responsibility for determining their needs, setting goals, identifying resources, implementing a plan to meet their goals, and evaluating outcomes. While the FDA Food Code and Conference for Food Protection have identified specific, science-based content and behaviors that management and employees are responsible for knowing and demonstrating, individuals may still determine the method of obtaining this information, the resources available, and evaluation of the learning.

Lastly, transformative learning can be described as learning that changes the way individuals think about themselves and their world that involves a shift of consciences. This is how behaviors and attitudes are changed, and corporate cultures are developed. Mezirow (2000) reported that as individuals reflect and discuss their assumptions about their world, a shift in their frame of reference or perspectives begins to emerge. However, Mezirow's theory has been criticized in that it does not account for the effect of an individual's race, class, gender, or the historical context in which learning occurs (Corley, 2003; Sheared and Johnson-Bailry, 2010; Taylor, 1998; Cervero and Wilson, 2001). Cultural differences have been identified as barriers to food safety practices (Cho et al., 2013).

5 CURENT FOOD SAFETY TRAINING AND EDUCATION

Trade organizations, such as the National Restaurant Association (NRA) and the Food Marketing Institute (FMI), as well as training development companies, have created educational programs in food safety training procedures tailored for the restaurant and retail food industries. Consequently, as more organizations create food safety training programs, managers are left to guess which program will most fit their needs (Frash et al., 2006). As the need for adequate food safety, training becomes a greater necessity in the restaurant and retail food industries, managers are also recognizing the training programs currently available lack job/task-specific training material (Egan et al., 2007).

The food safety training programs currently available are offered, both in a face-to-face classroom setting and online, in a self-guided format. Texts and manuals make up the majority of training tools used in the restaurant and retail food industry to train and educate employees (Harris and Bonn, 2000). However, the use of technology is revolutionizing the approach educators and trainers use to teach as well as the process students and trainees learn. Educators are increasing the use of various technologies, such as online video clips, computers, and data collection instruments, to teach more effectively and efficiently (Reid-Griffin and Carter, 2004). This will assist students in their learning process through creating a more interactive learning environment (Reid-Griffin and Carter, 2004).

Staying ahead of one's competition is a primary reason why organizations spend large amounts of money in the continual training of their employees. Properly trained employees are more confident, possess a more positive attitude, and become more effective and efficient employees (Aguinis and Kraiger, 2009). Read and Kleiner (1996) identified 10 learning methods with the highest effectiveness for knowledge retention. These include

- video,
- lectures,
- one-on-one instruction,
- role-plays,
- games/simulation,
- case studies,
- slides,
- computer-based training,
- audiotapes and film.

Each method can be as effective as the others; however, the **proper utilization of the method** is what determines its effectiveness on a trainee. For example, the effectiveness of videos relies heavily on whether or not there is interaction between the trainer and trainee. Ideally, the trainer would stop the video after major points to ask questions or encourage discussion between trainees. Similarly, lectures, as an individual learning method, may be ineffective; however, like videos, lectures can prove highly effective if interaction is incorporated and audience participation is encouraged (Read and Kleiner, 1996).

With the advancement of technology in classrooms, computer-based training is becoming more popular and can be divided into two separate groups: computer-managed instruction and computer-assisted instruction (Kulik and Kulik, 1991). A computer-managed training program evaluates a student's performance, guides the student through the instruction, and keeps track of the student's progress. A computer-assisted training program provides exercises and practice material; however, the material being taught is not new (Kulik and Kulik, 1991). With the use of computer-based training, it is easier to incorporate multiple learning methods while allowing the trainee to progress through the training program at his or her own pace (Read and Kleiner, 1996).

Many large foodservice operations have training departments and prefer to create their own food safety training material. One popular option is the use of e-learning technologies, which have been used to develop more effective training materials with enhanced learning efficiency (Bosco, 1986; Kulik et al., 1994; Parlangeli et al., 1999; O'Bryan et al., 2010). US Census data from 2003 indicated that approximately 83% of adult Americans age 18 years or older used the Internet at home (US Census Bureau, 2003). Convenience, efficiency, and affordability are some of the main features attracting participants to use online education courses (Santerre, 2005). While there are many benefits of using e-learning, e-learning designers must understand their target audience and how the learners will use the training tools created (Theofanos and Redish, 2003). One key element to consider when creating online material is usability testing, which evaluates the interactions between the user and the product being tested. O'Bryan et al. (2010) described how user-centered design (UCD) combined with usability testing to provide successful e-learning products and improve user satisfaction. Foodservice operations that are too small to have their own trainers or do not want to create their own material may choose to purchase existing food safety training material.

The opportunity to take an online food safety training module or class at the employee's own pace is a benefit for companies to offer. Santerre (2005) stated that individuals who work full-time like the option of Internet-based training. Another benefit for foodservice operations offering Internet-based training to employees is the frequency in which the modules can be offered. With turnover rates estimated at 200% in some markets, being able to offer employees food safety training when needed can eliminate gaps in training and can also be offered as continuing education for existing employees.

Despite the popularity of food safety training, it does not always translate into positive food handling practices (Clayton et al., 2002; Green et al., 2005; Powell et al., 1997; Riben et al., 1994). What is needed is a system that not only educates the employee but also changes their behavior. Therefore, before purchasing food safety materials, foodservice managers need to identify what needs to be included in a food safety training program. Seaman (2010) suggests using the food hygiene training model, which includes evaluation stages, managerial components, and overall performance measures. These components consider the planning and training of the program, managerial support needed to determine the type of training and support needed to facilitate it, and the overall performance measures needed to ensure that training transferred into changing behavior.

One of the challenges to having multiple online food safety training programs available is to evaluate the material to determine if it meets the needs of the individual company and if it is instructionally sound. Bertera (1990) identified the need for examination of service provision and program evaluation to ensure credibility and relevance in a study on the planning and implementation of workplace health promotion. The key to success is finding a tool that can be used to evaluate multiple food safety training programs consistently. Pisik (1997) recommends that the instrument should emphasize instructional soundness and include numeric data for comparison. While the evaluation tool should be thorough, it must also take into consideration the manager's time to complete it. Once the data has been analyzed, objective information can be obtained and save time and money.

When first introducing food handling safety, it is important to stick to the basics and avoid scare tactics. Scare tactics only work in the short term, but altering the employee's behavior and perception on food safety practices will prove more effective in the long term (Peregrin, 2001). The material in food safety training programs needs to stay within the confines of the five major risk factors: food from unsafe sources, inadequate cooking, improper holding temperatures, contaminated equipment, and poor personal hygiene (Medeiros et al., 2001). Having trainees that relate food safety practices to everyday life is helpful. For example, it is difficult for a person to remember that they should wash their hands for at least 20 s, but it is easier for them to remember singing "Happy Birthday" twice, beginning to end, will serve as a 20-s timer (Peregrin, 2001).

Through examining current training programs, one will notice how progressive the training process has become. The programs, which once consisted of groups watching filmstrips and face-to-face lectures, have now advanced to individuals using programs that contain a variety of interactive media, such as click-through computer programs, 30-s video clips, and online discussions (Medeiros et al., 2011). Training programs are continuing to progress with the discovery of more effective methods, advancement of technology, and new material.

Currently, most food safety training programs are fashioned after the Knowledge, Attitudes, and Practices (KAP) model, which implies that proper training leads to changes in behaviors (Ehiri et al., 1997). However, this model is criticized because of the assumption that once provided with the proper knowledge and material, individuals will proactively change their behaviors (Ehiri et al., 1997). Knowledge gained from food safety training programs alone is insufficient in addressing behaviors and by leaving these behaviors unaddressed, food workers are competent in the material but are continuing to use the same poor food safety practices (Hertzman and Barrash, 2007). Some of the behaviors needing attention are the use of gloves, personal hygiene, and storage of potentially hazardous foods (PHF) (Cho et al., 2013).

In an assessment to assess the current knowledge on food safety and best practices of caterers in Las Vegas, Hertzman and Barrash (2007) discovered most employees who were knowledgeable about food safety practices but found that they were neglectful of a number of safety precautions. The most common neglect was in the personal hygiene section of the observation, which included improper hand washing, eating/drinking in preparation areas, and improper restraint of hair (Hertzman and Barrash, 2007). This is consistent with the risk factors identified by the FDA.

6 ACTIVE MANAGERIAL CONTROL AND THE PERSON IN CHARGE

Active managerial control is the term used by the Food and Drug Administration (FDA) to describe the food industry's responsibility for the development and implementation of food safety management systems to reduce the risk of foodborne illnesses. The goal of active management control is to incorporate specific actions and procedures by foodservice managers into their operations to attain control over foodborne illness risk factors (FDA, 2004). With the implementation of active managerial control, emphasis has been placed on training foodservice workers by frontline supervisors, known as the person in charge (PIC). A similar trend has been observed in the United Kingdom (Seaman and Eves, 2010). The idea behind this emphasis is that, while many are responsible for food safety along the farm-to-fork continuum, the ultimate responsibility for food safety at the retail level lies with the retail and foodservice operators and their ability to develop and maintain effective food safety management systems (FDA, 2006, 2011). Therefore, FDA Food Code 2009 Chapter 2 states that the PIC must be able to demonstrate to the regulatory authority, upon request, knowledge of foodborne disease prevention, application of the Hazard Analysis Critical Control Points (HACCP) principles, and the requirements of the food code (FDA, 2011).

The food code requires that a PIC, who is food safety certified, be present on premises at all times. The PIC is expected to demonstrate and share his or her knowledge of food safety with entry-level or nonsupervisory employees. This mandate holds the PIC directly responsible for any food safety violations that may occur, and violations can lead to a fine for both the individual and the establishment. As a response to additional mandates and amendments, organizations and educational companies have begun creating food safety training programs to meet the state and/or local requirements to be a Certified Food Protection Manager (Frash et al., 2006).

7 BEHAVIOR-BASED TRAINING

While all foodservice employees need a general baseline of food safety education, at some point, it comes down to specific behaviors, and they need to take ownership of their actions. One of the limitations of traditional food safety training is its top-down control approach. Workplace barriers serve as an additional reason for foodservice workers not changing their behaviors. Several barriers that inhibit foodservice workers from altering their behavior are inadequate facilities, time restraints, understaffing, and insufficient information relating the cause to the effect (Pragle et al., 2007). Behavior-based training provides tools and procedures; employees can use to take personal control of food safety risks (Geller, 2005). Several models have been suggested on how to empower employees to take this next critical step.

Behavior-based safety (BBS) is not a novel concept, but its application to food safety is relatively new. Skinner helped to develop behavioral science, and behavior-based safety is founded in these principles (Geller, 2005; Skinner, 1938, 1953, 1974). Namely, the intervention process always targets specific behaviors for constructive change (Geller, 2005). In regards to food safety, despite training and food safety education, some employees still do not perform behaviors correctly or choose to take shortcuts as well as the barriers mentioned above. Identifying internal traits and attitudes of employees is difficult to identify objectively; therefore, it is more cost-effective to identify environmental conditions that influence behaviors and to address these factors when behavior change is needed (Geller, 2005). Sometimes this includes identifying management behaviors that encourage or promote food safety poor behavior; for example, not providing employees enough cleaning cloths in an effort to earn a bonus for reducing linen costs.

Typically, employees behave correctly or incorrectly and do what they do because of consequences they anticipate for the specific action or behavior. This is the basis for Skinner's foundation of this motivational principle. The key concept is that activators, or those signals preceding behavior, are only as powerful as the consequences supporting them (Geller, 2005). Activators tell people what to do to receive the reward or consequence, from the buzzing of a text message or knocking on a door to the instruction from a training seminar or one-on-one training session. This principle is known as the antecedent behavior consequence (ABC) model. Skinner was very concerned with people's feelings and attitudes and stressed that negative consequences should be reduced in order to increase perceptions of freedom and allow individuals the opportunity to make the right decision. Unfortunately, most of our metrics used to measure food safety compliance is the total number of food safety violations that puts retail and foodservice workers in the mind-set of avoiding failure rather than achieving success. The BBS approach to food safety provides proactive measures that employees set goals to achieve to reduce food safety risks and prevent unintentional injury to consumers (Geller, 2005).

The DO IT method can be used to provide feedback and cultivate food safety behavior improvement. DO IT stands for define, observe, intervene, and test (Geller, 2005; see Figure 1).

The DO IT process begins with *defining* task-specific target behaviors. These can be both behaviors that need to increase and behaviors that need to be corrected. Safe behaviors may need to be targeted to substitute particular at-risk behaviors. However, safe behaviors can be independent of at-risk behaviors. Safe behaviors may be as simple as wearing gloves or other methods of avoiding bare-handed contact with ready-to-eat foods. To develop precise definitions of a DO IT target, team members must consider the various categories of general operating procedures and then develop safe versus at-risk methods of performing a specific work task. When team members help develop a behavior checklist, they take ownership of the training process that can help improve human dynamics on both the outside (behaviors) and the inside (feelings and attitudes) of people (Geller, 2005; Yiannas, 2008).

Once individuals begin to observe one another for certain food safety at-risk behaviors, they quickly realize that everyone performs at-risk behavior, either intentionally or accidentally. This observation stage is not meant to be a faultfinding process but rather a fact-finding procedure to identify the behaviors and conditions that need to be continued, encouraged, reduced, or changed to prevent foodborne illness. This can be accomplished through a checklist that can be specific to a particular task, station, or team of employees, depending on the needs of the operation. Once the team has created the checklist, they can then begin to hold each other accountable. It is important that they not ask or approach other

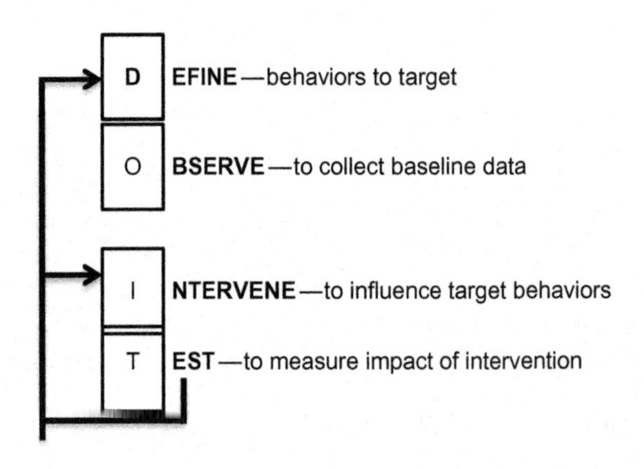

FIGURE 1

Behavior-based food safety as a continuous four-step improvement process (Geller, 2005).

workers to observe them, but when members of the work team look for the target behavior to occur (safe or at-risk) they can take out their checklist and complete it. Again, this is not a "gotcha" moment but rather a method to focus a worker's attention to specific food safety behaviors and identify how well the group performs this behavior.

During the *intervene* stage, interventions are devised and executed in an attempt to increase or decrease specific food safety behaviors. Interventions are aimed at changing the external conditions of the operation in order to make safe behaviors more likely than at-risk behaviors. This third step is critical because in order to change behaviors, motivating consequences occurs quickly, are is certain (expected), and sizeable (Geller, 2005). Once individuals or groups of employees see the results of a behavioral observation, they can receive the kind of information that enables them to improve their performance. In addition to behavioral feedback, safety researchers have reported multiple types of interventions such as worker-designed safety slogans, corrective-action reporting, safe behavior promise cards, as well as incentive and bonus programs (Geller, 1996, 1997, 1998, 2001a,b; McSween, 1995; Petersen, 1989). The final *test* phase provides work teams with data needed to refine or replace a behavior-change intervention. If the observations and checklists indicate that the desired behavior changes have not occurred, the work team can make adjustments or change the intervention method.

The gradual release of responsibility (GRR) model slowly shifts the teaching responsibly from the instructor to the learner. This model of training is typically used when "on-boarding" new employees to teach them the basic skills needed to perform specific tasks. This may or may not include food safety training; for example, proper assembling and using a deli slicer. In the GRR model, there are four phases of learning and transition: (1) focus lesson—the trainer is responsible for showing the trainee how tasks are done; (2) guided instruction—the trainer and the trainee complete the tasks together and questions are asked/answered; (3) collaborative work—the trainees work together in small groups, completing tasks and relying on each other for guidance; and (4) independent work—the trainee completes tasks entirely on his or her own (Fisher and Frey, 2008). Regardless of the varying views on a trainee's learning process, it is widely agreed that there is continuous engagement throughout the training program. While this model teaches specific tasks, it may or may not explain "why" things are done is a specific manner or specific order.

The health belief model (HBM) consists of four principles to direct an individual's behavior regarding health. The four principles are (1) perceived susceptibility, (2) perceived severity of a condition, (3) perceived benefits of treatment, and (4) perceived barriers to treatments (Cho et al., 2013). This model, tailored to food safety behavior, is called the food safety belief model (FSBM). The FSBM contains 10 principles focusing on food safety. Some of the principles are food safety knowledge, perceived benefits of food safety, cues to action, self-efficacy, and food safety behavior (Cho et al., 2013). The FSBM was tested on Latino foodservice workers, and the results revealed a positive correlation with an increase of food safety knowledge and food safety behavior. Additionally, Latino employees altered their food safety behaviors when they formed a link between improved food safety behavior and an increase in customer and manager satisfaction (Cho et al., 2013).

The "closed-loop" model combines supervisory observations with coaching to improve performance (Neal et al., 2014; Figure 2).

This model consists of the following six steps:

1. Breaking down a specific process into a sequence of smaller process or steps.
2. Determining the desired employee action/behavior at each step and identifying potential deficiencies.

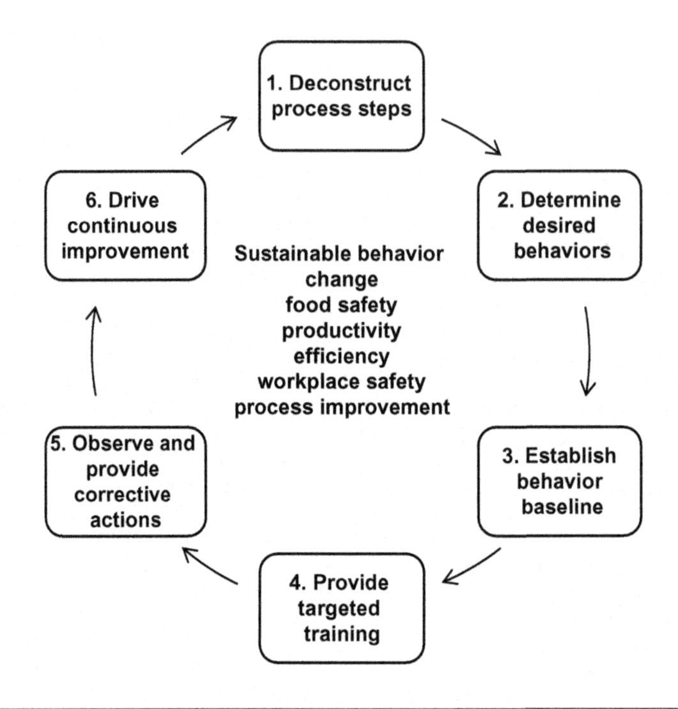

FIGURE 2

The closed-loop model of behavior-based food safety training.

3. Observing, measuring, and documenting a baseline level of employee behavior compliance following training.
4. Training frontline employees on what constitutes acceptable behaviors.
5. Empowering supervisors to make corrective observations of individual employees. Corrective actions should be nonjudgmental and nonpunitive in order to focus on continual improvements.
6. Repeating the process on a sustained basis to validate employee improvement.

Management should establish proper training procedures for all new employees and conduct periodic retraining sessions to ensure all employees continue to use proper hand hygiene and food handling practices. The importance of step 4, training frontline employees, cannot be stressed enough. There are multiple products and training materials available both in a classroom setting as well as for online training. Management should take into consideration the learning styles of their employees, which include auditory, kinesthetic, and visual. Alchemy Systems™ has developed a pedagogy based on significant visual metaphors to help hold trainees' attention, learning modules under 20 min, interactive involvement, group activities, and workplace training that is directly related to workplace experiences.

8 SUMMARY

Over the past 60 years, food safety has transitioned from "good manners and work rules" to risk management, epidemiology, and behavior science. This evolution has transitioned from workshops to certifications. The original courses focused on gaining knowledge about food-handling practices; proper food-handling practices for preparation, processing, handling, and storage; increasing knowledge of food safety regulations; and increasing knowledge of how handling practices affect an organization and the general population. Formal training was then established with the successful completion of a recognized training program to be administered either face-to-face or online with a passing score of 75 or higher. The content of the certification course focuses on the most common risk factors associated with foodborne illness: food from unsafe sources, inadequate cooking, improper holding temperatures, contaminated equipment, and poor personal hygiene. The FDA Trend Analysis Report suggests that specific control of several risk factors haves improved for most facility types; however, further improvement to adequately reduce or prevent foodborne illness outbreaks is still needed. By identifying foods commonly associated with foodborne illness, inspectors can focus their attention on the storage, preparation, holding, and serving of these items.

We are currently in our next transition from 2-day workshops to the development of integrated food safety systems composed of food safety education, continual monitoring and evaluation of employee behaviors, record keeping, and integration into a food safety culture. Researchers have focused their attention on developing models, assumptions and principles, theories and explanations to understand and explain the adult learning knowledge base. The major theories include androgony, self-directed learning, and transformational learning. Multiple models have been proposed as to improving employee behavior, but all include observations, monitoring, and corrective actions. In addition, multiple learning and delivery methods have been identified. While each method can be effective, the proper utilization of the method is what determines its effectiveness on an employee. Training programs are continuing to progress with the discovery of more effective methods, advancement of technology, and new material. Active managerial control is the term used by the Food and Drug Administration (FDA) to describe the food industry's responsibility for the development and implementation of food safety management systems to reduce the risk of foodborne illnesses. The goal of active management control is to incorporate specific actions and procedures by foodservice managers into their operations to attain control over foodborne illness risk factors. Proactive organizations will incorporate all of these recommendations into their organization and strive for continued improvement in both the flow of foods and employee actions.

REFERENCES

Aguinis, H., Kraiger, K., 2009. Benefits of training and development for individuals and teams, organizations, and society. Annu. Rev. Psychol. 60, 451–474.

Allos, B.M., Moore, M.R., Griffin, P.M., Tauxe, R.V., 2004. Surveillance for sporadic foodborne disease in the 21st century: the FoodNet perspective. Clin. Infect. Dis. 38 (3), S115–S120.

Almanza, B.A., Nesmith, M.S., 2004. Food safety certification regulations in the United States. J. Environ. Health 66 (9), 10–14 20.

American National Standards Institute – Conference for Food Protection, 2012. Standards for Accreditation of Food Protection Manager Certification Programs. http://www.foodprotect.org/media/managercert/CFP%20FPMC%20April%202012%20Standards.pdf (accessed 27 September 2014).

Bertera, R.L., 1990. Planning and implementing health promotion in the workplace: a cast study of the Du Pont company experience. Health Educ. Q. 17, 307–327.

Bosco, J., 1986. An analysis of evaluation of interactive video. Educ. Technol. 25, 7–16.

Bryan, F.L., 1988. Risks of practices, procedures and processes that lead to outbreaks of foodborne diseases. J. Food Prot. 51, 663–673.

Center for Disease Control, 2001. CDC estimates of food borne illness in the United States. Available at: http://www.cdc.gov/foodborneburden/2011-foodborne-estimates.html (accessed 27 September 2014).

Cervero, R.M., Wilson, A.L. (Eds.), 2001. Power in Practice: Adult Education and the Struggle for Knowledge and Power in Society. Jossy-Bass, San Francisco.

Cho, S., Hertzman, J., Erdem, M., Garriott, P., 2013. A food safety belief model for Latino(a) employees in foodservice. J. Hosp. Tour. Res. 37, 330–348.

Clayton, D.A., Griffith, C.J., Price, P., Peters, A.C., 2002. Food handlers' beliefs and self-reported practices. Int. J. Environ. Health Res. 12, 25–39.

Cogan, T., Slader, J., Bloomfield, S., Humphrey, T., 2002. Achieving hygiene in the domestic kitchen: the effectiveness of commonly used cleaning procedures. J. Appl. Microbiol. 92, 885–892.

Corley, M., 2003. Poverty, Racism, and Literacy (ERIC Digest No. 243). ERIC Clearinghouse on Adult, Career, and Vocational Education, Ohio State University, Columbus, OH.

Cotterchio, M., Gunn, J., Coffill, T., Tormey, P., Barry, M.A., 1998. Effect of a manager training program on sanitary conditions in restaurants. Public Health Rep. 113, 353.

Cross, K.P., 1981. Adults as Learners: Increasing Participation and Facilitating Learning. Jossey-Bass, San Francisco.

Dewaal, C.S., Hicks, G., Barlow, K., Alderton, L., Vegosen, L., 2006. Foods associated with foodborne illness outbreaks from 1990 through 2003. Food Protect. Trend. 26, 466–473.

Egan, M., Raats, M., Grubb, S., Eves, A., Lumbers, M., Dean, M., Adams, M., 2007. A review of food safety and food hygiene training studies in the commercial sector. Food Control 18, 1180–1190.

Ehiri, J., Morris, G., McEwen, J., 1997. Evaluation of a food hygiene training course in Scotland. Food Control 8, 137–147.

Fisher, D., Frey, N., 2008. Homework and the gradual release of responsibility: making "responsibility" possible. Engl. J. 98, 40–45.

Frash Jr., R., Binkley, M., Nelson, D., Almanza, B., 2006. Transfer of training efficacy in U.S. food safety accreditation. J.Culinary Sci. Technol. 4, 7–38.

Fritiz, B.R., Cohen, N.L., Evans, D.A., 1989. A food safety education program targeting food handlers in high risk settings. J. Nutr. Educ. 21, 284F–285F.

Geller, E.S., 1996. The truth about safety incentives. Prof. Saf. 41, 34.39.

Geller, E.S., 1997. Key processes for continuous safety improvement: behavior-based recognition and celebration. Prof. Saf. 42, 40–44.

Geller, E.S., 1998. Understanding Behavior-Based Safety: Step-by-Step Methods to Improve Your Workplace, second ed. J.J. Keller & Associates, Neenah, WI.

Geller, E.S., 2001a. The Psychology of Safety Handbook. CRC, Boca Raton, FL.

Geller, E.S., 2001b. Working Safe: How to Help People Actively Care for Health and Safety, second ed. Lewis, New York.

Geller, E.S., 2005. Behavior-based safety and occupational risk management. Behav. Modif. 29, 539–561.

Green, L., Selman, C., Banerjee, A., Marcus, R., Medus, C., Angulo, F.J., Radke, V., Buchanan, S., 2005. Food service workers' self-reported food preparation practices: and EHS-NET study. Int. J. Hyg. Environ. Health 208, 27–35.

Harris, K.J., Bonn, M.A., 2000. Training techniques and tools: evidence from the foodservice industry. J. Hosp. Tour. Res. 24, 320–335.

Hedberg, C.W., Smith, S.J., Kirkland, E., Radke, V., Jones, T.F., Selman, C.A., 2006. Systematic environmental evaluations to identify food safety differences between outbreak and non-outbreak restaurants. J. Food Prot. 69, 2697–2702.

Hertzman, J., Barrash, D., 2007. An assessment of food safety knowledge and practices of catering employees. Br. Food J. 109, 562–576.

Howes, M., McEwen, S., Griffiths, M., Harris, L., 1996. Food handler certification by home study: measuring changes in knowledge and behavior. Dairy Food Environ. Sanit. 16, 339–343.

Knowles, M., 1975. Self-Directed Learning: A Guide for Learners and Teachers. Follett Publishing Company, Chicago.

Knowles, M., 1980. The Modern Practice of Adult Education: Andragogy Versus Pedagogy, rev. and updated ed. Cambridge Adult Education, Englewood Cliffs, NJ.

Kulik, C.C., Kulik, J.A., 1991. Effectiveness of computer-based instruction: an updated analysis. Comput. Hum. Behav. 7, 75–94.

Kulik, C.C., Kulik, J.A., Shwalb, B.J., 1994. The effectiveness of computer applications: a meta-analysis. J. Res. Comput. Educ. 27, 48–61.

McSween, T.E., 1995. The Values-Based Safety Process: Improving Your Safety Culture with a Behavioral Approach. Van Nostrand Reinhold, New York.

Medeiros, L.C., Hillers, V.N., Kendall, P.A., Mason, A., 2001. Food safety education: what should we be teaching to consumers? J. Nutr. Educ. 33, 108–113.

Medeiros, C.O., Cavalli, S.B., Salay, E., Proença, R.P.C., 2011. Assessment of the methodological strategies adopted by food safety training programs for food service workers: a systematic review. Food Control 22, 1136–1144.

Mezirow, J., 2000. Learning to think like an adult: core concepts of transformation theory. In: Mezirow, J. (Ed.), Learning as Transformation. Critical Perspectives on a Theory Ion Progress. Jossey-Bass Publishers, San Francisco.

Neal, J.A., Dawson, M., Madera, J.M., 2011. Identifying food safety concerns when communication barriers exist. J. Food Sci. Educ. 10, 36–44.

Neal, J.A., O'Bryan, C.A., Crandall, P.G., 2014. A personal hygiene behavioral change study of a Midwestern cheese production plant. Agric. Food Anal. Bacteriol. 4, 13–19.

O'Bryan, C.A., Johnson, D.M., Shores-Ellis, K.D., Crandall, P.G., Marcy, J.A., Seideman, S.C., Ricke, S.C., 2010. Designing an affordable usability test for e-learning modules. J. Food Sci. Educ. 9, 6–10.

Olson, R.B., Carbone, E.T., 2011. Examining the exam: implications for participants and policy makers of the food manager certification exam. Food Protect. Trends 31, 93–103.

Parlangeli, O., Marchigiani, E., Bagnara, S., 1999. Multimedia systems in distance education: effects of usability on learning. Interact. Comput. 12, 37–49.

Peregrin, T., 2001. Teaching food-handling safety: stick to the basics. J. Am. Diet. Assoc. 101, 1339.

Petersen, D., 1989. Safe Behavior Reinforcement. Aloray, Goshen, NY.

Pisik, G.B., 1997. Is this course instructionally sound? A guide to evaluating online training courses. Educ. Technol. 7, 50–59.

Powell, S.C., Attwell, R.W., Massey, S.J., 1997. The impact of training on knowledge and standards of food hygiene – a pilot study. Int. J. Environ. Health Res. 7, 329–334.

Pragle, A.S., Harding, A.K., Mack, J.C., 2007. Food workers' perspectives on hand washing behaviors and barriers in the restaurant environment. J. Environ. Health 69, 27–31.

Read, C.W., Kleiner, B.H., 1996. Which training methods are effective? Manag. Dev. Rev. 9, 24–29.

Reid-Griffin, A., Carter, G., 2004. Technology as a tool: applying an instructional model to teach middle school students to use technology as a mediator of learning. J. Sci. Educ. Technol. 13, 495–504.

Riben, P.D., Mathias, R.G., Cambell, E., Wiens, M., 1994. The evaluation of the effectiveness of routine restaurant inspections and education of food handlers: critical appraisal of the literature. Can. J. Public Health 85 (1), S56–S60.

Santerre, C.R., 2005. X-train: teaching professionals remotely. J. Nutr. 135, 1248–1252.

Scallan, E., Hoekstra, R.M., Angulo, F.J., Tauxe, R.V., Widdonwson, M., Roy, S.L., Jones, J.L., Griffin, P.M., 2011. Foodborne illness acquired in the united states—major pathogens. Emerg. Infect. Dis. 17, 7–15.

Seaman, P., 2010. Food hygiene training: introducing the food hygiene training model. Food Control 21, 381–387.

Seaman, P., Eves, A., 2010. Efficacy of the theory of planned behavior model in predicting safe food handling practices. Food Control 21, 983–987.

Sheared, V., Johnson-Bailry, J., 2010. The Handbook of Race and Adult Education: A Resource for Dialogue on Racism. Wiley and Sons, San Francisco.
Skinner, B.F., 1938. The Behavior of Organisms. An Experimental Analysis. Copley, Acton, MA.
Skinner, B.F., 1953. Science and Human Behavior. Mac Million, New York.
Skinner, B.F., 1974. About Behaviorism. Alfred A. Knopf, New York.
Taylor, E., 1998. The Theory and Practice of Transformative Learning: A Critical Review. Center on Education and Training for Employment, Columbus, OH (ERIC document Reproduction Service No. ED423422).
Theofanos, M.F., Redish, J., 2003. Bridging the gap between accessibility and usability. Interactions 10, 36–51.
Tough, A., 1971. The Adult's Learning Projects. Ontario Institute for Studies in Education, Toronto.
US Department of Health and Human Services, Food and Drug Administration, 2009. Food Code 2009. http://www.fda.gov/downloads/Food/GuidanceRegulation/UCM189448.pdf (accessed 27 September 2014).
United States Census Bureau, 2003. Current Population Survey Report. http://www.census.gov/population/www/socdemo/computer/2003.html (accessed 27 September 2014).
United States Food and Drug Administration, 2004. Report of the FDA Retail Food Program Database of Foodborne Illness Risk Factors. http://vm.cfsan.fda.gov/~dms/retrsk.html (accessed 27 September 2014).
United States Food and Drug Administration, 2006. Managing Food Safety: A Regulator's Manual for Applying HACCP Principles to Risk-Based Retail and Food Service Inspections and Evaluating Voluntary Food Safety Management Systems. http://www.fda.gov/Food/FoodSafety/RetailFoodProtection/ManagingFoodSafetyHACCPPrinciples/Regulators/ucm078165.htm (accessed 27 September 2014).
United States Food and Drug Administration, 2011. FDA Trend Analysis Report on the Occurrence of Foodborne Illness Risk Factors in Selected Institutional Foodservice, Restaurant, and Retail Food Store Facility Types (1998–2008). http://www.fda.gov/Food/GuidanceRegulation/RetailFoodProtection/FoodborneIllnessRiskFactorReduction/ucm223293.htm (accessed 27 September 2014).
United States Food and Drug Administration, 2013. Recalls, Market Withdrawals and Safety Alerts Search. http://www.fda.gov (accessed 27 September 2014).
Yiannas, F., 2008. Food Safety Culture: Creating a Behavior-Based Food Safety Management System. Springer-Verlag, LLC, New York.

CHAPTER

17 APPROACHES TO FOOD SAFETY EDUCATION AMONG CRITICAL GROUPS

Armitra Jackson-Davis*, Sherrlyn Olsen†, Benjamin Chapman‡, Benjamin Raymond§, Ashley Chaifetz‡

Department of Food and Animal Sciences, Alabama A&M University, Huntsville, Alabama, USA, Department of Animal Science, Iowa State University, Ames, Iowa, USA†, Department of Youth, Family and Community Sciences, North Carolina State University, Raleigh, North Carolina, USA‡, Department of Food Bioprocessing and Nutrition Sciences, North Carolina State University, Raleigh, North Carolina, USA§*

1 INTRODUCTION

Communicating to food handlers, whether commercial or volunteers, about food safety risks targeting a specific audience and providing contextualized information, has been shown to be an effective strategy and has resulted in actual, observable behavior changes (Chapman et al., 2010). Putting food safety information into context for food handlers and encouraging a personal connection to the consequences of their food-handling actions have shown to effect change. Leventhal et al. (1965) demonstrated the importance of providing context in a study of risk communication materials. Schein (1993) found that creating dialog and generating discussion between a team is a necessary condition for teamwork and working toward shared values. Generating dialog has also been shown to promote collective learning and impact collective intentions within a group, which can be precursors to changing behavior (Ajzen, 1991; Bohm et al., 1991; Dignum et al, 2001). In the food handler context, Pragle et al. (2007) found that a lack of knowledge of consequences of not practicing specific food safety practices was a barrier to self-reported behavior change.

2 FOOD SAFETY EDUCATION PROGRAMS TAUGHT TO ENGLISH AND NON-ENGLISH ADULT LEARNERS

Beginning in 1862 with the Morrill Act, effective food safety education programs have continued to be developed in order to teach an increasingly diverse population (United States Department of Agriculture, 2009). Effective, safe food-handling practices for home and industry must be taught, especially due to the numbers of immigrants employed in the food and meat industries. Current estimates suggest that over 40 million Americans cannot read effectively, and many of these persons are immigrants who are

Food Safety. http://dx.doi.org/10.1016/B978-0-12-800245-2.00017-4

employed in the US workforce. Companies have started to implement programs that teach reading, writing, and basic math in their workers' native languages (Dutkowsky, 2011).

Challenges arise when teaching food safety to culturally diverse employees working in meatpacking and food-manufacturing industries. Recognizing that the numbers of non-English-speaking laborers employed in meatpacking and food-manufacturing industries will continue to grow (Po et al., 2011), it is evident that these individuals must learn safe food-handling behaviors and methods. The meatpacking industry in the United States relies heavily upon nondomestic workers who are typically non-English-speaking individuals. From 1980 to 2000, non-Hispanic Caucasian employees in meatpacking plants decreased from 74% to 49%, whereas the Hispanic workforce increased from 9% to 29%. In this same time span, the number of Hispanics born outside of the United States and entering this workforce increased from 50% to 82% of the Hispanic group. With this rapid influx of non-English-speaking persons into the food-processing sector, instruction and training in proper food handling have become especially problematic (Kandel, 2006).

Although food processors continue to develop and improve methods of safe food production (Jol et al., 2006), proper food-handling techniques can be difficult to teach to individuals when native language is not used, such as Spanish-speaking laborers in an English-speaking work environment (Nyachuba, 2008). Employee training programs must emphasize the importance of safe food, implement effective manufacturing practices, and enforce standard operating procedures in the plants so as to provide safe and wholesome products (Mikel et al., 2002). These programs must be designed using proven educational methods and meet the needs of the diverse cultures comprising the workforce of these plants. Attention must be given to the communication skills and educational levels of all employees, including the growing non-English-speaking population employed in these industries. A food safety curriculum designed to teach bilingual youth has been shown to provide successful training when the material was translated to the Spanish language (Hoover et al., 1996).

3 ADULT LEARNERS

According to Spencer (1998), educational programs delivered to adult learners require different methodology based upon their age. Spencer provides the following insights to adult learners:

- They wish to use their experiences in class and share those with other learners.
- Due to their individuality and maturity they expect to be treated with respect.
- Adult learners expect structure and distinct outcomes in the program as they have specific goals and wish to apply those to their professional and personal lives.
- Programs centered on active learning are the most appreciated as they allow interaction and search results to problem-solving activities.
- Adult learners want to be successful in the classroom and may be anxious about their competencies and abilities to learn the material.
- Due to the safe atmosphere, commitment of all participants and program coordinators, as well as the freedom to make choices throughout the course of the program, they seek out feedback, encouragement, and learning.
- Because they value their time dedicated to learning, they are critical of programs that are poorly planned, taught, and circumstances that interfere with their learning processes.

- In addition, they expect to be comfortable in the classroom with proper accommodations and breaks.
- Adult learners require balance between learning the material properly and evaluating how to best use the newly learned information.

Based upon the above information, educators are better equipped to provide meaningful instruction to adult learners employed in a variety of industries.

4 PREEMPLOYMENT FOOD SAFETY TRAINING PROGRAMS

Company trainers, extension specialists, and university personnel devoted significant time in a series of planning meetings for the development of preemployment food safety training programs that were taught in both English and Spanish languages for a major Midwest meat processing company. The program curriculum was developed based on a food safety handbook written by a nationally recognized corporation that provided program modules on allergens, bacteria, cross-contamination, foodborne illness, foreign material detection, personal hygiene, pest control, sanitation, security, and time and temperature. Six videos addressing cross-contamination, foodborne illnesses, microorganisms, personal hygiene, and sanitation were utilized for instruction. Adult learners participated in activities that taught proper hand-washing methods and differences that exist between poisons and sanitizers. Weekly training sessions were taught in a classroom at a plant on two mornings in succession. At the conclusion of the training, individuals considering employment were required to achieve a 70% score on a 65-question multiple-choice certification exam that measured their comprehension of the material. The exam questions were taken from quizzes included in the food safety handbook. Demographics of the 1265 participants that completed both the pre- and post-training assessments are shown in Table 1.

Table 2 contains results from pre- and post-training assessments for knowledge and behavior questions.

For knowledge questions, the Spanish-speaking group significantly ($P<0.05$) improved their mean scores by 50.56 percentage points, compared to an improvement of 13.38 percentage points ($P<0.05$) for the English-speaking group. Initially, the English-speaking group had a mean score of 83.5%, and the Spanish-speaking group had a mean score of 42.67%. The post-training assessments showed that the mean scores of the Spanish-speaking group (93.23%) were within 3.65 percentage points of the mean scores of the English-speaking group (96.88%).

For behavior questions, the Spanish-speaking group also improved their mean scores significantly ($P<0.05$) by 34.56 percentage points, compared to an improvement of 5.79 percentage points for the English-speaking group. On the pretraining assessments, the English-speaking group had a mean score of 91.41%, and the Spanish-speaking group had a mean score of 58.61%. The post-training assessments showed that the mean scores of the Spanish-speaking group (93.17%) were within 4.03 percentage points of the mean scores of the English-speaking group (97.20%).

A review of the mean scores for all questions provides revealing information. The English-speaking group achieved a mean score of 88.46% on the pretraining assessment, while the Spanish-speaking group mean score was 55.61%. The English-speaking group had a mean score that was 32.85 percentage points higher on the pretraining assessment.

However, the English-speaking group achieved a mean score of 98.68% on the post-training assessment while the Spanish-speaking group achieved a mean score of 96.60%. Following the training

Table 1 Demographic Information of Participants Completing Food Safety Pre- and Post-Training Assessments

Variable	Percent
Gender	
Male	58.81
Female	41.19
Native language	
Spanish	20.79
English	79.21
Reported age (years)	
Younger than 20	9.43
20-29	42.66
30-39	26.12
40-49	16.07
Equal to or greater than 50	5.72
Education-English	
11th Grade or less	11.10
High school diploma/general	
Education development (GED) certificate	51.10
Some college/vocational technical courses	23.90
Vocational technical certificate	5.10
Associate or advanced degree	8.50
No response	0.30
Education-Spanish	
Grades 1-6	34.99
Grades 7-9	31.94
Grades 10-12	19.77
Some college/vocational technical courses	4.18
College/vocational technical certificate	2.28
No response	6.84
Food service experience (years)	
None	22.92
Less than 1	20.16
1-5	38.18
Greater than 5	17.08
No response	1.66
Rehire for the company	
Yes	7.83
No	88.38
Current employee	1.50
No response	2.29

Table 2 Comparisons of Scores for Food Safety Pre- and Post-Training Assessments

Variable	Mean (SD) Pretraining	Mean (SD) Post-Training
Knowledge questions		
Native language		
Spanish	42.67[a] (27.5)	93.23[b] (19.4)
English	83.50[a] (17.0)	96.88[b] (13.3)
Total	74.21[a] (26.2)	96.05[b] (15.0)
Behavior questions		
Native language		
Spanish	58.61[b] (29.7)	93.17[b] (19.5)
English	91.41[a] (15.5)	97.20[b] (13.3)
Total	83.95[a] (23.9)	96.28[b] (15.0)
All questions		
Native language		
Spanish	55.61[a] (22.7)	96.60[b] (6.4)
English	88.46[a] (12.1)	98.68[b] (3.5)
Total	81.63[a] (20.0)	98.25[b] (4.3)

a,bMeans within a row with different superscripts are significantly different ($p < 0.05$).

program, the Spanish-speaking group had a mean score that was only 2.08 percentage points lower on the post-training assessment.

The high initial scores by the English-speaking group may reflect a greater familiarity with food safety issues among the English-speaking participants due to media coverage of these issues in the United States.

Impressive improvement was made by the Spanish-speaking group by training in the Spanish language. They increased their overall mean score 40.99 percentage points when pre- and post-training assessment scores were compared (compared to 10.22 percentage point improvement for the English-speaking group).

Extension specialists and trainers must communicate with the Hispanic population by using bilingual aides and Spanish-speaking educators and present materials in English and Spanish to be most effective (Farner et al., 2006). Results of the present study indicate that food safety training programs translated to the Spanish language and taught to adult Spanish-speaking learners were highly effective. This study addressed adult learner issues as follows: identify needs, define objectives, identify learning experiences to meet objectives, organize learning experiences into plans, and evaluate program outcomes (Brookfield, 1986).

5 FOOD SAFETY TRAINING PROGRAMS INFLUENCE BEHAVIORS

Although food and meat companies work to provide food safety education to all employees regardless of nationality, it is clearly noted that the most effective training programs must be coupled with behavioral changes. Campden BRI/Alchemy distributed information from *Lessons Learned From the 2014*

Global Food Safety Training Survey—data gathered to assist food processors and manufacturers document the effectiveness of their own training programs. It was concluded that employee training cannot be reactive to situations that occur in the plants and that these educational programs must be proactive and goal-oriented so as to achieve specific outcomes. Conclusions included that effective employee coaching is the most effective in assisting employees to implement lessons learned in training programs in manufacturing systems. Alchemy urges companies to utilize electronic coaching technologies to verify employee actions in the plant and by using this technology, the food safety training programs are more thorough.

6 SOCIAL MEDIA AUDIENCES

The popularity of social media presents an opportunity for disseminating food safety and health risk information broadly. Social media are primarily mobile and web-based applications that allow users to interact in various ways, such as sharing photos and video and instant messaging. Millions of people now participate in social media, such as Facebook, YouTube, Instagram, Pinterest, and Twitter. As of September 2013, an estimated 86% of adult Americans aged 18 years and older used the Internet (Duggan and Smith, 2013); of the adult Americans online, 73% report that they used a social networking site of some kind (Fox and Duggan, 2014). This is similar to the number of users in other developed countries.

Social media influences how individuals obtain information (Mangold and Faulds, 2009; Nielsen, 2009; Thackeray and Neiger, 2009), and the amount of time spent on social media applications is rising at a rate three times faster than the increase of time spent on the Internet overall (Nielsen, 2009). In addition, the increasing use of smartphones (Nielsen, 2009) and other mobile devices enables easy access to online resources, including social media.

Health information providers cannot ignore the changing behavior of information seekers. In 2012, 72% of Internet users said that they looked online for health information (Fox, 2012). In an earlier survey, 29% of Internet users reported that they look online for information about food safety or recalls. To help meet the needs of these online audiences, information providers must "go where the people are" (Centers for Disease Control and Prevention, 2011) and provide messages tailored to the environment of social media.

The potential use of social media for conveying risk communication related to foodborne illness has been contemplated, but very little research in the area has been conducted to date (Rutsaert et al., 2013). The variety of ways that social media could be used to inform and disseminate food safety risk information would be similar to those for other areas of public health.

In a 2013 survey of 1800 Americans, US communications firm Ketchum investigated the changing landscape of online discussions related to food quality, nutrition, and safety and identified a subset of the Internet public as food evangelists (Ketchum, 2013). According to Ketchum, this population generates up to 1.7 billion conversations about food weekly and do not see themselves as activists with entrenched beliefs around issues; they are an interested public. They have expectations that food companies work interactively to engage the eating public in dialog and share information proactively and transparently (Ketchum, 2013).

Therefore, there is evidence that people are already using online sources to seek out food information, so an audience for food safety risk communication through social media may already be established. In a study of domestic food practices by low-income women in Montreal (Canada), participants reported that the Internet was an important source of information for learning about food and finding recipes; in contrast, cooking shows were considered to be too complicated, and books or magazines were not often useful (Engler-Stringer, 2010). In a survey of college students during the 2011 *Escherichia coli* O104:H4 outbreak in Germany, van Velsen et al. (2012) found that the Internet was the most popular medium for passively receiving outbreak-related information.

Social media can be an asset to food safety risk communicators, and a hindrance as well. Benefits can be speed, accessibility, and interactive capacity when raising awareness about an issue or during crisis communications, but these may be countered by the lack of control on accurate information, low trust, the risk of information overload, and a communication preference for traditional media (Rutsaert et al., 2013).

Online discussion of risk may be susceptive to social amplification of risk, wherein risks assessed by technical experts as relatively minor elicit strong public concerns that result in substantial impacts on society and the economy (Kasperson et al., 1988). Misinformation and false assertions may be easily disseminated via social media with or without malicious intent and be widely believed (Chung, 2011). This effect was observed with the case of the lean, finely textured beef in the United States in 2012, where a blogger posted a petition on the website change.org that garnered a maelstrom of negative publicity and subsequently led to the removal of lean, finely textured beef from many food products and the closure of processing facilities (Gonzalez, 2012).

Another unexamined Internet group that is influential online and in direct care of a group vulnerable to foodborne illness is mom bloggers/online moms. Often targeted by marketers and marketing researchers, mom bloggers/online moms are a diverse group with an estimated 3,900,000 active bloggers in the United States alone (Piersall, 2014). As a population in caring for children, it is important to understand the knowledge level and concern of mom bloggers/online moms that may affect how norovirus and food safety information is shared and utilized. Finney et al. (2014) conducted a nationally representative survey of adults to determine if demographic characteristics influenced perceived severity and susceptibility of norovirus infections. The results of the current survey will be compared to the aforementioned survey for potential similarities and differences in attitude and knowledge levels between the different demographic groups. Cornell (2012) conducted a survey of 89 mom bloggers designed to assess whether they believed social media made the bloggers better parents. The majority of the mom bloggers (75%) believed use of social media did make them better parents. In addition, the average amount of time spent on social media per week was 4 h (Cornell, 2012), It is not currently known if food safety and norovirus specifically is a topic that is discussed or shared about in any way.

Mom bloggers obtain health and food safety information from a variety of sources and have varying knowledge levels and attitudes around norovirus (Figures 1 and 2). When asked about trusted sources of online health information, there were clusters of services that were trusted by a similar number of respondents. WebMD.com and mayoclinic.com were the most popular choices with 56 (70%) and 50 (62.5%),respectively, of respondents indicating them as trustworthy resources. These two were the most trusted resources for online health information, followed by the CDC, American Academy of Pediatrics, then NIH with 37 (46.3%), 34 (42.5%), and 32 (40%), respectively, of respondents selecting them as trusted resources. Online outlets for large news organizations such as FOX, NBC, CBS, MSNBC, and similar were not as trusted, with the most trusted of the cluster, CNN, at 14 (17.5%) respondents.

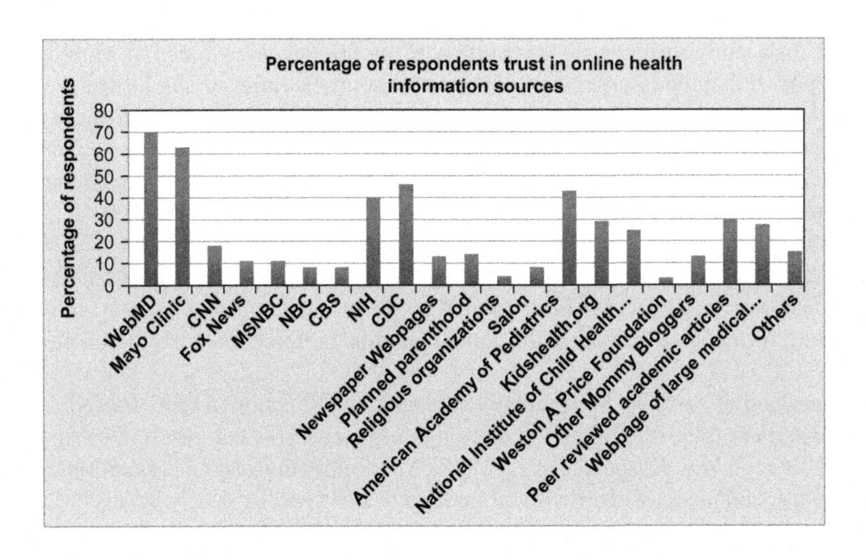

FIGURE 1

Mom bloggers trust in online sources.

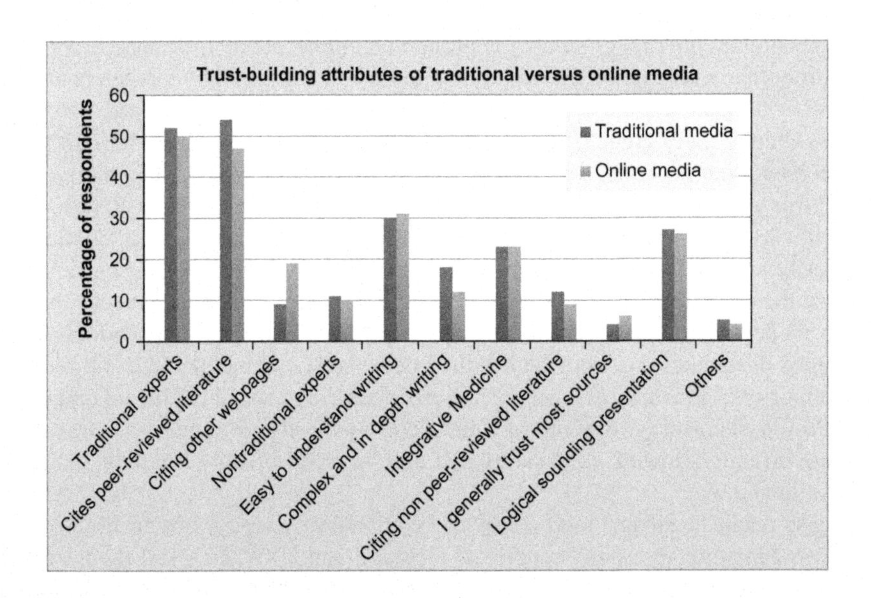

FIGURE 2

What builds trust within the mom blogger audience.

WebMD and the Mayo Clinic were selected as trustworthy sources of online health information by 70% and 62.5% of participants, respectively, while 40% and 42.3% found the NIH and CDC, respectively, to be reliable sources of online health information. Of the five major news' networks (Fox News, NBC, CBS, MSNBC, and CNN), none were selected as trustworthy by more than 17.5% of participants, with CNN being the most trusted. Online articles citing peer-reviewed literature were most selected attributes contributing to trust in traditional media forms of health information (67.5%) with traditional medical experts trusted by 65% of participants. Citing nonpeer-reviewed research was trusted by 15% of participants, while a logical sounding explanation and easy to understand terminology and writing were both found to be attributes that contributed to trust by over one-third of participants (33.8% and 37.5%). Comparing attributes of online versus traditional media that contributed to the mom bloggers' sense that it was reputable, reliable, and trustworthy did not yield any significant differences.

7 VOLUNTEERS

Neel (2010) surveyed the knowledge levels and attitudes of temporary foodservice establishment workers and found that personal hygiene and contamination control are strengths, but deficiencies existed in cooling, reheating, and temperature control. Workers had the highest agreement in questions about safe cross-contamination and personal hygiene practices. They agreed that there should be no smoking and that hand washing is important for general food safety and for preventing cross-contamination after handling raw meats. Neel (2010) suggests that this could be due to better training and/or increased awareness of high-profile foodborne illness outbreaks associated with beef.

On the other hand, workers held attitude deficiencies in cooling, reheating, and temperature control (Neel, 2010). This is likely correlated with poor knowledge of proper food-reheating procedure, concepts related to proper holding and cooling of cooked food, and the proper procedure for thawing frozen foods, as demonstrated by low knowledge scores in these areas. Lack of training or the nature of the facility could be responsible for deficient attitudes and poor knowledge. However, it could also be related to the nature of the temporary facility and the types of food handled.

Among their knowledge deficits were basic facts about conditions for pathogen growth, such as the fact that cooked food can be held for a period of time without temperature control, and that food should be held for service outside of the temperature danger zone. This contradiction may indicate that foodservice workers memorize facts but do not understand the concepts behind them (Neel, 2010). Focusing on why food temperature control is important and knowledge about the life cycles of organisms that can make people sick should be a component of food safety training.

Lee et al. (2010) found that food safety knowledge of vendors at fairs and festivals seemed to be related to age and attendance at an educational workshop, but not inspection violations. Training programs may help improve vendors' food safety knowledge and practices.

With fewer and fewer grocery stores throughout the United States, nonmetropolitan residents are clearly disadvantaged in their access to food, especially if they do not own vehicles or have physical disabilities (Blanchard and Matthews, 2007). Additionally, low-income neighborhoods have fewer instances of grocery stores but have access to convenience stores, with predominately black and Asian neighborhoods at lower rates than white neighborhoods (Powell et al., 2007). The lack of access is shown to lead individuals to alternative food sources, including the food pantries (Blanchard and Matthews, 2007; Smith and Morton, 2009). The food travels varying distances with varying levels of risk: hot backseats and trunks of cars, refrigerated vans, and so forth. Additionally, the food pantries often act as soup kitchens, backpack programs, and even serve community dinners, but more importantly, the food handlers are typically volunteers. The role of the food pantries and their volunteers in food provision is critical yet have been left out in studies related to food handling.

Without food safety studies available that concentrate on food pantries, studies that focused on other food-handling environments were reviewed. To begin, a 2007 study of food safety at home indicated that knowledge does not necessarily translate into behavior, and motivation, as well as habitual cooking practices, is needed for safer food outcomes (Fischer et al., 2007). While much of the research has revealed less-than-desirable practices by food handlers of various demographics, the field provides valuable insight into the goings-on of commercial and home kitchens. Qualitative and quantitative evidence and analysis of risky behaviors of young adults (Abbot et al., 2009; Dedonder et al., 2009), at home (Fischer et al., 2007; Shapiro et al., 2011), at work (Clayton et al., 2002; Green et al., 2005, 2006, 2007; Green and Selman, 2005; Pragle et al., 2007; Roberts et al., 2008), of WIC participants (Kwon et al., 2008), in families (Perry et al., 2012), and in student kitchens (Sharp and Walker, 2003) reveal that the intent to partake in safe behavior is not always manifested. While a comprehensive review of the studies show that they vary in the recommended practices analyzed and their prevalence, none of the aforementioned studies show all of the food handlers performing every best practice, either in observational or in interview-based research (or a combination). While they are likely interested in the well being of the food pantry clients, volunteers might not necessarily have the background or education in food handling that often comes alongside paid employment.

The organization of each food pantry is structured differently, in terms of paid employees versus unpaid volunteers, the number of people it takes to run the operation, length of a typical volunteer shift, the ages of the volunteers, required training, and volunteer requirements, as well as their agency within the organization; namely, who decides on quality and picks up the food (Table 3). In general, volunteers stock the shelves and refrigerators, pack bags of food, and assist clients in completing the requisite paperwork (either for the federal programs or specific to the food pantry). The interview questions focused on the structure of the food pantries themselves, from the descriptions of the managerial and volunteer activities and their respective responsibilities to the formalities of the organizations, including the provision of any guidelines or manuals that describe operations, details on types of food held at the food pantry and their storage facilities, the distribution of past-date food items and reasoning behind any rules, requirements for suppliers, on-site repackaging of food, knowledge of recalls and recall plans, and food distribution procedures. Organizational variables are presented in Table 3.

Table 3 Food Pantry Demographic and Organizational Characteristics ($N=105$)

Pantry Characteristics	No. of Food Pantries with Characteristic	Food Bank Partners	No. Food Bank
Paid manager	43 (40.95%)	35	8
Managers trained in managing volunteers	31 (32.29%)	27	4
Regular volunteers on each shift	81 (80.20%)	67	14
Use of sign-in sheet	48 (47.06%)	41	7
Volunteers responsible for intake of food	78 (77.23%)	66	12
Volunteers decide on quality of foods	91 (90.10%)	76	15
Volunteers drive own vehicle to pick up food	61 (64.89%)	50	11
Volunteers trained	42 (41.18%)	37	5
Volunteer requirements	29 (28.43%)	24	5

While food pantries describe the person in charge by varying titles, from coordinator to president, the most prevalent one was manager, which is used throughout this chapter for brevity and ease. Most (80.2%) food pantries have regular volunteers, defined as a person who repeatedly helps in the food pantry for no wage but also does not require training upon each visit. One food pantry did not have volunteers at all, just paid staff. Only 48 of 105 food pantries use a sign-in sheet to keep track of who is volunteering on any particular day. In more than 90% of the food pantries, the volunteers are responsible for deciding on the quality of the food, including its safety. In 77.2% of the food pantries, the volunteers are also responsible for the intake of the food; namely, how and where it is stored, but also how quickly it is put away. Similarly, only 41.2% of the volunteers go through any training at the food pantry, including but not limited to food safety. And while almost two-thirds of the food pantries have volunteers that use personal vehicles to pick up the food, only 28.8% of the food pantries have any sort of requirements for their volunteers to participate, cited as anything from a being member of the church to passing a formal background check.

8 IMPORTANCE OF K-12 FOOD SAFETY EDUCATION

Although many mechanisms are in place to control the spread of foodborne disease, the Centers for Disease Control and Prevention estimates that approximately one in every six Americans become ill each year as a result of consuming contaminated food products resulting in foodborne disease (CDC, 2014a). In 2013 alone, FoodNet surveillance identified 19,056 cases of culture-confirmed bacterial and laboratory-confirmed parasitic infections. *Salmonella* accounted for the most frequent infection (38%) followed by *Campylobacter* (35%) (CDC, 2014b). The information in these reported foodborne illnesses affect all age groups. Because children have immune systems that are not fully developed, they are more susceptible to becoming ill from consuming contaminated food. Additionally, many children today are in households where both parents must work to support the family unit. In many cases, parents working long hours results in children being left at home alone to prepare meals for themselves and/or their siblings. Because most children are not taught how to properly and safely prepare food, this unsupervised food preparation can potentially lead to foodborne illness due to mishandling of food. Therefore, it is crucial to equip individuals at an early age with the necessary tools and educational background needed to safely handle food. Food safety education is usually addressed in a college-level Food Microbiology class. By the time, an individual reaches adulthood, food safety habits are already formed. At this point, it becomes difficult to change habits that have been in place for many years. Therefore, it is critically important to address the topic of food safety education at the K-12 level. Patil et al. (2005) found that young adults (age 18-29) are more likely to participate in risky food-handling behaviors when compared to other age groups. Although some focus has been placed on food safety education at the high school level (Shearer et al., 2013), more attention is also needed at the elementary and junior high school levels.

9 OBSTACLES ASSOCIATED WITH INCLUDING FOOD SAFETY EDUCATION IN THE CLASSROOM

Testing and assessment represent a large component of today's educational system. For example, the state of Arkansas requires the administration of criterion-referenced tests (CRT) and norm-referenced tests (NRT) to all students. The Qualls Early Learning Inventory is administered to all kindergarten students at the beginning of the school year. This test is an assessment tool designed to identify student development. The Augmented Benchmark Exams (which represent the combination of the CRT and NRT) are administered at grades 3-8. In grades 9-12, the state mandated CRT include the following: (1) End-of-Course Algebra I, (2) End-of-Course Biology, (3) End-of-Course Geometry, and (4) End-of-Course Literacy Examination (Arkansas Department of Education, 2014). Because of state regulations regarding material that must be taught for testing purposes, teachers in the K-12 educational system are not left with much time to teach topics such as food safety. Furthermore, the majority of K-12 educational curriculums do not include food safety. Therefore, it becomes necessary to find creative ways to incorporate food safety education into existing curriculums.

10 OPERATION FOOD SAFETY

The state of Arkansas saw the need for the development of food safety material for the K-12 educational system. In the 1997 General Assembly, "Operation Food Safety" was developed as a result of Act 1274 (sponsored by former Arkansas Representative Perry Malone). To ensure that objectives were met, the Operation Food Safety Coalition was formed, and the manual was distributed to school districts in the state of Arkansas. Some of the organizations involved include Arkansas Department of Education Child Nutrition Program, University of Arkansas Center for Excellence for Poultry Science, Arkansas Department of Workforce Education, and University of Arkansas Cooperative Extension Service (Gurdon Times, 1998). "Operation Food Safety" consists of activities designed for different grade clusters (Pre-K-grade 4, grades 5-8, and grades 9-12). Each unit is accompanied by instructions for the teacher. These instructions provide a "road map" as the teacher gives instructions to the students. This is especially useful for teachers who may not have a background in food safety. These instructions include the lesson goal, specific objectives, materials needed, and procedures. Activities for each cluster are divided into the following units: hand washing, keeping things clean, and keeping food hot or cold. To make the parents aware of the fact that their child has been learning about food safety, a letter to the parents is included with each grade cluster. A Certificate of Participation is also included. Tables 4–6 provide example activities (adapted with permission from "Operation Food Safety") that can be incorporated into existing curricula.

11 CONCLUSIONS

As detailed by the case studies explored in this chapter, communicating food safety messages and providing education to individuals must be completed as a part of a targeted effort as opposed to a broad-brush, or one-size-fits-all general information provision strategy. As antecedents, experiences, knowledge level, attitudes, and capabilities vary across the spectrum of food handlers (from professionals, to volunteers, to moms, to children). Individuals need to be provided with specific information that meets their needs.

The goal of meat, poultry, and food manufacturers is to produce safe and wholesome products for their consumers, regardless of the languages that the employees in the industry speak and understand. Simply put, there are too many lives at stake. Developing food safety training programs in the learners' native language is of the utmost importance. In addition, due to age and experience, adult learners

Adapted with permission from "Operation Food Safety."

require special considerations when developing coursework and training programs. Research has shown that non-English-speaking individuals will continue to become employed in these industries. Providing food safety training programs for adult learners in their native languages is much more cost-effective than enduring the ordeals caused by major product recalls.

Delivering messages through trusted media that are already integrated into the lives of a target audience also increases the behavior-changing potential of messages. A still novel and successful model to

Table 5 Keeping Things Clean (Grades 1-2)

Find the Mistakes

Directions: Read the story below. Answer the questions

The Snack Mistakes

Juan petted his cat Felix. He wanted a snack. Juan did not wash his hands. He got an apple. Juan did not wash his apple. He took a bite. Juan was thirsty. Juan drank milk out of the carton. Felix jumped on the table. Felix drank from the carton too.

1. Juan did not wash his ____________________. This was the first mistake.
2. Circle the second mistake.

   Juan petted his cat Felix.   Juan did not wash his apple.
3. Write the third mistake.

   ____
4. Felix jumped on the ____________________. This was the fourth mistake.
5. Circle the last mistake.

   Felix drank from the cartoon.   Juan was thirsty.

Adapted with permission from "Operation Food Safety."

be used for addressing target audiences was created by Washington state researchers in the mid-1990s. In response to *Salmonella* outbreaks in Washington state associated with queso fresco, Bell et al. (1999) developed a program utilizing existing social networks of the affected community, which was largely Hispanic. The flow of information from Abeulas, a trusted source with compelling information, to the rest of the community is analogous to the social networks established through other settings.

The literature and case studies show that the most persuasive food safety messages provide direction and statements to which individuals can relate. Least persuasive messages contain information that is not related to individuals, use unfamiliar words, or contain unnecessary figures. Visual models connecting symbols and actions work well. An example of this would be an image of a thermometer and thoroughly cooked food. Regular repetition and reinforcement of the message also improves effectiveness.

Food safety education must be introduced in the K-12 system. Because most children will be placed in positions to prepare food for themselves and others at an early age, this is of utmost importance.

Table 6 Keeping Foods Hot or Cold (Grades 1-2)

Predict the Outcome

Instructions: Read the story. Answer the questions.

Ashley and the Pizza

Ashley ate a slice of pizza for lunch. She left the pizza on the table all afternoon. Ashley ate another piece of pizza before she went to bed.

1. Predict what will happen to Ashley.

____

____

2. Why?

____

____

3. What should Ashley have done with the pizza after eating a slice?

____

____

Adapted with permission from "Operation Food Safety."

12 OTHER RESOURCES

Suggest resources that may be used for incorporating food safety into existing curriculums:

1. *Eating Safe With Ace and Mace: The Birthday To Remember Forever* by Armitra Jackson (available on amazon.com),
2. Partnership for Food Safety Education (http://www.fightbac.org/),
3. American Society for Microbiology (http://www.microbeworld.org/washup),
4. National Coalition for Food Safe Schools (http://www.foodsafeschools.org/),
5. "Operation Food Safety,"
6. Centers for Disease Control and Prevention (http://www.cdc.gov/handwashing/training-education.html),
7. News for Food Safety Educators (http://www.foodsafety.gov/news/educators/index.html),

8. Science and Our Food Supply (http://www.fda.gov/Food/FoodScienceResearch/ToolsMaterials/ScienceandTheFoodSupply/default.htm),
9. *Journal of Food Science Education* (http://www.ift.org/knowledge-center/read-ift-publications/journal-of-food-science-education.aspx).

REFERENCES

Abbot, J.M., Byrd-Bredbenner, C., Schaffner, D., Bruhn, C.M., Blalock, L., 2009. Comparison of food safety cognitions and self-reported food-handling behaviors with observed food safety behaviors of young adults. Eur. J. Clin. Nutr. 63 (4), 572–579.

Ajzen, I., 1991. The theory of planned behavior. Org. Behav. Hum. Decis. Processes 50, 179–211.

Arkansas Department of Education, 2014. Student Assessment. http://www.arkansased.org/divisions/learning-services/student-assessment (assessed September 9, 2014).

Bell, R.A., Hillers, V.N., Thomas, T.A., 1999. The Abuela Project: safe cheese workshops to reduce the incidence of *Salmonella typhimurium* from consumption of raw-milk fresh cheese. Am. J. Public Health 89 (9), 1421–1424.

Blanchard, T.C., Matthews, T.L., 2007. Retail concentration, food desserts, and food-disadvantaged communities in rural America. In: Hinrichs, C.C., Lyson, T.A. (Eds.), Remarking the North American Food System: Strategies for Sustainability. University of Nebraska Press, Lincoln.

Bohm, D., Factor, D., Garrett, P., 1991. Dialogue—A Proposal. Available from: http://www.david-bohm.net/dialogue/dialogue_proposal.html (accessed October 11, 2014).

Brookfield, S., 1986. Understanding and Facilitating Adult learning, first ed. Jossey-Bass, San Francisco, California.

Campden BRI/Alchemy Systems, 2014. Lessons Learned from the 2014 Global Food Safety Training Survey. http://www.alchemysystems.com/files/7113/9903/7758/ALC_2014_Global_Food_Safety_Training_Study.pdf (accessed September 16, 2014).

Centers for Disease Control and Prevention, 2014a. Estimates of Foodborne Illness in the United States. http://www.cdc.gov/foodborneburden/ (accessed September 9, 2014).

Centers for Disease Control and Prevention, 2014b. Trends in Foodborne Illness in the United States. http://www.cdc.gov/foodborneburden/trends-in-foodborne-illness.html (accessed September 9, 2014).

Centers for Disease Control and Prevention (CDC), 2011. The Health Communicator's Social Media Toolkit, http://www.cdc.gov/socialmedia/tools/guidelines/pdf/socialmediatoolkit_bm.pdf (accessed October 11, 2014).

Chapman, B., Eversley, T., Filion, K., MacLaurin, T., Powell, D., 2010. Assessment of food safety practices of foodservice food handlers (risk assessment data): testing a communication intervention (evaluation of tools). J. Food Prot. 73, 1101–1107.

Chung, I.J., 2011. Social amplification of risk in the Internet environment. Risk Anal. 31, 1883–1896.

Clayton, D., Griffith, C., Price, P., Peters, A., 2002. Food handlers' beliefs and self-reported practices. Int. J. Environ. Health Res. 12 (1), 25–39.

Cornell, K.H., 2012. The Mommy Blogger Market: Social Media's Effect on Motherhood. East Carolina University, Greenville. http://www.katiecornelldesigns.com/ECUCourseWork/ICTN6855/The%20Mommy%20Blogger%20Market.pdf.

DeDonder, S., Jacob, C.J., Surgeoner, B.V., Chapman, B., Phebus, R., Powell, D.A., 2009. Self-reported and observed behavior of primary meal preparers and adolescents during preparation of frozen, uncooked, breaded chicken products. Br. Food J. 111 (9), 915–922.

Dignum, F., Dunin-Keplicz, B., Verbrugge, R., 2001. Creating collective intention through dialogue. Log. J. IGPL 9, 289–303.

Duggan, M., Smith, A., 2013. Social Media Update. Pew Research Center, January 2014. http://pewinternet.org/Reports/2013/Social-Media-Update.aspx (accessed October 11, 2014).

Dutkowsky, S., 2011. Trends in Training and Development—The New Economy, Training in US Companies, Who Does the Training in Corporations? http://careers.stateuniversity.com/pages/852/Trends-in-Training-Development.html. Trends in Training and Development – The New Economy, Training in US Corporations? (accessed December 17, 2011).

Engler-Stringer, R., 2010. The domestic foodscapes of young low-income women in Montreal: cooking practices in the context of an increasingly processed food supply. Health Educ. Behav. 37, 211–226.

Farner, S., Cutz, G., Farner, B., Seibold, S., Abuchar, V., 2006. Running successful extension camps for Hispanic children: from program planning to program delivery for a 1-week day camp. J. Ext. 44 (4). Available at: http://www.joe.org/joe/2006august/a4.php, Article 4FEA4.

Finney, M., Carney, M., Cates, S., Kosa, K., Brophy, J., Fraser, A., 2014. Need for Education About Noroviruses: Findings from a Nationally Representative Survey of U.S. Adults. In: Abstract Presented at International Association for Food Protection. Indianapolis, IN. August 5, 2014, submitted for publication. https://iafp.confex.com/iafp/2014/webprogram/Paper6526.html.

Fischer, A.R.H., De Jong, A.E.I., Van Asselt, E.D., De Jonge, R., Frewer, L.J., Nauta, M.J., 2007. Food safety in the domestic environment: an interdisciplinary investigation of microbial hazards during food preparation. Risk Anal. 27 (4), 1065–1082.

Fox, S., 2012. The Who, What, Where, When & Why of Health Care Social Media. Pew Research Internet Project. http://www.pewinternet.org/2013/10/18/the-who-what-where-when-why-of-health-care-social-media/ (accessed October 11, 2014).

Fox, S., Duggan, M., 2014. Health Online 2013. Pew Research Internet Project. http://www.pewinternet.org/2013/01/15/health-online-2013/ (accessed October 11, 2014).

Gonzalez, A., 2012. Lesson from Pink Slime Incident: Social Media is a Supply Chain Risk. Logistics Viewpoints. http://logisticsviewpoints.com/2012/03/28/lesson-from-pink-slime-incident-social-media-is-a-supply-chain-risk/ (accessed April 13, 2014).

Green, L.R., Selman, C.A., 2005. Factors impacting food workers' and managers' safe food preparation practices: a qualitative study. Food Prot. Trends 25 (12), 981–990.

Green, L., Selman, C., Banerjee, A., Marcus, R., Medus, C., Angulo, F.J., Radke, V., Buchanan, S., EHS-Net Working Group, 2005. Food service workers' self-reported food preparation practices: an EHS-Net study. Int. J. Hyg. Environ. Health 208, 1–2.

Green, L.R., Selman, C.A., Radke, V., Ripley, D., Mack, J.C., Reimann, D.W., Bushnell, L., 2006. Food worker hand washing practices: an observation study. J. Food Prot. 69 (10), 2417–2423.

Green, L.R., Radke, V., Mason, R., Bushnell, L., Reimann, D.W., Mack, J.C., Selman, C.A., 2007. Factors related to food worker hand hygiene practices. J. Food Prot. 70 (3), 661–666.

Gurdon Times, 1998. Safe Food Program Announced. http://www.newspapers.pcfa.org/index.php/ref/4624 (accessed September 30, 2014).

Hoover, T., Cooper, A., Tamplin, M., Osmond, J., Edgell, K., 1996. Exploring curriculum to meet the food safety needs of bilingual youth. J. Ext. 34 (3). Available at: http://www.joe.org/joe/1996june/a2.php, Article 3FEA2.

Jol, S., Kassianenko, A., Oggel, J., Wszol, K., 2006. A country-by-country look at regulations and best practices in the global cold chain. Food Saf. Mag. www.foodsafetymagazine.com/article.asp?id=561&sub (accessed June 23, 2009).

Kandel, W., 2006. Meat-processing firms attract Hispanic workers to rural America. In: Amber Waves—The Economics of Food, Farming, Natural Resources, and Rural America. www.ers.usda.gov/AmberWaves/June06/Features/MeatProcessing.htm (accessed April 16, 2009).

Kasperson, R.E., Renn, O., Slovic, P., 1988. The social amplification of risk: a conceptual framework. Risk Anal. 8, 177–187.

Ketchum, 2013. Food 2020: The Consumer as CEO. https://www.ketchum.com/food-2020-consumer-ceo (accessed April 13, 2014).

Kwon, J., Wilson, A.N., Bednar, C., Kennon, L., 2008. Food safety knowledge and behaviors of women, infant, and children (WIC) program participants in the United States. J. Food Prot. 71 (8), 1651–1658.

Lee, J., Almanza, B., Nelson, D., Cai, L., 2010. Food safety at fairs and festivals: vendor knowledge and violations at a regional festival. Event Manage. 14 (3), 215–223(9).

Leventhal, H., Singer, R., Jones, S., 1965. Effects on fear and specificity of recommendation upon attitudes and behavior. J. Pers. Soc. Psychol. 2, 20–29.

Mangold, W.G., Faulds, D.J., 2009. Social media: the new hybrid element of the promotion mix. Bus. Horiz. 52, 357–365.

Mikel, W.B., White, P., Senne, M., 2002. Findings of the National Pork Board Salmonella.

Neel, P.C., 2010. Temporary Public Eating Places: Food Safety Knowledge and Practices. Doctoral Dissertation. University of Missouri, Columbia.

The Nielsen Company. Global Faces and Networked Places, March 2009. http://www.nielsen.com/content/dam/corporate/us/en/newswire/uploads/2009/03/nielsen_globalfaces_mar09.pdf (accessed 13 April 2014).

Nyachuba, D., 2008. Project Plans FY 08 Food Safety. University of Massachusetts Amherst, Center for Agriculture. http://www.umassextension.org/index.php/public-issues/nutrition-health/project-plans-fy08/91-food-safety (accessed June 23, 2009).

Patil, S.R., Cates, S., Morales, R., 2005. Consumer food safety knowledge, practices, and demographic differences: findings from a meta-analysis. J. Food Prot. 68, 1884–1894.

Perry, C., Albrecht, J., Litchfield, R., Meysenburg, R.L., Er, I.N., Lum, A., Temple, J., 2012. The development of a food safety brochure for families: the use of formative evaluation and plain language strategies. J. Ext. 50 (1), 1RIB3.

Piersall, W., 2014. 16 Market research facts proving social media moms' influence. http://www.wendypiersall.com/.

Po, L.G., Bourquin, L.D., Occeña, L.G., Po, E.C., 2011. Food safety education for ethnic audiences. Food Saf. Mag. 17 (3), 26–31, 62.

Powell, L.M., Slater, S., Mirtcheva, D., Bao, Y., Chaloupka, F.J., 2007. Food store availability and neighborhood characteristics in the United States. Prev. Med. 44 (3), 189–195.

Pragle, A.S., Harding, A., Mack, J.C., 2007. Food workers' perspectives on handwashing behaviors and barriers in the restaurant environment. J. Environ. Health 69, 27–32.

Roberts, K.R., Barrett, B.B., Howells, A.D., Shanklin, C.W., Pilling, V.K., Brannon, L.A., 2008. Food Safety Training and Foodservice Employees' Knowledge and Behavior. http://krex.k-state.edu/dspace/handle/2097/806 (accessed October 11, 2014).

Rutsaert, P., Regan, Á., Pieniak, Z., et al., 2013. The use of social media in food risk and benefit communication. Trends Food Sci. Technol. 30, 84–91.

Schein, E.H., 1993. On dialogue, culture, and organizational learning. Organ. Dyn. 22, 40–51.

Sharp, K., Walker, H., 2003. A microbiological survey of communal kitchens used by undergraduate students. Int. J. Consum. Stud. 27 (1), 11–16.

Shapiro, M.A., Porticella, N., Jiang, L.C., Gravani, R.B., 2011. Predicting intentions to adopt safe home food handling practices. Applying the theory of planned behavior. Appetite 56 (1), 96–103.

Shearer, A., Snider, O., Kniel, K., 2013. Development, dissemination and preimplementation evaluation of food safety educational materials for secondary education. J. Food Sci. Educ. 12, 28–37.

Smith, C., Morton, L.W., 2009. Rural food deserts: low-income perspectives on food access in Minnesota and Iowa. J. Nutr. Educ. Behav. 41 (3), 176–187.

Spencer, K., 1998. Purposeful Teaching: Design and Instruction for Adult Learners. http://olms1.cte.jhu.edu/olms/data/resourse/5790/PURPOSEFUL%20TEACHING%20HANDOUT.pdf (accessed September 16, 2014).

Thackeray, R., Neiger, B.L., 2009. A multidirectional communication model: implications for social marketing practice. Health Promot. Pract. 10, 171–175.

United States Department of Agriculture Cooperative State Research, Education, and Extension Service, 2009. About Us—Extension. http://www.csrees.usda.gov/qlinks/extension.html (accessed April 13, 2009).

van Velsen, L., van Gemert-Pijnen, J.E., Beaujean, D.J., Wentzel, J., van Steenbergen, J.E., 2012. Should health organizations use web 2.0 media in times of an infectious disease crisis? An in-depth qualitative study of citizens' information behavior during an EHEC outbreak. J. Med. Internet Res. 14 (6), e181. doi:10.2196/jmir.2123. http://www.jmir.org/2012/6/e181/.

CHAPTER

18

THE ROLE OF TRAINING STRATEGIES IN FOOD SAFETY PERFORMANCE: KNOWLEDGE, BEHAVIOR, AND MANAGEMENT

Elke Stedefeldt, Laís Mariano Zanin, Diogo Thimoteo da Cunha, Veridiana Vera de Rosso, Vanessa Dias Capriles, Ana Lúcia de Freitas Saccol

1 INTRODUCTION

The food handler is a diamond, unique, and multifaceted that we need to know.

Food safety is a critical component of sustainable development, and problems that occur in one country may put others at risk. To make credible and sustainable legal and policy decisions, the decision-making process must be based on strong evidence. Considering the growing complexity of the food safety field, innovative approaches are required to improve prioritization, accounting for the overall available knowledge, and the need to integrate new scientific developments quickly (World Health Organization, 2013).

Foodborne diseases (FBD) are important health problems (Kafertein et al., 1997; Soares et al., 2013). The social and health impacts of FBD have been described and documented on all continents (World Health Organization, 2013). Every individual is threatened by FBD. This problem significantly contributes to mortality by diarrhea, with approximately 2.2 million deaths per year, mostly in children from developing countries (World Health Organization, 2008).

The Centers for Disease Control and Prevention (CDC) estimate that 31 of the most important known agents of FBD found in foods consumed in the United States cause 9.4 million illnesses, 55,961 hospitalizations, and 1351 deaths (Centers for Disease Control and Prevention, 2011) each year. In 2013, the CDC confirmed a total of 19,056 infections, 4200 hospitalizations, and 80 deaths by FBD in the United States (Crims et al., 2014). Data from FoodNet show that nearly 45% of the imported foods that caused outbreaks in the United States in 2010 came from Asia (Centers for Disease Control and Prevention, 2012). A total of 5648 foodborne outbreaks, including outbreaks for which there was both weak and strong evidence, were reported in Europe. In 2011 by the 25 reporting European Union member states. This finding represents an increase of 7.1% relative to 2010 (European Food Safety Authority, 2013).

Food Safety. http://dx.doi.org/10.1016/B978-0-12-800245-2.00018-6

In Latin America, studies show that the relative frequency of diarrheal diseases that are attributable to foodborne pathogens varies from 26% to 36% (Pan-American Health Organization, 2013). From 1993 to 2010, reports by the regional system of the Pan American Health Organization noted 9180 outbreaks for 22 countries in the region (Pan-American Health Organization, 2013). Among these countries, the Health Surveillance Office of Brazil indicated that there were 3351 FBD outbreaks from 2010 to September 2014 involving 64,936 individuals (Brazil Health Ministry, 2014).

In Africa, an unprecedented number of FBD outbreaks have been reported recently (World Health Organization, 2012). In the continent of Oceania, the OzFoodNet site reported 30,035 notifications of diseases or conditions that are commonly transmitted by food in Australia in 2010, which represents an increase relative to the average of 26,190 notifications per year for the last 5 years (2005–2009) (OzFoodNet, 2012). In New Zealand, there was a decrease in the number of reported outbreaks and associated cases in 2013 (652 outbreaks involving 7137 cases) compared with 2012 (719 outbreaks and 10,500 cases). A total of 113 outbreak-associated cases required hospitalization and there were four deaths (Public Health Surveillance, 2014).

These data show the magnitude of FBD even in countries with a complete monitoring system within their health surveillance programs.

However, there are no quantitative data for all FBD spectra on a global basis, primarily because foodborne illness is normally underreported (World Health Organization, 2013). For example, an estimated 38.4 million episodes of FBD occur annually in the United States, resulting in the death of approximately 1686 (90% confidence interval 369; 3338) individuals per year (Scallan et al., 2011), which is higher than reported by the CDC.

Reliable and science-based estimates of these diseases are necessary to support policy makers in their decisions and to mobilize resources at all levels; namely, local, national, and international (World Health Organization, 2013). Kafertein et al. (1997) concluded that epidemiological data provide a common field for meeting an international consensus on food safety questions.

In Brazil, the costs of hospitalization cases caused by FBD were approximately US$97 million from 1999 to 2004, with an average of US$16 million per year (Carmo et al., 2005). Hoffmann et al. (2012) estimated the annual cost of FBD (medical costs, productivity loss, and the valuation of premature mortality) at US$14.0 billion (ranging from US$4.4 billion to US$33.0 billion) in the United States. These costs demonstrate the negative economic impact caused by FBD.

The implementation of prerequisite programs (good manufacturing practices) improved food safety (Cenci-Goga et al., 2005). However, FBD outbreaks still occur (Soares et al., 2013). The Codex Alimentarius (2013) conceptualizes the food employee/food handler as "an individual working with unpackaged food, food equipment or utensils, or food-contact surfaces." In Brazil, the National Health Surveillance Agency (Brazil Health Ministry, 2004) recognizes food handlers as "anyone in the food service that comes into direct or indirect contact with food."

Among the negative behaviors of food handlers, who are often associated with outbreaks, are inadequate hand hygiene practices, inadequate hygiene of equipment and utensils, maintenance of ready-to-eat food at room temperature, preparation of meals in advance, insufficient cooking temperature, and inadequate thawing (Greig et al., 2007; Chan and Chan, 2008; Food and Drug Administration, 2009). Therefore, it has been suggested that these professionals may be responsible for up to 97% of FBD outbreaks (Egan et al., 2007).

Training food handlers in safe food handling is one of the most critical interventions in the prevention of FBD (World Health Organization, 2013). Food handlers can help to reduce FBD by either preventing food contamination or by preventing the growth or survival of bacteria (Clayton et al., 2002). Increasing the food safety standards implies food safety education (Kafertein et al., 1997).

Recommendations such as the Codex Alimentarius (Codex Alimentarius, 2013) and food safety laws and standards have been published to guide professionals and owners of foodservices, experts, and primarily food handlers on appropriate procedures to reduce the risk of FBD outbreaks. Individuals engaged in food operations that come directly or indirectly into contact with food should be trained (Codex Alimentarius, 2013). Resolution no. 216 of the Brazilian National Health Surveillance Agency (Brazil Health Ministry, 2004) establishes technical regulations for good manufacturing practices in food services. This legislation states that all food handlers should be periodically trained in food handling. The training program must be described and determined according to its hourly load and program content.

Karaman et al. (2012) studied food handler knowledge of dairy products from differently sized companies (small, medium, and large) in Turkey. He found that sites that had mandatory training were more effective in relation to good food-handling practices in comparison with companies that had a voluntary program.

According to ISO 22000:2005, the food security team and other people who perform activities that have an impact on food safety should be competent and have adequate education, training, skills, and experience. It is the responsibility of the organization to identify these necessary skills, provide training, or take actions to ensure that the staff has these competencies. The organization should also ensure that the people responsible for the monitoring, correction, and corrective action of the food safety management system are trained (International Organization for Standardization, 2005).

The Codex Alimentarius (2013) recommends that all people involved with food must be aware of their role and responsibility in food safety. Food handlers must have the necessary knowledge and skills to handle food hygienically. Organizations should establish a training program. This program should provide periodic evaluations of the effectiveness of the training, instructions that direct handlers, the supervision of work routines, and checks to ensure that the procedures are performed effectively.

Food operation managers and supervisors must have knowledge of food hygiene practices to assess the potential risks of FBD, and they should make decisions to prevent its occurrence. Training programs should be reviewed and updated regularly (Codex Alimentarius, 2013).

To support a credible system and build confidence with consumers and trading partners, food safety laws and regulations should be relevant and enforceable. In addition, the regulations should clearly define the mandate of the food safety authority and other relevant agencies to prevent and manage food safety issues. To the extent possible, modern food safety laws should not only confer the necessary legal powers, but also should allow them to build preventive approaches into the national food safety system (World Health Organization, 2013).

Food safety, food control systems, and the food safety education and training of workers must move their focus gradually from end product tests to process control throughout the food chain (World Health Organization, 2013).

This chapter is intended to discuss different strategies for diagnostics and for the training given to handlers in food chain production.

2 FOOD HANDLER BEHAVIOR, FEELINGS, AND PERCEPTIONS

Food Safety Management Systems highlights the importance of managing people to achieve their goals (International Organization for Standardization, 2005, 2009). The manager should, among other things, assess the effectiveness of the people management processes and training activities in relation to food safety and associated risks.

At present, ISO 31000:2009 (International Organization for Standardization, 2009) defines risk as the "effect of uncertainty on objectives, often expressed in terms of a combination of consequences of an event and the probability of expected occurrence." It is universally understood that the notion of risk involves some type of loss or damage. Yoe (2012) defines risk in a simplified manner by using the following equation:

$$\text{Risk} = \text{probability} \times \text{consequence} \tag{1}$$

The uncertainty addressed in the definition of ISO 31000:2009 is the possibility of an error, which may be caused by an erroneous measurement or calculation, a lack of available scientific data for decision making, and a measure of variability (Pan-American Health Organization, 2008).

The risk may be faced through three fundamental approaches: that is, as a risk analysis as defined in *Food Safety Management Systems*, which involves reason, science, and logic to address hazards; the second is risk as a feeling, which addresses our quick, intuitive, and instinctive responses to danger; and the third is of risk as a policy, which is based on the meeting perceptions/feelings with risk analysis (Slovic et al., 2004).

In addition to the evaluation of microbiological risks in food (Sant'ana and Franco, 2009), scientific research has been applied to a detailed study of risk as feeling. When risk is a feeling, the individual tends to consider his or her decision intuitively based on the risk that a decision or action leads to or aggravates a negative event, which is also called perceived risk (Slovic, 1987). By using heuristics as its basis, the risk perceptions (or perceptions of benefit) define human behaviors and lifestyles (Frewer et al., 1994). Risk judgments involve what people think and how they feel about this risk.

If the feelings are favorable, the risk is judged to be low; otherwise, there is a tendency to make the opposite judgment (Slovic et al., 2004; Dijk et al., 2011). Several factors, such as knowledge, experience, attitudes, and emotions, can influence thinking and individual judgment regarding the seriousness and acceptability of risk (Sjoberg, 2000; Wachinger et al., 2013).

When the perceived risk is high, individuals tend to take protective actions to prevent or reduce the risk (Brewer et al., 2007). There is a tendency to underestimate dangers, especially those involving personal risks such as those related to food or FBD (Weinstein, 1987).

When the individuals exercise (or identify) low control over the phenomenon, they tend to identify such situations as having a high risk to themselves and society. However, in the case of contamination of food by microorganisms, the risk is generally identified as low because the individual feels he or she has (real or illusory) control over these situations (Frewer et al., 1994).

Weinstein (1989) explains that the individual has the need to exert control over situations, but the belief that he or she has control leads to underestimating hazards, identifying them as low risk. In food handling, for example, the use of gloves generates feelings of comfort and safety in the handler, making him or her believe that the food will not be contaminated, because of the use of these gloves, regardless of other procedures or the proper hygiene of the gloves (Frewer et al., 1994; Weinstein, 1984).

This illusion leads the individual to believe that he or she is less susceptible to contamination than others, or that the food that he or she manipulates has lower risk than the food handled by others, under similar conditions or not. This phenomenon is called the optimism bias. The illusion of control is one concept that is intended to explain this phenomenon (Weinstein, 1989).

Another motivator of the optimism bias is a comparison with stereotypes risk (Weinstein, 1989). In considering that a person is more skilled or will have more luck than average, individuals believe that risk prevention campaigns are targeted at people who are less fortunate or those who exhibit clearly risky behavior (Horswill and McKenna, 1999; Weinstein, 1989).

The optimism bias is also related to individual experiences about each hazard, in which the handler ponders the personal risk by considering the occurrence referred to him or her from a negative event in the past or something that happened in his or her family and those close to him or her (Weinstein, 1987). In association with his or her knowledge, he or she can use heuristic methods to avoid risks. In terms of food safety, individuals who had had salmonellosis (or someone in their family had had salmonellosis) perceived themselves to be more susceptible to foodborne illness than individuals who had not had the illness (Parry et al., 2004).

Da Cunha et al. (2014a) observed that food handlers in different food businesses presented an optimistic bias (OB) toward the risk of being responsible for handling foods that can spread FBD to consumers, friends, and family members. The authors concluded that this positive outlook can emerge from environmental characteristics because food handlers from schools and hospitals presented higher OB levels than food handlers from street food vendors and restaurants.

It has also been shown that risk perception can directly interfere with the conduct of the individual (Janz and Becker, 1984) in that optimism bias can reduce the individual's precaution to minimize his or her own risk (Horswill and McKenna, 1999; Weinstein, 1984, 1987), and then engage in inadequate conduct. This assumption indicates that food handlers with a positive outlook on their (real or illusory) practices may neglect food protection attitudes.

Risk perception can establish an attitudinal ambivalence, mitigating the application of knowledge in the practices and attitudes of protection (Janz and Becker, 1984). Knowledge is an important attribute in the description of a risk. Determining the perception of risk is fundamental for determining the degree to which the individual knows that he or she is exposed to risk, how much the individual knows about the nature of the risk, and how much science he or she knows about the risk. When given some knowledge of the danger, the precision for estimating the risk of occurrence increases (Frewer et al., 1994). However, popular interpretations of a health threat are primarily based on the beliefs and convictions, not on the facts and scientific data (Slovic, 1987).

Risk perception is subjective and based on the experience, and understanding it is important to the success of food safety communication and the absorption of new technologies (Knox, 2000; Behrens et al., 2010).

Among tutors for food safety programs, there is the belief that only knowledge of microbiological hazards in foods can ensure safe practices. This understanding has a direct implication for internal risk communication, because food handlers can perceive low risks associated with other dangers and consequently neglect control measures applied in the best practices for food safety. One goal of risk communication is to promote the exchange of information among knowledge, attitudes, values, practices, and perceptions about risks to food safety (Food and Agriculture Organization of the United Nations, World Health Organization, 1999, 2006).

Evaluating risk perception may be an important diagnostic tool to improve understanding of food handler behavior at work and enable the use of different training strategies and interventions in this context (Nauta et al., 2008).

Knowledge is necessary, but it is not a condition for behavioral change (Ehiri et al., 1997; Askarian et al., 2004), and alone it has little direct effect on the intention of food handlers to engage in safe behavior. However, this behavior is compensated by social cognitive factors (Mullan et al., 2010).

It is important to determine gaps in the knowledge of food handlers to guide effective educational and behavioral interventions (Pichler et al., 2014). Accordingly, Jianu and Golet (2014) observed that, although results indicated a good level of knowledge and practice among meat handlers, some aspects, such as the identification of risks to food safety and hand hygiene, continued as problems that must be articulated in training programs.

This knowledge allied to attitudes is a precursor to behavioral change (Gilling et al, 2001; Mullan et al., 2010). Positive behavior in food handlers in terms of good manufacturing practices is the result of motivation; thus, people work to develop actions (Ellis et al., 2010).

To adapt this motivation to the workplace, Yiannas (2009) identifies some relevant actions, such as valuing people, recognizing the improvement of attitudes, encouraging initiatives, delegating authority, and conducting evaluations.

Lin (2007) emphasizes, "Efforts to foster the targeted reciprocal relationships and interpersonal interactions of employees are necessary for creating and maintaining a positive knowledge sharing culture in organizations."

Ellis et al. (2010) indicated that to increase in the probability that employees will be motivated to perform safe food-handling practices, supervisors should improve communication with employees and provide resources to them. All these functions can be controlled, implemented, and influenced by the supervisors, who need the required management skills to fulfill these roles in a way that influences the employees positively.

Food safety education, training, and motivation are factors that impact safe food preparation practices by food workers and managers (Green et al., 2005).

The psychosocial aspects of food safety, namely the beliefs, self-efficacy, locus of control, and stage of change, have been studied in the food safety field (Byrd-Bredbenner et al., 2007a,b).

In considering the above, researchers have studied ways to identify the nuances of behavior and their theories and to propose diagnostic strategies and training to improve food handler practices, in addition to the attending health legislation. From this perspective, carefully considering the pedagogical approach to adopt is fundamental. This approach allows for the guidance and development of a training program and assesses its impact. All educational activity is thought to obey the purposes of social, cultural, political, and economic development. Consequently, this activity responds to certain interests. It is therefore maintained by a philosophy of education in space and time (Bezerra et al., 2014).

3 THEORIES AND MODELS OF BEHAVIORAL CHANGE

Among the studies to evaluate food hygiene training, few used explicit theoretical models as their basis (Bezerra et al., 2014). Given the plurality of psychology theories, there is a question as to what would be the best model or theory to use in a training program for food handlers. The value of a psychological theory is not only judged by its explanatory and predictive power, but by its practical power to promote changes in human behavior (Bandura, 2004). Some theories and models that will be cited in diagnostic strategies and training are discussed below.

3.1 THE THEORY OF REASONED ACTION AND THE THEORY OF PLANNED BEHAVIOR

The theory of reasoned action (TRA) and its extension known as the theory of planned behavior (TPB) (Ajzen, 1991) include perceptions of control. These strategies have been successfully used in studies on food hygiene and food handler behavior (Clayton et al., 2002; Mari et al., 2012). In TRA/TPB, individual intentions are the primary factor in achieving certain behaviors (Figure 1) (Ajzen, 1991).

People's intentions and beliefs about food safety are influenced by two key factors: (1) the level of people's intent is greater if they have a positive attitude toward the behavior and (2) the level of intent will be greater if they are motivated to act in accordance with social norms (Yiannas, 2009).

The attitude toward a given behavior in the TPB and the perception of control are associated with the influence of subjective norms, also called social pressure or peer pressure. Peer pressure is often used to describe instances in which an individual feels indirectly pressured into changing his or her behavior to match that of his or her peers (Albarracín et al., 2001). In a study by Seaman and Eves (2010), it was noted that peer pressure was the factor that most influenced the intention to change hygiene and sanitation practices. Coworkers can positively or negatively influence the behavior of others quickly, underscoring the importance of involving all intervention strategies.

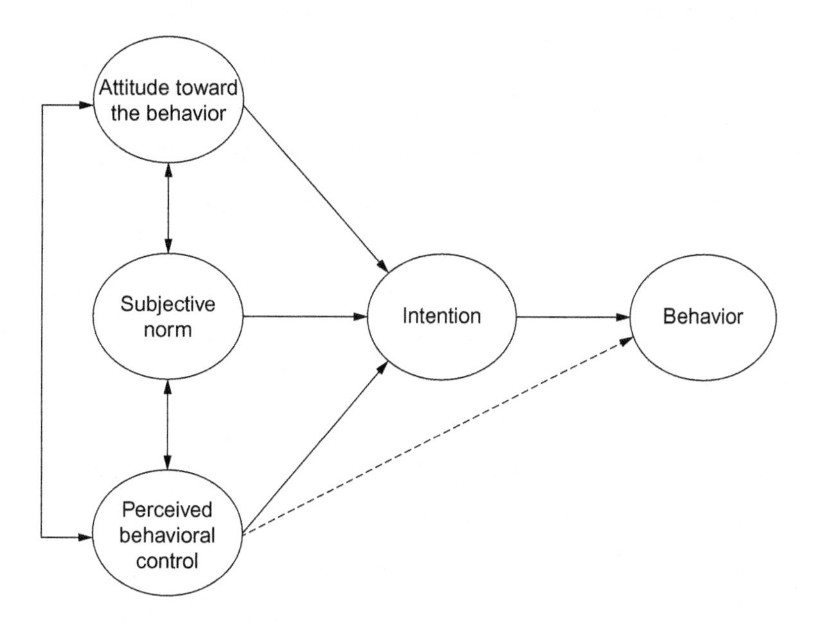

FIGURE 1

The theory of planned behavior model (TPB) (Ajzen, 1991).

3.2 THE HEALTH BELIEF MODEL

The health belief model (HBM) is intended to reveal the relations between what a person believes and how he or she acts. A person's health beliefs are the results of his or her ideas, beliefs, and attitudes about health and illness. These beliefs may be based on the information and misinformation, common ideas in a given community or family, or myths, lived realities, or misconceptions. Therefore, beliefs can both promote health and influence it negatively (Rosenstock, 1974).

The HBM is shown in Figure 2. Its application to the food safety field is presented in Tones' Health Action Model (THAM).

3.3 TONES' HEALTH ACTION MODEL

The THAM (Figure 3) is a combination of two other models: namely, the HBM and the TRA (Rennie, 1995) and contains several key points of the TRA/TPB such as the influence of norms and motivation to change behaviors.

The Tones' Health Model is divided into five constructs or systems as follows: (1) the knowledge system, initial knowledge about food safety; (2) the legal system, the laws and rules of the service; (3) the motivational system, the motivation elements of the service; (4) the belief system, the values and beliefs of the target audience; and (5) the environmental conditions of service (Nieto-Montenegro et al., 2008; Rennie, 1995). In this model, interventions that address all constructs/proposed systems are developed because all these factors are thought to affect the behavior of the food handler directly and not just knowledge.

3.4 SOCIAL COGNITIVE THEORY

Social cognitive theory (SCT), created in 1971 by Albert Bandura, a Canadian psychologist, argues that the individual's behavior, personal factors, and the environment mutually influence each other in a relation called triadic reciprocity. In this theory, self-development and change of human behavior are explained from the perspective of the agency. In a simplified way, being an agent means being able to

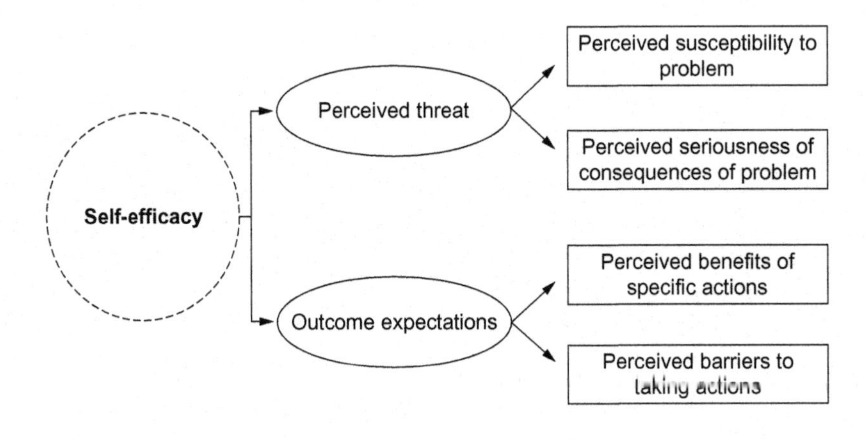

FIGURE 2

The health belief model (HBM).

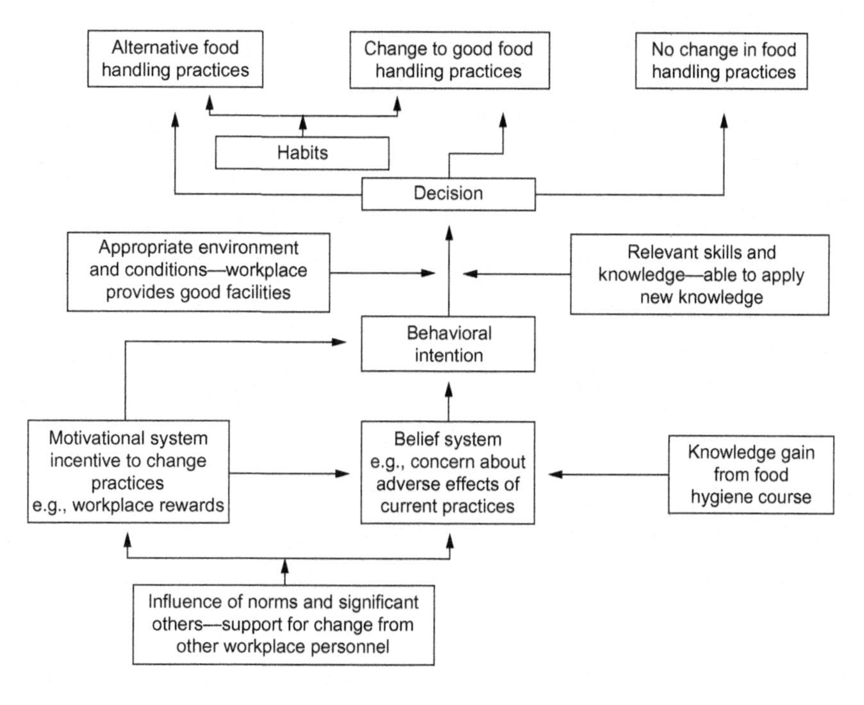

FIGURE 3

Tone's Health Action Model as applied to food safety (Rennie, 1995).

develop mechanisms of self-regulation that may determine the path to be followed. The individual can influence the course of events according to his or her interest. He is an active participant of the direction that his or her life will take, as he or she sets goals that will be reached through trajectories chosen by himself or herself (Bandura, 1977).

SCT emphasizes that behaviors are influenced by the environment and personal factors (Yiannas, 2009).

People are ambitious and proactive, not just reactive. The ability to predict allows people to exert control in advance, rather than simply react to the effects of their efforts. They are motivated and guided by predicting goals, not only by a retrospective view of limitations. SCT has been applied in studies about educational focuses on health promotion (Bandura, 2004).

3.5 THE TRANSTHEORETICAL MODEL

In the transtheoretical model (TM), behavioral change is conceptualized as "a process that unfolds over time and involves progression through a series of six stages: precontemplation, contemplation, preparation, action, maintenance, and termination. At each stage of change, different processes of change optimally produce progress" (Prochaska et al., 2002).

Based in this theory, specific interventions to influence behavioral change should correspond to the stage where an individual is or the degree of his or her willingness to change (Yiannas, 2009).

These models and theories should be seen as a way to understand the factors that influence decisions and individual behavior.

4 DIAGNOSTIC STRATEGIES

Diagnostic strategies are intended to support the development of training programs that are more appropriate to the specific needs of food handlers (Kaliyaperumal, 2004). Some diagnostic strategies will be presented and discussed below.

4.1 FOOD SAFETY KNOWLEDGE, ATTITUDES, AND PRACTICES

To introduce a discussion about knowledge, attitude, and practice, the following terms are defined:

Knowledge: "These are facts, information, and skills acquired by an individual through experience or education; it is the theoretical or practical understanding of a topic" (Pearsall, 2001).
Attitude: "A psychological tendency expressed in degree of favor or disfavor upon a particular entity" (Eagly and Chaiken, 2007); "an established way of thinking or feeling about something or someone, usually reflected in individual behavior" (Pearsall, 2001).
Practice: "Run or perform an activity or a particular method of habitual or regular manner" (Pearsall, 2001).

In relation to foods, practice can be defined as "Attitudes observed in an individual that can affect his nutrition or that of others, such as: eating, feeding, hand hygiene, cooking, and food choice" (Macías and Glasauer, 2014).

These definitions are simplified translations of what is generally understood to be knowledge, attitudes, and practices in food handling. The practice can be divided into self-reported and observed categories. The self-reported practice may be a bias in the research in that it considers the response of the manipulator, and the observed practice is a result of the researcher's observation.

Several studies have been devoted to assessing the knowledge, attitudes, and practices of food handlers (Angelillo et al., 2000; Ansari-Lari et al., 2010; Bas et al., 2006; McIntyre et al., 2013; Soares et al., 2012). Older and more recent studies indicate that the provision of knowledge is not enough to engender attitudinal change and the adoption of appropriate practices (Da Cunha et al., 2014b; Ansari-Lari et al., 2010; Ehiri and Morris, 1996; Ehiri et al., 1997; Iiu et al., 2015; Park et al., 2012; Rennie, 1995). This insufficiency occurs because individuals have attitudinal ambivalence (Newby-Clark et al., 2002) in several respects. This attitudinal ambivalence is defined by a concomitant degree of favor (positive attitude) and disfavor (negative attitude) toward a practice (Newby-Clark et al., 2002). Food handling researchers hypothesize some factors as generators of ambivalence; namely, OB (Miles et al., 1999), a low perception of risk in food handling activities (Da Cunha et al., 2012; Parry et al., 2004), an illusion of control (Horswill and McKenna, 1999), knowledge provision with low application in a practical context (Rennie, 1995), a lack of intention to change on the part of individuals (Ajzen, 1991), motivation (Kelly et al., 1991), and several behavioral aspects. It is noteworthy that attitudinal ambivalence has not yet been investigated in food handlers, and it is only a hypothesis of associate factors.

Attitude is an important factor in the application of knowledge, and it can influence behavior and practice, reducing the risk of FBD (Bas et al., 2006; Ko, 2013; Sani and Siow, 2014); thus, it is important that all food handlers have positive attitudes toward food safety. However, negative attitudes toward hand hygiene, the practice of safe storage, and the control of cross-contamination are observed among food handlers (Tan et al., 2013; Sani and Siow, 2014).

A way to evaluate knowledge, attitudes, and self-reported practices (KAP) is by using a structured questionnaire that can be created by using the method described by Bas et al. (2006) and Da Cunha et al. (2014b) and then administering it to food handlers.

The first part of the KAP questionnaire has the objective of evaluating food handler knowledge about food safety. Questions related to the daily practices of food handling are presented, and they address different aspects of food handling (e.g., personal hygiene, food hygiene, cross-contamination, temperature control, food thawing, and environmental hygiene). The response is generally given as one of three possible answers; that is, "yes," "no," and "I don't know." The order of "yes" and "no" as correct answers can be shuffled and should not follow a pattern. One point is assigned for each correct answer, and zero points are assigned for wrong or "I don't know" answers.

The second part of the KAP questionnaire includes affirmatives to assess food handler attitudes toward food safety, including affirmatives about the importance of hygienic procedures, their responsibility as food handlers in preventing FBD, and the importance of learning more about food safety. Food handlers can indicate their level of agreement by using a three-point rating scale of "agree," "not sure," and "disagree."

Self-reported practices are generally evaluated in the last part of the KAP questionnaire. Questions about daily practices can be included and preferably address the same topics as in the knowledge-related part. A five-point rating scale ranging from 1 = never to 5 = always can be used to evaluate each practice.

A checklist can be used to evaluate the observed food handler practices. The questions were retrieved from a food safety evaluation instrument that was created to consider the characteristics of food services and to evaluate the compliance and noncompliance resulting in a percentage of adequacy (Da Cunha et al., 2014b).

4.2 RISK PERCEPTION

Risk perception is a diagnostic strategy that can also be assessed by a questionnaire; for example, a structured questionnaire was designed to evaluate the risk perception of FBD as developed by Da Cunha et al. (2012). This instrument was based on a questionnaire used by Frewer et al. (1994). It contains seven questions about food storage and personal environmental food hygiene. Responses were given through a 10-cm structured scale that was anchored with the intensity descriptors ranging from "no risk" to "high risk." Risk perception can also be assessed by different scales such as a structured 7-point scale (with three descriptors; e.g., much less than most people, same as most people, much more than most people) (Sparks and Shepherd, 1994), a structured 9-point scale (1 = extremely unlikely to 9 = extremely likely) (Raats et al., 1999); a numeral scale from 0 (as defined as the certainty that the event will not happen) to 100 (certainty that it will happen) (Hoorens and Buunk, 1993); and others.

Thus, it is possible to evaluate different perceptions of risk as the perceived risk of specific practices (e.g., what is the risk of FBD for the consumer if you thaw food at room temperature?); the perceived risk of general practices (e.g., what is the risk of FBD for the consumer after eating a meal prepared by you?); or even to evaluate food handler willingness to change their practices (e.g., what is the risk of FBD for the consumer in the future if you don't change your food-handling practices?).

In food handling, the risk perception of food handlers may be the barrier between adopting and not adopting proper hygienic practices. If the food handler believes that a FBD will never occur during his or her work or after mishandling food, it may be more difficult to get him or her to adopt preventive measures, such as good handling practices. Perceived risk assessment can be used as a diagnostic tool and information can be used to demystify food-handling aspects perceived as low risk.

Other aspects of perceived risk assessment are described further in the next section.

4.3 OPTIMISTIC BIAS

OB, as described before, is defined as a "positive outlook regarding future events in which individuals consider themselves less likely than others to experience negative events" (Weinstein, 1987). OB can be identified by using risk perception with a direct or an indirect method. In the direct method, respondents are required to make a likelihood judgment form by themselves in relation to others on the same scale (Miles and Scaife, 2003) (e.g., compared with the average food handler of my sex and age, what is the likelihood of me being responsible for giving a FBD to the consumer?). In the indirect method, individuals separately indicate their own risk and their peers' risk of causing an FBD (Chock, 2011) (e.g., (a) What is the likelihood of me being responsible for causing a FBD in the consumer? (b) What is the likelihood of a food handler who is similar to me (same age and sex) being responsible for causing a FBD in the consumer?). The direct method tends to reveal a greater degree of OB, which may hinder the analysis of potential factors that influence this phenomenon (Helweg-Larsen and Shepperd, 2001; Miles and Scaife, 2003).

An average positive rating (when using a direct method) or a rating significantly greater than zero (when using an indirect method) is taken as evidence of an OB (Chock, 2011; Helweg-Larsen and Shepperd, 2001). Da Cunha et al. (2014a) assessed the OB of food handlers from different food businesses. In their assessment, the indirect method was used and significant differences between ratings were considered OB tendencies.

Food handlers can be asked the following questions as adjusted by Da Cunha et al. (2014a):

Question 1—"What is the consumers' likelihood of exhibiting abdominal pain and/or vomiting (FBD) after eating a meal or food in a restaurant (other than the one where you are working)?"
Question 2—"What is the consumers' likelihood of exhibiting abdominal pain and/or vomiting (FBD) after eating a meal or food prepared by you?"
Question 3—"What is the consumers' likelihood of exhibiting abdominal pain and/or vomiting (FBD) after eating a meal or food from another foodservice that is not a restaurant (such as hospitals, beach kiosks, street food kiosks, or school meal services)?"
Question 4—"What is the likelihood of your friends and family members exhibiting abdominal pain and/or vomiting (FBD) after eating a meal or food prepared by you?"
Question 5—"What is the likelihood of your friends and family members exhibiting abdominal pain and/or vomiting (FBD) after eating a meal or food prepared by a food handler other than you?"

The questions must be adjusted for each food business. Responses can be given on a scale similar the one used to assess risk perceptions, because OB is a phenomenon based on perceived risk.

To assess the food handlers' degree of OB, the score assigned to each perceived risk question must be compared with the scores assigned to the other questions. Five OB constructs were established, as follows: 1—consumer's risk (questions 1 and 2); 2—friend and family member risks (questions 4 and 5); 3—food business risk (questions 1 and 3); 4—consumer food business risk (questions 2 and 3); and

5—consumer family member risk (questions 2 and 4). Positive and statistically significant results indicated the presence of OB, and higher scores indicated a greater magnitude of OB (Chock, 2011; Da Cunha et al., 2014a; Helweg-Larsen and Shepperd, 2001).

4.4 BELIEFS

To reduce foodborne illness, it is crucial to gain an understanding about the interactions of prevailing food safety beliefs, knowledge, and practices of food handlers (World Health Organization, 1988).

Clayton et al. (2002) used elements of SCT to examine the beliefs of food handlers in relation to food safety, and they used a questionnaire to determine the self-reported practices of food handlers. The questionnaire first collected open-ended responses from the participants that were related to their beliefs about food safety, as follows:

- What are the important things that you can do, when preparing or handling food at work, in order to prevent food poisoning? (Please list as many things as you can.)
- Salient consequences: Please list any advantages or good things that would happen if you carried out these behaviors at every appropriate occasion during your working day. Please list any disadvantages or bad things that would happen if you carried out these behaviors at every appropriate occasion during your working day.
- Salient facilitators or barriers: What, if anything, might encourage you or make it easier for you to carry out these behaviors at every appropriate occasion during your working day? What, if anything, makes it difficult or prevents you from carrying out these behaviors at every appropriate occasion during your working day?

This questionnaire was based on the theory of planned behavior and the HBM. Both models require a pilot research stage in which remarkable beliefs of the target population about the specified behavior are determined by using open-ended questions. The questionnaire begins by asking participants to describe which behaviors could be performed to prevent food poisoning in their workplace. The specification of this food safety behavior allowed participants to answer the other questions based on their own definition of food safety behaviors.

The remaining open questions were designed to identify the primary consequences, meaning the advantages and disadvantages of performing food safety actions and to determine potential facilitators and barriers to engaging in food safety actions.

Consequences help to increase or decrease behaviors, meaning that they can be used to improve food safety performance. Positive and negative reinforcements are two behavioral consequences that increase the probability that a behavior occurs again. Historically, managers have focused excessively on creating negative consequences for performance below the ideal. To improve performance and results, consequences, or positive reinforcements should be used much more than negative consequences (Yiannas, 2009).

The second part of this questionnaire was designed to collect quantitative data in the form of closed questions. Self-reported practices by food handlers for food safety were verified by asking the respondents to designate which of four statements applied to themselves.

Statement A: I carry out all the food safety behaviors, which I know I should do, at every appropriate occasion during my working day.

Statement B: I carry out many of the food safety behaviors, which I know I should do, but sometimes I do not carry out all these food safety behaviors at every appropriate occasion during my working day.
Statement C: I carry out a few of the food safety behaviors, which I know I should do, but often I do not carry out all these food safety behaviors at every appropriate occasion during my working day.
Statement D: I do not carry out any of the food safety behaviors, which I know I should do, at the appropriate occasions during my working day.

To determine the level of the perceived control that food handlers had over performing food safety actions, closed questions were also used to assess their risk perceptions of someone who would contract food poisoning from their workplace, other businesses, and their home and to establish the type of food hygiene training participants had received.

4.4.1 The locus of control and self-efficacy

In personality psychology, the locus of control refers "to the extent to which individuals believe they can control events affecting them. A person's 'locus' (Latin for 'place' or 'location') is conceptualized as either internal (the person believes they can control their life) or external (meaning they believe their decisions and life are controlled by environmental factors which they cannot influence, or by chance or fate) (Figure 4)."

Self-efficacy is "the belief in one's capabilities to organize and execute the courses of action required to manage prospective situations." In 1977, Bandura developed a very important and influential publication that demonstrated how self-efficacy can impact everything from psychological states to behavior to motivation. Self-efficacy and skills are core concepts of SCT. For example, if a person perceives an incentive related to a specific behavior, it is important that he or she believes that he or she may be able to accomplish it (self-efficacy). Success in realizing this performance increases the likelihood that it will be performed again (Yiannas, 2009).

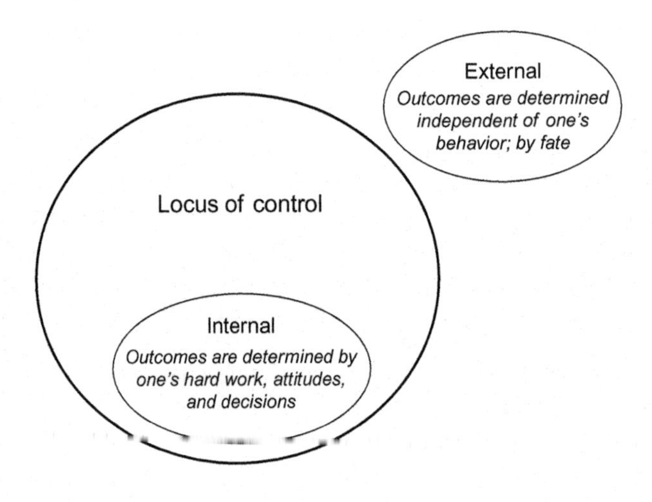

FIGURE 4

The locus of control.

The locus of control and generalized self-efficacy are significant predictors of both worker satisfaction and worker performance. Clayton and Griffith (2008) observed that self-efficacy was an important predictor of adequate food safety practices.

To initiate a study about the locus of control, we recommend Wallston et al. (1978) and Wallston (2005). With regards to self-efficacy, we recommend Abbot et al. (2009).

4.5 MOTIVATION

Motivation is "the driving force which gives purpose or direction to human and animal behaviors which operate at a conscious and subconscious level." Without motivation those behaviors would simply not occur. Another definition for motivation explains that, "it is a person's willingness to exert physical or mental effort in order to complete a goal or set aim."

Arendt and Sneed (2008) proposed a model for employee motivations to follow food safety practices. The motivators for this group included all external motivators that were controlled by the foodservice supervisor-leader.

Four motivation factors are part of the model. Factor 1 is communication. This extrinsic motivator refers to the verbal and nonverbal communication between the supervisor and foodservice worker in addition to communication by supervisors to employees through their role-modeling behaviors. Factor 2 is rewards and punishment. This extrinsic motivator refers to rewards (increased pay, "thank you," and time off) and punishments (reprimands) that are typically administered by the supervisor. Factor 3 is internal motivators. This category encompasses all intrinsic motivators that are not reliant on an external source. Examples in this area included pride and satisfaction when following safe food-handling practices and seeing benefits to the self. Factor 4 is resources. This extrinsic motivator included resources that were typically provided by the foodservice supervisor such as supplies and food safety training.

4.6 CHANGE STAGE

Garcia et al. (2013) developed a questionnaire known as The Scale of Attitude Change for Food Handlers (EMAMA) based on the TM by Prochaska.

The EMAMA was validated by judges and includes a semantic analysis of 31 statements that were prepared so that the food handler could choose an answer suited to his or her reality; in other words, what he or she performed at the restaurant in relation to good handling practices.

Responses were described on the basis of a Likert five-point scale, as follows: the answer "I do not think about it" or "I do not do it" was designated grade 1 and was classified as Precontemplation; the answer "I have been thinking in this way" was given grade 2 and was classified as Contemplation; the answer "I am determined to do it" was designated grade 3 and was classified as Decision; the answer "I have started doing it recently" was given grade 4 and was classified as Action; and the answer "I have been doing it for a long time" was designated grade 5 and was classified as Maintenance.

4.7 FOCUS GROUP

The perceived barriers to performing food safety practices were determined by Howells et al. (2008) by using the following three practices: (1) time/temperature control; (2) personal hygiene; and (3) cross-contamination. The investigators assessed restaurant employee perceptions of barriers to implementing the three food safety practices at work by using the following two series of focus groups: Group A, which was composed of restaurant employees who had not completed a food safety class prior to the focus groups; and Group B, which included employees who participated in focus group discussions immediately following a food safety class. After the focus groups, the researcher reviewed the data and developed barrier categories for each behavior. These data were managed by placing similar responses into categories.

The foodservice interventions were planned by using the results. Food safety posters were developed including phrases such as "how to" and "did you know."

Arendt et al. (2012) engaged in qualitative research that addressed the challenges encountered in foodservice processes and discussed strategies to overcome the identified challenges.

5 TRAINING STRATEGIES

Training is the primary strategy used to educate food handlers to perform the procedures established by the health legislation properly (Medeiros et al., 2011). Training, capacity building, and formation are generally treated as synonyms but have different definitions, as follows:

> Training "Making skillful, dexterous, able, by instruction, discipline or exercise; enable, train; perform (an activity) regularly; exercise, practice" (Pearsall, 2001).
> Capacity building: "Becoming able to; empower yourself; make understand or comprehend; persuade; have understanding; understand, finding out; be sure; convince yourself; persuade yourself" (Houaiss, 2001).
> Formation: "Act, effect or mode of forming; constitution, character; way by which constituted a mindset, a character or a professional knowledge. It is continuous and involves the knowledge and previous experiences" (Ferreira, 1986).

The purpose of these definitions is not to discuss the etymology of the three terms, although the term "training" has a connotation related to the reproduction of techniques by using regular practice. The term "capacity building" makes individuals capable, persuading them and making them understand something. In this sense, the term presents the individual as an object that must "be able," leaving aside the discussion of abilities and disabilities.

The term "formation" has a broader meaning linked to the gradual process of education. Formation is a term that best defines the actions (effective) of safe handling. It is derived from a bilateral action in the education process between educator and student, and it is based on exchange and continuity. In this text, the term training as used in the original quote was retained, and the term was used in the discussions among the authors.

Before the training, performance expectations should be created to achieve performance excellence in food safety. These expectations should be clear, achievable, and understood by all, based on risk. Without them, the actions and desired outcomes will not occur consistently (Yiannas, 2009).

Certainly, when creating performance expectations for food safety, they should be determined beyond adaptation to regulatory standards. The manager should think about all the things that food handlers should know about the risks associated with the food and clearly define what he or she wants them to do. The expectations should be documented so that they can be communicated consistently. When the manager relies excessively on negative effects, this demonstrates a lack of a comprehensive understanding of how to use expectations to generate improvements in performance. Food handlers will not be motivated to perform at full potential for fear of being punished or suffer a negative consequence. A work environment dominated by the fear of negative consequences is not a good working environment. Although negative consequences have their space in the food safety arena, they are not the ones that should be used (Yiannas, 2009).

Training has been established to promote an increased knowledge of food hygiene in food handlers (Hislop and Shaw, 2009; McIntyre et al., 2013; Osaili et al., 2013). The use of training is therefore motivated by the thought that knowledge can change attitudes and practices, for compliance with sanitary regulations and the low cost of this intervention.

The relation of training and adequate practice is complex. As mentioned, several authors have reported that training was not enough to change practices; however, other authors showed positive results in their research indicating that training can be transformative (Choudhury et al., 2011; McIntyre et al., 2013). In this sense, Nieto-Montenegro et al. (2008) reported that it is necessary to plan and use appropriate methods for the training to be effective.

Most food hygiene training uses a method of providing the information, following the KAP model, which has been widely criticized and has perceived limitations (Ehiri and Morris, 1996). A major flaw in this model is the fact that it is founded on the assumption that the information received by the workers is translated into behavior.

Therefore, there are variables/situations that were researched and considered as factors that affect training, such as the strategy used for intervention (Egan et al., 2007), the place where training is conducted (Rennie, 1995), the lowering of acquired knowledge applicability, mandatory participation against voluntary participation (Cotterchio et al., 1998), and factors that involve concepts of food safety culture such as leadership, communication, commitment, environment, perceptions of risk, and motivation (Ehiri et al., 1997; Ko, 2013).

Placing all bets on lectures is not associated with best practices by the food handler (Da Cunha et al., 2014b) because it ignores the perceptions of these professionals. A positive attitude is an important factor in the adoption of appropriate practices (Ko, 2013); thus, the handler must be motivated to learn and apply the acquired knowledge.

Food training programs based on practical activities and theoretical instruction have been shown to be important tools in which food handlers can translate information into practice (Soares et al., 2012). Soares et al. (2013) verified that the association of strategies based on practices enabled the food handler to translate the knowledge acquired in the theoretical lessons to practice and to reduce microbiological contamination successfully. Therefore, practical training should be implemented (Tan et al., 2013).

In 2010, Seaman proposed the food hygiene training model (Seaman, 2010). In this strategy, the following three evaluation stages are proposed: (1) the documented analysis of training needs, which consists of an assessment of the training needs of the food handler, why and when food handlers should be trained, analysis and previous records and controls, and established criteria for the success or failure of the training; (2) a knowledge test and/or evaluation of practical skill, which is a step to perform after training to carefully monitor and ensure the translation of knowledge into practice and to correct deficiencies in the practices of food handlers; and (3) an evaluation of the food handler training program, which is an assessment of the perceptions and reactions of food handlers in training, providing the opportunity to approve or disapprove certain aspects of training (Seaman, 2010).

After the evaluations, the food hygiene training model employs methods to increase knowledge and improve food-handling practices. These methods should be performed along with motivational strategies, adequate infrastructure, and the effects of social pressure, among others (Figure 5).

Training programs must be evaluated to ensure their effectiveness (Zain and Naing, 2002).

Based on the food hygiene training model and the concepts of knowledge, attitudes, and practices, an intervention model for food handlers in a school environment was tested (Da Cunha et al., 2013). This model uses different strategies to promote increases in knowledge and changes in practice and motivation (Figure 6). This model, which was created and tested in Brazil, adds points that are

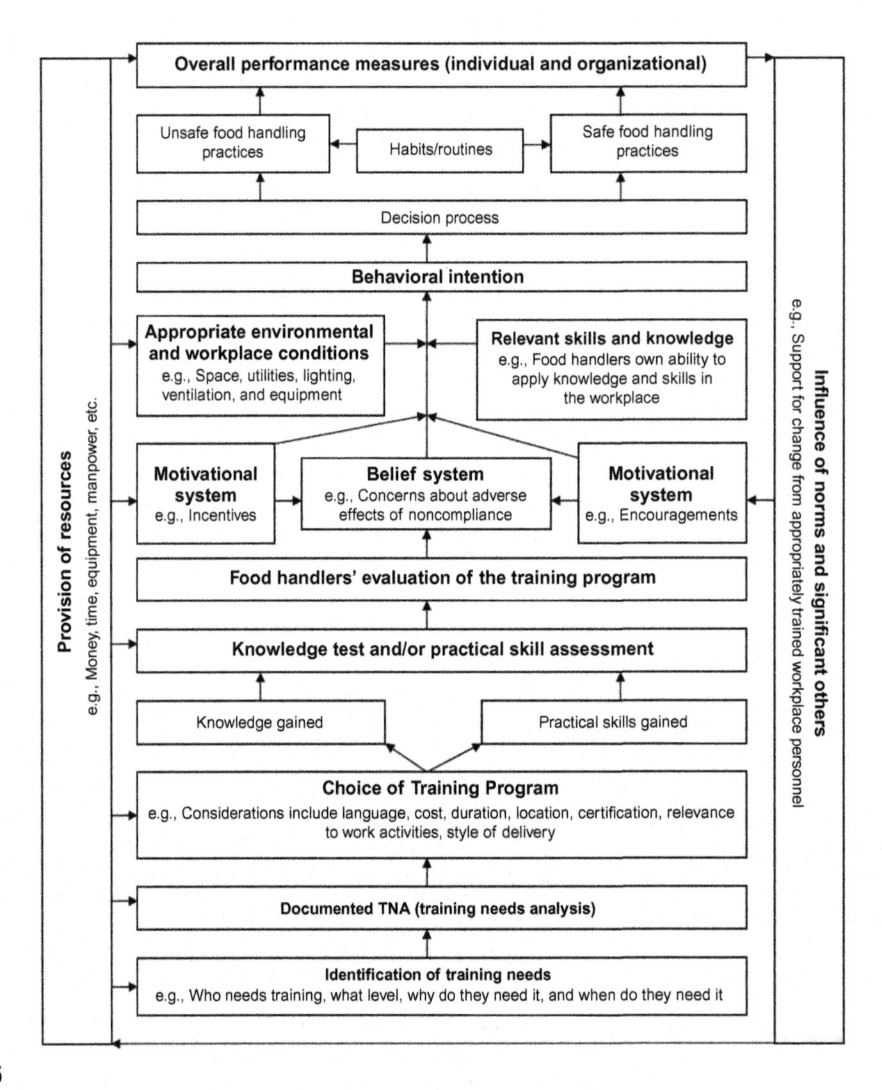

FIGURE 5

The food hygiene training model as proposed by Seaman (2010).

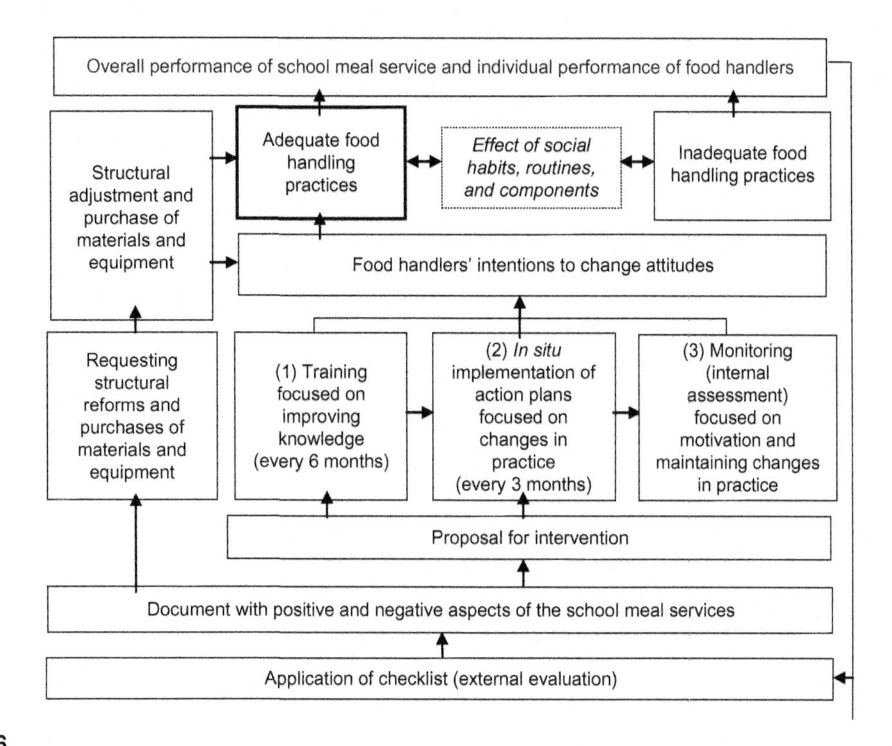

FIGURE 6

A systematic intervention model based on triad knowledge, attitude, and practice for school food services (Da Cunha et al., 2013).

infrequently discussed for intervention strategies with food handlers as a combination of internal and external evaluators, adequate infrastructure, and monitoring. When tested longitudinally, the proposed intervention improved school compliance with the health laws (Da Cunha et al., 2013).

Soares et al. (2013) assessed the influence of theoretical-practical food safety training based on the microbiological counts of food-contact surfaces, food equipment surfaces, food tools, and hand washing by handlers in canteens and cafes at a university campus in northern Portugal. A food safety training program was specially developed for all of the food establishment's handlers on the campus. Training was performed by only one trainer. The food safety training consisted in 9 h (three sessions of 3 h each) per group, and each session was divided into three parts as follows: (1) general concepts of hygiene and food safety, (2) practice, and (3) application of the acquired knowledge *in situ*.

The first part consisted of presenting food safety concepts in a theoretical approach. Training sessions were given to small groups with four food handlers belonging to canteens and groups and with two food handlers belonging to cafes to achieve a strict interaction between food handlers and trainers. The theoretical portion consisted of in-depth explanations of food safety practices.

The second part included practical training to translate the knowledge acquired in the first section. During the practical training, the practical aspect of each Hazard Analysis and Critical Control Points (HACCP) prerequisite was demonstrated to reinforce the food safety concepts that were previously discussed. Finally, the application of the previously acquired knowledge in situ included an audit by the food safety trainer to verify the procedures performed by the food handlers during the workday. After that, corrective measures were applied in cases of inadequate practices.

The results demonstrated that food handler training was essential to increasing the hygienic conditions of the food establishments in this study. After the food safety training program, the overall microbiological counts decreased and several differences related to the specific characteristics of the two categories of establishments were found. Thus, differences in the layout, hygienic design, equipment, foodservices, and food handlers in these establishments implied that the food safety training programs had been specially developed for each food sector and/or establishment size. Moreover, this study showed that food safety training is also important for ensuring safe foodstuffs, and for preventing the involvement of food handlers in FBD.

Gomes et al. (2014) evaluated a methodology for food handler training by using didactic material (booklet) developed for hotel sector training sessions and for evaluating microbiological variables. The microbiological analyses result in the need for constant vigilance to ensure safe food, in addition to providing examples to insert into teaching materials, with a focus on hotel food handlers, with the advantage of approaching examples of the routine in these establishments.

Food handler education in hotels should be a compulsory part of any food safety program. Gomes et al. (2014) demonstrated the development of a methodology that employs practical examples from the daily activities of the workers and the results of laboratory analysis. Improvements were observed after the implementation of the method and easy understanding was noted among all the food handlers. The methodology can be adopted without any extra cost, by any hotel foodservice.

Studies describing the effectiveness of the models/interventions mentioned here highlight the important points for understanding food handler behavior. However, all studies note that several factors were not investigated, because the behavioral context influenced the adoption of appropriate practices in food handling and few authors were motivated to identify these factors.

The training sessions, which focused on monitoring food safety practices, cause positive behavior to happen naturally (Da Cunha et al., 2013). According to Soon and Baines (2012), the recycling of training and the recurrent emphasis on the behavior of good handling practices has a positive effect on the hand-washing practice of handlers.

The effectiveness of food hygiene training is improved if the training is based on a more appropriate approach, and if it is designed on the basis of health education theories (Ehiri et al., 1997). Chapman et al. (2011) reasoned that the use of interactive media has a positive influence on food safety training.

Behavioral strategies open new ways of diagnosis and training; however, when applied in isolation, they become trapped or bound to the researcher. When greater involvement is intended, the approach of the research action can facilitate a diagnosis and a background in shared construction.

The goals of action research are represented in the following two ways: practical and theoretical. It is important that research addresses the proper relation and balance of both approaches. Good knowledge helps to achieve good practice. The purpose of knowledge is more focused on research and the willingness to expand the level of food handler awareness and produce knowledge. The practical goal is focused on the action itself, occurring as a concrete action, and an investigation of the situation and/or practices in food safety with an interaction between the researchers and workers.

When hosting all diagnostic strategies and training, this model also accepts others that may be created from the action research process. The meeting between the researcher and the handler can bring new solutions to food safety. It is important to know that reflective practice can be a useful precursor to action research. Reflective practice can lead to strategic action (McMahon, 1999).

The techniques used in training should be reviewed by minimizing theoretical concepts, a widely used feature (Medeiros et al., 2011). Studies conducted on this topic provide suggestions that may be fruitful for practical improvement or that can maximize the following intervention strategies:

(1) Train food handlers in the workplace to improve their understanding of procedures. In the workplace, a tutor can reinforce theoretical ideas in a practical manner (Da Cunha et al., 2013; Rennie, 1995).
(2) Involve the owner of the property or manager/supervisor in interventions regarding good practices. A supervisor with knowledge can assist in monitoring and correcting improper practices (Egan et al., 2007; Seaman and Eves, 2006).
(3) Provide a suitable environment, with the necessary resources, equipment, and utensils for the implementation of good manufacturing practices (Tannenbaum and Yukl, 1992; Tracey et al., 1995).
(4) Provide motivation. Feelings such as motivation and self-efficacy play important roles in enhancing training and the adoption of adequate practices (Seaman and Eves, 2006).
(5) Establish goals and present them to the group. A goal, when feasible, may motivate the fulfillment of tasks (Ray et al., 1997).
(6) Evaluate the impact of the acquired knowledge. The evaluation of results allows for the development of their own training methodologies (Garayoa et al., 2011; Soares et al., 2012).
(7) Select the content of the training. Irrelevant information reduces the interest of the handler (Abdul-Mutalib et al., 2012).

Some factors can negatively affect the impact of training and interventions performed in food services, such as hiring employees with a low socioeconomic status and low education, a high employee turnover, the cost of interventions for food safety, and dissatisfaction with pay (Mortlock et al., 2000; Seaman and Eves, 2006). Therefore, these factors must be evaluated with the intention to cause changes; for example, improving the socioeconomic situation and decreasing worker turnover.

In a study by Da Cunha et al. (2014b), the number of meals prepared per handler also had a negative effect on appropriate practices. The higher the workload is (i.e., the greater the number of meals produced by a food handler) the lower will be the number of correct food safety practices. The overloaded food handler may neglect correct practices, favoring practices that facilitate or accelerate the production of meals. In this study, the experience of food handlers positively affected the observed practices, independent of knowledge. The experienced food handler may be accustomed to techniques and procedures and know how to reproduce them. However, it is not possible to say whether the years of work that food handlers perform can change their attitudes and make them adopt appropriate practices in a voluntary manner.

Adequate buildings and facilities were also positively associated with the observed practices. Another significant factor associated with the observed practices was the presence of food safety leadership working in foodservice. With a specific curriculum addressing hygiene, food safety, and food microbiology, the nutritionist can perform the management and maintenance of good manufacturing practices in food services in Brazil (Da Cunha et al., 2013). In this sense, the nutritionist, and other professionals with the aforementioned knowledge, can monitor internal activities by acting as a food safety leader, which was already shown to influence food safety adequacy in a positive fashion (Tannenbaum and Yukl, 1992; Seaman and Eves, 2006; Egan et al., 2007; Da Cunha et al., 2013). For Griffith et al. (2010), food safety leadership is the ability of a leader to engage staff in hygiene/safety performance and encourage compliance in meeting goals and standards. In addition to monitoring, this professional can perform frequent evaluations. The assessment of food safety practices is itself a factor that enhances food safety in food handling (Bader et al., 1978).

6 VIRTUAL ENVIRONMENTS

Some countries such as Brazil and the United Kingdom are investing in training courses for food handlers via the Internet as a strategy to improve the knowledge of people interested in safe food handling.

In Brazil, ANVISA (National Health Surveillance Agency) launched a completely virtual 12-h course for food handlers with motivational and educational strategies in 2014. Videos and exercises were used for the assimilation of knowledge and for a dynamic, easy-to-understand presentation with a review of the content. The Brazilian course is divided into eight modules, and the participants can print a booklet with all the content and a certificate of participation. In the end, the course is assessed by the participants, facilitating the improvement of future versions.

7 CONSUMERS

A strategy that resonates with positive results for food establishments is the education of the population itself in matters related to proper food-handling practices.

This practice generates more critical and demanding consumers because they may also be food handlers. In a study performed by Deon et al. (2014), educational activities in the form of good practice programs for food preparation in households are of fundamental importance.

Such programs should be developed with consistent methodologies to meet their objectives, and they should consider the possible causes of failure. In addition, schools are the ideal environment for early interventions in hygiene education, assuming that childhood is the best time for learning. The dissemination of knowledge as early as possible in schools is a concrete method of risk communication because it builds trust and credibility.

In this sense, the U.S. Department of Agriculture (USDA) has directed its efforts to promote and educate the population about issues related to food safety, and it includes among its actions those related to the virtual environment. The USDA hotline responded to over 80,000 inquiries (calls, chats, and webmail) in fiscal year 2013. The USDA is also a major contributor of food safety consumer content for the site FoodSafety.gov, the primary initiative of the Working Group on Food Security. To engage consumers further on the importance of food safety at home, the USDA joined with the Ad Council to develop Families Safe Food Campaign, a national public education campaign (Fight BAC). The campaign encourages consumers to take four simple steps to protect themselves and their families from FBD. This program, which is in its fourth year, has been seen or heard by more than 1 billion people since its launch in June 2011 (United States Department of Agriculture, 2014).

8 FINAL CONSIDERATIONS

Scientific research has been directed at improving food-handling practices throughout the entire food chain, from primary production to the consumer's table. The legislation and regulations mention that the training of food handlers is critical for preventing FBD, and they recommend the need for continuous education actions for all involved, with records kept of these activities. Continuing training should be planned and enhanced to emphasize content directed at health risks and for improving the knowledge, attitudes, and practices of food handlers.

There is an observable scientific basis for improving food handler training in pursuit of updated theoretical and traditional models, and more studies are needed. In this context, the management steps used in the training of food handlers include a diagnosis of the demand, planning, and implementation of intervention activities and continuous assessment of the effectiveness of the program or training plan. This program should be documented, and all actions should be recorded to ensure execution.

Inspection practices remain important for the reduction of foodborne outbreaks. However, inspections alone do not reduce the risk of FBD. A combination of inspection results, mandatory training programs, and certifications can minimize this type of risk. Current evidence for the effectiveness of food hygiene training is limited. Studies with meaningful performance indicators are needed to elucidate the relations between training and adequate food-handling practices.

Educational programs, and not necessarily training in its traditional format or function, can be developed more effectively by considering the different variables related to food handlers. These variables can be observed in Figure 7.

Highlights for effective training in food safety:

- To establish performance expectations.
- To facilitate communication among employees, managers, and employers.
- To enroll people with skills in human resource management and give them knowledge of strategies.
- To provide encouragement for employers.
- To provide leadership for food safety.
- To elicit motivation, by stimulating feelings such as self-empowerment and self-efficacy.
- To enroll professionals with a curriculum including hygiene, food safety, and food microbiology disciplines.
- To establish and respect an adequate workload.
- To use languages according to food handler ethnicity.
- To train food handlers in their workplace.
- To provide an adequate infrastructure.
- To have similar workplace values.
- To provide a suitable environment, with the necessary resources, equipment, and utensils.
- To illustrate theoretical concepts with practical activities.
- To use a variety of diagnostic tools.
- To apply health educational theories.
- To achieve a consensus on food safety.
- To embody the concept of risk.
- To apply an action research approach.
- To establish goals and present them to the group.
- To work with consequences and incentives.
- To provide compelling, rapid, relevant, reliable, and repeated messages.
- To use interactive media.

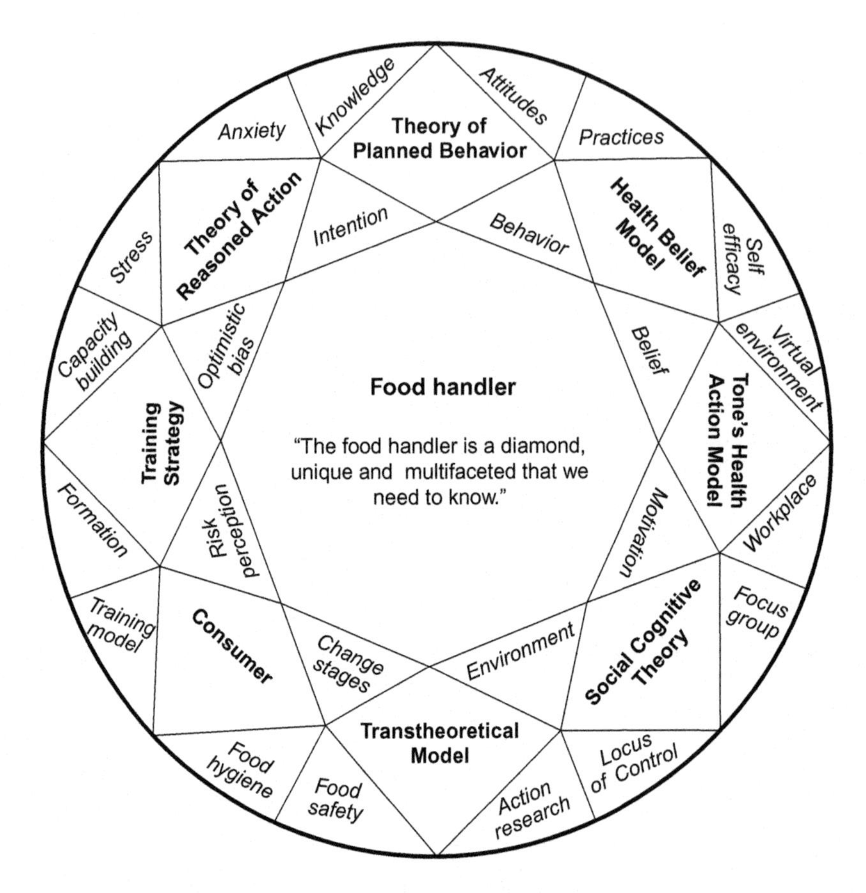

FIGURE 7

Different variables related to food handlers.

- To develop specific legislation.
- To perform continuous evaluation.
- To work toward purposes of social, cultural, economic, and political development.

The implementation of food safety practice requires adequate resources and an appropriate management culture. Food safety may be a field in which education and training can be organized to integrate the behavioral sciences with public health.

The creation of food safety within a business means supporting an environment in which food employees know about risks and risk management, and how to value consumer health. A food safety culture requires an application of the science with the best management and communication systems. This oversight includes compelling, relevant, rapid, reliable, and repeated messages.

Frank Yiannas proposed a systems approach entirely based on the scientific knowledge of human behavior, organizational culture, and food safety, called the Behavior-Based Food Safety Management System.

We have deconstructed the meaning of "to train" and "to intervene" here, by redirecting it into "to form." Punctual strategies tend to be forgotten quickly. To increase effectiveness, it is necessary to address the food handler as a whole person, in all his multifactorial integrality. The educational approach must contemplate, beyond formal training, the social relationships, personal perceptions, feelings, and experiences that shape their behavior. To ally such an understanding to a permanent work program valuation and food safety practice improvement could be the way to minimize inadequate practices and negligence.

ACKNOWLEDGMENT

Dr. Marcos Alberto Taddeo Cipullo—Psychologist.

REFERENCES

Abbot, J.M., Byrd-Bredbenner, C., Schaffner, D., Bruhn, C.M., Blalock, L., 2009. Comparison of food safety cognitions and self-reported food-handling behaviors with observed food safety behaviors of young adults. Eur. J. Clin. Nutr. 63, 572–579.

Abdul-Mutalib, N.A., Abdul-Rashid, M.F., Mustafa, S., Amin-Nordin, S., Hamat, R.A., Osmana, M., 2012. Knowledge, attitude and practices regarding food hygiene and sanitation of food handlers in Kuala Pilah, Malaysia. Food Control 27, 289–293.

Ajzen, I., 1991. The theory of planned behavior. Organ. Behav. Hum. Decis. Process. 50, 179–211.

Albarracín, D., Johnson, B.T., Fishbein, M., Muellerleile, P.A., 2001. Theories of reasoned action and planned behavior as models of condom use: a meta-analysis. Psychol. Bull. 127, 142–161.

Angelillo, I.F., Viggiani, N.M.A., Rizzo, L., Bianco, A., 2000. Food handlers and foodborne diseases: knowledge, attitudes, and reported behavior in Italy. J. Food Prot. 63, 381–385.

Ansari-Lari, M., Soodbakhsh, S., Lakzadeh, L., 2010. Knowledge, attitudes and practices of workers on food hygienic practices in meat processing plants in Fars, Iran. Food Control 21, 260–263.

Arendt, S.W., Sneed, J., 2008. Employee motivators for following food safety practices: pivotal role of supervision. Food Prot. Trends 28, 704–711.

Arendt, S.W., Roberts, K.R., Strohbehn, C., Ellis, J., Paez, P., Meyer, J., 2012. Use of qualitative research in foodservice organizations a review of challenges, strategies, and applications. Int. J. Contemp. Hosp. Manag. 24, 820–837.

Askarian, M., Kabir, G., Aminbaig, M., Memish, Z., Jafari, P., 2004. Knowledge, attitudes, and practices of food service staff regarding food hygiene in Shiraz, Iran. Infect. Control Hosp. Epidemiol. 25, 16–20.

Bader, M., Blonder, E., Henriksen, J., Strong, W., 1978. A study of food service establishment sanitation inspection frequency. Am. J. Public Health 68, 408–410.

Bandura, A., 1977. Self-efficacy: toward a unifying theory of behavioral change. Psychol. Rev. 84, 191–215.

Bandura, A., 2004. Health promotion by social cognitive means. Health Educ. Behav. 31, 143–164.

Bas, M., Ersun, A.S., Kivanç, G., 2006. The evaluation of food hygiene knowledge, attitudes, and practices of food handlers in food businesses in Turkey. Food Control 17, 317–322.

Behrens, J.H., Barcellos, M.N., Frewer, L.J., Nunes, T.P., Franco, B.D.G.M., Destro, M.T., Landgraf, M., 2010. Consumer purchase habits and views on food safety: a Brazilian study. Food Control 21, 963–969.

Bezerra, A.C.D., Mancuso, A.M.C., Heitz, S.J.J., 2014. Street food in the national agenda of food and nutrition security: an essay for sanitary qualification in Brazil. Cien. Saude Colet. 19, 1489–1494.

Brazil Health Ministry, 2004. Resolution no. 216 of the 15th of semptember 2004. Establishes the technical regulation of good manufacturing practices to foodservices.

Brazil Health Ministry, 2014. Information for Diseases Notification System (SINAN net).

Brewer, N.T., Chapman, G.B., Gibbons, F.X., Gerrard, M., McCaul, K.D., Weinstein, N.D., 2007. Meta-analysis of the relationship between risk perception and health behavior: the example of vaccination. Health Psychol. 26, 136–145.

Byrd-Bredbenner, C., Maurer, J., Wheatley, V., Schaffner, D., Bruhn, C., Blalock, L., 2007a. Food safety self-reported behaviors and cognitions of young adults: results of a national study. J. Food Prot. 70, 1917–1926.

Byrd-Bredbenner, C., Wheatley, V., Schaffner, D., Bruhn, C., Blalock, L., Maurer, J., 2007b. Development of food safety psychosocial questionnaires for young adults. J. Food Sci. Educ. 6, 30–37.

Carmo, G.M.I., Oliveira, A.A., Dimech, C.P., Santos, D.A., Almeida, M.G., Berto, L.H., Alves, R.M.S., Carmo, E.H., 2005. Vigilância epidemiológica das doenças transmitidas por alimentos no Brasil, 1999-2004. Boletim Eletrônico Epidemiológico 6, 1–7. Disponível em: http://portal.saude.gov.br/portal/arquivos/pdf/bol_epi_6_2005_corrigido.pdf (acesso em: 20 set. 2014).

Cenci-Goga, B.T., Ortenzi, R., Bartocci, E., Oliveira, A.C., Clementi, F., Vizzani, A., 2005. Effect of the implementation of HACCP on the microbiological quality of meals at a university restaurant. Foodborne Pathog. Dis. 2, 138–145.

Centers for Disease Control and Prevention, 2011. 2011 Estimates of foodborne illness in the United States. Available at: http://www.cdc.gov/Features/dsFoodborneEstimates (accessed 13.03.14).

Centers for Disease Control and Prevention, 2012. CDC research shows outbreaks linked to imported foods increasing. For Immediate Release, March 14, 2012. Available at: http://www.cdc.gov/media/releases/2012/p0314_foodborne.html (accessed 13.03.14).

Chan, S.F., Chan, Z.C.Y., 2008. A review of foodborne disease outbreaks from 1996 to 2005 in Hong Kong and its implications on food safety promotion. J. Food Saf. 28, 276–299.

Chapman, B., MacLaurin, T., Powell, D., 2011. Food safety infosheets: design and refinement of a narrative-based training intervention. Br. Food J. 113, 160–186.

Chock, T.M., 2011. The influence of body mass index, sex, and race on college students' optimistic bias for lifestyle healthfulness. J. Nutr. Educ. Behav. 43, 331–338.

Choudhury, M., Mahanta, L.B., Goswami, J.S., Mazumder, M.D., 2011. Will capacity building training interventions given to street food vendors give us safer food?: a cross-sectional study from India. Food Control 22, 1233–1239.

Clayton, D.A., Griffith, C.J., 2008. Efficacy of an extended theory of planned behaviour model for predicting caterers' hand hygiene practices. Int. J. Environ. Health Res. 18, 83–98.

Clayton, D.A., Griffith, C.J., Price, P., Peters, A.C., 2002. Food handlers' beliefs and self-reported practices. Int. J. Environ. Health Res. 12, 25–39.

Codex Alimentarius, 2013. Food Hygiene (Basic Texts), fifth ed. Food and Drug Administration of The United Nations, Rome.

Cotterchio, M., Gunn, J., Coffill, T., Tormey, P., Barry, M.A., 1998. Effect of a manager training program on sanitary conditions in restaurants. Public Health Rep. 113, 353–358.

Crims, S.M., et al., 2014. Incidence and trends of infection with pathogens transmitted commonly through food – foodborne diseases active surveillance network, 10 U.S. Sites, 2006–2013. Morb. Mortal. Wkly. Rep. 63, 328–332.

Da Cunha, D.T., Stedefeldt, E., De Rosso, V.V., 2012. Perceived risk of foodborne disease by school food handlers and principals: the influence of frequent training. J. Food Saf. 32, 219–225.

Da Cunha, D.T., Fiorotti, R.M., Baldasso, J.G., de Sousa, M., Fontanezi, N.M., Caivano, S., Stedefeldt, E., de Rosso, V.V., Camargo, M.C.R., 2013. Improvement of food safety in school meal service during a long-term intervention period: a strategy based on the knowledge, attitude and practice triad. Food Control 34, 662–667.

Da Cunha, D.T., Stedefeldt, E., de Rosso, V.V., 2014a. He is worse than I am: the positive outlook of food handlers about foodborne disease. Food Qual. Prefer. 35, 95–97.

Da Cunha, D.T., Stedefeldt, E., de Rosso, V.V., 2014b. The role of theoretical food safety training on Brazilian food handlers' knowledge, attitude and practice. Food Control 43, 167–174.

Deon, B.C., Medeiros, L.B., Saccol, A.L.F., Hecktheuer, L.H., Saccol, S., Naissinger, M., 2014. Good food preparation practices in households: a review. Trends Food Sci. Technol. 39, 40–46.

Dijk, H.V., Fischer, A.R.H., Frewer, L.J., 2011. Consumer responses to integrated risk-benefit information associated with the consumption of food. Risk Anal. 31, 429–439.

Eagly, A.H., Chaiken, S., 2007. The advantages of an inclusive definition of attitude. Soc. Cogn. 25, 582–602.

Egan, M.B., Raats, M.M., Grubb, S.M., Eves, A., Lumbers, M.L., Dean, M.S., Adams, M.R., 2007. A review of food safety and food hygiene training studies in the commercial sector. Food Control 18, 1180–1190.

Ehiri, J.E., Morris, G.P., 1996. Hygiene training and education of food handlers: does it work? Ecol. Food Nutr. 35, 243–251.

Ehiri, J.E., Morris, G.P., McEwen, J., 1997. Evaluation of a food hygiene training course in Scotland. Food Control 8, 137–147.

Ellis, J.D., Arendt, S.W., Strohbehn, C.H., Meyer, J., Pae, P., 2010. Varying influences of motivation factors on employees' likelihood to perform safe food handling practices because of demographic differences. J. Food Prot. 73, 2065–2071.

European Food Safety Authority, 2013. Technical report. Focal point activities 2013, Parma, Italy. Available at: http://www.efsa.europa.eu/fr/supporting/doc/570e.pdf (accessed 13.03.14).

Ferreira, A.B.H., 1986. Novo dicionário da Língua Portuguesa, second ed. Nova Fronteira, Rio de Janeiro.

Food and Agriculture Organization of the United Nations, World Health Organization, 1999. The application of risk communication to food standards and safety matters: report of a FAO/WHO Consultation, Food and Nutrition. Disponível em: ftp://ftp.fao.org/docrep/fao/005/x1271e/x1271e00.pdf.

Food and Agriculture Organization of the United Nations World Health Organization, 2006. Five keys to safer food manual. WHO Department of Food Safety, Zoonoses and Foodborne Diseases, Disponível em: http://www.who.int/foodsafety/publications/consumer/manual_keys.pdf.

Food and Drug Administration, 2009. FDA report on the occurrence of foodborne illness risk factors in selected institutional foodservice, restaurant, and retail food store facility types. In: FDA Retail Food Program Steering Committee (Ed.), United States of America.

Frewer, L.J., Shepherd, R., Sparks, P., 1994. The interrelationship between perceived knowledge, control and risk associated with a range of food-related hazards targeted at the individual, other people and society. J. Food Saf. 14, 19–40.

Garayoa, R., Vitas, A.I., Díez-Leturia, M., García-Jalón, I., 2011. Food safety and the contract catering companies: food handlers, facilities and HACCP evaluation. Food Control 22, 2006–2012.

Garcia, P.P.C., Akutsu, R.C.C.A., Oliveira K.E.S., Silva, I.C.R., 2013. The effectiveness of food handlers training: the Transtheoretical Model in focus. 161 f. Dissertation (Master in Human Nutrition) – Graduate Program in Human Nutrition, Faculty of Health Sciences, University of Brasilia, Brazil.

Gilling, S.J., Taylor, E.A., Kane, K., Taylor, J.Z., 2001. Successful hazard analysis critical control point implementation in the United Kingdom: understanding the barriers through the use of a behavioral adherence model. J. Food Prot. 64, 710–715.

Gomes, C.C.B., Lemos, G.F.C., Silva, M.C., Hora, I.M.C., Cruz, A.G., 2014. Training of food handlers in a hotel: tool for promotion of the food safety. J. Food Saf. 34, 218–223.

Green, L., Selman, C., Banerjee, A., Marcus, R., Medus, C., Angulo, F.J., Radke, V., Buchanan, S., 2005. Food service workers' self-reported food preparation practices: an EHS-Net study. Int. J. Environ. Health Res. 208, 27–35.

Greig, J.D., Todd, E.C.D., Bartleson, C.A., Michaels, B.S., 2007. Outbreaks where food workers have been implicated in the spread of foodborne disease. part 1. Description of the problem, methods, and agents involved. J. Food Prot. 70, 1752–1761.

Griffith, C.J., Livesey, K.M., Clayton, D., 2010. The assessment of food safety culture. Br. Food J. 112, 439–456.

Helweg-Larsen, M., Shepperd, J.A., 2001. Do moderators of the optimistic bias affect personal or target risk estimates? A review of the literature. Pers. Soc. Psychol. Rev. 5, 74–95.
Hislop, N., Shaw, K., 2009. Food safety knowledge retention study. J. Food Prot. 72, 431–435.
Hoffmann, S., Batz, M.B., Morris Jr., J.G., 2012. Annual cost of illness and quality-adjusted life year losses in the United States due to 14 foodborne pathogens. J. Food Prot. 7, 1184–1358.
Hoorens, V., Buunk, P.B., 1993. Social comparison of health risks: locus of control, the person-positivity bias, and unrealistic optimism. J. Appl. Soc. Psychol. 23, 291–302.
Horswill, M.S., McKenna, F.P., 1999. The effect of perceived control on risk taking. J. Appl. Soc. Psychol. 29, 377–391.
Houaiss, A., 2001. Dicionário Houaiss da Língua Portuguesa. Objetiva, Rio de Janeiro.
Howells, A.D., Roberts, K.R., Shanklin, C.W., Pilling, V.K., Brannon, L.A., Barrett, B.B., 2008. Restaurant employees' perceptions of barriers to three food safety practices. J. Am. Diet. Assoc. 108, 1345–1349.
Iiu, S., Iiu, Z., Zhang, H., Lu, L., Liang, J., Huang, Q., 2015. Knowledge, attitude and practices of food safety amongst food handlers in the coastal resort of Guangdong, China. Food Control 47, 457–461.
International Organization for Standardization, 2005. ISO 22000: Food Safety Management Systems-Requirements for Any Organization in the Food Chain. Organization for Standardization, International, 40 pp.
International Organization for Standardization, 2009. ISO 31010 – Risk management – Risk Assessment Techniques. ISO, Geneva, 96 pp.
Janz, N.K., Becker, M.H., 1984. The Health Belief Model – a decade later. Health Educ. Q. 11, 1–47.
Jianu, C., Golet, I., 2014. Knowledge of food safety and hygiene and personal hygiene practices among meat handlers operating in western Romania. Food Control 42, 214–219.
Kafertein, E.K., Motarjemi, Y., Bettcher, D.W., 1997. Foodborne disease control: a transnational challenge. Emerg. Infect. Dis. 3, 503–510.
Kaliyaperumal, K., 2004. Guideline for conducting a knowledge, attitude and practice (KAP) study. AECS Illum. 4, 7–9.
Karaman, A.D., Cobanoglu, F., Tunalioglu, R., Ova, G., 2012. Barriers and benefits of the implementation of food safety management systems among the Turkish dairy industry: a case study. Food Control 25, 732–739.
Kelly, R.B., Zyzanski, S.J., Alemagno, S.A., 1991. Prediction of motivation and behavior-change following health promotion – role of health beliefs, social support, and self-efficacy. Soc. Sci. Med. 32, 311–320.
Knox, B., 2000. Consumer perception and understanding of risk from food. Br. Med. Bull. 56, 97–109.
Ko, W.H., 2013. The relationship among food safety knowledge, attitudes and self-reported HACCP practices in restaurant employees. Food Control 29, 192–197.
Lin, H., 2007. Effects of extrinsic and intrinsic motivation on employee knowledge sharing intentions. J. Inf. Sci. 15, 1–15.
Macías, Y.F., Glasauer, P., 2014. Guidelines for Assessing Nutrition-Related Knowledge, Attitudes and Practices. Food and Agriculture Organization of the United Nations, Rome, Italy.
Mari, S., Tiozzo, B., Capozza, D., Ravarotto, L., 2012. Are you cooking your meat enough? The efficacy of the Theory of Planned Behavior in predicting a best practice to prevent salmonellosis. Food Res. Int. 45, 1175–1183.
McIntyre, L., Vallaster, L., Wilcott, L., Henderson, S.B., Kosatsky, T., 2013. Evaluation of food safety knowledge, attitudes and self-reported hand washing practices in FOODSAFE trained and untrained food handlers in British Columbia, Canada. Food Control 30, 150–156.
McMahon, T., 1999. Is reflective practice synonymous with action research? Educ. Action Res. 7, 163–169.
Medeiros, C.O., Cavalli, S.B., Salay, E., Proença, R.P.C., 2011. Assessment of the methodological strategies adopted by food safety training programmes for food service workers: a systematic review. Food Control 22, 1136–1144.
Miles, S., Scaife, V., 2003. Optimistic bias and food. Nutr. Res. Rev. 16, 3–19.
Miles, S., Braxton, D.S., Frewer, L.J., 1999. Public perceptions about microbiological hazards in food. Br. Food J. 101, 744–762.

Mortlock, M.P., Peters, A.C., Griffith, C.J., 2000. A national survey of food hygiene training and qualification levels in the UK food industry. Int. J. Environ. Health Res. 10, 111–123.
Mullan, B.A., Wong, C., Kothe, E.J., 2010. Predicting adolescents' safe food handling using an extended theory of planned behavior. Food Control 31, 454–460.
Nauta, M.J., Fischer, A.R.H., Asselt, E.D.V., Jong, A.E.I., Frewer, L.J., Jonge, R., 2008. Food safety in the domestic environment: the effect of consumer risk information on human disease risks. Risk Anal. 28, 179–192.
Newby-Clark, I.R., McGregor, I., Zanna, M.P., 2002. Thinking and caring about cognitive inconsistency: when and for whom does attitudinal ambivalence feel uncomfortable? J. Pers. Soc. Psychol. 82, 157–166.
Nieto-Montenegro, S., Brown, J.L., LaBorde, L.F., 2008. Development and assessment of pilot food safety educational materials and training strategies for Hispanic workers in the mushroom industry using the Health Action Model. Food Control 19, 616–633.
Osaili, T.M., Abu Jamous, D.O., Obeidat, B.A., Bawadi, H.A., Tayyem, R.F., Subih, H.S., 2013. Food safety knowledge among food workers in restaurants in Jordan. Food Control 31, 145–150.
Ozfoodnet, 2010. Monitoring the incidence and causes of diseases potentially transmitted by food in Australia: annual report. Network, 2010. Available at: http://www.health.gov.au/internet/main/publishing.nsf/Content/cda-cdi3603-pdf-cnt.htm/$FILE/cdi3603a.pdf. (accessed 13.03.14).
Pan-American Health Organization, 2008. PAHO Strategic Plan 2008–2012. Pan-American Health Organization, Washington, DC, USA.
Pan-American Health Organization, 2013. PAHO Strategic Plan 2014–2019. Pan-American Health Organization, Washington, DC, USA.
Park, S.H., Kwak, T.K., Chang, H.J., 2012. Evaluation of the food safety training for food handlers in restaurants operations. Nutr. Res. Pract. 4, 58–68.
Parry, S.M., Miles, S., Tridente, A., Palmer, S.R., South and East Wales Infectious Disease Group, 2004. Differences in perception of risk between people who have and have not experienced salmonella food poisoning. Risk Anal. 24, 289–299.
Pearsall, J., 2001. Concise Oxford, tenth ed. Oxford University Press, Oxford, UK.
Pichler, J., Ziegler, J., Aldrian, U., Allerberger, F., 2014. Evaluating levels of knowledge on food safety among food handlers from restaurants and various catering businesses in Vienna, Austria 2011/2012. Food Control 35, 33–40.
Prochaska, J., Redding, C., Evers, K., 2002. The transtheoretical model and stages of change. In: Glanz, K., Rimer, B., Lewis, F. (Eds.), Health Behavior and Health Education: Theory, Research, and Practice. Jossey-Bass, San Francisco, pp. 99–120.
Public Health Surveillance, 2014. Summary of Outbreaks in New Zealand. Institute of Environmental Science and Research Limited, Porirua, New Zealand.
Raats, M.M., Sparks, P., Shepherd, R., 1999. The effects of providing personalized dietary feedback a semi-computerized approach. Patient Educ. Couns. 37, 177–189.
Ray, P.S., Bishop, P.A., Wang, M.Q., 1997. Efficacy of the components of a behavioral safety program. Int. J. Ind. Ergon. 19, 19–29.
Rennie, D.M., 1995. Health-education models and food hygiene education. J. R. Soc. Health 115, 75–79.
Rosenstock, I.M., 1974. Historical origins of the health belief model. Health Educ. Monogr. 2, 328–335.
Sani, N.A., Siow, O.N., 2014. Knowledge, attitudes and practices of food handlers on food safety in food service operations at the Universiti Kebangsaan Malaysia. Food Control 37, 210–217.
Sant'ana, A.S., Franco, B.D.G.M., 2009. Microbial quantitative risk assessment of foods: concepts, systematics and applications. Braz. J. Food Technol. 12, 266–276.
Scallan, E., Griffin, P.M., Angulo, F.J., Tauxe, R.V., Hoekstra, R.M., 2011. Foodborne illness acquired in the United States—unspecified agents. Emerg. Infect. Dis. 17, 16–22.
Seaman, P., 2010. Food hygiene training: introducing the food hygiene training model. Food Control 21, 381–387.
Seaman, P., Eves, A., 2010. Efficacy of the theory of planned behaviour model in predicting safe food handling practices. Food Control 21, 983–987.

Seaman, P., Eves, A., 2006. The management of food safety the role of food hygiene training in the UK service sector. IJHM 25, 278–296.

Sjoberg, L., 2000. Factors in risk perception. Risk Anal. 20, 1–11.

Slovic, P., 1987. Perception of risk. Science 236, 280–285.

Slovic, P., Finucane, M.L., Peters, E., MacGregor, D.G., 2004. Risk as analysis and risk as feelings: some thoughts about affect, reason, risk, and rationality. Risk Anal. 24, 311–322.

Soares, L.S., Almeida, R.C.C., Cerqueira, E.S., Carvalho, J.S., Nunes, I.L., 2012. Knowledge, attitudes and practices in food safety and the presence of coagulase-positive staphylococci on hands of food handlers in the schools of Camacari, Brazil. Food Control 27, 206–213.

Soares, K., García-Díez, J., Esteves, A., Oliveira, I., Saraiva, C., 2013. Evaluation of food safety training on hygienic conditions in food establishments. Food Control 34, 613–618.

Soon, J.M., Baines, R.N., 2012. Food safety training and evaluation of handwashing intetion among fresh produce farm workes. Food Control 23, 437–448.

Sparks, P., Shepherd, R., 1994. Public perception of the potential hazards associated with food production and food consumption: an empirical study. Risk Anal. 14, 799–806.

Tan, S.L., Bakar, F.A., Karim, M.S.A., Lee, H.Y., Mahyudin, N.A., 2013. Hand hygiene knowledge, attitudes and practices among food handlers at primary schools in Hulu Langat district, Selangor (Malaysia). Food Control 34, 428–435.

Tannenbaum, S.I., Yukl, G.A., 1992. Training and development in work organizations. Annu. Rev. Psychol. 43, 399–441.

Tracey, J.B., Tannenbaum, S.I., Kavanagh, M.J., 1995. Applying trained skills on the job – the importance of the work-environment. J. Appl. Psychol. 80, 239–252.

United States Department Agriculture, 2014. Food Safety. Available at: http://www.usda.gov/wps/portal/usda/usdahome?navid=food-safety (accessed 13.03.14).

Wachinger, G., Renn, O., Begg, C., Kuhlicke, C., 2013. The risk perception paradox—implications for governance and communication of natural hazards. Risk Anal. 33, 1049–1065.

Wallston, K.A., 2005. The validity of the multidimensional health locus of control scales. J. Health Psychol. 10, 623–663.

Wallston, K.A., Wallston, B.S., DeVellis, R., 1978. Development of the multidimensional health locus of control (MHLC) scales. Health Educ. Monogr. 6, 160–170.

Weinstein, N.D., 1984. Why it wont happen to me – perceptions of risk-factors and susceptibility. Health Psychol. 3, 431–457.

Weinstein, N.D., 1987. Unrealistic optimism about susceptibility to health-problems – conclusions from a community-wide sample. J. Behav. Med. 10, 481–500.

Weinstein, N.D., 1989. Optimistic biases about personal risks. Science 246, 1232–1233.

World Health Organization, 1988. Health Education in Food Safety. WHO/EHE/FOS/88.7.

World Health Organization, 2008. The World Health Report 2008: Primary Health Care: Now More Than Ever. World Health Organization, Geneva, Switzerland.

World Health Organization, 2012. Manual for integrated foodborne disease surveillance in the WHO African region.

World Health Organization, 2013. Strategic plan for food safety including foodborne zoonoses 2013–2022.

Yiannas, F., 2009. Food Safety Culture: Creating a Behavior-Based Food Safety Management System. Springer Science, New York.

Yoe, C., 2012. Primer on Risk Analysis. CRC Press, United States of America.

Zain, M.M., Naing, N.N., 2002. Sociodemographic characteristics of food handlers and their knowledge, attitude and practice towards food sanitation: a preliminary report. Southeast Asian J. Trop. Med. Public Health 33, 410–417.

CHAPTER

19

ANAEROBIC MICROBIOLOGY LABORATORY TRAINING AND WRITING COMPREHENSION FOR FOOD SAFETY EDUCATION

Steven C. Ricke

Department of Food Science, Center for Food Safety, University of Arkansas, Fayetteville, Arkansas, USA

1 INTRODUCTION

Food safety represents a complex combination of scientific disciplines including not just the basic tenets of microbiology such as physiology and genetics, but also can incorporate seemingly unrelated concepts ranging from engineering and immunology to biochemistry, chemistry, and mathematics. Consequently, studying microorganisms associated with either food animals, vegetables in the field, or the resulting food products requires a high level of integration to develop an in-depth understanding of the entire spectrum of potentially interactive impact factors that influence the respective consortia of organisms present in a particular food production niche. While the involvement of microorganisms in food production is as ancient as the history of the cultivation and preparation of food itself, identifying the specific roles that the organisms play in this process is a much more recent phenomena. Certainly this is true of food fermentations and is becoming true for foodborne pathogens as well. This is in large part due to the explosion of knowledge evolving from the developments in genetics that led to a revolution in the number of applications relevant to food microbiology, including such technologies as rapid molecular-based assays for detection and quantitation of specific foodborne pathogens to in-depth sequence analyses of many of the important food fermentative microorganisms as well as the primary clinically important foodborne pathogens.

While this represents a technological boon to the science of food microbiology and food safety, it also represents somewhat of a difficult problem for food safety education. At the root of this education quandary is the risk that fundamental principles of food microbiology will be overcome by the race to more specialization, to the point that food microbiology training loses its integrative characteristics and, subsequently, an appreciation for the external impacts on organisms in food production. To some extent this may be even more true for food safety, as foodborne pathogens typically have highly involved life cycles in food production ranging from their gastrointestinal relationship with the multitude of animal hosts, both food animals as well as nonfood animals, to the occurrence on foods being

Food Safety. http://dx.doi.org/10.1016/B978-0-12-800245-2.00019-8

processed for retail. Designing educational curricula to accommodate these complexities is difficult not only from a background standpoint but also generating enough practical laboratory experience to handle a wide range of microorganisms is a daunting challenge. To achieve sufficient experience in the most efficient manner requires development of learning skills that incorporate experiential as well as integrative information. Developing an understanding of anaerobic microbiology techniques and principles in parallel with comprehensive writing exercises offers an opportunity to accomplish these objectives. Therefore, to illustrate this, this chapter will discuss the historical development of anaerobic microbiology and the roles it plays in food microbiology and food safety. This discussion will be followed by a description of instructional approaches involving anaerobic and food microbiology and teaching methods for integrating understanding through extensive writing projects.

2 HISTORICAL AND CURRENT STATUS OF MICROBIOLOGY TRAINING IN THE UNITED STATES

Over the past several decades, there has been a decline of college graduates with high-quality training in microbial physiology. There are several reasons for this, both at the undergraduate as well as the graduate level. Certainly, considerable efforts and advancements have been made to improve instructional approaches at the undergraduate educational level with incorporation of more group activities that encourage problem solving along with evidence-based teaching strategies (Suchman, 2015). Likewise, advantages have been identified with student-centered learning modes for science instruction at the elementary school level as well (Granger et al., 2012). Unfortunately, such efforts are impacted and in some cases offset by the rising costs of a college education and the ensuing stresses of student debt (Knisely, 2014; Price, 2014). This cost, in addition to increased flexibility and the ability to work at home, has made less hands-on educational approaches such as distance learning courses more popular (Braun, 2008; O'Bryan et al., 2010). As the popularity of online college course work continues to grow, the loss of integrated multifaceted training in laboratory settings to develop critical thinking will accelerate. This is important as such experiences are critical, not just for developing laboratory skills with actual hands-on activities, but also for the opportunity with laboratory reports to incorporate and introduce students to scientific writing that requires critical thinking and interpretation (Moskovitz and Kellogg, 2011).

At the graduate level, distance education programs are also growing. The attractiveness of being able to take online courses while working as a full-time employee has become an ongoing trend in fields such as food safety as a means to enhance professional credentials (Leake, 2014). More traditional microbiology-based graduate programs still retain considerable laboratory training both through advanced laboratory courses and, obviously, through graduate research. However, specialization has altered the approach and impacted the learning outcomes. For example, training of microbial physiologists in the United States has been neglected by academic microbiology departments and, in some centers, has been almost completely supplanted by molecular biology and genetics. Twenty years ago, this resulted in a serious shortfall of both PhDs and postdoctorals trained to address microbiology questions with physiological techniques and critical thinking (Hinman, 1992). In response to this shortfall, the American Society for Microbiology (Cassell, 1992) proposed increasing industrial contributions to the existing National Institute of General Medical Sciences biotechnology training grants as a means for reversing this trend. While theoretically a worthy cause, in practice this would be difficult to achieve.

In addition to the expense, revamping graduate educational emphasis to answer a current national problem is inevitably handicapped by the potentially slow turnaround response of the educational and administrative systems.

Currently, while the need for conventional microbiology is still prevalent, traditional microbiology still remains fairly neglected, resulting in a shortfall of basic microbiology trained graduates (ASM, 2007). This is unlikely to change given the current scientific research trends. The explosion in molecular biology has now evolved into the -omics era, where not only genomics but proteomics and metagenomics among others are viewed as the solutions to most of the questions being asked in microbiology and related fields (De Keersmaecker et al., 2006; Gilbert et al., 2011; Baker and Dick, 2013). Consequently, -omics has transcended into microbial ecology, industrial microbiology, gastrointestinal microbiology, food microbiology, and human nutrition, just to name a few of the research areas that have been influenced. The human gastrointestinal microbiome and the nutritional implications have been particularly impacted with a flurry of research funding, resulting in the publication of several seminal commentaries and comprehensive review papers (Walter and Ley, 2011; Balter, 2012; Costello et al., 2012; Gordon, 2012; Hood, 2012; Nicholson et al., 2012; Califf et al., 2014). As sequence data become less costly and more readily generated, the academic emphasis will probably continue to drift toward sophisticated bioinformatics interpretative tools (Kwon and Ricke, 2011). However, there are several problems with an exclusively large data set comparison as a primary research tool. Despite the insights gleaned from these interpretations, it has been pointed out in a summary report after an American Society for Microbiology workshop that actual experimental laboratory assessment is still required to not only confirm predictions based on sequence data but also to better understand network integration and potential modulation (ASM, 2007). There are also more long-term impacts as well. When the field of microbiology continues to evolve in the -omics direction, this has a number of long-term ramifications ranging from a philosophical change of emphasis on the type of faculty hired to more practical dilemmas such as the destiny of culture collections maintained by long-term faculty close to retirement (Enquist, 2015).

3 TRADITIONAL MICROBIOLOGY TRAINING: A ROLE FOR FOOD MICROBIOLOGY?

It is becoming more apparent that not only is the expertise drifting away from traditional microbiology physiology emphasis but also the philosophical focus of biology and microbiology departments is also moving in a different direction. The fundamental question is whether there is still a role for conventional microbiology education and training and, if so, where would this take place? Based on the needs of various commercial businesses involved in agriculture, food, or biological applications, it would appear that such training is still critical. However, rather than drastically change the educational infrastructure at the graduate level, this knowledge gap could potentially at least partially be filled by expanding the microbiology component of the food science undergraduate and graduate curricula into more physiology- and ecology-oriented studies, starting with an undergraduate food microbiology survey course that, in turn, could lead into more in-depth senior undergraduate level course(s) followed by specialized graduate level courses. Not only would this be more cost-effective, but also the food sciences are a logical grassroots source for well-trained microbial physiologists who can still take more molecular-oriented coursework in other departments when needed.

The evolvement of the food industry is already closely linked with the study and understanding of microbiological processes and the ecology of the organisms associated with particular foods (Boddy and Wimpenny, 1992; Gould, 1992; Mossel and Struijk, 1992; Rees et al., 1995). Although these advances in food technology may range anywhere from manipulation of fermentation pathways for food preservation to development of quality assurance strategies limiting the dispersion of foodborne pathogens, a fundamental knowledge of microbial physiology has always been required. In addition, the ready accessibility of recombinant DNA methodology has allowed food scientists to not only alter the microbial processes but also to use molecular biology techniques to study the physical and chemical structure of food sources and their further processed products (Kuipers, 1999; De Vos, 2005; O'Flaherty and Klaenhammer, 2011). This has naturally been led to the implementation of food biotechnology and the concept of direct alteration of food components. However, genetic construction of novel organisms is only the beginning; full industrial potential for expression and secretion of products of interest by these organisms is via optimization of physiological conditions. Hence, this training is not only important for biotechnology-oriented food industries but also for any industry seeking to utilize biotechnology for improving animal health care, agriculture, or environmental remediation (Hinman, 1992).

In food science and the field of food microbiology, the training area that may represent the most impact across a broad spectrum of biology disciplines beyond food science is the study of food fermentations. While food fermentations fundamentally describe microbial activities in a somewhat general sense, there has always been some understanding that anaerobic microbial metabolism was involved to some extent. Therefore, incorporating an anaerobic training slant to traditional microbiology education programs has more of a broad-spectrum appeal across multiple disciplines while still remaining easily and naturally identified with food science. Consequently, food fermentation/anaerobic microbiology has a readily identified education niche that, if taught in a comprehensive fashion, can not only teach anaerobic techniques but also encompass many of the same principles associated with microbial growth, physiology, and ecology that have currently been disappearing from modern microbiology curricula. The following sections describe the development of anaerobic microbiology in the understanding of food fermentations and the prominent role it is beginning to play with associated fields of study such as foodborne pathogen ecology and the emergence of a major commercial probiotic industry.

4 FOOD MICROBIAL FERMENTATIONS AND THE ROLE OF ANAEROBIC MICROBIOLOGY

Food microbial fermentations have had a prominent role in the long history of the development of cultivated food harvesting, storage, and preparation. Long before, it was understood that microorganisms were involved, methods for preservation and the prevention of spoilage were discovered and established (Ayres et al., 1980; Banwart, 1989; Jay, 2000; Montville and Matthews, 2005, 2008). Likewise, the art of indigenous food fermentations involving not only the production of alcoholic beverages but also fermented meats, vegetables, and dairy products also emerged early in the development of civilization and agriculture (Wood, 1985; Brock and Madigan, 1991; Nout and Rombouts, 1992; Steinkraus, 1995; Vaughan et al., 2005; Nout et al., 2007; Breidt et al., 2013; Campbell, 2013; Johnson and Steele, 2013; Parish and Fleet, 2013; Ricke et al., 2013). Certainly, a role still exists and perhaps the need is even expanding as locally produced food products, some of which require processing or some type of fermentation, are becoming more popular. Likewise, the concept of microbreweries with their ensuing need for classically trained microbiologists that understand industrial-scale fermentations continues to be relevant. Recently, such expertise has transcended into the biofuel industry from both a general need to understand the fermentation process and to develop effective means to control bacterial contaminations in these yeast-based processes (Muthaiyan et al., 2011; Limayen and Ricke, 2012). In addition, there has always been a demand for microbiologists in quality control laboratories for most of the major commercial food and protein processing companies that deal with food safety issues. Even though food safety government regulations are driving detection technologies toward more molecular-based methodologies, many laboratory routines still require the capabilities to isolate and cultivate microorganisms from food products.

This is particularly true for some of the anaerobic bacterial contaminants associated with certain foods and food ingredients. For example, the myriad of anaerobic microorganisms from food products, such as those contained in modified atmosphere food packaging with low levels of oxygen, require some degree of anaerobic methodology to do initial isolations for further identification and verification. Finally, as more becomes known about the life cycles of the primary foodborne pathogens, it is becoming better understood that they are capable of not only surviving but growing in fairly oxygen-limited environments including the gastrointestinal tract of most food animals (Dunkley et al., 2009; Horrocks et al., 2009; Lungu et al., 2009). This makes it more critical to conduct fundamental research studies under these conditions to better understand regulation of virulence and stress responses that may impact their dissemination and disease-causing capabilities.

Finally, as antibiotics are being phased out of agriculture and food animal systems due to public pressure and corresponding government regulatory responses, there has been a renaissance of interest in alternative products that have the potential to accomplish similar benefits to the animal when fed as well as to serve as potential barriers to foodborne pathogen colonization. These include a variety of biological applications, such as bacteriophages, bacteriocins, and other antimicrobials, but the primary interest has been on prebiotics and direct-fed microbials or probiotics (Fuller, 1992, 1997; Joerger, 2003; Patterson and Burkholder, 2003; Callaway and Ricke, 2012; Ricke et al., 2012; Montville and Chikindas, 2013). For prebiotics and probiotics, the focus has been on their activities when introduced to the gastrointestinal tract, which at least to some extent requires anaerobic microbiology capabilities to assess impact on the gut microbiome (Ricke and Pillai, 1999; Callaway and Ricke, 2012; Pfeiler and Klaenhammer, 2013).

There is a long history of development of anaerobic methods for food microbiology applications (Anderson and Fung, 1983; Barnes, 1985). Much of the early efforts were based on designing enclosed, self-contained incubation jars that allowed for removal of oxygen via a chemically based catalyst (Brown, 1921; Hall, 1929; Brewer, 1939; Brewer and Allgeier, 1966). Use of simple biological removal systems such as candle jars containing oats had also been examined (Vedamuthu and Reinbold, 1967). Over time these anaerobic jars became more commercialized with self-contained GasPak catalyst and gas generation systems that allowed for more routine use (Collee et al., 1972). Attempts to cultivate anaerobes outside of the anaerobic jar containment system were also attempted with modified tube systems for cultivation or the use of bacterial membrane fraction biocatalysts (Ogg et al., 1979; Tuitemwong et al., 1994). While most of these anaerobic systems were fairly effective for recovering facultative anaerobic organisms and spore formers, these systems suffered the limitation of still exposing isolates to oxygen during the time required to achieve complete oxygen removal and decrease oxidation-reduction potential of the atmosphere and cultivation media.

Unfortunately, more strict or obligate anaerobes such as those that inhabit the lower regions of the gastrointestinal tract would not survive even brief exposure to these elements and, thus, require much more strict anaerobic cultivation conditions. As discussed in the next section, it would take the dedication of focusing on these types of organisms to develop the appropriate methodology to ensure successful isolation and cultivation.

5 RUMEN MICROBIOLOGY AND EMERGENCE OF STRICT ANAEROBIC METHODOLOGY

Understanding the gastrointestinal microbial ecology of a variety of food animals has always been a very important component of nutritional research studies. Traditionally, much of the early research focused on ruminant animals and rumen microbiology, in part because of the interest in understanding how ruminants were able to subsist on cellulose. Although it was known for some time that ruminant animals could derive nutritional value from forages, it was unclear how that was accomplished. Research efforts beginning in the mid-twentieth century led to the isolation of many of the anaerobic microorganisms that were prominent in the rumen, including the primary cellulolytic bacteria and anaerobic protozoa (Bryant, 1959; Hungate, 1950, 1960, 1966, 1979; Schaechter, 2013). Once pure culture collections of the primary rumen bacterial isolates were established, this opened the door for numerous groups of researchers to enter the field of rumen microbiology and initiate studies to develop an in-depth understanding of individual organisms as well as metabolic interactions among groups of organisms. Consequently, rumen ecological concepts were identified and characterized such as the cross-feeding among cellulolytics and noncellulolytics during fiber degradation in the rumen, identifying organisms with multiple fermentation pathways to generate ATP for growth and metabolism, and discovering the ability of some organisms to both produce certain metabolites, such as lactic acid, and subsequently utilize these same metabolites as growth substrates under specific growth conditions (Hungate, 1966; Russell and Hespell, 1981; Hobson and Wallace, 1982a,b; Wolin and Miller, 1982, 1983; Ricke et al., 1996; Mackie, 2002; Weimer, 1992; Weimer et al., 2009). A key development was the identification of methanogens responsible for production of methane in the rumen, but perhaps more importantly from a rumen functionality standpoint, their ability to consume hydrogen generated by the saccharolytic population and drive overall fermentation to a more oxidized series of end products via a process referred to as interspecies hydrogen transfer (Hungate, 1966; Wolin and Miller, 1982, 1983; Wolin, 1979; Saengkerdsub and Ricke, 2014). As environmental interest in reducing rumen methane production became more important, competitors for hydrogen as a substrate such as acetogens were identified that had the capability of using hydrogen and carbon dioxide to form acetate (Sharak Genther and Bryant, 1987; Le Van et al., 1998; Boccazzi and Patterson, 2011, 2013, 2014; Jiang et al., 2012a,b; Pinder and Patterson, 2012, 2013).

To accomplish the tremendous amount of information on the rumen microbiota required the adaptation of traditional microbiology culture methods to accommodate strict anaerobes that were incapable of tolerating oxygen (Bryant, 1959, 1972; Hungate, 1960). However, routine anaerobic microbiology required a number of technological laboratory challenges to be overcome. Certainly first and foremost was developing media-handling techniques that limited initial oxygen exposure, removed residual oxygen, and retained oxygen-free liquid and headspace conditions during growth of strict anaerobes. The use of combinations of heating of media solutions and addition of chemical reductants allowed for poising of culture media at low oxidation-reduction levels for optimal growth (Bryant, 1972; Zehnder and Wuhrmann, 1976; Brock and O'Dea, 1977; Carlsson et al., 1979; Jones and Pickard, 1980; Moench and Zeikus, 1983; Wachenheim and Hespell, 1984; Ricke and Schaefer, 1990). Development of a roll tube technique with an inner layer of agar media that could be properly prepared anaerobically, oxygen-free gasses added, and sealed with oxygen-impermeable butyl rubber stoppers opened the door for isolation of individual colonies of strict anaerobic bacteria for further characterization (Hungate et al., 1966; Hungate, 1969; Holdeman and Moore, 1972). Utilization of serum bottles that could be sealed with stoppers via aluminum crimp sealing and inoculated by syringes allowed for pressurized anaerobic headspaces that revolutionized the cultivation and characterization of methanogens (Macy et al., 1972; Miller and Wolin, 1974; Balch and Wolfe, 1976; Balch et al., 1979). Characterizing strict anaerobic bacteria became more routine with the advent of anaerobic glove boxes that enabled manipulation of cultures using conventional plating techniques and subsequent enumeration in a carefully controlled anaerobic atmosphere and with some modifications that could also be used for growth of methanogens (Aranki et al., 1969; Aranki and Freter, 1972; Edwards and McBride, 1975; Cox and Herbert, 1978; Gill et al., 1978).

Once anaerobic methodology became well established, inevitably efforts to train the next generation of students became important. Consequently, rumen microbiology course work, particularly at the graduate level, was offered in several dairy and animal science departments of land grant institutions from the 1960s onward. One of the primary sources of material was the comprehensive monograph written by R.E. Hungate in 1966, which was still considered the seminal treatise on rumen microbiology 25 years later (Ling, 1991). Since the publication of Hungate's book, several monographs and edited books emerged to provide supplemental material on rumen theoretical aspects as well as protocols and descriptive assessments of the rumen microbial species and rumen function (Holdeman et al., 1977; Ogimoto and Imai, 1981; Van Soest, 1982; Hobson, 1988; Dehority, 1993; Russell, 2002). When laboratory principles developed for rumen microorganisms were applied to other animal gastrointestinal intestinal systems a series of book publications were generated that covered most of the major animal species as well as humans (Clark and Bauchop, 1977; Holdeman et al., 1977; Hentges, 1983; Drasar and Barrow, 1985; Woolcock, 1991; Ewing and Cole, 1994; Mackie and White, 1997a,b). Publications that focused on methodology protocols also emerged that proved useful (Holdeman et al., 1977; Levett, 1991). Along with gastrointestinal microbiology oriented material, complementary textbooks on fermentation biochemistry and microbial physiology could also be included for some of the fundamental tenets of anaerobic microbiology (Gottschalk, 1979, 1986; Ingraham et al., 1983; Gerhardt et al., 1994; Moat et al., 2002).

For the mid-twentieth century, anaerobic microbiology training and course work were considered the domain of food animal-oriented academic departments and, for the most part, primarily focused on rumen microbiology with some attempts to expand to other gastrointestinal systems. Consequently, the audience for this type of material was fairly limited and training remained confined to those dedicated to this academic discipline. However, this was soon to change dramatically and with these changes the target audiences for this type of training. While the fundamentals and principles would not necessarily be expected to change, the manner in which they might be taught would potentially have to be modified. To some extent, this also required a change in instructional philosophy as well. Some approaches and the underlying philosophies on how to develop and repackage effective delivery of this type of training are discussed in the following sections.

6 DEVELOPING ANAEROBIC MICROBIOLOGY TRAINING TO MEET CURRENT NEEDS

By the first decade of the 2000s, a new generation of researchers were expanding gastrointestinal microbiology in a multitude of directions and applications from probiotics and food safety to human nutrition (Ricke, 2010). This, in turn, increased the demand for training in handling the more obligate anaerobes associated with the respective gastrointestinal systems. In conjunction with this was the sudden flood of -omics approaches into many of these fields, which, while generating tremendous amounts of information, still needed to relate back to biological functionality within the corresponding ecosystem. In addition, the new political impetus to expand biofuel production from a variety of biomass sources initiated interest from traditional academic fields such as chemical engineering to gain a better understanding of fermentation and anaerobic metabolism to apply to biofuel production systems (Limayen and Ricke, 2012). However, developing the corresponding course work that could accommodate this myriad of interests and academic backgrounds remained a challenge and required efficient approaches to integrate instructional material in a fashion that met a fairly mixed audience's needs. Clearly, instruction in anaerobic microbiology would be considered essential but the question of how to package this in such a way as to appeal to such a mixed group of interests was the issue.

The other audience that is a prime target for anaerobic microbiology training is the food science student who eventually will become a professional employee either in the food industry, an allied industry, or a food-related academic field. With the increased commercial interest in functional foods and a healthy human gastrointestinal system along with probiotics and other dietary amendments such as prebiotics, understanding not only fundamental microbial concepts but microbial ecology in the gastrointestinal tract and the bacterial metabolism associated with these processes is important. Typically, food science students at most universities that have a food science program might be expected to take an introductory microbiology lecture and laboratory course and/or a food bacteriology course with an accompanying laboratory section. Depending on the course material and philosophy of the instructor, students from such programs typically receive a basic knowledge of prokaryote microbiology (morphology, physiology, genetics, taxonomy, and ecology) and specific knowledge of microbiology of human foods and accessory substances (raw and processed foods; physical, chemical, and biological phases of spoilage; and standard industry techniques of inspection and control).

However, given the current trends in the food industry, a well-qualified food microbiologist needs a more thorough understanding of the microorganism and its environment. Such curricula in the form of microbial ecology, physiology, and fermentation courses are not always available at the senior undergraduate or even at the graduate level at most universities and even if available may not be mandatory prerequisite courses for food science majors. Advanced experience in microbial physiology is a particularly important component of food microbiology because microbial processes in the food industry are closely linked with understanding the physiology of fermentative and pathogenic microorganisms. This has become more critical now that there is more focus on the development of novel, "green friendly" antimicrobials for limiting foodborne pathogens in food processing and preharvest applications. Being able to assess foodborne pathogen responses to antimicrobial agents under stressful environmental conditions is important in understanding the potential for variability in effectiveness. In preharvest intervention applications it is now becoming more evident that not only is it important to determine the antimicrobial efficacy on the foodborne pathogen growing in an oxygen-limited environment but also determining its impact on the indigenous microbiota is important as well to avoid deleterious outcomes due to upsetting the gastrointestinal microbiome balance with the host.

7 TEACHING ANAEROBIC MICROBIOLOGY: GENERAL APPROACHES AND PHILOSOPHY

Given the complex shifts in research priorities and commercial applications, it is critical that students graduating from programs such as food science possess a more integrated training experience in fields such as food microbiology. This requires a combination of training and teaching approaches. Certainly, classical lecture material is still an important component as is the corresponding laboratory training. However, while not novel, other learning tools, particularly critical reading of the scientific literature along with extensive analytical writing, are also important for achieving true integration. Regardless of a student's career trajectory, the ability to process and critically appraise written materials whether they be published research in peer reviews, trade magazine summaries, or written regulatory mandates, and in turn express this articulately both orally as well as in a written report is still considered an attribute. From a more fundamental aspect, any learning tool that amplifies the ability to critically think is a valuable teaching tool.

Interdisciplinary fields such as food science offer opportunities to incorporate multiple approaches to learning and food microbiology is particularly an ideal incubator for this philosophy. Although food science undergraduate and graduate students are usually required to take at least one food microbiology course as part of their curricula, access to more advanced laboratory training in microbiology may be limited. In addition to lecture material and writing exercises, development of student experiential laboratory learning projects directed toward research practicality offers opportunities to present a more integrated approach to training and teaching. Food microbiology laboratory exercises lend themselves quite well to multiple learning outcomes. For example, using fairly routine experiments such as microbial growth culture techniques to study microbial physiology of relevant food microorganisms still has tremendous utility for teaching complex scientific principles such as cell biology, genetics, and mathematics.

These types of approaches allow nonmicrobiology students in a hands-on environment to develop not only advanced bacteriology skills beyond the standard plating and identification strategies taught in introductory microbiology and food microbiology courses but also an appreciation for fundamental scientific principles as well. There are opportunities for development of such learning experiences courses both at the undergraduate and at the graduate level in food and anaerobic microbiology to accommodate students majoring in animal science, food science and technology, nutritional science or poultry science, along with other assorted disciplines. In the following sections, examples and approaches are provided based on my past experiences of delivering this type of teaching approach to these varied audiences.

8 TEACHING WRITING COMPREHENSION TO UNDERGRADUATE STUDENTS

Writing and composition are traditionally emphasized in academic curricula. This is critical for scientific training as well for careers in industry (Senauer, 1992; Summers, 1992; Pardue, 1997; Maciorowski and Ricke, 2000). Not only is the exercise of writing important for processing information but also is an efficient and cost-effective way to integrate otherwise seemingly disconnected sources of information. Precise measurements of effectiveness of writing exercises remain elusive; therefore, much of the perceptions presented here are based on my previous experiences in developing and evaluating writing exercises for undergraduate and graduate students who had various academic backgrounds and intellectual interests (Ricke, 1996a,b; Maciorowski and Ricke, 2000; Dittmar et al., 2006; Howard et al., 2006). While my undergraduate course focused primarily on foodborne pathogens, topics such as spoilage and food fermentations were addressed as well. My graduate course for the most part covered microorganisms present in the gastrointestinal tract and specific subject areas included methods for anaerobic bacterial isolation fermentation and growth, genetics and molecular biology of anaerobes, and metabolic disorders of microbial origin.

In the undergraduate food microbiology course, students were expected to become much more familiar with the scientific literature. This was accomplished by the requirement to write a definitive comprehensive review paper 14-15 pages long on a food safety topic that could be evaluated/graded during different stages of preparation (choice of topic, initial drafts, and final draft) and was set to be due in final form at the end of the semester. Instruction was provided during the first week of the semester to acquaint students with what was expected of them as well as how to conduct scientific literature electronic searches and critically assess the quality of the information retrieved from these sources (Maciorowski and Ricke, 2000). When students were surveyed, several issues were identified as ongoing concerns for effective teaching of written composition on a scientific topic (Maciorowski and Ricke, 2000). One primary area that appeared to be particularly problematic was the resistance of the students to limit procrastination, with nearly half admitting they did not start their written assignment until a week before it was due. To resolve the procrastination issue, several changes were implemented in the ensuing years including increasing the proportion of the total course grade being accounted for by the written paper assignment and being requested to submit an outline with at least three references as part of the student's grade early in the semester. In addition, an instructional writing exercise was developed as a part of a laboratory session early in the semester (Howard et al., 2006). In this laboratory exercise, students were not only taught how to utilize search engines effectively but also were taught proper methods for citing papers correctly. Students were also required to conduct a skim test using a series of focused questions on each section of the paper of a selected journal article as a class exercise to learn how to determine if a reference was appropriate for the subject of their chosen topic (Howard et al., 2006).

Despite these efforts, when the students were surveyed once again procrastination still remained an issue, although those with previous similar writing experiences did find teaching assistant input and evaluation of the rough draft helpful (Howard et al., 2006). The other issue that emerged was numerous students expressing difficulties in reading and understanding scientific literature despite the skim test exercise in their laboratory sections. Based on this, it is fairly apparent that undergraduate students not only need more exposure to scientific literature in general but also must be taught how to efficiently read papers and effectively interpret them. Certainly, more in-depth emphasis on technical writing is needed with more time devoted to these types of exercises in the classroom. At some universities there are classes that combine oral seminars in a journal club format where students are required to not only give a seminar but also present the seminar as a critique on a particular published scientific paper where data and other components are presented along with the student's interpretation. One wonders if this type of classroom activity were to be combined with writing a comprehensive review paper on a particular topic, that this might be a more effective way to teach and develop technical writing skills in students. Other opportunities such as involving students in translating scholarly publications for composing popular press articles have been suggested as a means to enhance student abilities to process information (Walker, 2011).

9 TEACHING WRITING COMPREHENSION TO GRADUATE STUDENTS

In the graduate course, writing instruction was accomplished in a couple of different ways (Ricke, 1996a). First of all, it was emphasized by the use of problem sets that provided hypothetical and analytical questions on key research papers. The published journal papers were chosen to be in synchrony with the lecture material being taught at that particular time and represented a more in-depth experimental background to support the general concepts being presented in class. The questions posed to the students for a particular scientific journal paper were both descriptive and interpretative in nature. Students typically might be asked to discuss specific experimental details, provide a rationale for why a particular method was chosen by the authors, or explain why the study might be important for providing evidence for a particular concept. To reinforce effort and development of interpretation and comprehension skills, students were advised that essay questions on the periodic course examinations would pose similar questions about the papers they had read in their problem set assignments.

In addition, because these were graduate students, they were also required to write a research proposal on a research problem that utilized information and concepts that corresponded to anaerobic and gastrointestinal microbiology topics taught in the course. The design of the proposal was to mirror what might be typically expected of a research grant for a governmental granting agency and included components such as a detailed budget and experimental protocols.

Written material was provided to the student at the beginning of the semester to describe how progress on the proposal would be defined and evaluated for the graduate course. An initial written proposal of three to five typed pages concerning the background and purpose of the project was required within the first month of the semester. The starting point for this was based on the initial discussion(s) with the instructor. The purpose of this initial proposal draft was to help the advisor gauge the student's level of understanding and help the student avoid pitfalls later in the project due to lack of clarification on a particular point. This instructional approach provided an effective means for graduate students to develop an understanding of gastrointestinal microbiology methodology and theory. The background and details of how these components were delivered and evaluated as instructional components are described in the following sections.

The proposals were designed to include five sections along with a budget:

(1) A justification section that discussed why the study should be considered important to those evaluating the proposal.

(2) A literature review, where initially the student would be given some literature as background information for the project by the instructor and/or the supervising teaching graduate student. This was either in the form of papers written on the subject or other material collected by the instructor. Students were evaluated on their ability to expand on this, with attention given to initiative, originality, and conciseness.

(3) An objective section was required that consisted of a short, clear statement of the question that was to be examined. The student was instructed to design the objective statement such that they pointed out the major subquestion(s) (anywhere from one to three) that were to be directly approached by the research effort. It was emphasized that a clear understanding of the objectives was necessary before presenting the design of experiments to make sure the experiments described actually answered the questions posed.

(4) Students were instructed to provide a methodology section that consisted of a general outline of how the objectives were to be met based on previous literature and discussions with the instructor. Students were advised that this section would be evaluated according to the feasibility of attaining the objectives with the proposed procedures during the remainder of the semester.

(5) A list of all references cited in text were required at the end of the proposal with a required style and format based on instructions provided at the beginning of the semester.

Once the proposals were completed they were evaluated by the instructor using similar criteria used by granting agencies. Students were divided into groups to serve on mock peer-review panels to evaluate the respective proposals of their counterparts in the other groups based on criteria discussed during class. These were designed and subsequently setup to closely resemble the type of grant peer-review panel used by government agencies to judge the merits of submitted proposals for potential funding by the corresponding government agency. Proposals to be evaluated were assigned essentially as a function of each group exchanging their respective proposals with the members of the other group and vice versa. The proposals were judged on the clarity of hypothesis presentation and the appropriateness of the experiential approaches in a prescribed set of benchmarks provided to the class early in the semester. Each student was required to review one proposal of another student from the other group. Once these individual reviews were completed, student panel groups were instructed to assemble into their respective groups and discuss individual evaluations of the corresponding proposals. The goal of the group discussion was to describe the merits and pitfalls of each proposal and decide on a ranking among the proposals of which would be most likely to be funded. Finally, they were asked as a group to give an oral presentation on their discussion and rankings of proposals. During this discussion the instructor, who had also read and evaluated the proposals independently of the student groups, provided evaluation input as well.

In general, the informal student feedback from this exercise was interesting as graduate students found the concept of providing constructive criticism on their peer's research ideas to be highly intimidating and to some extent one of the more difficult parts of the entire assignment. This was made more difficult in part because oftentimes they already knew each other fairly well. This may be an indication of the need to not only work with students at all levels on writing comprehension, but also focus on how to effectively develop their constructive criticism skills. This may be more important than is realized for career development, as students in their future employment may be asked to not only make judgment calls on a wide range of issues for their employer, but also provide detailed rationale for their assessment. Being able to deliver this in a constructive manner would be highly important, particularly in group settings.

10 TEACHING ANAEROBIC MICROBIOLOGY: STUDENT EXPERIENTIAL APPROACHES

The concept of student experiential microbiology laboratory research exercises is based on the general premise that such research-based laboratory projects must be manageable within the time frame of a semester but still be challenging at whatever level it is taught. For anaerobic microbiology training, students must initially become proficient at aseptic microbiological techniques, closely followed by training in anaerobic methods. This enables them to easily progress to more complex microbiology problems. As discussed previously, mastering anaerobic microbiology opens the door for conducting most types of microbiology laboratory work and prepares the student to handle complex laboratory manipulations. Once the work is completed the student is prepared to conduct and generate data from actual experiments. For anaerobic microbiology, conducting microbial growth studies represents an optimal teaching combination of fairly straightforward generation of data that offers experiments with sufficient variation to study most of the fundamental biology principles associated with microorganisms.

Fundamental concepts of growth physiology can be taught by conducting anaerobic batch culture growth experiments utilizing approaches described in numerous studies (Russell and Baldwin, 1978; Schaefer et al., 1980; Ricke and Schaefer, 1991, 1996b; Ricke et al., 1994; Jiang et al., 2012a; Pinder and Patterson, 2012, 2013). Essentially, batch culture represents a static culture in which all nutrients for growth are available upon inoculation and the respective microorganism is inoculated and allowed to run the course of growth phases (lag, exponential, and stationary) (Pirt, 1975; Brock and Madigan, 1991; Montville and Matthews, 2013). Measurements of growth (i.e., increase in population of bacterial cells) can be taken in a variety of ways, but probably the most common approach is via detection of changes in optical density/turbidity using a spectrophotometer (Kavanagh, 1963; Koch, 1994). Of course, the culture tubes need to be designed to contain reduced anaerobic media, either stoppered conventionally or as a serum-crimped top (Bryant, 1972; Balch and Wolfe, 1976). In addition to learning how to handle obligate anaerobes in liquid media systems, the student has the opportunity to design, construct, and prepare a variety of defined media with various nutrient amendments depending on the purpose of the study. In completing these tasks, the student learns the importance of precise calculations for molar concentrations of nutrients and buffer construction as well as more complicated procedures such as the introduction of a reductant to poise the medium anaerobically.

A number of different types of experimental outcomes can be achieved with batch culture beyond just monitoring growth responses. Certainly growth rate and doubling times can be estimated from standard growth responses and, depending upon the growth responses, kinetic parameters such as limiting nutrient affinity constants can be calculated from double reciprocal plots of different growth rates versus nutrient concentrations (Pirt, 1975; Russell and Baldwin, 1978; Schaefer et al., 1980; Ha et al., 1994; Froelich et al., 2002). The information gleaned from batch culture responses has several practical applications representative of concepts that illustrate fundamental principles of microbial physiology and genetics. For example, Monod (1949) used differences in growth responses to demonstrate the occurrence of diauxic growth, which eventually led to the concept of the *lac* operon and gene regulation. Assessment of affinity constants has been used to explain coexistence of multiple rumen bacterial species being able to coexist in the rumen via a hierarchy of substrate preferences based on different affinities for the same substrate (Russell, 1984). There are more biotechnology-oriented applications as well. Historically, generating differing maximum optical densities and constructing the corresponding standard curve of nutrient concentration versus maximum optical density has been used as a quantitative bioassay approach for estimating utilization of a wide range of amino acids and vitamins (Kavanagh, 1963). A more recent development has been to construct specific auxotrophic amino acid mutants in bacteria such as *E. coli* and insert bioluminescent or green fluorescent genes to allow for nonoptical density-based detection growth responses to amino acids such as lysine or methionine (Erickson et al., 2002; Froelich and Ricke, 2005; Chalova et al., 2009, 2010).

The availability of these modified whole cell sensor strains offers students the ability to not only develop a more in-depth understanding of genetic engineering, but also provides them with the opportunity to compare different methods for estimating microbial cell concentration as a function of potential differences in detection sensitivities.

Experiments in batch culture also help the student to link the theory of bacterial growth kinetics to the design of the appropriate test medium and generate the required preliminary data for the successful design of continuous culture experiments. Once the student has conceived an experimental objective and appropriate protocols for the allotted time available, then the continuous culture studies can be initiated. This gives the student an entirely new set of methodology problems to solve (e.g., how to efficiently scale up anaerobic techniques from milliliter volumes to liter volumes). Continuous culture also offers a teaching tool with massive experimental potential, not only in growth physiology but also in genetic, biochemical, and ecological studies (Bull and Slater, 1976). While numerous commercial continuous culture apparatuses are available, simple systems can be readily built that can accommodate anaerobic bacteria (Kafkewitz et al., 1973; Schaefer et al., 1980; Ricke and Schaefer, 1991, 1996a). Once set up and operational, nutrient levels, and dilution rates can be adjusted to assess microbial cell yields as a function of substrate utilization (Pirt, 1975). By varying dilution rates, affinity constants for the limiting nutrient can be assessed as well as maximum growth rates. Constructing and operating anaerobic continuous culture should provide sufficient understanding that will reinforce previously learned anaerobic microbiology principles. Once mastered, they can also be used to understand fundamental microbial ecology principles such as competition between microbial species in mixed cultures as well as setting up metabolic balances among multiple species in mixed multispecies bacterial-based continuous cultures (Veldkamp, 1972; Isaacson et al., 1975; Ushijima and Seto, 1991; Nisbet et al., 1996a,b).

11 TEACHING LABORATORY MICROBIOLOGY: INTEGRATED GROUP APPROACHES

Using laboratory exercises is important for mastering fundamental microbiology skills, but to gain true command of the entire discipline of microbiology requires development of multiple capabilities that allow for true integration. For such an integrated instructional approach to be successful requires combining extensive written reports, team-oriented projects, and oral presentations to communicate the results. A student's regular commitment to reporting and interpreting data nurtures critical thinking and develops the communication skills necessary to be competitive in today's job market. In industry for example, students who are to become employees must be articulate spokespersons for their area of expertise to speak to peer scientists, production personnel, and senior company management (Snetsinger, 1992; Urian et al., 2010). Finally, the design of this type of student experiential project requires the undergraduate student to work as part of a team to achieve mutual goals designed for group projects. This is a realistic research setting because the industry scientist must be a team player, as only rarely do complex research problems and technical support activities not involve integrated actions and approaches (Snetsinger, 1992). This means requiring interaction with other students at various stages of group projects, which can lead to additional learning experiences for the student.

To accomplish an integrated approach for training undergraduate students in food microbiology, we developed a group project exercise as part of the regular laboratory assignment (Dittmar et al., 2006). Essentially, the group laboratory project assignment involved students working as teams throughout the semester to design their own quality control microbiology laboratory for testing various foodborne pathogens. Both individual research papers and group project participation were required of students. While the individual student research reports entailed an in-depth background of the foodborne pathogen assigned to them, the team of students was responsible for delivering a project report describing and justifying a quality control laboratory design of an operating laboratory dedicated to that particular foodborne pathogen. The final report was delivered both as a written document as well as an oral presentation to the entire laboratory section. This meant that as a group they had to work within an assigned budget, incorporate regulatory requirements for that particular foodborne pathogen, and assign space allotment for physically setting up the laboratory and the equipment they chose to purchase. The purposes of these group student experiential projects were to (1) offer the student some training on group participation, management, and organization; (2) provide the student some experience planning, conducting, summarizing, and communicating research; and (3) increase exposure and awareness of contemporary issues of food safety. In order to accomplish these objectives students, once assigned to a research project, they were expected to develop the topic so that it contributed substantially to the group effort for designing the quality control laboratory.

The final project research report was evaluated for each individual student as the record of the scientific literature background information that particular student was able to generate. As such, it was made apparent to the student early in the semester that the report needed to be a clear, accurate, and thorough presentation of the project as the future quality control laboratory group project design would be based upon it. The format for the typed project report would also provide background information for the group oral presentation on their quality control laboratory design at the end of the semester. The student(s) were advised to outline their reports before they began writing, with the final report consisting of the following subsections. The abstract represented a brief synopsis section, which could be rapidly read to gain a superficial knowledge of what was to be described in the body of the research report. Therefore, in 100-200 words, it was expected to include a concise series of statements of the review paper objectives, description of the main take-home points, the summarized interpretations, and a conclusion. The introduction section was to be written such that it would briefly describe why this particular foodborne pathogen is important for the group project quality control laboratory (i.e., why should there be money invested in constructing a laboratory dedicated to detection of this pathogen?). The purpose of this section was to set the stage for a review of the existing information that followed in the rest of the report.

The bulk of the review was expected to be devoted to what is currently known in the scientific literature about the assigned foodborne pathogen along with the student's interpretations of the known literature and how it might fit for their respective group project. The literature review section was designed to include a thorough review of the scientific literature, which would be considered relevant to the experimental objectives. The discussion of pertinent references in this section was expected to deal with the particular methods, observations, or conclusions leading to the objectives for designing the quality control laboratory. Additionally, if there was disagreement in the literature over some method critical to the design of the quality control laboratory, a discussion concerning the relevant research on all sides would be expected. For example, a practical application of the published research may have already been described in the literature but some adaptation might be needed for making the technology feasible to the quality control laboratory being designed by the group. Conversely, investigation of this idea may reveal other approaches to solving an existing problem associated with the design of the quality control laboratory.

The group project report for proposing and designing the quality control laboratory was expected to incorporate what each member of the group learned from their respective background review papers on the particular foodborne pathogen that had been assigned to the group. The students were required to state their objective for what they hoped to accomplish. The materials and methods section was expected to include a detailed description of the equipment, cost of space allotment, and other details for the methods they would use for their proposed quality control laboratory. It was expected that the audience both for the written report and for the oral presentation should be able to understand not only the overall strategy for designing the laboratory but also the details and approaches that have been selected to accomplish the goals. Therefore, details on methods they anticipate employing for detection strategies on a particular pathogen would be expected to include information such as length of incubations for isolating and growing the pathogen, a description of equipment needed and its cost, along with information such as wavelength used for particular detection protocols, source of special reagents/substrates, along with other supplies that should be included. If the student group used some routine laboratory techniques, reference to a previous report on this was considered sufficient as long as this was made clear to the audience. However, if the student group modified the technique to fit their specific quality control laboratory needs, they were advised to state and justify the modifications. Students were reminded that organization and presentation of their proposed quality control laboratories was extremely important because their audience might be unable to interpret their overall justification for their particular design strategies if, for example, some of their descriptions of the methods were unintelligible.

Time management of the project was the responsibility of the students within each group. As the expectations of the laboratory instructor were specified to the students early in the semester, the students were advised to use this as a guide for time management during the semester. In practice, the students were encouraged to consult with the laboratory instructor at all stages of the project for guidance in experimental procedures and interpretation of methods for design of their quality control laboratories. In addition to the required written report, student groups were expected to present and discuss their results as part of the laboratory schedule at the end of the semester. Evaluation of students for assigning points from this group exercise was described in Dittmar et al. (2006) and will only be briefly discussed here. Obviously, group effort was somewhat difficult to gauge but teams were appraised by general consensus, albeit somewhat subjective; evaluations of their fellow group members based on each member's participation; and involvement in the activities associated with the project. Five criteria were used for evaluation and assigned points accordingly for grading purposes: (1) team member attending all group meetings; (2) attitude in terms of level of enthusiasm along with motivation toward placing the team first; (3) being a team player by cooperating, being helpful, and providing effective constructive criticism; (4) expressing a willingness to contribute to the workload of the project by actively being engaged in all aspects of the group project report such as writing, critiquing, and participating in the team presentations; and, finally, (5) the impression of the team on the overall effort made by each team member to the project.

Constructing meaningful learning experiences that engage groups of students and allow for meaningful evaluations of their individual efforts is a challenge. An attempt was made to assess effectiveness of the approach developed for the food microbiology students by directly surveying them with focused questions on their experiences with this exercise (Dittmar et al., 2006). When the students were surveyed for their own impressions of this approach to laboratory training, they identified finding time for all group members to meet and getting participation from all members of the team as being the most difficult part of the team component (Dittmar et al., 2006). Their conclusion is in line with suggestions by industry personnel that students need academic classes that cover resolving conflict, enhancing people skills, as well as holding effective meetings (Urian et al., 2010). While some formal instruction might be helpful, perhaps the more efficient approach might be to give students more practical experiences in group activities that require fairly high expectations from a performance standpoint. However, to implement this will also require accurate evaluation tools for objectively assessing that performance, which may be the more challenging aspect.

12 CONCLUSIONS AND FUTURE DIRECTIONS

Science-based undergraduate and graduate students to some extent represent somewhat different target audiences with diverse experiences and backgrounds and, in many cases, different career goals. However, for a scientific discipline such as food safety there are several common core educational principles and concepts that are needed for a student at any level to be successful in this career choice. This is true whether the person intends to go into academics in some capacity such as a researcher at a university or government laboratory or more along the lines as an instructor or extension person involved in teaching and outreach activities. The same holds true for careers in the private sector, whether it is more research and development focused or more directed toward marketing and sales. In all cases, the ability to integrate thinking and assimilate data from multiple sources, master fundamental microbiology laboratory skills, and work effectively in group settings, either as a manager or as a team member, appear to be critical for success.

In this chapter, it has been suggested that gaining laboratory experience specifically in anaerobic microbiology offers tremendous opportunities for the current generation of students to be completely trained in food safety. Not only do mastering anaerobic techniques represent an efficient training means for achieving a high-level skill with microbiological laboratory techniques, but several nonmicrobiology laboratory skills can be cultivated as well. Understanding anaerobic principles not only encompasses important concepts in microbiology but also provides background for other food science-related topics such as food fermentation processes and nonmicrobiology principles such as the concept of oxidation-reduction potential changes occurring in certain food processes. Once an understanding is achieved, a multitude of additional expertises can be developed for studying microbial physiology and growth kinetics, which are important considerations for either optimizing a food fermentation process or devising a more effective antimicrobial for application in a particular food matrix.

The field of anaerobic microbiology is also experiencing somewhat of a renaissance of interest and relevancy in the food and human nutrition academic fields and corresponding industries. This is not surprising for several reasons. Given the rapid advances being made in sequencing the gastrointestinal microbiome of multiple food animal species and humans, there is a pending demand for individuals skilled at recovering, handling, and characterizing these organisms from the corresponding ecosystems. This will be important because microbiome sequencing will only tell part of the story and, at some point, functional roles will need to be assigned to groups and individual organisms not only to confirm the microbiome data but also to establish ecological, metabolic, and physiological roles for the respective inhabitants. In addition, the explosion of probiotic and prebiotic products being marketed along with the claims being made will need to be examined more closely both to establish their role in the gastrointestinal tract as well as to establish metabolic activity and their relationship with the host. In addition, the activities of foodborne pathogens in the more oxygen-limited environments of the gastrointestinal tract need to be considered when devising preharvest intervention strategies for limiting their colonization.

The other issue with devising effective means for training the next generation of food safety professionals is the need to develop management and people skills in group settings. The personal skills of potential employees continue to be identified by the industry as somewhat lacking in graduates (Snetsinger, 1992; Urian et al., 2010). Likewise, this has also been identified as an important challenge for the academic careers of graduate students and professors as well (Barker, 2005; Tachibana, 2015). New faculty must learn how to manage a diverse set of individuals in their laboratories including undergraduate and graduate students, technicians, and postdoctorates. This represents motivational challenges due to the diverse training backgrounds and different personalities. As discussed in this review, perhaps the most effective way to deal with this issue from an academic curricula standpoint is to develop original course work that focuses on development of leadership skills and management. Other avenues that may also be helpful, particularly for undergraduate students, is participating in food industry internships where they not only receive experience in the "real world" but also get practical training on how to constructively interact and be productive in group settings.

Finally, food safety trained professionals need integration abilities to not only process the diverse and sometimes large data sets but also to interpret and capture the relevant concepts revealed by these data sets. Even though it is probably considered a traditionally academic exercise, comprehensive writing still remains a valuable tool for learning how to integrate information. The importance of learning how to effectively and efficiently write and compose comprehensive documents that are clearly written and coherent to a wide audience is truly an instructional challenge. For some aspects such as graduate student education, this is less of an obstacle because there are already considerable writing expectations in place along with several outlets for mentoring and advice. Obviously, publishing academic peer-reviewed manuscripts that document original scientific research is an expected task for the scientific community of professors, postdoctorates, and graduate students. Over the years, several sources have been published for providing advice on how to not only write scientific research and review manuscripts, but also to provide strategies for critically evaluating previously published literature as well as targeting and publishing in peer-reviewed journals (Day, 1983; Sargeant et al., 2006; Powell, 2010; Blow, 2013; Austin, 2014).

For undergraduate student education, there are less opportunities other than required technical writing courses and specific writing exercises within some courses, especially those with laboratory sections. However, particularly if these students do not continue on into a graduate program, there may not be many opportunities to develop more comprehensive writing skills before joining the workforce. Given the importance of developing communication skills in the workplace, academic institutions may need to determine effective ways to focus and develop more training on comprehensive writing and the associated skill sets such as efficient screening of literature database searches. Despite the technological advances made with electronic-based communications, analytical and critical discernment abilities are still very much needed in areas such as food safety. This is becoming more important as information is being publically shared now more than ever before.

REFERENCES

American Society for Microbiology, 2007. Basic research on bacteria – the essential frontier. Workshop report, American Society for Microbiology, Washington, DC.

Anderson, K.L., Fung, D.Y.C., 1983. Anaerobic methods, techniques and principles for food bacteriology: a review. J. Food Prot. 46, 811–822.

Aranki, A., Freter, R., 1972. Use of strictly anaerobic glove boxes for the cultivation of strictly anaerobic bacteria. Am. J. Clin. Nutr. 25, 1329–1334.

Aranki, A., Syed, S.A., Kenney, E.B., Freter, R., 1969. Isolation of anaerobic bacteria from human gingiva and mouse cecum by means of a simplified glove box procedure. Appl. Microbiol. 17, 568–576.

Austin, J., 2014. What it takes. Science 344, 1422.

Ayres, J.C., Mundt, J.O., Sandine, W.E., 1980. Microbiology of Foods. W. H. Freeman and Company, San Francisco, CA.

Baker, B.J., Dick, G.J., 2013. Omic approaches in microbial ecology: charting the unknown. Microbe 8, 353–360.

Balch, W.E., Wolfe, R.S., 1976. A new approach to the cultivation of methanogenic bacteria: 2-mercaptoethanesulfonic acid (HS-CoM)-dependent growth of *Methanobacterium ruminantium* in a pressurized atmosphere. Appl. Microbiol. 32, 781–791.

Balch, W.E., Fox, G.E., Magrum, L.J., Woese, C.R., Wolfe, R.S., 1979. Methanogens: reevaluation of a unique biological group. Microbiol. Rev. 42, 260–296.

Balter, M., 2012. Taking stock of the human microbiome and disease. Science 336, 1246–1247.

Banwart, G.J., 1989. Basic Food Microbiology, second ed. Chapman & Hall, Inc., New York, NY.

Barker, K., 2005. At the Bench – A Laboratory Navigator, Updated Version. Cold Spring Harbor Press, Cold Spring Harbor, NY.

Barnes, E.M., 1985. Isolation methods for anaerobes in food. Int. J. Food Microbiol. 2, 81–97.

Blow, N.S., 2013. The write way. Biotechniques 54, 235.

Boccazzi, P., Patterson, J.A., 2011. Using hydrogen-limited anaerobic continuous culture to isolate low hydrogen threshold ruminal acetogenic bacteria. Agric. Food. Anal. Bacteriol. 1, 33–44.

Boccazzi, P., Patterson, J.A., 2013. Isolation and initial characterization of acetogenic ruminal bacteria resistant to acidic conditions. Agric. Food. Anal. Bacteriol. 3, 129–144.

Boccazzi, P., Patterson, J.A., 2014. Batch culture characterization of acetogenesis in ruminal contents: influence of acetogen inocula concentration and addition of 2-bromoethanesulfonic acid. Agric. Food Anal. Bacteriol. 4, 177–194.

Boddy, L., Wimpenny, J.W.T., 1992. Ecological concepts in food microbiology. J. Appl. Bacteriol. Symp. Suppl. 73, 23S–38S.

Braun, T., 2008. The perceptions and attitudes of online graduate students. J. Technol. Teach. Educ. 16, 63–92.

Breidt, F., McFeeters, R.F., Perez-Diaz, I., Lee, C.-H., 2013. Fermented vegetables. In: Doyle, M.P., Buchanan, R.L. (Eds.), Food Microbiology—Fundamentals and Frontiers, fourth ed. American Society for Microbiology, Washington, DC, pp. 841–855 (Chapter 33).

Brewer, J.H., 1939. A modification of the Brown anaerobic jar. J. Lab. Clin. Med. 24, 1190–1192.

Brewer, J.H., Allgeier, D.L., 1966. Safe self-contained carbon dioxide-hydrogen anaerobic system. Appl. Microbiol. 14, 985–988.

Brock, T.D., Madigan, M.T., 1991. Biology of Microorganisms, sixth ed. Prentice Hall, Englewood Cliffs, New Jersey.

Brock, T.D., O'Dea, K.O., 1977. Amorphous ferrous sulfide as a reducing agent for culture of anaerobes. Appl. Environ. Microbiol. 33, 254–256.

Brown, J.H., 1921. An improved anaerobe jar. J. Exp. Med. 33, 677–681.

Bryant, M.P., 1959. Bacterial species of the rumen. Bacteriol. Rev. 23, 125–153.

Bryant, M.P., 1972. Commentary on the Hungate technique for culture of anaerobic bacteria. Am. J. Clin. Nutr. 25, 1324–1328.

Bull, A.T., Slater, J.H., 1976. The teaching of continuous culture. In: Dean, A.C.R., Ellwood, D.C., Evans, C.G.T., Melling, J. (Eds.), Continuous Culture 6: Applications and New Fields. Ellis Horwood Ltd., Sussex, England, pp. 49–68.

Califf, K., Gonzalez, A., Knight, R., Caporaso, J.G., 2014. The human microbiome: getting personal. Microbe 9, 410–415.

Callaway, T.R., Ricke, S.C. (Eds.), 2012. Direct Fed Microbials/Prebiotics for Animals: Science and Mechanisms of Action. Springer Science, New York, NY.

Campbell, I., 2013. Beer. In: Doyle, M.P., Buchanan, R.L. (Eds.), Food Microbiology—Fundamentals and Frontiers, fourth ed. American Society for Microbiology, Washington, DC, pp. 901–913 (Chapter 36).

Carlsson, J., Grandberg, G.P.D., Nyberg, G.K., Edlund, M.-J.K., 1979. Bactericidal effect of cysteine exposed to atmospheric oxygen. Appl. Environ. Microbiol. 37, 383–390.

Cassell, G., 1992. ASM comments on the NIH strategic plan. Am. Soc. Microbiol. News 58, 302–303.

Chalova, V.I., Sirsat, S., O'Bryan, C.A., Crandall, P.G., Ricke, S.C., 2009. *Escherichia coli*, an intestinal microorganism, as a biosensor for amino acid bioavailability quantification. Sensors 9, 7038–7057.

Chalova, V.I., Froelich, J., Ricke, S.C., 2010. Potential for development of an *Escherichia coli*-based biosensor for assessing bioavailable methionine: a review. Sensors 10, 3562–3584.

Clark, R.T.J., Bauchop, T., 1977. Microbial Ecology of the Gut. Academic Press, New York, NY.

Collee, J.G., Watt, B., Fowler, E.B., Brown, R., 1972. An evaluation of the GasPak system in the culture of anaerobic bacteria. J. Appl. Bacteriol. 35, 71–82.

Costello, E.K., Stagaman, K., Dethlefsen, L., Bohannan, B.J.M., Relman, D.A., 2012. The application of ecological theory toward an understanding of the human microbiome. Science 336, 1225–1262.

Cox, R.N., Herbert, S.D., 1978. An anaerobic glove box for the isolation and cultivation of methanogenic bacteria. J. Appl. Bacteriol. 45, 411–415.

Day, R.A., 1983. How to Write and Publish a Scientific Paper. ISI Press, Philadelphia, PA.

De Keersmaecker, S.C.J., Thijs, I.M.V., Vanderleyden, J., Marchal, K., 2006. Integration of omics data: how well does it work for bacteria? Mol. Microbiol. 62, 1239–1250.

De Vos, W.M., 2005. Frontiers in food biotechnology – fermentations and functionality. Curr. Opin. Biotechnol. 16, 187–189.

Dehority, B.A., 1993. Laboratory Manual for Classification and Morphology of Rumen Ciliate Protozoa. CRC Press, Boca Raton, FL, 128 pp.

Dittmar, R.S., Kundinger, M.M., Woodward, C.L., Donalson, L.M., Golbach, J.L., Kim, W.K., Chalova, V., Ricke, S.C., 2006. Quality control laboratory design project for poultry science undergraduate students enrolled in an advanced food microbiology course. J. Consum. Protect. Food Saf. 1, 77–82.

Drasar, B.S., Barrow, P.A., 1985. Intestinal Microbiology. Van Nostrand Reinhold, Wokingham, UK.

Dunkley, K.D., Callaway, T.R., Chalova, V.I., McReynolds, J.L., Hume, M.E., Dunkley, C.S., Kubena, L.F., Nisbet, D.J., Ricke, S.C., 2009. Foodborne *Salmonella* ecology in the avian gastrointestinal tract. Anaerobe 15, 26–35.

Edwards, T., McBride, B.C., 1975. New method for the isolation and identification of methanogenic bacteria. Appl. Microbiol. 29, 540–545.

Enquist, L.W., 2015. Passing the torch and the samples. Microbe 10, 46–47.

Erickson, A.M., Li, X., Zabala-Diaz, I.B., Ricke, S.C., 2002. Potential for measurement of lysine bioavailability in poultry feeds by microbiological assays—a review. J. Rapid Methods Autom. Microbiol. 10, 1–8.

Ewing, W.N., Cole, D.J.A., 1994. The living gut: an introduction to micro-organisms in nutrition. Context, N. Ireland.

Froelich Jr., C.A., Ricke, S.C., 2005. Rapid bacterial-based bioassays for quantifying methionine bioavailability in animal feeds: a review. J. Rapid Methods Automat. Microbiol. 13, 1–10.

Froelich, C.A., Zabala Diaz, I.B., Ricke, S.C., 2002. Methionine auxotroph *Escherichia coli* growth kinetics in antibiotic and antifungal amended media. J. Environ. Sci. Health B 37, 485–492.

Fuller, R. (Ed.), 1992. Probiotics – The Scientific Basis. Chapman & Hall, New York, NY, 398 pp.

Fuller, R. (Ed.), 1997. Probiotics 2 – Applications and Practical Aspects. Chapman & Hall, New York, NY.

Gerhardt, P., Murray, R.G.E., Wood, W.A., Krieg, N.R. (Eds.), 1994. Manual of Methods for General and Molecular Bacteriology. American Society for Microbiology, Washington, DC.

Gilbert, J.A., Laverock, B., Temperton, B., Thomas, S., Muhling, M., Hughes, M., 2011. Metagenomics. In: Kwon, Y.M., Ricke, S.C. (Eds.), High-Throughput Next Generation Sequencing: Methods and Applications, vol. 733. Springer Protocols, Humana Press, New York, NY, pp. 173–183 (Chapter 12).

Gill, V.J., Tipton, H.W., Gersch, S.M., 1978. Rapid entry port for an anaerobic glove box. J. Clin. Microbiol. 8, 736–739.

Gordon, J.I., 2012. Honor thy gut symbionts redux. Science 336, 1251–1253.

Gottschalk, G., 1979. Bacterial Metabolism. Springer-Verlag, Berlin, Germany.

Gottschalk, G., 1986. Bacterial Metabolism, second ed. Springer-Verlag, New York, NY.

Gould, G.W., 1992. Ecosystem approaches to food preservation. J. Appl. Bacteriol. Suppl. 73, 56S–68S.

Granger, E.M., Bevis, T.H., Saka, Y., Southerland, S.A., Sampson, V., Tate, R.L., 2012. The efficacy of student-centered instruction in supporting science learning. Science 338, 105–108.

Ha, S.D., Ricke, S.C., Nisbet, D.J., Corrier, D.E., DeLoach, J.R., 1994. Serine utilization as a potential competition mechanism between *Salmonella* and a chicken cecal bacterium. J. Food Prot. 57, 1074–1079.

Hall, I.C., 1929. A review of the development and application of physical and chemical principles in the cultivation of obligately anaerobic bacteria. J. Bacteriol. 17, 255–301.

Hentges, D.J. (Ed.), 1983. Human Intestinal Microflora in Health and Disease. Academic Press, New York, NY.

Hinman, R.L., 1992. The decline of microbial physiology training in the United States. Am. Soc. Microbiol. News 58, 62–63.

Hobson, P.N. (Ed.), 1988. The Rumen Microbial Ecosystem. Elsevier Applied Science, New York, NY.

Hobson, P.N., Wallace, R.J., 1982a. Microbial ecology and activities in the rumen: part I. Crit. Rev. Microbiol. 9, 165–225.

Hobson, P.N., Wallace, R.J., 1982b. Microbial ecology and activities in the rumen: part II. Crit. Rev. Microbiol. 9, 253–320.

Holdeman, L.V., Moore, W.E.C., 1972. Roll-tube techniques for anaerobic bacteria. Am. J. Clin. Nut. 25, 1314–1317.

Holdeman, L.V., Cato, E.P., Moore, W.E.C. (Eds.), 1977. Anaerobe Laboratory Manual, fourth ed. Virginia Polytechnic Institute and State University, Blacksburg, VA.

Hood, L., 2012. Tackling the microbiome. Science 336, 1209.

Horrocks, S.M., Anderson, R.C., Nisbet, D.J., Ricke, S.C., 2009. Incidence and ecology of *Campylobacter* in animals. Anaerobe 15, 18–25.

Howard, Z.R., Donalson, L.M., Kim, W.K., Li, X., Zabala Diaz, I.B., Landers, K.L., Maciorowski, K.G., Ricke, S.C., 2006. Development of research paper writing skills of poultry science undergraduate students taking food microbiology. Poult. Sci. 85, 352–358.

Hungate, R.E., 1950. The anaerobic mesophilic cellulolytic bacteria. Bacteriol. Rev. 14, 1–49.

Hungate, R.E., 1960. I. Microbial ecology of the rumen. Bacteriol. Rev. 24, 353–364.

Hungate, R.E., 1966. The Rumen and its Microbes. Academic Press, New York, NY.

Hungate, R.E., 1969. A roll tube method for cultivation of strict anaerobes. In: Norris, J.R., Ribbons, D.W. (Eds.), Methods in Microbiology, vol. 3B. Academic Press, New York, NY.

Hungate, R.E., 1979. Evolution of a microbial ecologist. Annu. Rev. Microbiol. 33, 1–20.

Hungate, R.E., Smith, W., Clarke, R.T.J., 1966. Suitability of butyl rubber stoppers for closing anaerobic roll culture tubes. J. Bacteriol. 91, 908–909.

Ingraham, J.L., Maaloe, O., Neidhardt, F.C., 1983. Growth of the Bacterial Cell. Sinauer Associates, Inc., Sunderland, MA.

Isaacson, H.R., Hinds, F.C., Bryant, M.P., Owens, F.N., 1975. Efficiency of energy utilization by mixed rumen bacteria in continuous culture. J. Dairy Sci. 58, 1645–1659.

Jay, J.M., 2000. Modern Food Microbiology, sixth ed. Aspen Publishers, Inc., Gaithersburg, MD.

Jiang, W., Pinder, R.S., Patterson, J.A., 2012a. Influence on growth conditions and sugar substrate on sugar phosphorylation activity in acetogenic bacteria. Agric. Food Anal. Bacteriol. 2, 94–102.

Jiang, W., Pinder, R.S., Patterson, J.A., Ricke, S.C., 2012b. Sugar phosphorylation activity in ruminal acetogens. J. Environ. Health Sci. Part A 47, 843–846.

Joerger, R.D., 2003. Alternatives to antibiotics: bacteriocins, antimicrobial peptides and bacteriophages. Poult. Sci. 82, 640–647.

Johnson, M.E., Steele, J.L., 2013. Fermented dairy products. In: Doyle, M.P., Buchanan, R.L. (Eds.), Food Microbiology—Fundamentals and Frontiers, fourth ed. American Society for Microbiology, Washington, DC, pp. 767–781 (Chapter 35).

Jones, G.A., Pickard, M.D., 1980. Effect of titanium (III) citrate as reducing agent on growth of rumen bacteria. Appl. Environ. Microbiol. 39, 1144–1147.

Kafkewitz, D., Iannotti, E.L., Wolin, M.J., Bryant, M.P., 1973. An anaerobic chemostat that permits the collection and measurement of fermentation gases. Appl. Microbiol. 25, 612–614.

Kavanagh, F., 1963. Analytical Microbiology. Academic Press, Inc., New York, NY, 707 pp.

Knisely, S., 2014. Are the kids really all right? On Wisconsin 115, 30–33.

Koch, A.L., 1994. Growth measurements. In: Gerhardt, P., Murray, R.G.E., Wood, W.A., Krieg, N.R. (Eds.), Manual of Methods for General and Molecular Bacteriology. American Society for Microbiology, Washington, DC, pp. 248–277 (Chapter 11).

Kuipers, O.P., 1999. Genomics for food biotechnology: prospects of the use of high-throughput technologies for the improvement of food microorganisms. Curr. Opin. Biotechnol. 10, 511–516.

Kwon, Y.M., Ricke, S.C. (Eds.), 2011. High-throughput next generation sequencing: methods and applications. Methods in Molecular Microbiology, vol. 733. Springer Protocols, Humana Press, New York, NY, 308 pp.

Le Van, R.D., Robinson, J.A., Ralph, J., Greening, R.C., Smolenski, W.J., Leedle, J.A., Schaefer, D.M., 1998. Assessment of reductive acetogenesis with indigenous ruminal bacterium populations and *Acetitomaculum ruminis*. Appl. Environ. Microbiol. 64, 3429–3436.

Leake, L.L., 2014. Cultivating new credentials in cyberspace. Food Qual. Saf. 21, 41–43.

Levett, P.N. (Ed.), 1991. Anaerobic Microbiology: A Practical Approach. Oxford University Press, New York, NY.

Limayen, A., Ricke, S.C., 2012. Lignocellulosic biomass for bioethanol production: current perspectives, potential issues and future prospects. Prog. Energy Combust. Sci. 38, 449–467.

Ling, J.R., 1991. Robert E. Hungate's *The Rumen and Its Microbes* after 25 years. Lett. Appl. Microbiol. 13, 179–181.

Lungu, B., Ricke, S.C., Johnson, M.G., 2009. Growth, survival, proliferation and pathogenesis of *L. monocytogenes* under low oxygen or anaerobic conditions: a review. Anaerobe 15, 7–17.

Maciorowski, K.G., Ricke, S.C., 2000. Improving content writing instruction in an undergraduate food bacteriology class. Dairy Food Environ. Sanit. 20, 196–204.

Mackie, R.I., 2002. Mutualistic fermentative digestion in the gastrointestinal tract: diversity and evolution. Integr. Comp. Biol. 42, 319–326.

Mackie, R.I., White, B.A. (Eds.), 1997a. Gastrointestinal Microbiology. Volume 1: Gastrointestinal Ecosystems and Fermentations. Chapman and Hall, New York.

Mackie, R.I., White, B.A. (Eds.), 1997b. Gastrointestinal Microbiology. Volume 2: Gastrointestinal Microbes and Host Interactions. Chapman and Hall, New York, NY.

Macy, J.M., Snellen, J.E., Hungate, R.E., 1972. Use of syringe methods for anaerobiosis. Am. J. Clin. Nut. 25, 1318–1323.

Miller, T.L., Wolin, M.J., 1974. A serum bottle modification of the Hungate technique for cultivating obligate anaerobes. Appl. Microbiol. 27, 985–987.

Moat, A.G., Foster, J.W., Spector, M.P., 2002. Microbial Physiology, fourth ed. Wiley-Liss, Inc., New York, NY.

Moench, T.T., Zeikus, J.G., 1983. An improved preparation method for a titanium (III) media reductant. J. Microbiol. Methods 1, 199–202.

Monod, J., 1949. The growth of the bacterial cultures. Annu. Rev. Microbiol. 3, 371–394.

Montville, T.J., Chikindas, M.L., 2013. Biological control of foodborne bacteria. In: Doyle, M.P., Buchanan, R.L. (Eds.), Food Microbiology—Fundamentals and Frontiers, fourth ed. American Society for Microbiology, Washington, DC, pp. 803–822 (Chapter 31).

Montville, T.J., Matthews, K.R., 2005. Food Microbiology. American Society for Microbiology, Washington, DC, 380 pp.

Montville, T.J., Matthews, K.R., 2008. Food Microbiology, second ed. American Society for Microbiology, Washington, DC, 428 pp.

Montville, T.J., Matthews, K.R., 2013. Physiology, growth, and inhibition of microbes in foods. In: Doyle, M.P., Buchanan, R.L. (Eds.), Food Microbiology—Fundamentals and Frontiers, fourth ed. American Society for Microbiology, Washington, DC, pp. 3–18 (Chapter 1).

Moskovitz, C., Kellogg, D., 2011. Inquiry-based writing in the laboratory course. Science 332, 919–920.

Mossel, D.A.A., Struijk, C.B., 1992. The contribution of microbial ecology to management and monitoring of the safety, quality and acceptability (SQA) of foods. J. Appl. Bacteriol. Symp. Suppl. 73, 1S–22S.

Muthaiyan, A., Limayen, A., Ricke, S.C., 2011. Antimicrobial strategies for limiting bacterial contaminants in fuel bioethanol fermentations. Prog. Energy Combust. Sci. 37, 351–370.

Nicholson, J.K., Holmes, E., Kinross, J., Burcelin, R., Gibson, G., Jia, W., Pettersson, S., 2012. Host-gut microbiota metabolic interactions. Science 336, 1262–1267.

Nisbet, D.J., Corrier, D.E., Ricke, S.C., Hume, M.E., Byrd II, J.A., DeLoach, J.R., 1996a. Cecal propionic acid as a biological indicator of the early establishment of a microbial ecosystem inhibitory to *Salmonella* in chicks. Anaerobe 2, 345–350.

Nisbet, D.J., Corrier, D.E., Ricke, S.C., Hume, M.E., Byrd II, J.A., DeLoach, J.R., 1996b. Maintenance of the biological efficacy in chicks of a cecal competitive-exclusion culture against *Salmonella* by continuous-flow fermentation. J. Food Prot. 59, 1279–1283.

Nout, M.J.R., Rombouts, F.M., 1992. Fermentative preservation of plant foods. J. Appl. Bacteriol. Symp. Suppl. 73, 136S–147S.

Nout, M.J.R., Sarkar, R.K., Beuchat, L.R., 2007. Indigenous fermented foods. In: Doyle, M.P., Beuchat, L.R. (Eds.), Food Microbiology—Fundamentals and Frontiers, third ed. American Society for Microbiology, Washington, DC, pp. 817–835 (Chapter 38).

O'Bryan, C.A., Dittmar, R.S., Chalova, V.I., Kundinger, M.M., Crandall, P.G., Ricke, S.C., 2010. Assessment of a food microbiology senior undergraduate course as a potential food safety distance education course for poultry science majors. Poult. Sci. 89, 2542–2545.

O'Flaherty, S., Klaenhammer, T.R., 2011. The impact of omic technologies on the study of food microbes. Annu. Rev. Food Sci. Technol. 2, 16.1–16.19.

Ogg, J.E., Lee, S.Y., Ogg, B.J., 1979. A modified tube method for the cultivation and enumeration of anaerobic bacteria. Can. J. Microbiol. 25, 987–990.

Ogimoto, K., Imai, S., 1981. Atlas of Rumen Microbiology. Japan Scientific Soc. Press, Tokyo, Japan.

Pardue, S.L., 1997. Educational opportunities and challenges in poultry science: impact of resource allocation and industry needs. Poult. Sci. 76, 938–943.

Parish, M.E., Fleet, G.H., 2013. Wine. In: Doyle, M.P., Buchanan, R.L. (Eds.), Food Microbiology—Fundamentals and Frontiers. fourth ed.. American Society for Microbiology, Washington, DC, pp. 915–947 (Chapter 37).

Patterson, J.A., Burkholder, K.M., 2003. Application of prebiotics and probiotics in poultry production. Poult. Sci. 82, 627–631.

Pfeiler, E.A., Klaenhammer, T.R., 2013. Probiotics and prebiotics. In: Doyle, M.P., Buchanan, R.L. (Eds.), Food Microbiology—Fundamentals and Frontiers. fourth ed.. American Society for Microbiology, Washington, DC, pp. 949–971 (Chapter 38).

Pinder, R.S., Patterson, J.A., 2012. Glucose and hydrogen utilization by an acetogenic bacterium isolated from ruminal contents. Agric. Food Anal. Bacteriol. 2, 253–274.

Pinder, R.S., Patterson, J.A., 2013. Growth of acetogenic bacteria in response to varying pH, acetate or carbohydrate concentration. Agric. Food Anal. Bacteriol. 3, 6–16.

Pirt, S.J., 1975. Principles of Microbe and Cell Cultivation. Blackwell Scientific Publications, Oxford, UK.

Powell, K., 2010. Publish like a pro. Nature 467, 873–875.

Price, J., 2014. The price is right. On Wisconsin 115, 24–29.

Rees, C.E.D., Dodd, C.E.R., Gibson, P.T., Booth, I.R., Stewart, G.S.A.B., 1995. The significance of bacteria in stationary phase to food microbiology. Int. J. Food Microbiol. 28, 263–275.

Ricke, S.C., 1996a. Development of a gastrointestinal microbiology graduate course in poultry science. Poult. Sci. 75 (Suppl. 1), 81.

Ricke, S.C., 1996b. Incorporation of microbial physiology concepts into advanced food microbiology training of undergraduate students. Poult. Sci. 75 (Suppl. 1), 82.

Ricke, S.C., 2010. Future prospects for advancing food safety research in food animals. In: Ricke, S.C., Jones, F.T. (Eds.), Perspectives on Food Safety Issues of Food Animal Derived Foods. University of Arkansas Press, Fayetteville, AR, pp. 335–350 (Chapter 24).

Ricke, S.C., Pillai, S.D., 1999. Conventional and molecular methods for understanding probiotic bacteria functionality in gastrointestinal tracts. Crit. Rev. Microbiol. 25, 19–38.

Ricke, S.C., Schaefer, D.M., 1990. An ascorbate-reduced medium for nitrogen metabolism studies with *Selenomonas ruminantium*. J. Microbiol. Methods 11, 219–227.

Ricke, S.C., Schaefer, D.M., 1991. Growth inhibition of the rumen bacterium *Selenomonas ruminantium* by ammonium salts. Appl. Microbiol. Biotechnol. 36, 394–399.

Ricke, S.C., Schaefer, D.M., 1996a. Growth and fermentation responses of *Selenomonas ruminantium* to limiting and non-limiting concentrations of ammonium chloride. Appl. Microbiol. Biotechnol. 46, 169–175.

Ricke, S.C., Schaefer, D.M., 1996b. Nitrogen-limited growth response of ruminal bacterium *Selenomonas ruminantium* strain D to methylamine addition in a minimal medium. J. Rapid Methods Automat. Microbiol. 4, 297–306.

Ricke, S.C., Schaefer, D.M., Chang, T.S., 1994. Influence of methylamine on anaerobic rumen bacterial growth and plant fiber digestion. Bioresour. Technol. 50, 253–257.

Ricke, S.C., Martin, S.A., Nisbet, D.J., 1996. Ecology, metabolism, and genetics of ruminal selenomonads. Crit. Rev. Microbiol. 22, 27–65.

Ricke, S.C., Hererra, P., Biswas, D., 2012. Bacteriophages for potential food safety applications in organic meat production. In: Ricke, S.C., Van Loo, E.J., Johnson, M.G., O'Bryan, C.A. (Eds.), Organic Meat Production and Processing. Wiley Scientific/IFT, New York, NY, pp. 407–424 (Chapter 23).

Ricke, S.C., Koo, O.K., Keeton, J.T., 2013. Fermented meat, poultry, and fish products. In: Doyle, M.P., Buchanan, R.L. (Eds.), Food Microbiology—Fundamentals and Frontiers, fourth ed. American Society for Microbiology, Washington, DC, pp. 857–880 (Chapter 34).

Russell, J.B., 1984. Factors influencing competition and composition of the rumen bacterial flora. In: Gilchrist, F.M.C., Mackie, R.I. (Eds.), Herbivore Nutrition in the Subtropics and Tropics. The Science Press (PTY) Ltd., Craighall, South Africa, pp. 313–345.

Russell, J.B., 2002. Rumen Microbiology and its Role in Animal Nutrition. James B. Russell, Ithaca, NY.

Russell, J.B., Baldwin, R.L., 1978. Substrate preferences in rumen bacteria: evidence of catabolite regulatory mechanisms. Appl. Environ. Microbiol. 36, 319–329.

Russell, J.B., Hespell, R.B., 1981. Microbial rumen fermentation. J. Dairy Sci. 64, 1153–1169.

Saengkerdsub, S., Ricke, S.C., 2014. Ecology and characteristics of methanogenic archaea in animals and humans. Crit. Rev. Microbiol. 40, 97–116.

Sargeant, J.M., Torrence, M.E., Rajic, A., O'Conner, A.M., Williams, J., 2006. Methodological quality assessment of review articles evaluating interventions to improve microbial food safety. Foodborne Pathog. Dis. 3, 447–456.

Schaechter, E., 2013. Small things considered – gut lecture. Microbe 8, 522.
Schaefer, D.M., Davis, C.L., Bryant, M.P., 1980. Ammonia saturation constants for predominant species of rumen bacteria. J. Dairy Sci. 63, 1248–1263.
Senauer, B., 1992. Preparing students for careers in food distribution and marketing: an opportunity for colleges of agriculture. J. Food Distribut. Res. 23, 1–7.
Sharak Genther, B.R., Bryant, M.P., 1987. Additional characterization of one-carbon compound utilization by *Eubacterium limosum* and *Acetobacterium woodii*. Appl. Environ. Microbiol. 53, 471–476.
Snetsinger, D.C., 1992. Poultry science training – what industry needs. Poult. Sci. 71, 1308–1312.
Steinkraus, K.H., 1995. Handbook of Indigenous Fermented Foods, second ed. Marcel Dekker, Inc., New York, NY, 792 pp.
Suchman, E., 2015. Encouraging microbiology students to think like scientists. Microbe 10, 68–72.
Summers, J.D., 1992. Will poultry science curricula meet projected needs of poultry and associated industries? Poult. Sci. 71, 1316–1318.
Tachibana, C., 2015. Faculty: making your research count. Science 347, 561–564.
Tuitemwong, J., Fung, D.Y.C., Tuitemwong, P., 1994. Acceleration of yogurt fermentation by bacterial membrane fraction biocatalysts. J. Rapid Methods Automat. Microbiol. 3, 127–139.
Urian, A.L., Yerkie, S.M., McCrea, B.A., 2010. Teaching panel on university and industry collaboration. J. Appl. Poultry Res. 19, 313–315.
Ushijima, T., Seto, A., 1991. Selected faecal bacteria and nutrients essential for antagonism of *Salmonella typhimurium* in anaerobic continuous flow cultures. J. Med. Microbiol. 35, 111–117.
Van Soest, P.J., 1982. Nutritional Ecology of the Ruminant. O and B Books, Inc., Corvallis, OR.
Vaughan, A., O'Sullivan, T., van Sindren, D., 2005. Enhancing the microbiological stability of malt and beer – a review. J. Inst. Brew. 111, 355–371.
Vedamuthu, E.R., Reinbold, G.W., 1967. The use of candle jar incubation for the enumeration, characterization, and taxonomic study of propionibacterium. Milchwissenchaft 22, 428–431.
Veldkamp, H., 1972. Mixed culture studies with the chemostat. J. Appl. Chem. Biotechnol. 22, 105–123.
Wachenheim, D.E., Hespell, R.B., 1984. Inhibitory effects of titanium (III) citrate on enumeration of bacteria from rumen contents. Appl. Environ. Microbiol. 48, 444–445.
Walker, E.L., 2011. Engaging agriculture students in the publication process through popular press magazines. North Am. Coll. Teach. Agric. 55, 53–58.
Walter, J., Ley, R., 2011. The human gut microbiome: ecology and recent evolutionary changes. Annu. Rev. Microbiol. 65, 411–425.
Weimer, P.J., 1992. Cellulose degradation by ruminal microorganisms. Crit. Rev. Biotechnol. 12, 189–223.
Weimer, P.J., Russell, J.B., Muck, R.E., 2009. Lessons from the cow: what the ruminant animal can teach us about consolidated bioprocessing of cellulosic biomass. Bioresour. Technol. 100, 5323–5331.
Wolin, M.J., 1979. The rumen fermentation: a model for microbial interactions in anaerobic ecosystems. Adv. Microb. Ecol. 3, 49–77.
Wolin, M.J., Miller, T.L., 1982. Interspecies hydrogen transfer: 15 years later. Am. Soc. Microbiol. News 48, 561–565.
Wolin, M.J., Miller, T.L., 1983. Interactions of microbial populations in cellulose fermentation. Fed. Proc. 42, 109–113.
Wood, B.J.B., 1985. Microbiology of Fermented Foods, vol. 1. Elsevier Science Publishing Co., Inc., New York, NY.
Woolcock, J.B., 1991. Microbiology of Animal and Animal Products. Elsevier Science, New York, NY.
Zehnder, A.J.B., Wuhrmann, K., 1976. Titanium (III) citrate as a nontoxic oxidation-reduction buffering system for the culture of obligate anaerobes. Science 194, 1165–1166.

CHAPTER

SYSTEMS-THINKING AND BEEF CATTLE PRODUCTION MEDICINE: ISSUES OF HEALTH AND PRODUCTION EFFICIENCY

20

Robert (Bob) L. Larson

Department of Clinical Sciences, College of Veterinary Medicine, Kansas State University, Manhattan, Kansas, USA

1 INTRODUCTION

A systems approach to food production and food safety recognizes the diversity of interacting components that impact producing, processing, delivering, and consuming foodstuffs and the need to recognize that the whole system is more than the sum of its parts (Ebersohn, 1976; Rountree, 1977). One of the key foodstuffs in many countries is beef. Beef cattle production in the United States is done within a system that typically includes grazing mature breeding cattle on large amounts of land per cow; long gestation and growth periods so that animals are sold for food 2-3 years from the time they were conceived; feeding young, growing cattle high-calorie diets for a few months in large populations immediately prior to slaughter; and, in most situations, having more than two changes of ownership from birth to being sold for food. Because of the long time lags and multiple changes in ownership between intervention decisions and health and economic outcomes, a systems approach is necessary to accurately evaluate numerous potential management interventions to optimize animal health and productivity of U.S. beef cattle production.

Veterinarians provide important services to cattle producers, beef consumers, and society at large within the beef cattle production system. As a profession, veterinarians work within wide boundaries that encompass much of the overall food system. Veterinary professionals work with producers to ensure the health, welfare, and productivity of breeding herds and growing and finishing herds. Veterinarians have integral roles within slaughter and processing facilities to ensure animal welfare and food safety expectations are met. Additionally, they serve within state and federal government research agencies to discover new methods to enhance animal health and welfare, improve food safety, and conserve specific resources.

Many factors have been influenced in the current system of beef production in the United States. While these influences extend beyond the boundaries of individual farms and ranches, they nevertheless impact the production and health efficiencies and risks that veterinarians face at the herd level. Because of the ruminant digestive system, beef cattle are able to utilize low-quality forages and a wide variety of feed types, allowing them to thrive in diverse environments. Beef cattle are raised in every U.S. state, and the diversity in latitude, altitude, rainfall, and soil types combine to create vastly

Food Safety. http://dx.doi.org/10.1016/B978-0-12-800245-2.00020-4

different forage productivity and quality in different cattle-raising areas (USDA NASS, 2014). Very few areas in the United States maintain high-quality and high-quantity forage production throughout the year, necessitating the use of stored forages such as hay or silage or the supplementation of standing dormant forages during periods of the year when available grazed forage is not sufficient for optimum productivity and health. Supplemental forages, grains, and other feeds differ greatly in availability and cost in different parts of the United States. Factors that influences the U.S. beef production systems to differ from production systems in other parts of the world include high land prices; high cost of labor; relatively low feed prices; an excellent system of rivers, railroads, and highways; the low cost and ready accessibility of technology; and a variety of climates. Because of the these structural influences, cow-calf production is predominant in areas with relatively low land prices and good forage production; feed production is predominant in areas with high land prices and abundant rainfall or irrigation; and feedlot production is predominant in areas with low rainfall, low land prices, and good access to feed grains. Because of these differences between regions of the United States as well as the excellent transportation infrastructure, different beef cattle production sectors are optimized in different geographic locations, which lead to multiple owners during each cattle's lifetime, and commingling of cattle from previous owners following transactions. From an animal health perspective, the U.S. system is challenged by frequency of commingling of cattle from multiple sources and transportation of cattle long distances between production segments. As individuals, veterinarians are likely to deal with system boundaries that are narrower than the food system as a whole, but that extend wider than the specific operations that they serve on a daily basis.

Veterinarians in clinical practice often serve a problem-solving role when called upon by cattle producers experiencing undesirable levels of cattle health or production. Within this role, they must assess the current situation, develop a rule-out list of plausible contributing factors, determine the best tests to narrow the rule-out list to the actual contributing factors, compare alternate interventions, implement one or more management changes, and monitor the outcome to determine if the problem has been solved or if the investigation process needs to continue.

Training veterinarians to approach beef cattle production medicine from a systems-thinking perspective builds upon existing strengths of veterinary education and practice, but expands traditional training to incorporate specific systems-thinking skills and methods. A systems approach to beef cattle production medicine will entail a veterinarian considering all facets of animal production and health when evaluating each potential change in cattle management. It addresses the question, "If I change this input or management, what is the effect on cattle health, human health, labor requirements, land utilization, water utilization, the environment (short-term and long-term horizons), many different types of biologic and economic risk, and the interaction of these considerations?"

2 DISEASE SYNDROMES WITH "SYSTEMS" CAUSES MUST BE SOLVED AT THE SYSTEM LEVEL

The health problems faced by U.S. cattle are typically syndromes that result from one or more of several infectious agents, toxins, or other insults. Rarely is one risk factor (such as a particular infectious agent) sufficient to cause disease in cattle. A combination of risk factors including pathogen factors, animal host factors, and environmental factors is necessary to elicit the disease syndromes commonly encountered in U.S. beef cattle production. Common syndromes addressed by U.S. veterinarians include reproductive losses, digestive tract disease, respiratory tract disease, and musculoskeletal system disease. These syndromes have complex causal webs that include the inciting agent, the host animal, and the cattle's environment. Because these problems are the result of complex systems, solutions must be found at the system level, not the individual animal level. For example, calves less than 1 month of age are susceptible to a number of bacterial, viral, protozoal, and parasitic agents that can result in severe diarrhea. These agents are all very common and, in fact, are present in nearly all beef cattle operations and in the digestive tract of most cattle; but clinical disease is only encountered when environmental factors such as poor sanitation allows large populations of the disease agents to concentrate, and animal factors such as age, innate immunity, and ability to repair tissue allow the animal to be susceptible to clinical disease. Given the characteristics of this syndrome, if a calf develops diarrhea and dies, a strategy to take tissue samples from the intestinal tract of the affected animal in order to identify which of several potential pathogens is present is not likely to provide necessary information to solve the disease problem. Because multiple pathogens can be contributors to the syndrome of diarrhea in calves, even if one could remove the pathogen associated with the most recent case from the population, other agents coupled with the remaining host and environment risk factors would leave the population at the same risk for diarrhea. In other words, when the disease is associated with a production system, addressing one infectious agent from a group of potential pathogens will not solve the disease problem; only by addressing the whole system can the outcome be changed as veterinarians move beyond the presenting problem to a study of the "big picture" (Hurd, 2011). Similar examples of failure to solve system-associated health problems by addressing specific infectious agents can be found with respiratory disease complex, musculoskeletal disease, and other common syndromes encountered by veterinarians serving U.S. beef-producing clients.

3 USING SYSTEM TOOLS TO IMPROVE DECISION ALGORITHMS

One of the key skills that veterinarians need to master is to move through a clinical decision algorithm quickly and efficiently. Without becoming proficient with this skill, the tremendous volume and detail of available information can be a hindrance rather than a benefit for problem solving. Veterinarians must learn which questions to ask, which physical examination findings to emphasize, and which additional diagnostic tests to order. Until this skill is mastered, a veterinarian may randomly ask questions, examine physical characteristics, and request diagnostic tests with the hope that some identifiable diagnosis appears. Efficient clinical decision-making saves the client time and money, reduces distress to the animal in situations of injury or disease, and correctly identifies the components of the problem and practical intervention alternatives. When a veterinarian is tasked with solving a production or health problem, systems-thinking concepts can help identify the best initial questions, most important physical examination findings, and most logical diagnostic tests to move through a problem to a logical intervention strategy. Submitting the client to random or exhaustive questions is not likely to capture valuable information, but strategic initial questions are likely to identify the most important examination findings at the animal level and the farm level, which can lead to identifying the most logical diagnostic tests so that time, money, and attention is not wasted on unnecessary tests.

The concept of stocks and flows and modeling the feed forward and feedback loops of a system can be adapted to veterinary decision making. An example of using these systems-thinking tools is when veterinarians investigate client concerns that too few cows are pregnant at a point in mid-gestation when the herd is evaluated for pregnancy status. A few key facts must be known to accurately model the flow of reproducing cows through several stages of a reproductive year. These facts include the expected length of bovine gestation is about 283 days; female cattle have 21-day estrous cycles such that approximately every 21 days, a female will ovulate a fertile oocyte (egg) and express behavior that initiates mating; the likelihood of a fertile mating resulting in pregnancy that can be detected 50 days later is 60-70% (BonDurant, 2007); and beef cattle have a period of time after calving, called postpartum anestrus, when they do not display the behavioral aspects of estrus necessary to initiate mating and they do not ovulate fertile eggs. This period of time is longer following the first pregnancy (80-100 days) (Berardinelle and Joshi, 2005; Ciccioli et al., 2003) compared to later pregnancies (50-80 days)(Cushman et al., 2007; Lents et al., 2008). Recognizing that there are 283 days of gestation and 365 days in a year means that there are 82 days from the time a cow calves to the time she needs to become pregnant again if she is going to calve about the same time each year. Although cattle do not have a seasonal pattern to breeding and fertility so that they are able to become pregnant throughout the year, cattle managers prefer that beef cows give birth during relatively short (2-3-months) predetermined times of the year that optimize the forage production of a region or to meet some other management or marketing goal.

From a reproductive standpoint, mature female cattle should pass through a series of "states" each year (Figure 1). Starting in late pregnancy, a cow moves from the "state of pregnancy" prior to calving

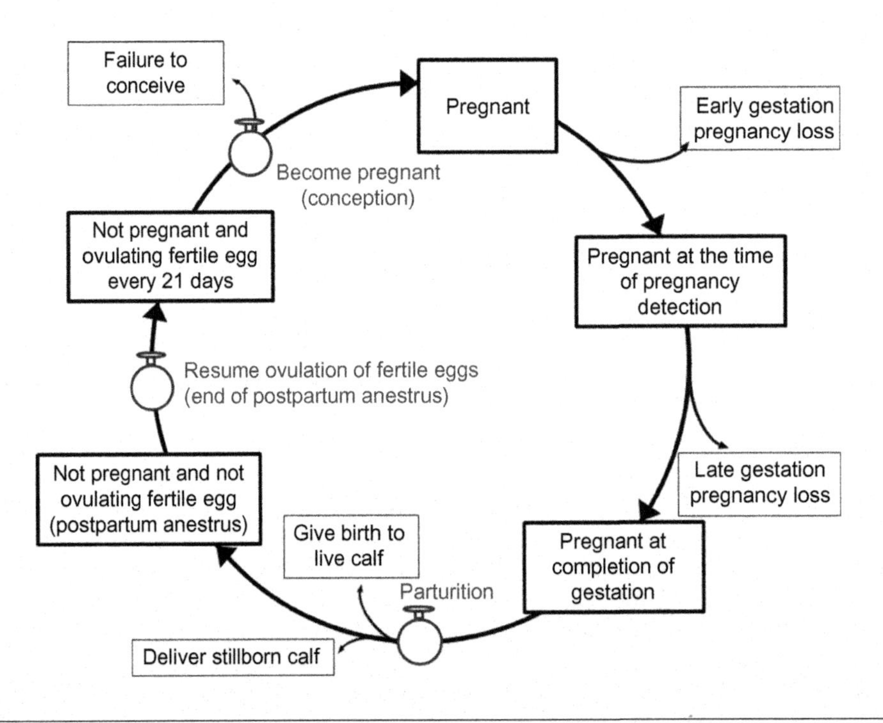

FIGURE 1

Beef cattle females flow through specific reproductive "states." Starting at late pregnancy, cows move from the state of being "pregnant"—to the state of being "not pregnant and not ovulating fertile eggs" at the time of parturition. Once the postpartum anestrous period has ended, cows move into the state of being "not pregnant and ovulating fertile eggs every 21 days" and upon mating with a fertile bull and subsequent conception, cows move to the state of being "pregnant" until the subsequent parturition. There are several alternative pathways in this system including: failure to conceive, early gestation pregnancy loss, late gestation pregnancy loss, and delivery of a stillborn calf.

to the "state of being nonpregnant but not having fertile ovulations" once parturition has completed. The cow will then resume ovulating fertile eggs every 21 days once the postpartum anestrus period is passed and she enters the "state of being nonpregnant and having fertile 21-day estrous cycles." If the cow is in the presence of a fertile bull, the subsequent mating has about a 60-70% probability of a pregnancy that can be detected 50 days later and, subsequently, a calf being born at the end of gestation. About 30% of the time following a fertile mating, even though fertilization takes place and an early embryo is formed, the embryo is imperfect and will die within the first 14 days. When an embryo is lost before 14 days, the cow will continue fertile estrous cycles and will ovulate another fertile egg about 21 days after the last ovulation and have another 60-70% chance of conceiving and maintaining a pregnancy. Cows that calve in the first 30-40 days of the calving season are likely to have completed the postpartum anestrous period and to be having fertile estrous cycles by the start of the subsequent breeding season, and they will have three opportunities to get bred in a 65-day breeding season. If a cow has three opportunities (each with a 65% chance of a successful pregnancy) during a 65-day breeding season, the cow has a 95% probability of giving birth to a calf at the end of gestation. In contrast, cows that give birth to a calf more than 40 days after the start of the calving season are not likely to have completed the postpartum anestrus period by the start of the subsequent breeding season and will not resume fertile estrous cycles until 3 weeks or later into the breeding season; therefore, these cows will only have two or fewer opportunities to become pregnant in a 65-day breeding season. If a cow has two opportunities to become pregnant, she will have about an 85-90% chance of maintaining a pregnancy to the end of gestation.

If a client is concerned that too few calves are born alive from the cows that entered the breeding season, a veterinarian looking at the flow of cows through the reproductive cycle will recognize that the potential pathways that result in calves not being born alive include failure to conceive, early gestation pregnancy loss, late gestation pregnancy loss, and birth of stillborn calves. Fortunately for this investigation, the most likely rule-outs for the various alternate pathways that result in cows that enter the breeding season failing to give birth to live calves are largely mutually exclusive with very little overlap. To begin the clinical decision algorithm, questions that specifically define the number of stillborn calves will determine whether excessive stillbirths is the problem (prompting follow-up investigation of the causes of stillbirths) or whether that alternate pathway can be ruled out in the current investigation. If stillbirths are not the problem, then specific questions about the number of cows that were pregnant at mid-gestation when pregnancy status was determined compared to the number of cows giving birth to calves will determine whether pregnancy loss occurred after mid-gestation and before parturition. If late gestation pregnancy loss is identified as the problem, a rule-out list of infectious agents or toxins that cause late gestation losses can be investigated with specific diagnostic tests for the most likely rule-outs. If late gestation pregnancy loss is not the problem, then physical examination findings and diagnostic tests specific for infectious, toxic, and physiologic insults that are associated with early gestation loss can be initiated. However, if early gestation pregnancy loss is not likely, then failure to conceive can be investigated by using physical examination findings and herd records to indicate whether a male infertility or female infertility problem is the most likely cause of the failure to conceive. Once the distinction of which sex is mostly likely associated with the conception failure has been made, specific diagnostic tests or examination of records can be used to investigate the principle male or female problem.

By starting with the flow of beef cows through the reproductive system, veterinarians can go through a diagnostic decision tree as efficiently as possible with the result of saving time and money, and improving animal health. Systems-thinking helps identify the best initial and follow-up questions, most important clinical exam findings, and most logical diagnostic tests to move through a clinical presentation to a logical intervention strategy.

4 SUMMARY

Systems-thinking has tremendous potential to improve the efficiency and effectiveness of veterinarians practicing beef cattle production medicine to enhance the health and production efficiency of U.S. beef herds. A process cannot be completely understood without relating it to the other processes with which it interacts (Rountree, 1977). Currently, veterinary education and practice emphasizes understanding the biologic systems that impact individual animal health and disease and epidemiologic principles that consider health and disease in populations of animals. These foundational skills of veterinary medicine can be enhanced by learning and adopting systems-thinking skills that extend the understanding of biologic systems beyond individual animals and groups of animals within a single herd, to industry, ecologic, and economic systems (Hurd, 2011).

REFERENCES

Berardinelle, J.G., Joshi, P.S., 2005. Introduction of bulls at different days. J. Anim. Sci. 83, 2106–2110.

BonDurant, R.H., 2007. Selected diseases and conditions associated with bovine conceptus loss in the first trimester. Theriogenology 68, 461–473.

Ciccioli, N.H., Wettemann, R.P., Spicer, L.J., Lents, C.A., White, F.J., Keisler, D.H., 2003. Influence of body condition at calving and postpartum nutrition on endocrine function and reproductive performance of primiparous beef cows. J. Anim. Sci. 81, 3107–3120.

Cushman, R.A., Allan, M.F., Thallman, R.M., Cundiff, L.V., 2007. Characterization of biological types of cattle (cycle VII): influence of postpartum interval and estrous cycle length on fertility. J. Anim. Sci. 85, 2156–2162.

Ebersohn, J.P., 1976. A commentary on systems studies in agriculture. Agr. Syst. 3 (1), 173–184.

Hurd, S.H., 2011. Food systems veterinary medicine. Anim. Health Res. Rev. 12 (2), 187–195.

Lents, C.A., White, F.J., Ciccioli, N.H., Wettemann, R.P., Spicer, L.J., Lalman, D.L., 2008. Effects of body condition score at parturition and postpartum protein supplementation on estrous behavior and size of the dominant follicle in beef cows. J. Anim. Sci. 86, 2549–2556.

Rountree, J.H., 1977. Systems thinking – some fundamental aspects. Agr. Syst. 4 (2), 247–254.

USDA NASS (National Agricultural Statistics Service), 2014. http://www.agcensus.usda.gov/ (accessed October, 2014).

CHAPTER

FOOD SAFETY TRAINING AND TEACHING IN THE UNITED KINGDOM AND EUROPE

21

J.S. Anil de Sequeira, Iain Haysom, Richard Marshall

Department of Science, School of Society, Enterprise and Environment, Bath Spa University, Newton St Loe, Bath, UK

1 INTRODUCTION

Food science and technology was developed as a field of study in the United Kingdom only in the early 1950s, with the establishment of the National College of Food Technology in 1952 and the development of food science degree courses in four UK universities and one polytechnic. Until then, the food industry drew on the expertise of chemists, microbiologists, physicists, and others who acquired knowledge of food and gradually took on the role of food technologists (Mounfield, 1966). In fact, lecturers in food science and technology do continue to have qualifications in chemistry, microbiology, biological sciences, and chemical engineering; though, lecturers with backgrounds in chemical engineering tend to teach food technology courses.

Food science and technology developed from the home economics curriculum. One can trace the origins of home economics to the start of the twentieth century when it was formed as a field of study in the United States by a group of women, most of whom were scientifically educated and reform-oriented, as well as men who were interested in applying philosophy to improving everyday life.

Students from the National College of Food Technology at Weybridge in Surrey, UK, which later became the Food Science and Technology Department at the University of Reading, UK, initiated the formation of the Institute of Food Technology, now the Institute of Food Science and Technology (IFST) (Mounfield, 1966). They felt that it was important for them to have something more than a paper qualification in order to ensure that they establish a rightful and clearly defined place for themselves in the food industry (Mounfield, 1966). The IFST is the qualifying body for food professionals in Europe and is concerned with all aspects of food science and technology.

Food safety education is delivered in the main as part of the curriculum of food science, food technology, and food and nutrition courses. In the United Kingdom, food safety training for industry is provided by registered training providers (e.g., the Chartered Institute of Environmental Health or Highfield Ltd.). Training for the industry in the United Kingdom is also provided by academic institutions as outreach work. In other parts of Europe, the majority of training for the industry is provided by academic institutions as outreach work.

Food Safety. http://dx.doi.org/10.1016/B978-0-12-800245-2.00021-6

In the United Kingdom, there are different levels of food safety training. Level 1 awards in food safety aim to raise awareness of key food safety issues and to provide employees with an introduction to food hygiene. On successful completion of Level 1 award, one can then obtain Level 2 qualification in food safety. Level 2 qualification provides the trainees with a firm grasp of the importance of food safety and knowledge of the systems, techniques, and procedures involved; understanding of how to control food safety risks (personal hygiene, food storage, cooking, and handling) and confidence and expertise to safely deliver quality food to customers. Level 3 qualification provides a thorough understanding of food safety procedures emphasizing the importance of monitoring staff and controls. Level 4 qualification meets the food industry's need for a high-level practical qualification with external accreditation. On successful completion of Level 4 qualification, one can deliver food safety training and become an accredited trainer (Chartered Institute of Environmental Health, 2015).

In the United Kingdom, some larger organizations have their own accredited food safety trainers that require those employees with a food safety permit and without an accredited professional food safety qualification to obtain the accredited professional qualification in-house.

Increasingly, higher-education institutions in the United Kingdom that deliver food safety as part of their curriculum are affording students the opportunity to obtain additional accredited professional qualifications.

For over 150 years a food microbiological approach has been taken with regard to the study of food safety (Griffith, 2006). Griffith (2006) rightly states that food safety is more than just food microbiology. Food safety is a multidisciplinary problem requiring new approaches (Mortlock et al., 2000). Companies have to address the complex mix between market, supply chain, and regulations in order to provide safer food (Hobbs et al., 2002).

The plethora of food safety initiatives in the United Kingdom were driven by the 1990 Food Safety Act and the bovine spongiform encephalopathy (BSE) (or "mad cow disease") crisis (Hobbs et al., 2002).

Foodborne illnesses have long been identified as one of the most widespread problems of the contemporary world (Wheelock, 1989; Notermans et al., 1994) with their notified incidence having increased significantly since the 1980s (Public Health England, 2015; Todd, 1989; Mossel, 1989). This has been attributed to the malpractice that occurs when food is prepared for human consumption in small food businesses, canteens, residential homes, and other places (Motrajemi and Käferstein, 1999). Therefore, it is important that adequate training is provided for the aforementioned food handlers, as a failure to provide adequate training and education within such premises could result in significant public health risks (Seaman and Eaves, 2006).

Seaman and Eaves (2006) reviewed the literature pertaining to the role of food hygiene training in a strategy to manage food safety. They state that one cannot assume that knowledge provided by traditional training translates into desired changes in behavior. In order for the training to be effective in terms of the actual change in behavior of the food handler to ensure that safe practices are carried out at all times, the trainee needs to be supported both pre- and post training by their organization and colleagues. Adequate resources also need to be made available (Seaman and Eaves, 2006).

The food industry faces the short-term challenge of providing safe, nutritious, and desirable food and the long-term challenge of providing for a growing world population when climate change will make food supply ever more critical. The industry accounts for 2% of European GDP and 13.5% of total employment in the EU manufacturing sector (European Commission, 2010). It plays a vital role in satisfying the needs of consumers and contributes annually more than €600 billion to the EU economy.

A number of high-profile product safety events and recalls have heightened public attention to the safety, provenance, and security of the products that they consume and use (Maruchecka et al., 2011); for example, the recent horse meat crisis. Supply chains have become increasingly complex, so traceability and the testing of products have become priorities. While the large companies in the supply chain have the knowledge and resources to cope with these challenges, staff in SMEs often have major gaps in their expertise.

We also need to meet the challenge of providing a sustainable and secure supply of safe, nutritious, and affordable high-quality food (Global Food Security, 2013). Developments in food science and technology have the potential to transform food security, safety, and quality. Yet the losses along the supply chain are immense, even in developed countries.

The costs of ingredients, equipment, energy, and waste disposal are continually increasing. Food businesses need to operate as efficiently as possible, reduce waste in all forms, and reuse and recycle materials whenever possible. There is a great need to understand the environmental impact of operations, including the carbon and water footprints, as this will give insights into the efficiency of those operations and where savings can be made.

2 PRINCIPLES OF SAFETY AND HYGIENE TRAINING

Against the background of a rapidly changing food system, increasing internationalization, new technologies in both food production and packaging, alongside the wider issue of population growth and the need for a sustainable and secure food system, the focus on food safety has perhaps never been greater. The move over the past few decades from a quality control-based approach to a risk analysis and safety management-based approach has led to an improvement in our academic understanding of food safety issues, but this has perhaps not always been translated into better practice on the front line of food preparation. The continuing high incidence of food poisoning cases and outbreaks in the United Kingdom and Europe would indicate that there is much progress still to be made. While a significant number of food poisoning outbreaks originate in the home due to poor hygiene practice by the domestic cook, improvements in hygiene practice in the food production and catering industries are needed to significantly reduce the burden of foodborne illness.

One key element in improving food hygiene is, of course, ensuring members of the staff are appropriately trained. It is a legal requirement (Regulation (EC) No. 852/2004 on the Hygiene of Foodstuffs, Chapter XII Annex II) that food workers are trained in food hygiene commensurate to their duties. There is no legal requirement in the United Kingdom that food workers hold an accredited food hygiene award, although many businesses do adopt this approach. Competence in food hygiene can be assessed in-house by a suitably qualified or trained individual, often via on the job training, by completing courses remotely at qualified training providers, or increasingly via online assessments.

Poor hygienic practice is at the root of the majority of food poisoning outbreaks, and often this poor practice is what could be considered to be elementary mistakes such as poor time and temperature control or inadequate separation of high- and low-risk activities and foods. Many studies investigating hygiene behaviors both commercially and domestically have shown that while knowledge of what constitutes good hygiene practice may be high, compliance with those hygiene practices is invariably lower (Redmond and Griffith, 2003; Clayton et al., 2002; Williamson et al., 1992).

This suggests that hygiene training based primarily on straightforward provision of information may not be effective in leading to a real improvement in food safety; hygiene training is not always converted into hygienic practice on the "shop floor," provision of information does not necessarily lead to behavior change, and therefore hygiene training needs to take a much broader approach to the issue to understand why this is.

The Global Food Safety Training Survey undertaken by Campden BRI in conjunction with Alchemy Systems, BRC Global Standards, SQF, and SGS in 2015, surveyed over 25,000 food manufacturing sites across the world. Over three-quarters of the respondents reported that finding the time for training was their greatest challenge. The surveys revealed that almost a quarter of the industry employees surveyed receive less than 4h of food safety training per year. The respondents recognized that improved food safety, better product quality, and a reduction in customer complaints are the greatest benefits of effective food safety training. One of the alarming findings was that one of the biggest deficiencies identified was incomplete training records, indicating that there is room for improvement with regard to food safety training in the food and drink industry (Global Food Safety Training Survey, 2014).

A review of 46 studies on commercial food safety and hygiene training programs from across the world highlighted that the majority of these studies did not investigate the attitude of food handlers or food managers, and there were limitations with methodologies where behaviors or practices were evaluated (Egan et al., 2007). The review concluded that a serious limitation to food hygiene training schemes was a lack of understanding of the key elements that would result in positive behavior change.

Effective food safety training programs need to do more than teach the fundamentals of microbiology, contamination, and good hygiene practice. Understanding human behaviors and attitudes is fundamental to designing food safety and hygiene training that does alter the behavior of food workers in a positive manner and, thus, reduce the risk of food poisoning outbreaks or food contamination events from occurring. Additionally, such programs need to take account of the wide variety of activities that a food worker may be involved with during his or her shift.

The apparent failure of traditional food hygiene training to significantly improve hygienic practice is evidenced by the continued high incidence of food poisoning outbreaks despite increasing numbers of food workers undergoing accredited training programs (Seaman, 2010). This has led to a focus on investigating the value of using social cognition models in an effort to understand what motivates food workers to comply or not with their training on a day-to-day basis. Such studies have tended to focus on adaptations of the health belief model (Becker, 1974), and the theory of planned behavior (Ajzen, 2005), which attempt to explain behaviors by understanding the drivers for and barriers against those behaviors.

The theory of planned behavior (TPB) states that an individual's intention to carry out a particular behavior is a result of three factors: (1) his or her attitude toward the behavior, (2) the social pressure to perform the behavior, and (3) his or her ability to perform the behavior (Ajzen, 2005). The first is regarded as a personal factor and reflects the individual's own positive or negative feelings toward carrying out the behavior. The second factor, the social or peer pressure to perform the behavior, is termed the subjective norm or normative influences and reflects the expected behavior in any particular scenario. Finally, the third factor is termed perceived behavioral control (PBC) and reflects the individual's perception regarding the ease or otherwise with which he or she can perform the behavior. Ultimately, an individual's intention to behave in a certain way is a function of his or her positive or negative attitude toward the behavior, the peer pressure the individual feels to carry out the behavior, and how easy it is for him or her to carry out the behavior. The more positive his or her attitude, the greater the peer pressure, and the easier the behavior is to perform, the more likely it is that the individual intends to act in that way. It should be noted that these three factors are key to understanding the intent to act in a certain manner, not necessarily that the behavior will be performed.

Research suggests that adopting a TPB approach can improve understanding of intent to follow a certain behavior, even if prediction of actual behavior is less precise (Connor and Norman (2005) cited in Mullan and Wong, 2009). However, there is evidence that another important factor relating to food handling behavior is past behavior; when behaviors are performed consistently they are important predictors of future behavior. Mullan and Wong (2009) investigated this idea using university students and food hygiene behavior and showed that while the TPB predicted 66% of variance in intention to prepare food hygienically, including past behavior in the model increased this to 69%. Subjective norms were significant at the $p<0.05$ level, but PBC and past behavior were significant at $p<0.01$. Intention was the only significant ($p<0.01$) predictor of behavior in the TPB model, but when the model included past behavior, only past behavior was significant ($p<0.01$). The authors highlighted the significant result for subjective norms and suggested that this was to be expected in scenarios when behaviors affect other people, as is the case with food handling behaviors. Despite the limitations of this study, it does suggest both the value of TPB and suggests that social pressure can be a positive influence on food hygiene behavior, and that training programs should aim to establish hygienic food handling as a habit.

Other studies have provided similar results in terms of the value of TPB at explaining intentions, and the importance of societal norms at influencing intentions (Mullan and Wong, 2010; Seaman and Eaves, 2010a). However, not all variance in intention is explained by such studies showing more research is needed, and factors such as the nature and intensity of the work can play a key role in determining whether intentions are translated to behavior. It has been shown that the level of hygiene behavior tends to correlate negatively with intensity and pressure of the work (O'Boyle et al., 2002).

Alongside the content of training courses themselves, the food industry must also consider the support systems that are in place to help those who have been trained to follow best practice and implement safe food handling practices. Without such support, any benefits of training in terms of more hygienic behaviors can be short lived (Seaman and Eaves, 2010b). Support and positive acknowledgement of training as well as ensuring that employees have the necessary resources to handle food safely are important elements that managers must ensure are in place to encourage continued correct food handling behaviors. Training in the workplace rather than in remote locations has been identified as being an important factor (Egan et al., 2007) as well as ensuring that employees recognize the different approaches to on the job training and support (Seaman and Eaves, 2010b).

3 TEACHING STYLES IN HIGHER EDUCATION ACROSS EUROPE

In all European universities where food science and technology is taught, food safety is a core part of the syllabus. Indeed, if it were not, the courses would not reach the basic level required for the subject. However, the way in which it is taught varies somewhat across the European countries. While the style of delivery is changing, it is still partly true that the farther east one goes, the more didactic the teaching. Universities in the more northerly countries have tended to use a more student-centered approach to teaching that gives students more autonomy in their learning than those in the east, who use more traditional methods (Osborn et al., 2003). In part, this difference is attributed to the lack of experience of faculty in institutions that are still undergoing major changes in their educational systems (Fernández-Díza et al., 2010).

Training is designed to build the knowledge and understanding of food operatives so that there is an overall culture of food safety (Powell et al., 2011). These authors review several incidents of failed food safety management in Europe, the United States, and Canada. They emphasize that improving food safety depends on creating cultural change within a food business. Just because people have the knowledge does not mean that they will apply it and change their practices. The essential way to bring about the necessary cultural change is to develop a sense of personal responsibility for food safety. Powell et al. (2011) believe that information sheets (Food Safety Infosheets) help develop the right culture (at the time of writing, the particular website is only providing links to training providers). The other area that can drive food safety is marketing. By selling not only product but also the claim to be "safer," food businesses can increase their market share. Powell et al. (2011) suggest that developing an open, transparent operation is the key to this. Thus, training in a food business requires a focus not only on managing food safety through production but also on publicizing the fact to customers.

A significant driver in the changes in the teaching of food safety, as part of the food science curriculum, are the series of thematic network projects that started in 1998 as Foodnet (Dumoulin, 2010). The Foodnet projects lasted until 2001 and were succeeded by four consecutive ISEKI projects, the last of which finished in September 2014. The partners were from European universities where food science is taught (ISEKI Food Association, 2014). In the projects, which were funded by the EU, data was collected on the structure of the food science curriculum, its delivery, and the needs of the food industry and consumers. The partners in these projects shared their experiences in teaching and developed a number of case studies for teaching food safety, as well as compiling databases of teaching resources that can be shared and used by the community.

The Foodnet and ISEKI projects collected information on the range of subjects within food studies courses in Europe and elsewhere. Members of both projects were invited at different times to submit examples of case study material used to teach various aspects of food safety.

4 EXAMPLES OF FOOD SAFETY CASE STUDY MATERIALS USED IN BSc AND MSc COURSES IN FOOD STUDIES

In their case study for students on using HACCP to control the microbiology of sliced ham, Tudorache and Goursaud (2001) set out the context for applying HACCP to this product, the legal situation in France, and the microbiological criteria for cooked presliced ham. Students are then guided through the development of the process flow diagram and the HACCP plan. This example shows clearly the importance of defining the reasons for using HACCP and its importance in managing food safety of the particular product. The legal aspects are built on the scientific principles that are used to manage food safety for this product, using other products as examples of existing controls. Although it is an excellent example of the application of HACCP to a real-world situation, the main limitation of this case study is that it does not give students clear learning objectives or outcomes.

A comprehensive case study on HACCP, designed for BSc students focused on the challenges facing small food businesses (Georgakopoulos, 2007), shows how HACCP can be applied in a business where scientific knowledge is limited, underlining that, in principle, it is a way of thinking about food safety as well as being a tool for managing food safety. Education of the staff is essential in order to implement it properly. The case study goes on to show how good communication and training are the key elements in implementation and that it is hard to teach the scientific principles of HACCP to non-scientific staff, be they managers or production-line operatives. Yet, food safety legislation requires all food businesses to have a food safety system based on HACCP principles. Therefore, creative, engaging ways of educating staff need to be developed.

A particularly engaging way of teaching food safety has been developed by partners in the ISEKI-Food 4 project and hosted on the ISEKI Food Association (IFA) webpages (https://www.iseki-food.net/students/fishcanninggame). This game is set in an imaginary fish cannery and players take on the investigation of a food poisoning incident, navigating around the cannery to isolate the cause. The designers have aimed the game not only at students in higher education but also those from industry who would like to practice their skills in a safe environment. The advantages of such an approach to teaching food safety are that the learners can go at their own pace and they can develop their understanding of the issues involved by themselves and in an environment that has some resemblance to reality. On the other hand, no virtual game can provide the nuances of the real world, and every such case may have subtle differences that will only become evident after considerable investigation.

IFA hosts a number of e-learning resources that have been developed for use in teaching and used extensively by ISEKI members and others over a number of years (https://www.iseki-food.net/training/e-learning). Topics include hygienic design and cleaning, canning, food packaging, and others. These courses are available free to IFA members and others on payment of a small fee. Some of these are credit bearing, with ECTS of 2-3. In each of these, students are provided with the study materials and short progress tests. In some, students are encouraged to add additional facts that they have about the topic. The study materials consist of text, diagrams, and animations that illustrate the topics. They are available online for stand-alone use or may be used as part of a short course delivered face-to-face.

The approach to microbiology and food safety teaching practiced at the authors' university follows that of UK institutions offering similar courses. In their first year, students receive introductory lectures on microbiology covering classification and diversity of microorganisms, growth, metabolism, and structure and function. This is developed in the second year to focus on the impact of the presence and growth of microorganisms in food and covers food poisoning, food spoilage, and food production, including fermentation reactions. These first 2 years provide students with a solid theoretical and practical understanding of the impacts that the presence and growth of microorganisms can have on the food industry and consumers. The final year builds on this and takes a risk analysis approach to apply the students' knowledge to real-world scenarios. The culmination of the module requires the students to prepare a HACCP plan appropriate to a specific short shelf life, chilled product. To prepare for this, students are guided through the process by working together as a group alongside the tutor to prepare a HACCP plan for a chicken-ready meal. This allows the students to see the HACCP approach in action and learn the necessary skills before tackling their own product. Use is also made of genuine HACCP plans from local businesses. The students' work is supported by a series of laboratory investigations on time-temperature abuse of food, stress responses of bacteria, shelf life evaluation, sampling techniques, food poisoning, and food contamination to give practical experience of issues relevant to food safety. This combination of lectures, practical investigations, worked examples, and real-world documentation has been very effective in educating students and enabling them to understand the principles of risk assessment and hazard analysis and apply them to food production. This combination of theory and practical application has also been shown to work well in industry (Soares et al., 2013).

5 FOOD SAFETY TRAINING FOR INDUSTRY

The importance of food safety and of training for the industry has been recognized for many years. The creation of the Food Standards Agency in the United Kingdom and similar bodies in other European countries was a response to several major food safety scares (Houghton et al., 2008). Consequently, it was recognized that the level of training needed to increase significantly and training providers began to offer a wide range of courses delivered face-to-face and, later, online. However, this provision may not be sufficient. There is still a great need to ensure proper training across the whole of the food supply chain following the principles of good nutritional practice (GNP) (Raspor, 2008). Whether this is really happening in 2015, is debatable.

In the United Kingdom, there are a number of organizations and businesses registered as training providers for the food industry; for example, the Chartered Institute of Environmental Health or Highfield Ltd. Some of these provide international services, especially where the courses are available online and in a suitable language. Most training for the industry across Europe is provided by academic institutions as outreach work. Some of this may be general training but in many cases it is via specialized courses aimed at particular sectors within the industry. It appears from a number of studies that training in the food industry is fairly patchy and somewhat inadequate in many cases. However, there is limited formal research in this area and some of the evidence is anecdotal.

A very few studies have been carried out on the effectiveness of online training in food safety, and none that we are aware of in the United Kingdom; Egan et al. (2007) looked at a wide range of courses but did not compare online and conventional ones. However, there is evidence from the United States that a combination of lecture and web-based delivery is more effective than either alone (Pintauro et al., 2005). A later study (Wallner et al., 2007) concluded that online food safety teaching is an effective method for the continuing education of health professionals. The recent development of Massive Open Online Courses (MOOCs), has increased the availability of online training and there a many examples of such courses that are aimed at food industry professionals. As with many online provisions, such resources can provide an easy way to improve or update employee knowledge, but despite many offering discussion boards and live chat facilities, they can lack the ability to tailor the course to the particular needs of individual companies. Many online training courses are offered by universities or training companies; it may be that if more food companies could be encouraged to get involved in the development and delivery of such courses, then they could be better tailored to the needs of the industry.

It would be expected that in Austria high standards of food safety training would be found in all parts of the industry because there is a legal requirement for training. Pichler et al. (2014) found that this is not the case. Nearly 25% of Austrian food handlers in catering have significant gaps in their knowledge of food safety and nearly as many do not undertake any training.

So even in Europe, food safety training and education may not be sufficient to reach small-scale, traditional producers. In the German-speaking areas of central Europe, there is a well-established traditional food industry producing cheeses, meat, and fish products; fermented vegetables; sour dough bread; and spiced biscuits (Lucke and Zangerl, 2014). In particular, it was found that there is limited knowledge of good manufacturing practice, and training in the principles of HACCP is needed. These authors stress that there should be more cooperation across the industry involving food scientists, food producers, enforcement agencies, and risk communicators.

In Slovenia, it was found that in food manufacturing and catering, employee satisfaction plays a significant role in implementing and maintaining the HACCP system (Jevšnik et al., 2008). There is often a lack of in-house knowledge about the reasons for particular actions, indicating that training is inadequate. These authors showed that training by in-house staff is the most efficient as it ensures that the training is relevant and thorough, but the trainers must be properly trained themselves, as giving misinformation could be more dangerous than giving none at all. They recognized that there are special problems in SMEs where the range of expertise is limited.

Similar research in Turkey (Baş et al., 2007), found that the main barriers to proper implementation of HACCP was the lack of motivation and the high turnover of employees. It was noticed that younger employees were more likely to see the benefits of training than older, long-term employees who saw no reason to change their behavior. In this study, it was found that more than two-thirds of employees lacked basic food hygiene training. Owners of these businesses had little background or experience in the food industry prior to establishing their company, and no training in hygiene. One of the key factors leading to high levels of foodborne illness is the lack of well-structured channels of communicating food safety. In order to improve matters, the Turkish government has instituted the Turkish Food Code under which better regulation of the industry can be established. Baş et al. (2007) recommended that the regulations include suitable training for food managers.

Meat is one of the highest-risk foods with respect to foodborne illness. Consequently, across Europe there are strict regulations concerning meat and meat products. Gomes-Neves et al. (2011) used a questionnaire to study the knowledge of meat handlers in a slaughterhouse in Portugal. Almost 25% of the respondents had received no formal food hygiene and safety training, and many had not even heard of HACCP. As with other studies, it was found that food hygiene courses were of limited benefit and that better understanding is gained through training in the workplace. However, management needs to have a positive attitude toward food safety and training in order to set the right environment. These authors refer to the review of Seaman (2010), who proposes a "food hygiene training model" that starts with training needs analysis that ensures training is targeted, then a testing stage, which is followed by evaluation of the food handlers' responses to the training. At the present time, there is no evidence that this model has been adopted across Europe, though it may form the basis of training in some areas.

Although training can be provided externally or internally, it is necessary to have some assessment of how effective that training is in ensuring and enhancing food safety. In Belgium, the food safety authority has encouraged food businesses to adopt a self-checking system certification (Jacxsens et al., 2015). In order to assess the effectiveness of this approach, 82 Belgium companies, of varying size and product lines, were assessed by a food safety management systems diagnostic instrument. All but one had an acceptable level of microbiological food safety. Thus, the Belgium self-checking system certification approach to food safety training appears to provide an effective basis for training.

Romania has a robust legal framework for food safety and training in food hygiene is mandatory (Jianu and Golet, 2014). However, the research shows that this is not always effective and that in the meat industry, staff do not tend to keep their professional qualifications up to date.

Within those universities where food science and technology is taught, the principles and application of food safety are covered comprehensively. However, many graduates do not go into the food industry, and there are many people working in the industry who have little or no scientific knowledge of food safety. Thus, there is a gap between academia and industry. Standard, predesigned, and specially designed courses can all provide useful training in food safety, but are the most effective when they are delivered in the workplace as part of the activities of all members of the staff. This suggests that teaching of food safety in universities should be closely linked with practical experience. It is well known that students who undertake work placements or sandwich years in industry have a better understanding of how the industry operates, food processing, and the practical application of HACCP and other systems. The situation in SMEs is different from larger enterprises in that they have much more limited resources, experience, and knowledge. The food industry is attractive to people who have no prior experience of the industry and no technical knowledge. This continues to present a challenge to national and international control of food safety. Regulatory authorities need to develop strategies that promote food safety and that include food enterprises of all sizes.

6 FOOD SAFETY TRAINING AND EDUCATION ELSEWHERE

As might be expected, the major industrialized countries, such as the United States, Australia, and the European countries, have well-developed and effective food safety control programs supported by many training providers. Many other countries have extensive legal frameworks that cover food safety and quality, but enforcement tends to be weak. China has introduced major reforms to its food safety laws, particularly in the run-up to the 2008 Olympics. In 2003, they created the State Food and Drug Administration of China (Jia and Jukes, 2014) to coordinate the activities of other regulatory bodies. The Ministry of Health has responsibility for developing the training requirements, inspection, testing, and for investigating breaches of food safety. Training is provided through agencies such as the United Kingdom's Chartered Institute of Environmental Health or by local universities, and through conferences such as the series of international workshops run by the South China University of Technology (2014).

A somewhat similar situation exists in India, where new regulatory bodies and laws have been brought in over the last decade or so to replace the previous ones that were based on the old colonial system. Responsibility for standards, certification, and product testing lies with the Bureau of Indian Standards (BIS). In-house and open training is offered by the National Institute for Training for Standardization, which is a subsidiary body under BIS (see http://www.bis.org.in/trg/train.htm). However, India has a vast informal economy across all sectors including food; hence, enforcement is weak. A survey by Choudhury et al. (2011) found that in the city of Guwahti, only about one-third of the street food vendors had any knowledge of food hygiene and only around 10% knew anything about sources of food contamination. Many of the international agencies provide training in food safety in India but with such a large informal economy, it is difficult to see how much improvement can be made in the short term.

Many other countries around the world have the legislation in place but lack the resources for enforcement. Without enforcement, there is little incentive for training; therefore, food safety is still an issue. This adversely affects the economy of these countries and limits their ability for international trade in food. The European Union and the United States impose strict conditions on imports from such countries, and meeting them is a major challenge. Some outreach work has been done, as for example under the ISEKI-Mundus projects (see https://www.iseki-food.eu/), but far more support is needed if the global food market is to expand. There is no doubt that the development of international supply chains will improve the standards of food safety through training (Kirezieva et al., 2015).

REFERENCES

Ajzen, I., 2005. Attitudes, Personality and Behavior, second ed. Open University Press, Maidenhead.

Baş, M., Yüksel, M., Çavuşoğlu, T., 2007. Difficulties and barriers for the implementing of HACCP and food safety systems in food businesses in Turkey. Food Control 18, 124–130.

Becker, M.H., 1974. The Health Belief Model and Personal Health Behaviour. CB Slack, New Jersey, USA.

Chartered Institute of Public Health, 2015. CIEH food safety training. Available from: http://www.cieh.org/training/food_safety.html (accessed 10.01.15).

Choudhury, M., Mahanta, L., Goswami, J., Mazumder, M., Pegoo, B., 2011. Socio-economic profile and food safety knowledge and practice of street food vendors in the city of Guwahati, Assam, India. Food Control 22, 196–203.

Clayton, D.A., Griffith, C.J., Price, P., Peters, A.C., 2002. Food handlers' beliefs and self reported practices. Int. J. Environ. Health Res. 12, 25–39.

Dumoulin, E., 2010. History of the food network before ISEKI food. ISEKI Food Association. Available from: https://www.iseki-food.eu/webfm_send/1005 (accessed 29.12.14).

Egan, M.B., Raats, M.M., Grubb, S.M., Eves, A., Lumbers, M.L., Dean, M.S., Adams, M.R., 2007. A review of food safety and food hygiene training studies in the commercial sector. Food Control 18, 1180–1190.

European Commission, 2010. High Level Forum to take on challenges of food sector. Available from: http://europa.eu/rapid/press-release_IP-10-1016_en.htm (accessed 13.02.15).

Fernández-Díza, M.J., Santaolalla, R.F., González, A.G., 2010. Faculty attitudes and training needs to respond to the new European higher education challenges. High. Educ. 60, 101–118.

Georgakopoulos, V., 2007. Application of HACCP in small food businesses. In: McElhatton, A., Marshall, R.J. (Eds.), Food Safety. A Practical and Case Study Approach. In: Kristbergsson, K. (Ed.), ISEKI Food Series, Vol. 1. Springer Science and Business Media, New York, pp. 239–252.

Global Food Safety Training Survey, 2014. Available from: http://www.campdenbri.co.uk/training/GFSI2014.pdf (accessed 10.02.15).

Global Food Security, 2013. Global food security strategic plan, Updated November 2013. Available from: http://www.foodsecurity.ac.uk/assets/pdfs/gfs-strategic-plan.pdf (accessed 15.01.15).

Gomes-Neves, E., Cardoso, C.S., Araújo, A.C., Correia da Costa, J.M., 2011. Meat handlers training in Portugal: a survey on knowledge and practice. Food Control 22, 501–507.

Griffith, C.J. 2006. Food safety: where from and where to? Br. Food J. 108 (1), 6–15.

Hobbs, J.E., Fearne, A., Spriggs, J., 2002. Incentive structures for food safety and quality assurance: an international comparison. Food Control 13, 77–81.

Houghton, J.R., Rowe, G., Van Frewer, L.J., Kleef, E., Chryssochoidis, G., Kehagia, O., Korzen-Bohr, S., Lassen, J., Pfenning, U., Strada, A., 2008. The quality of food risk management in Europe: perspectives and priorities. Food Policy 33, 13–26.

ISEKI Food Association, 2014. Available from: https://www.iseki-food.net/ (accessed 29.12.14).

Jacxsens, L., Kirezieva, K., Luning, P.A., Ingelrham, J., Diricks, H., Uyttendaele, M., 2015. Measuring microbial food safety output and comparing self-checking systems of food business operators in Belgium. Food Control 49, 59–69.

Jevšnik, M., Hlebec, V., Raspor, P., 2008. Food safety knowledge and practices among food handlers in Slovenia. Food Control 19, 1107–1118.

Jia, C., Jukes, D., 2014. The national food safety control system of China – a systematic review. Food Control 32, 236–245.

Jianu, C., Golet, I., 2014. Knowledge of food safety and hygiene and personal hygiene practices among meat handlers operating in western Romania. Food Control 42, 214–219.

Kirezieva, K., Luning, P.A., Jacxsens, L., Allende, A., Johannessen, G.S., Tondo, E.C., Rajkovic, A., Uyttendaele, M., van Boekel, M.A.J.S., 2015. Factors affecting the status of food safety management systems in the global fresh produce chain. Food Control 52, 85–97.

Lucke, F.-K., Zangerl, P., 2014. Food safety challenges associated with traditional foods in German-speaking regions. Food Control 43, 217–230.

Maruchecka, A., Greisb, N., Menac, C., Caid, L., 2011. Product safety and security in the global supply chain: issues, challenges and research opportunities. J. Oper. Manag. 29 (7–8), 707–720.

Mortlock, M.P., Peters, A.C., Griffith, C.J., Lloyd, D., 2000. Evaluating impacts of food safety control on retail buchers. In: N.H. Hooker, E.A. Murano, (Eds.), Interdisciplinary Food Safety Research, CRC Press, Boca Raton, FL.

Mossel, D.A.A., 1989. Food safety the need for public reassurance. Food Sci. Technol. Today 3, 2–10.

Motrajemi, Y., Käferstein, F., 1999. Food safety, hazard analysis and critical control point and the increase in food borne diseases: a paradox? Food Control 10 (4–5), 325–333.

Mounfield, J.D., 1966. Early history of the Institute of Food Science and Technology. J. Food Technol. 1, 1–8.

Mullan, B., Wong, C., 2009. Hygienic food handling behaviours. An application of the theory of planned behaviour. Appetite 52, 757–761.

Mullan, B., Wong, C., 2010. Using the theory of planned behaviour to design a food hygiene intervention. Food Control 21, 1524–1529.

Notermans, S., Zwietering, M.H., Mead, G.C., 1994. The HACCP concept: identifications of potentially hazardous micro-organisms. Food Microbiol. 11, 203–214.

O'Boyle, C.A., Henly, S.J., Larson, E., 2002. Understanding adherence to hand hygiene recommendations: the theory of planned behavior. Am. J. Infect. Control 29 (6), 352–360.

Osborn, M., Broadfoot, P., McNess, E., Planel, C., Ravn, B., Triggs, P., Cousin, O., Winther-Jensen, T., 2003. A World of Difference. Comparing Learners Across Europe. Open University Press/McGraw-Hill Education, Maidenhead.

Pichler, J., Ziegler, J., Aldrian, U., Allerberger, F., 2014. Evaluating levels of knowledge on food safety among food handlers from restaurants and various catering businesses in Vienna, Austria 2011/2012. Food Control 35, 33–40.

Pintauro, S.J., Krahl, A.G., Buzzell, P.R., Chamberlain, P.M., 2005. Food safety and regulation: evaluation of an online multimedia course. J. Food Sci. Educ. 4, 66–69.

Powell, D.A., Jacob, C.J., Chapman, B.J., 2011. Enhancing food safety culture to reduce rates of foodborne illness. Food Control 22, 817–822.

Public Health England, 2015. Notifiable diseases: annual totals from 1982 to 2013. Available from: https://www.gov.uk/government/uploads/system/uploads/attachment_data/file/335023/Noids_annual_totals_1982_to_2013_England_and_Wales.xls (accessed 13.02.15).

Raspor, P., 2008. Total food chain safety: how good practices can contribute? Trends Food Sci. Technol. 19, 405–412.

Redmond, E.C., Griffith, C.J., 2003. Consumer food handling in the home: a review of food safety studies. J. Food Prot. 66 (1), 130–161.

Seaman, P., 2010. Food hygiene training: introducing the food hygiene training model. Food Control 21, 381–387.

Seaman, P., Eaves, A., 2006. The management of food safety – the role of food hygiene training in the UK service sector. Hosp. Manage. 25, 278–296.

Seaman, P., Eaves, A., 2010a. Efficacy of the theory of planned behaviour model in predicting safe food handling practices. Food Control 21, 983–987.

Seaman, P., Eaves, A., 2010b. Perceptions of hygiene training amongst food handlers, managers and training providers – a qualitative study. Food Control 21, 1037–1041.

Soares, K., García-Díez, J., Esteves, A., Oliveira, I., Saraiva, C., 2013. Evaluation of food safety training on hygienic conditions in food establishments. Food Control 34, 613–618.

South China University of Technology, 2014. 2014 International training workshop on food safety assessment technology and food safety quality control. Available from: http://202.38.194.245/hgschool/indexfive/indexEducation.do?op=detail&id=1007&schoolFlagId=8 (accessed 05.02.15).

Todd, E.C.D., 1989. Costs of acute bacterial food-borne disease in Canada and USA. Int. J. Food Microbiol. 9, 313–326.

Tudorache, T., Goursaud, J., 2001. HACCP application on microbiological control of cooked pre-sliced ham. In: Mannino, S., Schleining, G. (Eds.), Practical Tasks for Food Quality Assurance. University of Milan, Department of, Food Science and Technology, Milan, pp. 74–80.

Wallner, S., Kendall, P., Hillers, V., Bradshaw, E., Medeiros, L.C., 2007. Online continuing education course enhances nutrition and health professionals' knowledge of food safety issues of high-risk populations. J. Am. Diet. Assoc. 107, 1333–1338.

Wheelock, V., 1989. Food safety in perspective. Br. Food J. 93 (8), 31–36.

Williamson, D.M., Gravani, R.B., Lawless, H.T., 1992. Correlating food safety knowledge with home food preparation practices. Food Technol. 46, 94–100.

Index

Note: Page numbers followed by *f* indicate figures and *t* indicate tables.

G

H

I

J

K

L

M

T

U

V

W

Printed in the United States
By Bookmasters